THIRD EDITION

Discrete Structures

S. B. SINGH

G.B. Pant University of Agriculture & Technology,
Pantnagar

JAI KISHORE

R. D. Engineering College,
Ghaziabad

EKATA

Krishna Institute of Engg. & Tech.,
Ghaziabad

KHANNA BOOK PUBLISHING CO. (P) LTD.

Publisher of Science, Technology and Engineering Books

4C/4344, Ansari Road, Darya Ganj, New Delhi-110002
Phone: 011-23244447-48 **Mobile:** +91-99109 09320
E-mail: contact@khannabooks.com
Website: www.khannabooks.com

Dear Readers,

To prevent the piracy, this book is secured with HIGH SECURITY HOLOGRAM on the front title cover. In case you don't find the hologram on the front cover title, please write us to at contact@khannabooks.com or whatsapp us at +91-99109 09320 and avail special gift voucher for yourself.

Specimen of Hologram on front Cover title:

Moreover, there is a SPECIAL DISCOUNT COUPON for you with EVERY HOLOGRAM.

How to avail this SPECIAL DISCOUNT:

Step 1: Scratch the hologram
Step 2: Under the scratch area, your "coupon code" is available
Step 3: Logon to www.khannabooks.com
Step 4: Use your "coupon code" in the shopping cart and get your copy at a special discount
Step 5: Enjoy your reading!

ISBN: 978-93-82609-40-7

Discrete Structures
by S.B. Singh, Jai Kishore, Ekata

First Edition: 2002
Second Edition: 2007
Third Edition: 2017
Reprint: 2025, **2026**

To view complete list of books, Please scan the QR Code:	 KPH

Published by:
Khanna Book Publishing Co. (P) Ltd.
CIN: U22110DL1998PTC095547

Visit us at: **www.khannabooks.com**
Write us at: **contact@khannabooks.com**

Printed in Bharat.

PREFACE TO THE PRESENT EDITION

The present volume is thoroughly revised and new objective type problems with their answers have been included. This edition offers a logical and lucid presentation of both the theory and problems and includes questions from various examinations of different universities.

We want to express our gratitude and appreciation to Mrs. Neelam Singh, Mrs. Jyotsna Tyagi and Dr. Neeraj Gupta for their constant support, patience and encouragement throughout the time of writing of this book. The authors are thankful to Mr. Puneet Khanna and Mr. Mukul Seth of Ms. Khanna Publishing Company (Pvt.) Ltd. and printer for publishing this work in the present nice form. The authors take the opportunity to thank all the faculty members who adopted/ referred this book for teaching.

Suggestions for improvement of the book will be gratefully acknowledged.

Authors

PREFACE TO THE FIRST EDITION

With the advancement of computer technology the study of Discrete Structures is regarded as most essential. A person can not be good Computer Engineer without having good knowledge of Discrete Mathematics. Keeping this fact in mind this book has been written to fulfill the growing needs of students of B.Tech., M.C.A., M.Sc. and B.C.A. etc. of various Universities specially U.P. Tech. Univ. Lucknow. A special care has been taken to present the book with an approach to explain the concepts in lucid and communicable language so that it should also be helpful to the students of M.C.A. and B.C.A. from arts and commerce background and also to students whose mathematical maturity is of high school standard. The book is divided into ten chapters. Numerous worked-out examples have been given which will help the readers in understanding various concepts easily. A problem set is given at the end of main topics and at the end of the chapters so that student may try the problems and access their confidence of answering the problem after studying the topics of the book.

The authors are indebted and obliged to Director, the faculty members, friends, and students of Dr. K.N. Modi Institute of Engineering and Technology, Modinagar for the suggestion and help in preparing this book. The authors are also thankful to many other authors whose books have been consulted during the preparation of this book. We want to express our gratitude and appreciation to Mrs. Neelam Singh and Mrs. Jyotsna Tyagi for their constant support, patience and encouragement throughout the time of writing of this book. The authors are thankful to Ms. Khanna Publishing Company (Pvt.) Ltd. and printer for publishing this work in present nice form.

Suggestions for improvement of the book will be gratefully acknowledged.

Authors

CONTENTS AT A GLANCE

CONTENTS

CHAPTER

1 Combinatorics

1.0 Introduction

In a department of a certain engineering college, there are 60 students. Director of the college wants to nominate three students to the department post of president, vice president and secretary. The Director first selects a group of three students, say Robert, John and Kamaljeet. Now, he wants to nominate them for the three department posts. He can do this in different ways as under:

President	Vice-President	Secretary
Robert	John	Kamaljeet
Robert	Kamaljeet	John
John	Robert	Kamaljeet
John	Kamaljeet	Robert
Kamaljeet	Robert	John
Kamaljeet	John	Robert

Here we can see that there are six possible ways in which Director can nominate after selecting a group of three students out of 60 students. We realize that the Director can face following two problems:

a. Selecting a group of three students in all possible ways.
b. Naming the selected students for different posts in all possible ways.

In Mathematics both the above problems are discussed under the heading Combinatorial Mathematics or Combinatorics, which deals with the problem of counting or enumeration. In particular number of ways of **combinations** deals with problems of selection of object and **permutations** deal with problems of number of ways of arrangements of objects.

1.1 Fundamental Principle of Counting

In the study of combinatorial mathematics, two types of problems related to addition and multiplication are of great importance. The problems are respectively based on addition and multiplication principles, which are stated below:

(a) Addition Principle

If an event can happen in m possible ways and another event can occur in n possible ways and both the events are mutually exclusive (i.e. only one event can occur at a time either first or second), then either of the two events can be occurred in $(m + n)$ ways. For example, suppose we want to find two digits even number that can be formed with the help of digits 1, 2, 3, 4 and 5. The possibility that a number of two digits will be even if the number formed has either 2 or 4 in the unit's place. The possibilities that the number ending with 2 are 12, 32,

42 and 52. Similarly the possibilities that the number ending with 4 are 14, 24, 34 and 54. Now the total possibilities that the two digits number will be even are 4 + 4 = 8 (4 numbers end with 2 and 4 numbers end with 4).

Example 1.1

There are four books on Discrete Mathematics and six books on Numerical Analysis in a certain show room. In how many ways can Ramesh buy a book either of Discrete Mathematics or of Numerical Analysis?

Solution

Book on Discrete Mathematics can be purchased in four ways and Numerical Analysis can be purchased in six ways. Hence Ramesh can purchase either a Discrete Mathematics or a Numerical Analysis book in 4 + 6 = 10 ways.

(b) Multiplication Principle

If an event can occur in m ways and corresponding to each way of occurring this event another event can occur in n ways independent of the first. Then the number of ways of happening both the events simultaneously is $m \times n$.

Example 1.2

Suppose there are 8 different courses offered in the morning and six different courses offered in the afternoon. Then there will be 7 × 5 choices for students who want to enroll one course in the morning and one in the afternoon.

Example 1.3

A carpenter has twelve patterns of chairs and five patterns of tables. In how many ways can he make a pair of chair and table?

Solution

For each ways of making a chair, a table can be made in five patterns, so for twelve ways of making chairs, the total number of ways of making a pair of chair and table by the multiplication principle are 12 × 5 = 60.

Example 1.4

In how many ways can we get a sum of 4 or of 8 when two distinguishable dice (say one is brown and the other is blue) are rolled? How many ways can we get an even sum?

Solution

Let us label the outcome of a 1 in brown dice and a 3 on the blue dice as the ordered pair (1, 3). The other possibilities of getting sum 4 are (2, 2) and (3, 1). Hence there are 3 ways to get sum 4.

Likewise, we get a sum 8 by the outcomes (2, 6), (3, 5), (4, 4), (5, 3) and (6, 2). Therefore, total number of possibilities in which sum is either 4 or 8 is 3 + 5 = 8.

Now, to get an even sum, the possibilities are 2, 4, 6, 8, 10 or 12. The possible ways to get the sum 2 is one, to get the sum 4 is three, to get the sum 6 is five, to get the sum 8 is five, to get the sum 10 is three and to get the sum 12 is one. Therefore, there are 1 + 3 + 5 + 5 + 3 + 1 = 18 ways to get an even sum.

When the dice are indistinguishable, then to get an even sum is possible in 1 + 2 + 3 + 3 + 2 + 1 = 12 *ways (In this case outcomes of the type* (6, 2) *and* (2, 6) *will be considered same).*

Example 1.5

In how many ways 100 voters can vote for three candidates contesting for the election of the post of president of their association?

Solution

Every member of the association can vote to either of the three candidates, thus by addition principle the number of ways that a member can vote is 3. Hence every member can vote in 3 ways. Thus by multiplication principle 100 members can vote in $3 \times 3 \times 3 \times \ldots \times 3 \times 3$ (100 times) i.e. 3^{100} ways.

Example 1.6

Suppose that license plates of a certain state require 3 English letters followed by 4 digits.

a. How many different plates can be manufactured if repetition of letters and digits are allowed?

b. How many plates are possible if only the letters can be repeated?

c. How many plates are possible if only the digits can be repeated?

d. How many plates are possible if no repetitions are allowed?

Solution

a. Since there are 26 possibilities for each of the 3 letters, thus it is possible in 26^3 ways and each of the 4 digits are possible in 10 ways. Thus the total possibilities for this case is 10^4. Therefore $26^3.10^4$ plates can be manufactured if repetition of letters and digits are allowed.

b. 26^3. 10.9.8.7 ways.

c. 26.25.24. 10^4 ways.

d. 26.25.24.10.9.8.7 ways.

Example 1.7

i. How many numbers are there between 100 and 1000 such that 7 is in the unit's place?

ii. How many numbers are there between 100 and 1000 such that at least one of their digits is 7?

iii. How many of them have exactly one of their digits as 7?

Solution

i. All the given numbers have three digits. We want to form the number in such a way that the 7 is placed at the unit's place. The middle digit can be any of the 10 integers between 0 and 9. The digit in hundred's place can be any one of the 9 digits between 1 and 9. Therefore, by the multiplication principle, there are $10 \times 9 = 90$ numbers.

ii. The numbers between 100 and 1000 such that at least one of their digits is 7 can be classified into three types. If the 7 is at the unit's place then possible ways as seen in (i) is 90. If the 7 is at middle place, but not in the unit's place, then possible ways are $9 \times 9 = 81$. If the 7 is at the hundred's place only, there are $9 \times 9 = 81$ possible ways.

   Totally, therefore, there are $90 + 81 + 81 = 252$ numbers are required.

iii. Here again in this case there are three kinds:

   a. The numbers that have 7 in the unit's place but not in any other place.

   b. The numbers that have 7 in the ten's place but not in any other place.

   c. The numbers that have 7 in the hundred's place but not in any other place.

The required number is $72 + 72 + 81 = 225$.

1.2 Factorials

Consider the product 1, 1.2, 1.2.3, 1.2.3.4, … We denote them respectively by 1!, 2!, 3!, 4! so on.

For each positive integer we define

$$n! = n(n-1)(n-2) \ldots 3.2.1 = \text{the product of all integers from 1 to } n.$$

$n!$ is read as "***n* factorial**". It is also denoted by the notation $\lfloor n$.

Also define $0! = 1$. Note that $1! = 1$

In general $n! = n[(n-1)!]$

In particular, $4! = 4\,(3!)$

$3! = 3\,(2!)$.

The factorial of proper fractions and negative integers are not defined. The factorial is defined only for whole numbers.

Example 1.8

Write each of the following in expanded form and compute its value.

(i) $\dfrac{9!}{5!}$ (ii) $\dfrac{10}{5!5!}$

Solution

(i) $\dfrac{9!}{5!} = \dfrac{9.8.7.6.5.4.3.2.1}{5.4.3.2.1} = 9.8.7.6 = 3024$ (ii) $\dfrac{10!}{5!5!} = \dfrac{10.9.8.7.6.5.4.3.2.1}{(5.4.3.2.1)(5.4.3.2.1)} = \dfrac{10.9.8.7.6}{5.4.3.2.1} = 252$

Example 1.9

Prove that $n!\,(n+2) = n! + (n+1)!$.

Solution

$$
\begin{aligned}
\text{R.H.S.} &= n! + (n+1)! \\
&= n! + n!(n+1) \\
&= n!\,[1 + (n+1)] \\
&= n!(n+2) \\
&= \text{L.H.S.}
\end{aligned}
$$

Example 1.10

If $(n+2)! = 2550.\,(n!)$, find n.

Solution

We know

$$(n+2)! = (n+2)\,(n+1).n \ldots 3.2.1$$
$$= (n+2)\,(n+1)n!$$

Its given $(n+2)! = 2550\, n!$

i.e. $(n+2)\,(n+1)\, n! = 2550\, n!$

or $(n+2)\,(n+1) = 2550$

or $n^2 + 3n - 2548 = 0.$

Which is a quadratic equation. Its roots are

$$n = \frac{-3 \pm \sqrt{9 + 4 \times 2548}}{2} = 49 \text{ or } -52.$$

Out of these two roots, the second root is ruled out because factorials of negative integers are not defined.

Exercise 1.1

1. How many two digit number is there in the set of natural numbers?
2. A lady wants to select one cotton saree and one polyester saree from a textile shop. If there are 10 cotton varieties and 12 polyester varieties, in how many ways can she choose the two sarees?
3. How many possible telephone numbers are there when there are seven digits, the first two of which are between 2 and 9 inclusive, the third digit between 1 and 9 inclusive, and each of the remaining may be between 0 and 9 inclusive?
4. i. How many ways are there to select 2 cards (without replacement) from a deck of 52?

   How many ways are there to select the two cards such that:

   ii. The first card is an ace and the second card is a king?

   iii. The first card is an ace and the second is not a king?

   iv. The first card is a heart and the second is a club?

   v. The first card is a heart and the second is a king?

   vi. The first card is a heart and the second is not a king?

   vii. Neither card is an ace?

   ix. At least one of the cards drawn is an ace?
5. In a cross-word puzzle there are two solutions to each of the three given places and three solutions for four other places. How many different solutions could be sent in to win a prize?
6. There are five different roads from city A to city B, three different roads from city B to city C, and three different roads that go directly from A to C.

   i. How many different ways are there to go from A to C via B?

   ii. How many different ways are there from A to C altogether?

   iii. How many different ways are there from A to C and then back to A?

   iv. How many different trips are there from A to C and back again to A that visit B both going and coming?

   v. How many different trips are there that go from A to C via B and return directly from C to A?

   vi. How many different trips are there that go directly from A to C and return to A via B?

   vii. How many different trips are there from A to C and back to A that visit B at least once?

   viii. Suppose that once a road is used it is closed and cannot be used again. Then how many different trips are there from A to C via B and back to A again via B?

   ix. Using the assumption in (viii) how many different trips are there from A to C and back to A again?
7. The students in a class are seated according to their marks in the previous examination. Once it so happens that four of these students get equal marks and therefore the same rank. To decide their seating arrangement, the teacher wants to write down all possible arrangements, one in each of separate bits of paper, in order to choose one of these by lots. How many bits of paper are required?

8. How many 5-letter words are there where the first and last letters
   i. are consonants?
   ii. are vowels?
   iii. are vowels and middle letters are consonants?
   iv. How many 5-letter words are there if vowels can only appear (if at all) as the first or last letter?
9. How many different four letter words can be formed out of the word 'LOGARITHMS' if repetition of letters is not allowed. (The word may or may not have a meaning)
10. In a certain programming language, an identifier is a sequence of a certain number of characters where the first character must be a letter of the English alphabet and remaining characters may be either a letter or a digit.
    i. How many identifiers are there of length 5?
    ii. In particular, in some implementations of Pascal an identifier is a sequence of from 1 upto 8 characters with the above restrictions. How many Pascal identifiers are there?
11. Each section in the first year of M.C.A. course has exactly 40 students. If there are 4 sections, in how many ways can a set of 4 students representatives be selected, one from each section?
12. A chain letter is sent to 10 people in the first week of the year. The next week each person who received a letter sends letter to 10 new people, and so on.
    i. How many people have received letters after 10 weeks?
    ii. At the end of the year?
13. A mother distributes 5 different apples among 8 children.
    i. How many ways can this be done if each child receives at most one apple?
    ii. How many ways can this be done if there is no restriction on the number of apples a child can receive?
14. If $(n+1)! = 12\,[(n-1)!]$, find n.
15. Prove that $1 + 2 + 3 + 4 + 5 + 6 = 21 = \frac{6}{2}(6+1)$
16. Find the total number of positive integers that can be formed from the digits 1, 2, 3, 4 and 5 if no digit is repeated in any one integer.
17. Find the sum of all 4-digit numbers that can be obtained by using (without repetition) the digits 2, 3, 5 and 7.
18. Prove that $(n+1)[n.n! + (2n-1)\overline{(n-1)}! + (n-1)\overline{(n-2)}!] = \overline{(n+2)}!$.
19. Compute (i) 15!/ 9!.6! (ii) 17!/ 12!.

Answers

1. 90
2. 120
3. $8 \cdot 8 \cdot 9 \cdot 10^4$
4. (i) 52·51 (ii) 4.4
   (iii) 4.47 (v) 1·48+12·47
   (vi) 48.47
   (vii) 52·51 – 48·47
5. 648
6. (i) 5.3 (ii) 15+3
   (iii) 182 (iv) 152
   (v) 15.3 (vi) 3.15

(vii) $15{\cdot}3 + 3{\cdot}15 + 152 = 182 - 32 = 15{\cdot}18 + 3{\cdot}15$

(viii) 15.8 (ix) 15.8+15.3 + 3.15 + 3.2

7. 24

8. (i) $21^2{\cdot}26^3$ (ii) $5^2{\cdot}26^3$

(iii) $5^2{\cdot}21^3$ (iv) $26^2{\cdot}\ 21^3 = 215 + 2{\cdot}5{\cdot}21^4 + 5^2{\cdot}21^3$

9. 5040

10. (i) In (i) identifiers are of length 5 and the first character can be any letter out of 26 letters, while the remaining 4 characters may be any of the 26 letters *A* through *Z*, or 10 digits 0 through 9. Thus, each of the remaining 4 characters can be filled in 36 ways.

Therefore, the length of identifiers of length 5 are $26{\cdot}36^4$.

(ii) $26 + 26{\cdot}36 + 26{\cdot}36^2 + \ldots + 26{\cdot}36^7$, Pascal identifiers.

11. 2560000

12. (i) $10 + 10^2 + 10^3 + \ldots + 10^{10}$

(ii) $10 + 10^2 + 10^3 + \ldots + 10^{52}$

14. 3

16. $5 + 5{\cdot}4 + 5{\cdot}4{\cdot}3 + 5{\cdot}4{\cdot}3{\cdot}2 + 5{\cdot}4{\cdot}3{\cdot}2{\cdot}1$

17. There are 4! such numbers. Every one of the 4 positions each digit occurs 6 times. The sum of the digits is 17. Hence the sum of these 24 numbers is (6)·(17)·(1111).

19. (i) 5005 (ii) 742560

1.3 Permutations

The different number of ways in which three books (i) Hindi (ii) English and (iii) Bengali can be arranged on a shelf are

(i)	Hindi	English	Bengali
(ii)	Hindi	Bengali	English
(iii)	Bengali	Hindi	English
(iv)	Bengali	English	Hindi
(v)	English	Bengali	Hindi
(vi)	English	Hindi	Bengali

The above six ways are total number of possible ways of arranging the books on a shelf in which no book is repeated, such types of arrangements are called **permutations**. We define the word permutation as under:

The ways in which a set of objects can be arranged without duplication of any of the object which can be made by taking some or all of the given number of distinct objects is called **permutation**. The total number of such arrangements is called the **number of permutation.**

The permutation of n objects taken r at a time is denoted by $P(n, r)$ or ${}^nP_r$.

Theorem 1.1

The number of possible permutations of r distinct elements taken from a set of n elements is given by

$$P(n,r) = n(n-1)(n-2)\ldots(n-r+1) = \frac{n!}{(n-r)!}$$

Proof

While arranging n objects taken r at a time, the position can be filled by any one of the n objects. This can be done in n ways. Any one of the remaining $n-1$ objects can fill the second position. This can be done in $n-1$ ways and so on. Finally, the rth position can be filled in $n-r+1$ ways, because after filling up the previous $r-1$ positions, exactly $n-(r-1) = n-r+1$ objects remain. After filling up the rth position, we are through in arranging r objects. By the fundamental principle of multiplication, the total number of ways of arranging r objects from n objects is $n(n-1)\ldots(n-r+1)$. Thus

$$P(n,r) = n(n-1)(n-2)\ldots(n-r+1)$$

$$= \frac{n(n-1)(n-2)\ldots(n-r+1)(n-r)!}{(n-r)!}$$

$$P(n,r) = \frac{n!}{(n-r)!}$$ [Here it is to be noted that $1 \le r \le n$].

Corollary

Out of n objects, if we take all at a time then the number of permutations will be

$$P(n,n) = n(n-1)(n-2)\ldots 3.2.1 = n!$$

In Theorem 1.1 we have got the number of permutations i.e. the total number of arrangements without repetition. In case an object is repeated certain number of times, then we call the total number of ways as arrangements and not permutations. In the following theorem we will discuss the total number of arrangements when certain object is repeated any number of times.

Theorem 1.2

Prove that the number of arrangements of n different objects taken r at a time, when each object may be repeated any number of times in one arrangement, is n^r (here r can be greater than n also).

Proof

Here we want to fill up r places with n different ways. Since any one of n objects can be placed at first place, hence first place can be filled up with n different ways. Now to fill up the second place, again we have n objects as the object placed at first place can be repeated. Thus there are altogether $n.n = n^2$ ways of filling first two places. The third place can again be filled in n ways and each of fillings in the third place may be associated with each of the n^2 ways of filling in the first two places. Therefore, first three places can be filled in n^3 ways. Generalizing the result we find that the filling in r places is n^r. Hence the required number of arrangements is n^r.

Example 1.11

Find the value of $P(4, 3)$.

Solution

We know $P(n, r) = n(n-1)\ldots(n-r+1)$

Here $n = 4, r = 3,$

$\therefore n-r+1 = 4-3+1 = 2.$

Hence $P(n, r) = P(4, 3) = 4\cdot3\cdot2 = 24$

Aliter: We know $P(n, r) = \dfrac{n!}{(n-r)!}$

Here $n = 4, r = 3$

$$P(4,3) = \frac{4!}{(4-3)!} = \frac{4!}{1!} = 4 \times 3 \times 2 = 24$$

Example 1.12

How many five different letter words can be formed out of the word "LOGARITHMS"?

Solution

The total letters in the word "LOGARITHMS" is 10 i.e. $n = 10$ and we have to form five letter words thus $r = 5$.

The required number of words is.

$$P(10,5) = \frac{10!}{(10-5)!} = \frac{10!}{5!}$$

$$= \frac{10 \times 9 \times 8 \times 7 \times 6 \times 5!}{5!}$$

$$= 10 \times 9 \times 8 \times 7 \times 6 = 30240$$

Example 1.13

How many different license plates are there that involve 1, 2 or 3 letters followed by 4 digits?

Solution

Out of total 26 letters, 3 letter words can be formed with no repeated letters in P (26, 3) ways. Complete yourself.

Example 1.14

Find n if $P(n, 4) = 42.\ P(n, 2)$.

Solution

$$P(n, 4) = n(n-1)(n-2)(n-3) \quad \text{... (1)}$$

and $$P(n, 2) = n(n-1) \quad \text{... (2)}$$

Given $$P(n, 4) = 42 \cdot P(n, 2) \quad \text{...(3)}$$

Substituting values (1) and (2) in (3), we get

$$n(n-1)(n-2)(n-3) = 42\ n(n-1)$$

or $$(n-2)(n-3) = 42$$

or $$n^2 - 5n + 6 = 42$$

or $$n^2 - 5n - 36 = 0$$

or $$(n-9)(n+4) = 0$$

or $$n = 9 \text{ or } n = -4,$$

Out of these two answers, $n = -4$ is not possible because for negative numbers permutation is not defined, hence $n = 9$ is the required answer.

Example 1.15

It is required to sit 5 men and 4 women in a row so that the women occupy the even places. How many such arrangements are possible?

Solution

Here, there are exactly four even places and exactly four women to occupy them. Hence, these even positions can be filled in $P(4, 4) = 4!$ ways. Now we are left with 5 positions and we have exactly 5 men. These 5 positions can be filled in $P(5, 5) = 5!$ ways. Therefore, the required number of ways by the fundamental principle of multiplication is

$$P(4, 4) \cdot P(5, 5) = 4!.5! = 24.120 = 2880.$$

Example 1.16

In how many ways can the letters of English alphabet be arranged so that there are exactly 5 letters between the letters a and b?

Solution

5 letters between a and b can be arranged in $P(24, 5)$ ways. a and b can be arranged in two ways and then 20! ways to arrange any 7 letter word treated as one unit along with the remaining 19 letters. Hence the total numbers of possible ways are $P(24, 5) \cdot (20!) \cdot 2$.

Example 1.17

There are four prizes to be distributed among 6 students. In how many ways can it be done when (i) no student gets more than one prize; (ii) there is no restriction as to the number of prizes any student gets; (iii) no student gets all the prizes.

Solution

i. Number of students = 6

Number of prizes = 4

When no student to get more than one prize. This can be done in

$$P(6,4) = \frac{6!}{(6-4)!} = \frac{6!}{2!} = 6 \times 5 \times 4 \times 3 = 360 \text{ ways}$$

ii. As there is no restriction to the number of prizes to be won by the students, the first prize can be distributed in 6 ways. Corresponding to each way of distribution of the prize there are again 6 ways of distributing the second prize since it can be awarded to the same student who has won the first prize. Therefore, the first two prizes can be distributed in $6 \times 6 = 6^2$ ways. Similarly each of the third and fourth prizes can be awarded in 6 different ways.

$$\text{The required number} = 6 \times 6 \times 6 \times 6 = 6^4 = 1296.$$

ii. Total number of ways a prize can be awarded = 6 different ways. Therefore, the total number of ways in which no student gets all the prizes $6^4 - 6 = 1290$.

Example 1.18

How many three-digit numbers are there with no digit repeated?

Solution

Since total number of digits are 10. The number of possible ways that a three-digit number can be formed is $P(10, 3)$. But to form the three digits number 0 should not be in the left-most position. For such position total number of permutation is $P(9, 2)$. Hence, $P(10, 3) - P(9, 2)$ is the required number of ways a three-digit number can be formed with no digit repeated.

Theorem 1.3

Prove that $$P(n, r) = P(n-1, r) + r \cdot P(n-1, r-1).$$

Proof

$$\text{R. H. S} = P(n-1,r) + r \times P(n-1,r-1)$$

$$= \frac{(n-1)!}{(n-1-r)!} + \frac{r.(n-1)!}{(n-1-r-1)!}$$

$$= (n-1)!\left\{\frac{1}{(n-r-1)!} + \frac{r}{(n-r)!}\right\}$$

$$= (n-1)!\left\{\frac{n-r+r}{(n-r)!}\right\} = \frac{(n-1)!n}{(n-r)!}$$

$$= \frac{n!}{(n-r)!} = P(n,r) = \text{L.H.S.}$$

1.3-1 Circular Permutations

The permutations we have considered so far are more properly called **linear permutations** for the objects are being arranged in a line. If instead of arranging objects in a line, we arrange them in a cycle, then the permutation is said to be **circular permutations.**

Let us arrange 5 integers 1, 2, 3, 4, and 5 around a circular table. The five possible arrangements of the integers are 12345, 23451, 34512, 45123, and 51234.

Here we can see that all above five arrangements are different if they are to be placed in a straight line. But if the arrangements are made in the cyclic order then they are not different as seen from the above Fig.1.1. We can note that in a circular arrangement, there is neither a beginning nor an end. In this situation, we have to check the relative position of the given object. The circular permutations are different only when the relative order of the object is changed, otherwise these are the same.

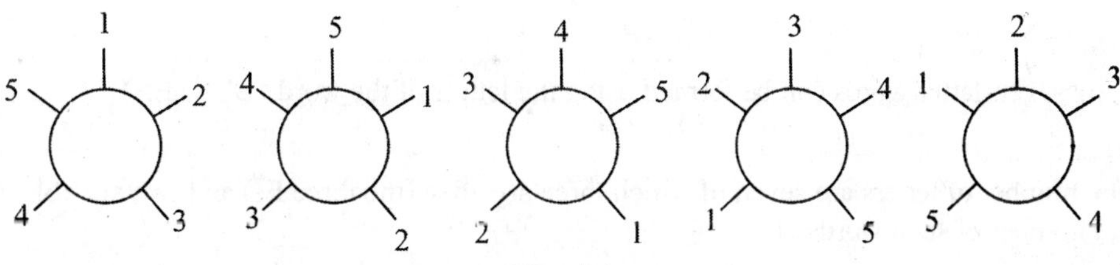

Fig. 1.1

As the circular position depends upon the relative position of the objects, we fix the position of one object and then arrange the remaining $(n-1)$ objects in all possible ways. Thus the number of circular arrangements of five objects 1, 2, 3, 4, 5 will be $(5-1)! = 4! = 24$.

Example 1.19

In how many ways can a party of seven gentlemen and six ladies be seated at a round table so that no two ladies are seated together?

Solution

As there are 7 gentlemen and they can be arranged around the table in $(7-1)!$ ways. Now when the gentlemen have been seated, there are seven places for ladies one each between two gentlemen. Hence the ladies can be seated in $P(7, 6)$ ways. Therefore, the required number of ways this arrangements can be made by the fundamental principle of multiplication is $P(7, 6) \times 6! = 362880$.

1.3-2 Permutations with Repetitions

Theorem 1.4

The number of arrangements of a multi-set objects some of which are alike say n_1 are alike of the first kind, n_2 are alike of the second kind, … n_r are alike the rth kind, is given by

$$P(n; n_1, n_2, ..., n_r) = \frac{n!}{n_1!n_2! .... n_r!}.$$

where $$n_1 + n_2 + n_3 + ... + n_r \le n.$$

Proof

Let n objects be denoted by n letters, n_1 of them being alike each be denoted by p, n_2 being alike each be denoted by q and so on.

Let t be the required number of arrangements. Let us replace one of these t arrangements by n_1 alike letters 'p' by unlike letters, say p_1, p_2, p_3,……, P_{n_1} different from any of the rest and arrange these n_1 new letters amongst themselves without changing the positions of the remaining letters. Then, we have $n_1!$ new arrangements and if we change each of t arrangements, the total number of arrangements will be $t \cdot n_1!$.

Similarly for n_2 alike objects, 'q' be replaced by n_2 new letters different from each other and different from the rest, the total number of arrangements would be $t \cdot n_1! \cdot n_2$.

Similarly, if we replace all the alike objects by unlike objects then the total number of arrangements will be $t \cdot n_1! \cdot n_2! \cdot n_3! \; .... \; n_r!$.

But here these n objects all are different and admit $n!$ arrangements.

Hence $x \cdot n_1! \cdot n_2! .. .... n_r! = n!$.

or $$x = \frac{n!}{n_1!n_2!n_3! \; ....n_r!}$$

Example 1.20

How many seven-letter words can be formed using the letters of the word "BENZENE"?

Solution

Here total number of letters are seven of which three are alike (the three Es) and two are alike (the two Ns). Thus the total number of such words is

$$P(7,3,2) = \frac{7!}{3!2!} = 420$$

Example 1.21

How many number greater than one million can be formed with the digits 4, 6, 6, 0, 3, 6, 3?

Solution

For numbers to be greater than one million, it must consist of 7 or more digits. Here, we are given with 7 digits in all. Out of these 7 digits, there are 3 sixes, 2 threes and the rest all digits are different. Therefore,

$$\text{Total number of possible numbers} = \frac{7!}{2! \cdot 3!} = 420$$

Out of these numbers, some begin with zero and are thus less than one million, which must be rejected. The members beginning with zero

$$= \frac{6!}{2! \cdot 3!} = 60$$

∴ The required numbers greater than one million

$$= 420 - 60 = 360.$$

Example 1.22

How many different signals, each consisting of eight flags hung in a vertical line, can be formed from a set of four indistinguishable red flags, three indistinguishable white flags, and a blue flag?

Solution

As there are four indistinguishable red flags, three indistinguishable white flags. Therefore, the required numbers of permutations of eight objects are

$$P(8;4,3) = \frac{8!}{4!.3!} = 280$$

Example 1.23

In how many ways can four mathematics books, three history books, three chemistry books, and two sociology books be arranged on a shelf so that all books of the same subject are together?

Solution

There are four mathematics books that can be arranged in 4! ways, the history books can be arranged in 3! ways, the chemistry books in 3! ways, and sociology books in 2! ways. Thus, altogether, there are $4! \cdot 4! \cdot 3! \cdot 3! \cdot 2! = 41472$ ways.

1.3-3 Restricted Permutation

Theorem 1.5

The numbers of permutations of n different objects take r at a time in which m particular objects are:

a. never included will be $P(n - m, r)$.

b. always included will be $P(n - m, r - m)\, P(r, m)$, where $(n \geq m, r \geq m)$.

Proof

a. When we are not interested in including m particular objects, then arrange r objects out of remaining $n - m$ objects. This can be done in $P(n - m, r)$ ways.

b. When we are interested in including m particular objects always, we keep them aside and form the permutation of the remaining $(n - m)$ objects taking $(r - m)$ at a time. This can be done in $P(n - m, r - m)$ ways.

Now we take the case of m particular objects kept aside one by one. Out of total m objects kept aside, the first object can be occupied by $(r - m + 1)$ places as $(r - m)$ objects are already arranged. Thus the first can be introduced in $(r - m + 1)$ ways. Similarly second object can be introduced in $(r - m + 2)$ ways and proceeding in the same ways the mth object can be introduced in $(r - m + m) = r$ ways. Therefore, m objects can be arranged in

$$(r - m + 1)(r - m + 2) \cdot \cdots (r) = P(r, m) \text{ ways.}$$

∴ The required number of permutations are

$$P(n - m, r - m) \cdot P(r, m).$$

Example 1.24

How many words of four letters can be formed with the letters a, b, c, d, e, f, g, h and i when

i. e and f are not to be included and.
ii. e and f are to be included.

Solution

i. We are given with eight letters out of which we are to take four letters after keeping aside 2. After leaving 2 letters we are left with 6 letters, now the 4 letters out of 6 can be arranged in $P(6, 4) = 360$ ways i.e. there are 360 four lettered word which do not contain e and f.
ii. When two letters e and f are included 2 letters e and f for 2 places out of 4 places of 4 lettered words and the remaining 2 letters are to be taken out of the remaining 6 letters a, b, c, d, e, g, h, i. Thus the number of words of 4 letters in this case will be

$$P(6, 2) \cdot P(4, 2) = (6 \cdot 5) \cdot (4 \cdot 3) = 360.$$

Example 1.25

There are 15 points in a plane, no three of which are collinear. Find the number of straight lines formed by joining them.

Solution

A straight line can be formed whenever two points are joined. Now selection of two points out of 15 points can be made, i.e. $C(15, 2) = 105$.

Exercise 1.2

i. Find n if
   i. $P(n, 2) = 72$
   ii. $P(n, 3) + 3 \cdot P(n, 2) + P(n, 1) = n^3$
   iii. $2P(n, 2) + 50 = P(2n, 2)$.
2. In how many ways the letters of word (i) LIBERTY (ii) RATION (iii) REPUBLIC can be arranged?
3. In how many ways can the letter of word 'PENCIL' be arranged so that N is always next to E?
4. In how many ways 10 boys can be arranged in a line if two particular boys are (i) always together (ii) never together?
5. The Director wants to arrange 5 students on the platform such that the boy Salim occupies the second position and such that the girl Gita is always adjacent to the girl Rita. How many such arrangements are possible?
6. An electric network contains 14 switches such that each switch may have three possible positions. How many different switching are there?
7. a. How many automobile license plates can be made if each plate contains two different letters followed by three different digits?
   b. Solve the problem if first digit cannot be 0.
8. Ten students are participating in a race. In how many ways the first three prizes are won?
9. In how many ways can we arrange 15 books on a shelf; 7 on mathematics, 6 on Numerical Analysis, and 2 on Operations Research, when no distinction is to be made in the books of the same subject?
10. If there are six periods in each working day of an institute, in how many ways can one arrange 5 subjects such that each subject is allowed at least one period?

11. i. Find the number of ways in which five persons can sit in a row.
    ii. How many ways are there if two of the persons insist on sitting next to one another?
    iii. Solve part (i) assuming they sit around a circular table.
    iv. Solve part (ii) assuming they sit around a circular table.
12. In how many ways can 7 persons address a meeting so that out of these, three persons A, B, C are to address in such a way that A will speak before B and B before C?
13. Four alphabets E, K, S and V, one each, were purchased from a plastic warehouse. How many ordered pair of alphabets, to be used as initials, can be formed from them?
14. In how many ways can 7 boys and 5 girls be seated together in a row, so that no two girls are together?
15. In an examination hall, there are four rows of chairs. Each row has 8 chairs one behind the other. There are two classes sitting for the examination, with 16 students in each class. It is desired that in each row all students belong to the same class. In how many ways can these 32 students be seated?
16. Find the number of arrangements, which can be made out of the letters of the word "ALGEBRA" without altering the relative positions of vowels and consonants.
17. Among the 36 teachers in a college, one principal and one vice-principal are to be appointed. In how many ways can this be done?
18. Five boys and five girls are to be seated together. In how many ways can they be seated if:
    i. All boys must be seated in the five leftmost seats?
    ii. No two boys can be seated together?
    iii. John and Mary must be seated together?
19. i. In how many ways can three boys and two girls sit in a row?
    ii. In how many ways can they sit in a row if the boys and girls are to sit together?
    iii. In how many ways can they sit in a row if just the girls are to sit together?
20. How many words of four letters can be formed with the letters of the words "RATION" if:
    i. R and O are not to be included.
    ii. R and O are to be included.
21. When a group photograph is taken, all the seven teachers should be in the first row and all the twenty students should be in the second row. If the two corners of the second row are reserved for the two tallest students, interchangeable only between them, and if the middle seat of the front row is reserved for the principal, how many arrangements are possible?
22. Find the number of natural number less than 1000 and divisible by 5 which can be formed with the digits 0, 1, 2, 3, 4, 5, 6, 7, 8, 9 each digit not occurring more than once in each number.
23. Find the number of ways that a party of seven persons can arrange themselves: (i) in a row of seven-chair (ii) around a circular table.
24. In how many ways can 9 cabinet ministers occupy their seats at a round table if one of the seats is to be reserved for the Prime Minister and there are eight other seats?
25. Find the number of ways of forming a ring with the help of 10 girls.
26. Three boys and three girls are to be seated around a table in a circle. Among them, the boy X doesn't want any girl neighbour and the girl Y doesn't want any boy neighbour. How many such arrangements are possible?

27. If the different permutations of the word EXAMINATION are listed as in a dictionary, how many items are there in this list before the first word starting with E?
28. How many numbers lying between 3000 and 4000 and divisible by 5 can be formed with the digits 3, 4, 5, 6, 7 and 8 repetition of the digits are not allowed?

Answers

1. (i) 9 (ii) all natural numbers (iii) 5
2. (i) 5040 (ii) 720 (iii) 40320
3. 120
4. i. $2 \cdot 9!$ (ii) $8 \cdot 9!$
5. 8
6. 3^{14}
7. (i) $26 \cdot 25 \cdot 10 \cdot 9 \cdot 8 = 468000$
8. 720
9. 180180
10. 3600
11. (i) 120 (ii) 48 (iii) 24 (iv) 12
12. 840
13. 12
14. 3386880
15. $2 \cdot (16!)^2$
16. 72
17. 1260
19. (i) 120 (ii) 24 (iii) 48
20. (i) 24 (ii) 144
21. $(1440) \cdot 18!$
22. 154
23. (i) 7! (ii) 6!
24. 40320
25. 9!
26. 4
27. 907200
28. 12

1.4 Combinations

On many occasion we are not interested in the ordering or arrangements but only in selecting r objects from n objects. For example if a company wants to recruit 6 of the applicants for a position; a student may want to choose 4 books from a library at a time.

Suppose we want to select two balls out of four balls marked A, B, C and D. We may choose either A, B or A, C or A, D or B, C or B, D or C, D. Note that we do not list B, A separately, because it is the same as the choice A, B. Thus there are six ways of selecting 2 persons out of 4 persons. Here, we can see that in this selection, we didn't care the order but only two different balls. We denote the above fact by writing $C(4, 2) = 6$.

Consider a finite collection of n objects. A combination of these n objects taken r at a time is, any selection of r objects where order does not count. The number of such selection is called the **number of combination**.

The combination of r objects out of n given objects is denoted by ${}^nC_r$ or $C(n, r)$ or $C_{n,r}$ or C_r^n or $\binom{n}{r}$.

1.4-1 Difference between a Permutation and a Combination

In a combination only selection is made whereas in permutation not only a selection is made but also an arrangement in a definite order is also taken into consideration.

In a combination, the ordering of the selected objects is immaterial. In a permutation, this ordering is essential. For example A, B and B, A are same as combinations but different as permutations.

Example 1.26

Find the number of combinations of the four objects A, B, C, and D taken three at a time.

Solution

Here total number of given objects are 4 and we are interested in finding combination consisting of three objects which is 3! = 6 permutations of the objects in the combinations as shown in the table below.

Here we can see that number of combinations multiplied by 3! equals the number of permutations, i.e.

$$C(4, 3) \cdot 3! = P(4, 3)$$

or
$$C(4, 3) = \frac{P(4,3)}{3!}$$

Combination	Permutations
ABC	ABC, ACB, BAC, BCA, CAB, CBA
ABD	ABD, ADB, BAD, BDA, DAB, DBA
ACD	ACD, ADC, CAD, CDA, DAC, DCA
BCD	BCD, BDC, CBD, CDB, DBC, DCB
Total No. 4	24

Since $P(4, 3) = 4.3.2 = 24$ and $3! = 6$.

Therefore $C(4, 3) = 4$

We can observe from the above example that the number of permutations exceeds the number of combinations which is usually true except for some trivial cases.

In higher mathematics, the combinations $C(n, r)$ occur more frequently then the permutations.

1.4-2 Combination of *n* Different Objects

Theorem 1.6

The number of combination of n distinct objects taken r at a time is given by

$$C(n,r) = \frac{n!}{r!(n-r)!} = \frac{P(n,r)}{r!}$$

Proof

As we have seen in the above example the selection of r objects from n objects, the selected r objects can be arranged in $P(r, r) = r!$ ways.

Thus for each combination counted in $C(n, r)$, there are $r!$ permutations counted in $P(n, r)$. Therefore, we conclude that

$$P(n, r) = r!\, C(n, r)$$

or
$$C(n,r) = \frac{P(n,r)}{r!} = \frac{n!}{r!(n-r)!}$$

COROLLARY 1 $C(n, 0) = 1$

$$C(n,0) = \frac{n!}{0!(n-0)!} = \frac{n!}{1.n!} = 1$$

COROLLARY 2 $C(n, n) = 1$

$$C(n,n) = \frac{n!}{n!(n-n)!} = \frac{n!}{n!.0!} = 1$$

Theorem 1.7

Prove that $C(n, r) = C(n, n-r)$.

Proof

$$\text{L.H.S.} = C(n,r) = \frac{n!}{r!(n-r)!}$$

$$\text{R.H.S.} = C(n, n-r) = \frac{n!}{(n-r)!(n-n+r)!}$$

Hence $C(n, r) = C(n, n-r)$

C (n, r) and C (n, n – r) are called ***complementary combinations.***

Theorem 1.8

Prove that $C(n+1, r) = C(n, r-1) + C(n, r)$.

Proof

$$\text{R.H.S.} = C(n, r-1) + C(n,r)$$

$$= \frac{n!}{(r-1)!(n-r+1)!} + \frac{n!}{r!(n-r)!}$$

$$= n!\left\{\frac{1}{(n-r+1)(n-r)!(r-1)!} + \frac{1}{r(r-1)!(n-r)!}\right\}$$

$$= \frac{n!}{(n-r)!(r-1)!}\left\{\frac{1}{(n-r)!(r-1)!} + \frac{1}{r}\right\}$$

$$= \frac{n!}{(n-r)!(r-1)!}\left\{\frac{n+1}{(r)(n-r+1)}\right\}$$

$$= \frac{(n+1)!}{r!(n-r+1)!}$$

$$= C(n+1, r) = \text{L.H.S.}$$

Above identities are combinatorial identities. They can alternatively be proved by combinatorial arguments.

Example 1.27

Compute $C(13, 5)$.

Solution

$$C(13,5) = \frac{13!}{5!(13-5)!} = \frac{13!}{5!.8!}$$
$$= \frac{13.12.11.10.9.8!}{5.4.6.1.8!}$$
$$= \frac{13.12.11.10.9}{5.4.3.2.1} = 13.11.9$$
$$= 1287.$$

Example 1.28

Find n, if $C(n, 7) = C(n, 5)$.

Solution

Given $C(n, 7) = C(n, 5)$

$$\frac{n!}{7!(n-7)!} = \frac{n!}{5!(n-5)!}$$

or $\quad (n!)(5!)(n-5)! = (n!)(7!)(n-7)!$

or $\quad (n-6)(n-5) = 6 \times 7$

or $\quad n^2 - 11n - 12 = 0$

or $\quad n = 12, -1$

In this problem n should be non-negative. Therefore, $n = 12$ is the only solution.

Example 1.29

i. In how many ways can a committee of 5 is chosen from 9 people?
ii. How many committees of 5 or more can be chosen from 9 people?
iii. In how many ways can a committee of 5 teachers and 4 students be chosen from 9 teachers and 15 students?
iv. In how many ways can the committee in (iii) be formed if teacher A refuses to serve if student B is in the committee?

Solution

i. $C(9, 5)$ ways.
ii. $C(9, 5) + C(9, 6) + C(9, 7) + C(9, 8) + C(9, 9)$.
iii. The 4 students can be chosen out of 15 students in $C(15, 4)$ ways and the teachers can be selected in $C(9, 5)$ ways. Hence the committee can be formed is $C(9, 5) \times C(15, 4)$ ways.
iv. First we count the number of committees where both A and B are on the committee. Now, we are left with 8 teachers from whom 4 teachers are to be chosen. Similarly we are left with 14 students from whom 3 more students are to be chosen. There are $C(8, 4)\cdot C(14, 3)$ committees containing both A and B, hence there are

$$C(9, 5)\cdot C(15, 4) - C(8, 4)\cdot C(14, 3)$$

committees that do not have both A and B in the committee.

Example 1.30

In how many ways can a student choose a programme of 5 courses if 9 courses are available and two courses are compulsory for all the students?

Solution

Here two courses are given to be compulsory, the students are required to choose 3 out of 7 courses and this can be done in $C(7,3) = \dfrac{7!}{3!\cdot 4!} = 35$.

Example 1.31

The question paper of mathematics contains two questions divided into two groups of five questions each. In how many ways can an examinee answer six questions taking at least two questions from each group?

Solution

The examinee can answer questions from two groups in following ways:

i. 2 from first group and 4 from the second group.

ii. 3 from first group and 3 from the second group.

iii. 4 from the first group and 2 from the second group.

Now, the number of ways of selecting the questions in case (i) is

$$= C(5,2)\cdot C(5,4) = 10\times 5 = 50$$

Likewise the number of ways of selecting the question in case (ii) is

$$= C(5,3)\cdot C(5,3) = 10\times 10 = 100$$

and the number of ways of selecting the question in case (iii) is

$$= C(5,4)\cdot C(5,2) = 5\times 10 = 50$$

Hence the examinee can answer the questions in either of the three ways.

Therefore, the required number of ways

$$= 50 + 100 + 50$$
$$= 200.$$

Example 1.32

i. How many 5-card hands consist only of hearts?

Solution

Out of total 52 cards we have 13 hearts. Since there are 13 hearts to choose from, each such hand is a 5 combination of 13 objects. Therefore,

$$C(13,5) = \frac{13!}{5!(8)!} = \frac{13!}{5!.8!} = 1287$$

ii. How many 5-card hands consist of cards from a single suit?

Solution

Out of each 4 suits, spades, hearts, diamonds or clubs there are $5\cdot C(13, 5)$ cards in hands. Hence, there are a total of $4\cdot C(13, 5)$ such hands.

Example 1.33

A farmer buys 3 hens, 2 pigs and 4 cows from a man who has 6 hens, 5 pigs and 8 cows. How many choices does the farmer have?

Solution

The farmer can choose the hens in C (6, 3) ways, the pigs in C (5, 2) ways, and hens in C (8, 4) ways. Hence by the fundamental principle of multiplication, required number of ways is

$$C(6,3)\cdot C(5,2)\cdot C(8,4) = \frac{6\cdot 5\cdot 4\cdot 5\cdot 4\cdot 8\cdot 7\cdot 6\cdot 5}{1\cdot 2\cdot 3\cdot 1\cdot 2\cdot 1\cdot 2\cdot 3\cdot 4} = 20\cdot 10\cdot 70$$

$$= 14{,}000 \text{ ways.}$$

Example 1.34

If there are 12 persons in a party, and if each two of them shake hands with each other, how many handshakes happen in the party?

Solution

When two persons shake hands, it is counted as one handshake, not two. Therefore, the total number of handshakes is the same as number of ways of selecting 2 persons from among 12 persons. This is given by

$$C(12,2) = \frac{12\times 11}{1\times 2} = 66$$

Example 1.35

A bag contains six white marbles and five red marbles. Find the number of ways four marbles can be drawn from the bag if (i) they can be any colour, (ii) two must be white and two red (iii) they must all be of same colour.

Try yourself.

1.4-3 Restricted Combinations

Theorem 1.9

The number of combinations of n object taken r at a time in which m particular objects are

i. always included is $C(n-m, r-m)$ ways and

ii. never included is $C(n-m, r)$ ways, where $(n \geq m)$.

Proof

i. When we are to include m particular objects compulsorily then for our selection we keep them aside and selection of $(r-m)$ out of remaining $(n-m)$ objects is made. This can be done in $C(n-m, r-m)$ ways.

ii. When we are to exclude m particular objects compulsorily then for selections we should ignore them. Thus we shall be selecting r out of remaining $(n-m)$ objects, which can be done in $C(n-m, r)$ ways.

Example 1.36

In how many ways a football eleven can be chosen out of 17 players when (i) five particular players are to be always included (ii) two particular players are to be always excluded?

Solution

i. Here, first we include five particular players. Then, we are left with 12 players. Out of which we are to select remaining 6 players and this can be done in C (12, 6) ways, i.e.

$$\frac{12}{6!.6!} = \frac{12 \times 11 \times 10 \times 9 \times 8 \times 7!}{6!.6!}$$

$$= \frac{12 \times 11 \times 10 \times 9 \times 8 \times 7!}{6 \times 5 \times 4 \times 3 \times 2 \times 1} = 924 \text{ ways}$$

ii. Here, we wish to exclude two particular players. Hence, we are left with 15 players. Out of which we are to select football eleven, which can be done in $C(15, 11)$ ways.

1.4-4 Combinations with Repetitions

Theorem 1.10

The total number of combinations of $(n_1 + n_2 + n_3 \ldots)$ objects where n_1 are alike, n_2 are alike, and so on taken any number at a time is

$$(n_1 + 1)(n_2 + 1)(n_3 + 1) \ldots -1.$$

Proof

First of all consider n_1 like objects. These n_1 like objects can be dealt in $n_1 + 1$ ways for we may take 1, 2, 3, … , n_1 or none in any selection. Similarly n_2 like objects can be dealt in $(n_2 + 1)$ ways and so on. By the principle of mathematical induction, the total number of selection is $(n_1 + 1)(n_2 + 1)(n_3 + 1) \ldots$ ways. But these selections include one case where in all objects has been left out. Therefore, the total number of ways in which at least one object has been selected $(n_1 + 1)(n_2 + 1)(n_3 + 1) \ldots -1$.

Example 1.37

How many different choices can be made of selections out of 15 Maths, 10 Numerical Analysis and 12 Operations Research books when at least one book is to be selected?

Solution

Total number of ways mathematics books can be dealt is (15 + 1) ways including one of not selecting. Similarly books on Numerical Analysis and Operations Research can be dealt in (10 + 1) and (12 + 1) ways respectively. Then the total numbers of ways of dealing with these books are $(15 + 1). (10 + 1)\cdot(12 + 1) = 16\cdot 11\cdot 13 = 2288$. But there is only one way in which none of these books are selected. Therefore, the required number of ways of selecting at least one of them is $2288 - 1 = 2287$.

Example 1.38

Prove that the total number of ways in which it is possible to make a selection by taking some or all of the given n different objects, is $2^n - 1$.

Proof

As each object can be disposed of in two different ways, i.e. it may be accepted or it may be rejected. And each way of disposing of one object may be associated with each way of disposing of any other object. Therefore, the total number of ways in which all objects may be disposed of is $2 \times 2 \times 2 \times \ldots n$ times $= 2^n$ ways.

But there is one way in which nothing is accepted.

Hence the required number of ways $= 2^n - 1$.

Example 1.39

In how many ways Gaurav can invite eight of his friends to a party by inviting at least one of them?

Solution

There are two ways in which Gaurav can deal with his friends i.e. he may invite him or he may not invite him. Therefore, he can deal with eight friends 2^8 ways, but this includes the case when he does not invite any one of them which can be possible only in one way. Therefore, the required number of ways he can invites his friends $2^8 - 1 = 255$ ways.

1.4-5 Combinatorial Identities and Agreements

Here is the list of certain identities or equations involving the symbols $P(n, r)$ and $C(n, r)$ including the ones that we have already seen:

1. $P(n, n) = P(n, n-1)$
2. $P(n, n) = n\,P(n-1, n-1)$
3. $(n+1)\,P(n, r) = (n-r+1)\,P(n+1, r)$
4. $P(n, n) = r!\,P(n, n-r)$
5. $P(n, r) = P(n-1, r) + r\,P(n-1, r-1)$
6. $P(\text{n, r}) = \text{r! C (n, r)}$
7. $C(n, r) + C(n, r-1) = C(n+1, r)$
8. $r\,C(n, r) = n\,C(n-1, r-1)$
9. $C(n, r) - C(n, n-r)$
10. $C(n, 0) + C(n, 1) + C(n, 2) \ldots + C(n, n) = 2^n$

All these can be proved using the formula $C(n, r)$ and $P(n, r)$ or other agreements. The readers are advised to try each of the above identities/ equations.

Exercise 1.3

1. Evaluate (i) $C(14, 2)$ (ii) $C(20, 17)$ (iii) $C(52, 4)$ (iv) $C(10, 2)$.
2. If $C(n, 8) = C(n, 6)$. Find $C(n, 2)$.
3. Find n, if $C(n, 6): C(n-3, 3) = 91: 4$.
4. Prove that $C(2, 1) + C(3, 1) + C(5, 1) = C(3, 2) + C(4, 2)$.
5. Show that
   i. $C(n, 0) + C(n, 1) + C(n, 2) + \ldots + C(n, n) = 2^n$.
   ii. $C(n, 0) - C(n, 1) + C(n, 2) - C(n, 3) + \ldots + C(n, n) = 0$.
6. Prove the relation
$$C(n, r) = C(n-1, r) + C(n-1, r-1).$$
7. How many committees can be formed consisting of 2 professors, 3 Asst. professors and 5 lecturers in a college having 9 professors, 12 Asst. professors and 20 lecturers?
8. A woman has 11 close friends.
   i. In how many ways can she invite five of them to lunch?
   ii. In how many ways if two of the friends are married and will not attend separately?
   iii. In how many ways if two of them are not on speaking terms and will not attend together?
9. A collection of 100 light bulbs contains 8 defective ones.
   i. In how many ways can a sample of 10 bulbs be selected?
   ii. In how many ways can a sample of 10 bulbs be selected which contain 6 good bulbs and 4 defective ones?
   iii. In how many ways can the sample contains either the sample contains 6 good ones and 4 defective ones or 5 good ones and 5 defective ones?
10. A question paper has two parts, part A and part B, each containing 10 questions. If the student has to choose 8 from part A and 5 from part B, in how many ways can he choose the questions?

11. A student is to answer 10 out of 13 question on an examination.
    i. How many choices have he?
    ii. How many if he must answer the first two questions?
    iii. How many if he must answer the first or second question but not both?
    iv. How many if he must answer exactly three out of the first five questions?
    v. How many if he must answer at least three of the first five questions?
12. A man has 15 close friends, 6 of them are women.
    i. In how many ways can he invite 3 or more of his friends to a party?
    ii. In how many ways can he invite 3 or more of his friends if he wants the same number of men (including him) as woman?
13. A woman has 11 colleagues in her office, of which 8 are men. She could like to invite some of her colleagues to dinner. Find the number of her choices if she decides to invite:
    i. at least 9 of them,
    ii. All her women colleagues and sufficient men colleagues to make the number of women and men equal.
14. In how many ways can we select a cricket eleven from 17 players in which 5 players can bowl? Each cricket team must have at least two bowlers.
15. In how many ways can 5 days be chosen from each of the 12 months of an ordinary year of 365 days.
16. Suppose there are 15 red balls and 5 white balls. Assume that the balls are distinguishable and that a sample of 5 balls is to be selected.
    i. How many samples of 5 balls are there?
    ii. How many samples contain all red balls?
    iii. How many samples contain 3 red balls and 2-white balls?
    iv. How many samples contain at least four red balls?
17. In how many ways can a committee consisting of 5 members be formed each containing 3 consonants and 2 vowels?
18. From a class of 25 students, 10 are to be chosen for an excursion party. There are three students who decide that either all of them will join or none of them will join. In how many ways can they be chosen?
19. A committee of 8 is to be selected out of 6 males and 8 females. How many ways can it be done so that male may not be out numbered?
20. A group of 8 scientists is composed of 5 physicists and 3 mathematicians:
    i. In how many ways a committee of 5 be formed.
    ii. In how many ways can a committee of 5 be formed that has 3 physicist and 2 mathematicians?
21. Shyam invited a party of 10 students to a dinner and places half of them on one side of a long table and half on the other side. If 3 particular men desires to sit one side and two on the other. Find the number of ways in which he can arrange the guests?
22. There are 16 points in a plane, no three of which are collinear.
    i. Find the number of straight lines.
    ii. Number of triangles formed by joining them.
23. How many triangles can be formed by joining 10 points, 5 of which are in the same straight line.

24. In a town, there are 87 families, of which 52 families have atmost 2 children. In a rural development programme, 20 families are to be helped chosen for assistance, of which at least 18 families must have at most 2 children. In how many ways can the choice are made?
25. A committee of eight is to be selected out consisting of five gentlemen and three ladies out of eleven gentlemen and nine ladies with the condition that Miss X refuses to serve on the committee when Mr. Y is selected. In how many ways can the committees are formed.
26. Compute the number of 6-letter combination of the letters of the English alphabet if no letter is to appear in a combination more than two times.
27. For the posts of 5 teachers, there are 23 applicants. 2 posts are reserved for SC candidates and there are 7 SC candidates among the applicants. In how many ways can the selection are made?
28. A candidate is required to answer 6 out of 10 questions which are divided into two groups each containing five questions and he is not permitted to attempt more than four from any group. In how many ways can he make up his choice?
29. A committee of 5 is to be formed out of 6 gentlemen and 4 ladies. In how many ways can this be done if
    a. At least two ladies are to be included?
    b. At most two ladies are to be included?
30. A businessman buys 3 computers, 8 mobile phones, and 12 washing machines from a man who has 9 computers, 25 mobile phones, and 100 washing machines. How many choices does the farmer have?
31. Find the number of diagonals of a polygon of 80 sides.
32. A teacher intends to take eight students to a factory taking three of them at a time. How many times will the teacher go and how many times each student will go?
33. A woman has 11 close friends of whom six are also women.
    i. In how many ways can she invite three or more to a party?
    ii. In how many ways can she invite three or more of them if she wants the same number of men as women (including herself)?

Answers

1. (i) 91 (ii) 1140
   (iii) 270725 (iv) 45
2. 91
3. 15
7. 122791680
8. (i) 462 (ii) 210
   (iii) 252
9 (i) C(100,10) (ii) C(92, 6)·C(8, 4)
   (iii) $C(92, 6)\cdot C(8, 4)+C(92,5)\cdot C(8, 5)$
10. 11340
11 (i) 286 (ii) 165
   (iii) 110 (iv) 80
   (v) 276
12 (i) $2^{15}-1-C(15, 1)-(15, 2)=C(15, 3)+C(15, 4)+\ldots+C(15, 4)$
   (ii) $C(9, 1)\cdot C(6, 2)+C(9, 2)\cdot C(6, 3)+C(9, 3)\cdot C(6, 4)+C(9, 4)\cdot C(6, 5)+C(9, 5)\cdot C(6, 6)$

14. 50050 15. $\{C(30,5)\}^4 \{C(31,5)\}^7 \cdot C(28,5)$
16. (i) $C(20,5)$ (ii) $C(15,5)$
(iii) $C(15,3)\cdot C(5,2)$
17. (i) 56 (ii) 56
18. $C(12,4)\cdot C(8,1)\cdot C(14,2)$ 19. 1414
20 (i) $C(8,5)$ (ii) $C(5,3)\cdot C(3,2)$
21. 288000
22 (i) 120 (ii) 56
24. $C(52,18)\cdot C(35,2) + C(52,19)\cdot C(35,1) + C(52,20)$
25. 32928
26. There are $C(26,6)$ combinations with all 6 letters distinct; $C(26,1)$ $C(25,4)$ combinations with 1 pair of letters and 4 other letters etc. Total: $C(26,6) + C(26,1)\cdot C(25,4) + C(26,3) + C(26,2)\cdot C(24,2)$.
27. There are $C(26,6)$ combinations with all 6 letters distinct; $C(26,1)$ $C(25,4)$ combinations with 1 pair of letters and 4 other letters etc. Total: $C(26,6) + C(26,1)\cdot C(25,4) + C(26,3) + C(26,2)\cdot C(24,2)$.
27. $C(7,2)\cdot C(16,3) = 11760$ 28. 200
29 (i) 186 (iii) 126
30. $C(9,3)\cdot C(25,8)\cdot C(100,12)$ 31. 3080
32. (i) 56 (ii) 21
33 (i) $2^{11}-1 - C(11,2) - C(11,2) = 1981$ or $C(11,3) + C(11,4) + \ldots + C(11,11) = 1981$
(ii) $C(5,5)\cdot C(6,4) + C(5,4)\cdot C(6,3) + C(5,3)\cdot C(6,2) + C(5,2)\cdot C(6,1) = 325$.

1.6 Recurrence Relations and Generating Functions

1.6-1 Polynomial Expression

An nth degree polynomial ($n \geq 0$) is an expression of the form

$$a_0x^n + a_1x^{n-1} + a_2x^{n-2} + \ldots.a_n, \quad a_0 \neq 0$$

where $a_0, a_1, \ldots\ldots, a_n$ are called set of coefficients.

This is a **non-recursive** definition of polynomial.

1.6-2 Zeroth Degree Polynomial

Polynomials of zeroth degree are called constant polynomials and are simply elements of set of coefficients.

1.6-3 Evaluation of Polynomials (non-recursively)

Let $$f(x) = 5x^4 + 12x^3 - 6x^2 + x + 6 \quad \ldots(1)$$

be a polynomial of degree four.

To evaluate it for $x = 2$, put it in the polynomial (1).

$$f(2) = 5(2)^4 + 12(2)^3 - 6(2)^2 + 2 + 6 = 160$$

Here we can see that to evaluate this polynomial we have to perform seven multiplication and four additions/subtractions operations.

1.6-4 Polynomial Expression (Recursive)

Let A be a set of coefficients, then

i. A polynomial expression of degree zero (zeroth degree) is an element of set A.

ii. An nth degree polynomial expression for $n \geq 1$ is an expression of the form $[\varphi(x)]\, x + a$, where $\varphi(x)$ is an $(n-1)^{th}$ degree polynomial expression and $a \in A$.

1.6-5 Evaluation of Polynomials (Recursive)

The polynomial expression (1) can be expressed in recursive form as

$$\begin{aligned} f(x) &= 5x^4 + 12x^3 - 6x^2 + x + 6 \\ &= (5x^3 + 12x^2 - 6x + 1) + 6 \\ &= ((5x^2 + 12x - 6)x + 1)x + 6 \\ &= (((5x + 12)x - 6)x + 1)x + 6 \\ &= ((((5)x + 12)x - 6)x + 1) \end{aligned} \quad \ldots (2)$$

Now, we can see in the above polynomial (5) is a zeroth degree polynomial $(5)x + 12$ **is a first degree** polynomial, $((5)x + 12)\, x - 6$ is a second degree polynomial $(((5)x + 12)\, x - 6)x + 1$ is a third degree polynomial, $((((5)x + 12)x - 6)\, x + 1)x + 6$ is a fourth degree polynomial.

Putting $x = 2$ in (2), we have

$$f(2) = ((((5)\, 2 + 12)2 - 6)\, 2 + 1)\, 2 + 6 = 160.$$

In this form we can see that only four multiplication and four additions/subtractions are used. We can see that operations required to evaluate (1) with the help of recursive form are less than that of non-recursive form polynomial.

1.6-6 Telescopic Form

The final expression (2) of the polynomial (1) is called **telescoping form** of (1).

Method of evaluating a polynomial, after converting into telescoping form is called **Horners Method.**

Example 1.40

i. The telescoping form of $f(x) = 3x^4 - 5x^3 - 2x^2 + 7x + 8$ is

$$((((3)\, x - 5)x - 2)x + 7)x + 8.$$

ii. The telescoping from of $f(x) = x^5 + 3x^4 + 3x^3 - 2x^2 + x$ is

$$f(x) = (((((-1)\, x - 5)\, x + 3)\, x + 3)\, x + 2)\, x + 1)\, x + 0.$$

1.7 Recursion

We give the name recursion to the technique of defining a function, a set, or an algorithm in terms of itself, where it is generally understood that the definition will be in terms of previous values or simpler objects. Describing an object in a simple definition form is usually called the **basis** for recursion.

Example 1.41

Out of several recursively-defined functions, let us take one very familiar recursively defined function $f_n = n!$ (factorial function).

Solution

(i) $f_0 = 1$ (ii) $f_{n+1} = (n+1) f_n$ for all $n \geq 0$.

Here $f(0) = f_0$ is called the initial condition or basis. When $n = 0, 1, 2$.

(ii) becomes

$$f_1 = 1.f_0 = 1.1 = 1, f_2 = (1+1) f_1 = 2.1 = 2, f_3 = (2+1) f_2 = 3\cdot 2 = 6 \text{ and so on.}$$

1.7-1 Recursion Theorem

Let A be a given function from a set A into A. Let $a_0 \in A$ and N denote the set of positive integers. Then there exist a unique function $f: N \to A$ such that

i. $f(0) = a_0$ and

ii. $f(n+1) = F(f(n)) \; \forall \; n \in N$.

i.e $f(n+1)$ in a function of $f(n)$.

The condition (i) is called the initial condition and condition (ii) is called **recurrence relation** or generating rule, which will be defined later on in this chapter. The proof of the theorem is beyond the scope of this book.

Example 1.42

Consider the famous **Fibonacci sequence**, which is defined as

$f_{n+1} = f_n + f_{n-1}$, for all integers $n \geq 1$, $f_0 = 1 = f_1$. Here f_0 and f_1 are called basis or initial condition.

Now, if we want to compute f_3 then,

$$f_3 = f_2 + f_1 = (f_1 + f_0) + f_1 = (1+1) + 1 = 3$$

One can see here that we have moved backward from the third number to second and first which is closer to the basis. Similarly, we can compute f_5,

$$\begin{aligned} f_5 &= f_4 + f_3 \\ &= (f_2 + f_3) + (f_1 + f_2) \\ &= f_1 + 2f_2 + f_3 = f_1 + 2f_2 + f_1 + f_2 \\ &= 2f_1 + 3f_2 \quad = 2f_1 + 3(f_0 + f_1) \\ &= (3f_0 + 5f_1) \quad = 3.1 + 5.1 = 8. \end{aligned}$$

1.8 Iteration

Iteration is a process, in which with the help of initial value(s) we can find the values of higher order terms or we can say forward values.

For example, if we want to compute f_5 of Fibonacci with the help of iteration, then

$$\begin{aligned} f_2 &= f_1 + f_0 = 1 + 1 = 2 \\ f_3 &= f_2 + f_1 = 2 + 1 = 2 \\ f_4 &= f_3 + f_2 = 3 + 2 = 5 \\ f_5 &= f_4 + f_3 = 5 + 3 = 8 \end{aligned}$$

Here, we can see that iterative computation is usually faster than the computations that apply recursion. Therefore, if it is possible better to transform recursive function into an equivalent iterative process for computation or a faster method if possible.

The drawback of recursion is that usually the execution of a recursive program is slower than that of its iterative counterparts, and a recursive program is more difficult to debug than a corresponding iterative one.

1.9 Mathematical Induction

Induction is a process in which we guess and conclude from careful observations of a number of special examples. There may emerge a pattern or a rule, which may found to be common to all the observations. We then formulate a statement in general terms, which applies to all the individual cases under consideration. Consider the following cases of the sums of natural numbers.

$$1 + 2 + 3 + 4 = 10 = \frac{4}{2}(4 + 1)$$

$$1 + 2 + 3 + 4 + 5 = 15 = \frac{5}{2}(5 + 1)$$

$$1 + 2 + 3 + 4 + 5 + 6 = 21 = \frac{6}{2}(6 + 1)$$

We may conclude that the sum of first n natural numbers is $n(n + 1) / 2$. But is this statement true for all n? As seen above it is true for all cases from $n = 1$ to $n = 6$. To verify that this statement is always true, we take it to be true for any arbitrary n and then show it to be true for the successor.

Sum of first n natural number $n(n + 1) / 2$ to be true, so that

$$1 + 2 + 3 + 4 + \ldots + n = n(n + 1) / 2$$

Add its successor i.e. $(n + 1)$ to both the sides, we get

$$\begin{aligned} 1 + 2 + 3 + 4 + \ldots + n + (n + 1) &= \frac{n}{2}(n + 1) + (n + 1) \\ &= \left(\frac{n}{2} + 1\right)(n + 1) = \frac{1}{2}(n + 2)(n + 1) \\ &= \frac{n + 1}{2}\left(\overline{n + 1} + 1\right) \end{aligned}$$

Therefore $1 + 2 + 3 + 4 + \ldots + n = \frac{n}{2}(n + 1)$

is also true statement when n is replaced by $n + 1$. The above is an illustration of principle of mathematical induction. We state the principle as follows:

Let $P(n)$ be a statement such that

a. $P(1)$ is true.

b. $P(r + 1)$ is true whenever $P(r)$ is true where $n = r$.

Then $P(n)$ is true for all natural numbers n.

Example 1.43

Show that $1^2 + 2^2 + 3^2 + \ldots + n^2 = \frac{n(n + 1)(2n + 1)}{6}$ by principle of mathematical induction.

Proof

Let $P(n)$ be the statement such that

$$P(n) = 1^2 + 2^2 + 3^2 + ... + n^2 = \frac{n(n+1)(2n+1)}{6}$$

Then $$P(1) = \frac{1(1+1)(2.1+1)}{6} = 1$$

Hence, $P(1)$ is true

If $P(n)$ is true for an arbitrary r, then

$$1^2 + 2^2 + 3^2 + ... + r^2 = \frac{r(r+1)(2r+1)}{6}$$

Add $(r+1)^2$ to both the sides, we have

$$1^2 + 2^2 + 3^2 + ... + r^2 + (r+1)^2 = \frac{r(r+1)(2r+1)}{6} + (r+1)^2$$

$$= (r+1)\left(\frac{r(2r+1)}{6} + (r+1)\right)$$

$$= (r+1)\left(\frac{(2r^2+7r+6)}{6}\right)$$

$$P(r+1) = \frac{(r+1)(r+2)(2r+3)}{6}$$

$\therefore P(r+1)$ is true. Hence the given statement is true for all $n \in N$.

The formula that we have just proved for $P(n)$ is called a **closed form expression**. It involves no recursion.

1.10 Closed form Expression

Let $F = F(x_1, x_2, ..., x_n)$ be an algebraic expression in which variables x_i (i = 1, 2, ... n) can take any values from some fixed set. F is said to be in closed form expression if there exist a number D such that evaluation of F will take no more than D operations for any allowed values of x_i (i = 1, 2,. . ., n).

Example 1.44

Closed form expression for sum of first n natural number, i.e.

$$\sum_{m=1}^{n} m = 1 + 2 + 3 + 4 + ... + n$$

$$= \frac{n}{2}(n+1)$$

Some more examples of closed form expressions are given below.

Expression	Closed form
$\sum_{m=1}^{n} 2m = 2 + 4 + 6 + ... + 2n$	$n(n+1)$
$\sum_{m=1}^{n} 2m - 1 = 1 + 3 + 5 + ... + (2n-1)$	n^2

Exercise 1.4

1. Write the telescoping form of
   i. $f(x) = 5x^4 + 12x^3 - 6x^2 + x + 6$
   ii. $f(x) = x^5 + 3x^4 - 15x^3 + x - 10$ and calculate the value of $f(3)$.
2. Prove that function $f(n) = 2(3^n) - 5$ is the unique function defined by
   $$f(n) = 4f(n-1) - 3f(n-2) \text{ for } n \geq 2, \text{ where } f(0) = -3, f(1) = 1.$$
3. Use mathematical induction to prove that each of the following statements is true for all positive integers n.
   i. $1^3 + 3^3 + 5^3 + \ldots + (2n-1)^3 = n^2(2n^2 - 1)$
   ii. $1 \cdot 2 + 2 \cdot 3 + 3 \cdot 4 + \ldots\ldots + n \cdot (n+1) = \dfrac{n(n+1)(n+2)}{3}$.
4. Prove by induction that $g(k) = 3k + 2$, $k \geq 0$ is a closed form expression for the g defined recursively by $g(0) = 2$ and $g(k) = g(k-1) + 3$ for $k \geq 1$.
5. Prove by induction that $g(n) = (3)5^n + (7)\,2^n$ is a closed form expression for the sequence g defined recursively by $g(0) = 10$, $g(1) = 29$ and $g(n+1) = 7g(n) - 10g(n-1)$ for $n \geq 1$.
6. Prove by induction that $g_n = 2^n - 1$ is a closed form expression for the sequence g defined recursively by $g_0 = 0, g_1 = 1$ and $g_n = 3g_{n-1} - 2g_{n-2}$ for $n \geq 2$.
7. Using Recursion theorem verify that the definition $g(n+1) = 3g(n)^2 + 7$ for $n \geq 0$, $g(0) = 1$.
8. What is wrong in the definition $f: R \rightarrow R$? $f(0) = 1$ and $f(x) = f(x/2)/2$?

Answers

1. a. $((((5)x + 12)x - 6)x + 1)x + 6$
   b. $(((((1)x + 3)x - 15)x)\,x + 1)x - 10$.
8. The basis is not reached in a finite number of steps if you try to compute $f(x)$ for x not zero.

1.11 Sequences or Discrete Functions

Let us consider the ordered collection of numbers 2, 4. 6, 8, 10,…. and assign the order amongst its members. We observe that 2 is the first member, 4 is the second member, 6 is the third member and so on. These respective numbers are called the terms of the sequence, i.e the first member is called the I[st] **term**, the second member the 2nd term and so on. We in general denote the terms of the sequence by u_1, u_2, u_3, ..., u_n, or a_1, a_2, $a_3 \ldots, a_k$, or t_1, t_2, t_3, ..., t_n, the subscript denoting the place (or position) of the term in the sequence.

We define a sequence in a nonempty set X to be a map $f : N \rightarrow X$. If $X = R$ (real number), we call the sequence f a **real sequence** and if $X = C$ (complex number.) we call f a **complex sequence.** If domain of f is finite N (set of Natural numbers), the sequence f is said to be **finite sequence** and if domain of $f = N$, it is called an **infinite sequence**.

A **sequence of integers** is a function from the set of natural numbers into the set of integers, i.e. if f is a sequence then $f : N \rightarrow Z$. The image of any $k \in N$ can be written as $f(k)$ or f_k.

A sequence is often called a **discrete function.**

Example 1.45

i. The sequence $f(n) = 2n^2 + 3n$, $n \geq 0$ is a sequence of integers.

ii. The sequence $f(n) = 2n^3 + 4n^2 - 3n$, $n \geq 0$ is a sequence of integers.

iii. $f(n) = n^2 - 1, n \in N$.

Example 1.46

The sequence t defined recursively by $t_1 = 1, t_2 = 1$ and $t_n = t_{n-1} + 2t_{n-2}, n \geq 3$ is a sequence of integers. Various terms of t can be computed either by applying recursion formula or by iteration.

By recursion

$$\begin{aligned} t_5 &= t_4 + 2t_3 \\ &= (t_3 + 2t_1) + 2t_3 \\ &= ((t_2 + 2t_1 + 2t_1) + 2(t_2 + 2t_1)) \\ &= t_2 + 4t_1 + 2(t_2 + 2t_1) \\ &= 1 + 4 \cdot 1 + 2 \cdot (1 + 2) = 5 + 6 = 11 \end{aligned}$$

By iteration

$$\begin{aligned} t_3 &= t_2 + 2t_1 = 1 + 2 \cdot 1 = 3 \\ t_4 &= t_3 + 2t_2 = 3 + 2 \cdot 1 = 5 \\ t_5 &= t_4 + 2t_3 = 5 + 2 \cdot 3 = 11 \end{aligned}$$

1.12 Recurrence Relation

For a numeric function $(a_0, a_1, a_2, \ldots, a_r, \ldots)$ or for sequence of numbers an equation relating a_r, for any r, to one or more of the a_i's, $i < r$, is called **recurrence relation**. Recurrence relations are also called **difference equations**, because they can be written in terms of the difference between the consecutive terms of a sequence.

We will see that there is a striking resemblance between recurrence relations of Discrete Mathematics and differential equation of continuous mathematics. Although there is some difference between the two, but many concepts and results about recurrence relations and their solutions are similar to the concepts and solutions of differential equations.

Examples of recurrence relations

1. If a_n denotes the sum of first n terms then $a_n = n + a_{n-1}$.
2. If t_n denotes the nth term of a geometric progression with common ratio r, then $t_n = r\, t_{n-1}$.
3. $a_n = 3a_{n-1} - 2a_{n-2}, n \geq 2$.
4. $a_n = (n-1)\, a_{n-1} - (n-1)\, a_{n-2}, n \geq 2$.
5. $a_n = 9a_{n-1} + 26\, a_{n-2} - 24\, a_{n-3} + 5n, n \geq 3$.
6. $a_n = 2\,(a_{n-1})^2 - n\, a_{n-3}, n \geq 3$.
7. $a_n^2 = -(a_{n-1})^2 - 1, n \geq 1$.

In above examples, relations (1) and (2) are first order recurrence relations because a_n and t_n depend upon only one previous term of a and t respectively. Examples (3) and (4) are second order recurrence relation as they depend upon previous two terms. Examples (5) and (6) are third order recurrence relation as they depend upon previous three terms a_0, a_1 and a_2.

Example 1.47

Consider the Fibonacci sequence

$$F_k = F_{k-2} + F_{k-1}, F_0 = F_1 = 1, k \geq 2.$$

This is a second order recurrence relation.

1.12-1 Solutions of Recurrence Relations

In elementary algebra solving a quadratic equation means finding the values of x which satisfy the quadratic equation. Here we see that $x = 2, 3$ are two values which satisfies the $x^2 - 6x + 5 = 0$ and hence they are the solutions.

Sequences are often most easily defined with a recurrence relation. It is clear that according to the recurrence relation, we can carry out step-by-step computation to determine a_r from $a_{r-1}, a_{r-2},\ldots$. The value of the function at one or more points is given so that computation can be initiated, however the calculation of terms by directly applying a recurrence relation can be time consuming. The process of finding a closed form expression for the terms of sequence from its recurrence relation is called **solving the relation.**

There are some recurrence relations whose solutions do not exist. For example recurrence relation $a_n^2 + (a_{n-1})^2 = -1$ have no solution since there are no real valued function which satisfy the relation, as the squares of real numbers are always nonnegative. Unfortunately, no general method or algorithm of solution for handling all recurrence relations is known. In the following section, we shall study a class of recurrence relations known as linear recurrence relations with constant coefficients.

1.13 n^th^ Order Recurrence Relations with Constant Coefficients

Let S be a sequence of numbers with domain $k \geq 0$. Then a recurrence relation of the form

$$S(k) + c_1 S(k-1) + \ldots + c_n S(k-n) = f(k) \qquad \ldots (1)$$

where c_i's are constant and f is a numeric function defined for $k \geq n$ is called a **linear recurrence** relation of order n with constant coefficients.

Example 1.48

(i) $2\,S(k) + 3\,S(k-1) = 2^k$ is a first order recurrence relation with constant coefficients. And $3a_r - 5a_{r-1} + 2a_{r-2} = r^2 + 5$ and $a_r + 2a_{r-3} = r^2$ are 2nd are 3rd order recurrence relation with constant coefficients respectively.

If b_1 and b_2 are arbitrary constants then $a_n = b_1\,2^n + b_2 5^n$ satisfies the recurrence relation

$$a_n - 7a_{n-1} + 10a_{n-2} = 0,\ n \geq 2. \qquad \ldots (1)$$

Substituting the value of a_n in (1), we have

$$b_1 2^n + b_2 5^n - 7\left(b_1 2^{n-1} + b_2 5^{n-1}\right) + 10\left(b_1 2^{n-2} + b_2 5^{n-2}\right)$$

$$= b_1 2^n + b_2 5^n - 7b_1 2^{n-1} - 7b_2 5^{n-1} + 10b_1 2^{n-2} + 10b_2 5^{n-2}$$

$$= b_1 2^n + b_2 5^n - 7b_1 2^{n-1} - 7\,b_2 5^{n-1} + 5.2\,b_1 2^{n-2} + 5.2b_1 5^{n-2}$$

$$= b_1 2^n + b_2 5^n - 7b_1 2^{n-1} - 7b_2 5^{n-1} + 7b_1 2^{n-1} + 5b_1 5^{n-1}$$

For a nth order linear recurrence relation with constant coefficients, fewer than n values of the numeric function will not be sufficient to determine the numeric function uniquely. As we know for a given set of boundary conditions the solution to a linear recurrence relation with constant coefficients is unique, as long as we are able to find a numeric function that satisfies the recurrence relation as well as initial conditions or boundary conditions. Before giving an algorithm for solving finite order linear relations, we will examine some recurrence relations that arise from some closed form expressions.

Example 1.49

Solve the recurrence relation $\quad T(k) = 2T(k-1),\ T(0) = 1.$

Solution

Given $\quad T(k) = 2T(k-1),\ T(0) = 1$

$$= 2 \cdot 2\, T(k-2)$$
$$= 2^2\, T(k-2)$$
$$= 2^2\, 2\, T(k-3)$$
$$\dots\dots\dots\dots$$
$$\dots\dots\dots\dots$$
$$= 2^k\, T(k-k) \qquad \text{[By Mathematical induction]}$$
$$= 2^k\, \mathrm{T}(0) = 2^k.1 \qquad \text{[Using initial condition]}$$
$$= 2^k\ \forall\, k \geq 1$$

which is the required solution.

Aliter:

The same problem we can solve with the help of mathematical induction

$$T(k) = 2\, T(k-1).$$
$$k = 1,\ T(1) = 2.T(0) = 2.1 = 2$$
$$k = 2,\ T(2) = 2.T(1) = 2.2 = 2^2$$
$$k = 3,\ T(3) = 2.T(2) = 2.\ 2^2 = 2^3$$
$$k = 4,\ T(4) = 2^4$$

We can see that $T(k) = 2^k,\ k \geq 1.$

Example 1.50

Solve the recurrence relation $a_k = (1.08)\ a_{k-1}$ for $k \geq 1$.

Solution

Given $a_k = (1.08)\ a_{k-1}$

When

$$k = 1,\ a_1 = (1.08)\ a_0 = 1.08\ (100)$$
$$K = 2,\ a_2 = (1.08)\ a_1 = (1.08)^2\ (100)$$
$$K = 3,\ a_3 = (1.08)^3\ (100)$$

By mathematical induction, we have

$$a_k = (100)\ (1.08\ )^k$$

This is the required solution.

Aliter:

Given recurrence relation is

$$a_k = (1.08)\ a_{k-1} = (1.08)\ (1.08)\ a_{k-2}$$
$$= (1.08)^2\ a_{k-2} = (1.08)^3\ a_{k-3}$$
$$\dots\dots$$
$$= (1.08)^k\ a_0 = 100\ (1.08)^k$$

Example 1.51

Obtain recurrence relation for the solution $D(k) = k^2 - k$.

Solution

Given

$$D(k) = k^2 - k \qquad \dots(1)$$
$$D(k-1) = (k-1)^2 - (k-1)$$
$$= k^2 - 2k + 1 - k + 1$$

$$= k^2 - 3k + 2 \quad \ldots (2)$$

Similarly $$D(k-2) = (k-2)^2 - (k-2)$$

$$= k^2 - 5k + 6 \quad \ldots (3)$$

Multiplying (2) by 2 and subtracting from (1) and then adding (3), we get

$$D(k) - 2D(k-1) + D(k-2) = k^2 - k - 2k^2 + 6k - 4 + k^2 - 5k + 6 = 2$$

Putting $k = 0$, $k = 1$ in (1), we get

$$D(0) = 0, D(1) = 0.$$

∴ The recurrence relation is

$$D(k) - 2D(k-1) + D(k-2) = 2$$

with initial conditions $D(0) = 0$, $D(1) = 0$.

Example 1.52

If $S(k) = 3^{k-1} + 2^{k+1} + k$, $k \geq 0$. Find the recurrence relation.

Solution

Given $$S(k) = 3^{k-1} + 2^{k+1} + k \quad \ldots(1)$$

$$S(k-1) = 3^{k-2} + 2^k + (k-1) \quad \ldots(2)$$

Multiplying (2) by 3 and subtracting from (1), we get

$$S(k) - 3S(k-1) = 3^{k-1} - 3\cdot 3^{k-2} + 2^{k+1} - 3.\ 2^k + k - 3(k-1)$$

$$S(k) - 3S(k-1) = -2^k - 2k + 3 \quad \text{(First order recurrence relation)}$$

Similarly, we can get the second order recurrence relation

$$S(k) - 3\,S(k-1) + 6S(k-2) = 2k - 7$$

With initial conditions

$$C(6) = 7/3 \text{ and } C(1) = 5 \text{ for } k \geq 2.$$

Example 1.53

Find the recurrence relation of $a_n = A3^n + B(-4)^n$.

Solution

Given $$a_n = A\cdot 3^n + B(-4)^n \quad \ldots (1)$$

∴ $$a_{n+1} = A.3_{n+1} + B.(-4)^{n+1} \quad \ldots (2)$$

and $$a_{n+2} = A\cdot 3^{n+2} + B(-4)^{n+2} \quad \ldots (3)$$

Now adding (2) and (3) and subtracting (1) from it, we get

$$a_{n+2} + a_{n+1} - a_n = 9A.3^n + 3A.3^n - A.3^n + 16B.(-4)^n + 4B.(-4)^n - B.(-4)^n$$

$$a_{n+2} + a_{n+1} - 12a_n = 0$$

This is the required recurrence relation.

Exercise 1.5

Find the recurrence relations for the closed form expressions:

1. $B(k) = 5.2^k$
2. $C(k) = 2k^2 + 1$
3. $a_n = \frac{5}{2}.2^n + \frac{3}{2}n.2^n$
4. $a_n = \frac{2}{5}4^n + \frac{3}{5}(-1)^n$

5. $D(k) = 2.4^k - 5(-3)^k$ 6. $E(k) = (3 + k)(2)^k$

Answers

1. $B(k) - 2B(k-1) = 0$ 2. $C(k) - 2C(k-1) + C(k-2) = 4$
3. $a_n - 4a_{n-1} + 4a_{n-2} = 0$ for $n \geq 2$ $a_0 = 5/2$, $a_1 = 8$.
4. $a_n - 3a_{n-1} - 4a_{n-2} = 0$ for $n \geq 2$ and $a_0 = a_1 = 1$.
5. $D(k) - D(k-1) + 12D(k-2)$ for $k \geq 2$.
   $D(0) = -3$, $D(1) = 23$.
6. $E(k) - 4E(k-1) + 4E(k-2) = 0$ $k \geq 2$, $E(0) = 3$, $E(1) = 8$.

1.14 Homogeneous Recurrence Relation

Suppose n and k are nonnegative integers. A linear recurrence relation of the form

$$S(k) + C_1 S(k-1) + \dots + C_n S(k-n) = f(k)$$

or

$$S_k + C_1 S_{k-1} + \dots + C_n S_{k-n} = f(k)$$

said to be **homogeneous** if $f(k)$ is identically zero i.e.

$$S(k) + C_1 S(k-1) + \dots + C_n S(k-n) = 0$$

or

$$S_k + C_1 S_{k-1} + \dots + C_n S_{k-n} = 0$$

otherwise it is said to be **inhomogeneous.**

Example 1.54

i. $D(k) - 3D(k-1) + 2D(k-2) = 0$ is a second order homogeneous recurrence relation.

ii. $a_n - 2a_{n-1} = 0$ is a first order homogeneous recurrence relation with $a_0 = 5$.

1.14-1 Homogeneous Solutions

The total solution of a linear recurrence relation (or linear difference equation) with constant coefficient consist of two parts, the **homogeneous solution** and **particular solution.** The homogeneous solution satisfies the recurrence relation when R.H.S. is set to zero. We will discuss the particular solution later.

Let $a^{(h)} = (a_0, a_1, \dots, a_r, \dots)$ denote the homogeneous solution of

$$S_k + c_1 S_{k-1} + \dots + c_n S_{k-n} = 0 \quad \dots(1)$$

then

$$a_r + c_1 a_{r-1} + \dots + c_n a_{r-n} = 0 \quad \dots(2)$$

A homogeneous solution of a linear recurrence relation with constant coefficients is of the form $A\alpha^r$, where α is called a **characteristic root** and A is a constant to be determined by the initial boundary conditions. Putting $S_k = A\alpha^r$ in (1), we get

$$A\alpha^r + c_1 A\alpha^{r-1} + c_2 A\alpha^{r-2} + \dots + c_n A\alpha^{r-n} = 0$$

$$A\alpha^{r-n}\left[\alpha^n + c_1\alpha^{n-1} + c_2\alpha^{n-2} + \dots + c_n\right] = 0$$

$$\alpha^n + c_1\alpha^{n-1} + c_2\alpha^{n-2} + \dots + c_n = 0 \quad \dots(3)$$

(3) is called the **characteristic equation** of the recurrence relation (difference equation). If α_1 is one of the roots (characteristic roots) of the characteristic equation then $A\alpha_1^r$ will be a homogeneous solution to the recurrence relation.

Example 1.55

Find the characteristic equation of

$$a_n - 7a_{n-1} + 12a_{n-2} = 0,\ n \geq 2 \qquad \ldots(1)$$

Putting, $a_n = A \cdot \alpha^n$ in (1), we get

$$A(\alpha^n - 7\alpha^{n-1} + 12\alpha^{n-2}) = 0$$

or $$A\alpha^{n-2}[\alpha^2 - 7\alpha + 12] = 0$$

or $$\alpha^2 - 7\alpha + 12 = 0$$ is the characteristic equation.

Example 1.56

Characteristic equation of $S(k) + 2S(k-1) - 3S(k-2) - 6S(k-4) = 0$ is given by

$$\alpha^4 + 2\alpha^3 - 3\alpha^2 - 6 = 0.$$

1.14-2 Algorithm for solving nth order Homogeneous Linear Recurrence Relation

1. First write down the characteristic equation of the relation

$$S_k + c_1 S_{k-1} + c_2 S_{k-2} + \ldots\ldots + c_n S_{k-n} = 0$$

as $$\alpha^n + c_1\alpha^{n-1} + c_2\alpha^{n-2} + \ldots\ldots + c_n = 0.$$

2. Find all the characteristic roots of characteristic equations. Let it be $\alpha_1,\ \alpha_2, \ldots\ \alpha_n$.
3. If all the roots are distinct and real, i.e. $\alpha_1 \neq \alpha_2 \neq \alpha_3 \neq \ldots.. \neq \alpha_n$, then the general solution is

   $A_1\alpha_1^k + A_2\alpha_2^k + A_3\alpha_3^k + \ldots\ldots + A_n\alpha_n^k = 0$, where $A_1, A_2 \ldots\ldots, A_n$ are constants which are to be determined by initial conditions.
4. If there is a multiple roots, say $\alpha_1 = \alpha_2, \alpha_3, \alpha_4, \ldots, \alpha_n$,

   then $S_k = (A_1 + A_2 k)\alpha_1^k + A_3\alpha_3^k + \ldots\ldots + A_n a_n^k$ is a general solution.

   Here α_1 is a double root.
5. Find the values of the constants $A_1, A_2, \ldots\ldots, A_n$ by substituting it into initial conditions.
6. To get the required solution, substitute the values of $A_1, A_2, \ldots\ldots, A_n$ in the general solution.
7. When the characteristic equation has imaginary roots.

   Let $\alpha + i\beta$ is a root of the characteristic equation. Since imaginary roots always occur in pair, hence $\alpha - i\beta$ will also be the root of the characteristic equation.

   The solution corresponding to these roots will be given by

$$S_k = A_1(\alpha + i\beta)^k + A_2(\alpha - i\beta)^k \qquad \ldots(1)$$

Let $$\alpha + i\beta = r\cos\theta + ir\sin\theta \qquad \ldots(2)$$

Then $$\alpha - i\beta = r\cos\theta - ir\sin\theta \qquad \ldots(3)$$

Now putting (2) and (3) in (1) and using DeMoivre's theorem, we have

$$S_k = A_1(r\cos\theta + ir\sin\theta)^k + A_2(r\cos\theta - ir\sin\theta)^k$$

$$= r^k\,[A_1\cos k\theta + iA_1\sin k\theta + A_2\cos k\theta - iA_1\sin k\theta]$$

Let $B = A_1 + A_2$ and $C = A_1 - A_2$, then

$$S_k = r^k\,[B\cos k\theta + C\sin k\theta]$$

where $r = \sqrt{\alpha^2 + \beta^2}$ and $\phi = \tan^{-1}\dfrac{\beta}{\alpha}$

If the pair of complex roots $\alpha \pm i\beta$ occur twice, then the solution is given by

$$S_k = r^k\,[(B_1 + B_2 k)\cos k\theta + (C_1 + C_2 k)\sin k\theta]$$

REMARK

If the coefficients of characteristic equation are all integers and constant term equal to *m*, then the only possible rational characteristic roots are divisors of *m*, both positive and negative.

Example 1.57

Solve $a_n - 5a_{n-1} + 6a_{n-2} = 0$, where $a_0 = 2$, and $a_1 = 5$.

Solution

Given $$a_n - 5a_{n-1} + 6a_{n-2} = 0 \qquad \ldots(1)$$

The characteristic equation is

$$\alpha^2 - 5\alpha + 6 = 0$$

or $$(\alpha - 3)(\alpha - 2) = 0$$

or $\alpha = 3, 2$ are the roots, hence the general solution

$$a_n = B_1 3^n + B_2 2^n \qquad \ldots(2)$$

The given initial conditions are $a_0 = 2$ and $a_1 = 5$. Substituting these values in (2), we get

$$a_0 = B_1 3^0 + B_2 2^0$$

or $$2 = B_1 + B_2 \qquad \ldots(3)$$

and $$5 = 3B_1 + 2B_2 \qquad \ldots(4)$$

Solving (3) and (4) for B_1 and B_2, we get $B_1 = 1$ and $B_2 = 1$.

Therefore, the required solution is $a_n = 3^n + 2^n \;\; \forall\, n \geq 0$.

Example 1.58

Solve $T(k) - 7T(k-2) + 6T(k-3) = 0$, where $T(0) = 8$, $T(1) = 6$ and $T(2) = 22$.

Solution

Given relation is

$$T(k) - 7\,T(k-2) + 6T(k-3) = 0 \qquad \ldots(1)$$

The characteristic equation is

$$\alpha^3 - 7\alpha^2 + 6 = 0$$

The roots of the characteristic equation are $\alpha = 1, 2, -3$.

Therefore, the general solution is

$$A_1.1^k + A_2.2^k + A_3(-3)^k = 0 \qquad \dots (2)$$

The given initial conditions are

$$T(0) = 8,\ T(1) = 6 \text{ and } T(2) = 22$$

Putting the initial conditions in (2), we get

$$T(0) = A_1 + A_2 + A_3 = 8 \qquad \dots (3)$$

$$T(1) = A_1 + 2A_2 - 3A_3 = 6 \qquad \dots (4)$$

$$T(2) = A_1 + 4A_2 + 9A_3 = 22 \qquad \dots (5)$$

Solving (3), (4) and (5) for the values of A_1, A_2 and A_3, we get

$$A_1 = 5, A_2 = 2, A_3 = 1$$

Putting the values in (2), one can get the required solution

$$T(k) = 5 + 2^{k+1} + (-3)^k.$$

Example 1.59

Solve $$a_r + 6a_{r-1} + 12a_{r-2} + 8a_{r-3} = 0.$$

Solution

Given $$a_r + 6a_{r-1} + 12a_{r-2} + 8a_{r-3} = 0.$$

The characteristic equation is

$$\alpha^3 + 6\alpha^2 + 12\alpha + 8 = 0$$

The roots of the equation is $\alpha = -2, -2, -2$ (triple characteristic root)

Therefore, the general solution is

$$a_r = (A_1 + A_2 r + A_3 r^2)(-2)^r.$$

Example 1.60

Solve $$4t_r - 20t_{r-1} + 17t_{r-2} - 4t_{r-3} = 0.$$

Solution

The characteristic root of the given recurrence relation is

$$4\alpha^3 - 20\alpha^2 + 17\alpha - 4 = 0$$

The characteristic roots are 1/2, 1/2, 4.

The general solution is given by

$$t_r = (A_1 + A_2 r)\left(\frac{1}{2}\right)^r + A_3(4^r).$$

Example 1.61

Solve the recurrence relation $a_r + a_{r-1} + a_{r-2} = 0$.

Solution

Given $a_r + a_{r-1} + a_{r-2} = 0$

The characteristic equation is

$$\alpha^2 + \alpha + 1 = 0$$

or

$$\alpha = \frac{-1 \pm i\sqrt{3}}{2}$$

The solution is given by

$$a_r = A_1\left(\frac{-1 + i\sqrt{3}}{2}\right)^r + A_2\left(\frac{-1 - i\sqrt{3}}{2}\right)^r.$$

Example 1.62

Solve the recurrence relation $a_{r+4} + 8a_{r+2} + 16a_r = 0$.

Solution

Given $a_{r+4} + 8a_{r+2} + 16a_r = 0$.

The characteristic equation is

$$\alpha^4 + 8\alpha^2 + 16 = 0$$

or

$$(\alpha^2 + 4)^2 = 0$$

The roots of the equation is

$$\alpha = \pm 2i, \pm 2i$$

Now comparing the roots with polar form, we have

$$\alpha = 0 \pm 2i = r\cos\theta + ir\sin\theta$$

$$\Rightarrow r\cos\theta = 0 \text{ and } r\sin\theta = 2$$

Therefore $r = 2$ and $\theta = \pi/2$

Hence the complementary function is given by

$$a_r = 2^r\left[(B_1 + B_2 r)\cos\left(k\frac{\pi}{2}\right) + (C_1 + C_2 r)\sin\left(k\frac{\pi}{2}\right)\right]$$

Exercise 1.6

Solve the following recurrence relations:

1. $a_r - 3a_{r-1} - 4a_{r-2} = 0$ for $r \geq 2$; $a_0 = a_1 = 1$.
2. $a_r - 4a_{r-1} + 4a_{r-2} = 0$ for $r \geq 2$; $a_0 = 5/2, a_1 = 8$.
3. $a_r - 7a_{r-1} + 10a_{r-2} = 0$; $a_0 = 0, a_1 = 3$.
4. $a_r + 5a_{r-1} + 5a_{r-2} = 0$; $a_0 = 0, a_1 = 2\sqrt{5}$.
5. $a_r - 2a_{r-1} + 2a_{r-2} - a_{r-3} = 0$; $a_0 = 2, a_1 = 1$ and $a_2 = 1$.
6. $a_r - 20a_{r-1} + 100a_{r-2} = 0$; $a_0 = 2, a_1 = 30$.
7. $a_r - a_{r-1} - 6a_{r-2} = 0$; $a_0 = 12, a_1 = -1$.
8. $a_r + a_{r-1} - 5a_{r-2} + 3a_{r-3} = 0$; $a_0 = 0, a_1 = 1$ and $a_2 = 2$.
9. $a_r - 4a_{r-1} - 11a_{r-2} + 30a_{r-3} = 0$, $a_0 = 0, a_1 = -35, a_2 = -85$.

10. $a_r - 9a_{r-1} + 18a_{r-2} = 0$, $a_0 = 0, a_1 = 4$.

11. $a_r - 7a_{r-2} + 10a_{r-4} = 0$; $a_0 = 7, a_1 = 8\sqrt{2} + 5\sqrt{5}$, $a_2 = 41$, $a_3 = 16\sqrt{2} + 25\sqrt{5}$.

12. $a_r - 6a_{r-1} + 12a_{r-2} - 8a_{r-3} = 0$.

13. $a_r - 7a_{r-1} + 16a_{r-2} - 12a_{r-3} = 0$, $a_0=1, a_1=4$ and $a_2=8$.

14. $2a_r - 7a_{r-1} + 3a_{r-2} = 0$; $a_0 = 1$ and $a_1 = 1$.

15. $6a_r - 19a_{r-1} + 15a_{r-2} = 0$; $a_0 = 0$, $a_1 = -1/6$.

16. $a_n = 2a_{n-1} - a_{n-2}$; $a_1 = 1.5$ and $a_2 = 3$.

Answers

1. $a_r = \frac{2}{5}4^r + \frac{3}{5}(-1)^r$

2. $a_r = \frac{5}{2}2^n + \frac{3}{2}r.2^n$

3. $a_r = 5^r - 2^r$

4. $a_r = 2\left(\frac{-5+\sqrt{5}}{2}\right)^r - 2\left(\frac{-5-\sqrt{5}}{2}\right)^r$

6. $(2+r)10^r$

9. $a_r = 4(-3)^r + 2^r - 5^{r+1}$

10. $a_r = \left(\frac{2}{3}\right)3^r + \left(\frac{1}{3}\right)6^r$

11. $a_r = 3\left(\sqrt{2}\right)^r - 5\left(-\sqrt{2}\right)^r + 7\left(\sqrt{5}\right)^r + 2\left(-\sqrt{5}\right)^r$

12. $a_r = \left(A_1 + A_2 r + A_3 r^2\right)2^r$

13. $a_r = 5.2^r + 3.n2^r - 4.3^r$

1.14-3 Solution of Inhomogeneous Recurrence Relation

As discussed earlier the total solution of a difference equation consist of two parts i.e. homogeneous solution and particular solution. **Particular solution** is a solution which satisfies the difference equation with $f(k)$ on the R.H.S., i.e.

$$S_k + c_1 S_{k-1} + c_2 S_{k-2} + \dots\dots + c_n S_{k-n} = f(\text{k})$$

Let $a^{(h)} = \left(a_0^{(h)}, a_1^{(h)}, \dots\dots, a_r^{(h)}\right)$ denote the homogeneous solution

and $a^{(p)} = \left(a_0^{(p)}, a_1^{(p)}, \dots\dots, a_r^{(p)}\right)$ denote the particular solution, then

$$a_r^{(h)} + c_1 a_{r-1}^{(h)} + c_2 a_{r-2}^{(h)} + \dots\dots + c_n a_{r-n}^{(h)} = 0$$

and
$$a_r^{(p)} + c_1 a_{r-1}^{(p)} + c_2 a_{r-2}^{(p)} + \dots\dots + c_n a_{r-n}^{(p)} = f(k)$$

and complete solution will be

$$a_r^{(h)} + c_1 a_{r-1}^{(h)} + c_2 a_{r-2}^{(h)} + \dots\dots + c_n a_{r-n}^{(h)} + a_r^{(p)} + c_1 a_{r-1}^{(p)} + c_2 a_{r-2}^{(p)} + \dots\dots + c_n a_{r-n}^{(p)} = f(k)$$

i.e. $S = a^{(h)} + a^{(p)}$

Here one may recall that many discrete numeric function satisfy the given difference equation but only one of them satisfies the boundary conditions at the same time. Particular solution alone in general does not satisfy the boundary conditions but we can adjust the homogeneous solution so that the total solutions will satisfy the difference equation as well as boundary conditions.

There is no general method for determining the particular solution of a difference equation. In simple cases, however it can be determined by the method of inspection or by taking an educational guess. **Following is the algorithm for solving inhomogeneous finite order linear recurrence relation.**

Let the recurrence solution is $S_k + c_1 S_{k-1} + \ldots\ldots + c_n S_{k-n} = f(k)$

i. Find the homogeneous solution by putting $f(k) = 0$.
ii. To find the particular solution of the relation, take an educated guess in the form of a particular solution.
iii. Substitute your educated guess from step (ii) in the recurrence relation. If your guess is correct you will be able to determine the unknown coefficient of your guess. If your guess is wrong, it should be apparent from the result of this substitution, so go back to step (ii).
iv. The general solution of the recurrence relation is the sum of homogeneous and particular solutions.
v. The solution obtained in the step (iv) is over if no initial conditions are given. If n initial conditions are given, obtain n linear equations in n unknowns and solve the system, and get a complete solution.

TABLE 1.1: Form of particular solutions for given R.H.S. of the recurrence relation

S. No.	R.H.S. of Recurrence relation	Form of particular Solution
1	D, a constant	A, a constant
2	$D\,a^n$ and if a is not a root of characteristic polynomial	$A\,a^n$
3	$D\,a^n$ and if a is a root of multiplicity m	$A\,n^m\,a^n$
4	D_0+D_1k, a linear function	$A_0+A_1\,k$,a linear function
5	$D_0+D_1k+D_2k^2+\ldots\ldots+D_mk^m$an mth degree polynomial an mth degree polynomial	$A_0+A_1k+A_2k^2+\ldots.+A_mk^m$
6	$D\,n^s\,a^n$, and if a is not a root of characteristic polynomial	$\left(A_0+A_1n+\ldots..A_sn^s\right)a^n$
7	$D\,n^s\,a^n$, if a is a root of multiplicity m	$n^m\left(A_0+A_1n+\ldots..+A_sn^s\right)a^n$
	$D\,n^s$ and if 1 is not a root	$A_0+A_1n+\ldots..A_sn^s$
	$D\,n^s$ and if 1 is a root of multiplicity m	$n^m\left(A_0+\ldots\ldots.A_sn^s\right)$

Example 1.63

Solve $T(k) + 5T(k-1) = 9$, $T(0) = 6$.

Solution

Associated homogeneous relation is

$$T(k) + 5T(k-1) = 0$$

And its characteristic equation is $\alpha + 5 = 0$ or $\alpha = -5$.

$\therefore$ The homogeneous solution is

$$T^{(h)}(k) = A_1(-5)^k$$

To find the particular solution, we assume

$$T^{(p)}(k) = A \text{ as R.H.S.} = 9, \text{ a constant}$$

Substituting it into the given relation we get

$$A + 5A = 9 \Rightarrow A = 1\cdot 5 \Rightarrow T^{(p)}(k) = 1.5$$

∴ The general solution is

$$T(k) = T^{(h)}(k) + T^{(p)}(k)$$

or $$T(k) = A_1(-5)^k + 1.5 \quad ...(1)$$

Putting initial condition $T(0) = 6$ in (1), we get

$$T(0) = A_1 + 1.5 = 6$$

or $$A_1 = 4.5$$

∴ The required solution is

$$T(k) = 4.5\,(^-5)^k + 1.5.$$

Example 1.64

Solve $a_n - 6a_{n-1} + 8a_{n-2} = 3^n$, where $a_0 = 3$ and $a_1 = 7$.

Solution

The associated homogeneous relation of the given relation is

$$a_n - 6a_{n-1} + 8a_{n-2} = 0$$

and its characteristic equation is

$$\alpha^2 - 6\alpha + 8 = 0 \Rightarrow \alpha = 2, 4.$$

$$a_n^{(h)} = A_1(2)^n + A_2(4)^n$$

Now, to find the particular solution, as R.H.S. $= 3^n$, let us assume $a_n^{(p)} = A \cdot 3^n$.

Putting it into the given recurrence relation, we get

$$A \cdot 3^n - 6 \cdot 3^{n-1}A + 8 \cdot 3^{n-2}A = 3^n$$

or $$3^{n-2}(3^2 - (6)^3 + 8)A = 3^n$$

or $$[9 - 18 + 8]\,\mathrm{A} = 3^2$$

or $$-\mathrm{A} = 9 \text{ or } \mathrm{A} = -9$$

∴ $$a_n^{(p)} = (-9)3^n$$

The complete solution is

$$a_n = (-9)3^n + A_1(2)^n + A_2(4)^n \quad ...(1)$$

Given initial conditions are $a_0 = 3$ and $a_1 = 7$

Putting these in (1), we get

$$a_0 = (-9)\,3^0 + A_1(2)^0 + A_2(4)^0 = 3$$

or $$-9 + A_1 + A_2 = 3$$

or $$A_1 + A_2 = 12 \quad ...(2)$$

and $$a_1 = (-9)\,3^1 + A_1(2)^1 + A_2(4)^1 = 7$$

or $$-27 + 2A_1 + 4A_2 = 7$$

or $$2A_1 + 4A_2 = 34$$

or $$A_1 + 2A_2 = 17 \quad ...(3)$$

Solving for the values of A_1 and A_2 with the help of equation (2) and (3), we get

$$A_1 = 7, A_2 = 5$$

$\therefore$ The complete solution is $a_n = (-9)3^n + 5(4)^n + 7(2^n)$.

Example 1.65

Solve $S(k) - 7S(k-1) + 10S(k-2) = 6 + 8k$.

Solution

The associated homogeneous relation is

$$S(k) - 7S(k-1) + 10S(k-2) = 0$$

Its characteristic equation is $\alpha^2 - 7\alpha + 10 = 0 \Rightarrow \alpha = 2, 5$

$\therefore$ The homogeneous solution is $S^{(p)}(k) = B_1 2^k + B_2 5^k$

Now to find the particular solution, we assume

$$S^{(p)}(k) = A_0 + A_1 k$$

Substituting this in given recurrence solution, we get

$$(A_0 + A_1 k) - 7[A_0 + A_1(k-1)] + 10[A_0 + A_1(k-2)] = 6 + 8k$$

or $$4A_0 + 4A_1 k - 13A_1 = 6 + 8k$$

or $$(4A_0 - 13A_1) + 4A_1 k = 6 + 8k$$

Two polynomials are equal only if their coefficients are equal

$$4A_0 - 13A_1 = 6 \text{ and } 4A_1 = 8$$

$\Rightarrow$ $$A_1 = 2 \text{ and } A_0 = 8$$

$$S^{(p)}(k) = 8 + 2k$$

$\therefore$

$\therefore$ The complete solution is

$$S(k) = B_1 2^k + B_2 5^k + 8 + 2k \quad \text{... (1)}$$

Given initial conditions are $S(0) = 1$ and $S(1) = 2$,
Substituting these in (1), we get

$$S(0) = B1 + B2 + 8 = 1 \Rightarrow B1 + B2 = -7$$

and $$S(1) = 2\,B1 + 5\,B2 + 8 + 2 = 2$$

$\Rightarrow$ $$2B_1 + 5B_2 = -8$$

Solving these, we get

$$B_1 = -9 \text{ and } B_2 = 2$$

$\therefore$ The complete solution is

$$S(k) = -9.2^k + 2.5^k + 8 + 2k.$$

Example 1.66

Solve $a_n - 6a_{n-1} + 8a_{n-2} = n4^n$ with initial conditions $a_0 = 8$ and $a_1 = 22$.

Solution

Characteristic equation of the given equation is

$$\alpha^2 - 6\alpha + 8 = 0 \Rightarrow \alpha = 2, 4$$

$$\therefore \quad a_n^{(h)} = B_1(2)^n + B_2(4)^n$$

To find particular solution let us assume $a_n^{(p)} = n(A_0 + A_1 n)4^n$ as R.H.S. involves an exponential function (i.e. $n4^n$) with a base that equals characteristic equation [Refer Table 1.1, S.No.7].

Substituting $a_n^{(p)}$ in the given recurrence relations, we get

$$n(A_0 + A_1 n)4^n - 6(n-1)(A_0 + A_1(n-1))4^{n-1} + 8(n-2)(A_0 + A_1(n-2)4^{n-2} = n4^n$$

or $4^{n-2}[16n(A_0 + A_1 n) - 24(n,-1)((A_0 + A_1)n - 1) + 8(n-2)(A_0 + A_1)(n-2)] = n4^{n-2} \cdot 4^2$

or $16n(A_0 + A_1 n)4^n - 24(n-1)(A_0 + A_1(n-1)) + 8(n-2)(A_0 + A_1(n-2) = 16n$

Simplifying, we get $a_0=1$, $a_1=-7$,

$$\therefore \quad a_n^{(p)} = n(n-7)4^n$$

Hence, general solution is

$$a_n = B_1 2^n + B_2 4^n + n(n-7)4^n.$$

Example 1.67

Solve the difference equation $a_r + 5a_{r-1} + 6a_{r-2} = 3r^2 - 2r + 1$.

Solution

Given difference equation is

$$a_r + 5a_{r-1} + 6a_{r-2} = 3r^2 - 2r + 1 \quad \ldots (1)$$

The associated homogeneous difference equation is

$$a_r + 5a_{r-1} + 6a_{r-2} = 0$$

and its characteristic equation is

$$\alpha^2 + 5\alpha + 6 = 0 \Rightarrow \alpha = -2, -3.$$

Hence the homogeneous solution is

$$a_r^{(h)} = B_1(-2)^r + B_2(-3)^r$$

For finding particular solution, let us assume

$$a_r^{(p)} = A_1 r^2 + A_2 r + A_3$$

Substituting this in (1), we get

$$A_1 r^2 + A_2 r + A_3 + 5A_1(r-1)^2 + 5A_2(r-1) + 5A_3 + 6A_1(r-2)^2 + 6A_2(r-2) + 6A_3 = 3r^2 - 2r + 1$$

Simplifying, we get

$$12A_1 r^2 - (34A_1 - 12A_2)r + (29A_1 - 17A_2 + 12A_3) = 3r^2 - 2r + 1$$

Equating the coefficient of same degree variables, we get

$$12A_1 = 3 \Rightarrow A_1 = 4$$

$$34A_1 - 12A_2 = 2 \Rightarrow A_2 = \frac{13}{24}$$

$$29A_1 - 17A_2 + 12A_3 = 1 \Rightarrow A_3 = \frac{71}{288}$$

∴ The particular solution is

$$a_r^{(p)} = \frac{1}{4}r^2 + e\,\frac{13}{24}r + \frac{71}{288}$$

And the complete solution is

$$a_r = B_1(-2)^r + B_2(-3)^r + \frac{1}{4}r^2 + \frac{13}{24}r + \frac{71}{288}.$$

Example 1.68

Find the general solution of $S_r - 4S_{r-1} + 4S_{r-2} = 2^r + r2^r$.

Solution

Given difference equation is

$$S_r - 4S_{r-1} + 4S_{r-2} = 2^r + r2^r \quad ...(1)$$

The associated homogeneous difference equation is

$$S_r - 4S_{r-1} + 4S_{r-2} = 0$$

Its characteristic equation is

$$\alpha^2 - 4\alpha + 4 = 0 \Rightarrow (\alpha - 2)^2 = 0$$

or $\alpha = 2, 2.$

∴ The homogeneous solution is

$$S_r^{(h)} = (B_1 + B_2 r)2^r$$

Now to find the particulars solution let us assume

$$S_r^{(p)} = r^2(A_1 r + A_2)2^r \quad ...(2)$$

(Since 2 is a double characteristic root).

Substituting (2) in (1), we get

$$r^2(A_1 r + A_2)\,2^r - 4(r-1)^2\{A_1(r-1) + A_2\}\,2^{r-1} + 4(r-1)^2\{A_1(r-2) + A_2\}\,2^{r-2} = (r+1)\,2^r$$

After simplification, we obtain

$$6A_1 2^r r = 2^r r$$

and $(-6A_1 + 2A_2)\,2^r = 2^r$

which gives $A_1 = \frac{1}{6}$ and $A_2 = 1$

∴ The particular solution is

$$S_r^{(p)} = r^2\left(\frac{r}{6} + 1\right)2^r$$

The complete solution is given by

$$S_r = (B_1 + B_2 r)\,2^r + r^2\left(\frac{r}{6} + 1\right)2^r$$

Example 1.69

Find the general solution of $a_r - a_{r-1} - 12a_{r-2} = (-3)^r + 6 \cdot 4^r$.

Solution

The characteristic roots are –3 and 4, therefore homogeneous solution is

$$a_r^{(h)} = B_1(3)^r + B_2(4)^r$$

And the particular solution will be of the form

$$a_r^{(p)} = A_1 r(-3)^r + A_2 r 4^r$$

Try general solution yourself.

Example1.70

Solve the recurrence relation $a_{r+2} + a_r = r^2 + r + 1$

Solution

Given recurrence relation is

$$a_{r+2} + a_r = r^2 + r + 1 \quad \ldots (1)$$

The associated homogeneous recurrence relation is

$$a_{r+2} + a_r = 0$$

and its characteristic equation is given by

$$\alpha^2 + 1 = 0 \quad \Rightarrow \quad \alpha = \pm i$$

Now, comparing the values with the polar form, we get

$$0 \pm i = r\cos\theta \pm ir\sin\theta$$

$$\Rightarrow \quad r\cos\theta = 0, \quad r\sin\theta = 1$$

$$\Rightarrow \quad r = \sqrt{0+1} = 1, \theta = \tan^{-1}\frac{1}{0} = \frac{\pi}{2}$$

Hence the homogeneous solution is

$$a_r^{(h)} = 1^r\left[B_1\cos\left(r\frac{\pi}{2}\right) + B_2\cos\left(r\frac{\pi}{2}\right)\right]$$

For finding particular solution, let us assume

$$a_r^{(p)} = \mathrm{A}_1 r^2 + \mathrm{A}_2 r + \mathrm{A}_3$$

Substituting this in (1), we get

$$(\mathrm{A}_1(r+2)^2 + \mathrm{A}_2(r+2) + \mathrm{A}_3) + \mathrm{A}_1 r^2 + \mathrm{A}_2 r + \mathrm{A}_3 = r^2 + r + 1$$

or $$\mathrm{A}_1(r^2 + 4r + 4) + \mathrm{A}_2 r + 2\mathrm{A}_2 + \mathrm{A}_3 + \mathrm{A}_1 r^2 + \mathrm{A}_2 r + \mathrm{A}_3 = r^2 + r + 1$$

or $$2\mathrm{A}_1 r^2 + 4\mathrm{A}_1 r + 2\mathrm{A}_2 r + 4\mathrm{A}_1 + 2\mathrm{A}_2 + 2\mathrm{A}_3 = r^2 + r + 1$$

Equating the coefficients of same degree variables, we get

$$2A_1 = 0 \Rightarrow A_1 = 1/2$$

$$4A_1 + 2A_2 = 1 \Rightarrow A_2 = -12$$

$$4A_1 + 2A_2 + 2A_3 = 1 \Rightarrow A_3 = 0$$

Therefore the particular solution is given by

$$a_r^{(p)} = \frac{1}{2}r^2 - \frac{1}{2}r + 0 = \frac{1}{2}r(r-1)$$

And the complete solution is given by

$$a_r = B_1 \cos(r\frac{\pi}{2}) + B_2 \sin(r\frac{\pi}{2}) + \frac{1}{2}r(r-1).$$

Example1.71

Solve the difference equation $a_{r+1} - 3a_r = 2^r \cos(r\frac{\pi}{2})$.

Solution

Given difference equation is

$$a_{r+1} - 3a_r = 2^r \cos(r\frac{\pi}{2}) \quad \text{... (1)}$$

The associated homogeneous difference equation is

$$a_{r+1} - 3a_r = 0$$

and its characteristic equation is

$$\alpha - 3 = 0 \text{ or } \alpha = 3$$

hence the homogeneous solution is given by

$$a_r^{(h)} = B3^r$$

For finding particular solution, let us assume

$$a_r^{(h)} = 2^r\left[A_1 \cos\left(r\frac{\pi}{2}\right) + A_2 \sin\left(r\frac{\pi}{2}\right)\right]$$

Substituting this in (1), we get

$$2^{r+1}\left(A_1 \cos(r+1)\frac{\pi}{2} + A_2 \sin(r+1)\frac{\pi}{2}\right) - 3.2^r\left[A_1 \cos(r\frac{\pi}{2}) + A_2 \sin(r\frac{\pi}{2})\right] = 2^r \cos(r\frac{\pi}{2})$$

$$2^r\left(2A_1 \cos\left(r\frac{\pi}{2} + \frac{\pi}{2}\right) + 2A_2 \sin\left(r\frac{\pi}{2} + \frac{\pi}{2}\right)\right) - 3A_1 \cos(r\frac{\pi}{2}) - 3A_2 \sin(r\frac{\pi}{2}) = 2^r \cos(r\frac{\pi}{2})$$

or

$$2A_1 \sin(r\frac{\pi}{2}) + 2A_2 \cos(r\frac{\pi}{2}) - 3A_1 \cos(r\frac{\pi}{2}) - 3A_2 \sin(r\frac{\pi}{2}) = \cos(r\frac{\pi}{2})$$

or

$$(2A_2 - 3A_1)\cos(r\frac{\pi}{2}) - (2A_1 + 3A_2)\sin(r\frac{\pi}{2}) = \cos(r\frac{\pi}{2})$$

Equating the coefficients of $\cos(r\frac{\pi}{2})$ and $\sin(r\frac{\pi}{2})$ on both sides, we get

$$2A_2 - 3A_1 = 1 \qquad \text{... (2)}$$

$$-2A_1 - 3A_2 = 0 \qquad \text{... (3)}$$

From (2) $\qquad A_1 = -\frac{3}{2}A_2$

Substituting this in (3), we have

$$2A_2 - 3\left(-\frac{3}{2}A_2\right) = 1$$

or $\qquad 4A_2 + 9A_2 = 2 \Rightarrow A_2 = \frac{2}{3}$

Therefore $\qquad A_1 = -\frac{3}{2}\cdot\frac{2}{3} = -\frac{3}{13}$

Hence the particular solution is

$$a_r^{(p)} = 2^r\left(-\frac{3}{13}\cos\left(r\frac{\pi}{2}\right) + \frac{2}{13}\sin\left(r\frac{\pi}{2}\right)\right)$$

And the complete solution is given by

$$a_r = B3^r + 2^r\left(-\frac{3}{13}\cos\left(r\frac{\pi}{2}\right) + 2\sin\left(r\frac{\pi}{2}\right)\right).$$

1.14-4 Method of operator

Consider the difference equation

$$S_{k+n} + C_1S_{k+n-1} + C_2S_{k+n-2} + ... + C_nS_n = f(k)$$

which can be written in symbolic form as

$$\phi(E)S_k = f(k)$$

where $\qquad \phi(E) = E^n + C_1E^{n-1} + ... + C_n$

and E is the shifting operator, defined as

$$S_{k+n} = E^nS_k$$

Then the particular solution is given by

$$S^{(p)} = \frac{1}{\phi(E)}f(k)$$

Now, let us take the following particular cases:

Case (1): When right hand side of difference equation is a^k, i.e.

$$f(k) = a^k$$

Then particular solution

$$S^{(p)} = \frac{1}{\phi(E)}a^k$$

Putting $E = a$, we have

$$S^{(p)} = \frac{1}{\phi(a)} a^k \qquad \text{(provided } \phi(a) \neq 0)$$

If $\phi(a) = 0$, then

a. $(E - a)S_k = a^k$

$$S^{(p)} = \frac{1}{(E-a)} a^k = k\,a^{k-1}$$

b. $(E - a)^2 S_k = a^k$

$$S^{(p)} = \frac{1}{(E-a)^2} a^k = \frac{k(k-1)}{2!} a^{k-2}$$

c. $(E - a)^3 S_k = a^k$

$$S^{(p)} = \frac{1}{(E-a)^3} a^k = \frac{k(k-1)(k-2)}{3!} a^{k-3}$$

and in general

d. $(E - a)^k S_k = a^k$

$$S^{(p)} = \frac{1}{(E-a)^n} a^k = \frac{k(k-1)(k-2)...(k-\overline{n-1})}{n!} a^{k-n}$$

When $f(k) = \sin ck$, where c is a constant then particular integral

$$S^{(p)} = \frac{1}{\phi(E)} \sin ck$$

$$= \frac{1}{\phi(E)} \left[\frac{e^{ick} - e^{-ick}}{2i} \right]$$

$$= \frac{1}{2i} \left[\frac{1}{\phi(E)} e^{ick} - \frac{1}{\phi(E)} e^{-ick} \right]$$

Now apply $\frac{a^k}{\phi(E)}$

Case (2): When $f(k) = \cos ck$, where c is a constant.

Then particular integral is given by

$$S^{(p)} = \frac{1}{\phi(E)} . \cos ck = \frac{1}{\phi(E)} \left[\frac{e^{ick} + e^{-ick}}{2} \right]$$

$$= \frac{1}{2} \left[\frac{1}{\phi(E)} e^{ick} + \frac{1}{\phi(E)} e^{-ick} \right]$$

Case (3): When $f(k) = \psi(k)$.

Then particular integral is given by

$$S^{(p)} = \frac{1}{\phi(E)}.\Psi(k)$$

$$= \frac{1}{\phi(1+\Delta)}\Psi(k) \qquad \text{[since } E = 1 + \Delta\text{]}$$

Expanding by Binomial theorem, we have

$$S^{(p)} = \left(C_0 + C_1\Delta + \ldots + C_n\Delta^n\right)\Psi(k)$$

It is to be noted here that the difference of order higher than that of polynomial will always be zero, i.e. $\Delta^m\psi(k) = 0$, if $m >$ degree of $\psi(k)$.

Case (4): When $f(k) - C^k f(k)$, where $f(k)$ is a polynomial.

Then the particular integral is given by

$$S^{(p)} = \frac{1}{\phi(E)}.C^k f(k)$$

or

$$S^{(p)} = C^k \frac{1}{\phi(CE)}.f(k)$$

Now use case (3).

Example1.72

Solve the following difference equation $a_{k+2} - 7a_{k+1} + 6a_k = 2 \cdot 3^k$.

Solution

Given difference equation is

$$a_{k+2} - 7a_{k+1} + 6a_k = 2 \cdot 3^k$$

The associated homogeneous difference equation can be written as

$$E^2 a_k - 7Ea_k + 6a_k = 2 \cdot 3^k$$

or

$$(E^2 - 7E + 6)a_k = 2 \cdot 3^k$$

And its characteristic equation is given by

$$\alpha^2 - 7\alpha + 6 = 0$$

or

$$\alpha = 1, 6$$

Therefore, the homogeneous solution is

$$a_k^{(h)} = B_1 1^k + B_2 6^k$$

Now, particular solution is obtained by

$$a_k^{(p)} = \frac{2 \cdot 3^k}{E^2 - 7E + 6} = 2 \cdot \frac{3^k}{E^2 - 7E + 6}$$

$$= 2 \cdot \frac{3^k}{3^2 - 7 \cdot 3 + 6} = 2 \cdot \frac{3^k}{-6} = -\frac{1 \cdot 3^k}{3}$$

Hence the complete solution is

$$a_k = B_1 + B_2 6^k - \frac{1}{3} \cdot 3^k .$$

Example 1.73

Solve the recurrence relation $t_{r+2} + p^2 t_r = \sin pr$.

Solution

Associated homogeneous difference equation is

$$t_{r+2} + p^2 t_r = 0$$

which can be written with the help of shifting operator as

$$\left(E^2 + p^2\right) t_r = 0$$

The characteristic equation is given by

$$\alpha^2 + p^2 = 0 \text{ or } \alpha = 0 \pm ip$$

This can be written in polar form as

$$r\cos\theta \pm ir\sin\theta = 0 \pm ip$$

$$r\cos\theta = 0, \quad r\sin\theta = p$$

i.e.

$$\theta = \tan^{-1}\frac{p}{0} = \tan^{-1}\infty = \frac{\pi}{2}$$

$$r^2 = p^2 \Rightarrow r = p$$

Hence the complementary function (homogeneous solution) is given by

$$t^{(h)} = p^r \left[B_1 \cos r\frac{\pi}{2} + B_2 \sin r\frac{\pi}{2} \right]$$

Now to find the particular solution

$$t^{(p)} = \frac{\sin pr}{E^2 + p^2}$$

Use Demoiver's theorem to find the particular integral. As we know

$$e^{i\theta} = \cos\theta + i\sin\theta$$

Since here we have to find the particular integral corresponding to sine function, hence after writing with the help of exponential function, we find particular integral for imaginary part (I.P.).

Now, $$t_r^{(p)} = \text{I.P. of } \frac{e^{ipr}}{E^2 + p^2}$$

$$= \text{I.P. of } \frac{e^{ipr}}{(e^{ip})^2 + p^2}$$

$$= \text{I.P. of } \frac{e^{ipr}(e^{-2ip} + p^2)}{(e^{2ip} + p^2)(e^{-2ip} + p^2)}$$

$$= \text{I.P. of } \frac{e^{ip(r-2)} + p^2 e^{ipr}}{1 + p^2(e^{2ip} + e^{-2ip}) + p^4}$$

$$= \text{I.P. of} \frac{[\cos p(r-2) + i\sin p(r-2) + p^2(\cos pr + i\sin pr)]}{1 + 2p^2\cos 2p + p^4}$$

Hence $\quad t_r^{(p)} = \dfrac{\sin p(r-2) + p^2 \sin pr}{1 + 2p^2\cos 2p + p^4}$

and the complete solution is given by

$$t_r = p^r\left[B_1\cos(r\frac{\pi}{2}) + B_2\,\sin(r\frac{\pi}{2})\right] + \frac{p^2\sin pr + \sin p(r-2)}{1 + 2p^2\cos 2p + p^4}$$

Example1.74

Solve the difference equation $t_r + 4t_{r-1} + 4t_{r-2} = r^2 - 3r + 5$.

Solution

The associated homogeneous equation of the given equation is

$$t_r + 4t_{r-1} + 4t_{r-2} = 0$$

Which can be written with the help of shifting operator as

$$(1 + 4E^{-1} + 4E^{-2})t_r = 0$$

The corresponding characteristic equation is

$$1 + 4\alpha^{-1} + 4\alpha^{-2} = 0$$

or $\quad \alpha^2 + 4\alpha + 4 = 0$

or $\quad \alpha = -2, -2$

Hence the homogenous solution is

$$t_r^{(h)} = (B_1 + B_2 r)(-2)^r$$

Now to find the particular solution

$$t_r^{(p)} = \frac{1}{(E^2 + 4E + 4)}(r^2 - 3r + 5)$$

$$= \frac{1}{(E+2)^2}(r^2 - 3r + 5)$$

$$= \frac{1}{(1 + \Delta + 2)^2}(r^2 - 3r + 5) \qquad [\text{Since } E = 1 + \Delta]$$

$$= \frac{1}{(3 + \Delta)^2}(r^2 - 3r + 5)$$

$$= \frac{1}{9\left(1 + \frac{\Delta}{3}\right)^2}(r^2 - 2r + 5)$$

Let us express r in factorial notation $r^{(n)} = r(r-1)(r-2)...(r - \overline{n-1})$ and $r^1 = r$.

$$t_r^{(p)} = \frac{1}{9}\left(1+\frac{\Delta}{3}\right)^{-2}(r^2-2r+5)$$

$$= \frac{1}{9}\left(1-\frac{2\Delta}{3}+\frac{3\Delta^2}{9}+...\right)(r^2-2r+5)$$

$$= \frac{1}{9}\left[r^2-2r+5-\frac{2\Delta}{3}(r^2-2r+5)+\frac{3\Delta^2}{9}(r^2-2r+5)+...\right]$$

$$= \frac{1}{9}\left[r^2-2r+5-\frac{2}{3}(2r-2)+\frac{3}{9}(2)\right] \quad [\because \Delta r^2 = 2r, \Delta^2 r^2 = 2]$$

$$= \frac{1}{9}\left[r^2-2r-\frac{4r}{3}+5+\frac{4}{3}+\frac{2}{3}\right]$$

$$= \frac{1}{9}\left[r^2-\frac{10r}{3}+\frac{21}{3}\right]$$

$$= \frac{1}{9}\left[r^2-\frac{10r}{3}+7\right]$$

Hence the complete solution is

$$t_r = (B_1+B_2r)(-2)^r + \frac{1}{9}\left[r^2-\frac{10r}{3}+7\right].$$

Example 1.75

Solve the recurrence relation $t_{r+2} + 4t_{r+1} - 12t_r = 2^r(r-1)$.

Solution

The given equation with the help of characteristic equation can be written as

$$(E^2+4E-12)t_r = 2^r(r-1)$$

The characteristic equation is given by

$$\alpha^2 + 4\alpha - 12 = 0$$

$$\Rightarrow \alpha = 2, \alpha = -6$$

Hence the homogenous solution is

$$t_r^{(h)} = B_1 2^r + B_2(-6)^r$$

Now, particular solution is obtained by

$$t_r^{(p)} = \frac{2^r(r-1)}{E^2+4E-12} = \frac{2^r(r-1)}{(E-2)(E+6)}$$

$$= \frac{1}{(E-2)}\left[\frac{1}{(E+6)}2^r\right](r-1)$$

$$= \frac{1}{(E-2)}\left[\frac{2^r}{(2+6)}\right](r-1)$$

$$= \frac{1}{8} \cdot \frac{2^r \cdot (r-1)}{(E-2)}$$

$$= \frac{1}{8} \cdot \frac{2^r}{(E-2)}(r-1)$$

$$= \frac{1.2^r}{16} \cdot \frac{1}{(E-1)}(r-1)$$

$$= \frac{2^r}{16} \cdot \frac{1}{\Delta}(r-1) \qquad \text{[since } E = 1 + \Delta\text{]}$$

$$= \frac{2^r}{16} \cdot \frac{1}{\Delta}(r-1)$$

$$= \frac{2^r}{16} \cdot \Delta^{-1}(r^{(1)} - r^{(0)}) \qquad \text{[since } r^{(0)} = 1\text{]}$$

$$= \frac{2^r}{16}\left[r^{(2)} - r^{(1)}\right]$$

$$= \frac{2^r}{16}\left[\frac{r(r-1)}{2} - r\right] \qquad \text{[Operator } \Delta^{-1} \text{ is analogous with integral]}$$

$$= \frac{2^r}{16}\frac{r}{2}(r-1-2)$$

$$= \frac{2^r}{32}r(r-3)$$

Hence the complete solution is given by

$$t_r = B_1 2^r + B_2(-6)^r + \frac{2^r}{32}r(r-3)$$

Exercise 1.7

Find the solution of following recurrence relations:

1. $a_r - 2a_{r-1} + a_{r-2} = 2$, $a_0 = 25$ and $a_1 = 16$.
2. $a_r + 6a_{r-1} + 9a_{r-2} = 3$, $a_0 = 0$ and $a_1 = 1$.
3. $a_r - 5a_{r-1} + 6a_{r-2} = 2$, $a_0 = 1$ and $a_1 = -1$.
4. $a_r \, . \, a_{r-1} = 7$.
5. $a_r - a_{r-1} - 6a_{r-2} = 30$, $a_0 = 20$ and $a_1 = -5$.
6. $a_r - 5a_{r-1} + 6a_{r-2} = 1$.
7. $a_r - 4a_{r-1} + 4a_{r-2} = 2^r$.
8. $a_r - 5a_{r-1} = 5^r$, $a_0 = 3$.
9. $a_r - 6a_{r-1} + 8a_{r-2} = 4^r$ $a_0 = 8$ and $a_1 = 22$.
10. $a_r - 9a_{r-1} + 20a_{r-2} = 2.5^r$.
11. $a_r - 7a_{r-1} + 10a_{r-2} = 3^r$, $a_0 = 1$ and $a_1 = 1$.
12. $a_r - 10a_{r-1} + 25a_{r-2} = 2^r$, $a_0 = 2/3$ and $a_1 = 1$.
13. $a_r + 2a_{r-1} = r + 3$ for $r \geq 1$ and $a_0 = 3$.
14. $a_r - 4a_{r-1} + 4a_{r-2} = 3r + 2^r$, $a_0 = 1$ and $a_1 = 1$.
15. $a_r = 1.08a_{r-1} + 2^r$, $a_0 = 1$.
16. $a_r - 2a_{r-1} = 2^r$.
17. $a_r - 7a_{r-1} + 12a_{r-2} = r\,4^r$.
18. $a_r - 7a_{r-1} + 12a_{r-2} = 3^r$.
19. $a_r - 2a_{r-1} + a_{r-2} = 4$.
20. $a_r + 3a_{r-1} - 10a_{r-2} = r^2 + r + 1$.
21. $a_r - 3a_{r-1} + 2a_{r-2} = 5r + 3$.

22. Given $a_0 = 0$, $a_1 = 1$, $a_2 = 4$ and $a_3 = 12$ which satisfy the recurrence relation

$$a_r + c_1 a_{r-1} + c_2 a_{r-2} = 0 \text{, determine } a_r.$$

23. Solve $a_r - a_{r-1} = 7r$. Find also the solution when $f(r) = 7r^2$.

24. Solve the recurrence relation $a_r + 3a_{r-1} + 2a_{r-2} = f(r)$, given

$$f(r) = \begin{cases} 1, r=2 \\ 0, \text{otehrwise with boundary condition } a_0 = a_1 = 0 \end{cases}$$

25. $a_{r+2} - 2a_{r+1} + a_r = r^2 2^r$.

26. $a_{r+1} - 3a_r = 2^r \cos(\pi / 2)r$.

27. $a_{r+2} - 7a_{r+1} + 12a_r = \cos r$.

28. $a_{r+2} - 6a_{r+1} + 8a_r = r^2 + 3^r$.

29. $a_{r+2} + a_r = \sin\frac{\pi}{2} r$.

30. $a_{r+2} + k^2 a_r = \cos kr$.

31. $a_{r+2} - 3a_{r+1} + 2a_r = \cos 2r$.

32. $a_{r+2} + a_{r+1} + a_r = r^2 + r + 1$.

33. $a_{r+3} + 2a_{r+2} + a_{r+1} = 2^r + r^2 + r$.

34. $t_{r+1} - at_r = a^2(2r+1)$ with $t_0 = 1/a$.

Answers

1. $a_r = r^2 - 10r + 25$

3. $a_r = -2 \cdot 3^r + 2^{r+1} + 1$

4. $a_r = B \cdot (1)^r + 7r$ (since 1 is a characteristic root $a_r^{(p)} = pr$)

5. $a_r = 11(-2)^r + 4 \cdot 3^r + 5$

6. $a_r = B_1 2^2 + B_2 3^2 + \frac{1}{2}$

7. $a_r = \frac{r^2}{2} \cdot 2^r + A_1 2^r + A_2 r 2^n$

8. $a_r = (3+r)5^r$

9. $a_r = r(r-1)4^r + 3(4^r) + 5.2^r$

12. $a_r = \frac{2}{9} \cdot 5^r + \frac{1}{5} r 5^r + \frac{2^{r+2}}{9}$

13. $a_r = \frac{16}{9} \cdot (-2)^r + \frac{r}{3} + \frac{11}{9}$

14. $a_r = 12 + 3r + (r^2 + 7r + 22).2^{r-1}$

15. $a_r = (-1.174)(1.08)^r + (2.174).2^r$

16. $a_r = B2^r + Ar2^r$ (since 2 is a charateristic root)

17. $a_r = B_1 4^r + B_2 3^r + r(A_1 + A_2 r)4^r$

18. $a_r = B_1 4^r + B_2 3^r + Ar3^r$ (since 3 is a characteistic root)

19. $a_r = (B_1 + B_2 r).1^r + Ar^2$ (since 1 is a charateristic root (find the value of A))

25. $a_r = B_1 + B_2 r + 2^r(r^2 - 8r + 20)$

26. $a_r = B\,3^r + \frac{2^r}{13}\left(-3\cos r\frac{\pi}{2} + 2\sin r\frac{\pi}{2}\right)$

27. $a_r = B_1 3^r + B_2 4^r + \frac{\cos(r-2) - 7\cos(r-1) + 12\cos r}{194 + 12\cos 2 - 91\cos 1}$

28. $a_r = B_1 2^r + B_2 4^r + \frac{1}{3}\left(r^2 + \frac{8}{3}r + \frac{38}{9}\right) - 3^r$

29. $a_r = B_1 \cos r\frac{\pi}{2} + B_2 \sin r\frac{\pi}{2} - \frac{1}{2}\sin r\frac{\pi}{2}$

30. $a_r = k^r\left(B_1 \cos r\frac{\pi}{2} + B_2 \sin r\frac{\pi}{2}\right) + \frac{k^2 \cos k' \cos(k(r-2))}{1 + 2k^2\cos 2k + k^4}$

31. $a_r = B_1 + B_2 2^r + \frac{\cos\{2(r-2)\} - 3\cos\{2(r-1)\} + 2\cos 2r}{14 - 18\cos 2k + 4\cos 4}$

32. $a_r = B_1 \cos\frac{2\pi r}{3} + B_2 \sin\frac{2\pi r}{3} + \frac{r^2}{3} - \frac{r}{3} + \frac{1}{9}$

33. $a_r = (-B_1 + B_2 - B_3 r)(-1)^r + \frac{2^r}{18} + \frac{1}{9}\left(r^2 - 3r - \frac{1}{2}\right)$

34. $a_r = (1/a)\cdot a^r + r^2 a^{r-1}$.

1.15 Generating Functions

As we know for a Computer science and Engineering student notion of alternative representation is very important. As decimal numbers can alternatively be represented with the help of binary numbers. And if we want to add, subtract, multiply or divide decimal numbers, we represent them as binary numbers, use a computer to carry out the arithmetic operations on binary numbers, and then obtain results of our computation by converting the results in binary numbers into decimal numbers. It can be noted here that suitably chosen alternatively form leads to the easiness and efficiency of operations.

We introduce here an alternative way to represent a sequence of numbers called **Generating function**, and it is one of very useful tools for solving the recurrence relation.

For a sequence S with terms $\langle S_0, S_1, S_2 \ldots\ldots, S_n\rangle$ we define an infinite series

$$G(S, z) = S_0 + S_1 z + S_2 z^2 + S_3 z^3 + \ldots\ldots$$

or

$$G(S, z) = \sum_{n=0}^{\infty} S_n z^n \text{ , where z is a variable.}$$

Example 1.76

Generating function of the sequence $\langle 1, 3, 5, 7\ldots\rangle$ is $G(S,z) = 1 + 3z + 5z^2 + 7z^3 + \ldots\ldots.$

Example 1.77

Generating function of the sequence $\langle 2^0, 2^1, 2^2, 2^3, 2^4, \ldots\rangle$ is

$$G(S,z) = 1 + 2z + 2^2 z^2 + 2^3 z^3 + ...$$
$$= 1 + 2z + (2z)^2 + (2z)^3 + ...$$
$$= \sum_{n=0}^{\infty} 2^n z^n$$

1.15-1 Closed form Expression for Generating Functions

As we know closed form expression is a compact way to represent a sequence, we see that from the closed form expression of a generating function a sequence can be manipulated conveniently.

Example 1.78

Find the closed form expression of the generating function for the sequence $\langle 3^0, 3^1, 3^2, 3^3, 3^4, 3^5, \ldots \rangle$.

Solution

The generating function is given by

$$G(S, z) = 1 + 3z + 3^2 z^2 + 3^3 z^3 + ...$$
$$= 1 + 3z + (3z)^2 + (3z)^3 + ...$$
$$= (1 - 3z)^{-1} = \frac{1}{(1 - 3z)}$$

is the required closed form expression.

Example 1.79

Obtain partial fraction composition and identify the sequence having the expression $\frac{5 + 2z}{1 - 4z^2}$ as a generating function.

Solution

Given $$G(S,z) = \frac{5 + 2z}{1 - 4z^2}$$
$$= \frac{5 + 2z}{(1 + 2z)(1 + 2z)} = \frac{2}{1 + 2z} + \frac{3}{1 - 2z}$$
$$= 2(1 + 2z)^{-1} + 3(1 - 2z)^{-1}$$

Expanding with the help of binomial theorem, we get

$$= 2\left(1 - 2z + (2z)^2 - (2z)^3 + ...\right) + 3\left(1 + 2z + (2z)^2 - (2z)^3 + ...\right)$$
$$= 2\sum_{n=0}^{\infty}(-1)^n 2^n z^n + 3\sum_{n=0}^{\infty} 2^n Z^n$$

∴ The required sequence is

$$S(r) = 2(-2)^r + 3.2^r.$$

Example 1.80

Find the closed form expression for the Fibonacci sequence defined by $F_r = F_{r-1} + F_{r-2}$, $r \geq 2$, $F_0 = 0$ and $F_1 = 1$.

Solution

Given Fibonacci sequence is

$$F_r = F_{r-1} + F_{r-2}\ ,\ r \geq 2 \text{ and },\ F_0 = 0 \text{ and } F_1 = 1 \qquad \ldots (1)$$

And the generating function is given by

$$F(S,z) = F_0 + F_1 z + F_2 z^2 + F_3 z^3 + \ldots..$$

$$= \sum_{n=0}^{\infty} F_n z^n \qquad \ldots (2)$$

Multiplying both sides of (1) by z^r and sum over $r \geq 2$, we get

$$\sum_{r\geq 2} F_r z^r = \sum_{r\geq 2} F_{r-1} z^r + \sum_{r\geq 2} F_{r-2} z^r$$

$$F(S,z) - F_0 - F_1 = z \sum_{r\geq 2} F_{r-1} z^{r-1} + z^2 \sum_{r\geq 2} F_{r-2} z^{r-2}$$

$$= z\,(F(S,z) - F_0) + z^2\,(F(S,z)) \qquad \text{[by (2)]}$$

$$\therefore \quad F(S,z) - z\,F(S,z) - z^2 F(S,z) = F_0 + F_1 z - z\,F_0$$

Using initial condition, $F_0 = 0$ and $F_1 = 1$, we get

$$\left(1 - z - z^2\right) F(S,z) = z$$

$$\therefore \quad F(S,z) = \frac{z}{1 - z - z^2}$$

This is the closed form expression for the Fibonacci sequence as a rational function.

1. *Finite Geometric Series*:

$$a + ar^2 + \ldots.. + ar^n = a\frac{(1 - r^{n-1})}{(1 - r)}.$$

2. *Infinite Geometric Series*

$$a + ar + ar^2 + \ldots\ldots = \frac{a}{1 - r}.$$

3. *Arithmetic Series*

$$a + (a + d) + (a + 2d) + \ldots + (a + (n-1)d) = \frac{n}{2}\left\{2a + \overline{(n-1)d}\right\}.$$

Closed form expressions of the generating function for some sequences

	Sequence	**Generating function**
1.	1	$\frac{1}{1 - z}$
2.	$(-1)^n$	$\frac{1}{1 + z}$
3.	a^n	$\frac{1}{1 - az}$
4.	$(-a)^n$	$\frac{1}{1 + az}$

5.	$C(k, n)$	$(1+z)^n$
6	$(C(k-1+n, n))\, a^n$	$\dfrac{1}{(1-z)^k}$
7.	n	$\dfrac{z}{(1-z)^2}$
8.	$n+1$	$\dfrac{1}{(1-z)^2}$
9.	n^2	$\dfrac{z(1+z)}{(1-z)^3}$
10.	n^3	$\dfrac{z(1+4z+z^2)}{(1-z)^4}$
11.	na^n	$\dfrac{az}{(1-az)^2}$
12.	$(n+1)a^n$	$\dfrac{1}{(1-az)^2}$
13.	ba^n	$\dfrac{b}{(1-az)}$
14.	n^2a^n	$\dfrac{az(1+az)}{(1-az)^3}$
15.	$(n+1)n$	$\dfrac{2z}{(1-z)^3}$
16.	$\dfrac{1}{n!}$	e^z
17.	bna^n	$\dfrac{abz}{(1-az)^2}$

Exercise 1.8

Find the closed form expression for the generating function of the sequence given below

1. $T(k) = 4, k \geq 0.$
2. $T(n) = S(n) + P(n) + U(n)$
   $= 9 \cdot 5^n + 4 + 3(-1)^n.$
   [**Hint:** $G(T, z) = G(S, z) + G(P, z) + G(U, z)$].
3. $a_n = 3^n.$
4. $a_n = 3^n + 5^n.$
5. $a_n = 7(n+3)(n+2)(n+1).$
6. $a_n = n^2 3^n.$
7. $a_n = 9.5^n,\ n \geq 0.$
8. $a_n - 6a_{n-1} + 5a_{n-2} = 0, a_0 = 2, a_1 = 2.$

9. Obtain fractional decomposition and identify the sequence having the expression as a generating function:

i. $\dfrac{32 - 22z}{2 \quad 3z \mid z^2}$ ii. $\dfrac{1}{5 \quad 6z \mid z^2}$

iii. $\dfrac{6 - 29z}{1 - 11z + 30z^2}$ iv. $\dfrac{7z^2}{(1 - 2z)(1 + 3z)}$

v. $\dfrac{(1 + z)^2}{(1 - z)^4}$ vi. $\dfrac{3 - 5z}{1 - 2z - 3z^2}$

vii. $\dfrac{z^5}{5 - 6z + z^2}$ viii. $(1 + z)^n + (1 + z)^n$.

Answers

1. $G(T, z) = \dfrac{4}{1 - z}$
2. $\dfrac{16 - 34z - 11z^2}{(1 - 5z - z^2 + 5z^3)}$
3. $\dfrac{1}{1 - 3z}$
7. $\dfrac{9}{1 - 5z}$
8. $a_n = 2, n \geq 0;\ G(a, z) = \dfrac{2}{1 - z}$
9. i. $G(S, z) = \dfrac{10}{1 - z} + \dfrac{12}{2 - z},\ S(k) = 10 + 6\left(\dfrac{1}{2}\right)^k$

   iii. $G(S, z) = \dfrac{-1}{1 - 5z} + \dfrac{7}{1 - 6z},\ S(k) = 7.6^k - 5^k$

   vi. $G(S, z) = \dfrac{1}{1 - 3z} + \dfrac{2}{(1 + z)},\ S(k) = 3^k + 2.(-1)^k$.

1.15-2 Properties of Generating Function

Let S and T be two sequences and their generating functions are $G(S, z)$ and $G(T, z)$ respectively, then $G(S, z) = \sum_{n=0}^{\infty} S_n z^n$ and $G(S, z) = \sum_{n=0}^{\infty} T_n z^n$

1. Uniqueness $G(S, z) = G(T, z)$ if and only if $S_n = T_n \forall n$.
2. Sum of two generating functions is again a generating function.

$$G(S, z) + B(S, z) = \sum_{n=0}^{\infty} S_n z^n + \sum_{n=0}^{\infty} T_n z^n$$

$$= \sum_{n=0}^{\infty} (S_n + T_n) z^n$$

$$= \sum_{n=0}^{\infty} C_n z^n = G(C, z)$$

(Here C_n is sum of series S_n and T_n)

3. Let C be a number, then scalar product $C.G(S, z)$ is again a generating function.

$$C.G(S,z) = C.\sum_{n=0}^{\infty} S_n z^n = \sum_{n=0}^{\infty} (C.S_n) z^n$$

4. The product of two generating functions is again a generating function.
Let $G(S, z)$ and $G(T, z)$ be two sequences, then

$$G(S,z) \cdot G(T,z) = \sum_{n=0}^{\infty} S_n z^n \sum_{n=0}^{\infty} T_n z^n$$

$$= \sum_{n=0}^{\infty} C_n z^n \text{, where } C_n = \sum_{n=0}^{\infty} S_i T_{n-i}$$

5. The monomial product $z^n G(S,z)$, $n \geq 1$ is again a generating function.

$$z^n G(S,z) = z^n \sum_{m=0}^{\infty} S_m z^m$$

$$= \sum_{m=0}^{\infty} S_m z^{m+n}$$

$$= \sum_{m=n}^{\infty} S_{m-n} z^m$$

6. Differentiation
Let $G(S, z)$ be an infinitely differentiable function, then

$$G(S, z) = S_o + S_1 z + S_2 z^2 + \ldots\ldots = \sum_{n=0}^{\infty} S_n z^n$$

Its first differential coefficient is given by

$$G'(S, z) = S_1 + 2S_2 z + 3S_3 z^2 + \ldots\ldots = \sum_{n=1}^{\infty} n S_n z^{n-1}$$

Similarly, second differential coefficient is given by

$$G''(S, z) = 2S_2 + 6S_3 z + \ldots\ldots$$

$$= \sum_{n=2}^{\infty} n(n-1) S_n z^{n-2}$$

By mathematical induction,

$$G^m(S, z) = \sum_{n=1}^{\infty} n(n-1)(n-2)\ldots(n-m+1) S_n z^{n-m}.$$

1.15-3 Solution of a Recurrence Relation using Generating Function

Outline of the method of generating functions.

To solve the recurrence relation with constant coefficients of degree n;

$$S_k + C_1 S_{k-1} + \ldots\ldots\ldots + C_n S_{k-n} = 0,$$

where $C_1, C_2, C_3, \ldots\ldots, C_n$ are constants $C_n \neq 0$ and $k \geq n$.

Let $$S(z) = \sum_{k=0}^{\infty} a_k z^k,$$

i. Multiply each term of the recurrence relation by z^n, sum from n to ∞, and replace all infinite sums by equivalent expressions, then transform the recurrence relation into an algebraic equation.

$$S(z) = \frac{P(z)}{Q(z)},$$

where $P(z) = S_0 + (S_1 + C_1S_0)z + (S_2 + C_1S_1 + C_2S_0)z^2 + \ldots + (S_{n-1} + C_1S_{n-2} + \ldots + C_{n-1}S_0)z^{n-1}$

and $Q(z) = 1 + C_1z + C_2z^2 + \ldots + C_nz^n$

ii. After getting the values of $P(z)$ and $Q(z)$, transform $S(z)$ back to obtain the coefficients of Sn. This can be done in following two ways:

a. Factorize $Q(z)$ and then use partial fractions to get $S(z)$ as a sum of familiar series and hence that S_k is the sum of the coefficients of known series.

b. If the initial conditions are given then, we can find as many coefficients of $S(z)$ as we desire even if we cannot factorize $Q(z)$ by long division of $Q(z)$ into $P(z)$ [or by finding the multiplicative inverse of $Q(z)$ since the constant term of $Q(z)$ is non zero].

Example 1.81

Solve, $a_n - 2a_{n-1} - 3a_{n-2} = 0,\ n \geq 2$ with $a_0 = 3$, and $a_1 = 1$.

Solution

Given recurrence relation is

$$a_n - 2a_{n-1} - 3a_{n-2} = 0 \qquad \ldots (1)$$

The generating function is given by

$$G(a, z) = \sum_{n=0}^{\infty} a_n z^n = a_0 + a_1z + a_2z^2 + \ldots . \qquad \ldots (2)$$

Multiplying (1) by z^n and sum over all $n \geq 2$, we get

$$\sum_{n\geq 2} a_nz^n - 2\sum_{n\geq 2} a_{n-1}z^n - 3\sum_{n\geq 2} a_{n-2}z^n = 0$$

$$(G(a, z))a_0 - a_1z) - 2z\sum_{n\geq 2} a_{n-1}z^{n-1} - 3z^2\sum_{n\geq 2} a_{n-2}z^{n-2} = 0 \qquad \text{[Using (2)]}$$

Now using initial conditions $a_0 = 3$ and $a_1 = 1$, we get

$$\{G(a,z) - 3 - z\} - 2z\{G(a,z) - a_0\} - 3z^2G(a,z) = 0$$

or

$$G(a,z) - 3 - z - 2z\{G(a,z) - 3\} - 3z^2G(a,z) = 0$$

or

$$(1 - 2z - 3z^2)G(a,z) = 3 - 5z$$

$\therefore$

$$G(a,z) = \frac{3 - 5z}{1 - 2z - 3z^2}$$

$$= \frac{3 - 5z}{(1 - 3z)(1 + z)}$$

$$= \frac{A}{1 - 3z} + \frac{B}{1 + z}$$

∴ The required solution is

$$a_n = 1.3^n + 2(-1)^n = 3^n + 2(-1)^n.$$

Example 1.82

Solve the following recurrence relation $a_n - 9a_{n-1} + 26a_{n-2} - 24a_{n-3} = 0$ for $n \geq 3$ with $a_0 = 0$, $a_1 = 1$ and $a_2 = 10$.

Solution

Given recurrence relation is

$$a_n - 9a_{n-1} + 26a_{n-2} - 24a_{n-3} = 0 \qquad \ldots(1)$$

The generating function is given by

$$G(a, z) = \sum_{n=0}^{\infty} a_n z^n = a_0 + a_1 z + a_2 z^2 + \ldots. \qquad \ldots(2)$$

Multiplying throughout (1) by z^n and sum from 3 to∞. (Since $n \geq 3$), we get

$$\sum_{n=3}^{\infty} a_n z^n - 9\sum_{n=3}^{\infty} a_{n-1} z^n + 26\sum_{n=3}^{\infty} a_{n-2} z^n - 24\sum_{n=3}^{\infty} a_{n-3} z^n = 0$$

Replacing the infinite sum by equivalent expressions, we get

$$-(G(a, z) = a_0 - a_1 z - a_2 z^2) - 9z(G(a,z)) - a_0 - a_1 z) + 26z^2(G(a,z)) - a_0) \; 24z^3 G(a,z)(a,z) = 0.$$

Simplifying, we get

$$G(a,z)(1 - 9z + 26)z^2 - 24z^3) = a_0 + a_1 z + a_2 z^2 - 9a_0 z - 9a_1 z^2 + 26a_0 z^2$$

or

$$G(a,z) = \frac{a_0 + (a_1 - 9a_0)z + (a_2 - 9a_1 + 26a_0)z^2}{1 - 9z + 26z^2 - 24z^3}$$

Now, using initial conditions $a_0 = 0, a_1 = 1$ and $a_2 = 10$, we get

$$G(a,z) = \frac{z + z^2}{(1-2z)(1-3z)(1-4z)}$$

$$= \frac{A_1}{(1-2z)} + \frac{A_2}{(1-3z)} + \frac{A_3}{(1-4z)}$$

$$= \frac{3}{2(1-2z)} - \frac{4}{1-3z} + \frac{5}{2(1-4z)}$$

∴

$$G(a, z) = \frac{3}{2}(2^n) - 4(3^n) + \frac{5}{2}(4^n)$$

Example 1.83

Solve $a_n - 8a_{n-1} + 21a_{n-2} - 18a_{n-3} = 0$ for $n \geq 3$.

Solution

The given recurrence relation is

$$a_n - 8a_{n-1} + 21a_{n-2} - 18a_{n-3} = 0 \quad \text{... (1)}$$

The generating function is given by

$$G(a,z) = \sum_{n=0}^{\infty} a_n z^n = a_0 + a_1 z + a_2 z^2 + \ldots. \quad \text{... (2)}$$

Multiplying (1) by z^n and sum over all $n \geq 3$, we get

$$\sum_{n=3}^{\infty} a_n z^n - 6\sum_{n=3}^{\infty} a_{n-1} z^n + 12\sum_{n=3}^{\infty} a_{n-2} z^n - 8\sum_{n=3}^{\infty} a_{n-3} z^n = 0$$

Replacing the infinite sum by equivalent expression and simplifying, we get

$$G(a,z) = \frac{a_0 + (a_1 - 6a_0)z + (a_2 - 6a_1 + 12a_0)z^2}{1 - 6z + 12z^2 - 8z^3}$$

$$= \frac{a_0 + (a_1 - 6a_0)z + (a_2 - 6a_1 + 12a_0)z^2}{(1-2z)^3}$$

$$G(a,z) = \frac{A_1}{1-2z} + \frac{A_2}{(1-2z)^2} + \frac{A_3}{(1-2z)^3}$$

$$\Rightarrow \qquad G(a,z) = A_1 2^n + A_2(n+1)2^n + A_3\frac{(n+2)(n+1)}{2}2^n$$

Example 1.84

Solve the recurrence relation $a_{n+2} - 5a_{n+1} + 6a_n = 2$, $a_0 = 1$, $a_0 = 2$, by the method of generating function.

Solution

Given recurrence relation is

$$a_{n+2} - 5a_{n+1} + 6a_n = 2 \quad \text{... (1)}$$

The generating function is given by

$$G(a,z) = \sum_{n=0}^{\infty} a_n z^n = a_0 + a_1 z + a_2 z^2 + \ldots \quad \text{... (2)}$$

Multiplying (1) by z^n and sum from $r = 0$ to ∞, we have

$$\sum_{n=0}^{\infty} a_{n+2} z^n - 5\sum_{n=0}^{\infty} a_{n+1} z^n + 6\sum_{n=0}^{\infty} a_n z^n = \sum_{n=0}^{\infty} 2z^n$$

or $$\left(a_2 + a_3 z + a_4 z^2 + \ldots\right) - 5\left(a_1 + a_2 z + a_3 z^2 + \ldots\right) + 6G(a,z) = 2\left(1 + z + z^2 + \ldots\right)$$

or $$\frac{\left(a_2 z^2 + a_3 z^3 + \ldots\right)}{z^2} - 5\frac{\left(a_1 z + a_2 z^2 + \ldots\right)}{z} + 6G(a,z) = 2(1-z)^{-1}$$

$$\frac{\left(a_0 + a_1 z + a_2 z^2 + \ldots\right) - a_0 - a_1 z}{z^2} - 5\frac{\left(a_0 + a_1 z + a_2 z^2 + \ldots\right) - a_0}{z^2} + 6G(a,z) = \frac{2}{1-z}$$

or $$\frac{G(a,z) - a_a - a_1 z}{z^2} - 5\frac{G(a,z) - a_0}{z} + 6G(a,z) = \frac{2}{1-z}$$

Now using initial condition $a_0 = 1$ and $a_1 = 2$, we have

$$\frac{G(a,z)-1-3z}{z^2} - 5\frac{G(a,z)-1}{z} + 6G(a,z) = \frac{2}{1-z}$$

or
$$(1-5z+6z^2)G(a,z) = \frac{2z^2}{1-z} + 1 - 3z$$

or
$$G(a,z) = \frac{1-4z+5z^2}{(1-5z+6z^2)(1-z)}$$

$$G(a,z) = \frac{A}{1-z} + \frac{B}{1-2z} + \frac{C}{1-3z}$$

$$= \frac{1}{1-z} - \frac{1}{1-2z} + \frac{1}{1-3z}$$

Hence the solution after inverse transformation is given by

$$a_n = 1 - 2^n + 3^n.$$

Example 1.85

Solve $4a_{n+2} - 4a_{n+1} + a_n = 2^{-n}$, $a_0 = 0$, $a_1 = 1$, by the method of generating function.

Solution

Given recurrence relation is

$$4a_{n+2} - 4a_{n+1} + a_n = 2^{-n} \qquad \dots (1)$$

The generating function is given by

$$G(a,z) = \sum_{n=0}^{\infty} a_n z^n = a_0 + a_{1z} + a_2 z^2 + \dots \qquad \dots (2)$$

Multiplying (1) by z^n and summing from $n = 0$ to $n = \infty$, we have

$$4\sum_{n=0}^{\infty} a_{n+2} z^n - 4\sum_{n=0}^{\infty} a_{n+1} z^n + \sum_{n=0}^{\infty} a_n z^n = \sum_{n=0}^{\infty} 2^{-n} z^n$$

or
$$4(a_2 + a_3 z + a_4 z^2 + \dots) - 4(a_1 + a_2 z + \dots) + G(a,z) = 1 + \frac{z}{2} + \left(\frac{z}{2}\right)^2 + \dots$$

or
$$4\frac{(a_2 z^2 + a_3 z^3 + \dots)}{z^2} - 4\frac{(a_1 z + a_2 z^2 + \dots)}{z} + G(a,z) = \frac{1}{1-\frac{z}{2}}$$

$$4\frac{(a_0 + a_1 z + a_2 z^2 + a_3 z^3 \dots) - a_0 - a_1 z}{z^2} - 4\frac{(a_0 + a_1 z + a_2 z^2 + \dots) - a_0}{z} + G(a,z) = \frac{2}{2-z}$$

or
$$4[G(a,z) - a_0 - a_1 z] - 4[G(a,z) - a_0] + G(a,z) = \frac{2}{2-z}$$

Using initial condition $a_0 = 1$ and $a_1 = 1$, we have

$$4[G(a,z) - 0 - 1.z] - 4z[G(a,z) - 0] + z^2 G(a,z) = \frac{2z^2}{2-z}$$

$$\text{or} \qquad 4G(a,z) - 4z - 4zG(a,z) + z^2G(a,z) = \frac{2z^2}{2-z}$$

$$\text{or} \qquad (4 - 4z + z^2)G(a,z) = 4z + \frac{2z^2}{2-z}$$

$$\text{or} \qquad (2-z)^2 G(a,z) = \frac{8z - 2z^2}{2-z}$$

$$\text{or} \qquad G(a,z) = \frac{8z - 2z^2}{(2-z)^3} = \frac{2z(4-z^2)}{(2-z)^3}$$

To resolve into partial fraction, let us put $2 - z = t$, or $z = 2 - t$, then

$$\frac{8z - 2z^2}{(2-z)^3} = \frac{8(2-t) - 2(2-t)^2}{t^3} = \frac{8 - 2t^2}{t^3}$$

$$= \frac{8}{t^3} - \frac{2}{t} = \frac{8}{(2-z)^3} - \frac{2}{2-z}$$

$$\text{or} \qquad G(a,z) = \frac{8}{(2-z)^3} - \frac{2}{2-z}$$

$$= \frac{8}{8\left(1 - \frac{z}{2}\right)^3} - \frac{2}{2\left(1 - \frac{z}{2}\right)}$$

$$= \left(1 - \frac{z}{2}\right)^{-3} - \left(1 - \frac{z}{2}\right)^{-1}$$

$$\text{or} \qquad \sum_{r=0}^{\infty} a_n z^n = \sum_{n=0}^{\infty} \frac{(n+1)(n+2)}{2}\left(\frac{z}{2}\right)^n - \sum_{n=0}^{\infty}\left(\frac{z}{2}\right)^n$$

Equating the coefficient of z^n on both sides, we have

$$a_n = \frac{(n+1)(n+2)}{2} \cdot \frac{1}{2^n} - \frac{1}{2^n}$$

$$= \frac{n^2 + 2n + 2 - 2}{2^{n+1}}$$

$$\text{or} \qquad a_n = (n^2 + 3n) \cdot 2^{-n-1}.$$

Example 1.86

Solve the recurrence relation $t_{r+1} - t_r = r$ with $t_0=1$ by using method of generating functions.

Solution

The given recurrence relation is

$$t_{r+1} - t_r = r \qquad \dots (1)$$

The generating function is given by

$$G(a,z) = \sum_{r=0}^{\infty} t_r z^r = t_0 + t_1 z + t_2 z^2 + \ldots \qquad \ldots (2)$$

Multiplying (1) by z^r and sum over $r = 0$ to $r = \infty$, we have

$$\sum_{r=0}^{\infty} t_{r+1} z^r - \sum_{r=0}^{\infty} t_r z^n = \sum_{r=0}^{\infty} r \cdot z^r$$

$$\sum_{r=0}^{\infty} t_{r+1} z^r - \sum_{r=0}^{\infty} t_r z^r = \sum_{r=0}^{\infty} r \cdot z^r$$

$$(t_1 + t_2 z + t_3 z^2 + \ldots) - G(a,z) = z + 2z^2 + 3z^2 + \ldots$$

$$\frac{(t_1 z + t_2 z + t_3 z^2 + \ldots)}{z} - G(a,z) = z(1 + 2z + 3z^2 + \ldots)$$

or $$\frac{(t_0 + t_1 z + t_2 z^2 + \ldots) - t_0}{z} - G(a,z) = Z(1-z)^{-1}$$

Since $t_0 = 1$, we have

$$G(a,z) - 1 - zG(a,z) = z^2(1-z)^{-1}$$

or $$(1-z)\,G(a,z) = 1 + z^2(1-z)^{-2}$$

or $$G(a,z) = 1 + z^2(1-z)^{-2}$$

$$G(a,z) = (1-z)^{-1} + z^2(1-z)^{-3}$$

or $$\sum_{r=0}^{\infty} t_r z^r = \sum_{r=0}^{\infty} z^r + z^2 \sum_{r=0}^{\infty} \frac{(r+1)(r+2)}{2} z^r$$

or $$\sum_{r=0}^{\infty} t_r z^r = \sum_{r=0}^{\infty} z^r + z^2 \sum_{r=0}^{\infty} \frac{(r+1)(r+2)}{2} z^{r+2}$$

Equating coefficients of z^r on both sides, we get

$$t_r = 1 + \frac{1}{2}(r-1)r.$$

1.15-4 Some More Applications of Generating Functions

1. Generating function is one of the very important tools to solve not only the recurrence relations but many other problems in discrete mathematics.
2. With the help of generating function, one can find the coefficient of n^r from the expression of the type $(1 + x^2 + x^3)(x^2 + x^3)(x + 4)$.
3. It can be used to find the number of committees having the particular size that can be formed from a number of delegates from different countries subject to some restrictions.

Exercise 1.9

Solve the following recurrence relation using generating functions:

1. $a_n - 9a_{n-1} + 20a_{n-2} = 0$ for $n \geq 2, a_0 = -3$ and $a_1 = -10$.

2. $a_n - a_{n-1} - 9a_{n-2} + 9a_{n-3} = 0$ for $n \geq 3; a_0 = 0, a_1 = 1$ and $a_2 = 2$.

3. $a_n - 7a_{n-1} + 10a_{n-2} = 0, a_0 = 0$ and $a_1 = 3$.

4. $a_n - 5a_{n-1} + 6a_{n-2} = 0$ for $n \geq 2$, $a_0 = 1$ and $a_1 = -2$.

5. $a_n - 3a_{n-2} + 2a_{n-3} = 0$ for $n \geq 3$ and $a_0 = 1, a_1 = 0, a_2 = 0$.

6. $a_n = a_{n-1} + a_{n-2}$ for $n \geq 3, a_1 = 2$ and $a_2 = 3$.

7. $a_n - 4a_{n-2} = 0$ for $n \geq 2, a_0 = 0$ and $a_1 = 1$.

8. $a_n - 10a_{n-1} + 33a_{n-2} - 36a_{n-3} = 0$ for $n \geq 3, a_0 = 1, a_1 = 1$ and $a_2 = -23$

9. $a_{n+2} + a_{n+1} - 6a_n = 0, a_0 = 3, a_1 = 1$.

10. $a_n - 5a_{n-1} + 6a_{n-2} = 2^n + n,\ n \geq 2\ a_0 = 1$ and $a_1 = 1$.

11. Show that generating function for $a_n = \binom{2n}{n}$ is $(1 - 4z)^{-1/2}$.

12. Solve the recurrence relation using generating function

$$a_{n+2} - 4a_{n+1} + 3a_n = 0, \text{ given } a_0 = 2, a_1 = 4.$$

13. Find a simple expression for the generating function of the numeric function

$$1, -2, 3, -4, 5, -6.$$

14. Solve the recurrence relation $2a_{r+1} - a_r = 4$.

Answers

1. $a_n = 2.5^n - 5.4^n$

2. $a_n = \{-3 + 4.3^n - (-3)^n\} / 12$

3. $a_n = 2^{n+1} + (-3)^n$

4. $a_n = 5.2^n - 4.3^n$

5. $a_n = \frac{8}{9} - \frac{6}{9}n + \frac{1}{9}(-2)^n$

6. $a_n = \left(\frac{5 + 3\sqrt{5}}{10}\right)\left(\frac{1 + \sqrt{5}}{2}\right)^n + \left(\frac{5 - 3\sqrt{5}}{10}\right)\left(\frac{1 - \sqrt{5}}{2}\right)^n$

7. $a_n = 0$ if n is even, $a_n = 2^{n-1}$ if n is odd

8. $a_n = 15.3^n + 6(n + 1)3^n - 20.4^n$

9. $a_n = 2^{n+1} + (-3)^n$

11. $1 + 3^n$

14. $a_r = 4 - \frac{1}{2^r}$.

1.16 Discrete Numeric Functions

As we know that a function is a binary relation which associates to each element in the domain a unique element in the co-domain. In this section we shall study those functions whose domain is the set N of natural numbers and whose range is the set R of real numbers. This type of functions are known as discrete numeric functions or, in short, numeric functions. These functions are useful in digital computations to solve the counting problems involving non-negative integers.

Definition

A function whose domain is the set N of natural numbers and whose range is the set R of real numbers is said to be **discrete numeric functions** or in brief, **numeric functions**.

From the definition, it is clear that a numeric function associates to each non-negative integer r, a unique real number. We shall use bold face lower case letters to denote numeric functions. For a numeric function $\boldsymbol{a}$ we use $a_0, a_1, a_2, \ldots, a_r, \ldots$ to denote the values of the function at 0, 1, 2, …, r instead of $\boldsymbol{a}(0), \boldsymbol{a}(1), \boldsymbol{a}(2), \ldots, \boldsymbol{a}(r), \ldots$

1.16-1 Representation of Numeric Functions

We can represent a numeric function is several different ways:

i. By listing in order $(a_0, a_1, a_2, \ldots, a_r, \ldots)$ the values of a numeric function. For example $(2, 3, 10, 29, \ldots, r^3 + 2, \ldots)$ is the numeric function a whose r^{th} term a_r is $r^3 + 2$.

ii. By using a formula for its rth term. For example, the numeric function $(2, 3, 10, 29, .., r^3 + 2, \ldots)$ can be represented as $a_r = r^3 + 2, r \geq 0$.

The numeric function (0, 1, 2, 3, 17, 26, 37, …) can be represented as

$$a_r = \begin{cases} r, & 0 \leq r \leq 3 \\ r^2 + 1, & r \geq 4 \end{cases}$$

iii. A numeric function $(a_0, a_1, a_2, \ldots, a_r, \ldots)$ can also be represented by recurrence relation. For example, $a_0 = 2$ and $a_r = 2r\, a_{r-1}, r \geq 1$ describe the numeric function (2, 4, 16, 96, …).

Example 1.87

A ball is dropped to the floor from a height 100 m. Suppose that the ball always rebounds to reach three fourth of the height from it falls. Find the numeric function $\boldsymbol{a}$, where a_r, denote the height it reaches in the rth rebound.

Solution

The numeric function so formed is given by $a_0 = 100$ and $a_r = \frac{3}{4} a_{r-1}, r \geq 1$

or
$$a_r = 100\left(\frac{3}{4}\right)^r, \quad r \geq 1$$

Example 1.88

In a process control system, a monitoring device measures the temperature inside a chemical reaction chamber once every 30 Sec. Let a_r denote the rth reading in degrees in centigrade. Find an expression for a_r if it is known that the temperature rises from 100 to 120° at a constant rate in the first 300 Sec and stays at 120° from then on.

Solution

We have

$$a_r = \begin{cases} 100 + \dfrac{r}{15}, & 0 \leq r \leq 300 \\ 120, & r \geq 330 \end{cases}$$

1.16-2 Manipulation of Numeric Functions: Sum of Numeric Functions

Let $\boldsymbol{a}$ and $\boldsymbol{b}$ be any two numeric functions. The sum $\boldsymbol{a} + \boldsymbol{b}$ is another numeric function $\boldsymbol{c}$ whose value at r is equal to the sum of the values of $\boldsymbol{a}$ and $\boldsymbol{b}$ at r, i.e. $c_r = a_r + b_r$.

Example 1.89

Consider two numeric functions $\boldsymbol{a}$ and $\boldsymbol{b}$,

where
$$a_r = \begin{cases} 2r, & 0 \le r \le 2 \\ 3^r - 1, & r \ge 3 \end{cases}$$

and
$$b_r = \begin{cases} 0, & 0 \le r \le 1 \\ r + 5, & r \ge 2 \end{cases}$$

Let $\boldsymbol{c}$ be the sum of $\boldsymbol{a}$ and $\boldsymbol{b}$ denoted by $\boldsymbol{a} + \boldsymbol{b}$. Then

$$c_r = a_r + b_r$$

$$= \begin{cases} a_0 + a_1 + b_0 + b_1 \ , & 0 \le r \le 1 \\ a_2 + b_2, & r = 2 \\ a_r + b_r, & r \ge 3 \end{cases}$$

$$= \begin{cases} 2\ , & 0 \le r \le 1 \\ 11, & r = 2 \\ 3^r + r + 4, & r \ge 3 \end{cases}$$

1.16-3 Scalar Multiplication of a Numeric function

Let $\boldsymbol{a}$ be any numeric function and λ be a real number (scalar). Then $\lambda\boldsymbol{a}$ be a numeric function whose value at r is λ times the value of a_r. This numeric function $\lambda\boldsymbol{a}$ is known as a scaled version of $\boldsymbol{a}$ with scaling factor λ.

Example 1.90

Consider a numeric function $\boldsymbol{a}$, where

$$a_r = \begin{cases} r\ , & 0 \le r \le 3 \\ 3^r - 5, & r \ge 4 \end{cases}$$

then $5\boldsymbol{a}$ is a numeric function whose value at r is $5a_r$, i.e.

$$5a_r = \begin{cases} 5r\ , & 0 \le r \le 3 \\ 5(3^r - 5), & r \ge 4 \end{cases}$$

1.16-4 Product of Numeric Functions

Let $\boldsymbol{P}$ be the product of two numeric functions $\boldsymbol{a}$ and $\boldsymbol{b}$, denoted by $\boldsymbol{ab}$. The value of $\boldsymbol{P}$ is equal to the product of the values of two numeric function $\boldsymbol{a}$ and $\boldsymbol{b}$ at r, i.e.

$$P_r = a_r b_r$$

Example 1.91

Let
$$a_r = \begin{cases} r\ , & 0 \le r \le 3 \\ 3^r - 5, & r \ge 4 \end{cases}$$

and
$$b_r = \begin{cases} 2r, & 0 \le r \le 2 \\ r+3, & r \ge 3 \end{cases}$$

Then
$$P_r = a_r b_r = \begin{cases} r \cdot 2r, \ 0 \le r \le 2 \\ 3 \cdot (3+3), \ r = 3 \\ (3^r - 5)(r+3), \ r \ge 4 \end{cases}$$

$$= \begin{cases} 2r^2, & 0 \le r \le 2 \\ 18, & r = 3 \\ r3^r - 5r + 3^{r+1} - 15, & r \ge 4 \end{cases}$$

Verification

i. Here $a_0 = 0, b_0 = 0 \Rightarrow P_0 = a_0 \cdot b_0 = 0.0 = 0$

ii. $a_1 = 1, b_1 = 2 \Rightarrow p_1 = a_1 \cdot b_1 = 1 \cdot 2 = 2$

iii. $a_2 = 2, b_2 = 4 \Rightarrow p_2 = a_2 \cdot b_2 = 2 \cdot 4 = 8$

iv. $a_3 = 3, b_3 = 6 \Rightarrow p_3 = a_3 \cdot b_3 = 3 \cdot 6 = 18$

v. $a_4 = 76, b_4 = 7 \Rightarrow p_4 = a_4 \cdot b_4 = 76 \cdot 7 = 532$

All the values are verified.

1.16-5 Numeric Functions (special)

1. $S^n\boldsymbol{a}$: Let $\boldsymbol{a}$ be a numeric function and n be a positive integer. A numeric function denoted by $S^n\boldsymbol{a}$ is a function whose value at r is equal to 0 for $r = 0, 1, 2, \ldots, n-1$ and is a_{r-n}, for $r \ge n$. In other words,

$$S^n a = \begin{cases} 0, & 0 \le r \le n-1 \\ a_{r-n}, & r \ge n \end{cases}$$

Example 1.92

Let
$$a_r = \begin{cases} 0, & 0 \le r \le 7 \\ 2^{-r} + 5, & r \ge 8 \end{cases}$$

Find S^7a.

Solution

By definition of S^na.

$$S^7 a = \begin{cases} 0, & 0 \le r \le 7-1 \\ a_{r-7}, & r \ge 7 \end{cases}$$

$$= \begin{cases} 0, & 0 \le r \le 6 \\ a_{r-7}, & r \ge 7 \end{cases}$$

Here
$$a_r = \begin{cases} 0, & 0 \le r \le 7 \\ 2^{-r} + 5, & r \ge 8 \end{cases}$$

Replacing r by $r - 7$, we get

$$a_{r-7} = \begin{cases} 0, & 0 \le r-7 \le 7 \\ 2^{-(r-7)} + 5, & r-7 \ge 8 \end{cases}$$

$$= \begin{cases} 0, & 7 \le r \le 14 \\ 2^{-(r-7)} + 5, & r \ge 15 \end{cases}$$

Therefore,

$$S^7 a = \begin{cases} 0, & 0 \le r \le 6 \\ 0, & 7 \le r \le 14 \\ 2^{-r+7} + 5, & r \ge 15 \end{cases}$$

$$= \begin{cases} 0, & 0 \le r \le 14 \\ 2^{-r+7} + 5, & r \ge 15 \end{cases}$$

Example 1.93

Let $\boldsymbol{a}$ be a numeric function such that

$$a_r = \begin{cases} 2r, & 0 \le r \le 3 \\ 3^r \cdot r, & r \ge 4 \end{cases}$$

Find $S^2 a$.

Solution

By definition,

$$S^2 a = \begin{cases} 0, & 0 \le r \le 2-1 \\ a_{r-2}, & r \ge 2 \end{cases}$$

$$= \begin{cases} 0, & 0 \le r \le 1 \\ a_{r-2}, & r \ge 2 \end{cases}$$

Here

$$a_r = \begin{cases} 2r, & 0 \le r \le 3 \\ 3^r \cdot r, & r \ge 4 \end{cases}$$

Replace r by $r - 2$, we get

$$a_{r-2} = \begin{cases} 2(r-2), & 0 \le r-2 \le 3 \\ 3^{r-2}(r-2), & r-2 \ge 4 \end{cases}$$

$$= \begin{cases} 2r-4, & 2 \le r \le 5 \\ 3^{r-2}(r-2), & r \ge 6 \end{cases}$$

Therefore

$$S^2 a = \begin{cases} 0, & 0 \le r \le 1 \\ 2r-4, & 2 \le r \le 5 \\ 3^{r-2}(r-2), & r \ge 6 \end{cases}$$

Let $\boldsymbol{a}$ *be a numeric function and* $\boldsymbol{n}$ *be a positive integer. A numeric function denoted by* $S^{-n}\boldsymbol{a}$ *is a function whose value at r is equal to* a_{r+n} *for* $r \ge 0$. *In other words,*

$$S^{-n}\boldsymbol{a} = a_{r+n},\ r \ge 0.$$

Example 1.94

Let $$a_r = \begin{cases} 1, & 0 \le r \le 10 \\ 2, & r \ge 11 \end{cases}$$

Find $S^{-5}a$.

Solution

By definition, $$S^{-5}a = a_{r+5},\ r \ge 0$$

Here $$a_r = \begin{cases} 1, & 0 \le r \le 10 \\ 2, & r \ge 11 \end{cases}$$

Replacing r by $r + 5$ in a_r, we get

$$a_{r+5} = \begin{cases} 1, & 0 \le r + 5 \le 10 \\ 2, & r + 5 \ge 11 \end{cases}$$

$$= \begin{cases} 1, & -5 \le r \le 5 \\ 2, & r \ge 6 \end{cases}$$

But we know that r is a non-negative integer. Thus

$$a_{r+5} = \begin{cases} 1, & 0 \le r \le 5 \\ 2, & r \ge 6 \end{cases}$$

Hence $$S^{-5}a = \begin{cases} 1, & 0 \le r \le 5 \\ 2, & r \ge 6 \end{cases}.$$

1.16-6 Accumulated Sum of Numeric Functions

The accumulated sum of a numeric function $\boldsymbol{a}$ is defined as $a_0 + a_1 + a_2 + ... + a_r$, then

$$b_r = \sum_{n=0}^{r} a_n,\ r \ge 0.$$

Example 1.95

A person deposits Rs. 200 each year in a saving account at an yearly compound interest rate of 6 percent. What is his total amount after r years.

Solution

$$\text{The total amount} = \text{sum}\left(1 + \frac{\text{rate}}{100}\right)^{\text{time}}$$

Let $\boldsymbol{a}$ be a numeric function which gives the amount in the saving account at the end of rth year, then

$$a_r = 200\left(1 + \frac{6}{100}\right)^r$$

$$= 200(1.06)^r,\ r \ge 0$$

Clearly, the total amount in the account after r years is equal to the accumulated sum of the numeric function $\boldsymbol{a}$ i.e.

$$b_r = \sum_{n=0}^{r} a_n = \sum_{n=0}^{r} 200(1.06)^n$$

$$= 200\left(1 + (1.06) + (1.06)^2 + ... + (1.06)^r\right)$$

$$= \frac{20000}{6}\left[(1.06)^{r+1} - 1\right],\ r \geq 0.$$

1.16-7 Forward Difference

Let a be a numeric function. The **forward difference** of a is a numeric function, whose value at r is equal to $a_{r+1} - a_r$. It is denoted by Δa. If b be the forward difference of a, i.e. $b = \Delta a$; then

$$b_r = a_{r+1} - a_r.$$

Example 1.96

Let

$$a_r = \begin{cases} 2, & 0 \leq r \leq 3 \\ 2^{-r} + 5, & r \geq 4 \end{cases}$$

Find Δa.

Solution

Here

$$a_r = \begin{cases} 2, & 0 \leq r \leq 3 \\ 2^{-r} + 5, & r \geq 4 \end{cases}$$

$$a_{r+1} = \begin{cases} 2, & 0 \leq r+1 \leq 3 \\ 2^{-(r+1)} + 5, & r+1 \geq 4 \end{cases}$$

$$= \begin{cases} 2, & 0 \leq r \leq 2 \\ 2^{-r-1} + 5, & r \geq 3 \end{cases}$$

Let $b = \Delta a$, then

$$b_r = a_{r+1} - a_r$$

$$= \begin{cases} 0, & 0 \leq r \leq 2 \\ \frac{49}{16}, & r = 3 \\ -2^{-r-1}, & r \geq 4 \end{cases}$$

Verification

i. $b_0 = a_1 - a_0 = 2 - 2 = 0$

ii. $b_1 = a_2 - a_1 = 2 - 2 = 0$

iii. $b_2 = a_3 - a_2 = 2 - 2 = 0$

iv. $b_3 = a_4 - a_3 = \frac{81}{16} - 2 = \frac{49}{16}$

v. $b_4 = a_5 - a_4 = (2^{-5} + 5) - (2^{-4} + 5)$
$= 2^{-5} - 2^{-4} = -2^{-4-1}$

vi. $b_r = a_{r+1} - a_r = -2^{-r-1}, r \geq 4$

Example 1.97

Show that $S^{-1}(\nabla a) = \Delta a$.

Solution

Let a be any numeric function such that $b = \nabla a$ and $c = \Delta a$, then, $b_r = a_r - a_{r-1}$, and $c_r = a_{r+1} - a_r$.

Let $d = S^{-1}(\nabla a)$, then

$$d_r = b_{r+1}, r \geq 0$$

$$= a_{r+1} - a_r = c_r$$

Hence, $$S^{-1}(\nabla \boldsymbol{a}) = \Delta \boldsymbol{a}.$$

1.16-8 Backward Difference

Let $\boldsymbol{a}$ be a numeric function. The **backward difference** of $\boldsymbol{a}$ is a numeric function, whose value at r is equal to $a_r - a_{r-1}$ at $r \geq 1$ and denoted by $\nabla \boldsymbol{a}$.

If $\boldsymbol{c}$ is the forward difference of $\boldsymbol{a}$ i.e. $\boldsymbol{c} = \Delta \boldsymbol{a}$; then

$$c_r = a_r - a_{r-1}.$$

Example 1.98

Let $$a_r = \begin{cases} 2, & 0 \leq r \leq 3 \\ 2^{-r} + 5, & r \geq 4 \end{cases}$$

Find $\nabla \boldsymbol{a}$.

Solution

Here $$a_r = \begin{cases} 2, & 0 \leq r \leq 3 \\ 2^{-r} + 5, & r \geq 4 \end{cases}$$

$$a_{r-1} = \begin{cases} 2, & 0 \leq r - 1 \leq 3 \\ 2^{-r+1} + 5, & r - 1 \geq 4 \end{cases}$$

$$= \begin{cases} 2, & 1 \leq r \leq 4 \\ 2^{-r+1} + 5, & r \geq 5 \end{cases}$$

Let $\boldsymbol{c} = \nabla \boldsymbol{a}$, then

$$c_r = a_r - a_{r-1}$$

$$= \begin{cases} 0, & 0 \leq r \leq 3 \\ \dfrac{49}{16}, & r = 4 \\ -2^{-r}, & r \geq 5 \end{cases}$$

Example 1.99

Let $$a_r = 2^r, r \geq 0 \text{ and } b_r = 5^r, r \geq 0$$

then, for $\boldsymbol{c} = \boldsymbol{a} * \boldsymbol{b}$, $$c_r = \sum_{i=0}^{r} 2^i \cdot 5^{r-i}, \quad r \geq 0.$$

Example 1.100

Show that $S^{-1}(\nabla a) = \Delta a$.

Solution

Let $\boldsymbol{a}$ be any numeric function such that $\boldsymbol{b} = \nabla \boldsymbol{a}$ and $\boldsymbol{c} = \Delta a$, then, $b_r = a_r - a_{r-1}$, and $c_r = a_{r+1} - a_r$. Let $\boldsymbol{d} = S(\nabla \boldsymbol{a})$, then

$$d_r = b_{r+1}, r \geq 0$$

$$= a_{r+1} - a_r$$
$$= c_r$$

Hence, $S^{-1}(\nabla \boldsymbol{a}) = \Delta \boldsymbol{a}.$

Example 1.101

Determine $a * b$ if

$a_r = 2^r$ for all r and $b_r = \begin{cases} 0, & 0 \le r \le 2 \\ 2^r, & r \ge 3 \end{cases}.$

Solution

Let $c = a * b$. Then

$$c_0 = a_0 b_0 = 2^0 \cdot 0 = 0$$
$$c_1 = a_0 b_1 + a_1 b_0 = 2^0 \cdot 0 + 2^1 \cdot 0 = 0$$
$$c_2 = a_0 b_2 + a_1 b_1 + a_2 b_0 = 2^0 \cdot 0 + 2^1 \cdot 0 + 2^2 \cdot 0 = 0$$
$$c_3 = a_0 b_3 + a_1 b_2 + a_2 b_1 + a_3 b_0 = 2^0 \cdot 2^3 + 2^1 \cdot 0 + 2^2 \cdot 0 + 2^3 \cdot 0$$
$$= 1.2^3$$
$$c_4 = a_0 b_4 + a_1 b_3 + a_2 b_2 + a_3 b_1 + a_4 b_0$$
$$= 2^0 \cdot 2^4 + 2^1 \cdot 2^3 + 2^2 \cdot 0 + 2^3 \cdot 0 + 2^4 \cdot 0$$
$$= 2^4 + 2^4$$
$$= 2.2^4$$
$$c_5 = a_0 b_5 + a_1 b_4 + a_2 b_3 + a_3 b_2 + a_4 b_1 + a_5 b_0$$
$$= 2^0 \cdot 2^5 + 2^1 \cdot 2^4 + 2^2 \cdot 2^3 + 2^3 \cdot 0 + 2^4 \cdot 0 + 2^5 \cdot 0$$
$$= 1 \cdot 2^5 + 2^5 + 2^5$$
$$= 3.2^5$$

....

$$c_r = a_0 b_r + a_1 b_{r-1} + \ldots + a_r b_0$$
$$= (r-2)2^r$$

Thus, $c = a * b$ is given by $c_r = \begin{cases} 0, & 0 \le r \le 2 \\ (r-2)2^r, & r \ge 3 \end{cases}.$

1.16-9 Convolution of Numeric Function

The convolution of two numeric functions ***a*** and ***b***, denoted by ***a * b***, is a numeric function ***c*** such that

$$c_r = \sum_{i=0}^{r} a_i b_{r-i}$$
$$= a_0 b_r + a_1 b_{r-1} + a_2 b_{r-2} + \ldots + a_{r-1} b_1 + a_r b_0$$

Exercise 1.10

1. A ball is dropped to the floor from a height 40 m. Suppose that the ball always rebounds to reach half of the height from it falls. Find the numeric function ***a***, where a_r denotes the height it reaches in the rth rebound.
2. Let a_r denotes the altitude of an aircraft, in thousands of feet, at the rth minute. Suppose the aircraft takes off after spending 10 minutes on the ground, climbs up at a uniform speed to a cruising altitude of 30,000 feet in 10 minutes, starts to descend uniformly after 110 minutes of flying time, and lands 10 minutes later. Find the numeric function *a*.
3. Let *a* and *b* be any two numeric functions where:

$$a_r = \begin{cases} 0, & 0 \le r \le 2 \\ 2^{-r} + 5, & r \ge 3 \end{cases} \text{ and } b_r = \begin{cases} 3 - 2^r, & 0 \le r \le 1 \\ r + 2, & r \ge 2 \end{cases}$$

Find

(i) Sum of *a* and *b*.

(ii) Product of *a* and *b*.

(iii) Δa and ∇a.

4. Let *a* be a numeric function such that

$$a_r = \begin{cases} 1, & 0 \le r \le 10 \\ 2, & r \ge 11 \end{cases}$$

Find (i) $S^5(a)$ (ii) $S^{-7}(a)$.

5. Determine the convolution of *a* and *b*, where

$$a_r = 3^r, \quad r \ge 0 \text{ and } b_r = 2^r, \quad r \ge 0.$$

6. Let *a*, *b* and *c* be numeric functions such that $a*b = c$. Given that

$$a_r = \begin{cases} 1, & r = 0 \\ 2, & r = 1 \\ 0, & r \ge 2 \end{cases} \text{ and } c_r = \begin{cases} 1, & r = 0 \\ 0, & r \ge 1 \end{cases}$$

determine ***b***.

7. Let $a_r = \begin{cases} 1, & r = 0 \\ 3, & r = 1 \\ 2, & r = 2 \\ 0, & r \ge 3 \end{cases}$ and $c_r = 5^r \ \forall r$.

Determine *b* such that $a * b = c$.

8. A person deposit Rs. 200 in a saving account at an interest rate 6% per year compounded annually. Obtain the amount in this account at the end of *r* year. Find the numeric function.
9. If *a* be a numeric function then find (i) $S^{11}(a)$ (ii) $S^{-11}(a)$, where

$$a_r = \begin{cases} 1, & 0 \le r \le 20 \\ 2, & r \ge 21 \end{cases}$$

Answers

1. $a_r = 40\left(\frac{1}{2}\right)^r, \ r \geq 0$

2. $a_r = \begin{cases} 0, & 0 \leq r \leq 10 \\ 3(r-10), & 11 \leq r \leq 20 \\ 30, & 21 \leq r \leq 120 \\ 3(130-r), & 121 \leq r \leq 130 \\ 0, & r \leq 131 \end{cases}$

3. (i) $c_r = \begin{cases} 3 - 2^r, & 0 \leq r \leq 1 \\ 4, & r = 2 \\ 2^{-r} + r + 7, & r \geq 3 \end{cases}$ (ii) Product $= \begin{cases} 0, & 0 \leq r \leq 2 \\ r \cdot 2^{-r} + 2^{-r+1} + 5r + 10, & r \geq 3 \end{cases}$

(iii) $\Delta a = \begin{cases} 0, & 0 \leq r \leq 1 \\ \frac{41}{8}, & r = 2 \\ -2^{-r-1}, & r \geq 3 \end{cases}$ (iv) $\Delta a = \begin{cases} 0, & 0 \leq r \leq 2 \\ \frac{41}{8}, & r = 3 \\ -2^{-r}, & r \geq 4 \end{cases}$

4. (i) $S^5(a) = \begin{cases} 0, & 0 \leq r \leq 4 \\ 1, & 5 \leq r \leq 15 \\ 2, & r \geq 16 \end{cases}$ (ii) $S^{-7}(a) = \begin{cases} 1, & 0 \leq r \leq 3 \\ 2, & r \geq 4 \end{cases}$

5. $a * b = \sum_{i=0}^{r} 3^i 2^{r-i}, \ r \geq 0$

6. $b_r = (-2)^r, \ r \geq 0$

8. $a_r = 200(1.06)^r, \ r \geq 0$

9. (i) $S^{11}(a) = \begin{cases} 0, & 0 \leq r \leq 10 \\ 1, & 11 \leq r \leq 31 \\ 2, & r \geq 32 \end{cases}$ (ii) $S^{-11}(a) = \begin{cases} 1, & 0 \leq r \leq 9 \\ 2, & r \geq 10 \end{cases}$.

1.17 Pigeonhole Principle

The **pigeonhole principle** is also known as **Dirichlet Drawer Principle** or the **Shoe Box Principle**. This principle states that if a flock of pigeons flies into a set of pigeonholes to roost and the number of pigeons are more than the number of pigeonholes then there must be at least one pigeonhole with at least two pigeons in it. In practical life situation this principle applies to other objects also. The pigeonhole principle tells us nothing about how to locate the pigeons that contain two or more pigeons.

Theorem 1.11 Pigeonhole Principle

If k pigeons fly into n pigeonholes and $(k > n)$ then at least one pigeonhole contains two or more pigeons.

Proof

If none of n pigeonholes contains more than one pigeon. Then the total number of pigeons would be at most n. But this is a contradiction, since there are $(k > n)$ pigeons.

Example 1.102

If there is a group of 27 English words, there must exist at least two that begin with the same letter, since there are 26 letters in the English alphabet.

Example 1.103

Determine the minimum number of elements that one needs to take from the set $A = \{1, 2, 3, 4, 5, \ldots, 9\}$ to be sure that two of the numbers add up to 10.

Solution

We have five pigeonholes here namely {1, 9}, {2, 8}, {3, 7}, {4, 6}, {5, 5}. One can observe from here that any choice of six elements (pigeons) of A will guarantee that two of the numbers add upto ten.

Example 1.104

Let there are five men and three women at a function. Show that if these people are lined up in a row, at least two men will be next to each other.

Solution

Do yourself.

Example 1.105

Determine the number of students must be in a B.Tech class to guarantee that at least two students receive the same score in the final exam, if the exam is graded on a scale from 0 to 100 points.

Solution

There are 101 possible scores on the final exam. From the pigeonhole principle it is clear that among any 102 students there must be at least 2 students with the same score.

1.17-1 Generalized Pigeonhole Principle

If there exist n pigeonholes and more than $2n$ pigeons, than three or more pigeons will have to be assigned to at least one of the pigeonholes. In general, if the number of pigeons is much larger than the number of pigeonholes. For this situation theorem 1 can be restated as follows.

Theorem 1.12

If n pigeons are to occupy m pigeonholes, then there is at least one pigeonhole containing at least $\lceil n/m \rceil$* objects.

Proof

Let none of the pigeonholes are occupied by more than $\lceil n/m \rceil - 1$ pigeons. Then total number of pigeons is at most

$$m(\lceil n / m \rceil) - 1 < m((n / m + 1) - 1) = n$$

Where the inequality

$$\lceil n / m \rceil < (n / m) + 1$$

But this is a contradiction as there are only n objects.

Example 1.106

If 30 books in a school contain a total of 61,327 pages, then show that one of the books must have at least 2045 pages.

---

* The ceiling function at x is denoted by $\lceil x \rceil$ and assigns to the real number x the smallest integer that is greater than or equal to x. For example $\lceil 1/2 \rceil = 1, \lceil 4{\cdot}1 \rceil = 5, \lceil 9 \rceil = 9$.

Solution

From the extended pigeonhole principle we see that if we assign each page (pigeon) to the books (pigeonholes), then one book must contain at least

$$\left\lceil \frac{61{,}327}{30} \right\rceil = \lceil 2044.2 \rceil = 2045 .$$

Example 1.107

Find the minimum number of persons in a joint family to be sure that 4 of them born in the same month.

Solution

Total number of months (pigeonholes) is 12. Complete yourself.

Example 1.108

What is the minimum number of students required in a particular class to be sure that at least six students will receive the same division if there are five possible divisions.

Solution

The least number of students required to make sure that at least 6 student get the same grade in the smallest integer n such that $\left\lceil \frac{4}{5} \right\rceil = 6$. The smallest such integer is $n = 5.5 + 1 = 26$.

$\Rightarrow$ the minimum number of students required to be sure that at least 6 students will get the same division.

Exercise 1.11

1. Show that among any group of five (not necessarily consecutive) integers, there are two with the same remainder when divided by 4.
2. A box contains 10 black balls and 10 green balls. Shyam selects balls at random without looking at them.
   i. How many balls must Shyam select to be sure of having at least three balls of the same color?
   ii. How many balls must Shyam select to be sure of having at least three green balls?
3. If thirteen people are assembled in a get together party. Show that at least two of them must have their birthday in the same month.
4. Consider a tournament in which each of n players plays against every other player and each player wins at least once. Show that at least two players having the same number of wins.
5. How many students each of whom comes from one of the 50 districts must be enrolled in an engineering college to guarantee that there are at least 100 who come from the same state?
6. There are 51 electricity poles on a street. Each pole has a number between 1000 and 1099 inclusive. Show that at least two poles have numbers that are consecutive integers.
7. There are seven colours, which are to be used to paint 50 cars, show that at least eight bicycle will be of same colour.
8. Prove that if any 14 numbers from 1 to 25 are chosen, then one of them is multiple of another.
9. Show that if five integers are selected from the first eight positive integers, there must be a pair of these integers with a sum equal to 9.
9. If there are 38 different time periods during which classes at an engineering college can be scheduled. If there are 677 different classes, how many different rooms will be needed?
10. Show that there are at least four people in Modinagar (population: 25 million) with the same three initials who were born on the same days of the year.

11. How many friends you must have to guarantee at least five of them will have birthdays in the same month?

Answers

1. Total number of possible remainder when an integer is divided by 4 is four, hence from the pigeonhole principle there are at least two have the same remainder.
3. Let the birth month is like pigeons and the calendar months, the pigeonhole i.e. there are 13 pigeons and 12 pigeonholes. Hence from the pigeonhole principle at least two people were born in the same month.
4. One can observe that the number of wins for a player is at least 1 and at most $n-1$. Here, this $n-1$ numbers correspond to $n-1$ pigeonholes which can not accommodate n player (pigeons). Thus, at least two players will have the same number of wins.
5. 4951.
7. Use extended pigeonhole principle.
8. By labeling the pigeonholes with 1, 3, 5, ..., 25 (inclusive both). If we assign each of the selected 14 numbers to the pigeonhole labelled with its odd part. But there exist only 13 pigeonholes, so two numbers must have the same odd part. One is a multiple of other.
9. Grouping the first eight positive integers of each pair which are four in number 9 : {1, 8}, {2, 7}, {3, 6} and {4, 5}. If we select first five integers from the first eight positive integers. By the pigeonhole principle at least two of them must come from the same subset. Two such integers have a sum of 9, as required.
10. 18.
11. There are 6,432,816 possibilities for the three initials and a birthday. Hence from the generalized pigeonhole principle, there are at least [25,000,000/6,432,816] = 4 people who share the same initials and birthday.
12. At least 49 friends.

Chapter Objective Type Questions

A. Multiple choice questions

1. The number of distinguishable permutations of the letters in the word BANANA are,
   (A) 20 (B) 10
   (C) 60 (D) 36
2. In a Mathematics examination there are 15 questions of type True or False. How many sequences of answers are possible?
   (A) 215 (B) 205
   (C) 251 (D) none of these
3. Suppose there are 12 students, among whom three students are, A, B and C. A delegation of four students (chosen from the 12 students) is to be sent to a conference. How many ways can this be done so that the delegation includes exactly two students from {A, B, C}?
   (A) 144 (B) 88
   (C) 64 (D) 108
4. In how many ways 5 students out of a class of 20 line up for a game?
   (A) ${}^{20}C_5$ (B) ${}^{20}C_4$
   (C) ${}^{20}P_5$ (D) ${}^{20}P_4$

5. How many different rearrangements are there of the letters in the word TATARS if the two A's are never adjacent?
   (A) 180 (B) 144
   (C) 120 (D) 220
6. In a survey of 85 people it is found that 43 like to drink mango shake, 39 like pineapple juice and 31 like orange juice. Also 13 like both orange juice and pineapple juice, 15 like orange juice and mango shake, 20 like pineapple juice and mango shake and 12 like none of the three drinks. Find the number of people who like all the three drinks.
   (A) 8 (B) 73
   (C) 20 (D) 13
7. The permutations of {a,b,c,d,e,f,g} are listed in lex order. What permutations are just before and just after bacdefg?
   (A) Before: agfedcb, After: bacdegf (B) Before: agfedbc, After: bacdfge
   (C) Before: agfedcb, After: bacdfge (D) Before: agfebcd, After: bacedgf
8. Suppose there are ten balls in an urn, four black, four violet, and two yellow. The balls are also numbered 1 to 10. How many ways are there to select an ordered sample of four balls without replacement such that the number $B \geq 0$ of black balls, the number $V \geq 0$ of violet balls, and the number $Y \geq 0$ of yellow balls are all different?
   (A) 2144 (B) 1446
   (C) 1152 (D) 864
9. How many words can be formed out of the letters of the word 'PECULIAR' beginning with P and ending with R?
   (A) 720 (B) 125
   (C) 150 (D) 100
10. The minimum number of cards to be dealt from an arbitrarily shuffled deck of 52 cards to guarantee that three cards are from some same suit is?
    (A) 9 (B) 12
    (C) 3 (D) 8
11. In how many ways can a chairman and deputy chairman be chosen from a set of 30 candidates?
    (A) 850 (B) 820
    (C) 870 (D) 880
12. How many different rearrangements are there of the letters in the word BUBBLE?
    (A) 80 (B) 120
    (C) 70 (D) 40
13. In how many ways can 5 balls be chosen so that 2 are blue and 3 are green?
    (A) 910 (B) 970
    (C) 990 (D) 980
14. The number of ways to seat 3 girls and 2 boys in a row if each boy must sit next to at least one boy is:
    (A) 36 (B) 148
    (C) 48 (D) 248

15. How many 5-cards consist only of diamonds?

(A) 1127 (B) 1287
(C) 1487 (D) 1687

16. Suppose there are ten balls in an urn, four black, four yellow, and two violet. The balls are also numbered 1 to 10. How many ways are there to select an ordered sample of four balls without replacement such that there are two black balls and two yellow balls in the sample?

(A) 446 (B) 664
(C) 864 (D) 256

17. In how many ways can a party of 7 students arrange themselves around a circular table?

(A) 5! (B) 6!
(C) 7 (D) 7!

18. The English alphabet has 26 letters of which 5 are vowels (a, e, i, o, u). How many seven letter words, with all letters distinct, can be formed that start with b, end with the letters es, and have exactly three vowels?

(A) $24 \times 34 \times 19$ (B) $24 \times 33 \times 17$
(C) $24 \times 33 \times 19$ (D) $23 \times 34 \times 17$

19. In how many ways can a hungry person choose 3 toppings for his prize from a list of 10 delicious possibilities?

(A) 100 (B) 110
(C) 120 (D) 150

20. Teams *X* and *Y* play in a cricket tournament. The first team to win two games in a row or a total of three games wins the tournament. What is the number of ways the tournament can occur?

(A) 11 (B) 10
(C) 9 (D) 12

21. A quiz team consists of 3 girls and 2 boys. Find the number of ways they can sit in a row?

(A) 12 (B) 24
(C) 120 (D) 720

22. The number of four letter words that can be formed from the letters in BUBBLE (each letter occurring at most as many times as it occurs in BUBBLE) is:

(A) 78 (B) 80
(C) 74 (D) 72

23. The number of four letter words that can be formed from the letters in SASSABY (each letter occurring at most as many times as it occurs in SASSABY) is:

(A) 120 (B) 108
(C) 90 (D) 114

24. For the sequence defined by the recurrence relation $S_n = nS_{n-1}$ with initial condition $S_1 = 7$, the explicit formula for S_n is:

(A) $S_n = n7^{n-1}$ (B) $S_n = 7n!$
(C) $S_n = \dfrac{n!}{7}$ (D) $S_n = n! - 7$

25. Pigeonhole principle states that $f: A \to B$ and $|A| > |B|$ then:
    (A) f is not onto
    (B) f is not one-one
    (C) f is neither one-one nor onto
    (D) f may be one-one

    This problem may be attempted after studying chapter 4.

B. Fill in the blanks.

1. The number of combination of n distinct objects taken r at a time is given by _____.
2. Method of evaluating a polynomial, after converting into telescoping form is called ________.
3. The total solution of a linear recurrence relation (or linear difference equation) with constant coefficient consist of two parts, the homogeneous solution and ________ .
4. Closed form expression is a compact way to represent a ________.
5. The convolution of two numeric functions a and b is given by ________.
6. If k pigeons fly into n pigeonholes and $(k > n)$ then at least one pigeonhole contains ___________ pigeons.
7. Let a be a numeric function. The forward difference of a is a numeric function, whose value at r is equal to________.
8. The accumulated sum of a numeric function a is defined as ______.

C. Which of the following statements is True (T)/ False (F)?

1. In a combination only selection is made whereas in permutation not only a selection is made but also an arrangement in a definite order is also taken into consideration.
2. For a nth order linear recurrence relation with constant coefficients, fewer than n values of the numeric function will not be sufficient to determine the numeric function uniquely.
3. $S_k + C_1 S_{k-1} + \ldots\ldots + C_n S_{k-n} = 0$ is a inhomogeneous linear recurrence relation.
4. If the coefficients of characteristic equation are all integers and constant term equal to m, then the only possible rational characteristic roots are divisors of m, both positive and negative.
5. The product of two generating functions need not be a generating function. (*T/F*)
6. The pigeonhole principle tells us everything about how to locate the pigeons that contain two or more pigeons.
7. Let ***a*** be a numeric function. The backward difference of ***a*** is a numeric function, whose value at r is equal to $a_r - a_{r-1}$ at $r \geq 1$.
8. Let ***P*** be the product of two numeric functions ***a*** and ***b***, denoted by ***ab***. The value of ***P*** is equal to the product of the values of two numeric function ***a*** and ***b*** at r, i.e $P_r = a_r b_r$.

Answers (Objective Type Questions)

A.

1	(C)	2	(A)	3	(D)	4	(C)
5	(C)	6	(A)	7	(A)	8	(C)
9	(A)	10	(D)	11	(C)	12	(B)
13	(C)	14	(A)	15	(B)	16	(C)
17	(B)	18	(C)	19	(C)	20	(B)
21	(C)	22	(D)	23	(D)	24	(B)

B.

1. $C(n,r)$	2. Horners Method	3. particular solution	4. Sequence
5. $c_r = \sum_{i=0}^{r} a_i b_{r-i}$	6. Two or more	7. $a_{r+1} - a_r$	8. $\sum_{n=0}^{r} a_n,\ r \geq 0$

C.

1. T	2. T	3. F	4. T
5. F	6. F	7. T	8. T

CHAPTER

2 Set Theory

2.0 Introduction

The theory of sets was originated in the year 1895 by the German mathematician G. Cantor. The concept of set is most basic to computer science and mathematics and hence we should have a very clear definition of sets. But because a set is such a simple notion, one may be surprised to learn that it is one of the most difficult concepts for mathematicians to define to their own liking.

In this chapter we will cover some of the basic concepts, notations and definitions of set theory. Later on we introduce Venn diagrams to have good understanding of the operations on sets. In this chapter an understanding of sets is developed from experience using computers and computer programs.

2.1 Set

A **set** may be defined as the well-defined collection of objects. The objects in a set are also called **elements** or **members** of the set. A set denoted by capital letters $A, B, C, D, X, Y, \ldots$ and elements of a set usually by lowercase letters $a, b, c, d, x, y, \ldots$.

2.1-1 Set Membership

Let A be any set and x be an element of A, then we write $x \in A$ and read it as "x belongs to the set A" or equivalently "x is an element of set A" or "x is a member of set A"

If x is not an element of A, then we write $x \notin A$ and read it as "x does not belong to the set A".

2.1-2 Specifying Sets

There are essentially following two ways by which sets can be denoted.

(i) Tabular form of a set (ii) Set builder form.

(i) Tabular form of a Set

The set can be represented by listing all the members which belong to it, if possible. In this listing the elements are separated by commas and enclosed in braces {}. For example the set A of odd positive integers less than 10 can be written as

$$A = \{1, 3, 5, 7, 9\}.$$

Example 2.1

The set A of all vowels in English alphabet can be written as

$$A = \{a, e, i, o, u\}.$$

Example 2.2

The set of positive integers less than 50 is written as

$$A = \{1, 2, 3, 4, \ldots, 49\}$$

The three consecutive dots "..." are called **ellipses** and are used when the general pattern of the elements is clear.

Example 2.3

Write the following in tabular form.

i. $A = \{x \mid x \text{ is a prime number less than } 15\}$

ii. $B = \{x \mid x \text{ is a letter in the word 'induction'}\}$

iii. $C = \{x \mid x \in N, 3 < x < 12\}$.

Solution

i. $A = \{2, 3, 5, 7, 11, 13\}$

ii. $B = \{i, n, d, u, c, t, i, o, n\}$

iii. $C = \{4, 5, 6, 7, 8, 9, 10, 11\}$.

(ii) Set Builder Form

A set can be described by stating the property or properties, shared by all the elements of the set. The standard notation for a set so described is $\{x: \ldots\}$ or $\{x| \ldots\}$. The space between ':' (or |) and '}' is to be filled by the statement that x has the property in question. Here x is a dummy symbol and could have been replaced by any other dummy symbol. The ': ' or '|' is read as such that or 'where'.

For example, set of natural number is denoted by $N = \{x \mid x \text{ is a natural number}\}$. We read it as N is the set of all those elements x, which is a natural number or N is the set of x such that x is a natural number.

Generally we use set builder form of expression to represent sets when it is difficult or impossible to list all the elements of a set.

In the set builder form:

i. *All commas are read as "and".*

ii. $x, y \in R$ is an abbreviated way of saying $x \in R$ and $y \in R$.

Example 2.4

Following are some of the examples of set builder form.

i. If A is a set consisting of the letters a, e, i, o, u, then we write

$$A = \{x: x \text{ is a vowel}\}$$

ii. If A is a set of English alphabets, then we write

$$A = \{x: x \text{ is a letter of English alphabet}\}$$

iii. If $B = \{1, 3, 5, 7, 9, 11\}$, then we write

$$B = \{x \mid x \text{ is odd positive integer}, x \leq 11\}$$

iv. The set of Americans who visited India in 2002, we write as

$$B = \{x \mid x \text{ is a American who visited India in 2002}\}$$

v. Let $A = \{x \mid x^2 + x - 6 = 0\}$. Here, we can say that A consists of those elements which are the solutions of the equation $x^2 \quad x \quad 6 \quad 0$, sometime this is called solution set of the given equation. As the solution of the equation are 2 and –3, we can also write $A = \{2, -3\}$.

A few of the other sets of numbers that we use frequently are:

N = the set of natural numbers = $\{1, 2, 3, 4, \ldots\}$

Z = the set of integers = $\{\ldots, -2, -1, 1, 2, 3, \ldots\}$

Q = the set of rational numbers

= $\{x \mid x = p/q$, where p and q are integers, $q \neq 0\}$

R = the set of real numbers

C = the set of complex numbers

= $\{x: x = a + ib$, where a and b are real numbers and $i^2 = -1\}$.

Example 2.5

Write the following sets in set builder form when $N = \{1, 2, 3, 4, \ldots\}$

i. $A = \{1, 2, 3, \ldots, 8\}$

ii. $B = \{1, 4, 9, 16, 25, 36\}$

iii. $C = \{1, 3, 5, 7, 9\}$

iv. $E \quad 1, 1/2, 1/3, 1/4, 1/5$.

Solution

i. $A = \{x: x \in N \text{ and } 1 \leq x \leq 8\}$

ii. $B = \{x: x \in N \text{ and } x = n^2, \text{ where } 1 \leq n \leq 6\}$

iii. $C = \{x: x \in N \text{ and } x = 2n + 1, \text{ where } n = 0, 1, 2, 3, 4\}$

iv. $E = \{x: x = 1/n, \text{ where } n = 1, 2, 3, 4, 5\}$.

It is to be noted here that even if we can list the elements of a set, it may not always be practical to do so. For example if we want to list the number of rats born in the world during the year 2002, it is not possible, though theoretically it is possible to compile such a list. Here, we mean to say that we describe a set by listing its members only if the set contains a few elements, otherwise we describe a set in terms of the property by which we characterize its elements. This fact of describing a set in terms of the property is formally stated as the ***principle of abstraction****.*

Principle of Abstraction

Consider a set S and any property P in it, then there is a set A such that the members of A are exactly those elements of S which have the property P.

2.2 Finite and Infinite Set

A set is said to be **finite** if it has finite number of elements and if the numbers of elements are infinite then the set is said to be **infinite**. For example

$S = \{a, e, i, o, u\}$ is a finite set while A = {Number of stars in the sky} is an infinite set.

2.2-1 Cardinality

The number of different elements in a finite set is called **cardinality** of a set and denoted by #*A*. Sometimes cardinality of set *A* is also called as the **order** of a set and is denoted by *n*(*A*). For example, set *A* = {2, 4, 6, 8, 10} is a finite set and its cardinality or order is 5.

Some texts use cardinality as | A | or card (A) instead of # A or n (A).

Example 2.6

Let set *A* = {Delhi, Mumbai, Kolkata, Chennai}. The set *A* is finite and its cardinality is four.

REMARK

One must keep in mind that while listing the elements of a set, an element may occur more than once. But to write the set that element may not be repeated, since in a set each element is thought of as occurring only once. For example let us have collection {*a*, *a*, *e*, *u*, *i*, *o*, *i*, *u*, *u*, *a*} then the set corresponding to these letters is written as {*a*, *e*, *i*, *o*, *u*}.

Here we note that in listing the elements of the set, order of the elements is immaterial.

2.3 Empty Set or Null Set

A set with no element is called an **empty set** or a **null set** or a **void set** or a **vacuous** and is denoted by ∅. Thus ∅ = {}.

Also by definition, the null set is regarded as a finite set of order zero.

i. {0} is not a null set, since it contains 0 as its member.
ii. {∅} is not a null set. It is a set having empty set.

Example 2.7

Some examples of empty sets are as follows:

i. { *x*: *x* is a year with 380 days}
ii. { *y*: *y* is a real number satisfying $y^2 = -3$}
iii. Set of fish, that cannot swim in water
iv. { *x*: *x* is an integer satisfying $3x + 5 = 6$}.

Example 2.8

Which of the following sets are empty sets?

i. {*x*: *x* is an imaginary number and $x^2 - 1 = 0$}
ii. {*x*: *x* is a real number and $x^2 = -4$}
iii. {*x*: *x* is a real number and $x = x + 1$}
iv. {*x*: *x* is a real number and $x = 3x + 1$}.

Solution

i. Since *x* is an imaginary number and $x^2 - 1 = 0$ gives $x = \pm 1$ which is real, hence the given set is empty
ii. empty

iii. empty
iv. not empty.

2.4 Singleton

A set consisting of a single element is called a **singleton set**. For example $A = \{3\}$ is a singleton, as it contains only one element, i.e. 3.

2.5 Subset

Let A and B be two sets. The set A is called **subset** of set B iff every element of A is an element of B. This relationship is written as

$$A \subseteq B \text{ or } B \supseteq A$$

And read as A is **contained** in B or B contains A. So, a set B is said to be **superset** of set A iff set B contains all the elements of set A. If A is not a subset of B, it means there exist at least one element of A that does not belong to B, we write $A \not\subseteq B$.

The words "subfamily" and "sub collection" are synonymous with subset.

Example 2.9

i. If $A = \{2, 4, 6\}$ and $B = \{1, 2, 3, 4, 5, 6\}$ then $A \subseteq B$.
ii. If $S = \{a, e, i, o, u\}$ and $T = \{a, b, e, i, o, u\}$ then $S \subseteq T$.
iii. $N \subseteq Z \subseteq R \subseteq C$.
iv. If $A = \{3, 4, 5\}$ and $B = \{1, 2, 4, 5, 6\}$ then $A \not\subseteq B$ because $3 \in A$ but $3 \notin B$.

2.5-1 Proper Subset

Let A and B be two sets. The set A is called the **proper subset** of B if every element of A is the element of B and there exist at least one element of B which is not an element of the set A. We denote proper subset by "$\subset$". For example $A = \{2, 4, 6\}$ is a proper subset of $B = \{2, 4, 6, 8\}$. Here element 8 of B is not contained in A.

i. The larger set is always kept towards the open end of the sign $\subset$ or $\supset$. The symbol $\subset$ or $\supset$ resembles in some ways the sign $<$ or $>$.

ii. "$\subseteq$" denotes subset whereas "$\subset$" denotes proper subset.

Example 2.10

Rewrite the following statements using set notations:

i. 1 is an element of S
ii. 2 is not an element of S
iii. A is a subset of S
iv. B is not a subset of A
v. S is a superset of A.

Solution

i. $1 \in S$
ii. $2 \notin S$
iii. $A \subseteq S$
iv. $B \not\subseteq A$
v. $S \supseteq A$.

2.6 Equality

Two sets S and T are said to be **equal** if every element of S is an element of T and every element of T is also an element of S. Equality of two sets is expressed by the notation "=" i.e. $S = T$.

$S = T$ can equivalently be said as $S \subseteq T$ and $T \subseteq S$ i.e. each set is a subset of the other. In other words $S = T$ implies whenever $x \in S$, then $x \in T$ and whenever $x \in T$, then $x \in S$.

Example 2.11

i. If $A = \{a, b, c, d\}$, $B = \{c, a, b, d\}$ then $A = B$.
ii. If $A = \{1, 3, 5, 7\}$, $B = \{1, 7, 5, 3\}$ then $A = B$.
iii. If $X = \{x, y, z, i, m\}$ and $Y = \{y, i, z, n, r\}$ then $X \neq Y$.
iv. $\{2, 8, 10\} = \{2, 8, 8, 10\}$.
v. If $X = \{\{2, 4\}, 6\}$ and $Y = \{2, 4, 6\}$ then $X \neq Y$.
vi. If $P = \{x: (x\text{-}1)(x\text{-}2) = 0\}$ and $Q = \{1, 2\}$, then $P = Q$.

To show equality of two sets say $S = T$, it is sufficient to show that $S \subseteq T$ and $T \subseteq S$.

Example 2.12

Which of the following sets are equal $\{p, q, r\}$, $\{r, p, q, r\}$, $\{p, q, r, q\}$.

Solution

All the three sets are equal, since order and repetition do not change a set.

2.7 Universal Set

The **universal** set or the universe is the set of elements under discussion for possible membership in a set. It is usually denoted by the letter U. For example if we are discussing about human population, then the universal set consists of all the people in the world. Similarly, if we have a set S = {Mars, Jupiter, Pluto, Earth}, then the universal set U for S will be set of all planets in the universe.

2.8 Properties of a Set

i. Every set S is a subset of the universal set U, since all the elements of S belongs to set U.
ii. Every set S is a subset of itself since all the elements of S belong to S.
iii. Empty set $\varnothing$ is a proper subset of every set except possibly $\varnothing$ itself.

Proof

Let us consider a set A. Suppose $\varnothing$ is not a proper subset of A. If it is so, then there exists at least one element say x such that $x \in \varnothing$ and $x \notin A$. But for all x, $x \notin \varnothing$ since $\varnothing$ is the empty set. This implies that $x \in \varnothing$ and $x \notin \varnothing$ which is a contradiction hence our assumption that $\varnothing$ is not a proper subset of A is wrong therefore, empty set $\varnothing$ is a proper subset of every set except $\varnothing$ itself.

iv. If every element of A belongs to set B, and every element of B belongs to a set C, then every element of A belongs to C

or

If A is a subset of B and B is a subset of C then A is a subset of C, i.e. if $A \subseteq B$ and $B \subseteq C$ then $A \subseteq C$.

Proof

Let x be any element of set A. Since $A \subseteq B$, therefore $x \in A \Rightarrow x \in B$. As $B \subseteq C$, therefore $x \in B \Rightarrow x \in C$. Thus it follows that $x \in A \Rightarrow x \in C$. Therefore by the definition of the subset we have $A \subset C$.

It is important to have a clear-cut distinction between membership and inclusion. The difference can be represented with the help of following example.

Let $A = \{2, 4, 6\}$, $B = \{2, 4\}$, $C = \{2, 6\}$ and $D = \{6\}$, then we can see that $B \subseteq A$, $C \subseteq A$ and $D \subseteq A$ or $\{2, 4\} \subseteq \{2, 4, 6\}$, $\{2, 6\} \subseteq \{2, 4, 6\}$ and $\{6\} \subseteq \{2, 4, 6\}$.

But $2 \in \{2, 4, 6\}$, and 2 is not included in $\{2, 4, 6\}$ though $\{2\} \subseteq \{2, 4, 6\}$. It is to be noted that a set can be included in but an element can be members of a set. But there are many occasions in which a set may have some other sets as elements. For example $A = \{\{2\}, 4, 6\}$, then

$$\{2\} \in A;\ \{\{2\}, 4\} \subseteq A;\ \{\{2\}\} \subseteq A;\ 4 \in A;\ \{2, 6\} \subseteq A$$

v. If $A \subseteq B$ and $B \subseteq A$, then A and B have same elements, i.e. $A = B$. Conversely, if $A = B$ then $A \subseteq B$ and $B \subseteq A$.

Proof

If $A \subseteq B$, then each element of A is an element of B and similarly if $B \subseteq A$, then each element of B is an element of the set A. This implies that each element of A is an element of B and each element of B is an element of A, i.e. $A = B$.

2.9 Comparability of Sets

Two sets A and B are said to be **comparable** if either of following happens:

i. $A \subset B$

ii. $B \subset A$

iii. $A = B$.

If neither of above three exist i.e. $A \not\subset B$, $B \not\subset A$ and $A \neq B$, then A and B are said to be **incomparable**.

Example 2.13

Let $A = \{2, 4, 6\}$ and $B = \{2, 4\}$. Here $B \subseteq A$, hence set A and B are comparable. But $A = \{1, 3, 5, 7\}$ and $B = \{2, 4, 6\}$ are incomparable as any of the three required property for the comparability is not satisfied.

2.10 Family of a Set

If the objects constituting the set are sets themselves then the set formed is called the **Family of sets** or the **class of sets** or the **set of sets**. For example if $\{a\}$, $\{b\}$, $\{a, b\}$, $\{a, b, c\}$ are the sets then the set A of these sets or family A of sets is

$$A = \{\{a\}, \{b\}, \{a, b\}, \{a, b, c\}\}$$

But it is to be noted that when a set A is given and we are interested in its particular subsets then to avoid confusion we speak of **class of sets** or **collection of sets** rather then set of sets.

Example 2.14

Consider $P = \{5, 10, 15, 20, 25\}$. Let Q be the class of subsets of P which contain exactly three elements of P, then

$$P = \{\{5, 10, 15\}, \{10, 15, 20\}, \{5, 15, 20\}, \{5, 10, 20\}\}.$$

The elements of P are $\{5, 10, 15\}$, $\{10, 15, 20\}$, $\{5, 15, 20\}$ and $\{5, 10, 20\}$.

2.11 Power Set

If S is a set, then the set of all subsets (proper or not) of S is called the **power set** of S and is denoted by Power(S) or $P(S)$. The number of elements in $P(S)$ is 2 raised to the power of number of elements of S, i.e.

$n\,P(s) = 2^{n(s)}$, where $n\,P(s)$ is the number of elements of power set and $n(s)$ is the number of element of set S.

The set S itself and the null set ∅ are both members of power set.

Example 2.15

i. If $S = \{1, 2\}$, then $P(S) = \{\{1\}, \{2\}, \{1, 2\}, \varnothing\}$. Total number of elements $= 2^2 = 4$.

ii. If $S = \{a, b, c\}$, then $P(S) = \{\{a\}, \{b\}, \{c\}, \{a, b\}, \{a, c\}, \{b, c\}, \{a, b, c\}, \varnothing\}$. Total number of elements $= 2^3 = 8$.

iii. If $S = \varnothing$, then $P(S) = \{\varnothing\}$.

Theorem 2.1

If a finite set A consists of n elements then there are 2^n subsets of the set A.

Proof

Let $A = \{a_1, a_2, a_3 \cdots, a_n\}$. The subsets each of which has exactly one element out of n element of A will be of the type $\{a_1\}, \{a_2\}, \{a_3\}, \ldots, \{a_n\}$, and the number of subsets is $nC1$. Similarly, the subsets each of which has exactly two elements out of n elements of A are of the type $\{a_1, a_2\}, \{a_1, a_3\}, a_1, a_4\}, \ldots$ and the number of such subsets is nC_2. The number of those subsets which have exactly r elements of A is nCr. Similarly, the number of subsets each of which has exactly n elements of A is nCn, Finally the number of those subsets which have no element at all is $n\,C_0$. Hence the total number of subsets of A

$$^nC_0 \quad ^nC_1 \quad ^nC_2 \quad \cdots\cdots \quad ^nC_r \quad \cdots\cdots \quad ^nC_n$$

$$1 \quad ^nC_1 \quad ^nC_2 \quad \cdots\cdots \quad ^nC_r \quad \cdots\cdots \quad 1$$

$$(1 \quad 1)^n \quad 2^n$$

$$[\therefore (1+x)^n = 1 + {}^nC_1x + {}^nC_2x^2 + \ldots + {}^nC_rx^r + \ldots + {}^nC_nx^n]$$

Hence, the number of elements in a power set of a finite set of order n will be having 2^n elements.

2.12 Index Set

Let $I = \{s_1, s_2, s_3 \cdots\}$ and A be a family of sets $A = \{A_{s1}, A_{s2}, A_{s3}, \ldots\}$ such that for any $s_i \in I$ there corresponds a set $A_{si} \in A$, and also $A_{si} = A_{sj}$ if and only if $s_i = s_j$, then A is called **index set** and any subscript such as s_i in A_{si} is called an **index**.

We can also write index family of set as $A = \{A_i\}_{i \in I}$.

Example 2.16

Identify which of the following are true and which are false, when $A = \{1, 2, 3, x, y, z\}$.

i. $1 \in A$	ii. $x \subset A$
iii. $\{y\} \in A$	iv. $\varnothing \in A$
v. $A \subset A$	vi. $A \in A$.

Solution

i. Since 1 is an element of the set A, therefore the statement is true.

ii. x is an element of A, not a subset of A, therefore the statement is false.

iii. $\{y\}$ is a set and set A is not a set of sets, therefore the statement is false.

iv. False.

v. A is not a proper subset of itself, hence false

vi. False.

Example 2.17

Assume A, B and C are arbitrary sets. State whether the following statements are True or False.

1. $\{a, \varphi\} \in \{a, \{a, \varphi\}\}$.
2. If $A \in B$ and $B \subseteq C$, then $A \in C$.
3. If $A \in B$ and $B \subseteq C$, then $A\ C \subseteq C$.
4. If $A \in B$ and $B \subseteq C$, then $A\ C \in C$.
5. If $A \subseteq B$ and $B \in C$, then $A\ C \subseteq C$.

Solution

1. True
2. True
3. False
4. False
5. False.

Exercise 2.1

1. Represent the following set in tabular form.

   i. $B = \{x: x$ is a letter in the word 'MATHEMATICS'$\}$

   ii. $A = \{x: x \in Z$ and $x^2 < 16\}$

   iii. $C = \{x: x$ is a positive integer and $x \leq 5\}$.

2. Represent the following set in set builder form.

   i. $A = \{a, e, i, o, u\}$

   ii. $B = \{3, 5, 7, 9, 11, \ldots, 77, 79\}$

   iii. $C =$ the rational numbers that are strictly between -2 and 2.

3. Determine which of the following sets are finite.

   i. $A = \{$days in a week$\}$

   ii. $B = \{$dogs living in India$\}$

   iii. $C - \{$number of books at Modinagar$\}$

   iv. $D = \{$negative integers greater than $-1\}$

   v. $E = \{$even integers$\}$

   vi. $F = \{$human beings of age less than 25 years$\}$.

4. Which of the following sets are equal .

   i. $A = \{x: x^2\text{-}7x + 12 = 0\}$, $B = \{x: x \in N, x < 7\}$, $C = \{3, 4\}$, $D = \{3, 4, 3\}$

   ii. $P = \{x: x^2\text{-}\ 5x + 6\}$, $Q = \{2, 3\}$, $R = \{x: x \in N$ and $1 < x < 4\}$,
   $S = \{x: x \in N$ and $x \leq 3\}$.

   iii. $A = \{x \mid x$ is a letter in the word "title"$\}$,
   $B = \{x \mid x$ is a letter in the word "little"$\}$,
   $C = \{e, i, l, t, k\}$.

5. Let $A = \{1, 2, 3, 4\}$, $B = \{2, 3\}$, $D = \{4, 1, 2, 3, 2\}$ and $C = \{1, 5, 7\}$. Identify each of the following as true or false. Give reasons for your answer.
   i. $A \subseteq B$ ii. $4 \in A$
   iii. $\{5\} \in C$ iv. $A = D$
   v. $B \subseteq A$ vi. $C \not\subseteq A$
   vii. $\varnothing \in D$ viii. $\varnothing \subseteq B$.
6. Given $S = \{2, a, \{3\}, 4\}$ and $R = \{\{a\}, 3, 4, 1\}$. Answer whether the following statements are true or false:
   i. $\{a\} \in S$ ii. $\{a\} \in R$
   iii. $\{a, 4, \{3\}\} \subseteq S$ iv. $\{\{a\}, 1, 3, 4\} \subset R$
   v. $R = S$ vi. $\varnothing \subset R$.

Answers

1. i. B = {M, A, T, H, E, M, A, T, I, C, S} ii. A = {-1, -2, -3, 1, 2, 3}
   iii. C = {1, 2, 3, 4, 5}.
2. i. A = {x: x is a vowel of English alphabet} ii. $B = 2n + 1 : n \in Z, 2 \le n \le 39\}$
   iii. $C = \{x \in Q : -2 < x < 2\}$.
3. i. Finite ii. Finite
   iii. Finite iv. Finite as there is no negative integer greater than 1
   v. infinite vi. Finite
4. i. $A = C = D$ ii. $P = Q = R$
   iii. $A = B$
5. i. False ii. True
   iii. False iv. True
   v. True vi. True
   vii. False vii. True
6. i. False ii. True
   iii. True iv. False
   v. False vi. False

2.13 Venn Diagrams

It is often very useful to represent a situation under consideration with the help of a picture or diagram or graph. Sets can also be represented graphically using Venn diagrams, named after the British logician John Venn who started their use in 1881. It can formally be defined as follows

A **Venn diagram** is a pictorial representation of sets in which enclosed areas in the plane represent sets. The universal set U is represented by the interior of a rectangle and set by the disks lying within the rectangle. For example if $A \subseteq B$, then it can be represented by Venn diagram as shown in Fig.2.1 (a). And similarly $A \not\subseteq B$ can be represented as shown in Fig. 2.1 (b)

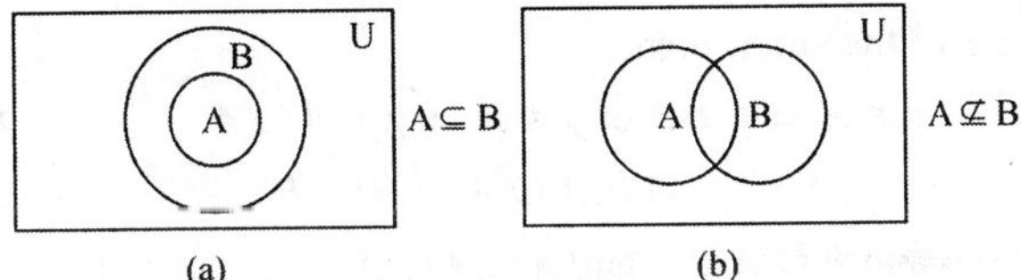

Fig. 2.1

2.14 Set Operations

We are very much familiar with the operations on real numbers like addition, multiplication etc. Here, we would discuss system of operations on the set, some of which resemble with the algebraic properties satisfied by real numbers and their operations. There are many operations on sets, the two most commonly used are union and intersection of sets. These operations play a key role in many applications and ideas that follow.

2.15 Union of Sets

Let A and B be two sets. The **union** also called **Join** of sets A and B is a new set consists of all elements that belong to either A or B (or both) and denoted by $A \cup B$. Thus

$$A \cup B = \{x: x \in A \text{ or } x \in B\}$$

or

$$A \cup B = \{x: x \in A \vee x \in B\}$$

Fig. 2.2(a) and (b) are Venn diagrams in which $A \cup B$ is shaded. Here one can note that in Fig. 2.2(a) no element is common to A and B but in Fig.2.2 (b) some elements are common.

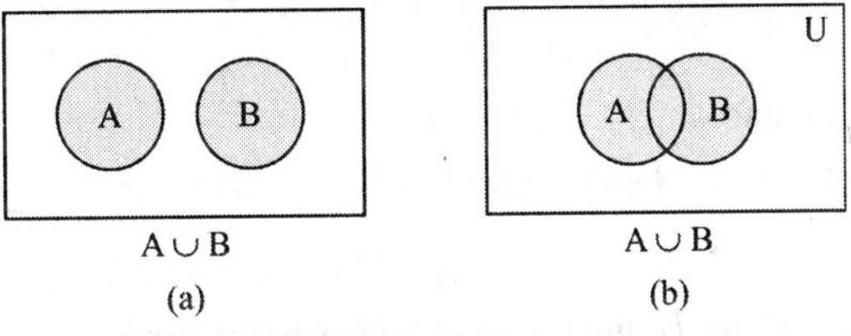

Fig. 2.2

REMARK

"$A \cup B$" is read as "A union B" or "A cup B". Here "or" is used in the sense of "and/or" i.e. "inclusive or".

Example 2.18

Let $A = \{a, b, c, d\}$, $B = \{b, d, p, q\}$ and $C = \{a, e, i, o, u\}$, then

$$A \cup B = \{a, b, c, d\} \cup \{b, d, p, q\} = \{a, b, c, d, p, q\}$$

$$A \cup C = \{a, b, c, d\} \cup \{a, e, i, o, u\} = \{a, b, c, d, e, i, o, u\}$$

$$B \cup C = \{b, d, p, q\} \cup \{a, e, i, o, u\} = \{a, b, d, i, o, p, q, u\}$$

Theorem 2.2 [Idempotent Law]

The union of sets is idempotent i.e. $A \cup A = A$.

Proof

Let x be any element of $A \cup A$, then

$$x \in A \cup A \Rightarrow x \in A \text{ or } x \in A \qquad \text{[by definition of union]}$$

$$\therefore \quad A \cup A \subseteq A \qquad ...(1)$$

Further, let y be any element of the set A, then

$$y \in A \Rightarrow y \in A \text{ or } y \in A \Rightarrow y \in A \cup A$$

$$\therefore \quad A \subseteq A \cup A \quad \ldots(2)$$

Now from (1) and (2) we see that $A \cup A \subseteq A$ and $A \subseteq A \cup A \Rightarrow A \cup A = A$

To prove that the set $A = B$, we must prove $A \subseteq B$ and $B \subseteq A$.

Theorem 2.3

Identity Law $A \cup \varnothing = A$, where A is any set.

Proof

Let x be an element of the set $A \cup \varnothing$, then

$$x \in A \cup \varnothing \Rightarrow x \in A \text{ or } x \in \varnothing \quad \text{[by definition of union]}$$

$$\Rightarrow x \in A \quad [\text{Since } \varnothing \text{ is the null set, } x \notin \varnothing]$$

$$\therefore \quad A \cup \varnothing \subseteq A \quad \ldots(1)$$

Further if y be any element of the set A, then

$$y \in A \Rightarrow y \in A \text{ or } y \in \varnothing$$

$$\Rightarrow y \in A \cup \varnothing$$

$$\therefore \quad A \subseteq A \cup \varnothing \quad \ldots(2)$$

From (1) and (2), we conclude that

$$A \cup \varnothing \subseteq A \text{ and } A \subseteq A \cup \varnothing \Rightarrow A \cup \varnothing = A.$$

Theorem 2.4

If A is any subset of the universal set U, then $A \cup U = U$ (**Bound law**).

Proof

Let x be any element of the universal set U, then

$$x \in U \Rightarrow x \in U \text{ or } x \in A$$

$$\Rightarrow x \in A \cup U$$

$$\therefore \quad U \subseteq A \cup U \quad \ldots(1)$$

Further, $\quad x \in A \cup U \Rightarrow x \in A \text{ or } x \in U \Rightarrow x \in U \quad [\because A \subseteq U]$

Hence $\quad A \cup U \subseteq U \quad \ldots(2)$

Therefore, from (1) and (2) we conclude that $A \cup U = U$.

Theorem 2.5

Union of two sets is **commutative**, i.e. if A and B are two sets, then

$$A \cup B = B \cup A.$$

Proof

Let x be any element of the set $A \cup B$, then

$$x \in A \cup B \Rightarrow x \in A \text{ or } x \in B$$

$$\Rightarrow x \in B \text{ or } x \in A$$
$$\Rightarrow x \in B \cup A$$

Hencc $A \cup B \subseteq B \cup A$ ... (1)

Further if y be any element of the set $B \cup A$, then

$$y \in B \cup A \Rightarrow y \in B \text{ or } y \in A$$
$$\Rightarrow y \in A \text{ or } y \in B$$
$$\Rightarrow y \in A \cup B$$

Hence $B \cup A \subseteq A \cup B$ ... (2)

Therefore, from (1) and (2) we conclude that $A \cup B = B \cup A$. Hence union of two sets is commutative.

Theorem 2.6

The union of sets is **associative,** i.e. if A, B and C are any sets, then

$$(A \cup B) \cup C = A \cup (B \cup C).$$

Proof

Let x be any element of the set $(A \cup B) \cup C$, then

$$x \in (A \cup B) \cup C \Rightarrow x \in (A \cup B) \text{ or } x \in C$$
$$\Rightarrow [x \in A \text{ or } x \in B] \text{ or } [x \in C]$$
$$\Rightarrow [x \in A] \text{ or } [x \in B \text{ or } x \in C]$$
$$\Rightarrow [x \in A] \text{ or } [x \in B \cup C]$$
$$\Rightarrow x \in A \cup (B \cup C)$$

This shows that if x is any element of the set $(A \cup B) \cup C$, then x is also an element of $A \cup (B \cup C)$.

$\therefore$ $(A \cup B) \cup C \subseteq A \cup (B \cup C)$ ... (1)

Further, if y be any element of the set $A \cup (B \cup C)$, then

$$y \in A \cup (B \cup C) \Rightarrow [y \in A] \text{ or } [y \in B \cup C]$$
$$\Rightarrow [y \in A] \text{ or } [y \in B \text{ or } y \in C]$$
$$\Rightarrow [y \in A \text{ or } y \in B] \text{ or } [y \in C]$$
$$\Rightarrow [y \in A \cup B] \text{ or } [y \in C]$$
$$\Rightarrow y \in (A \cup B) \cup C$$

Hence $A \cup (B \cup C) \subseteq (A \cup B) \cup C$ ...(2)

Therefore, from (1) and (2) we conclude that

$$(A \cup B) \cup C = A \cup (B \cup C).$$

2.16 Intersection of Sets

Let A and B be two nonempty sets. The **intersection** sometimes called **meet** of A and B, is a new set consists of all elements that belong to both A and B. It is denoted by $\cap$. Thus

$$A \cap B = \{x : x \in A \text{ and } x \in B\}.$$

Example 2.19

Let $A = \{1, 2, 3, 4, 5\}$, $B = \{2, 4, 6, 8\}$ and $C = \{1, 3, 5, 6, 7\}$, then

$$A \cap B = \{1, 2, 3, 4, 5\} \cap \{2, 4, 6, 8\} = \{2, 4\}$$

$$A \cap C = \{1, 2, 3, 4, 5\} \cap \{1, 3, 5, 6, 7\} = \{1, 3, 5\}$$

and $$B \cap C = \{2, 4, 6, 8\} \cap \{1, 3, 5, 6, 7\} = \{6\}.$$

Theorem 2.7

i. Intersection of sets is idempotent, i.e. $A \cap A = A$.

ii. $A \cap \varnothing = \varnothing$ (Bound law).

iii. **Identity Law,** $A \cap U = A$, where A is any subset of the universal set U.

iv. Intersection of sets is **commutative** i.e. $A \cap B = B \cap A$.

iv. Intersection of sets is **associative** i.e. $(A \cap B) \cap C = A \cap (B \cap C)$.

v. **Distributive laws**

a. $A \cap (B \cup C) = (A \cap B) \cup (A \cap C)$ (Distributive law of intersection over union)

b. $A \cup (B \cap C) = (A \cup B) \cap (A \cup C)$ (Distributive law of union over intersection)

These distributive laws can be easily understood by the following Venn diagrams.

a.

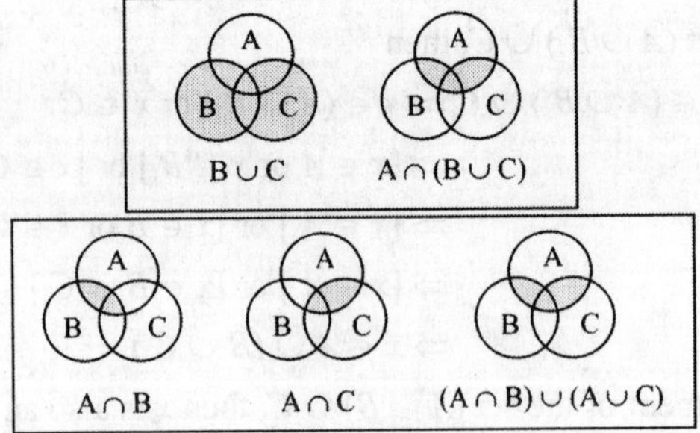

Fig. 2.3

b.

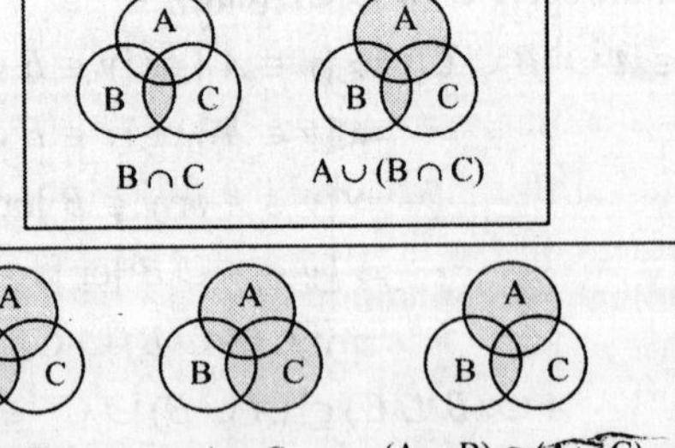

Fig. 2.4

All the above theorems (laws) can easily be proved in the same way as we have proved for union of sets.

Example 2.20

Given, $A = \{a, b, c, d, e\}$, $B = \{a, c, e, g\}$ and $C = \{b, e, f, g\}$, prove that

$$(A \cup B) \cap C = (A \cap C) \cup (B \cap C)$$

Solution

$$A \cup B = \{a, b, c, d, e\} \cup \{a, c, e, g\} = \{a, b, c, d, e, g\}$$

$$\Rightarrow \quad (A \cap B) \cap C = \{a, b, c, d, e, g\} \cap \{b, e, f, g\} = \{b, e, g\} \quad \ldots (1)$$

and $\quad A \cap C = \{a, b, c, d, e\} \cap \{b, e, f, g\} = \{b, e\},$

$B \cap C = \{a, c, e, g\} \cap \{b, e, f, g\} = \{e, g\}$

$\Rightarrow \quad (A \cap C) \cup (B \cap C) = \{b, e\} \cup \{e, g\} = \{b, e, g\}.$ ... (2)

$\Rightarrow$ (1) and (2) are equal.

2.17 Disjoint Sets

Two sets A and B are called **disjoint** if they have no common element i.e. $A \cap B = \varnothing$. For example if $A = \{2, 4, 6\}$ and $B = \{1, 3, 5\}$, then A and B are disjoint as they have no element in common. Venn diagram for disjoint sets is shown in Fig. 2.5.

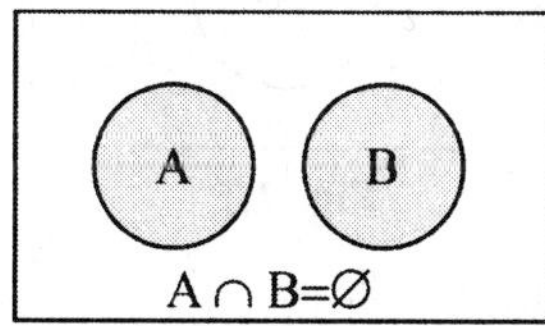

Fig. 2.5

A collection of sets is called a **disjoint collection** if for every pair of sets in the collection the two sets are disjoint. The elements of a disjoint collection are said to be **mutually disjoint**.

Consider an index set $A = (A_i)_{i \in j}$. The set A is disjoint collection if $A_i \cap A_j = \varnothing \ \forall \ i, j \in J, i \neq j$.

For example if $A_1 = \{\{a, b\}, \{c\}\}$, $A_2 = \{\{a\}, \{b, c\}\}$, and $A_3 = \{\{a, b, c\}\}$, then A_1, A_2 and A_3 are mutually disjoint because

$$A_1 \cap A_2 = \varnothing, A_1 \cap A_3 = \varnothing, \text{ and } A_2 \cap A_3 = \varnothing.$$

2.18 Relative Complement

Let A and B are two sets. The complement of A relative to B (or of A with respect to B or relative complement of A in B) written as $B - A$ or $B \sim A$ is the set of elements that are in B and not in A, i.e.

$$B - A = \{x: x \in B \text{ and } x \notin A\}.$$

The **relative complement** of A in B is also called the **difference** of B and A. If U be the universal set, then for any set A, the relative complement of A with respect to U i.e. $U - A$ is written as A^c or A' or $\sim A$ and is called the **absolute complement** or simply the **complement** of A.

Example 2.21

Let $B = \{1, 2, 3, 4, 5, 6, 7, 8\}$ and $A = \{1, 3, 5, 7\}$.

Then $B - A = \{2, 4, 6, 8\}$ and $A - B = \varnothing$. This shows $A - B \neq B - A$.

Example 2.22

i. If $U = Q$ (rational number), then Q'= the irrational numbers

ii. $U' = \varnothing$ and $\varnothing' = U$.

Example 2.23

Represent (i) $A - B$, (ii) A' and (iii) If $A \subseteq B$ then $B - A$ by Venn diagram.

Solution

i. Venn diagram for $A - B$

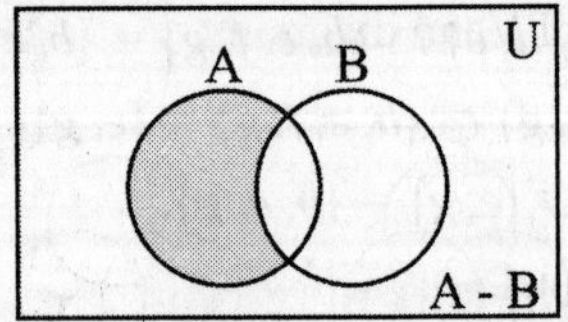

Fig. 2.6

ii. Venn diagram for A'

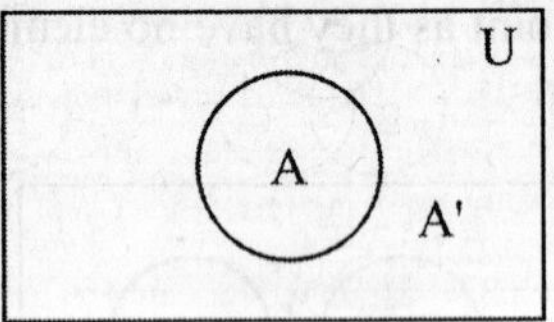

Fig. 2.7

iii. Venn diagram for $B - A$ when $A \subseteq B$

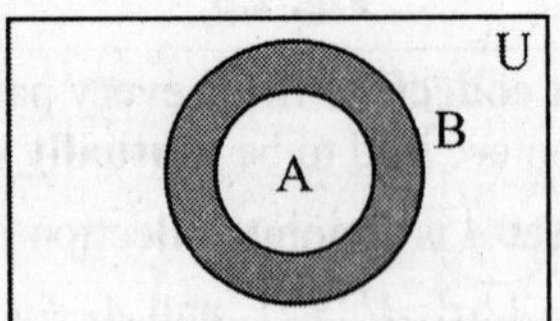

Fig. 2.8

For any three sets A, B, C.

i. $A - (B \cup C) = (A - B) \cap (A - C)$

ii. $A - (B \cap C) = (A - B) \cup (A - C)$.

Proof

Let $x \in A - (B \cup C)$

$\Rightarrow x \in A$ and $x \notin (B \cup C)$

$\Rightarrow (x \in A)$ and $(x \notin B$ and $x \notin C)$

$\Rightarrow (x \in A$ and $x \notin B)$ and $(x \in A$ and $x \notin C)$

$\Rightarrow (x \in (A - B))$ and $(x \in (A - C))$

$\Rightarrow x \in [(A - B) \cap (A - C)]$

$\Rightarrow A - (B \cup C) \subseteq (A - B) \cap (A - C)$ ... (1)

Let y be any element of $(A - B) \cap (A - C)$, then

$y \in (A - B) \cap (A - C) \Rightarrow [y \in (A - B)]$ and $[y \in (A - C)]$

$\Rightarrow [y \in A$ and $y \notin B]$ and $[y \in A$ and $y \notin C]$

$\Rightarrow y \in A$ and $[y \notin B$ and $y \notin C]$

$\Rightarrow y \in A$ and $[y \notin B \cup C]$

$\Rightarrow y \in A - (B \cup C)$

$\Rightarrow (A - B) \cap (A - C) \subseteq A - (B \cup C)$ ... (2)

From (1) and (2) we conclude that $A - (B \cup C) = (A - B) \cap (A - C)$

We can also prove this with the help of Venn diagrams.

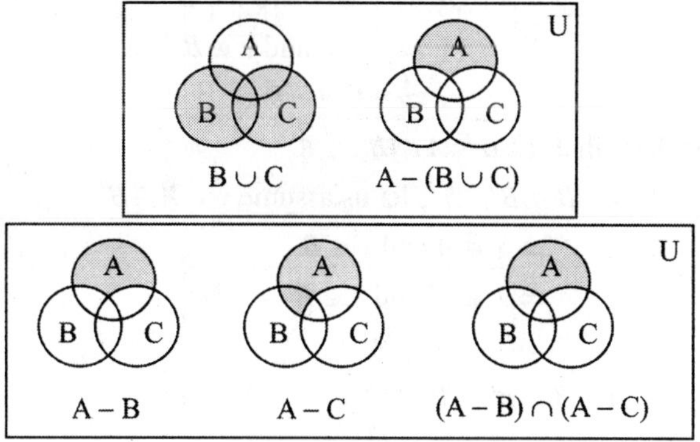

Fig. 2.9

ii. $A - (B \cap C) = (A - B) \cup (A - C)$ can be proved easily as we have proved for (i). The same can be represented with the help of Venn diagram as shown below

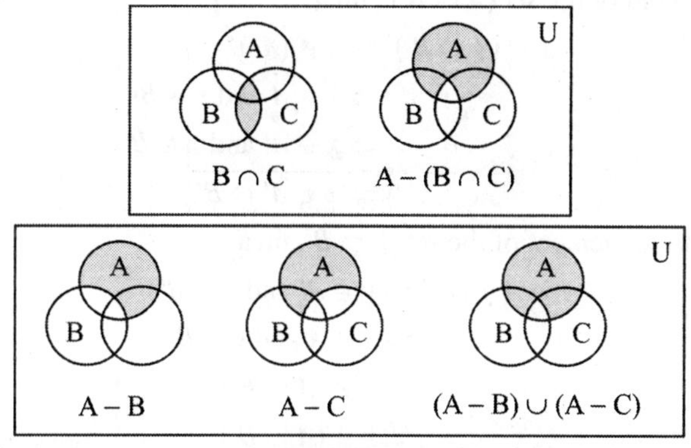

Fig. 2.10

Example 2.24

Prove that if A and B are two sets then $A - B \subseteq A$.

Proof

Consider an element x of the set A - B, then

$$x \in A - B \Rightarrow x \in A \text{ and } x \notin B \Rightarrow x \in A$$

Hence, $$A - B \subseteq A.$$

Example 2.25

Prove that $A - B = A \cap B' = B' - A'$.

Solution

Let x be an arbitrary element of the set A - B, then

$$\begin{aligned} x \in A - B &\Rightarrow x \in A \text{ and } x \notin B \\ &\Rightarrow x \in A \text{ and } x \in B' \\ &\Rightarrow x \in A \cap B' \end{aligned} \quad \ldots (1)$$

Further, let y be an arbitrary element of $A \cap B'$, then

$$y \in A \cap B' \Rightarrow y \in A \text{ and } y \in B'$$
$$\Rightarrow y \in A \text{ and } y \notin B$$
$$\Rightarrow y \in A - B \quad \text{...(2)}$$

From (1) and (2), we conclude that $A - B = A \cap B'$.

Now to show $A - B = B' - A'$, let us assume $x \in A - B$

$$\Rightarrow x \in A \text{ and } x \notin B$$
$$\Rightarrow x \notin A' \text{ and } x \in B'$$
$$\Rightarrow x \in B' - A'$$

Therefore $A - B = B' - A'$.

Example 2.26

Prove that $(A \cup B)' = A' \cap B'$.

Proof

Let x be an arbitrary element of the set $(A \cup B)'$, then

$$x \in (A \cup B)' \Rightarrow x \notin (A \cup B)$$
$$\Rightarrow x \notin A \text{ and } x \notin B$$
$$\Rightarrow x \in A' \text{ and } x \in B'$$
$$\Rightarrow x \in A' \cap B' \quad \text{... (1)}$$

Further, let y be an arbitrary element of the set $A' \cap B'$, then

$$y \in A' \cap B' \Rightarrow y \in A' \text{ and } y \in B'$$
$$\Rightarrow y \notin A \text{ and } y \notin B$$
$$\Rightarrow y \notin (A \cup B)$$
$$\Rightarrow y \in (A \cup B)' \quad \text{... (2)}$$

$\therefore$ From (1) and (2), we conclude that $(A \cup B) = A' \cap B'$.

Example 2.27

If A and B are sets, prove that $A \cup B = (A - B) \cup B$.

Proof

Let $x \in A \cup B$

$$\Rightarrow x \in A \text{ or } x \in B$$
$$\Rightarrow [x \in B \text{ or } x \in A] \text{ and } [x \in B \text{ or } x \notin B]$$

$[\because x \in B$ or $x \notin B$ is always true$]$

$$\Rightarrow [x \in B \text{ or } x \in A \text{ and } x \notin B]$$
$$\Rightarrow x \in B \text{ or } x \in (A - B)$$
$$\Rightarrow x \in (A - B) \text{ or } x \in B$$
$$\Rightarrow x \in (A - B) \cup \text{B} \quad \text{... (i)}$$

Further, let $y \in (A - B) \cup B$

$$\Rightarrow y \in B \text{ or } (y \in A - B)$$
$$\Rightarrow y \in B \text{ or } (y \in A \text{ and } y \notin B)$$
$$\Rightarrow (y \in B \text{ or } y \in A) \text{ and } (y \in B \text{ or } y \notin B)$$
$$\Rightarrow (y \in A \text{ or } y \in B) \text{ and } (y \in B \text{ or } y \notin B)$$
$$\Rightarrow y \in A \text{ or } y \in B$$
$$\Rightarrow y \in A \cup B$$
$$\Rightarrow (A - B) \cup B \subseteq A \cup B \qquad \text{... (ii)}$$

Therefore (i) and (ii) implies that $A \cup B = (A - B) \cup B$.

Example 2.28

Prove that $A \quad B - A \Rightarrow A \cap B - \varnothing$.

Solution

Given $A - B = A$

Let $x \in A \cap B$, then

$$x \in A \cap B \Rightarrow x \in A \text{ and } x \in B$$
$$\Rightarrow x \in (A - B) \text{ and } x \in B \qquad [\because A = A - B]$$
$$\Rightarrow (x \in A \text{ and } x \notin B) \text{ and } x \in B$$
$$\Rightarrow x \in A \text{ and } (x \notin B \text{ and } x \in B)$$
$$\Rightarrow x \in A \text{ and } x \in \varnothing$$
$$\Rightarrow x \in \varnothing$$
$$A - B = A \Rightarrow A \cap B = \varnothing. \text{ But } \varnothing \subseteq A \cap B$$
$$\Rightarrow A - B = A \Rightarrow A \cap B = \varnothing$$

Conversely, let us suppose $A \cap B = \varnothing$ and let $y \in A$, then

$$y \in A \Rightarrow y \in A \text{ and } (y \in B \text{ or } y \notin B)$$
$$\Rightarrow (y \in A \text{ and } y \in B) \text{ or } (y \in A \text{ and } y \notin B)$$
$$\Rightarrow y \in (A \cap B) \text{ or } y \in (A - B)$$
$$\Rightarrow y \in \varnothing \text{ or } y \in (A - B)$$
$$\Rightarrow y \in (A - B) \qquad [\because y \notin \varnothing]$$

$\therefore$ $\quad A \cap B = \varnothing \Rightarrow A \subseteq A - B.$

But $\quad A - B \subseteq A$

Hence $\quad A \cap B = \varnothing \Rightarrow A - B = A.$

2.19 Symmetric Difference

Let A and B are any two sets. The **symmetric difference** (or **Boolean Sum**) of A and B is denoted by $A \Delta B$ and defined by

$$A \Delta B = (A - B) \cup (B - A)$$

or $\quad A \Delta B = \{x: [x \in A \text{ and } x \notin B] \text{ or } [x \in B \text{ and } x \notin A]\}.$

Boolean sum is also expressed by $A \oplus B$.

One can observe here that the symmetric difference of two sets is the set of all elements which are not common to both the given sets. It can be represented with the help of Venn diagram as shown in Fig.2.11.

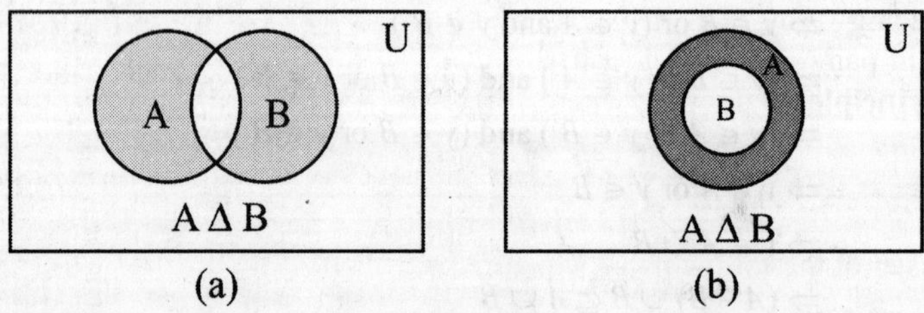

Fig. 2.11

Example 2.29

Let $A = \{a, b, c\}$, $B = \{c, d, e\}$ then $A \Delta B = \{a, b, d, e\}$.

2.20 Basic Set Identities

The identities which we have discussed earlier can be proved by assuming certain other identities also. Here we will list certain identities (laws) with their name in the Table 2.1.

TABLE 2.1

Idempotent Laws	
1(a) $A \cup A = A$	1(b) $A \cap A = A$
Commutative Laws	
2(a) $A \cup B = B \cup A$	2(b) $A \cap B = B \cap A$
Associative Laws	
3(a) $(A \cup B) \cup C = A \cup (B \cup C)$	3(b) $(A \cap B) \cap C = A \cap (B \cap C)$
Distributive Laws	
4(a) $A \cup (B \cap C) = (A \cup B) \cap (A \cup C)$	4(b) $A \cap (B \cup C) = (A \cap B) \cup (A \cap C)$
Identity Laws	
5(a) $A \cup \varnothing = A$	5(b) $A \cap U = A$
Bound Laws	
6(a) $A \cup U = U$	6(b) $A \cap \varnothing = \varnothing$
Absorption Laws	
7(a) $A \cup (A \cap B) = A$	7(b) $A \cap (A \cup B) = A$
Involution Laws	
8(a) $A \cup A' = U$	8(b) $A'' = A$
Complement Laws	
9(a) $A \cup A' = U$	9(b) $A \cap A' = \varnothing$
0/1 Laws	
10(a) $U' = \varnothing$	10(b) $\varnothing' = U$
De Morgan's Laws	
11(a) $(A \cup B)' = A' \cap B'$	11(b) $(A \cap B)' = A' \cup B'$

2.21 Duality

Two formulas/ identities E and E * in set algebra are said to be **dual** of each other if either one can be obtained from the other by replacing $\cup$ by $\cap$, $\cap$ by $\cup$, U by $\varnothing$ and $\varnothing$ by U. As one can observe from Table 2.1, the identities are arranged in pairs where each pair is a dual of other. For example 4(a) and 4(b). This fact of set algebra is called the **principle of duality**, where if any equation E is an identity, then its dual E^* is also an identity.

Example 2.30

Write the dual of each of the following set equation

i. $(A \cap U) \cap (\varnothing \cup A') = \varnothing$

ii. $(A \cup B \cup C)' = (A \cup C)' \cap (A \cup B)'$.

Solution

Duals can be obtained by interchanging $\cup$ by $\cap$ and vice-versa also U by $\varnothing$ and vice-versa in each set equation

i. $(A \cup \varnothing) \cup (U \cap A') = U$

ii. $(A \cap B \cap C)' = (A \cap C)' \cup (A \cap B)'$.

2.22 Counting Principle

Let A and B are finite subsets of a universal set U. If A and B are disjoint set (i.e. $A \cap B = \varnothing$) then the cardinality of union of A and B is given

$$|A \cup B| = |A| + |B|$$

If A and B overlap as shown in Fig. 2.12, then elements in $A \cap B$ belong to both sets, and the sum $|A| + |B|$ counts these elements twice. Hence to correct for this double counting we subtract $|A \cap B|$. Keeping these facts in mind we have the following theorem called addition principle. Because of above Fig. 2.12 this is also called the inclusion-exclusion principle.

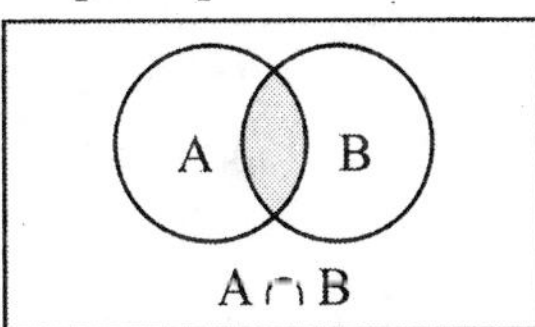

Fig. 2.12

Theorem 2.9

a. If A and B are finite sets, then $|A \cup B| = |A| + |B| - |A \cap B|$.

b. If A and B are disjoint then $A \cap B = \varnothing$ and $|A \cup B| = |A| + |B|$.

Corollary

If A, B, and C are finite sets, then so is $A \cup B \cup C$, then

$$|A \cup B \cup C| = |A| + |B| + |C| - |A \cap B| - |A \cap B| - |B \cap C| + |A \cap B \cap C|$$

This corollary can be generalized for more than three sets.

Example 2.31

Verify above corollary for the sets $A = \{1, 2, 3, 4, 5\}$, $B = \{1, 2, 5, 7, 8\}$ and $C = \{2, 4, 5, 7, 10, 11, 13, 15\}$.

Solution

We have $A \cup B \cup C = \{1, 2, 3, 4, 5, 7, 8, 10, 11, 13, 15\}$,

And $A \cap B = \{1, 2, 5\}, A \cap C = \{2, 4, 5\}, B \cap C = \{2, 5, 7\}, A \cap B \cap C = \{2, 5\}$.

So $|A| = 5, |B| = 5, |C| = 8, |A \cap B \cap C| = 2, |A \cap B| = 3,$
$|A \cap C| = 3, |B \cap C| = 3$, and $|A \cup B \cup C| = 11$

Since $|A \cup B \cup C| = |A| + |B| + |C| - |A \cap B| - |B \cap C| - |A \cap C| + |A \cap B \cap C|$

$\therefore$
$$11 = 5 + 5 + 8 - 3 - 3 - 3 + 2$$
$$11 = 11,$$

Hence the theorem.

Example 2.32

An institute wants to hire 20 computer personnel for computer-related work and 30 electronics personnel for performing certain electronics job. Of those hired eight can perform both jobs of types. How many personnel must be hired?

Solution

Let C be the set of computer personnel and E be the set of electronics personnel hired. Hence, $|C| = 20$, $|E| = 30$, $|C \cap E| = 8$. The number of personnel that must be hired is $|C \cup E|$.

But
$$|C \cup E| = |C| + |E| - |C \cap E|$$
$$= 20 + 30 - 8 = 22$$

So the institute must hire 22 personnel.

2.23 Generalized Set Operations

Earlier we have discussed the operations of union and intersection of two sets. Here we will extend the operations on any number of sets say $S_1, S_2, S_3, \ldots, S_n$.

Let S be the collection of sets. The union and the intersection of these sets can be defined respectively by

$$S_1 \cup S_2 \cup \ldots \cup S_m = \bigcup_{i=1}^{m} S_i$$
$$= \{ y : y \in S_i \text{ for some } S_i \}$$

$$S_1 \cap S_2 \cap \ldots \cap S_m = \bigcap_{i=1}^{m} S_i$$
$$= \{ y : y \in S_i \text{ for every } S_i \}.$$

Example 2.33

Consider the sets

$$S_1 = \{a, b, c, d, e, f, g, \ldots\} = A, S_2 = \{b, c, d, e, \ldots\},$$
$$S_3 = \{c, d, e, f, g, \ldots\}, S_4 = \{d, e, f, g, \ldots\},$$
$$\cup (S_n : n \in 1, 2, 3, 4\} = A \text{ and } \cap (S_n : n \in 1, 2, 3, 4\} = \varnothing.$$

Example 2.34

Let $S_n = \{ n, n+1, n+2, \ldots \}$. Then

$$\bigcup_{n=1}^{m} S_n = \bigcup_{n=1}^{m} \{n, n+1, n+2, \ldots\} = \{1, 2, 3, \ldots\}$$

and

$$\bigcap_{n=1}^{m} S_n = \bigcap_{n=1}^{m} \{n, n+1, n+2, ...\} = \{m, m+1, m+2, ...\}.$$

2.24 Applications of Set Theory

Why we study set theory? Here the applications of the same will make us clear about it. We know that the number of distinct elements of a finite set A is called cardinal number, denoted by $n(A)$.

Let A and B are non-empty sets.

Then,

i. $n(A \cup B) = n(A) + n(B) - n(A \cap B)$

ii. $n(A) = n(A - B) + n(A \cap B)$

iii. $n(B) = n(B - A) + n(A \cap B)$

iv. $n(A \cup B) = n(A - B) + n(B - A) + n(A \cap B)$

v. $n(A \cup B \cup C) = n(A) + n(B) + n(C) - n(A \cap B) - n(B - C) - n(B - C) + n(A \cap B \cap C)$.

Example 2.35

In a class of 25 students, 13 have taken Mathematics. 9 have taken Mathematics but not Biology. Find the number of students who have taken Mathematics and Biology and those who have taken Biology but not Mathematics.

Solution

Let A and B be sets of students who have taken Mathematics and Biology respectively. Then $(A - B)$ is the set of students who have taken Mathematics but not Biology. So $n(A) = B$, $n(A \cap B) = 25$, $n(A - B) = 9$.

We know that $n(A) = n(A - B) + n(A \cap B)$

$$13 = 9 + n(A \cap B)$$

$$\Rightarrow n(B) = 16$$

Thus, 4 students have taken both Mathematics and Biology.

Again, $n(A \cup B) = n(A) + n(B) - n(A \cap B)$

$$25 = B + n(B) - 4$$

$$\Rightarrow n(B) = 16$$

Now $n(B - A) + n(A \cap B) = n(B)$

$$\Rightarrow n(B - A) = n(B) - n(A \cap B) = 16 - 4 = 12$$

Hence, 12 students have taken Biology but not Mathematics.

2.25 Ordered Pairs and *n* Tuples

An ordered pair consists of two objects in a given fixed order. If x and y be two objects, then $\langle x, y \rangle$ or (x, y) is called their ordered pair, where x is called the first member and y the second member. It is to be noted that two objects need not be distinct, and the ordered pair $\langle x, y \rangle$ and $\langle y, x \rangle$ are quite different while the set $\{x, y\}$ and $\{y, x\}$ are equal.

A very familiar example of an ordered pair is the representation of a point in a two-dimensional plane in Cartesian coordinates. Accordingly, the ordered pair $\langle 1, 7 \rangle$, $\langle 2, 6 \rangle$, $\langle 4, 4 \rangle$ and $\langle 6, 2 \rangle$ are different points.

2.25-1 Equality of two ordered pairs

Two ordered pair $\langle x, y\rangle$ and $\langle p, q\rangle$ are said to be equal if $x = p$ and $y = q$.

2.25-2 Ordered *n* Tuple

The idea of an ordered pair can be extended to define an ordered triple, and more generally, an n-tuple.

An ordered triple can be defined as an ordered pair whose first element is itself an ordered pair. Hence we can write an ordered triple as $\langle\langle a, b\rangle, c\rangle$. Now from the definition of equality of an ordered pair, we can write the equality of $\langle\langle a, b\rangle, c\rangle$ and $\langle\langle x, y\rangle, z\rangle$ as

$$\langle a, b\rangle = \langle x, y\rangle \text{ and } c = z$$

But $\langle a, b\rangle = \langle x, y\rangle$ if $a = x$ and $b = y$ hence we say

$$\langle\langle a, b\rangle, c\rangle \text{ and } \langle\langle x, y\rangle, z\rangle \Leftrightarrow a = x \text{ and } b = y \text{ and } c = z$$

Hence we can write ordered triple as $\langle a, b, c\rangle$ with an understanding that $\langle a, b, c\rangle$ actually stands for $\langle\langle a, b\rangle, c\rangle$.

Here it is to be noted that $\langle a, b, c\rangle \neq \langle b, a, c\rangle \neq \langle a, c, b\rangle$

Continuing this process an ordered n tuple can be defined as an ordered pair whose first member is an ordered n -1 tuple. The ordered n tuple is written as

$$\langle\langle a_1, a_2, \ldots, a_{n-1}\rangle, a_n\rangle$$

which may also be written as

$$\langle a_1, a_2, a_3, \ldots, a_{n-1}, a_n\rangle$$

2.26 Cartesian Product

If A and B are two nonempty sets then we define the product set or Cartesian product $A \times B$ (read A cross B) as the set of all ordered pairs such that the first member of the ordered pair is an element of A and the second member is an element of B.

Example 2.36

If $A = \{x, y\}$ and $B = \{1, 2, 3\}$, then

$$A \times B = \{\langle x, 1\rangle, \langle x, 2\rangle, \langle x, 3\rangle, \langle y, 1\rangle, \langle y, 2\rangle, \langle y, 3\rangle\}$$

$$B \times A = \{\langle 1, x\rangle, \langle 2, x\rangle, \langle 3, x\rangle, \langle 1, y\rangle, \langle 2, y\rangle, \langle 3, y\rangle\}$$

$$A \times A = \{\langle x, x\rangle, \langle x, y\rangle, \langle y, x\rangle, \langle y, y\rangle\}$$

$$B \times B = \{\langle 1, 1\rangle, \langle 1, 2\rangle, \langle 1, 3\rangle, \langle 2, 1\rangle, \langle 2, 2\rangle, \langle 2, 3\rangle, \langle 3, 1\rangle, \langle 3, 2\rangle, \langle 3, 3\rangle\}$$

Note that $(A \times B) \cap (B \times A) = \varnothing$.

Example 2.37

A person have four shirts say p = white shirt, q = brown shirt, r = blue shirt, s = grey shirt and three paints say b = black paint, r = red paint, y = yellow paint. If we let $A = \{p, q, r, s\}$ and $B = \{b, r, y\}$ then the cartesian product of $A \times B$ lists 12 possible combinations of paint and shirt a person can wear.

If either of A or B is a null set then $A \times B = \varnothing$. If either of A or B is infinite and the other is nonempty, then $A \times B$ is infinite. Show yourself.

Example 2.38

Show that $(A \times B) \times C \neq A \times (B \times C)$.

Solution

Let us consider the Cartesian product of three sets A, B and C. If $A = \{a, b, c\}$, $B = \{p, q\}$, $C = \{x, y, z\}$, then

$A \times B \times C = \{ \langle a, p, x\rangle, \langle a, p, y\rangle, \langle a, p, z\rangle, \langle a, q, x\rangle, \langle a, q, y\rangle, \langle a, q, z\rangle, \langle b, p, x\rangle, \langle b, p, y\rangle, \langle b, p, z\rangle, \langle b, q, x\rangle, \langle b, q, y\rangle, \langle b, q, z\rangle, \langle c, p, x\rangle, \langle c, p, y\rangle, \langle c, p, z\rangle, \langle c, q, x\rangle, \langle c, q, y\rangle, \langle c, q, z\rangle \}$.

Now consider the expressions $(A \times B) \times C$ and $A \times (B \times C)$. From the definition

$$(A \times B) \times C = \{\langle\langle a, b\rangle, c\rangle \mid \langle a, b\rangle \in A \times B \text{ and } c \in C\}$$
$$= \{\langle a, b, c\rangle \mid a \in A \text{ and } b \in B \text{ and } c \in C\}$$

and
$$A \times (B \times C) = \{\langle a, \langle b, c\rangle\rangle \mid a \in A \text{ and } \langle b, c\rangle \in B \times C\}$$

It is to be noted here that $\langle a, \langle b, c\rangle\rangle$ is not an ordered triple. If we consider $(A \times B) \times C$ as an ordered pair, then the first member is an ordered pair and second member is an element of C. But $A \times (B \times C)$ is an ordered pair in which the first member is an element of A whereas second member is an ordered pair. This gives

$$(A \times B) \times C \neq A \times (B \times C).$$

Example 2.39

If $A = \{x, y\}$, $B = \{1, 2\}$ and $C = \{2, 3\}$. Evaluate

i. $A \times (B \cup C)$

ii. $A \times (B \cap C)$.

Solution

i. $B \cup C = \{1, 2, 3\}$

$\Rightarrow A \times (B \cup C) = \{x, y\} \times \{1, 2, 3\}$

$= \{\langle x, 1\rangle, \langle x, 2\rangle, \langle x, 3\rangle, \langle y, 1\rangle, \langle y, 2\rangle, \langle y, 3\rangle\}$

ii. $B \cap C = \{2\}$

$\Rightarrow A \times (B \cap C) = \{x, y\} \times \{2\}$

$= \{\langle x, 2\rangle, \langle y, 2\rangle\}$.

Example 2.40

If A, B and C are sets, prove that $A \times (B \cup C) = (A \times B) \cup (A \times C)$.

Solution

Let $\langle b, c\rangle \in A \times (B \cup C)$

$\Leftrightarrow b \in A$ and $c \in B \cup C$

$\Leftrightarrow b \in A$ and $(c \in B$ or $c \in C)$

$\Leftrightarrow (b \in A$ and $c \in B)$ or $(b \in A$ and $c \in C)$

$\Leftrightarrow \langle b, c\rangle \in A \times B$ or $\langle b, c\rangle \in A \times C$

$\Leftrightarrow \langle b, c\rangle \in (A \times B) \cup (A \times C)$

$\therefore A \times (B \cup C) = (A \times B) \cup (A \times C)$.

Example 2.41

Prove that $(A \times B) \cap (R \times S) = (A \cap R) \times (B \cap S)$.

Solution

Let $\langle b, c \rangle \in (A \times B) \cap (R \times S)$

$$\Leftrightarrow \langle b, c \rangle \in A \times B \text{ and } \langle b, c \rangle \in R \times S$$
$$\Leftrightarrow (b \in A \text{ and } c \in B) \text{ and } (b \in B \text{ and } c \in S)$$
$$\Leftrightarrow (b \in A \text{ and } b \in R) \text{ and } (c \in B \text{ and } c \in S)$$
$$\Leftrightarrow b \in A \cap R \text{ and } c \in B \cap S$$
$$\Leftrightarrow \langle b, c \rangle \in (A \cap R) \times (B \times S)$$

$\therefore (A \times B) \cap (R \times S) = (A \cap R) \times (B \cap S)$.

Example 2.42

Prove or provide a counter example that $A \cup B \subseteq A \cap B \Rightarrow A = B$.

Solution

Let $A = \{a, b, c\}, B = \{b, c, d\}$

Now $A \cup B = \{a, b, c, d\}$

$A \cap B = \{b, c\}$

Clearly, here $A \neq B \Rightarrow A \cup B \subseteq A \cap B$

Again, Let $A = \{a, b, c\}, B = \{a, b, c\}$

$A \cap B = \{a, b, c\}, A \cup B = \{a, b, c\}$

$\text{A} = \text{B} \Rightarrow A \cup B \subseteq A \cap B$.

Exercise 2.2

1. Let $A = \{2, 5, 8, 10\}$, $B = \{1, 2, 3, 4, 5\}$, $C = \{2, 4, 6, 8\}$ and $U = \{1, 2, 3, \ldots 10\}$. List the elements of following set

i. $A \cup B$

ii. $A \cap C$

iii. $B \cap C$

iv. $A - B$

v. $B - A$

vi. $(A \cup B) - C$

vii. A

viii. U

ix. $B \cap C$

x. $A \cap (B \cup C)$

xi. $(A \cap B) - C$

xii. $(A \cap B)' \cup C$

xiii. $A \cup B - (C - B)$

xiv. $A \oplus B$

xv. $A \oplus C$

xvi. $A \cup B \cup C$

xvii. $A \cap B \cap C$

xviii. $(A \cap B) - (B \cap C)$ (xix) $(A \cup B) - (C - B)$

xx. $(A \cap B)' \cup C$

xxi. $A \cup U$

xxii. $(A \cup B)'$

xxiii. $A \cap A'$

xxiv. $B \cup B'$.

2. Let A and B be two sets, then show that
   i. $A \cap U = A$, where A is any subset of set U.
   ii. $A \cap \varnothing = \varnothing$
   iii $(A \cap B) \subseteq A$
   iv. $A \subseteq (A \cup B)$
   v. $A \cap A = A$
   vi. $A - \varnothing = A$
   vii. $\varnothing - A = \varnothing$
   viii. $A - B \subseteq A$
   ix. $A \cup (B - A) = A \cup B$.
3. If A and B are two sets, prove that $A - B = A \cap B'$.
4. If $A \cup B = A \cup C$, is it imply $B = C$? Explain.
5. Show that if A, B and C are sets, then

$$(A \cap B \cap C)' \equiv A' \cup B' \cup C'$$

   i. by showing each side is a subset of the other side
   ii. using membership table.
6. Prove that A û $(A - B) \subseteq B$.
7. Suppose that $A \oplus B = A \oplus C$. Does this guarantee that $B = C$? Justify your answer.
8. Let C be a circle and let D be the set of all diameters of C, then is $C \cap D$
9. For each condition in the following exercises (i) to (iv), what relation must hold between sets A and B?
   i. $A \cap B = A$
   ii. $A \cup B = A$
   iii. $A' \cap U = \varnothing$
   iv. $(A \cap B)' = B'$.
10. What can you say about the sets X and Y if the following are true?
    i. $X \cup Y = X$
    ii. $X \cap Y = X$
    iii. $X - Y = X$
    iv. $X \cap Y = Y \cap X$
    v. $X - Y = Y - X$.
11. Can you conclude that $X = Y$ if X, Y and Z are sets such that
    i. $X \cup Z = Y \cup Z$
    ii. $X \cap Z = Y \cap Z$.
12. Draw the Venn diagrams for each of the following combination of sets A, B and C
    i. $A \cap (B \cup C)$
    ii. $A' \cap B' \cap C'$.
    iii. $(A - B) \cup (A - C) \cup (B - C)$
    iv. $A' \cap (B \cup C)$.
13. In a survey of 600 television viewers given the following information: 385 watch cricket matches, 295 watch hockey matches, 215 watch football matches, 145 watch cricket and football matches, 170 watch cricket and hockey matches, 150 watch hockey and football matches and 150 do not watch any of the three kind of matches.
    i. How many people in survey watch all three kinds of matches?
    ii. How many people watch exactly one of the sports?
14. A survey of a sample of 50 new cars being sold at a local auto dealer was conducted to see which of three popular options, air conditioning, radio, and power windows, were already installed. The survey found.
    a. 40 had air-conditioning
    b. 37 had radio
    c. 36 had power windows
    d. 30 had air-conditioning and power windows
    e. 34 had air-conditioning and radio
    f. 29 had radio and power windows
    g. 28 had all three options.

    Find the number of cars that had:

i. Only power windows
ii. Only air conditioning
iii. Only radio
iv. Radio and power windows but not air-conditioning
v. Air-conditioning and radio, but not power windows
vi. Only one of the options
vii. At least one options
viii. None of these options.

15. If A, B and C are three sets, then prove that
    i. $A \oplus (B \oplus C) = (A \oplus B) \oplus C$ [Associative law]
    ii. $A \oplus B = (B \oplus A)$ [Commutative law]
    iii. If $A \oplus B = A \oplus C$, then $B = C$ [Cancellation law]
    iv. $A \cap (B \oplus C) = (A \cap B) \oplus (A \cap C)$. [Distributive law]
16. (a) Show that
    i. $A \oplus B = A \cup B - (A \cap B)$
    ii. $A \oplus B = (A - B) \cup (B - A)$.

    (b) Show that if A is a subset of a universal set U, then
    i. $A \oplus A = \varnothing$
    ii. $A \oplus \varnothing = A$
    iii. $A \oplus U = \bar{A}$
    iv. $A \oplus \bar{A} = U$

    (c) If A, B, C and D are sets, does it mean that
    $$(A \oplus B) \oplus (C \oplus D) = (A \oplus D) \oplus (B \oplus C)?$$
    (d) If A, B, C and D are sets does it mean that
    $$(A \oplus B) \oplus (C \oplus D) = (A \oplus D) \oplus (B \oplus D)?$$
17. Prove that $(A \times B) \cap (A \times C) = A \times (B \cap C)$.
    [Hint $(A \times B) \cap (A \times C)$
    $$= \{(x, y) \mid (x, y) \in A \times B \text{ and } (x, y) \in A \times C\}$$
    $$= \{(x, y) \mid x \in A, y \in B \text{ and } x \in A, y \in C\}.$$
18. If $A = \{1\}$, $B = \{a, b\}$ and $C = \{2, 3\}$. Write
    i. $A \times B$
    ii. $A \times B \times C$
    iii. B^2
    iv. A^3
    v. $B^2 \times A$
19. Show by means of an example $A \times B \neq B \times A$ and $(A \times B) \times C \neq A \times (B \times C)$.
20. Show that $A \times (B \cap C) = (A \times B) \cap (A \times B)$.
21. Suppose 63% persons like oranges whereas 76 % like apples. What can be said about the percentage of persons who like both oranges and apple.
22. In a group of 1000 people, there are 750, who can speak Hindi and 400, who can speak Bengali. How many can speak Hindi only? How many can speak Bengali only?

Answers

1. i. { 1, 2, 3, 4, 5, 8, 10}
   ii. { 2, 8}
   iii. { 2, 4}
   iv. { 8, 10}
   v. {1, 3, 4}
   vi. { 1, 3, 5, 10}
   vii. { 1, 3, 4, 6, 7, 9}
   viii. $\varnothing$
   ix. { 1, 2, 3, 4, 5, 6, 8}
   x. { 2, 5, 8}

xi. { 5}
xii. { 1, 2, 3,4, 6, 7, 8, 9, 10}
xiii. { 1, 2, 3, 4, 5, 10}
xv. { 5, 4, 6, 10}
xvi. { 1, 2, 3, 4, 5, 6, 8, 10}
xvii. { 2}
xviii. { 5}
xix. { 1, 2, 3, 4, 5, 10}
xx. { 1, 2, 3, 4, 6, 7, 8, 9, 10}
xxi. U
xxii. { 6, 7, 9}
xxiii. $\varnothing$
xxiv. { 1, 2, 3, 4, 5, . . ., 9, 10}

10. i. $Y \subseteq X$
ii. $X \subseteq Y$
iii. $X \cap Y = \varnothing$
iv. nothing, since this is always true
v. $X = Y$

12. (i)

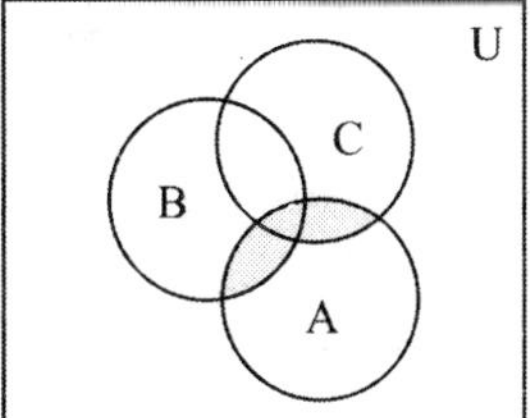

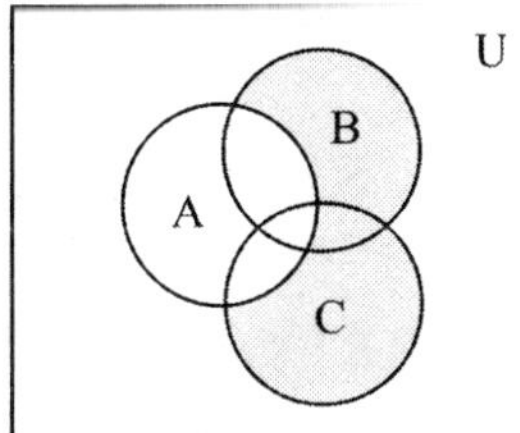

21. 39%
22. 600, 250.

Chapter Objective Type Questions

A. Multiple choice questions.

1. If A = {1, 3, 5, 7, 9, 11, 13, 15, 17}, B = {2, 4, …, 18} and N is the universal set, then $A' \cup (A \cup B) \cap B')$ is:
(A) A
(B) N
(C) B
(D) none of these
2. If A and B are two sets, then $A \cap (A \cup B)$ equals to:
(A) A
(B) B
(C) $\varnothing$
(D) none of these
3. If A is a finite set having n elements, then $P(A)$ has:
(A) $2n$
(B) 2^n elements
(C) n elements
(D) none of these
4. If $A = \{ \varnothing, \{\varnothing\}\}$, then the power set of A is-
(A) A
(B) $\{\varnothing, \{\varnothing\}, A\}$
(C) $\{\varnothing, \{\varnothing\}, \{\varnothing\}\}, A\}$
(D) none of these
5. Which of the following is an empty set?
(A) $\{x| x$ is a real number and $x^2 - 1 = 0\}$
(B) $\{x| x$ is a real number and $x^2 + 1 = 0\}$
(C) $\{x| x$ is a real number $- 9 = 0\}$
(D) $\{x| x$ is a real and x^2 number and $x^2 = x + 2\}$
6. Let A = {1, 2, 3}, B = {3, 4}, C = {4, 5, 6}. Then A' $(B \cap C)$ is-
(A) {3}
(B) {1, 2, 3, 4}
(C) {1, 2, 5, 6}
(D) {1, 2, 3, 4, 5, 6}

7. If X and Y are two sets, then $X \cap (Y \cup X)'$ equals-
   (A) X　　(B) Y
   (C) $\varnothing$　　(D) none of these
8. Let A and B be two disjoint subsets of a universal set U. Then $(A \cup B) \cap B' =$
   (A) A　　(B) B
   (C) $\varnothing$　　(D) none of these
9. If X and Y are two sets, then $X \cap (X \cup Y)$ equals to:
   (A) X　　(B) Y
   (C) $\varnothing$　　(D) none of these
10. $A - (B \cup C) =$
   (A) $(A-B) \cup (A-C)$　　(B) $(A \cup C)$
   (C) $(A-B) \cap (A-C)$　　(D) $(A \cap B)\,(A \cap C)$
11. $(A \cup B) - C =$
   (A) $(A-C) \cap (B-C)$　　(B) $(A-C) \cup (B-C)$
   (C) $(A-B) \cup (A-C)$　　(D) none of these
12. $A = \{x: x \text{ is a multiple of } 4\}$ and $B = \{x : x \text{ is a multiple of } 6\}$, then $A \cap B$ consists of all multiples of:
   (A) 2　　(B) 24
   (C) 12　　(D) 4
13. In a group of 1000 persons, 750 can speak English and 400 can speak Hindi. How many persons can speak Hindi only?
   (A) 150　　(B) 25
   (C) 350　　(D) 600
14. The symmetric difference of two sets A and b is:
   (A) $(A-C) \cap (B-A)$　　(B) $(A-B) \cup (B-A)$
   (C) $(A \cup B) - (A \cap B)$　　(D) none of these
15. For any sets A and B, $(A - B) =$
   (A) B　　(B) $A-B$
   (C) $A \cap B$　　(D) $A' \cap B'$
16. Two finite sets have m and n elements. The total number of subsets of the first set is 56 more than the total number of subsets of the second set. The values of m and n are:
   (A) 7, 6　　(B) 6, 3
   (C) 5, 1　　(D) 8, 7

B. Fill in the blanks.

1. The number of different elements in a finite set is called ________ of a set.
2. A set B is said to be ______ of set A iff set B contains all the elements of set A.
3. If $A \subseteq B$ and $B \subseteq A$, then A and B have ________ elements.
4. $A \not\subset B$, $B \not\subset A$ and $A \neq B$, then A and B are said to be ________.
5. The elements of a disjoint collection are said to be ________.
6. The relative complement of A in B is also called the ________ of B and A.

7. The symmetric difference $A \Delta B =$ __________.
8. If A and B are finite sets, then $|A \cup B| =$ __________.

C. Which of the following statements Is True (T)/ False (F)?

1. Sometimes cardinality of set A is also called as the order of a set.
2. {5} is a null set.
3. The set A is called the proper subset of B if every element of A is the element of B and there exist at least one element of B which is not an element of the set A.
4. If A is a subset of B and B is a subset of C then A is a subset of C.
5. A null set $\varnothing$ is not a member of power set.
6. $A \cap (B \cup C) = (A \cap B) \cup (A \cap C)$.
7. If A and B are two sets then $A - B \subseteq A$.
8. If A and B are two sets then $(A \cup B)' = A' \cup B'$.
9. If A, B and C are three sets then $(A \times B) \times C = A \times (B \times C)$.

Answers (Objective Type)

A.

1. B	2. A	3. B	4. C
5. B	6. B	7. C	8. A
9. A	10. C	11. B	12. C
13. B	14. B	15. C	16. B

B.

1. Cardinality	2. Superset	3. Same	4. Incomparable
5. Mutually disjoint	6. Difference	7. $(A - B) \cup (B - A)$	
8. $\lvert A \rvert + \lvert B \rvert - \lvert A \cap B \rvert$			

C.

1. T	2. F	3. T	4. T
5. F	6. T	7. T	8. F
9. F			

CHAPTER

3 Relations

3.0 Introduction

In our daily life we often use sentences like (i) Rakhi is the sister of Karan (ii) 8 is greater than 5 (iii) Line A is parallel to line B. Here "is the sister of ", "is greater than", or "is parallel to" describe a relationship between two objects one which proceeds it and other which follows it.

In this chapter, we shall consider relation, called binary relation between a pair of objects. It is to be noted that a relation between two objects can be defined by listing the two objects as an ordered pair. A set of all such ordered pairs in each of which the first member has some definite relationship to the second, describes a particular relationship.

3.1 Relation

Let A and B be two sets. A (binary) **relation** from A to B is a subset of the Cartesian product $A \times B$. If (x, y) is a member of relation R, then we say that "x is in the relation R to y" or "x is R-related to y". Symbolically it is written as (x, y) or $\langle x, y \rangle \in R \Leftrightarrow x\, R\, y$.

If R is a relation from A to B, then R is a set of ordered pairs where each first element will be from A and each second element from B. We can conclude from above that either of the following will be true.

a. $(x, y) \in R$, i.e. x is R-related to y.

b. $(x, y) \notin R$ i.e. x is not R-related to y, written $x \not R\, y$.

A relation is also called a **correspondence**. In this book we will only concern with binary relation, hence by a relation, we shall always mean binary relation.

Example 3.1

Let $A = \{1, 2, 3\}$ and $B = \{x, y\}$. Then $\{(1, x), (2, x), (3, x), (2, y)\}$ is a relation from A into B. It is worth noticing that there are many more relations from A into B we could describe, 64 i.e. 2^6 to be precise.

If A and B are two finite sets, then the number of elements in set S i.e. n (S) is given by

$$n(A \times B) = n(A).\, n(B),$$

where n(A) and n(B) are number of elements of A and B respectively.

Example 3.2

Let A = {milk, corn, eggs} and B = {buffaloes, cows, birds}. Here, we can define a relation R from set A to set B by $(a, b) \in R$ if a is produced by b, i.e. R = {(milk, buffaloes), (milk, cows), (eggs, birds)}and the relation can be expressed as milk R buffalo, eggs R birds etc.

Example 3.3

Let $X = R$, the set of real numbers. We define the following relation R on X.

$a\ R\ b$ iff a and b satisfy the equation $\frac{x^2}{16}+\frac{y^2}{4}=1$. The set R consist of all points on the ellipse shown below.

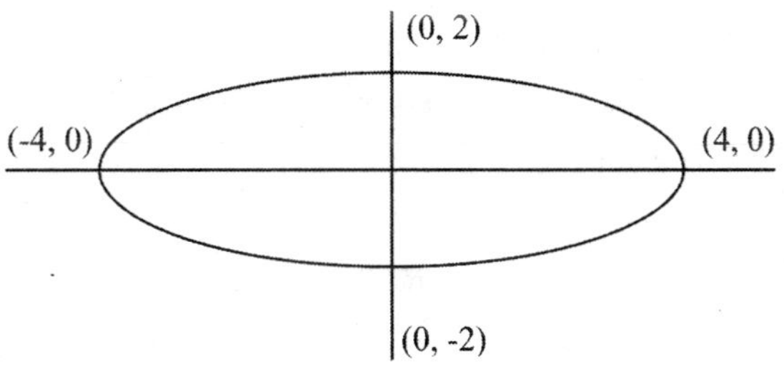

Fig. 3.1

3.2 Relation on a set

A relation from a set X into itself i.e. if R is a subset of $X^2 = X \times X$, then we say that R is a relation on X.

3.3 Domain and Range of a Relation

If R is a relation defined from set X to Y, then the **domain** D of the relation R is defined as the set of all first elements of the ordered pair which belong to R, i.e. $D = \{x : x \in X \text{ and } (x, y) \in R\}$.

Domain of R is also denoted by Dom(R). The **range** of the relation R is defined as the set of all second elements of the ordered pairs which belong to R. It is denoted by Ran (R).

$$\text{Ran}(R) = \{y : y \in Y \text{ and } (x, y) \in R\}$$

One can observe here very easily that $D \subseteq X$ and Ran(R) $\subseteq Y$ and elements of X that are not in Dom(R) are not involved in the relation R in any way. This is also true for elements of Y that are not in Ran (R).

Example 3.4

Let $X = \{1, 2, 3\}$ and $Y = \{p, q\}$. Then $R = \{(1, p), (2, q), (3, p)\}$ is a relation from X to Y. Here Dom (R) = X and Ran (R) = Y.

Example 3.5

Let $A = \{4, 5, 6\}$ and $B = \{5, 6, 7, 8, 9\}$. If we define a relation R from A to B by $(a, b) \in R$ if a divides b with zero remainder, we get $R = \{(4, 8), (5, 5), (6, 6)\}$.

Then the dom (R) = $\{4, 5, 6\}$ and Ran (R) = $\{5, 6, 8\}$.

Theorem 3.1

Let R be a relation from set X to set Y and let X_1 and X_2 be subsets of X. Then

i. If $X_1 \subseteq X_2$, then $R(X_1) \subseteq R(X_2)$

ii. $R(X_1 \cup X_2) = R(X_1) \cup R(X_2)$

iii. $R(X_1 \cap X_2) \subseteq R(X_1) \cap R(X_2)$.

Proof

i. If $b \in R(X_1)$, then $a\ R\ b$ for some $a \in X_1$.

$$\text{As, } X_1 \subseteq X_2,\ a \in X_2 \Rightarrow b \in R(X_2).$$

ii. Let $b \in R(X_1 \cup X_2)$, then from the definition of $a\,R\,b$ there will be some a in $X_1 \cup X_2$ such that $b \in R(X_1 \cup X_2)$. But, if $a \in X_1$, then since $a\,R\,b$ we have $b \in R(X_1)$. By the same argument, if $a \in X_2$, then $b \in R(X_2)$.

In either case, $b \in R(X_1) \cup R(X_2)$.

Thus $R(X_1 \cup X_2) \subseteq R(X_1) \cup R(X_2)$.

Conversely, as $X_1 \subseteq X_1 \cup X_2$, from part (i), $R(X_1) \subseteq R(X_1 \cup X_2)$.

And, $R(X_2) \subseteq R(X_1 \cup X_2)$.

$\Rightarrow$ $R(X_1) \cup R(X_2) \subseteq R(X_1 \cup X_2)$.

Hence $R(X_1 \cup X_2) = R(X_1) \cup R(X_2)$.

iii. Try yourself.

Example 3.6

Consider a set $A = Z$ and relation R on the set is "$\leq$". Let $A_1 = \{0, 1, 2\}$ and $A_2 = \{11, 13\}$. Then $R(A_1)$ contains all integers n s.t. $0 \leq n$ or $1 \leq n$, or $2 \leq n \Rightarrow R(A_1) = \{0, 1, 2, 3\ldots\}$.

Similarly, $R(A_2) = \{11, 12, 13\ldots.\}$. Hence $R(A_1) \cap R(A_2) = \{11, 12, 13,\ldots\}$.

But $A_1 \cap A_2 = \varnothing$, hence $R(A_1 \cap A_2) = \varnothing$. This shows that containment of the theorem 3.1 (iii) is not always equality.

3.4 Pictorial representations of Relations

The pictorial representation of the relation is sometimes called the **graph** of the relation. For example consider the relation R taken in example 3.5. A picture of R can be drawn as shown in Fig. 3.2.

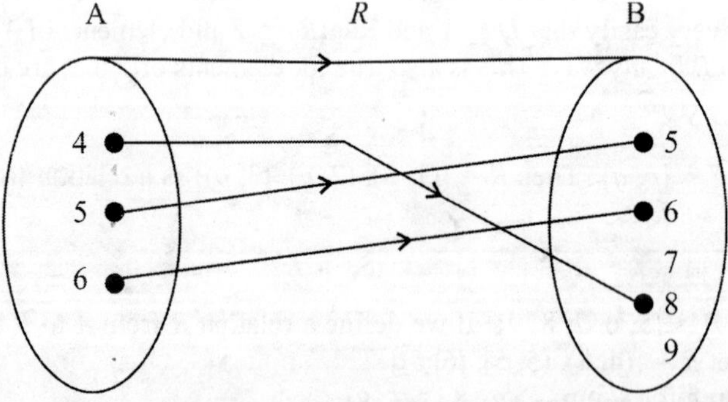

Fig. 3.2

Where the arrow indicates that 4 is related to 8 under R, 5 is related to 5 and 6 is related to 6 while the upper arrow denotes that R is a relation from the whole set A into the set B.

3.4-1 Total Number of distinct relations from set *A* to a set *B*

Let the number of elements of A and B be m and n respectively. Then the number of elements of $A \times B$ is mn. Therefore, the number of elements of the power set of $A \times B$ is 2^{mn}. Thus $A \times B$ has 2^{mn} different subsets. Now every subset of $A \times B$ is a relation from A to B. Hence the number of different relations A to B is 2^{mn}.

3.5 Inverse Relation

Let R be a relation from set A to set B, then the inverse relation of R, denoted by R^{-1} is the relation from B to A and is defined as

$$R^{-1} = \{(y, x) : y \in B, x \in A, (x, y) \in R\}$$

Thus $(x, y) \in R \Leftrightarrow (y, x) \in R^{-1}$ i.e. R^{-1} is the relation from B to A which consists of those ordered pairs which when reversed belong to R.

Example 3.7

Let $X = \{x, y, z\}$, $Y = \{2, 4, 6\}$ and relation $R = \{(x, 2), (x, 6), (y, 4), (y, 6), (z, 6)\}$ then

$$R^{-1} = \{(2, x), (6, x), (4, y), (6, y), (6, z)\}.$$

Theorem 3.2

If R is a relation from A to B, then the domain of R is identical with the range of $R^{-1}$and range of R is identical with the domain of R^{-1}.

Proof

Let $b \in$ domain of R^{-1}, then there exist $a \in A$ and $b \in B : (b, a) \in R^{-1}$.

But $\quad (b, a) \in R^{-1} \Rightarrow (a,b) \in R \Rightarrow b \in$ Range of R

$\therefore \quad b \in$ domain of $R^{-1} \Rightarrow b \in$ range of R.

It means domain of $R^{-1} \subseteq$ range of R.

In the same way we can prove that range of $R \subseteq$ domain of R^{-1}. We can conclude from here that domain of R^{-1} = Range of R. Similarly, we can prove that domain of R = Range of R^{-1}.

Theorem 3.3

Let R and S be relations from X to Y. Then

i. If $R \subseteq S$ then $R^{-1} \subseteq S^{-1}$

ii. $(R \cup S)^{-1} = R^{-1} \cup S^{-1}$

iii. $(R \cap S)^{-1} = R^{-1} \cap S^{-1}$.

Proof

i. Let $R \subseteq S$ and $(x, y) \in R^{-1}$, then $(y, x) \in R$ and also $(y, x) \in S$ since $R \subseteq S$. Again if $(y, x) \in S \Rightarrow (x, y) \in S^{-1}$. Therefore $R^{-1} \subseteq S^{-1}$.

ii. Let $(x, y) \in (R \cup S)^{-1}$. Then $(y, x) \in R \cup S$. This implies $(y, x) \in R$ or $(y, x) \in S$. Hence $(x, y) \in R^{-1}$ or $(x, y) \in S^{-1}$. Therefore $(x, y) \in R^{-1} \cup S^{-1}$.

Hence, $\quad (R \cup S)^{-1} \subseteq R^{-1} \cup S^{-1}. \quad \ldots (1)$

Conversely, let $(x, y) \in R^{-1} \cup S^{-1}$. This implies $(x, y) \in R^{-1}$ or $(x, y) \in S^{-1}$. Hence $(y, x) \in R$ or $(y, x) \in S$. Therefore $(y, x) \in R \cup S$.

Hence $\quad (x, y) \in (R \cup S)^{-1}$.

This implies $\quad R^{-1} \cup S^{-1} \subseteq (R \cup S)^{-1} \quad \ldots (2)$

From (1) and (2)

$$(R \cup S)^{-1} = R^{-1} \cup S^{-1}.$$

iii. Let $(x, y) \in (R \cap S)^{-1}$. Then $(y, x) \in R \cup S$. This implies $(y, x) \in R$ and $(y, x) \in S$.

Hence $\quad (x, y) \in R^{-1}$ and $(x, y) \in R^{-1} \cup S^{-1}$.

Therefore $\quad (x, y) \in R^{-1} \cap S^{-1}$.

Hence $\quad (R \cap S)^{-1} \subseteq R^{-1} \cap S^{-1} \quad \ldots (1)$

Conversely, let $(x, y) \in R^{-1} \cap S^{-1}$. This implies $(x, y) \in R^{-1}$ and $(x, y) \in S^{-1}$.
Hence, $(x, y) \in R$ and $(x, y) \in S$.

Therefore $\qquad (y, x) \in R \cap S.$

Hence $\qquad (x, y) \in (R \cap S)^{-1}.$

This implies $\qquad R^{-1} \cap S^{-1} \subseteq (R \cap S)^{-1}$ ... (2)

From (1) and (2), we have

$$(R \cap S)^{-1} = R^{-1} \cap S^{-1}.$$

Example 3.8

A relation R defined on the set of integers Z in the following way:

$$(y, x) \in R \Leftrightarrow x^2 + y^2 = 25.$$

Find R and R^{-1} as the sets of ordered pairs and hence find their respective domains.

Solution

Given, $\qquad (y, x) \in R \Leftrightarrow x^2 + y^2 = 25 \Leftrightarrow y = \sqrt{(25 - x^2)}$

When $\qquad x = 0 \Rightarrow y = \pm 5$

$\therefore \qquad (0, 5)$ and $(0, -5) \in R$

When $\qquad x = \pm 3, y = \sqrt{25-9} = \pm 4$

$\therefore (3, 4) \in R, (-3, 4) \in R, (3, -4) \in R$, and $(-3, -4) \in R$,

When $\qquad x = \pm 4, y = \sqrt{(25 - 16)} = \pm 3$

$\therefore (4, 3) \in R, (-4, 3) \in R, (4, -3) \in R$, and $(-4, -3) \in R$,

When $\qquad x = \pm 5, y = \sqrt{(25 - 25)} = 0$

$\therefore \quad (5, 0) \in R, (-5, 0) \in R.$

We can see that for any other integral value of x, y is not an integer.

$\therefore R = \{(0, 5), (0, -5), (3, 4), (-3, 4), (3, -4), (-3, -4), (4, 3), (-4, 3), (4, -3), (-4, -3), (5, 0), (-5, 0)\}$

and $R^{-1} = \{(5, 0), (-5, 0), (4, 3), (4, -3), (-4, 3), (-4, -3), (3, 4), (3, -4), (-3, 4), (-3, -4), (0, 5), (0, -5)\}$

Domain $(R) = \{0, 3, -3, 4, -4, 5, -5\}$

Domain $(R^{-1}) = \{0, 3, -3, -4, -4, 5, -5\}$.

3.6 Directed Graphs of Relations on Sets

In this section we will discuss a brief explanation of procedures for graphing a relation. As it is always easy to recognise a situation or to have a clear idea of the situation under consideration with the help of a graph. Here we investigate two methods of representation of relation by graph namely directed graph and Hasse diagram other then the pictures of relation discussed previously.

Let A is a finite set and R is a relation on A, then this relation R can be represented in a picture as follows:

The elements of A are called **vertices** of the graph. Draw a small circle for each element (vertex) of A and label the circle with the corresponding element of A. Draw an arrow called an **edge** from vertex a to vertex b if and only if $a\ R\ b$. This type of pictorial representation of R is called a **directed graph** or **digraph** of R.

Now it is to be noted here that if R is a relation of set A, the edges in the digraph of R correspond exactly to the pairs in R, and the vertices correspond exactly to the elements of the set A.

Now consider an example,

$$A = \{a, b, c, d\}$$

$$R = \{(a, a), (a, b), (b, a), (b, b), (b, c), (b, d), (c, d), (d, a)\}.$$

The diagraph of R is shown in the Fig. 3.3.

A collection of vertices with edges between some of the vertices determine a relation in a natural manner. In the digraph the actual location of the vertices is immaterial. The main emphasis is on the placement of the vertices, i.e. to place the vertices in such a way that the graph is easy to read. Obviously, after a rough-draft graph of relation, we may decide to relocate and/or order the vertices so that the final result will be clear.

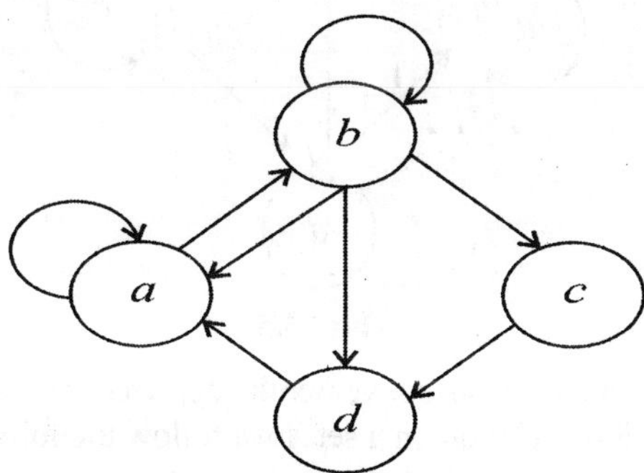

Fig. 3.3

1. *A vertex of a graph is also called a* ***node****, a* ***point*** *or a* ***junction****. An edge of a graph is also referred to as an* ***arc****, a* ***line*** *or a* ***branch****.*
2. *Two graphs of a given relation may look different.*

Example 3.9

Find the relation detrmined by the following figure.

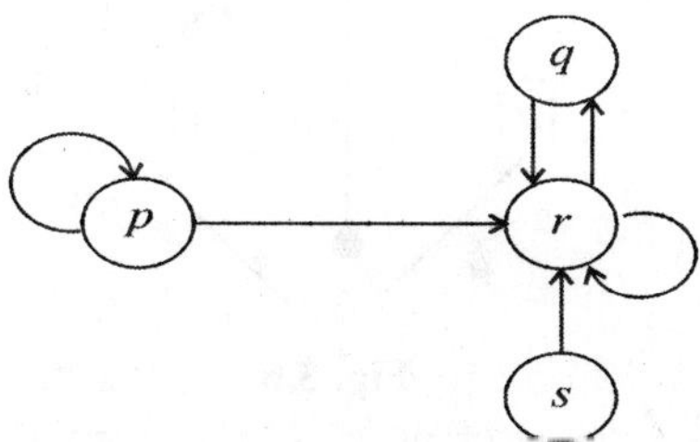

Fig. 3.4

Solution

As $a\ R\ b$ if there is an edge from a to b. Hence from the Fig. 3.4, we will have following relation.

$$R = \{(p, p), (p, r), (q, r), (r, q), (r, r), (s, r)\}.$$

3.6-1 Hasse Diagram

Let $X = \{1, 2\}$ and let $Y = P(X) = \{\varnothing, \{1\}, \{2\}, \{1, 2\}\}$. Then $\subseteq$ is a relation on X whose digraph is shown in the Fig. 3.5.

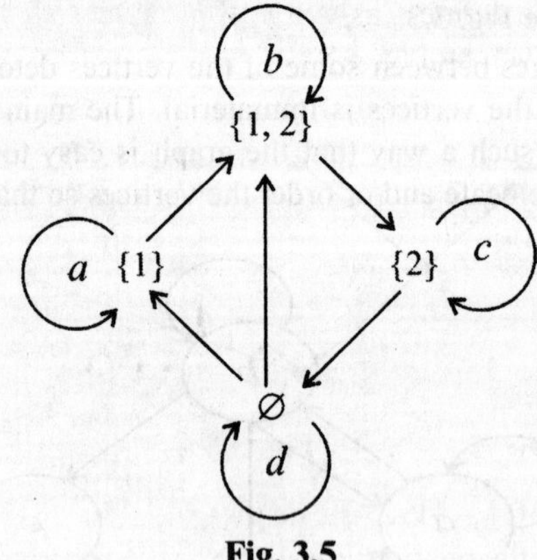

Fig. 3.5

Some relations such as this can also be conveniently depicted by what is called **Hasse** or **Ordering** diagram. To read a Hasse diagram for a relation on a set A we follow the following rules:

a. Each vertex of A must be related to itself, so the arrows from a vertex to itself are not necessary.

b. If a vertex b appears above vertex a and if vertex a is connected to vertex b by an edge, then $a\ R\ b$, so direction arrows are not necessary.

c. If vertex c is above vertex a and if c is connected to a by a sequence of edges then $a\ R\ c$.

d. The vertices are denoted by point rather then circles.

Now, following above rules the Hasse diagram of Fig. 3.5 can be drawn as follows:

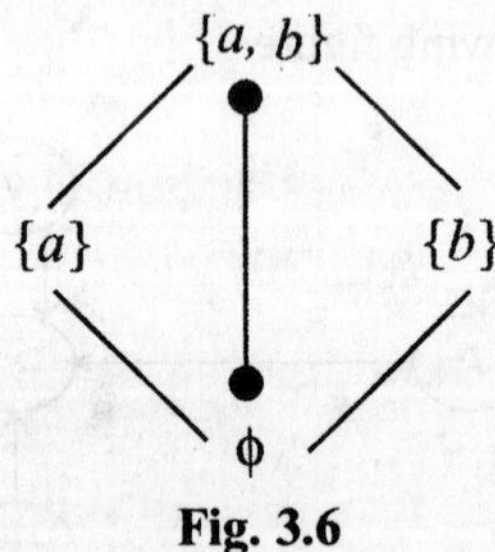

Fig. 3.6

Example 3.10

i. Let $A = \{1, 2, 3, 6\}$. Draw Hasse diagram for 'divides'.

Solution

Given $A = \{1, 2, 3, 6\}$. Then the Hasse diagram for 'divides' is as follows:

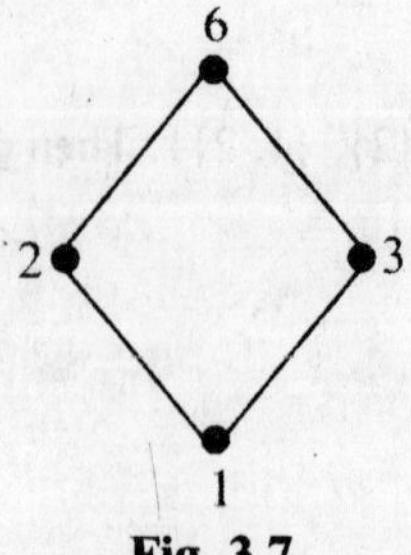

Fig. 3.7

ii. Let $S = \{1, 2, 3, 4, 5, 6\}$. We define $\leq$ as $m \leq n$ if 'm divides n', i.e. n is an integer multiple of m. Then the Hasse diagram of $\leq$ is given by Fig. 3.8.

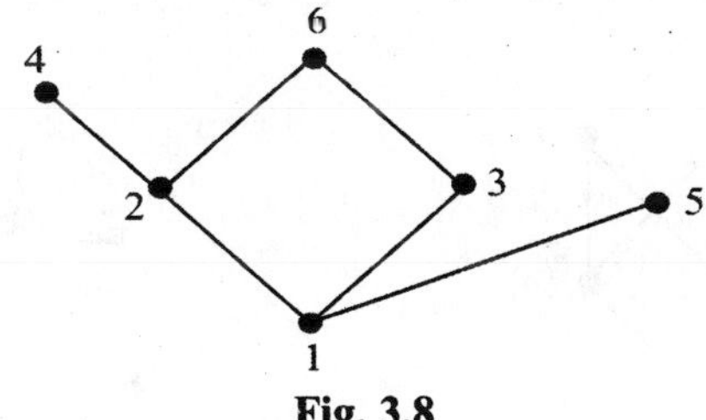

Fig. 3.8

iii. Let $B = P(A)$ be power set of $A = \{1, 2, 3\}$, i.e. $B = \{\varnothing, \{1\}, \{2\}, \{3\}, \{1, 2\}, \{1, 3\}, \{2, 3\}, \{1, 2, 3\}\}$. Then the Hasse diagram for the relation R is given in Fig.3.9. Here $X \leq Y$ if $X \subseteq Y$ for $X, Y \in R$.

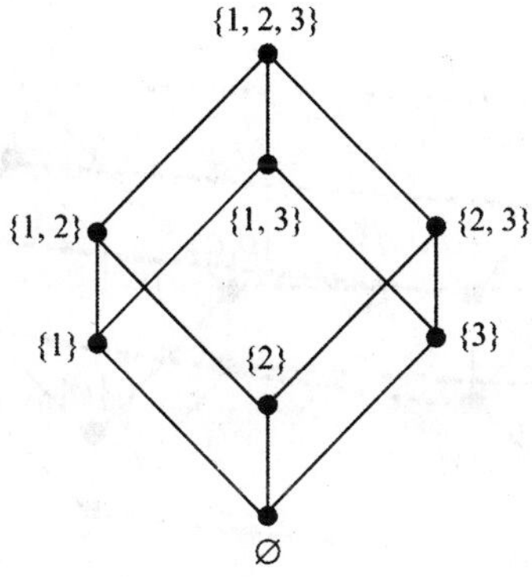

Fig. 3.9

iv. Consider a finite set A and its power set $P(A)$. The relation R defined on $P(A)$ is inclusion relation $\subseteq$. Draw Hasse diagram of R for

i. $A = \{a\}$ ii. $A = \{a, b\}$

iii. $A = \{a, b, c, d\}$.

Solution

i.

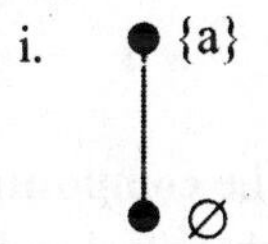

ii.

{a, b}

{a} {b} or {b} {a}

∅

{a, b}

∅

iii.

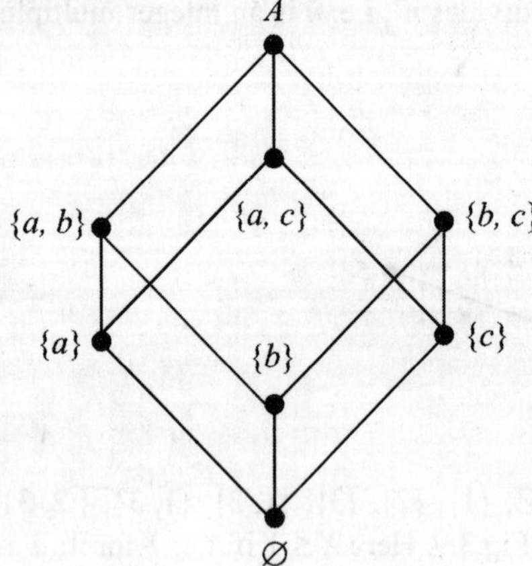

iv.

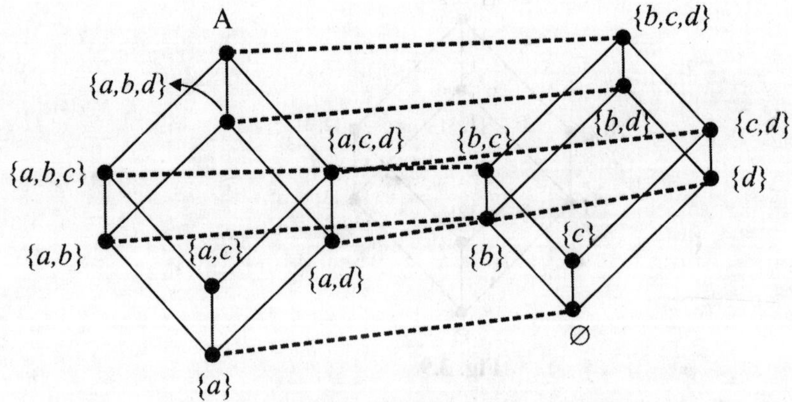

Fig. 3.10

Hasse diagram of a relation need not be unique as one can see from the example 3.10(*ii*).

3.7 Composition of Relations $R \circ S$

Let A, B, and C be sets. Let R be a relation from A to B and let S be a relation from B to C. The **composition** of R and S, denoted by $R \circ S$, (also by RS) defined as the relation having ordered pair (a, c) such that $a \in A$ and $c \in C$, and for which there exists an element $b \in A$ and $c \in C$, and for which there exists an element $b \in B$ such that $a \, R \, b$ and $b \, S \, c$. Thus,

$$R \circ S = \{(a, c) \in A \times C : \text{for some } b \in B,\ a \, R \, b \text{ and } b \, S \, c\}.$$

Example 3.11

Let R and S be the following relations on $A = \{a, b, c, d\}$ defined as

$R = \{(a, a), (a, c), (c, b), (c, d), (d, b)\}$

$S = \{(b, a), (c, c), (c, d), (d, a)\}$

Find (i) $R \circ S$ (ii) $S \circ R$ (iii) $R \circ R$ (iv) $S \circ S$.

Solution

i. $R \circ S = \{(a, c), (a, d), (c, a), (d, a)\}$

ii. $S \circ R = \{(b, a), (b, c), (c, b), (c, d), (c, b), (d, a), (d, c)\}$

iii. $R \circ R = \{(a, a), (a, c), (a, b), (a, d), (c, b)\}$

iv. $S \circ S = \{(c, c), (c, d), (c, a)\}$

The composition of a relation R with itself is denoted by a power of relation R. Let R be relation on the set A. Then RoR is the composition of R with itself and denoted by R^2. Similarly, $R^3 = R^2 \circ R = R \circ R \circ R$ and so on.

Theorem 3.4

If R is a relation from A to B and S is a relation form B to C. Show that $(R \circ S)^{-1} = S^{-1} \circ R^{-1}$.

Proof

Since, $R : A \to B \Rightarrow R^{-1} : B \to A$

And $S : B \to C \Rightarrow S^{-1} : C \to B$

Now, $R \circ S : A \to C \Rightarrow (R \circ S)^{-1} : C \to A$ ... (1)

Also, $S^{-1} \circ R^{-1} : C \to A$ ... (2)

From (1) and (2) $(R \circ S)^{-1} = S^{-1} \circ R^{-1}$.

Example 3.12

Let $A = \{2, 3, 4, 5\}$. The relation R and S on A is defined by

$R = \{(2, 2), (2, 3), (2, 4), (2, 5), (3, 4), (3, 5), (4, 5), (5, 3)\}$

$S = \{(2, 3), (2, 5), (3, 4), (3, 5), (4, 2), (4, 3), (4, 5), (5, 2), (5, 5)\}$

Verify $(R \circ S)^{-1} = S^{-1} \circ R^{-1}$.

Solution

$R \circ S = \{(2, 3), (2, 5), (2, 4), (2, 5), (2, 2), (3, 2), (3, 3), (3, 5), (4, 2), (4, 5), (5, 4), (5, 5)\}$

$(R \circ S)^{-1} = \{(3, 2), (5, 2), (4, 2), (2, 2), (2, 3), (3, 3), (5, 3), (2, 4), (5, 4), (4, 5), (4, 5), (5, 5)\}$ ... (1)

Also, $S^{-1} - \{(3, 2), (5, 2), (4, 3), (5, 3), (2, 4), (3, 4), (5, 4), (2, 5), (5, 5)\}$

and $R^{-1} = \{(2, 2), (3, 2), (4, 2), (5, 2), (4, 3), (5, 3), (5, 4), (3, 5)\}$

$S^{-1} \circ R^{-1} = \{(3, 2), (5, 2), (4, 2), (4, 5), (5, 5), (2, 2), (2, 3), (3, 3), (5, 3), (2, 4), (5, 4)\}$ ... (2)

From (1) and (2), $(R \circ S)^{-1} = S^{-1} \circ R^{-1}$.

1. *The readers who are familiar with the composition of functions note that traditional way of describing a composition of two relations is RS (or $R \circ S$) where R is the first relation and S is the second. Actually, function composition is traditionally expressed "backwards" i.e. as SR (or $S \circ R$), where R is the first relation and S is the second relation. Many texts denote the composition of relations R and S by $S \circ R$ rather than $R \circ S$. This is done in order to confirm with the usual use of $g \circ f$ to denote the composition of f and g where f and g are functions. (The readers who are not familiar with composition of functions disregard this).*
2. *$R \circ S$ is empty if the intersection of the range of R and domain of S is empty.*
3. *For the relation $R \circ S$, the domain is a subset of A and the range is a subset of C.*

Example 3.13

Consider the relations on {1, 3, 5}. Let R be the relation $x\ R\ y$ if $y = x + 2$ and let S be the relation $x\ S\ y$, if $x \leq y$. Find (i) RS (ii) SR.

Solution

$R = \{(1, 3), (3, 5)\}$, $S = \{(1, 3), (3, 5), (1, 5), (1, 1), (5, 3), (5, 5)\}$

i. $RS = \{(1, 5), (1, 3), (3, 5)\}$

ii. $SR = \{(1, 5), (1, 3), (3, 5)\}$.

Example 3.14

Let $A = \{4, 5, 6, 7\}$, $B = \{p, q, r, s\}$, $C = \{a, b, c\}$ and let consider the relations.

$R = \{(4, p), \{(5, s), (6, p), (6, q), (6, s)\}$ and

$S = \{(q, x)\ (q, z), (r, y), (s, z)\}$.

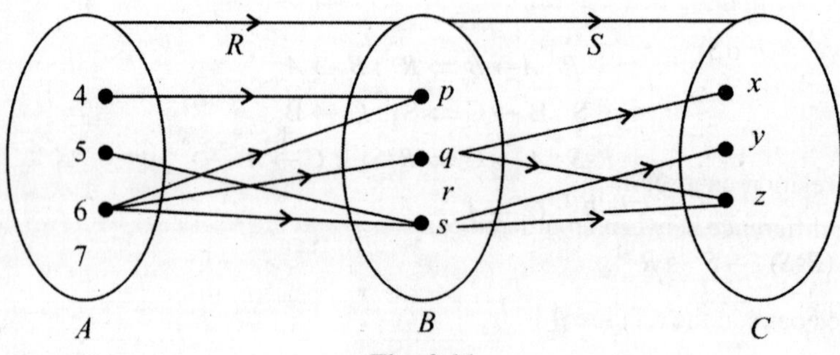

Fig. 3.11

Now see the Fig.3.11 and consider the arrow diagram of R and S. Here, one can observe that there is an arrow from 5 to s which is followed by an arrow from s to z. These two arrows can be viewed as a path which connects the element $5 \in A$ to the element $z \in C$. Thus

$$5(R \circ S)\ z \text{ since } 5\ R\ s \text{ and } s\ S\ z.$$

Similarly, one can observe that there is a path from 6 to x and a path from 6 to z. Hence $6\ (R \circ S)\ x$ and $6(R \circ S)\ z$.

Further note that no other element of A is connected to an element of C. Therefore $R \circ S = \{(5, z), (6, x), (6, z)\}$.

Example 3.15

Let $A = \{1, 2, 3, 4, 5\}$ and define R on A by $x\ R\ y$ if and only if $x + 1 = y$. Find (i) R, (ii) R^2, (iii) R^3.

Solution

Relation R is defined by $x\ R\ y$ iff $x + 1 = y$ and $A = \{1, 2, 3, 4, 5\}$. We can see that when $x = 1$, $y = 2$, both are in set A. Similarly when $x = 2$, $y = 3$ both are in A. Similarly for $x = 3$, 4 and $y = 4$, 5. Hence,

i. $R = \{(1, 2), (2, 3), (3, 4), (4, 5)\}$

ii. $R^2 = \{(1, 3), (2, 4), (3, 5)\}$

iii. $R^3 = \{(1, 4), (2, 5)\}$.

Example 3.16

Let R_1, R_2 and R_3 be relations on any set A. Prove that if $R_1 \subset R_2$, then $R_1 \circ R_3 \subseteq R_2 \circ R_3$.

Solution

Let $(x, z) \in R_1 \circ R_3$. Then there exist a 'y' in such a way that $(x, y) \in R_1$ and $(y, z) \in R_3$.

Since $R_1 \subset R_2 \Rightarrow (x, y) \in R_2$.

Again $(x, y) \in R_2$ and $(y, z) \in R_3 \Rightarrow (x, z) \in R_2 \circ R_3$

But $(x, z) \in R_1 \circ R_3 \Rightarrow R_1 \circ R_3 \subseteq R_2 \circ R_3$.

Theorem 3.5

If R is a relation from X to Y and S is a relation from Y to Z then $(R \circ S)^{-1} = S^{-1} \circ R^{-1}$.

Proof

Given R is a relation from X to Y hence R^{-1} is a relation from Y to X.

Similarly, S^{-1} is a relation from Z to Y. This implies relation $S^{-1} \circ R^{-1}$ is from Z to X.

Further, if $x \, R \, y$ and $y \, S \, z$, then $x(R \circ S) \, z$ hence if $z(R \circ S)^{-1}x$.

But $z \, S^{-1}y$ and $y \, R^{-1}x$ implies $z(S^{-1} \circ R^{-1}) \, x$.

This is possible when $(R \circ S)^{-1} = S^{-1} \circ R^{-1}$ for any $x \in X$ and $z \in Z$.

Exercise 3.1

1. What is input by relation on a set A.
2. Explain the main difference between an ordered pair (a, b) and the set $\{a, b\}$ with two elements.
3. Let $A = \{x, y, z\}$ and $B = \{1, 2\}$. Find (i) $A \times B$ (ii) $B \times A$.
4. List all the ordered pairs in the relation R from $A = \{0, 1, 2, 3, 4\}$ to $\{0, 1, 2, 3\}$, where $(x, y) \in R$ iff
   i. $x + y = 4$
   ii. $x = y$
   iii. $x > y$.
   iv. x/y
   v. LCM $(x, y) = 2$
   vi. GCD $(x, y) = 1$.
5. Let $X = Z^+$, the positive integers, and R be the relations defined by $x \, R \, y$ iff $2x \leq y + 1$. Which of the ordered pairs belong to R?
6. Let $X = \{1, 2, 3, 4, 6\}$ and R be the relation on X defined by a/b. Write R as a set of ordered pair.
7. Let $R_1 = \{(1, 2), (2, 3), (3, 4)\}$ and $R_2 = \{(1, 1), (1, 2), (2, 1), (2, 2), (2, 3), (3, 1), (3, 2), (3, 3), (3, 4)\}$ be relations from $A = \{1, 2, 3\}$ to $B = \{1, 2, 3, 4\}$. Find
   i. $R_1 \cup R_2$
   ii. $R_1 \cap R_2$
   iii. $R_1 - R_2$
   iv. $R_2 - R_1$.
8. Let R be a relation from a set X to Y. The **complementary Relation** $\bar{R}$ is the set of ordered pairs $\{(a, b) \mid (a, b) \notin R\}$.
   Let R be the relation $R = \{(x, y) \mid x < y\}$ on the set of integers. Find (i) R^{-1} (ii) $\bar{R}$.
9. Let R be a relation defined on the set $\{1, 2, 3, 4, 5\}$ by the rule $(x, y) \in R$ if 3 divides $x - y$.
   i. List the elements of R^{-1}.
   ii. Find the domain of R.
   iii. Find the range of R.
   iv. Find the domain of R^{-1}.
   v. Find the range of R^{-1}.
10. Let $A = \{p, q, r\}$, $B = \{q, r, s\}$ and $C = \{p, s\}$. Construct the tree diagram of $A \times B \times C$ and find $A \times B \times C$.
11. Write the ranges of the relations

    $R = \{(x, x^2) \mid x \in N\}$

and $S = \{ (x, 2x) \mid x \in N\}$

where $N = \{0, 1, 2, \ldots\}$. Also find $R \cup S$ and $R \cap S$.

12. Let R be the relation on the set $\{a, b, c, d, e\}$ containing the ordered pairs (a, a), (a, b), (a, c), (b, c), (b, d), (c, a), (c, d), (c, e), (d, b), (d, e), (e, a), (e, b) and (e, d). Find (i) R (ii) R^3 (iii) R^4 (iv) R^5.
13. Given a set $A = \{1, 2, 3, 4\}$. Define a relation R "a is less than b". Find the R^{-1}.
14. Consider the relation R defined on the set N of positive integers defined by the equation $x + 4y = 14$, i.e. $R = \{(x, y): x + 4y = 14\}$.
15. Let R be the relation on the set N of positive integers defined by the equation $x^2 + 2y = 100$. Find the domain of R. Find R^{-1} and describe R^{-1} by an equation.
16. Let $X = \{1, 2, 3, 4, 5\}$ and $Y = \{1, 2, 3\}$. Graph the relation $R = \{(x, y) \mid x > y\}$ from a subset of X into Y. Given the domain and range of the relation.
17. Draw the directed graph of the relation R defined by

$$R = \{(2, 3), (2, 9), (3, 2)\ (3, 4), (4, 3), (4, 9), (9, 2), (9, 4)\}.$$

18. Let $A = \{a, b, c, d\}$, $B = \{1, 2, 3, 4\}$ and $C = \{p, q, r\}$. Find the composite relation $R°S$, where relation R is defined from A to B and S from B to C.

$$R = \{(a, 1), (b, 4), (c, 1), (d, 2), (d, 4)\},$$
$$S = \{(2, p), (2, r), (3, q), (4, r)\}.$$

19. Let $X = \{w, x, y, z\}$, $Y = \{1, 2, 3\}$, and $Z = \{a, b, c, d\}$. Find the composite relation $R°S$ for the relation R from X to Y and S from Y to Z defined by

$$R = \{(w, 3), (x, 3), (y, 1), (y, 3), (z, 2)\} \text{ and } S = \{(1, b), (2, c), (2, d)\}.$$

20. Describe the Hasse diagram of following diagraph.

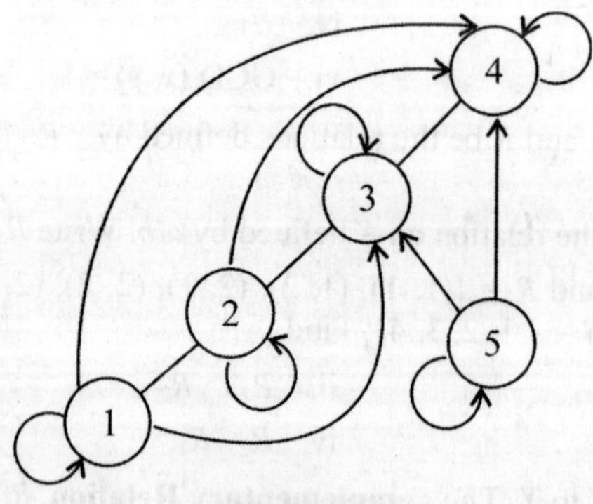

Fig. 3.12

3.8 Types of Relation

3.8-1 Identity Relation

A relation R in a set is said to be **identity relation**, generally denoted by I_A if

$$I_A = \{(x, x): x \in A\}.$$

Example 3.17

Let $A = \{3, 4, 5\}$ then $I_A = \{(3, 3), (4, 4), (5, 5)\}$ is an identity relation in A.

3.8-2 Universal Relation

A relation R in a set A is said to be **universal relation** if R is equal to $A \times A$, i.e. if $R = A \times A$.

Example 3.18

Let $A = \{a, b, c\}$, then $R = A \times A = \{(a, a), (a, b), (a, c), (b, a), (b, b), (b, c), (c, a), (c, b), (c, c)\}$ is a universal relation.

3.8-3 Void or null Relation

A relation R in a set A is said to be **void or null relation** if R is a null set, i.e. if $R = \varnothing$.

Example 3.19

Let $A = \{3, 4, 7\}$ and relation R be defined by 'aRb' if and only if 'a divides b' then we observe that $R = \varnothing$ is a void relation. One can observe that $R = \varnothing \subset A \times A$.

3.8-4 Diagonal Relation

Consider a set A. An important relation on A is that of equality, i.e. $\{(a, a) \mid a \in A\}$, which is ordered pair (a, b) of $A \times A$ for which $a = b$.

This is usually denoted by "=". The relation is also called the **identity** or **diagonal relation** on A. It is also denoted by Δ_A or simply Δ.

3.9 Properties of Relations

Here in this section we will discuss some of the important properties that are used to classify relations on a set.

3.9-1 Reflexive Relation

A relation R defined in a set A is said to be **reflexive** if every element of X is related to itself. So R is reflexive if $x R x \; \forall \, x \in X$

i.e. $$(x, x) \in R \; \forall \, x \in X.$$

If R is **not reflexive** then there exist at least one element $x \in A$ such that x is not related to x i.e. $(x, x) \notin R$.

If R is a reflexive relation in a set A, then $I_A \subset R$. Thus every identity relation of a set is also reflexive. But every reflexive relation in a set is not necessarily an identity relation.

Example 3.20

Consider a relation R on set $A = \{2, 4, 6, 8\}$ defined by $(a, b) \in R$ if $a \leq b$, $\forall \, a, b \in A$. Then

$$R = \{(2, 2), (2, 4), (2, 6), (2, 8), (4, 4), (4, 6), (4, 8), (6, 6), (6, 8), (8, 8)\}.$$

The relation R on A is reflexive since for each element $a \in A$, $(a, a) \in R$. To be specific (2, 2), (4, 4), (6, 6) and (8, 8) are each in R.

Example 3.21

Consider a set Z and let $a, b \in Z$ defined by a relation R by $x R y = x - y$ is a multiple of 3. As $x - x = 0$ is a multiple of 3. It follows that $x R x \; \forall \, x \in Z$. Hence R is reflexive.

Example 3.22

Consider the relation R in the set N of natural numbers defined by 'a divides b'. Then R is reflexive since every natural number divides itself.

Example 3.23

Consider the set of all straight lines say L in a plane. The relation "is parallel to" in L is reflexive, since every line is parallel to itself.

Example 3.24

Consider the following relations.

i. Relation of set inclusion in a family of set.

ii. Relation 'is congruent' in the set of all triangles in a plane.

iii. Relation $\perp$ (perpendicular) on the set L of lines in the plane.

iv. Set inclusion $\subseteq$ on a collection C of sets.

Determine which of the above relations are reflexive.

Solution

All relations except for (iii) are reflexive.

Theorem 3.6

Let R be a relation on X. Then if R is reflexive, so is R^{-1}.

Proof

Suppose R is reflexive. Then

$$(x, x) \in R \;\forall\, x \in X$$

$$\Rightarrow (x, x) \in R^{-1} \;\forall\, x \in X$$

Therefore, R^{-1} is reflexive.

3.9-2 Anti Reflexive or Irreflexive Relations

Let R be a relation in a set A. Then R is called an **anti-reflexive relation** if $(a, a) \notin R$ for every $a \in A$ i.e. $a \not R a$ $\forall\, a \in A$.

Example 3.25

i. The relation in the set of natural number N defined by $a < b$ is anti-reflexive, since x is not less than x for any natural number x, i.e. $x \not R x \;\forall\, x \in N$.

ii. Let $A = \{1, 2, 3, 4\}$. Then the relation $R = \{(1, 1), (2, 3), (3, 4)\}$ is not anti reflexive since, $(1, 1) \in R$.

3.9-3 Symmetric Relations

A relation R defined in a set X, is said to be symmetric if for any elements $x, y \in X$ whenever 'x' is related to 'y' then 'y' is also related to 'x'.

So R is symmetric if $a R b \Rightarrow b R a$ i.e. if $(a, b) \in R \Rightarrow (b, a) \in R$.

It is to be noted that $(b, a) \in R^{-1}$. So R is symmetric if $R = R^{-1}$.

Further R is not symmetric or asymmetric if there exist at least two distinct elements $x, y \in X$ such that $(x, y) \in R$ but $(y, x) \notin R$.

Example 3.26

Which of the following relation(s) is/are symmetric?

i. $R_1 = \{(1, 1), (2, 2), (3, 3)\}$ defined on set $A = \{1, 2, 3\}$.

ii. $R_2 = \{(1, 2), (2, 3), (1, 3)\}$ defined on set $A = \{1, 2, 3\}$.

iii. $R_3 = \{(2, 1), (3, 1), (3, 2), (4, 1), (4, 2), (4, 3)\}\}$ defined on set $A = \{1, 2, 3\}$.

iv. The relation 'is parallel to' in the set of all straight lines in a plane.

Solution

i. R_1 is symmetric.

ii. R_2 is symmetric.

iii. R_3 is not symmetric.

iv. Symmetric.

Example 3.27

Consider the set of natural numbers N and the relation R in N 'a divisor of b' is not symmetric because 2 divides 4 but 4 does not divide 2 i.e. $2 R 4$ but $4 \not R 2$.

Example 3.28

Let A be the set of members of a family and the relation R is given by 'a is a sister of b' then aRb does not necessarily imply bRa as b may be the brother of a. Hence R is not symmetric.

Theorem 3.7

If R be a relation on X then R is symmetric if and only if $R = R^{-1}$.

Proof

Let R is symmetric and $(a, b) \in R^{-1}$. This implies $(b, a) \in R$ and hence $(a, b) \in R$ is symmetric. Therefore $R^{-1} \subseteq R$. Similarly we can show that $R^{-1} \subseteq R$. Hence $R = R^{-1}$.

Conversely suppose that $R = R^{-1}$. Let $(a, b) \in R$, then $(a, b) \in R^{-1}$. This implies $(b, a) \in R$. Hence R is symmetric. Therefore, R is symmetric if and only if $R = R^{-1}$.

3.9-4 Anti-Symmetric Relations

A relation R defined in a set X is said to be antisymmetric if for any two elements $x, y \subset X$, $x R y$ and $y R x \Rightarrow x = y$

On the other hand a relation R in a set is not antisymmetric if there are at least two elements x and y in R such that $x \neq y$ but $x R y$ and $y R x$.

The terms symmetric and antisymmetric are not opposite. It is possible to have a relation which is both symmetric and antisymmetric. Further note that negative of antisymmetric doesn't mean not symmetric.

Example 3.29

Let N be the set of natural numbers and the relation R in N is defined by "a divides b". Then R is antisymmetric relation in N. Because for any two natural numbers a and b, if a divides b and b divides a then $a = b$.

One can note here that this relation is not antisymmetric in the set Z of integers, because we find that for any nonzero integer a, $a R (-a)$ and $(-a) R a$ but $a \neq (-a)$.

Example 3.30

Consider the relation R of 'set inclusion' in a family of sets. The relation R is antisymmetric as for any two sets A and B of the given family

$$A \subset B \text{ and } B \subset A \Rightarrow A = B.$$

Theorem 3.8

Let R be a relation on X_3 then R is anti-symmetric if and only if $R \cap R^{-1} \subseteq I_x$ (i.e. identity relation on X).

Proof

Suppose relation R is anti-symmetric on X. Let $(x, y) \in R \cap R^{-1}$. Then $(x, y) \in R$ and $(x, y) \in R^{-1}$ implies $(y, x) \in R$. Hence $(x, y) \in R$ and $(y, x) \in R$. But R is antisymmetric, thus $x = y$. Hence, every element of $R \cap R^{-1}$ is of the form (x, x) for $x \in X$.

Therefore $$R \cap R^{-1} \subseteq I_x.$$

Conversely, suppose $R \cap R^{-1} \subseteq I_p$. Let $(x, y) \in X \times X$ in such a way that $(x, y) \in R$ and $(y, x) \in R$. This implies. $(x, y) \in R$ and $(x, y) \in R^{-1}$. Hence $(x, y) \in R \cap R^{-1}$. But $R \cap R^{-1} \subseteq I_x$. This is only possible when $x = y$. Hence R is anti symmetric.

3.9-5 Transitive Relation

A relation R defined in a set X is said to be **transitive** if for any $x, y, z \in X$ whenever x is related to y and y is related to z then 'x' is related to 'z'. So

$$R \text{ is transitive if } x\,R\,y \text{ and } y\,R\,z \Rightarrow x\,R\,z.$$

or we can say if $$(x, y) \in R \text{ and } (y, z) \in R \Rightarrow (x, z) \in R.$$

The relation R in a set X is not transitive if there exist elements x, y and $z \in X$ not necessarily distinct and

$$(x, y) \in R \text{ and } (y, z) \in R \text{ but } (x, z) \notin R.$$

The following table summaries the above properties:

S. No.	Property	Definition
1.	Reflexivity	$x\,R\,x \;\forall\, x \in A$
2.	Anti reflexivity	$x \not R\, x \;\forall\, x \in A$
3.	Symmetry	$x\,R\,y \Rightarrow y\,R\,x \;\forall\, x, y \in A$
4.	Asymmetry	$x \not R\, y \Rightarrow y\,R\,x \;\forall\, x, y \in A$
5.	Anti symmetry	i. If $x\,R\,y$ and $y\,R\,x \Rightarrow x = y$ ii. If $x \neq y \Rightarrow$ either $x\,R\,y$ or $y\,R\,x \;\forall\, x, y \in A$
6.	Transitivity	$x\,R\,y$ and $y\,R\,z \Rightarrow xRz \;\forall\, x, y \in A$

Example 3.31

Let L be the set of all straight lines in a plane and R be the relation in L defined by 'a is parallel to b. If x is parallel to y and y is parallel to z then obviously x is parallel to z. Thus, $(x, y) \in R$ and $(y, z) \in R \Rightarrow (x, z) \in R$. Hence R is transitive.

Example 3.32

Consider the set N of natural numbers and let R be the relation in N defined by 'a is less than b'. Is it transitive?

Solution

As we know $x < y$ and $y < z \Rightarrow x < z$, therefore R is transitive.

Example 3.33

In the set of natural numbers N, the relation R defined by $a \; R \; b \Leftrightarrow$ 'a divides b' is transitive, as for any a, b, $c \in N$ if a divides b and b divides c then there exist $p, q \in N$ such that

$$b = pa \text{ and } c = qb$$
$$\Rightarrow c = q\,(pa) = (qp)\,a$$
$$\Rightarrow a \text{ divides } c$$

Therefore a divides b and b divides $c \Rightarrow a$ divides c.

In other words we can say that $a \; R \; b$ and $b \; R \; c \Rightarrow a \; R \; c$.

Example 3.34

Let A be the set of all Americans. Consider a relation R in A defined by 'a hates b' if x hates y and y hates z, it does not necessarily follow that x hates z. So, R is not a transitive relation.

Example 3.35

If a relation R is transitive, then R^{-1} is also transitive.

Solution

Let (x, y) and $(y, z) \in R^{-1}$. This implies $(y, x) \in R$ and $(z, y) \in R$. But R is transitive, hence $(z, x) \in R$. Therefore $(x, y) \in R^{-1}$. Hence R^{-1} is also transitive.

Example 3.36

Let $A = \{1, 2, 3, 4, 6\}$ and let R be the relation on A defined by 'x divides y' indicated as x/y iff there exists an integer z such that $x\,z = y$.

i. Write R as a set of ordered pairs.

ii. Find the inverse relation R^{-1} and describe R^{-1} in words.

Solution

i. $R = \{(1, 1), (1, 2), (1, 3), (1, 4), (1, 6), (2, 2), \{(2, 4), (2, 6), (3, 3), (3, 6), (4, 4), (6, 6)\}$.

ii. $R^{-1} = \{(1, 1), (2, 1), (3, 1), (4, 1), (6, 1), (2, 2), \{(4, 2), (6, 2), (3, 3), (6, 3), (4, 4), (6, 6)\}$.

R^{-1} is the relation in which y divides x.

Example 3.37

Prove that if a relation R on set A is transitive and irreflexive, then it is asymmetric (not symmetric).

Solution

Let R is not asymmetric then bRa whenever aRb. Since R is transitive, aRb and bRa implies aRa. This shows that R is reflexive, which is not true because R is irreflexive. Hence our assumption is wrong. So, R is asymmetric.

Example 3.38

How many reflexive relation and symmetric relations are there on a set with n elements.

Solution

We know that a relation R on set A is a subset of $A \times A$. Thus a relation is determined by specifying whether each of the n^2 ordered pairs in $A \times A$ is in R. If R is reflexive, each of the other $n\,(n-1)$ ordered pairs (a, a) for

$a \in A$ must be in R. Each of the other $n(n-1)$ ordered pairs of the form may or may not be in R. By product rule of counting, there are $2^{n(n-1)}$ reflexive relations.

If R is symmetric, each of the n ordered pairs (a, a) for $a \in A$ must be in R and set of pairs of elements of the form (a, b) and (b, a) for $a, b \in A$ must also be in R. There are $n(n-1)/2$ such pairs. By product rule of counting, there are $2^n \cdot 2^{n(n-1)/2} = 2^{n(n+1)/2}$ symmetric relations.

Example 3.39

Let R and S are relations on a set A. Prove that

i. If R and S are reflexive, then $R \cup S$ and $R \cap S$ are reflexive.
ii. If R and S are symmetric, then $R \cup S$ and $R \cap S$ are symmetric.
iii. If R and S are transitive, then $R \cup S$ is not transitive but $R \cap S$ is transitive.

Solution

i. Given, R and S are reflexive implies $a\ R\ a$ and $a\ S\ a$ $\forall a \in A$. Thus, $a\ R \cup S\ a$ and a $R \cap S\ a$. Hence, $R \cup S$ and $R \cap S$ are reflexive.
ii. Given, R and S are symmetric, we know that if R is symmetric then $R = R^{-1}$. Similarly $S = S^{-1}$. Also $(R \cup S)^{-1} = R^{-1} \cup S^{-1} = R \cup S$ and $(R \cap S)^{-1} = R^{-1} \cap S^{-1} = R \cap S$. Hence, $R \cup S$ and $R \cap S$ are symmetric.
iii. Given, R and S are transitive. Let $a\ R \cap S\ b$ and $b\ R \cap S\ c$, then $a\ R\ b$, $a\ S\ b$, $b\ R\ c$ and $b\ S\ c$. Now $a\ R\ b$, and $b\ R\ c \Rightarrow a\ R\ c$, and $a\ S\ b$, and $b\ S\ c \Rightarrow a\ S\ c$. Therefore, $a\ R \cap S\ c$. Thus, $R \cap S$ is transitive.

   Again, if R and S are transitive relations on A. Then $R \cup S$ need not be a transitive relation on A. For example, if $R = \{(1, 1), (1, 2)\}$ and $S = \{(2, 2), (3, 3)\}, (2, 3)\}$ on $A = \{1, 2, 3\}$, then R and S are transitive but $R \cup S = \{(1, 1), (1, 2), (2, 2), (3, 3), (2, 3)\}$ is not transitive because (1, 2) and (3, 3) are is $R \cup S$ but (1, 3) is not in $R \cup S$.

Example 3.40

Let R be the binary relation defined as:

$$R = \{(a, b) \in R^2 \mid a - b \leq 3\}$$

Determine whether R is reflexive, symmetric, anti–symmetric or transitive.

Solution

Reflexive: $(a, a) \in R^2 \because a - a = 0 < 3$

$\therefore R$ is reflexive.

Symmetric: Let $(a, b) \in R^2 \Rightarrow a - b \leq 3 \Rightarrow b - a \leq 3$

$(a, b) \in R^2 \Rightarrow R$ is symmetric.

Anti-Symmetric: Let $(a, b) \in R^2$ and $(b, a) \in R^2, \Rightarrow a - b \leq 3$ and $b - a \leq 3 \not\Rightarrow a = b$

R is not anti-symmetric.

Transitive: Let $(a, b) \in R^2$ and $(b, c) \in R^2$

$\Rightarrow a - b \leq 3$ and $b - c \leq 3$

Now $a - c = (a - b) + (b - c)$

This is not necessarily less than or equal to 3.

$\Rightarrow R$ is not transitive.

Exercise 3.2

1. The relations R_1, R_2 and R_3 are defined in $A = \{1, 2, 3\}$ as follows

$$R_1 = \{(1, 1), (1, 2), (1, 3), (2, 2), (2, 3), (3, 1), (3, 2), (3, 3)\}$$
$$R_2 = \{(1, 2), (2, 1), (1, 3), (3, 1)\}$$
$$R_3 = \{(1, 2), (2, 3), (3, 1)\}.$$

Find whether each of R_1, R_2 and R_3 is

a. Reflexive
b. Symmetric
c. Transitive.

2. Consider the following relations on set $A = \{x, y, z\}$

$$R_1 = \{(x, x), (x, y), (x, z), (z, z)\}$$
$$R_2 = \{(x, x), (x, y), (y, x), (y, y), (z, z)\}$$
$$R_3 = \{(x, x), (x, y), (y, y), (y, z)\}$$
$$\varnothing = \text{empty relation}$$
$$R_4 = A \times A.$$

Determine which of the relations are

i. reflexive
ii. symmetric
iii. transitive
iv. antisymmetric

3. Determine which of the following relation defined on the set $A = \{a, b, c, d\}$ is irreflexive.

i. $\{(a, c), (a, d), (b, c), (b, d), (c, a), (c, d)\}$
ii. $\{(a, b), (b, c), (c, d)\}$
iii. $\{(b, b), (b, c), (b, d), (c, b), (c, c), (c, d)\}$
.iv. $\{(a, a), (b, b), (c, c), (d, d)\}$.

4. Can a relation on a set X be neither reflexive nor irreflexive?

5. Give an example of a relation which is

i. Symmetric but not reflexive or transitive.
ii. Transitive but not reflexive or symmetric.
iii. Reflexive but not symmetric or transitive.
iv. Reflexive and transitive but not symmetric.
v. Reflexive and symmetric but not transitive.
vi. Symmetric and transitive but not reflexive.

6. Give an example of a relation on a set that is

i. Symmetric and antisymmetric.
ii. Neither symmetric nor antisymmetic.

7. Determine which of the following relation defined on the set of positive integers is reflexive, symmetric, antisymmetric and transitive.

i. $(a, b) \in R$ if $a = b^2$.
ii. $(a, b) \in R$ if $a = b$.

8. Determine whether the relation R whose diagraph is given is reflexive, irreflexive, symmetric, asymmetic*, antisymmetic or transitive.

---

* A relation R is called asymmetric if $(a, b) \in R \Rightarrow (b, a) \notin R$.

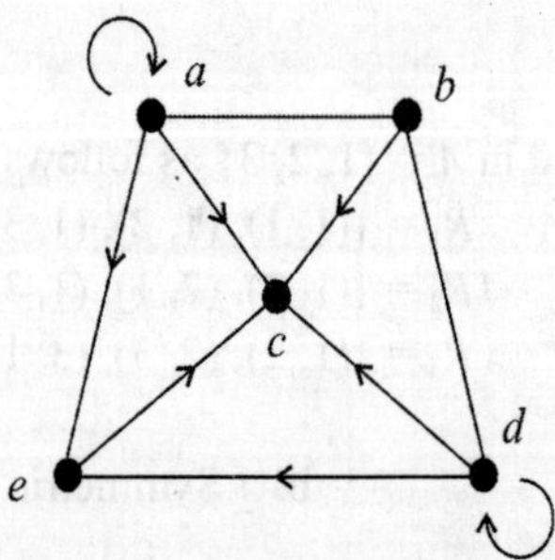

Fig. 3.13

9. Determine whether the relation R on the set X is reflexive, irreflexive, symmetric, asymmetric, antisymmetric or transitive.
   i. $X = Z^+, x\,R\,y$ iff $x = y^k$ for some $k \in Z^+$.
   ii. $X = Z, x\,R\,y$ iff $x \le y + 1$.
   iii. $X = Z, x\,R\,y$ iff $|x - y| = 2$.
10. Prove that if a relation R on a set A is symmetric then the relation R^2 is also symmetric.
11. Each of the following open sentences defines a relation in the set of all people:
   i. a is a husband of b.
   ii. a is a father of b
   iii. a is a brother of b
   iv. a lives in the same house in which b lives.
   v. a is a friend of b.

   Investigate whether the relation is reflexive, symmetric and transitive.
12. i. Show that the relation R on a set A is symmetric if and only if $R = R^{-1}$ where R^{-1} is the inverse relation.
   ii. Show that the relation R on a set A is reflexive iff inverse relation is reflexive.
   iii. Suppose that the relation R is reflexive. Is R^2 necessarily irreflexive? Give a reason for your answer.
13. Let R be a relation of set X to Y. The **complementary relation** R is the set of ordered pairs $\{(x, y) | (x, y) \notin R\}$. Let R be the relation $R = \{(x, y) | x \text{ divides } y\}$ on the set of positive integers. Find R.
14. Let R and S be reflexive relations on a set X. Show that
   i. $R \cap S$ is reflexive
   ii. $R \cap S$ is symmetric
   iii. $R \cap S$ is transitive.
15. Let R be a relation of a set X. Show that $R \cup R^{-1}$ is symmetric.
16. Show that the relation R on a set X is reflexive iff its complementary relation R^{-1} is reflexive.
17. Use the matrix of relation to test for transitivity for the relation
   i. $R = \{(a, b), (b, c), (c, d), (d, e)\}$ ordering of $A : a, b, c, d, e$
   ii. $R = \{(a, b) | a < b\}$ ordering of A : 1, 2, 3, 4.
18. Let X be a family of sets and R be a relation in X defined by 'X is disjoint from Y'. State whether or not R is
   i. reflexive
   ii. Symmetric
   iii. transitive.
19. Prove or disprove that if a relation on a set A is transitive and irreflexive, then it is asymmetric.

20. Consider a nonempty relation R on a set X. Suppose that R is symmetric and transitive. Show that R is not irreflexive.

21. Let A be any set. Define the diagonal relation on A, can you give another name to this relation?

Answers

1. R_1 is reflexive but it is neither symmetric nor transitive.

   R_2 is symmetric but it is neither reflexive nor transitive

   R_3 is neither reflexive nor symmetric nor transitive.

2. R_2 and R_4 are reflexive.

   R_2, R_4 and $\varnothing$ are symmetric.

   R_1, R_2, R_4 and $\varnothing$ are transitive.

   R_1, R_3 and $\varnothing$ are antisymmetic.

3. (i) and (ii).

4. Yes, for instance $\{2, 2\}$ on $\{2, 3\}$.

7. (i) Antisymmetric (ii) Reflexive, symmetric, antisymmetric and transitive.

8. Antisymmetric and transitive

9. (i) Reflexive, antisymmetric and transitive. (ii) Reflexive.

iii. Irreflexive, symmetric.

14. $\{(x, y) \mid x \text{ does not divide } y\}$.

19. R is symmetric but it is neither reflexive nor symmetric.

3.10 Representing Relations Using Matrices

We can represent a relation R from a finite set X to a finite set Y with the help of a matrix called the **relation matrix** of R or **matrix of the relation R** or **adjacency matrix** or **Boolean Matrix.** Such a representation are used in computer programming to analyze a relation.

Let $A = \{a_1, a_2, a_3, ..., a_m\}$, $B = \{b_1, b_2, b_3, ..., b_n\}$ are finite sets containing m and n elements respectively. The relation matrix of R can be obtained by labeling the rows with the elements of A and column with the element of B (with in some arbitrary order). We then set the ith row and j^{th} column to 1 if $a_i\ R\ b_j$ other wise 0, i.e. if $a_k \not R\ b_r$, then we enter a 0 in the kth row and r^{th} column. The relation R can be represented by matrix.

$$R_{ij} = \begin{cases} 1 & \text{if } a_i R\ b_j \\ 0 & \text{otherwise} \end{cases}$$

where R_{ij} is the element in the ith row and j^{th} column. Matrix obtained in this way is called **relation matrix**. If A has m elements and B has n elements then the relation matrix is an $m \times n$ matrix.

The use of matrices to represent relations is illustrated in the following examples.

Example 3.41

Let $X = \{1, 2, 3\}$ and R be the relation $\{(1, 1), (1, 2), (2, 3), (3, 3)\}$ on X. Since R is a relation from X into the same set X. Let us have $a_1 = 1$, $a_2 = 2$, $a_3 = 3$ and $b_1 = 1$, $b_2 = 2$, and $b_3 = 3$.

Since $1\ R\ 1$, we have $R_{11} = 1$

$1\ R\ 2$, we have $R_{12} = 1$

$2\ R\ 3$, we have $R_{23} = 1$

and $3\ R\ 3$, we have $R_{33} = 1$

And all other entries of R are zero. Relation R is represented by the matrix

$$\begin{array}{c} \\ 1 \\ 2 \\ 3 \end{array}\begin{array}{c} \begin{matrix} 1 & 2 & 3 \end{matrix} \\ \begin{bmatrix} 1 & 0 & 0 \\ 1 & 1 & 0 \\ 0 & 0 & 0 \end{bmatrix} \end{array}$$

Example 3.42

Consider the set $A = \{2, 4, 6, 8\}$ and $B = \{p, q, r, s\}$.
The matrix of relations

$$R = \{(2, q), (2, s), (4, r), (6, r), (6, q), (8, p)\}$$

from A to B relative to the orderings 2, 4, 6, 8 and p, q, r, s is

$$\begin{array}{c} \\ 2 \\ 4 \\ 6 \\ 8 \end{array}\begin{array}{c} \begin{matrix} p & q & r & s \end{matrix} \\ \begin{bmatrix} 0 & 1 & 0 & 1 \\ 0 & 0 & 1 & 0 \\ 0 & 1 & 1 & 0 \\ 1 & 0 & 0 & 0 \end{bmatrix} \end{array}$$

If we find the matrix of relation R relative to the ordering 4, 6, 8, 2 and s, q, p, r it is given by

$$\begin{array}{c} \\ 4 \\ 6 \\ 8 \\ 2 \end{array}\begin{array}{c} \begin{matrix} s & q & p & r \end{matrix} \\ \begin{bmatrix} 0 & 0 & 0 & 1 \\ 0 & 1 & 0 & 1 \\ 0 & 0 & 1 & 0 \\ 1 & 1 & 0 & 0 \end{bmatrix} \end{array}$$

Here, one can observe that the matrix of relation from A to B is dependent on the orderings of A to B i.e. changes with respect to ordering.

Example 3.43

The matrix of relation R from {3, 4, 5} to {5, 6, 8, 10} relative to the orderings 3, 4, 5 and 5, 6, 8, 10 defined by $a\ R\ b$ if a divides b is given by

$$\begin{array}{c} \\ 3 \\ 4 \\ 5 \end{array}\begin{array}{c} \begin{matrix} 5 & 6 & 8 & 10 \end{matrix} \\ \begin{bmatrix} 0 & 1 & 0 & 0 \\ 0 & 0 & 1 & 0 \\ 1 & 0 & 0 & 1 \end{bmatrix} \end{array}$$

When we write the matrix of a relation R on a set A (i.e. from A to A), the ordering of rows and columns should be same.

3.11 Determination of Nature of relation using relation matrix

By examining the relation matrix we can quickly determine the nature of the relation in following way:

(1) Reflexive Relation

A relation R is reflexive if all the elements of main diagonal of the relation matrix M_R is 1, i.e. for $M_R = [m_{ij}]$, all $m_{ii} = 1$.

Hence $$M_R = \begin{bmatrix} 1 & 0 & 0 \\ 1 & 1 & 0 \\ 0 & 1 & 1 \end{bmatrix}$$ represents a reflexive relation.

If all the elements of main diagonal of relation matrix M_R is 0 then the relation R is irreflexive.

Reflexivity and irreflexivity depend only on the main diagonal of the relation matrix.

2. Symmetric Relation

A relation R is symmetric if the relation matrix M_R is symmetric about the main diagonal. That is for

$$M_R = [\, m_{ij}],\ m_{ij} = m_{ij}\ \forall\ i \text{ and } j.$$

For example:

$$M_R = \begin{bmatrix} 1 & 0 & 1 \\ 0 & 0 & 0 \\ 1 & 0 & 1 \end{bmatrix}$$ represents a symmetric relation.

But a relation R is antisymmetric if for any $m_{ji} = 1$, $m_{ij} = 0$.

For example:

$$M_R = \begin{bmatrix} 1 & 1 & 0 \\ 0 & 1 & 0 \\ 1 & 0 & 0 \end{bmatrix}$$ represents a antisymmetric.

3. Transitive Relation

We show later how composition of relation matrix be used to test the transitivity.

3.12 Composition of Relations and Matrices

Let R is a relation from X to Y and S is a relation from Y to Z. Let M_R and M_S represent the matrix of relation R and S respectively. Then to obtain the composition $R \circ S$ in matrix form, follow following procedure:

i. Choose the ordering X, Y and Z.
ii. Find the product of M_R and M_S.
iii. Replace all nonzero elements by 1 in the matrix product M_R. M_S.

For example:

Let $X = \{a, b, c, d\}$, $Y = \{1, 2, 3, 4\}$, $Z = \{x, y, z\}$ and let $R = \{ (a, 1), (b, 4), (c, 1), (c, 2), (c, 4)\}$ and $S = \{(2, x), (2, z), (3, y), (4, z)\}$. Let M_R and M_S denote respectively the matrices of the relations R and S.

$$M_R = \begin{array}{c} \\ a \\ b \\ c \\ d \end{array}\begin{array}{c} \begin{array}{cccc} 1 & 2 & 3 & 4 \end{array} \\ \begin{bmatrix} 1 & 0 & 0 & 0 \\ 0 & 0 & 0 & 1 \\ 1 & 1 & 0 & 1 \\ 0 & 0 & 0 & 0 \end{bmatrix} \end{array} \text{ and } M_S = \begin{array}{c} \\ 1 \\ 2 \\ 3 \\ 4 \end{array}\begin{array}{c} \begin{array}{ccc} x & y & z \end{array} \\ \begin{bmatrix} 0 & 0 & 0 \\ 1 & 0 & 1 \\ 0 & 1 & 0 \\ 0 & 0 & 1 \end{bmatrix} \end{array}$$

Multiplying M_R and M_S, we get

$$M = M_R \circ M_S = \begin{array}{c} \\ a \\ b \\ c \\ d \end{array}\begin{array}{c} \begin{array}{ccc} x & y & z \end{array} \\ \begin{bmatrix} 0 & 0 & 0 \\ 0 & 0 & 1 \\ 1 & 0 & 2 \\ 0 & 0 & 0 \end{bmatrix} \end{array} \approx \begin{array}{c} \\ a \\ b \\ c \\ d \end{array}\begin{array}{c} \begin{array}{ccc} x & y & z \end{array} \\ \begin{bmatrix} 0 & 0 & 0 \\ 0 & 0 & 1 \\ 1 & 0 & 1 \\ 0 & 0 & 0 \end{bmatrix} \end{array}$$

By checking the elements of matrix M, we can see the elements related by $R \circ S$. We see that the relation $R \circ S$ is given by

$$R \circ S = \{(b, z), (c, x), (c, z)\}.$$

and

$$M_{RoS} = \begin{array}{c} \\ a \\ b \\ c \\ d \end{array}\begin{array}{c} \begin{array}{ccc} x & y & z \end{array} \\ \begin{bmatrix} 0 & 0 & 0 \\ 0 & 0 & 1 \\ 1 & 0 & 1 \\ 0 & 0 & 0 \end{bmatrix} \end{array}$$

$M = M_R \,.\, M_S$ and M_{RoS} have same entries, i.e. $M_{RoS} = M_R \,.\, M_S$.

Aliter: The matrix of composition of relation can also be obtained using Boolean product of matrices. Let the sets X, Y, and Z have m, n and p elements respectively. Let $M_R = [a_{ij}]_{m\times n}$, $M_S = [b_{ij}]_{m\times p}$, and $M_{RoS} = [c_{ij}]_{m\times p}$ are the relation matrices of relation R, S and $R \circ S$ respectively. Then by the Boolean product $[c_{ij}]_{m\times p}$ denoted by $M_R \circ M_S$ is obtained in an analogous way to the ordinary product of these matrices with addition replaced by the operation $\vee$ * (which sense 'OR') and multiplication replaced by the operation* (which sense 'AND').

Hence the $(i, j)^{th}$ entry of C_{ij} is given by

$$C_{ij} = (a_{i1} \wedge b_{1j}) \vee (a_{i2} \wedge b_{2j}) \vee \ldots \vee (a_{im} \wedge b_{mj})$$

For example:

Let $X = \{a, b, c\}$, $Y = \{1, 2\}$ and $Z = \{x, y, z\}$,

and $R = \{(a,1), (b,2), (c,2)\}$, $S = \{(1,y), (1,z), (2,x), (2,z)\}$

Then

$$M_R = \begin{array}{c} \\ a \\ b \\ c \end{array}\begin{array}{c} \begin{array}{cc} 1 & 2 \end{array} \\ \begin{bmatrix} 1 & 0 \\ 0 & 1 \\ 0 & 1 \end{bmatrix} \end{array}, \quad M_S = \begin{array}{c} \\ 1 \\ 2 \end{array}\begin{array}{c} \begin{array}{ccc} x & y & z \end{array} \\ \begin{bmatrix} 0 & 1 & 1 \\ 1 & 0 & 1 \end{bmatrix} \end{array},$$

The Boolean product of M_R and M_S is given by

$$M_R \circ M_S = \begin{bmatrix} (1\wedge 0)\vee(0\wedge 1) & (1\wedge 1)\vee(1\wedge 0) & (1\wedge 1)\vee(0\wedge 1) \\ (0\wedge 0)\vee(1\wedge 1) & (0\wedge 1)\vee(1\wedge 0) & (0\wedge 1)\vee(1\wedge 1) \\ (0\wedge 0)\vee(1\wedge 1) & (0\wedge 1)\vee(1\wedge 0) & (0\wedge 1)\vee(1\wedge 1) \end{bmatrix}$$

$$= \begin{bmatrix} 0\text{V}0 & 1\text{V}0 & 1\text{V}0 \\ 0\text{V}0 & 0\text{V}0 & 0\text{V}1 \\ 0\text{V}0 & 0\text{V}0 & 0\text{V}1 \end{bmatrix} = \begin{bmatrix} 0 & 1 & 1 \\ 1 & 0 & 1 \\ 0 & 0 & 1 \end{bmatrix}$$

* $1 \vee 0 = 1$, i.e. 1 or 0 = 1; $0 \vee 1 = 1$, i.e. 0 or 1 = 1, $1 \vee 1 = 1$, i.e. 1 or 1 = 1; $0 \vee 0 = 0$, i.e. 0 or 0 = 0; $1 \wedge 0 = 0$, i.e. 1 and 0 = 1; $0 \wedge 1 = 0$, i.e. 0 and 1 = 0; $1 \wedge 1 = 1$, i.e. 1 and 1 = 1; $0 \wedge 0 = 0$, i.e. 0 and 0 = 0.

Consider a relation R. Let M_R be the matrix of relation R. Then the relation R is transitive if and only if the $(i, j)^{th}$ nonzero element of $M_R^2 = M_R \cdot M_R$ is equal to the $(i, j)^{th}$ element of M_R. This is because of the fact that $(i, j)^{th}$ element is nonzero (i.e.1) if and only if there exist (i, k) and (k, j) nonzero elements in R. But (i, j) is in R if and only if element (i, j) in M_R is nonzero.

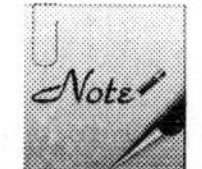

1. *Boolean matrix multiplication is not commutative in general and hence the composition of relation is not commutative in general.*
2. *If M_R is relation matrix of relation R, then R^{-1} = Transpose of M_R.*

Example 3.44

Consider the matrix $M = \begin{bmatrix} 0 & 1 & 1 & 0 \\ 1 & 0 & 0 & 1 \\ 0 & 1 & 0 & 1 \end{bmatrix}$, since M is a 3×4 matrix, if we assume $X = \{x_1, x_2, x_3\}$ and $Y = \{y_1, y_2, y_3, y_4\}$. Then $(x_i, y_j) \in R$ iff $m_{ij} = 1$.

$\Rightarrow R = \{(x_1, y_2), (x_1, y_3), (x_2, y_1), (x_2, y_4), (x_3, y_2), (x_3, y_4)\}$.

Example 3.45

Let $A = \{p, q, r, s, t, u\}$ and B = {quit, player, sun, rat} and consider the relation R from A to B where $(a,–b) \in R$ if a is a letter in the word B.

Find the matrix M which represents R.

Solution

To find the relation matrix M let us order the elements of A and B. Here, the matrix M will have six rows, labelled by the elements of A and four columns labeled by the elements of B. Then put 1 in the row a and column b if a is a letter in B and 0 otherwise. Thus

$$M = \begin{array}{c} \\ p \\ q \\ r \\ s \\ t \\ u \end{array}\begin{array}{c} \text{quit} \quad \text{player} \quad \text{sun} \quad \text{rat} \\ \begin{bmatrix} 0 & 1 & 0 & 0 \\ 1 & 0 & 0 & 0 \\ 0 & 1 & 0 & 1 \\ 0 & 0 & 1 & 0 \\ 1 & 0 & 0 & 1 \\ 1 & 0 & 1 & 0 \end{bmatrix} \end{array}$$

Exercise 3.3

1. Let $A = \{a, b, c\}$ be a set. Represent each of the following relations with matrix
   i. $\{(a, a), (a, b), (a, c), (b, c)\}$
   ii. $\{(a, a), (a, b), (a, c), (b, b), (b, c), (c, c)\}$
   iii. $\{(a, c), (c, a)\}$
   iv. The relation S on $A = \{1, 2, 3, 4\}$ where $S = \{(1, 1), (2, 3), (3, 2), (4, 4)\}$.
2. Find the matrix of the relation $R = \{(p, 1), (p, 2), (q, 1), (r, 2), (s, 1)\}$, defined on the set $A = \{p, q, r, s\}$ and $B = \{1, 2, 3\}$.
3. Find the matrix of relation R on A relative to the ordering given $R = \{(a, b), (b, c), (c, d), (d, e)\}$ ordering $A = a, b, c, d, e$.

4. Find the matrix of relation R from A to B relative to the ordering given
   i. $R = \{(a, \delta), (b, \alpha), (b, \Sigma), (c, \beta), (c, \Sigma)\}$. Relative to the ordering of A: a, b, c; ordering of B: $\alpha, \beta, \Sigma, \delta$.
   ii. $R = \{(a, 1), (a, 3), (b, 1), (b, 2), (c, 4)\}$, ordering of A: a, b, c; ordering of B : 1, 2, 3, 4.
5. i. Let $X = \{x_1, x_2, x_3\}$ and $Y = (y_1, y_2, y_3, y_4, y_5\}$. Which ordered pairs are in the relation R represented by the matrix

$$M_R = \begin{bmatrix} 0 & 1 & 0 & 0 & 0 \\ 1 & 0 & 1 & 1 & 0 \\ 1 & 0 & 1 & 0 & 1 \end{bmatrix}$$

   ii. Write the relation R, given by the matrix as a set of ordered pair

$$\text{(a)} \quad \begin{array}{c} \\ 1 \\ 2 \\ 3 \\ 4 \end{array} \begin{array}{c} \begin{array}{cccc} a & b & c & d \end{array} \\ \begin{bmatrix} 1 & 0 & 1 & 0 \\ 0 & 0 & 0 & 0 \\ 0 & 0 & 1 & 0 \\ 1 & 1 & 1 & 1 \end{bmatrix} \end{array} \qquad \text{(b)} \quad \begin{array}{c} \\ a \\ b \\ c \\ d \end{array} \begin{array}{c} \begin{array}{cccc} w & x & y & z \end{array} \\ \begin{bmatrix} 1 & 0 & 1 & 0 \\ 0 & 1 & 0 & 0 \\ 1 & 0 & 1 & 0 \\ 0 & 0 & 0 & 1 \end{bmatrix} \end{array}.$$

6. How can the matrix for the inverse of the relation R^{-1} be found from the matrix representing R, where R is a relation on a finite set.
7. Find the inverse relation R^{-1} of the relation R in the set $A = \{a, b, c, d\}$ to set $B = \{0, 1\}$ and find the matrix M which represents R^{-1}.
8. (i) Let $X = \{a, b, c, d\}$ and

$$M_R = \begin{bmatrix} 1 & 1 & 0 & 1 \\ 0 & 1 & 1 & 0 \\ 0 & 0 & 1 & 1 \\ 1 & 0 & 0 & 0 \end{bmatrix}.$$ Find the relation R and its digraph defined on X.

   ii. If M_R be a relation matrix of a set $\{x, y, z\}$. Find the relation matrix of

$$R^2 = R \circ R, R^3 = R \circ R \circ R$$

$$M_R = \begin{bmatrix} 1 & 0 & 1 \\ 1 & 1 & 0 \\ 1 & 1 & 1 \end{bmatrix}$$

9. Find the relation determined by the digraph and give its matrix.

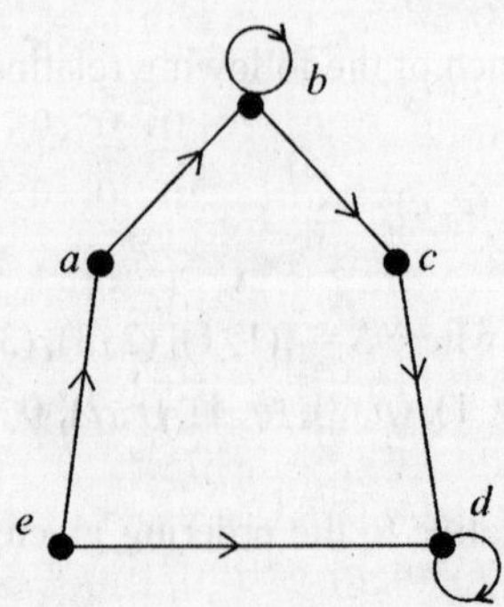

10. Let $M_R = \begin{bmatrix} 0 & 1 & 1 \\ 1 & 1 & 0 \\ 1 & 0 & 1 \end{bmatrix}$ be the matrix of the relation R, Find (i) R^2 (ii) R^{-1}.

11. Suppose R_1 and R_2 are relations on a set X represented by the matrices M_{R1} and M_{R2} respectively. The matrix representing the union of these relations has 1 in the positions where either M_{R1} or M_{R2} has a 1. Thus matrices representing the union and intersection of these relations are $M_{R_1 \cup R_2} = M_{R_1} \vee M_{R_2}$ and $M_{R_1 \cap R_2} = M_{R_1} \wedge M_{R_2}$ respectively, where '$\vee$' and '$\wedge$' denotes 'or' and 'and' respectively.

If $M_{R_1} = \begin{bmatrix} 1 & 0 & 1 \\ 1 & 0 & 0 \\ 0 & 1 & 0 \end{bmatrix}$ and $M_{R_2} = \begin{bmatrix} 1 & 0 & 1 \\ 0 & 1 & 1 \\ 1 & 0 & 0 \end{bmatrix}$

What are the matrices representing $R_1 \cup R_2$ and $R_1 \cap R_2$?

12. Suppose that the matrix of the relation R_1 on $\{a, b, c\}$ is $\begin{bmatrix} 1 & 0 & 0 \\ 0 & 1 & 1 \\ 1 & 0 & 1 \end{bmatrix}$ relative to the ordering a, b, c and that the matrix of the relation R_2 on $\{ a, b, c \}$ is $\begin{bmatrix} 0 & 1 & 0 \\ 0 & 1 & 0 \\ 1 & 0 & 1 \end{bmatrix}$ relative to the ordering a, b, c . Find the matrix of relation $R_1 \cup R_2$ relative to the ordering a, b, c.

Answers

1. i. $\begin{bmatrix} 1 & 1 & 1 \\ 0 & 0 & 0 \\ 0 & 0 & 0 \end{bmatrix}$ ii. $\begin{bmatrix} 1 & 1 & 1 \\ 0 & 1 & 1 \\ 0 & 0 & 1 \end{bmatrix}$

   iii. $\begin{bmatrix} 0 & 0 & 1 \\ 0 & 0 & 0 \\ 1 & 0 & 0 \end{bmatrix}$ iv. $\begin{bmatrix} 1 & 0 & 0 & 0 \\ 0 & 0 & 1 & 0 \\ 0 & 1 & 0 & 0 \\ 0 & 0 & 0 & 1 \end{bmatrix}$

2. $\begin{bmatrix} 1 & 1 & 0 \\ 1 & 0 & 0 \\ 0 & 1 & 0 \\ 1 & 0 & 1 \end{bmatrix}$

3. $\begin{bmatrix} 0 & 1 & 0 & 0 & 0 \\ 0 & 0 & 1 & 0 & 0 \\ 0 & 0 & 0 & 1 & 0 \\ 0 & 0 & 0 & 0 & 1 \\ 0 & 0 & 0 & 0 & 0 \end{bmatrix}$

4 i. $\begin{bmatrix} 0 & 0 & 0 & 1 \\ 1 & 0 & 1 & 0 \\ 0 & 1 & 1 & 0 \end{bmatrix}$ ii. $\begin{bmatrix} 1 & 0 & 1 & 0 \\ 1 & 1 & 0 & 0 \\ 0 & 0 & 0 & 1 \end{bmatrix}$

5. i. $R = \{(x_1, y_2), (x_2, y_1), (x_2, y_3), (x_2, y_4), (x_3, y_1), (x_3, y_3), (x_3, y_5)\}$

   ii. (a) $\{(1, a), (1, c), (3, c), (4, a), (4, b), (4, c), (4, d)\}$

(b) $\{(a, w), (a, y), (b, x), (c, w), (c, y), (d, z)\}$.

7. $M_R = \begin{bmatrix} 0 & 1 \\ 1 & 0 \\ 1 & 0 \\ 0 & 1 \end{bmatrix}$

8. i. $R = \{(a, a), (a, b), (a, d), (b, b), (b, c), (c, c), (c, d), (d, a)\}$

ii.

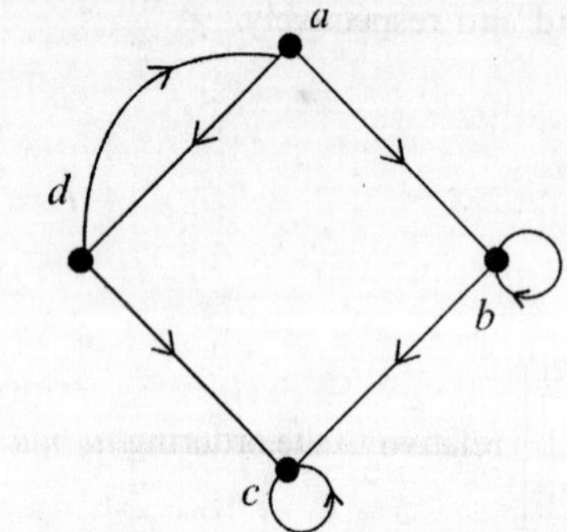

9. $R = \{(a, b), (b, b), (b, c), (c, d), (d, d), (e, a), (e, d)\}$.

$$\begin{bmatrix} 1 & 0 & 0 & 0 & 0 \\ 0 & 1 & 1 & 0 & 0 \\ 0 & 0 & 0 & 1 & 0 \\ 1 & 0 & 0 & 1 & 0 \end{bmatrix}$$

12. $M_{R_1 \cup R_2} = \begin{bmatrix} 1 & 0 & 1 \\ 1 & 1 & 1 \\ 1 & 1 & 0 \end{bmatrix}$, $M_{R_1 \cap R_2} = \begin{bmatrix} 1 & 0 & 1 \\ 0 & 0 & 0 \\ 0 & 0 & 0 \end{bmatrix}$

3.13 Equivalence Relations

Let R be a relation in a set A. Then R is an **equivalence relation** in A if and only if

i. R is reflexive, i.e., $(a, a) \in R \; \forall a \in A$.

ii. R is symmetric i.e. $(a, b) \in R \Rightarrow (b, a) \in R$.

iii. R is transitive i.e. $(a, b) \in R$ and $(b, c) \in R \Rightarrow (a, c) \in R$.

The relation '=' on R is an example of equivalence relation. In fact, the term equivalence relation is used because the relations which satisfy this definition behave like the '=' relation.

Example 3.46

Let R be a relation in the real numbers defined by '$a \leq b$' then

i. $a \leq a$ i.e. $(a, a) \in R$ i.e. R is reflexive.

ii. Let $(a, b) \in R \Rightarrow a \leq b \Rightarrow b \nleq a \Rightarrow (b, a) \notin R$ i.e. R is not symmetric.

Thus R is not an equivalence relation.

Example 3.47

Let R be a relation in the set of all lines in a plane defined by $a\ R\ b$ if line a is parallel to line b'. Then R is an equivalence relation.

Since for any three lines x, y, z of the given set,

i. $x \parallel x$ for each line x.

$\Rightarrow$ $x\ R\ x$ for each line x

$\therefore$ R is reflexive.

ii. Since $x \parallel y \Rightarrow y \parallel x$

$\therefore$ $x\ R\ y \Rightarrow y\ R\ x$

$\therefore$ R in symmetric.

iii. Since $x \parallel y$ and $y \parallel z \Rightarrow x \parallel z$

$\therefore$ $x\ R\ y$ and $y\ R\ z \Rightarrow x\ R\ z$

$\therefore$ R is transitive.

Hence, R being reflexive, symmetric and transitive, is an equivalence relation.

Example 3.48

Prove that the relation R in the set N of natural numbers defined by $a\ R\ b \Leftrightarrow (a - b)$ is divisible by n, where $n \in N$, is an equivalence relation.

Solution

To show that the relation R is an equivalence relation, let us show that the R is reflexive, symmetric and transitive.

i. **Reflexive**

Let $x \in N$, then $x - x = 0 = 0 \times n \Rightarrow (x - x)$ is divisible by $n \Rightarrow x\ R\ x$ for all $x \in N$. Therefore, R is reflexive.

ii. **Symmetric**

Let $x\ R\ y$, then

$$x\ R\ y \Rightarrow (x - y) \text{ is divisible by } n.$$
$$\Rightarrow (x - y) = n \times p \qquad [p \in Z]$$
$$\Rightarrow y - x = n \times (-p) \qquad [-p \in Z]$$
$$\Rightarrow (y - x) \text{ is divisible by } n$$
$$\Rightarrow y\ \mathrm{R}\ x$$

Therefore, R is symmetric

iii. **Transitive**

Let $x\ R\ y$ and $y\ R\ z$, then

$$x\ R\ y \Rightarrow (x - y) \text{ is divisible by } n.$$
$$\Rightarrow x - y = np \qquad [p \in Z]$$

and

$$y\ R\ z \Rightarrow (y - z) \text{ is divisible by } n$$
$$\Rightarrow y - z = nq \qquad [q \in Z]$$

Now $x\ R\ y$ and $y\ R\ z$

$$\Rightarrow x - y = np \text{ and } y - z = nq$$

$$\Rightarrow (x-y)+(y-z)=np+nq$$
$$\Rightarrow x-z=n\,(p+q) \qquad [p+q \in Z]$$
$$\Rightarrow x-z \text{ is divisible by } n$$
$$\Rightarrow x\,R\,z$$

Therefore, R is transitive.

Hence R being reflexive, symmetric and transitive is, an equivalence relation in N.

Example 3.49

Consider a relation R in the set of real numbers defined as $x\,R\,y$ if and only if $1 + xy > 0$. Show that this relation is reflexive and symmetric but not transitive.

Solution

i. ***R* is reflexive**

$$1+xx = 1+x^2 > 0 \text{ for all } x \in R \text{ (real number)}$$
$$\Rightarrow x\,R\,x, \text{ for all } x \in R$$
$$\Rightarrow R \text{ is reflexive.}$$

ii. ***R* is symmetric**

$$x\,R\,y \Rightarrow 1+xy>0$$
$$\Rightarrow 1+yx>0 \qquad [\text{since } xy = yx \Rightarrow y\,R\,x,\ \forall\, x, y \in R]$$
$$\Rightarrow R \text{ is symmetric.}$$

iii. ***R* is not transitive**. Let us consider three real numbers, 1, –2/3 and – 4. Then $1+1\left(\frac{-2}{3}\right)=\frac{1}{3}>0$, hence $1\,R\left(\frac{-2}{3}\right)$.

Further $$1+\frac{2}{3}.4=\frac{11}{3}>0 \Rightarrow \left(\frac{-2}{3}\right)R(-4).$$

But $1+(-4) = -3$ which is not greater than zero.

i.e. 1 is not related to –4. Therefore, R is not transitive.

Example 3.50

Prove that the relation R in the set Z of integers defined by $a\,R\,b \Leftrightarrow$ '$(a - b)$ is even' is an equivalence relation.

Solution

Reflexive Let $x \in Z$, then $x-x=0=0.2 \Rightarrow (x-x)$ is even $\Rightarrow x\,R\,x\ \forall\, x \in Z$.

Therefore, R is reflexive.

Symmetric If $x\,R\,y$, then

$$x\,R\,y \Rightarrow (x-y) \text{ is even.}$$
$$\Rightarrow x-y=2p \qquad [\text{for some } p \in Z]$$
$$\Rightarrow y-x=2.(-p) \qquad [-p \in Z]$$
$$\Rightarrow y-x \text{ is even}$$
$$\Rightarrow y\,R\,x$$

Therefore, R is symmetric.

Transitive, Let $x\,R\,y$ and $y\,R\,z$, then

$$x\,R\,y \Rightarrow (x-y) \text{ is even}$$

$$\Rightarrow x - y = 2.p \qquad \text{[for some } p \in Z\text{]}$$

and $$y\,R\,z \Rightarrow (y-z) \text{ is even} \qquad \text{[for some } q \in Z\text{]}$$

$$\Rightarrow y - z = 2.q,$$

$\therefore$ $$x\,R\,y \text{ and } y\,R\,z \Rightarrow x - y = 2.p \text{ and } y - z = 2.q$$

$$\Rightarrow (x-y) + (y-z) = 2.p + 2.q$$

$$\Rightarrow x - z = 2.(p+q)$$

$$\Rightarrow x - z \text{ is even}$$

$$\Rightarrow x\,\text{R}\,\text{z}$$

Therefore, R is reflexive, symmetric and transitive and hence is an equivalence relation in Z.

Example 3.51

Let A be a set of natural numbers and let R be a relation on $A \times A$ defined by $(a, b)\,R\,(c, d)$ whenever $ad = bc$. Prove that R is an equivalence relation.

Solution

Reflexive

Let $$(a, b) \in N \times N, \text{ then}$$

$$(a, b) \in A \times A\ \forall\ a, b \in A$$

$$\Rightarrow ab = ba \qquad \text{[by commutativity in } A\text{]}$$

$$\Rightarrow (a, b)\,R\,(a, b)$$

Therefore, R is reflexive.

Symmetric

Let $(a, b)\,R\,(c, d)$, then $$(a, b)\,R\,(c, d) \Rightarrow ad = bc$$

$$\Rightarrow bc = ad$$

$$\Rightarrow cb = da \qquad \text{[by commutativity in } A\text{]}$$

$$\Rightarrow (c, d)\,R\,(a, b)$$

Therefore, R is symmetric.

Transitive

Let $(a, b)\,R\,(c, d)$ and $(c, d)\,R\,(e, f)$.

Then $$(a, b)\,R\,(c, d) \Rightarrow ad = bc \qquad \ldots (1)$$

and $$(c, d)\,R\,(e, f) \Rightarrow cf = de \qquad \ldots (2)$$

(1) and (2) $$\Rightarrow (ad)(cf) = (bc)\,de$$

$$\Rightarrow af = \text{be}$$

$$(a, b)\,R\,(e, f)$$

$\therefore$ R is transitive.

Hence, R is an equivalence relation in $A \times A$.

Example 3.52 (Congruence modulo)

Let Z be the set of integers. Let $m > 1$ be any fixed integer. Then we say a is congruent to another integer b modulo if $a - b$ is divisible by m. Equivalently we can say if there exists an integer k such that $a - b = k.m$, and it is symbolically written as

$$a \cong b \pmod{m}.$$

Show that this defines an equivalence relation.

Solution

Reflexive

Let $x \in Z$, then $x \cong x \pmod{m}$ because $x - x = 0.\ m$, where $k = 0$

Thus, $x \cong x \pmod{m}\ \forall\ x \in Z,$

$\Rightarrow x\ R\ x$, hence congruence modulo m is reflexive.

Symmetric

Let $x\ R\ y$, then $x \cong y \pmod{m}$

$\Rightarrow x - y = km$ for $k \in Z$

$\Rightarrow y - x = (-k)m$ $\quad [\because k \in Z \Rightarrow -k \in Z]$

Thus $y - x$ is divisible by m.

$\Rightarrow y \cong x\ (mod\ m)$

i.e. $x\ R\ y \Rightarrow y\ R\ x$.

Hence, congruence modulo m is symmetric on z.

Let $x\ R\ y$ and $y\ R\ z$, then

Transitive

Let $xR\ y$ and $yR\ z$, then

$x \cong y \pmod{m}$ and $y \cong z \pmod{m}$,

But $\quad x \cong y \pmod{m} \Rightarrow (x - y) = k_1\, m$ for some $k_1 \in Z$

and, $\quad y \cong z \pmod{m} \Rightarrow (y - z) = k_2\, m$ for some $k_2 \in Z$

Therefore $\quad x - z = (x - y) + (y - z) = k_1\, m + k_2\, m$

$= (k_1 + k_2)\, m$ for some $k_1 + k_2 \in Z$

i.e. $x \cong z \pmod{m}$. Hence congruence modulo m is transitive on Z.

Thus R being reflexive, symmetric and transitive, hence is an equivalence relation in Z.

Example 3.53

If R is an equivalence relation in a set X, then prove that R^{-1} is also an equivalence relation in X.

Solution

Since R is a relation in X.

$\Rightarrow R \subset X \times X$

$\Rightarrow R^{-1} \subset X \times X$

$\Rightarrow R^{-1}$ is also a relation in X.

Let us now prove that R^{-1} is an equivalence relation.

Reflexive

Let $x \in X$, then

$$x \in X \Rightarrow (x, x) \in R \qquad [\because R \text{ is reflexive}]$$

$$\Rightarrow (x, x) \in R^{-1} \text{ for all } (x, x) \in X \times X$$

Therefore, R^{-1} is reflexive.

Symmetric

Let $(x, y) \in R^{-1}$, then $(y, x) \in R$

$$\Rightarrow (x, y) \in R \qquad [\because R \text{ is symmetric}]$$

$$\Rightarrow (y, x) \in R^{-1}$$

Therefore, R^{-1} is symmetric.

Transitive

Let $(x, y) \in R^{-1}$ and $(y, z) \in R^{-1}$, then $(y, x) \in R$ and $(z, y) \in R$.

$$\Rightarrow (z, y) \in R \text{ and } (y, x) \in R$$

$$\Rightarrow (z, x) \subset R \qquad [\because R \text{ is Transitive}]$$

$$\Rightarrow (x, z) \in R^{-1}$$

Therefore, R^{-1} is transitive.

Therefore, R^{-1} is also an equivalence relation in X.

Example 3.54

If R and S be two equivalence relations in a set X, then prove that $R \cap S$ is also an equivalence relation in X.

Solution

As R and S are equivalence relations in X.

$$\Rightarrow R \subset X \times X \text{ and } S \subset X \times X$$

$$\Rightarrow R \cap S \subset X \times X$$

$$\Rightarrow R \cap S \text{ is also a relation in } X.$$

Let us now show that $R \cap S$ is an equivalence relation in X.

Reflexive:

Let $x \in X$, then

$$x \in X \Rightarrow (x, x) \in R \text{ and } (x, x) \in S \qquad [\because R \text{ and } S \text{ are reflexive}]$$

$$\Rightarrow (x, x) \in R \cap S \text{ for all } (x, x) \in X \times X$$

Therefore, $R \cap S$ is reflexive.

Symmetric:

Let $(x, y) \in R \cap S$, then

$$\Rightarrow (x, y) \in R \text{ and } (x, y) \in S$$

$$\Rightarrow (y, x) \in R \text{ and } (y, x) \in S \qquad [\because R \text{ and } S \text{ are symmetric}]$$

$$\Rightarrow (y, x) \in R \cap S$$

Therefore, $R \cap S$ is symmetric.

Transitive:

Let $(x, y) \in R \cap S$ and $(y, z) \in R \cap S$, then

$$(x, y) \in R \cap S \Rightarrow (x, y) \in R \text{ and } (x, y) \in S$$

$$(y, z) \in R \cap S \Rightarrow (y, z) \in R \text{ and } (y, z) \in S$$

As R and S are transitive

$$(x, y) \in R \text{ and } (y, z) \in R \Rightarrow (x, z) \in R$$

$$(x, y) \in R \text{ and } (y, z) \in S \Rightarrow (x, z) \in S$$

$$\Rightarrow (x, z) \in R \cap S$$

Therefore $(x, y) \in R \cap S$ and $(y, z) \in R \cap S \Rightarrow (x, z) \in R \cap S$

Hence $R \cap S$ is transitive.

$\Rightarrow R \cap S$ is an equivalence relation in X.

Example 3.55

Consider the set of ordered pair of natural numbers $N \times N$ defined by

$$(a, b)\, R\, (c, d) \Leftrightarrow a + d = b + c.$$

Prove that R is an equivalence relation.

Solution

Reflexive:

Let $(a, b) \in N \times N \Rightarrow a + b = b + a$ i.e. $(a, b)\, R\, (a, b)$, hence R is reflexive.

Symmetric:

As $(a, b)\, R\, (c, d) \Rightarrow a + d = b + c$

$\Rightarrow c + b = d + a$ [$\because$ addition is commutative]

$\Rightarrow (c, d)\, R\, (a, b)$

$\Rightarrow R$ is symmetric.

Transitive:

If $(a, b)\, R\, (c, d)$ and $(c, d)\, R\, (e, f)$ ... (1)

$\Rightarrow a + d = b + c$

and $c + f = d + e$ ... (2)

Adding (1) and (2), we get $(a + d) + (c + f) = (b + c) + (d + e)$

$\Rightarrow a + f = b + e \Rightarrow (a, b)\, R\, (e, f)$

$\Rightarrow R$ is transitive.

Hence R is an equivalence relation on $N \times N$.

3.13-1 Equivalence Class

Let R be an equivalence relation on a set X. The **equivalence class** denoted by $[a]$ or a of each element a of A is defined as $[a] = \{b \in X \mid a\, R\, b\}$, i.e. the set of all elements that are related to the element a.

If $b \in [a]$, b is called a **representative** of this equivalence class.

When we have only one relation under consideration then the equivalence class of a is written as $[a]$, however if we have more relations under consideration, then equivalence class of a with respect to R is denoted by $[a]_R$.

3.13-2 Properties of Equivalence Classes

Let R be an equivalence relation defined on a set A. Let a and b be arbitrary elements in A. Then

i. $a \in [a]$

ii. $b \in [a] \Rightarrow [b] = [a]$

iii. $[a] = [b] \Leftrightarrow (a, b) \in R$ if $a\,R\,b$

iv. Any two equivalence classes are either identical or disjoint, i.e. either $[a] = [b]$ or $[a] \cap [b] = \varnothing$.

Proof

i. Since R is reflexive, i.e. $a\,R\,a$. Hence for any $x \in A$, $[a] = \{x: x \in A \text{ and } x\,R\,a\}$.

i.e. $a\,R\,a \Rightarrow a \in [a]$.

ii. If $b \in [a]$ then $b\,R\,a$. Let $x \in A$ be any other element such that $x \in [b] \Rightarrow x\,R\,b$. Because the relation R is transitive so

$$x\,R\,b \text{ and } b\,R\,a \Rightarrow x\,R\,a \Rightarrow x \in [a]$$

Therefore $\quad x \in [b] \Rightarrow x \in [a] \Rightarrow [b] \subseteq [a] \quad \ldots (1)$

Let there be an element $y \in [a] \Rightarrow y\,R\,a$. But by symmetry of relation R, we have $b\,R\,a \Rightarrow a\,R\,b$. Thus $y\,R\,a$ and $a\,R\,b \Rightarrow y\,R\,b$, because the relation R is transitive. From this we see that $y \in [b]$.

Therefore $\quad y \in [a] \Rightarrow y \in [b]$

or $\quad [a] \subseteq [b] \quad \ldots (2)$

Hence from (1) and (2) $[a] = [b]$.

iii. Given $[a] = [b]$. Since the relation R is reflexive therefore for any a, $a\,R\,a$.

Further $\quad a\,R\,a \Rightarrow a \in [a] \Rightarrow a \in [b] \quad [\because [a] = [b]]$

$\Rightarrow a\,R\,b$

Therefore $\quad [a] = [b] \Rightarrow a\,R\,b$.

Conversely let $a\,R\,b$ is given. Let $x \in [a]$, then $x\,R\,a$. But $a\,R\,b$, so from the transitivity of relation R, $x\,R\,b$. Hence $x \in [b]$.

Therefore $\quad x \in [a] \Rightarrow x \in [b] \Rightarrow [a] \subseteq [b] \quad \ldots (3)$

Consider any arbitrary element y of $[b]$, then

$$y \in [b] \Rightarrow y\,R\,b$$

But R is symmetric i.e. $\quad a\,R\,b \Rightarrow b\,R\,a$.

Therefore $\quad y\,R\,b$ and $b\,R\,a \Rightarrow y\,R\,a \quad [\because R$ is transitive$]$

Hence $\quad y \in [a]$

Therefore $\quad y \in [b] \Rightarrow y \in [a]$ i.e. $[b] \subseteq [a] \quad \ldots (4)$

Hence from (3) and (4), we conclude

$$[a] = [b]$$

Therefore $\quad [a] = [b] \Leftrightarrow a\,R\,b$.

iv. If $[a] \cap [b] = \varnothing$, then there is nothing to prove.

Here, let us assume $[a] \cap [b] \neq \varnothing$

$\Rightarrow$ there exist $x \in A$ such that $x \in [a] \cap [b]$.

But $\quad x \in [a] \cap [b] \Rightarrow x \in [a]$ and $x \in [b]$

$\Rightarrow x\,R\,a$ and $x\,R\,b$

$\Rightarrow a\,R\,x$ and $x\,R\,b \quad [\because R$ is symmetry$]$

$\Rightarrow a\,R\,b \quad$ [as R is transitive]

$$\Rightarrow [a] = [b] \quad \text{[using (iii)]}$$

Therefore $[a] \cap [b] \neq \varnothing \Rightarrow [a] = [b]$.

3.14 Partitions

Consider a non empty set A and let $P = \{A_1, A_2, A_3, \ldots\}$, where $A_1, A_2, A_3, \ldots, An$ are nonempty subsets of A. Then the set P is called **partition** of A, if following two conditions are satisfied:

i. The set A is the union of subsets $A_1, A_2, A_3, \ldots$, i.e.

$$A_1 \cup A_2 \cup A_3 \cup \ldots = A.$$

ii. The intersection of every pair of distinct subsets of $A \in P$, is the null set i.e. if A_1 and $A_2 \in P$ then either $A_1 = A_2$ or $A_1 \cap A_2 = \varnothing$.

The following figure illustrates the concept of partition

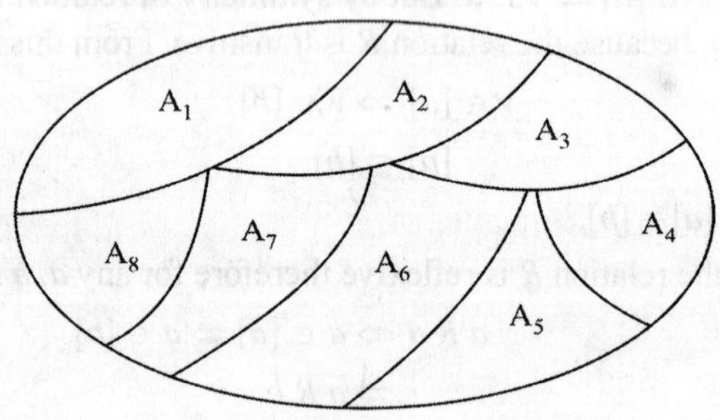

Fig. 3.14

Example 3.56

Write down the sets in the partition of integers Z arising from the congruent modulo 4.

Solution

There are four congruence classes, corresponding to $[0]_4$, $[1]_4$, $[2]_4$, and $[3]_4$. And the sets are

$$[0]_4 = \{\ \ldots, -8, -4, 0, 4, 8 \ldots\}$$
$$[1]_4 = \{\ \ldots, -7, -3, 1, 5, 9 \ldots\}$$
$$[2]_4 = \{\ \ldots, -6, -2, 2, 6, 10 \ldots\}$$
$$[3]_4 = \{\ \ldots, -5, -1, 3, 7, 10 \ldots\}$$

It is to be noted here that these congruence classes are disjoint and every integer is in exactly one of them.

Theorem 3.10

For any positive integer m, the relation $\cong$ (mod m) is an equivalence relation on the integers and partitions the integer into m distinct equivalent classes: $[0], [1], \ldots, [m-1]$.

Proof

We have already proved in exp. 3.52 that $\cong$ mod (m) is an equivalence relation.

Now, if a is any integer, then from the division algorithm $a = mp + q$, where p and q are integers and $0 \leq q < m$.

Hence $a \cong q \text{ mod } m$ and $[a] = [q]$.

Therefore each equivalence class for this relation is one of the classes $[0], [1], \ldots, [m-1]$.

Further, if $[a] = [b]$, where $a \leq b \leq m-1$, Then $b = m\,x + a$ for some integer x.

$$0 \leq b - a = mx < m \Rightarrow x = 0 \text{ and } a = b.$$

3.15 Relation Induced by the Partition of a Set

Given any partition of a set A, we can define a relation R in A by the requirement that $x\ R\ y$ if and only if x and y belong to the same subset of A belonging to the partition. The relation R is then said to be induced by the partition.

For example consider the subsets

$$A_1 = \{3, 6, 9, \ldots, 24\}, A_2 = \{1, 4, 7, \ldots, 25\},$$

$$A_3 = \{2, 5, 8, \ldots, 23\}, \text{ of the set } A = \{1, 2, 3, \ldots, 25\}$$

One can conclude from here that $A_1 \cup A_2 \cup A_3 = A$

and $$A_1 \cap A_2 = A_1 \cap A_3 = A_2 \cap A_3 = \varnothing.$$

Therefore $\{A_1, A_2, A_3\}$ is the partition of A. Further, if R is the relation induced by the above partition, then we have $x\ R\ y$ iff x and y belong to the same subset A_1, A_2, A_3.

Theorem 3.11 (Fundamental theorem on Equivalence Relation)

An equivalence relation R in a non-empty set A partitions A and conversely a partition of A defines an equivalence relation in A.

Proof

Consider an equivalence relation R defined in a nonempty set A and let P be the set of equivalence classes of A with respect to R.

Since R is reflexive hence $P = \{[x]: x \in A\}$ where $[x] = \{a: a \in A \text{ and } a\ R\ x\}$

$x\ R\ x$, and thus by definition of equivalent class

$$x \in [x] \text{ i.e. } [x] \neq \varnothing \quad \ldots (1)$$

Further every element x of A is an element of the equivalence class

$$[x] \text{ in } P, \text{ so } A = \cup[x] \quad \ldots (2)$$

where x is an element of A. As we know that two equivalence classes are either identical or disjoint, i.e.

$$[x] = [y] \text{ or } [x] \cap [y] = \varnothing \quad \ldots (3)$$

Therefore, from (1), (2) and (3), we conclude P is a partition of A,

Now, to prove the second part let us assume

$B = \{ S_a, S_b, S_c, \ldots\}$ be any portion of A and let $m, n \in A$. Let us define a relation R in A by $m\ R\ n$ if there exist a S_i in the partition such that $m, n \in S_i$.

But, $A = S_a \cup S_b \cup S_c \ldots$, so for all $a \in A$, there exist $S_i \in B$, such that $a \in S_i$.

i.e. $a \in S_i \Rightarrow a\ R\ a$.

Thus for all $a \in A$, $a\ R\ a$ and thus R is reflexive.

Again let for any $a\ R\ b$ there exist $S_i \in P : a \in S_i$ and $b \in S_i$.

This implies that $b \in S_i$ and $a \in S_i$ i.e. bRa.

Hence R is symmetric.

Finally let $a\ R\ b$ and $b\ R\ c$. Then from the definition of R, there exist subsets say S_j and S_k (may or may not be distinct) such that $a, b \in S_j$ and $b, c \in S_k$. But $b \in S_j$ and also $b \in S_k$ imply that $S_j \cap S_k \neq \varnothing$. But S_j and S_k belong to a partition of A, so that $S_j \cap S_k \neq \varnothing \Rightarrow S_j = S_k$. $S_j = S_k$, implies $a, c \in S_j$ and therefore $a\ R\ c \Rightarrow R$ is transitive.

Thus R is reflexive, symmetric and transitive and hence R is an equivalence relation.

Example 3.57

Find all partitions of
i. {1, 2, 3} ii. {1, 2, 3, 4}.

Solution

i. [{1}, {2},{3}], [{1, 2, 3}], [{1, 2}, {3}], [{1, 3}, {2}], [{2, 3},{1}], i.e. total 5.
ii. Total: 15, complete yourself.

Example 3.58

Find the equivalence relations induced by the partitions in the set $A = \{a,b,c,d\}$.

Solution

i. $R = \{ (a, a), (b, b), (c, c), (d, d)\}$.
ii. $R = \{ (a, a), (b, b), (c, c), (d, d), (a, c), (c, a), (b, d), (d, b)\}$.

Example 3.59

Consider the equivalence relation R defined on $X \times X$ where $X = \{1, 2, 3, \ldots, 14, 15\}$ and defined by (a, b) R (c, d) if $a + d = b + c$. Find the equivalence class of (2, 11).

Solution

Let us find all (m, n) such that $(2, 11)$ R (m, n) such that

$$2 + n = 11 + m \text{ or } n = 9 + m$$

Set $m = 1, 2, ..,$ to obtain

$$[(2, 11)] = \{(1, 10), (2, 11), (3, 12), (4, 13), (5, 14), (6, 15) \}.$$

Example 3.60

Prove that the relation R of similarity in the set of all triangles in a plane is an equivalence relation.

Solution

Consider the set of all triangles X in a plane and let R be the relation in X defined by $a\ R\ b$ if and only if triangle a is similar to the triangle b for all a and $b \in$ X.

Reflexive

Since every triangle is similar to itself, therefore by the definition of R, we have aRa.

Symmetry

$A\ R\ b \Rightarrow$ triangle a is similar to triangle b
$\Rightarrow$ triangle b is similar to triangle a
$\Rightarrow$ b R a

i.e. $a\ R\ b \Rightarrow b\ R\ a$.

Therefore, R is symmetric.

Transitive

$a\ R\ b$ and $b\ R\ c \Rightarrow a\ R\ c$
$a\ R\ b$ and $b\ R\ c$.
$\Rightarrow$ triangle a is similar to triangle b and triangle b is similar to triangle c.
$\Rightarrow$ triangle a is similar to triangle c.
$\Rightarrow c\ R\ b$.
$\Rightarrow R$ is reflexive, symmetric and transitive and hence R is an equivalence relation.

Example 3.61

Let R be the equivalence relation defined on a set $X = \{1, 2, 3, 4, 5, 6\}$ and defined by
$R = \{(1, 1), (1,5), (2, 2), (2, 3), (2, 6), (3, 2), (3, 3), (3, 6), (4, 4), (5, 1), (5, 5), (6, 2), (6, 3), (6, 6)\}$.
Find the partition of X induced by R i.e. find the equivalence classes of R.

Solution

From the relation we can see that the elements related to 1 are 1 and 5, hence $[1] = \{1, 5\}$. And we can pick an element not belong to [1], say 2. The elements related to 2 are 2, 3 and 6 and hence $[2] = \{2, 3, 6\}$. Further, the only element which does not belong to [1] or [2] is 4. And the only element related to 4 is 4. Thus $[4] = \{4\}$. Therefore, $[\{1, 5\}, \{2, 3, 6\}, \{4\}]$ is the partition of A induced by R.

Example 3.62

Let $A = R \times R$ (R is the set of real numbers) and define the following relation on A:

$$(a, b)\, R\, (c, d) \text{ if } a^2 + b^2 = c^2 + d^2.$$

i. Verify that (A, R) is an equivalence relation.
ii. Describe geometrically what the equivalence classes are for this relation (Justify).

Solution

Reflexive: Let $(a,b) \in R \times R \Rightarrow a^2 + b^2 = a^2 + b^2$

$$\Rightarrow (a, b)\, R\, (a, b) \Rightarrow R \text{ is reflexive.}$$

Symmetric: Let $(a,b)R(c,d) \Rightarrow a^2 + b^2 = c^2 + d^2$

$$\Rightarrow c^2 + d^2 = a^2 + b^2 \Rightarrow (c,d)R(a,b)$$

$$\Rightarrow R \text{ is symmetric.}$$

Transitive: Let $(a, b)\, R\, (c, d)$ and $(c, d)\, R\, (e, f)$

$$\Rightarrow a^2 + b^2 = c^2 + d^2 \quad \ldots (1)$$

and $$c^2 + d^2 = e^2 + f^2 \quad \ldots (2)$$

From (1) and (2), we get $a^2 + b^2 = e^2 + f^2$

$$\Rightarrow (a, b)\, R\, (e, f) \Rightarrow R \text{ is transitive}$$

Hence R is an equivalence relation.

ii. Do yourself.

Example 3.63

Let $A = \{1, 2, 3, \ldots, 9\}$ and let $\sim$ be the relation of $A \times A$ defined by

$$(a, b) \sim (c, d) \text{ if } a + d = b + c.$$

i. Prove that $\sim$ is an equivalence relation.
ii. Find $[(2, 5)]$, the equivalence class of $(2, 5)$.

Solution

i. See example 3.55
ii. We see that

$(2, 5)\, R\, (1, 4), (2, 5)\, R\, (2, 5), (2, 5)\, R\, (3, 6), (2, 5)\, R\, (4, 7), (2, 5)\, R\, (5, 8), (2, 5)\, R\, (6, 9)$
So $[(2, 5)] = \{(1, 4), (2, 5), (3, 6), (4, 7), (5, 8), (6, 9)\}$.

3.16 Quotient Set

If R be an equivalence relation on a nonempty set A then the set of all pair-wise disjoint equivalence classes of A with respect to R is called **quotient set** of A with respect to relation R. This quotient set is written as A/R.

Example 3.64

Consider the set of words: A = {Sun, room, tree, summer, and animal}.

Find A/R where R is the equivalence relation on A defined by either (i) begins with the same letter (ii) has the same number of letters.

Solution

i. Those words beginning with the same letter belong to the same cell, hence

A/R = {{sun, summer}, {room}, {tree}, {and, animal}}.

ii. Those words beginning with the same number of letters belong to the same cell, hence

A/R = {{sun, and}, {summer, animal}, {room, tree}}.

Exercise 3.4

1. Consider a relation R "is a blood relative of" on the set A of people. Is R an equivalence relation?
2. Let $A = \{a, b, c, d\}$. Give two example of a set of ordered pairs having equivalence relation.
3. Which of the following relations on $\{a, b, c, d\}$ are equivalence relations? Determine the equivalence relation that the others lack.
   i. $\{ (a, a), (b, b), (b, d), (c, c), (c, d), (d, b), (d, c), (d, d)\}$
   ii. $\{ (a, a), (b, b), (b, c), (c, b), (c, c), (d, d)\}$
   iii. $\{ (a, a), (b, b), (c, c), (d, d)$.
4. Let A = $\{a, b, c, d, e\}$, $B = \{c, d\}$. Define the relation R on power set of A i.e. $P(A)$, as

   $$X R Y \text{ if } X \cup B = Y \cup B.$$

   Show that R is an equivalence relation.
5. Determine whether the following relation R on the set X is an equivalence relation
   i. $X = \{a, b, c, d\}$
      $R = \{(a, a), (a, b), (b, a), (b, b), (c, a), (c, c), (a, c), (d, a), (d, d)\}$
   ii. $X = \{1, 2, 3, 4\}$
      $R = \{(1, 1), (2, 1), (2, 2), (3, 3), (4, 4), (4, 3)\}$.
6. i. Let $A = \{1, 2, 3, 4, 5\}$ and $X = A \times A$. Define the following relation R on

   $X: (x, y)\, R\, (x', y')$ if $x y' = x' y$. Show that R is an equivalence relation.

   ii. Let $A = \{1, 2, 3, 4\}$ and $X = A \times A$. Define the following relation R on

   $$X: (x, y)\, R\, (x', y') \text{ if } x + y = x' + y'.$$

   Show that R is an equivalence relation.
7. Let R be the relation on $A = \{1, 2, 3, 4\}$ defined by $(a, b) \in R$ if $a \le b$; $a, b \in A$. Is R an equivalence relation?
8. Show that if R is an equivalence relation on A, then

   $$\text{domain } R = \text{range } R = A.$$
9. Show that the relation $\subseteq$ of set inclusion is not an equivalence relation on the subsets of N.

10. Show that if R_1 and R_2 are equivalence relations on X, then $R_1 \cap R_2$ and $R_1 \cup R_2$ are equivalence relations.
11. How many equivalence relations are there on the set $\{a, b, c\}$?
12. Determine whether the relation with directed graphs is an equivalence relation.

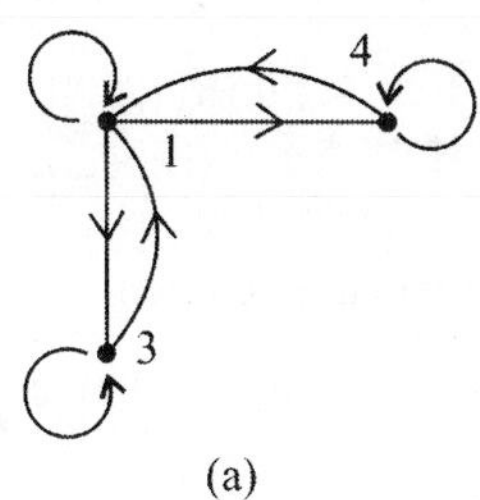

(a)

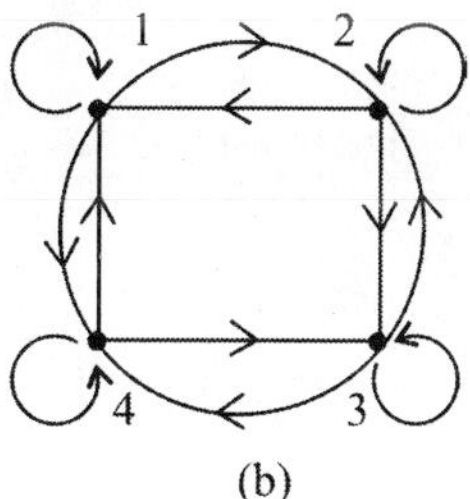

(b)

13. Consider the equivalence relation

    $R = \{(1, 1), (1, 3), (1, 5), (2, 2), (2, 4), (3, 1), (3, 3), (3, 5), (4, 2), (4, 4), (5, 1), (5, 3), (5, 5)\}$.

    Determine the equivalence classes.
14. Let R_1 and R_2 be equivalence relation of A. Describe the equivalence classes $R_1 \cap R_2$ in terms of the equivalence classes of R_1 and equivalence classes of R_2.
15. Let $A = \{1, 2, 3, 4,\ldots,, 10\}$ be a set. Define a relation R on $A \times A$ by $(a, b)\ R\ (c, d)$ if $ad = bc$. Show that R is an equivalence relation on $A \times A$. List one member of each equivalence class of $A \times A$.
16. Let R be a relation defined on a set $X = \{1, 2, 3, \ldots, 13, 14, 15\}$. Let R be the relation on X defined by the congruence modulo 4. Find the equivalence classes determined by R.
17. Consider the equivalence relation $R = \{(a, a), (a, b), (b, a), (c, c)\}$ defined on the set $A = \{a, b, c\}$. Find the quotient set A/R.
18. Let R be a relation such that $R = \{(a, b) \mid a - b \text{ is an integer}\}$
    i. Find the equivalence class of 1 for this equivalence relation?
    ii. Find the equivalence class of 1/2 for this equivalence relation?
19. Determine the number of different equivalence relations on a set with three elements by listing them.
20. Which of the following collections of subsets are partitions of $\{a, b, c, d, e, f\}$.
    i. $\{a\}, \{b, c, f\}, \{d\}, \{e\}$.
    ii. $\{a, d, e\}, \{b, e\}$.
    iii. $\{a, b\}, \{b, c, d\}, \{d, e, f\}$.
    iv. $\{b, d, f\}, \{a, c, e\}$.
21. Consider an equivalence relation R defined on a set $A = \{a, b, c, d, e, f\}$ defined as $R = \{(a, a), (a, e), (b, b), (b, c), (b, f), (c, b), (c, c), (c, e), (d, d), (e, a), (e, e), (f, b), (f, c), (f, f)\}$

    Find the partition of A induced by R i.e. find the equivalence classes of R.
22. Determine whether the relations represented by the following matrices are equivalence relations.

    i. $\begin{bmatrix} 1 & 1 & 1 & 0 \\ 1 & 1 & 1 & 0 \\ 1 & 1 & 1 & 0 \\ 0 & 0 & 0 & 1 \end{bmatrix}$ ii. $\begin{bmatrix} 1 & 1 & 1 \\ 0 & 1 & 1 \\ 1 & 1 & 1 \end{bmatrix}$
23. Determine whether each of the following is a partition of $A = \{1, 2, 3, \ldots, 8, 9\}$.
    i. $[\{2, 4, 5, 8\}, \{1, 9\}, \{3, 6, 7\}]$
    ii. $[\{1, 2, 7\}, \{3, 5\}, \{4, 6, 8, 9\}, \{3, 5\}]$

iii. [{1, 5, 7}, {2, 4, 8, 9}, {3, 5, 6}].

24. Show that the composition of two congruence relations on a set is necessarily a congruence relation.

25. Let $I = \{\ldots -3, -2, 0, 1, 2, 3, \ldots\}$ and R be an equivalence relation in I defined by $x \cong y \pmod 5$, then find the quotient set I/R.

Answers

1. No.

3. i. Not transitive ii. Equivalence relation.
   iii. Equivalence relation.

5. i. No. ii. No.

6. i. $(x, y)\, R\, (x, y)$ because $xy = yx \Rightarrow R$ is reflexive. If $(x, y)\, R\, (x', y')$ then $x y' = y x' \Rightarrow x' y = y' x \Rightarrow (x', y')$ $R\, (x, y) \Rightarrow R$ is symmetric. Further suppose that $(x, y)\, R\, (x', y')$ and $(x', y')\, R\, (x'', y'')$.
   Then $xy' = yx'$ and $x' y'' = y' x''$.

$$xy'' = x \cdot \frac{y'x''}{x'} = xy' \frac{x''}{x'} = yx' \frac{x''}{x'} = yx'' \Rightarrow (x, y)\, R\, (x'', y'') \text{ and } R \text{ is transitive.}$$

7. No as it is not symmetric.

12. i. No. ii. No.

13. [1] = [3] = [5] = [1, 3, 5], [2] = [4] = [2, 4].

16. [1] = {1, 5, 9, 13}, [2] = {2, 6, 10, 14},
    [3] = {3, 7, 11, 14}, [4] = {4, 8, 12}.

17. $[a] = \{a, b\}$, $[b] = \{a, b\}$ and $[c] = \{c\} \Rightarrow [a] = [b]$. Hence $A/R = \{[a], [b]\}$.

18. i. Z ii. $\left\{\left(n + \frac{1}{2}\right) : n \in Z\right\}$.

20. i. Yes ii. No
    iii. No iv. Yes.

22. $[\{a, e\}, \{b, c, f\}, \{d\}]$ is the partition of A induced by R.

23. i. Yes ii. Yes
    iii. No.

3.17 Partial Ordering

A relation R on a set P is called a **partial ordering** or **partial order relation** if it is reflex-ive, antisymmetric and transitive. A set P together with partial ordering R is called a **partially ordered set** or **poset** and is denoted by (P, R).

Here throughout in this book we will denote partial ordering relation by the symbol $\preceq$. This symbol should not be confused with "less than or equal to". As the relation is reflexive, henceforth we call it a relation on a set. Some author use symbol '$\leq$' for poset.

Posets $(P, \preceq)$ are traditionally represented in a more economical way by Hasse (or poset) diagrams.

Example 3.65

Consider a power set P of a set X. Is the relation of set inclusion $\subseteq$ is a partial order on P?

Solution

The relation 'inclusion' is partial ordered on P. Since for any subset A of a set X, $A \subseteq A$, hence $\subseteq$ is reflexive. It is antisymmetric since for any two subsets A and B of X, $A \subseteq B$ and $B \subseteq A \Rightarrow A = B$. Finally, $\subseteq$ is transitive since for any subsets A, B and C of X. $A \subseteq B$ and $B \subseteq C$ imply that $A \subseteq C$. Hence '$\subseteq$' is a partial ordering of P and thus $(P, \subseteq)$ is a poset.

Example 3.66

Consider the relation of divisibility on the set Z of integers. Is this relation a partial order of Z?

Solution

This relation is not partially ordered since the relation is not antisymmetric. For example: 3 divides –3 and –3 divides 3 but $3 \neq -3$.

3.17-1 Special Elements in Posets

Let $(P, \preceq)$ be a poset and let Y be a subset of X. Then

1. If $y \leq x \; \forall \, x \in Y$, then the element $y \in Y$ is called the **least element** of Y. The set Y can have at most one least element.

   An element $y \in Y$ is called **greatest element** if Y of $x \leq y \; \forall \, x \in Y$. Greatest element of a poset is often called the **unit element** and donated by I and the least element is often called **zero element** and denoted by O.
2. An element $y \in Y$ is a **minimal (maximal)** element of Y of $x < y$ $(x > y)$ for no x in X. In other words we can say an element of poset is minimal (maximal) if it is not greater (less) than any element of the set. It is to be noted here that if the set Y contains least element y, then y will be the only minimal element. But if the set Y contains a minimal element it need not be the only minimal element of Y. The minimal and maximal elements are easy to find using Hasse diagram. They are the "bottom" and the "top" elements in the diagram.
3. An element $y \in X$ is called a **lower (upper) bound** of Y if $y \leq x$ $(y \geq x)$ $\forall \, x \in Y$.
4. The element y is called the **least upper bound (lub)** of subset Y if y is the least element in the set of upper bounds of Y. Similarly, the element y is called the **greatest lower bound (glb)** of subset Y if y is the greatest element in the set of lower bounds of Y. The greatest lower bound and least upper bound of subset Y are denoted by glb(Y) and lub (Y) respectively.

Example 3.67

Find the least upper bound and greatest lower bounds of subsets of P. (*i*) $A_1 = \{1, 2\}$ (*ii*) $A_2 = \{3, 4, 5\}$ where poset $P = \{1, 2, 3, 4, 5, 6, 7, 8\}$. Hasse diagram is shown in the Fig.3.15, find the glb and lub of this Hasse diagram.

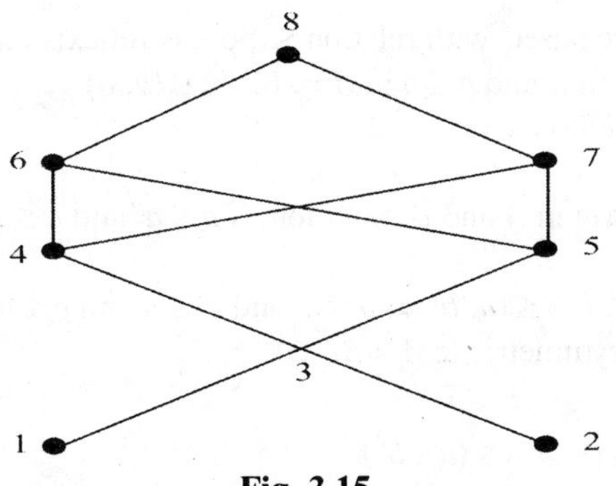

Fig. 3.15

Solution

i. From the Hasse diagram we can see that A_1 has no lower bounds but its upper bounds are 3, 4, 5, 6, 7 and 8.

ii. The upper bounds of A_2 are 6, 7 and 8; and its lower bounds are 3, 1 and 2.

Since A_1 has no lower bounds hence it has no greatest lower bounds. But lub $(A_1) = 3$. Further the lower bounds of A_2 are 3, 1 and 2 we see that glb $(A_2) = 3$. The upper bounds of A_2 are 6, 7 and 8. It is to be noted here that 6 and 7 are not comparable hence A_2 has no least upper bound (lub).

Example 3.68

Let $A \subseteq Z$ and $f: A \to N$ a one to one function. (Z is set of integers; N is set of natural numbers) We define a relation R on a as follows:

For all $x, y \in A$, $(x, y) \in R$, if and only if $f(y) = k\,f(x)$ for some $k \in N$

i. Prove that R is a partial order relation on A.

ii. Let $A = \{4, 5, 6, 7, 8, 9\}$ and suppose $f(x) = 10 - x$.

Draw the Hasse diagram for the poset (A, R).

Solution

i. $\because f(x) = 1 - f(x)$ for $| \in N$

$\Rightarrow (x, x) \in R \Rightarrow R$ is reflexive.

Anti-Symmetry: Let $x \neq y$ then either $f(y) = k\,f(x)$ or $f(x) = k\,f(y)$ for some $k \in N \Rightarrow R$ is anti-symmetric.

Transitive: Let $(x, y) \in R \Rightarrow f(y) = k_1\,f(x)$ for some $k_1 \in N$

and $(y, z) \in R \Rightarrow f(z) = k_2\,f(y)$ for some $k_2 \in N$

Now $f(z) = k_2\,f(y) = k_2\,k_1\,f(x) = k_3\,f(x)$ for some $k_3 \in N$

$(x, z) \in R \Rightarrow R$ is transitive

So, R is partial order relation on A.

ii. Here $A = \{4, 5, 6, 7, 8, 9\}$

$f(x) = 10 - x$

$R = \{(4, 6), (5, 5), (6, 4), (7, 3), (8, 2), (9, 1)\}$.

Example 3.69

Show that if $(A, \leq)$ and $(B, \leq)$ are posets, then $(A \times B, \leq)$ is a poset, with partial order $\leq$ defined by $(a, b) \leq (a', b')$ if $a \leq a'$ in A and $b \leq b'$ in B.

Solution

Reflexive: Because A and B are posets with relation $\leq$. So $\leq$ is reflexive in A and B

$\Rightarrow a \leq a$ in A and $b \leq b$ in $B \Rightarrow (a, b) \leq (a, b)$

$\Rightarrow \leq$ is reflexive in $A \times B$.

Anti-Symmetric:

We know that $\leq$ is antisymmetric in A and B. Solution, if $a \leq a'$ and $a \leq a$ in $A \Rightarrow a = a'$ and iff $b \leq b'$ in B. $\Rightarrow b = b^{\prime}$.

Now let $(a, b) \leq (a', b')$ and $(a', b') \leq (a, b) \Rightarrow a \leq a'$ and $a' \leq a$ and $b \leq b'$ and $b' \leq b \Rightarrow a = a'$ and $b = b'$. $\Rightarrow (a, b) = (a', b') \Rightarrow \leq$ is antisymmetric in $A \times B$.

Transitive:

Let $(a, b) \leq (a', b')$ and $(a, b) \leq (a', b') \leq (a'', b'')$

$$\Rightarrow a \leq a'' \text{ in } A \text{ and } b \leq b'' \text{ in } A$$

$$\Rightarrow (a, b) \leq a'', b'') \Rightarrow \leq \text{ is transitive.}$$

Hence $A \times B, \leq)$ is a poset.

Example 3.70

Draw the Hasse diagram representing the partial ordering $\{(x, y) | x \text{ divides } y\}$ on $\{1, 2, 3, 4, 6, 8, 12\}$.

Solution

Let us first draw the digraph of partial order which is as shown in the Fig. 3.16. Hasse diagram corresponding to the relation is shown in the Fig. 3.17.

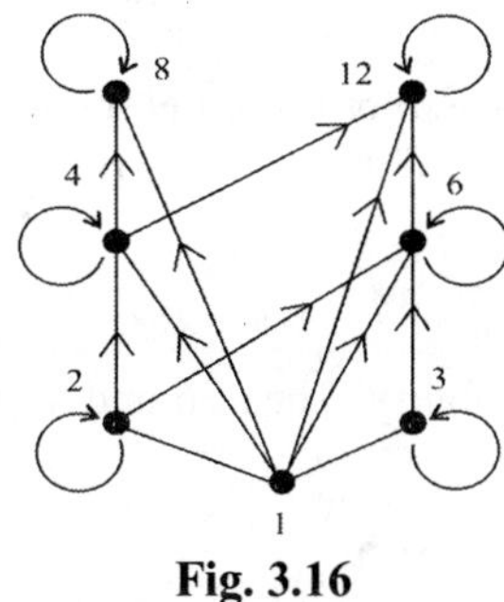

Fig. 3.16

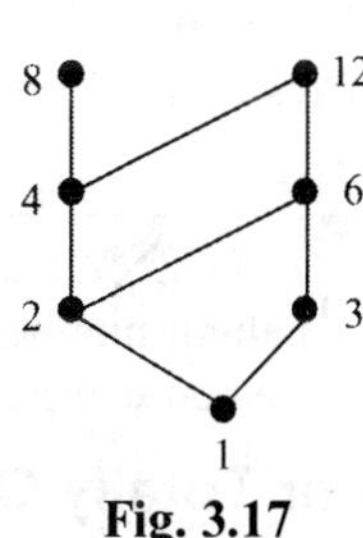

Fig. 3.17

Example 3.71

Draw the Hasse diagram for the partial ordering $\{(X, Y) | X \subseteq Y\}$ on the power set $P(A)$, where $A = \{a, b, c\}$.

Solution

The Hasse diagram for the partial ordering is given in the Fig.3.18.

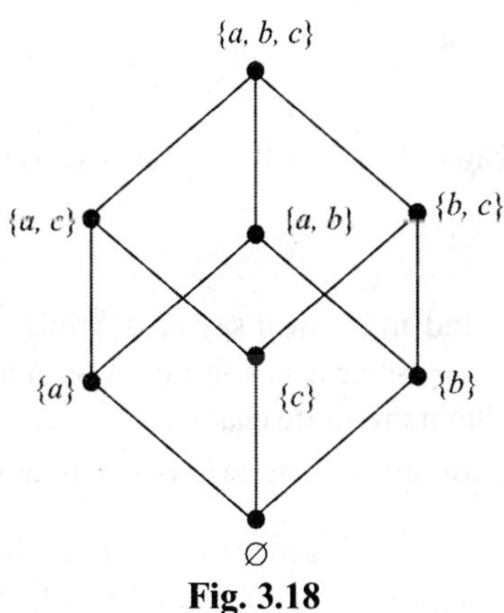

Fig. 3.18

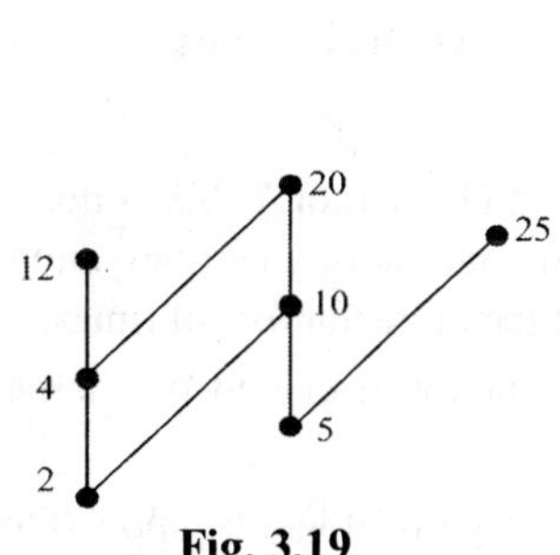

Fig. 3.19

Example 3.72

Find the maximal and minimal elements of the poset $(P, |)$ of the set $P = \{2, 4, 5, 10, 12, 20, 25\}$.

Solution

The Hasse diagram for the poset $(\{2, 4, 5, 10, 12, 20, 25\}, | \}$ is shown in the Fig.3.19.

Example 3.73

Show that

i. "greater than or equal to" relation '$\geq$'.

ii. "less than or equal to" relation $\leq$ are partial ordering in set of real number R.

Solution

i. Since $x \geq x$ for every x hence '$\geq$' is reflexive. If $x \geq y$ and $y \geq x$, then $x = y$. Hence, $\geq$ is antisymmetry. Further, $\geq$ is transitive since $x \geq y$ and $y \geq z \Rightarrow x \geq z \Rightarrow (P, \geq)$ is a poset.

ii. Try yourself.

3.17-2 Comparability

The elements a and b of a poset $(P, \preceq)$ are called **comparable** if either $a \preceq b$ or $b \preceq a$ i.e. if one of the element precedes the other. Thus a and b are called **non comparable** if neither $a \preceq b$ nor $b \preceq a$ (i.e. a and b are not related).

Example 3.74

Consider a set of natural number N. Let N is ordered pair by divisibility. Then 36 and 4 are comparable since 4/36 but 3 and 7 are non-comparable since neither 3/7 nor 7/3.

3.17-3 Linearly or Totally Ordered

If every pair of elements in a poset A is comparable, we say that A is **linearly ordered** or **simply ordered** set and the partial order is called **linear order**. We can also say that A is a **chain**.

3.17-4 Well Ordered Sets

A total order $\preceq$ on a set X is **well order** if every nonempty subset Y of X contains a least element. Then, $(X, \preceq)$ is said to be **well ordered**.

Note that as X is a totally ordered set therefore we can say that if a set Y contains a minimal element, then Y contains only one minimal element and this element is a least element of Y as well.

Theorem 3.12

Consider a finite nonempty poset X with partial ordering $\preceq$. Then $(X, \preceq)$ will have at least one maximal element and one minimal element.

Proof

Let x be an element of X. If x is not a maximal element, we can find an element say $x_1 \in X$ such that $x < x_1$. But if x_1 is not maximal we can find another say $x_2 \in X$ such that $x_1 < x_2$. Since X is a finite set we may continue the argument for finite number of times. In this way we eventually obtain the finite chain $x < x_1 < \ldots < x_{i-1} < x_i$.

And this cannot be extended. Therefore, we cannot have $x_i < y$ for any $y \in A$, so x_i is maximal element of $(X, \preceq)$.

The same argument will be applied to the **dual*** **poset** $(X, \preceq)$.

Theorem 3.13

A poset has at most one greatest element and one least element.

---

*If (P, R) is a poset then (P, R^{-1}) is also a poset. The poset (P, R^{-1}) is called the dual of the poset (P, R) and the partial order is called the dual of the partial order R.

Proof

Let us assume that there exist two greatest elements x and y of poset X. As y is the greatest element, we have $x \leq y$ but x is also greatest element, hence we also have $y \leq x \Rightarrow x = y$ by the antisymmetry property.

This implies that if the poset has a greatest element it has only one such element.

Example 3.75

Consider poset P of nonnegative real numbers with partial ordered "less than or equal to" i.e. ' $\leq$ '. In this set there exists a minimal element of P. But there is no maximal element in the poset P.

Example 3.76

Consider the set of integers Z with partial order $(P, \leq)$. In this poset there is no maximal and no minimal elements.

Example 3.77

Consider the Hasse diagram of the poset P shown in the Fig. 3.19. Write down the minimal and maximal elements of the poset P.

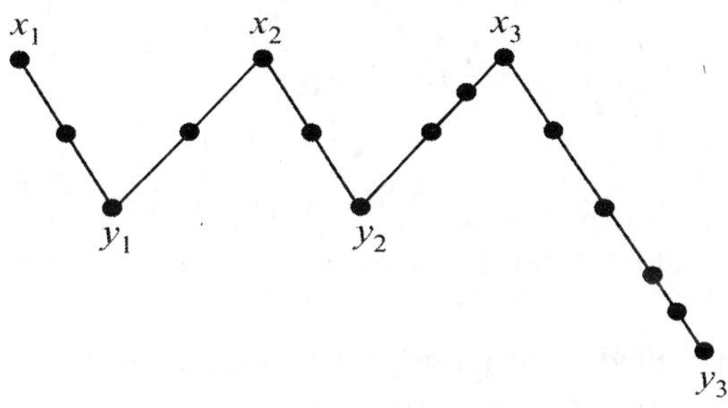

Fig. 3.20

Solution

From the Hasse diagram one can observe that x_1, x_2 and x_3 are the maximal elements of P while y_1, y_2 and y_3 are minimal elements of $(P, \leq)$.

Theorem 3.14

Consider a poset $(P, \preceq)$. Then the subset A of P has at most one lub and at most one glb.

Proof

The proof is similar to proof of theorem 3.13.

Exercise 3.5

1. Determine whether the relation R is a partial order on the set X.

   i. $X = Z$ and $x R y$ iff $x = 2y$ ii. $X = Z$ and $x R y$ iff y^2/x.

2. Consider the set Z of integers and relation R is divisibility on Z. Is this relation partial ordering of Z?

3. Determine whether the relation R is a partial ordering on the set Z defined by $x R y$ iff $x = y^r$ for some $r \in Z^+$.

4. The relation R on the set $A = \{a, b, c, d, e\}$ defined by the rule $(a, b) \in R$ if $a = b - 1$. Determine whether the relation is partial order.

5. Determine the relation R defined on the set of positive integers and defined by rule $(a, b) \in R$ if $a = b$ is reflexive, symmetric antisymmetric, transitive, partial order.
6. Define the usual order on positive integers N. One of the most familiar order relation, called the **usual order** is the relation $\leq$ ("less than or equal") on N (or we can say on any subset of R). This is the reason why, a partial order relation is denoted by $\preceq$ and $x \preceq y$ is read "x precedes y". Determine whether the relation R is a linear order on the set A, where $A = R \times R$ and $(a, b)\, R\, (c, d)$ iff $a \leq c$ and $b \leq d$ and $\leq$ is usual partial order on R.
7. Consider the set $X = \{x, y, z\}$. Find all partial orders $\leq$ in which $a \leq b$.
8. Determine the Hasse diagram of the relation on set $A = \{a, b, c, d, e\}$ whose matrix is given below.

$$\begin{bmatrix} 1 & 1 & 1 & 1 & 1 \\ 0 & 1 & 1 & 1 & 1 \\ 0 & 0 & 1 & 1 & 1 \\ 0 & 0 & 0 & 1 & 1 \\ 0 & 0 & 0 & 0 & 1 \end{bmatrix}$$

9. Determine whether the following relational matrix are partial order

i. $\begin{bmatrix} 1 & 0 & 0 \\ 0 & 1 & 0 \\ 1 & 0 & 1 \end{bmatrix}$ ii. $\begin{bmatrix} 1 & 0 & 1 & 0 \\ 0 & 1 & 1 & 0 \\ 0 & 0 & 1 & 1 \\ 1 & 1 & 0 & 1 \end{bmatrix}$

10. Let (P, R) be a poset. Show that (P, R^{-1}) is also a poset, where R^{-1} is the inverse of R. The poset (P, R^{-1}) is called the **dual** of (P, R).
11. Consider N with usual order $\leq$. Find all the linearly ordered subsets of N.
12. Let $X = \{1, 2, 3\}$ and the set P is the set of all subsets of power set. Draw the Hasse diagram for the elements of P under the partial order of inclusion.
13. Determine whether the relation with following directed graph is partial order.

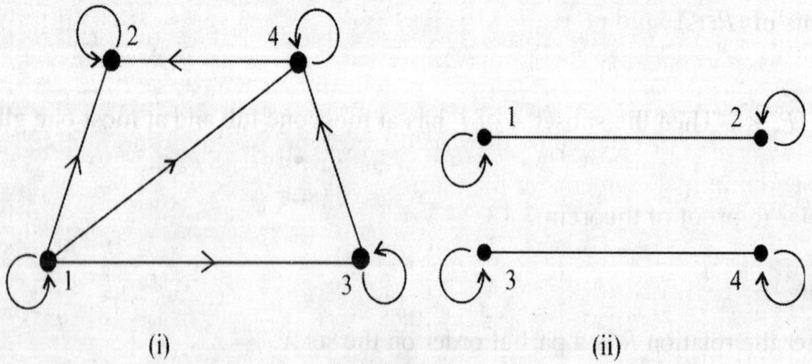

Fig. 3.21

14. Find the duals of the following posets:

i. $(Z, \geq)$ ii. $(\{0, 1, 2\}, \leq)$.

iii. $(Z^+, /)$.

15. Consider a partial order relation R_i on A_i, $i = 1, 2$. Show that R is a partial order on $A_1 \times A_2$ if we define $(a_1, a_2)\, R\, (a'_1, a'_2)$ if $a_1 R_1 a'_1$ and $a_2 R_2 a'_2$.

16. i. Find the pairs of elements that are incomparable in the following posets.

    a. $(P\{0, 1, 2\}, \subseteq)$ b. $(\{1, 2, 4, 6, 8\}, /)$.

17. Give a poset that has

    i. a minimal element but no maximal element.

    ii. a maximal element but no minimal element.

    iii. neither a maximal nor a minimal element.

18. Consider the following Hasse diagram and answer the questions

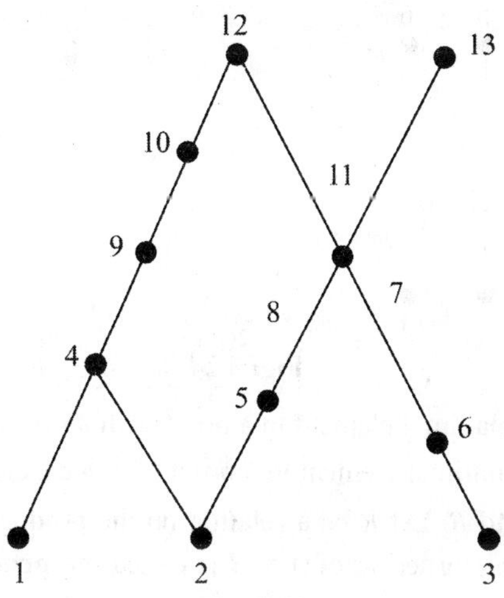

Fig. 3.22

i. Find the minimal elements.

ii. Find the maximal elements.

iii. Is there a least element?

iv. Is there a greatest element?

v. Find the lower bounds of $\{6, 7, 8\}$

vi. Find the upper bounds of $\{1, 2, 3\}$.

vii. Find the least upper bound of $\{1, 2, 3\}$, if it exists.

viii. Find the greatest lower bound of $\{6, 7, 8\}$ if it exists.

19. Find the maximal and minimal elements of the set $A = \{2, 3, 5, 7, 11, 13, \ldots\}$ of prime numbers ordered by divisibility.

20. Consider a set $A = \{2, 3, 4, 16\}$ ordered by divisibility. Find the maximal and minimal elements of A.

21. Find the maximal and minimal elements of the partially ordered set shown in the following Hasse diagram.

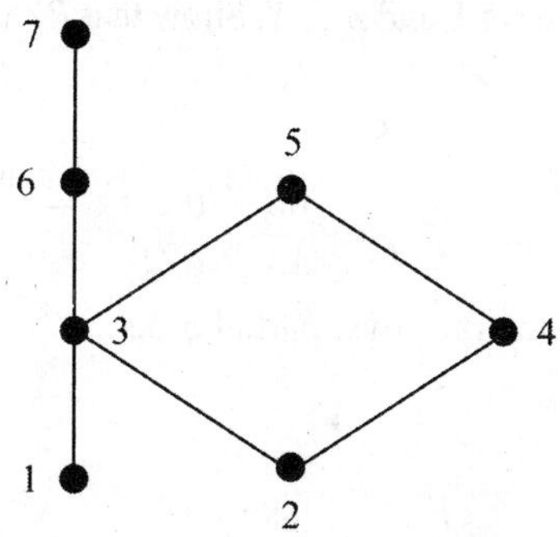

Fig. 3.23

22. Consider a poset $\{(\{1\}, \{2\}, \{4\}, \{1,2\}, \{1,4\}, \{2,4\}, \{3,4\}, \{1,3,4\}, \{2,3,4\}), \subseteq\}$. Answer the following questions:

i. Find the maximal elements.
ii. Find the minimal elements.
iii. Is there a greatest element.
iii. Is there a least element?
iv. Find all the upper bounds of $\{\{2\}, \{4\}\}$
v. Find the lub of $\{\{2\}, \{4\}\}$, if it exists.
vi. Find the glb of $\{\{1, 3, 4\}, \{2, 3, 4\}\}$, if it exists.

23. List all the ordered pairs in the partial ordering with the accompanying Hasse diagram.

(i)

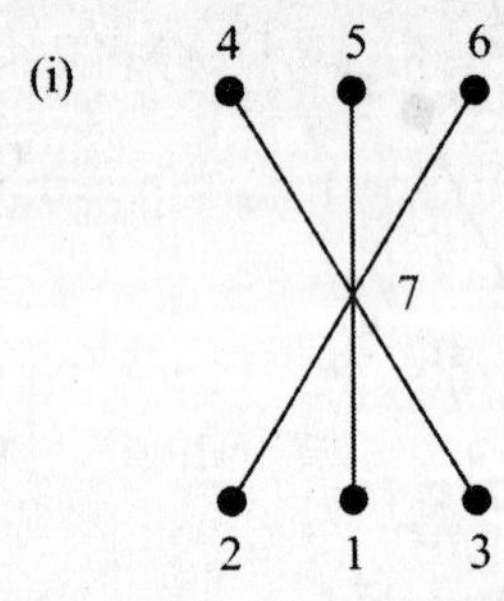

(ii)

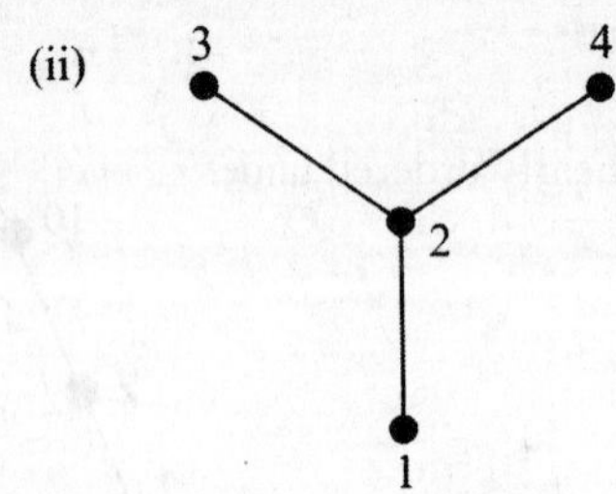

Fig. 3.24

24. Show that there is exactly one maximal element in a poset with a greatest element.
25. Show that there is exactly one minimal element in a poset with least element.
26. Consider two ordered sets A and B. Let R be a relation on the product set $A \times B$ defined by $(x, y)\ R\ (x', y')$ whenever $x \preceq x'$ and $y \preceq y'$. This ordering of $A \times B$ is called the **product order**. Show that R is a partial order.
27. Consider two posets $(A, \preceq)$ and $(B, \preceq)$. Then the **lexicographic ordering** or **dictionary order** on A×B is defined as

$$(a, b) \preceq (a', b') \text{ if } a \prec a' \text{ and } b \leq b'.$$

The ordering of the elements in the first coordinate dominates except in case when attention passes to the second coordinate.

Show that lexicographic order is a partial ordering on the Cartesian product of two posets.

28. Let $N = \{1, 2, 3,\ldots\}$ and $A = \{a, b, c, \ldots, z\}$ are ordered pairs with usual orders. Consider the following subset of $N \times A$ such that

$$X = \{(2, a), (1, c), (2, c), (4, b), (4, z), (3, b)\}.$$

If we assume $N \times A$ to be lexicographically ordered, place the elements of X in the appropriate order.

29. If R is a partial ordering relation on a set X and $A \subseteq X$, Show that $R \cap (A \times A)$ is a partial ordering relation on A.

Answers

1. i. No. ii. No.
2. No. 4. No.
5. Reflexive, symmetric, antisymmetric, transitive, partial order.
6. Yes.

8.

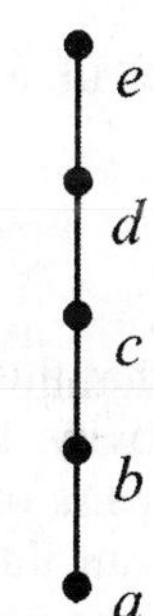

11. Since N itself is linearly ordered under $\leq$, every subset of N will also be linearly ordered.

12.

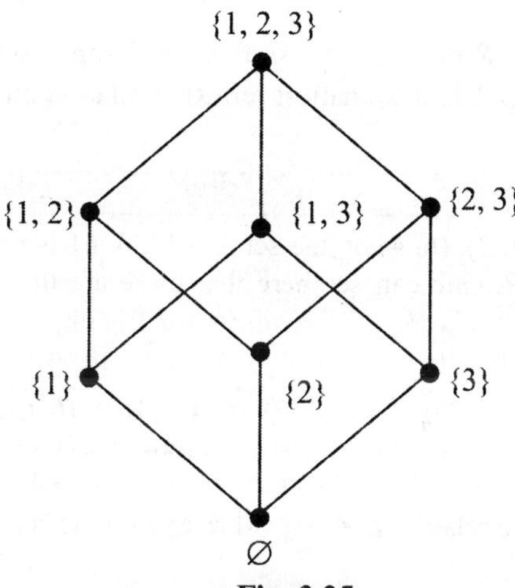

Fig. 3.25

13. ii. Yes.

14. i. $(Z, \leq)$ ii. {(0, 0), (1, 0), ((1, 1), (2, 0), (2, 1), (2, 2)}.

iii. { Z^+, "is a multiple of"}.

16. i. {0} and {1} ii. 4 and 6.

19. Each prime number is divisible only by itself and 1 and the given set does not contain 1. Further any two numbers in the infinite set A are non-comparable. Hence, every element of A is both maximal and minimal element.

20. 12 and 30 are maximal elements, and 2 is minimal element.

21. Minimal element 1; 2; maximal element 5, 7.

22. i. {1, 2}, {1, 3, 4}, {2, 3, 4}, ii. {1}, {2}, {4}

iii. No iv. No

v. {2, 4}, {2, 3, 4} vi. {2, 4}

vii. {3, 4}, { 4} viii. {3, 4}.

23. i. (1, 1), (1, 7), (1, 4), (1, 5), (1, 6), (2, 2), (2, 7), (2, 4), (2, 5),(2, 6), (3, 3), (3, 7), (3, 4), (3, 5), (3, 6), (7, 4), (7, 5), (7, 6), (7, 7), (4, 4), (5, 5), (6, 6).

ii. (1, 2), (1, 3), (1, 4), (2, 3), (2, 4), (1, 1), (2, 2), (3, 3), (4, 4).

28. (1, *c*), (2, *a*), (2, *c*), (3, *b*), (4, *b*), (4, *z*).

3.18 Closures

Let R be a relation on a set X, it is quite possible that this may lack some of important relational properties discussed earlier i.e. reflexivity, symmetry and transitivity. There may occur a situation where R does not have a particular property but with the addition of certain pairs of R, we get a relation with that property. With the addition of minimum number of ordered pairs we can find a smallest relation R_1 on X that contains R and possesses the property we desired. But a relation of the type R_1 doesn't always exist. If there exist a relation such as R_1 then we call it the **closure** of R with respect to the property in question.

3.18-1 Reflexive Closure

Consider a non reflexive relation R on a set X. As this can happen only when some pairs of diagonal relation* Δ are not in R. Thus $R_1 = R \cup \Delta$ is the smallest reflexive relation on X containing R; i.e. the **reflexive closure** of R is $R \cup \Delta$.

Example 3.78

The relation $R = \{(2, 2), (2, 4), (4, 2), (6, 4)$ on the set $X = \{2, 4, 6\}$ is not reflexive. Here by adding (4, 4) and (6, 6), it can be made reflexive R. One can see here that these are the only pairs of the form (a, a) [i.e. diagonal relation] that are not in R.

Therefore the reflexive closure of R is

$$R \cup \Delta_X = \{(2, 2), (2, 4), (4, 2), (6, 4), (4, 4), (6, 6)\}.$$

Example 3.79

What is the reflexive closure of the relation $R = \{(x, y) \mid x < y\}$ on the set of integers?

Solution

The reflexive closure of R is

$$R \cup \Delta = \{(x, y) \mid x < y\} \cup \{(x, x) \mid x \in Z\}$$
$$= \{(x, y) \mid x \le y\}.$$

3.18-2 Symmetric Closure

Consider a relation R on a set X and let it is not symmetric. Then there must exist a pair (a, b) in R such that (b, a) is not in R. As we know $(b, a) \in R^{-1}$. From here one can conclude that if R is to be symmetric we have to add all pairs from R^{-1} i.e. we must enlarge R to $R \cup R^{-1}$.

As $(R \cup R^{-1})^{-1} = R \cup R^{-1}$, hence $R \cup R^{-1}$ is the smallest symmetric relation containing R i.e. $R \cup R^{-1}$ is the **symmetric closure** of R. In other words we can define a relation R_1 to be symmetric closure of R if R_1 is the smallest relation containing R which is symmetric.

---

* A relation Δ is diagonal relation where $\Delta = \{(a, a)$: $a \in X$. Some times diagonal relation Δ w.r.t set A is also written as Δ_A.

Example 3.80

If $X = \{1, 2, 3, 4\}$ and the relation R is defined as $R = \{(1, 2), (2, 3), (1, 3), (3, 4)\}$, then $R^{-1} = \{(2, 1), (3, 2), (3, 1), (4, 3)\}$, hence the symmetric closure of R is $R \cup R^{-1} = \{(1, 2), (2, 1), (2, 3), (3, 2), (1, 3), (3, 4), (3, 1), (4, 3)\}$.

Symmetric closures of relation R can be easily visualized geometrically. The graph of symmetric closure of R i.e. $R \cup R^{-1}$ *will be a digraph with all edges made bidirectional.*

Example 3.81

Let R be the relation on $X = \{a, b, c\}$ defined by $R=\{(a, a), (a, b), (b, c)\}$. Find the symmetric closure of R.

Solution

$R \cup R^{-1} = \{(a, a), (a, b), (b, a), (b, c), (c, b)\}$ is symmetric closure of R.

3.18-3 Transitive Closure

A relation R is the **transitive closure** of R if R is the smallest relation containing R which is transitive. The transitive closure is usually denoted by R^* or R^+.

Theorem 3.15

If R_1 is transitive closure relation of R on a set X and if X is a finite set with n elements. Then

$$R^* = R \cup R^2 \cup \ldots \cup R^n.$$

Proof

Let x and y be the elements of set X and suppose that $x, a_1, a_2, \ldots, a_m, y$ is a path* from x to y in R. This means $(x, a_1), (a_1, a_2), \ldots, (a_m, y)$ are all in R. If a_i and a_j are same vertex say $i < j$ then this path can be divided into three sections $(a, x_i), (x_i, x_j)$ and (x_j, b). But the middle path is a cycle* since $x_i = x_j$, thus we can leave it out and keep the remaining two together. Hence we get a shorter path from x to y as shown in the Fig. 3.26.

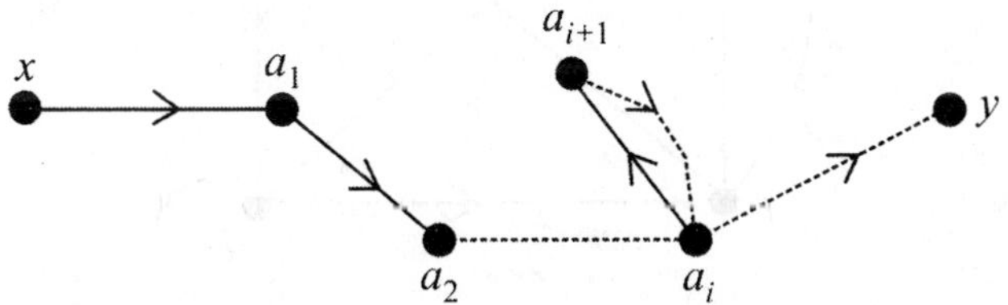

Fig. 3.26

Let us suppose $x, a_1, a_2, \ldots, a_k, y$ be the shortest path from x to y. Further if $x \neq y$ then vertices $x, a_1, a_2, \ldots, a_k, b$ are distinct otherwise from the preceding discussion we can find a shorter path. This implies the length of the path is at most $n - 1$ (since $|X| = n$).

Further, $x = y \Rightarrow$ the vertices $x, a_1, a_2, \ldots, a_k$, are distinct (for similar reasons).

$\Rightarrow$ The length of the path is at most n. Hence we can say if $x\, R_1\, y$, then $x\, R^k\, y$, for some, $1 \leq k \leq n$.

$\Rightarrow R_1 = R \cup R^2 \cup R^3 \cup \ldots \cup R^n$.

Example 3.82

Consider the relation R on $X = \{2, 4, 6\}$ such that $R = \{(2, 4), (4, 6), (6, 6)\}$. Find the transitive closure.

Solution

As we know

$$R^2 = R \circ R = \{(2, 6), (4, 6), (6, 6)\} \text{ and } R^3 = R^2 \circ R = \{(2, 6), (4, 6), (6, 6)\}.$$

Therefore, transitive closure $(R^*) = R \cup R^2 \cup R^3 = \{(2, 6), (4, 6), (6, 6)\}$.

Example 3.83

Find the transitive closure R^* of the relation R on $X = \{a, b, c, d\}$ defined by the following directed graph.

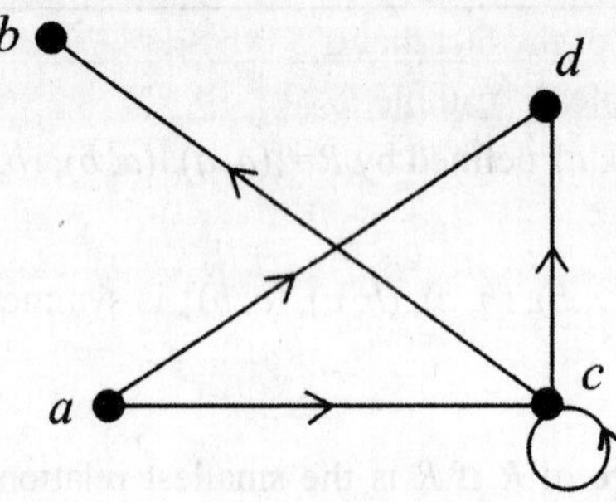

Fig. 3.27

Solution

From the Fig. 3.27 we can see that there is a path from vertex a to vertices b, c and d and there is no path from b to any other vertex. Further there is a path from vertex c to the vertices b, c and d. Finally, there is no path from d to any other vertex.

Therefore $\quad R^* = \{(a, b), (a, c), (a, d), (c, b), (c, c), (c, d)\}$.

Fig. 3.28

Example 3.84

Find the transitive closure R^* of the relation R on set $X = \{a, b, c, d\}$ defined by the following directed graph.

Solution

From the Fig. 3.28 we can see that relation R is

$R^* = \{(a, b), (b, b), (b, d), (c, b), (c, d), (d, a), (d, c)\}$. i.e. there is a path from every vertex in X to every other vertex in X. Further there is also a path from each vertex to itself. Therefore $R^* = X \times X$, the universal relation.

3.18-4 Warshall's Algorithm

Consider the relation matrix R defined by the relation r on the set $\{1, 2, 3, 4, \ldots, n\}$. The matrix of the transitive closure R^*, can be computed by the equation $R^* - R + R^2 + R^3 + \ldots + R^n$. We can compute R^* with $(n-1)$ matrix multiplication by using ordinary polynomial evaluation methods.

$$R^* = R\,(I + R\,(I + R + \ldots + R\,(I + R)))$$

For example if $n = 3$,

$$R^* = R\,(I + R\,(I + R\,))$$

Warshall's algorithm makes use of the fact that if S is a relation matrix,

$S + S = S$ as $1 + 1 = 1$ in Boolean Arithmetic.

Let $\qquad T_k = R + R^2 + \ldots + R^k$, then $R = T_1$.

$$T_1\,(I + T_1) = R\,(I + R) = R + R^2 = T_2$$

$$T_2\,(I + T_2) = (R + R^2)\,(I + R + R^2)$$

$$= (R + R^2) + (R^2 + R^3) + (R^3 + R^4)$$

$$= R + R^2 + R^3 + R^4 = T_4.$$

Similarly $\qquad T_4\,(I + T_4) = T_8.$

Warshall's algorithm can be expressed as follows:

Warshall's Algorithm (version – 1)

Let R^* be a transitive closure matrix of a relational matrix R. R^* can be computed as

1.0 $\quad S := R$

2.0 $\quad$ Repeat

2.1 $\quad T = S$

2.2 $\quad S : T = (I + T)$ $\quad$ (using Boolean Arithmetic)

$\quad$ Until $S = T$

3.0 $\quad$ Terminate with $R^* = S$.

Improved version (version II)

Let R^* be a transitive closure matrix of a relational matrix R. R^* can be computed as

1.0 $\quad S := R$

2.0 $\quad$ For $k := 1$ to n Do

$\quad$ For $i : 1$ to n Do

$\quad$ For $j : 1$ to n Do

$\quad S(i, j) = S\,(i, j) + (S\,(i, k) + S\,(k, j))$

Terminate with $R^* = S$.

Exercise 3.6

1. Let $X = \{w, x, y, z)$ and $R = \{(x, w), (x, y), (y, x), (y, y), (x, x\,), (z, x\,)\}$. Find the reflexive and symmetric closure of R.
2. Let R be a relation on a set X. Define, reflexive, symmetric and transitive closure of R.
3. Consider a relation R on a set $A = \{a, b, c, d\}$ and let $R = \{(a, b), (b, b), (b, c), (c, a), (c, c), (d, a)\}$. Find
   i. reflexive closure of R.
   ii. symmetric closure of R.
4. Explain why the concept of closure is not applicable for irreflexivity or antisymmetry.
5. Consider a relation $R = \{(a, b) \mid a \text{ divides } b\}$ on the set of integers. What is the symmetric closure of R?

6. Find the (i) reflexive closure of R. (ii) symmetric closure of R defined on the relation matrix M_R.

$$\begin{bmatrix} 1 & 0 & 0 & 1 & 1 \\ 0 & 0 & 1 & 0 & 1 \\ 1 & 1 & 1 & 0 & 0 \\ 0 & 1 & 1 & 0 & 0 \\ 0 & 0 & 1 & 0 & 1 \end{bmatrix}$$

7. Draw the directed graph of the reflexive closure of the relations with the directed graph shown in Fig. 3.29(i) and 3.29(ii).

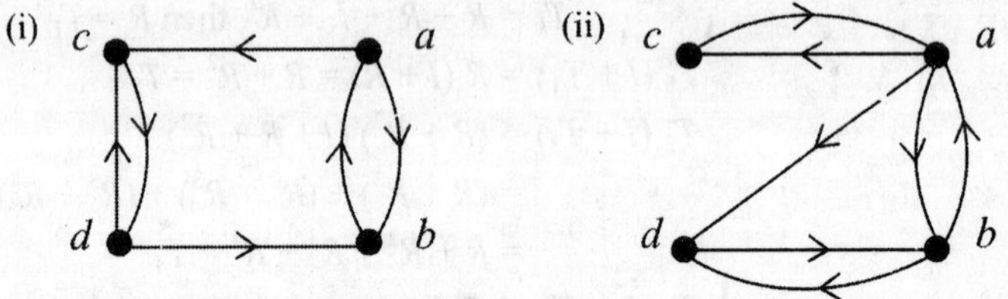

Fig. 3.29

8. Find the transitive closure R^* of the relation R on $X=\{a, b, c, d\}$ defined by the following directed graphs.

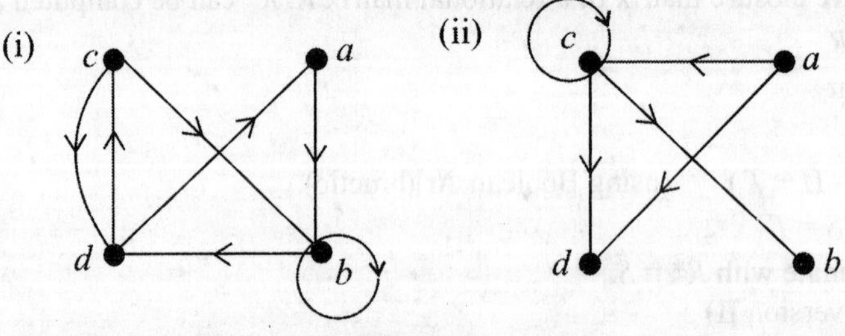

Fig. 3.30

9. Let R be the relation represented by the digraph shown in the Fig. 3.31.
   i. Find transitive closure
   ii. Determine the digraphs of transitive closure R^* directly from the digraph of R.

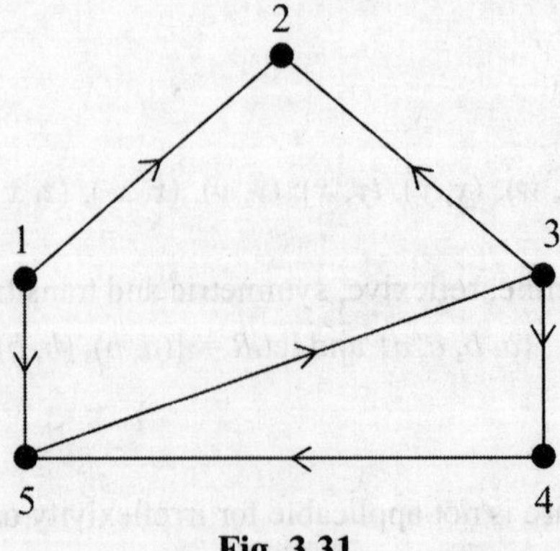

Fig. 3.31

10. Is the transitive closure of a relation symmetric and reflexive? Explain.

11. What common relation on integers Z are transitive closures of the following relations?
    i. $x\,R\,y$ iff $x + 1 = y$.
    ii. $x\,R\,y$ iff $|x - y| = 2$.
12. Use Warshal's algorithm to find the transitive closures of the following relations on $A = \{1, 2, 3, 4, 5\}$.
    i. $\{(1, 3), (2, 4), (3, 1), (4, 2), (5, 4)\}$.
    ii. $\{(1, 5), (2, 1), (2, 4), (3, 4), (4, 1), (4, 3), (5, 1), (5, 2), (5, 3), (5, 5)\}$.
    iii. $\{(1, 2), (1, 3), (1, 5), (2, 1), (2, 3), (3, 1), (3, 2), (4, 1), (5, 4)\}$.
13. Use Warshal's algorithm to find the transitive closure of following relation on $X = \{a, b, c, d\}$.
    i. $\{(b, a), (b, c), (c, a), (c, d), (d, a), (d, c)\}$.
    ii. $\{(a, a), (a, d), (b, a), (b, c), (c, a), (c, b), (c, d), (d, b)\}$.
    iii. $\{(a, b), (b, a), (b, c), (c, d), (d, a)\}$.
14. Show that the closure of a relation R with respect to a property P, if it exists is the intersection of all the relations with property P that contains R.

Answers

3. i. $\{(a, a), (a, b), (b, b), (b, c), (c, a), (c, c), (d, a), (d, d)\}$.
   ii. $(a, b), (a, c), (a, d), (b, a), (b, b), (b, c), (c, a), (c, b), (c, c), (d, a)\}$.
5. $\{(a, b) \mid a$ divides b or b divides $a\}$.
6. i. $\{(a, a), (a, d), (a, e), (b, b), (b, c), (b, e), (c, a), (c, b), (c, c), (d, b), (d, c), (d, d), (e, c), (e, e)\}$.
   ii. $\{(a, a), (a, d), (d, a), (a, e), (e, a), (b, c), (c, b), (b, e), (e, b), (e, a), (a, c), (c, c), (d, b), (b, d), (d, c), (c, d), (e, c), (c, e), (e, e)\}$.
7. (i) 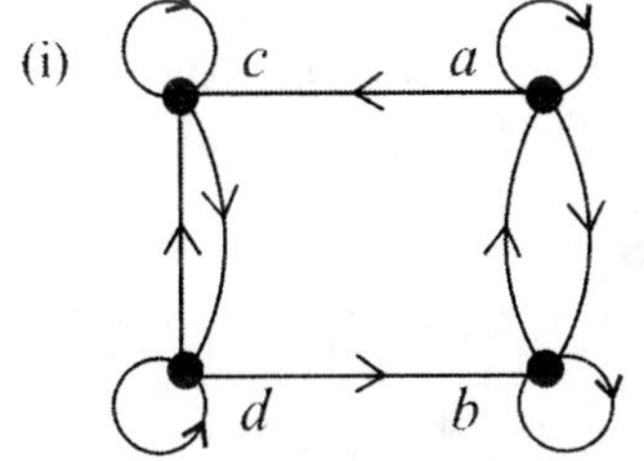

(ii) 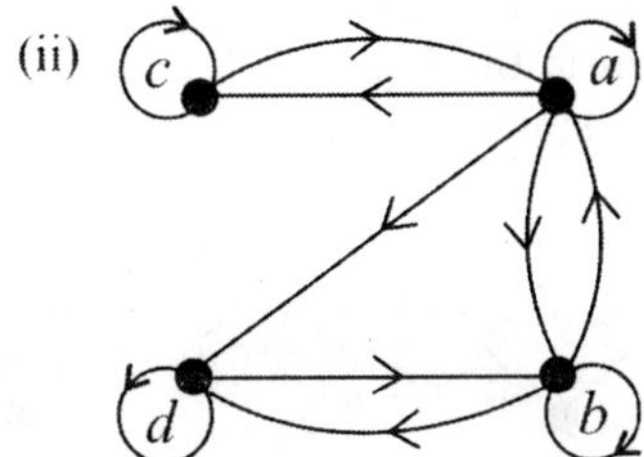

8. i. $R^* = A \times A$.
   ii. $R^* = \{(a, b), (a, c), (a, d), (c, b), (c, c), (c, d)\}$.
9. i. $R^* = [(1, 2), (1, 5), (1, 3), (1, 4), (5, 3), (5, 4), (5, 2), (5, 5), (3, 4), (3, 2), (3, 5), (3, 3), (4, 5), (4, 3), (4, 4), (4, 2)\}$.
11. i. $x\,R^*\,y$ iff $x < y$
    ii. $x\,R^*\,y$ iff $x - y$ is even
13. i. $\begin{bmatrix} 0 & 0 & 0 & 0 \\ 1 & 0 & 1 & 1 \\ 1 & 0 & 1 & 1 \\ 1 & 0 & 1 & 1 \end{bmatrix}$
    ii. $\begin{bmatrix} 1 & 1 & 1 & 1 \\ 1 & 1 & 1 & 1 \\ 1 & 1 & 1 & 1 \\ 1 & 1 & 1 & 1 \end{bmatrix}$
    iii. $\begin{bmatrix} 1 & 1 & 1 & 1 \\ 1 & 1 & 1 & 1 \\ 1 & 1 & 1 & 1 \\ 1 & 1 & 1 & 1 \end{bmatrix}$

3.19 Isomorphic (Similar) Ordered Sets

Consider two partially ordered set A and B. An injective (one to one) function $f : A \to B$ is called a similarity mapping from A into B if f preserves the order relation. In other words we can say $f : A \to B$ is called **similarity mapping** if the following two conditions hold for any pair x and x' in X:

i. If $x \preceq x'$ then $f(x) \preceq f(x')$

ii. If $x \parallel x'$ (non comparable), then $f(x) \parallel f(x')$.

Accordingly if A and B are linearly ordered, then only (i) is required for f to be similarity mapping.

Two ordered set A and B are said to be **isomorphic** or **similar**, written $A \cong B$ if there exists a one-to-one correspondence (bijective mapping) $f : A \to B$ which preserves the order relation, i.e. which is a similarity mapping.

Example 3.85

Consider a poset A isomorphic to a poset B and $f: X \to Y$ is a similarity mapping. Are the following statements true?

i. An element $x \in X$ immediately precedes an element $x' \in X$ i.e. $x \ll x'$, iff $f(x) \ll f(x')$.

ii. An element $x \in X$ has r immediate successors in X iff $f(x)$ has r immediate successors in Y.

iii. An element $x \in X$ is a first (last, minimal, or maximal) element of X iff $f(x)$ is a first (last, minimal, or maximal) element of Y.

Solution

All the statements are true, the order structure of X is the same as the order structure of Y.

Example 3.86

Consider an ordered set X as given in the Fig.3.32. Suppose $X = \{1, 2, 3, 4, 5\}$ is isomorphic to X and $f = \{(p, 1), (q, 3), (r, 5), (s, 2), (t, 4)\}$ is a similarity mapping from X to Y. Draw the Hasse diagram of X.

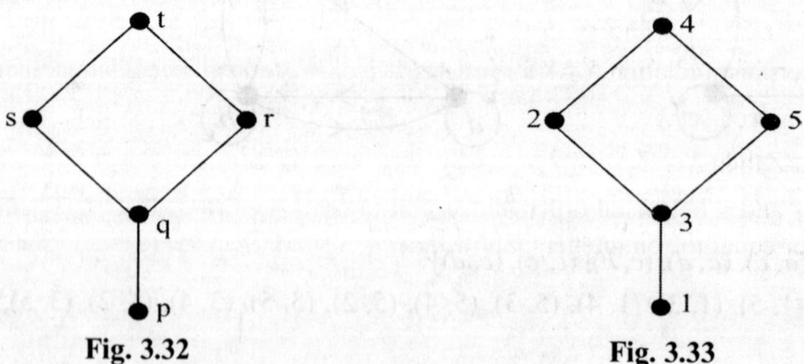

Fig. 3.32 Fig. 3.33

Solution

Since the similarity mapping f preserves the order structure of X and hence f may be seen simply as a relabeling of the vertices in the diagram of X. The Hasse diagram of Y is given in the Fig. 3.33.

Example 3.87

Give an example of a finite nonlinearly ordered set $A = (X, R)$ which is isomorphic to $B = (X, R^{-1})$, the set X with the inverse order.

Solution

Let R be the partial ordering of $X = \{x, y, z, u, v\}$ as shown in Fig. 3.34(a). Fig. 3.34(b) shows X with inverse order R^{-1} (it is obtained by simply turning upside down to R). One can observe here that the two diagrams are identical except for the labeling.

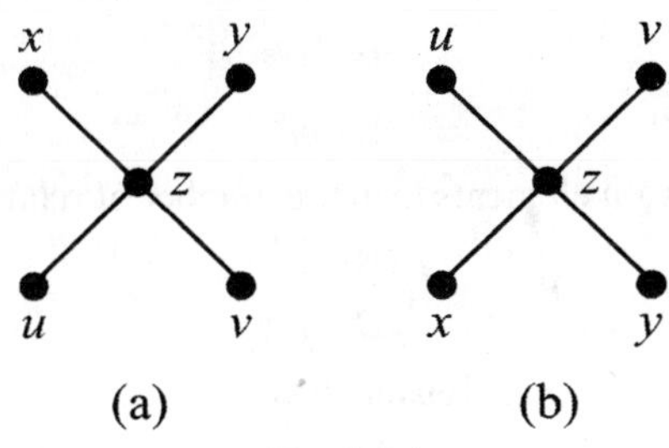

Fig. 3.34

Exercise 3.7

1. Find the number of non–isomorphic posets with three elements x, y, z and draw their diagram.
2. Let P be the ordered set in Fig. 3.35. Suppose $X = \{1, 2, 3, 4, 5\}$ is isomorphic to P and the following is a similarity mapping from P into A: $f = \{(x, 1), (y, 4), (z, 5), (u, 2), (v, 3)\}$. Draw the Hasse diagram of A.

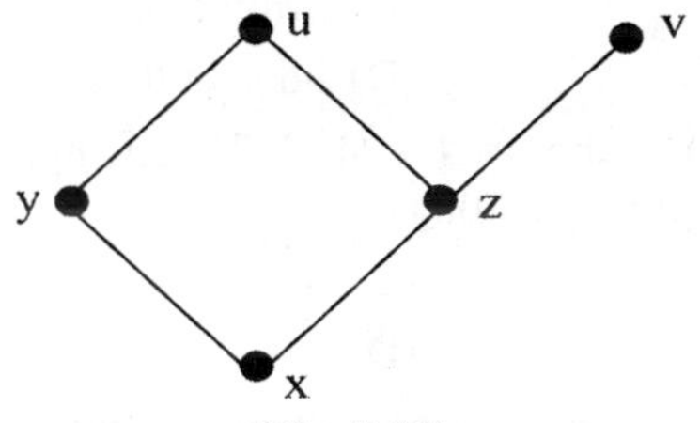

Fig. 3.35

3. Show that the isomorphism relation $X \cong Y$ for ordered sets is an equivalence relation, that is
   i. $X \cong X$ for any ordered set X
   ii. If $X \cong Y$ then $Y \cong X$
   iii. If $X \cong Y$ and $Y \cong Z$ then $X \cong Z$.
4. Let $A = \{1, 2, 4, 8\}$ and $\leq$ be partial order of divisibility on A. Let $A' = \{0, 1, 2, 3\}$ and let $\leq'$ be the usual relation "less than or equal to" on integers. Show that $(A, \leq)$ and $(A', \leq')$ are isomorphic posets.

Answers

1. Four (i) x, y, z (ii) $x, y \preceq z$ (iii) $x \preceq y, x \preceq z$ (iv) $x \preceq y \preceq z$.
2.

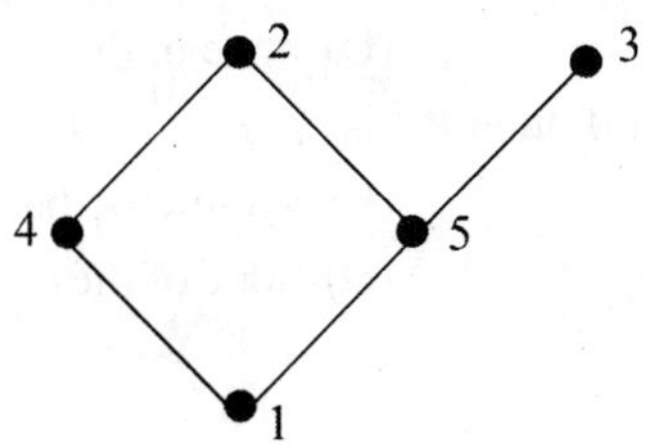

Chapter Objective Type Questions

A. Multiple choice questions

1. If R is a relation from a finite set A having m elements to a finite set B having n elements, then the number of relations from A to B is:
   (A) 2^{mn} (B) $2^{mn} - 1$
   (C) $2mn$ (D) m^n
2. If R is relation on a finite set having n elements then the number of relation A is:
   (A) 2^n (B) 2^{n2}
   (C) n^2 (D) n^n
3. The relation R defined in N as $a\,R\,b \Leftrightarrow b$ is divisible by a is:
   (A) reflexive but not symmetric (B) symmetric but not transitive
   (B) symmetric and transitive (D) none of these
4. Let R be a reflexive relation on set A and I be the identity relation on A. Then-
   (A) $R \subset I$ (B) $I \subset R$
   (C) $R = I$ (D) none of these
5. Let $A = \{1, 2, 3, 4\}$ and let $R = \{(2, 2), (3, 3), (4, 4), (1, 2)\}$ be a relation on A, then R is:
   (A) reflexive (B) symmetric
   (C) transitive (D) none of these
6. Given the relation $R = (1, 2), (2, 3)$ on the set $A = \{1, 2, 3\}$. The minimum number of ordered pairs which when added to R make it an equivalence relation is-
   (A) 5 (B) 6
   (C) 7 (D) 8
7. Let R be a reflexive relation of a finite set A having n-elements, and let there be m ordered pairs in R. Then-
   (A) $m \geq n$ (B) $m \leq n$
   (C) $m = n$ (D) none of these
8. Let $P = \{(x, y) \mid x^2 + y^2 = 1, x, y \, J \, R\}$, then P is:
   (A) reflexive (B) symmetric
   (C) transitive (D) antisymmetric
9. Let R and S be two equivalence relations on a set A, then-
   (A) $R \cup S$ is an equivalence relation on A (B) $R \cap$ S is an equivalence relation on A
   (C) $R - S$ is an equivalence relation on A (D) none of these
10. The relation of subset on the power set $P(A)$ of a set A is:
    (A) symmetric (B) anti-symmetric
    (C) equivalence relation (D) none of these
11. If R is an equivalence relation on a set A, then R^{-1} is:
    (A) reflexive only (B) symmetric but not transitive
    (C) equivalence (D) none of these

12. In the set $A = \{1, 2, 3, 4, 5\}$, a relation R is defined by $R = \{(x, y) \mid x, y \leq R \text{ and } x < y\}$. Then R is:
   (A) reflexive
   (B) symmetric
   (C) transitive
   (D) none of these
13. Let R and S be two relations on a set A. Then which of the statements is not correct?
   (A) R and S are transitive, then $R \cup S$ is also transitive
   (B) R and S are transitive, then $R \cap S$ is also transitive
   (C) R and S are transitive, then $R \cap S$ is also reflexive
   (D) R and S are symmetric, then $R \cup S$ is also symmetric
14. If A and B have n elements in common, then the number of elements common to $(A \times B)$ and $(B \times A)$ is :
   (A) n
   (B) $(n/2)$
   (C) n^2
   (D) $n!$
15. The relation $\geq$ on the set I of all integers is:
   (A) symmetric
   (B) antisymmetric
   (C) equivalence
   (D) none of these
16. Let S be the set of all real umbers. Then the relation $R = \{(a, b): 1 + ab > 0\}$ on S is:
   (A) Reflexive and symmetric but not transitive
   (B) Reflexive and transitive but not symmetric
   (C) Symmetric and transitive but
   (D) none of these not reflexive
17. The union of two equivalence relations is:
   (A) always an equivalence relation
   (B) always reflexive and symmetric but need not be transitive
   (C) always reflexive and transitive but need not be symmetric
   (D) none of these
18. The intersection of two equivalence relation is:
   (A) always an equivalence relation
   (B) always reflexive and symmetric but need not be transitive
   (C) always reflexive and transitive but need not be symmetric
   (D) none of these
19. Let S be the set of all real numbers. A relation R on S has been defined by $aRb \Leftrightarrow |a - b| \leq 1$. Then R is:
   (A) reflexive and symmetric but not transitive
   (B) reflexive and transitive but not symmetric
   (C) symmetric and transitive but not reflexive
   (D) is an equivalence relation
20. A partial ordered relation is transitive, reflexive and-
   (A) antisymmetric
   (B) bisymmetric
   (C) antireflexive
   (D) asymmetric

21. Let $N = \{1, 2, 3, ...\}$ be ordered by divisibility, which of the following subset is totally ordered?
    (A) (2, 9, 16) (B) (4, 15, 30)
    (C) (2, 6, 24) (D) (3, 5, 15)
22. Let R be a symmetric and transitive relation on a set A. Then:
    (A) R is reflexive and hence a partial order
    (B) R is reflexive and hence an equivalence relation
    (C) R is not reflexive and hence not an equivalence relation
    (D) none of above
23. The relation { (1,2), (1,3), (3,1), (1,1), (3,3), (3,2), (1,4), (4,2), (3,4)} is:
    (A) Symmetric (B) Asymmetric
    (C) Transitive (D) Reflexive
24. The number of distinct relations on a set of 3 elements is:
    (A) 512 (B) 18
    (C) 9 (D) 8
25. If R is a relation "Less than" from $A = \{1, 2, 3, 4\}$ to $B = \{1, 3, 5\}$, then $R \circ R^{-1}$ is:
    (A) {(1,3), (1,5), (2,3), (2,5), (3,5), (4,5)} (B) {(3,1), (5,1), (3,2), (5,2), (5,3), (5,4)}
    (C) {(3,3), (3,5), (5,3), (5,5)} (D) {(3,3), (3,4), (3,5)}
26. A relation that is reflexive, anti-symmetric and transitive is a–
    (A) partial order (B) equivalence relation
    (C) Function (D) none of these

B. Fill in the blanks.

1. The range of the relation R is defined as the set of all ________ elements of the ordered pairs which belong to R.
2. Let R and S be relations from X to Y then $(R \cap S)^{-1} =$ ________.
3. $R \circ S$ is empty if the intersection of the range of R and domain of S is ________.
4. A relation R in a set A is said to be universal relation if R is equal to ________
5. A relation R defined in a set X is said to be transitive if for any $x, y, z \in X$ whenever x is related to y and y is related to z then 'x' is related to ______
6. If R and S are symmetric, then $R \cup S$ and $R \cap S$ are ________.
7. An element $y \in Y$ is called ________ if Y of $x \leq y \; \forall \; x \in Y$.
8. The least element is often called ________.
9. An element $y \in X$ is called a ________ of Y if $y \leq x \; \forall \; x \in Y$.
10. The element y is called the glb of subset Y if y is the greatest element in the set of________ of Y.
11. Posets $(P, \preceq)$ are traditionally represented in a more economical way by ________
12. A total order $\preceq$ on a set X is well order if every nonempty subset Y of X contains a________.
13. The reflexive closure of R is ______ .
14. Given $A = \{1 2, 3\}$, $B = \{3, 4\}$, $C = \{4, 5, 6\}$, then $(A \times B) \cap (B \times C) =$ ________.

C. Which of the following statements is True (T)/ False (F)?

1. The number of elements of the power set of $A \times B$ is 2^{mn}.
2. Let R and S be relations from X to Y then if $R \subseteq S$ then $S^{-1} \subseteq R^{-1}$.
3. For the relation $R°S$, the domain is a subset of A and the range is a subset of C (where relation R is defined from A to B and S from B to C).
4. A relation R defined in a set A is said to be reflexive if some element of X is related to itself.
5. If R be a relation on X then R is symmetric if and only if $R = R^{-1}$.
6. Let R and S are relations on a set A then if R and S are reflexive, then $R \cap S$ is reflexive.
7. If R and S are transitive, then $R \cup S$ and $R \cap S$ are not transitive.
8. Boolean matrix multiplication is not commutative in general and hence the composition of relation is not commutative in general.
9. Greatest element of a poset is often called the unit element.
10. The element y is called the lub of subset Y if y is the least element in the set of lower bounds of Y.
11. If every pair of elements in a poset A is comparable then we say that A is linearly ordered.
12. $R \cap R^{-1}$ is the symmetric closure of R.
13. An injective function $f: A \to B$ is called a similarity mapping from A into B if f preserves the order relation.

Answers Objective Type

A.

1.	A	2.	B	3.	A	4.	B
5.	C	6.	C	7.	A	8.	B
9.	B	10.	B	11.	C	12.	C
13.	B	14.	C	15.	B	16.	A
17.	A	18.	A	19.	A	20.	A
21.	C	22.	D	23.	C	24.	A
25.	C	26.	A				

B.

1.	second	2.	$R^{-1} \cap S^{-}1$	3.	empty	4.	$A \times A$
5.	z	6.	symmetric	7.	greatest element	8.	Zero element
9.	lower bound	10.	lower bounds	11.	Hasse diagrams	12.	least element
13.	$R \cup \Delta$	14.	{3, 4}				

C.

1.	T	2.	F	3.	T	4.	F
5.	T	6.	T	7.	F	8.	T
9.	T	10.	F	11.	T	12.	F
13.	T						

CHAPTER

4 Functions

4.0 Introduction

In this chapter we will study special type of relations which are called *functions*. A function plays very important role in almost all branches of mathematics. Primarily, we are concerned with discrete function, which transform a finite set into another finite set. Here, we discuss the basic properties of functions, several types and their applications. Computer science is an area where a number of applications of function can be seen.

4.1 Function or Mapping

Suppose P and Q are any two non-empty sets and let there exists a rule denoted by 'f' by which each element of P is associated with a unique (well-defined) element of Q. Then the rule f is said to be a **function** or **mapping** from P to Q and denoted by $f: P \to Q$ which reads 'f is a function from P to Q' or 'f maps P to Q'.

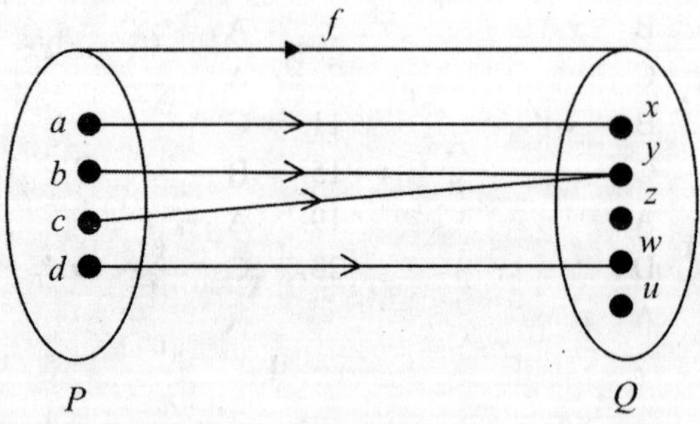

Fig. 4.1

Let $P = \{a, b, c, d\}$ and $Q = \{x, y, z, w, u\}$ be two non-empty sets. Mapping f from P to Q is defined in Fig.4.1. Here a is associated with an element x, b and c are associated with y and d is associated with w. The set $\{(a, x), (b, y), (c, y), (d, w)\}$ is called a **function** or a **mapping** from P to Q.

1. *It must be noted that there may be some elements of the set Q which are not associated to any element of the set P.*
2. *Each element of the set P must be associated to one and only one element of the set Q.*

4.2 Image of a Subset

Let f be a function from P to Q and let X be any subset of P, then the set $Y = \{y : y \in Q \text{ and } y = f(x) \text{ for some } x \in X\}$ is said to be image of X under f and denoted by $f(X) = Y$.

4.3 Function as a Set of Ordered Pairs

Consider two non-empty sets P and Q given in Fig. 4.1. Then the function f from P to Q is a set of ordered pairs $(a, f(a))$, where $a \in P$ and $f(a) \in Q$.

Example 4.1

Let $P = \{a, b, c, d\}$; $Q = \{1, 2, 3, 4, 5\}$ be two sets, then $f = \{(a, 1), (b, 1), (c, 3), (d, 4)\}$ is a function from P to Q, represented by Fig. 4.2.

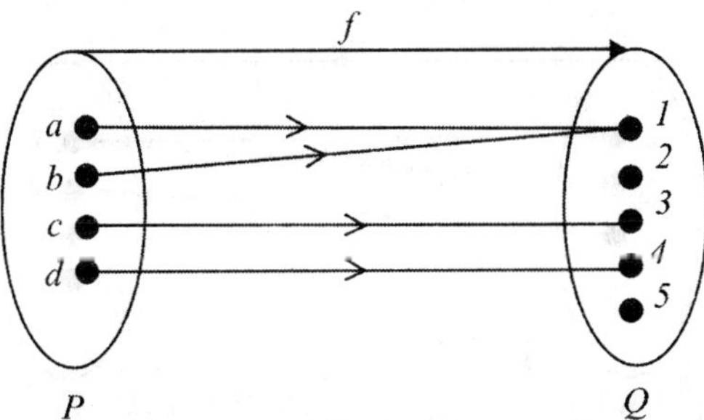

Fig. 4.2

Here $f(a) = 1, f(b) = 1, f(c) = 3, f(d) = 4$.

Example 4.2

Let $P = \{a, b, c\}$ and $Q = \{b, c, d, e\}$ be two sets, then
$f = \{(a, b), (a, c), (b, c)\}$ is not a function from P to Q. Since $f(a) = b$
and $f(a) = c$ is not possible because element a of P is related to two different elements b and c of Q.

Example 4.3

Let $P = \{-1, -2, 2, 3\}$ and $Q = \{1, 4, 6, 9\}$ be any two non empty sets, then the mapping f which associates each element of P to its square in Q is a function from P to Q. Thus $f: P \to Q$. and $f = \{(a, a^2): a \in P \text{ and } a^2 \in Q\} = \{(-1, 1), (-2, 4), (2, 4), (3, 9)\}$.

Diagrammatically it can be represented as

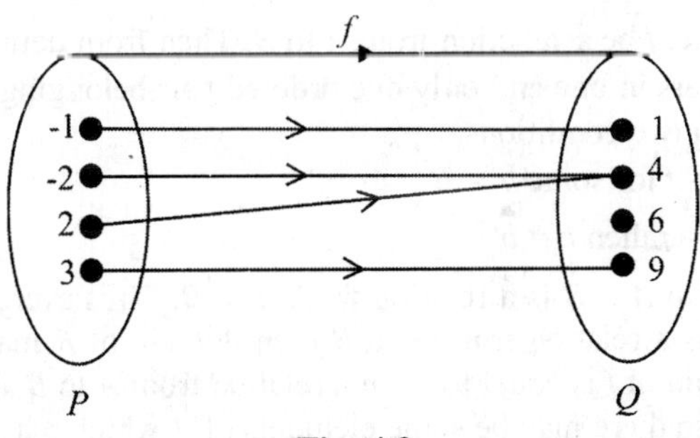

Fig. 4.3

4.4 Domain, Co-domain and Range of a Function

If f is a function from P to Q, then the set P is said to be the **domain** of f and the set Q is said to be the **co-domain** of the function f. Further if an element $x \in P$ is associated with an element $y \in Q$, then y is called "the **image of x under f**" and x is called **pre-image** of y or **argument** of y.

The set of all f - **images** of elements of P is said to be the **range** of f or **image** set of P under f denoted by $f(P)$. Therefore $f(P) = \{f(x): x \in P\}$. Clearly $f(P) \subseteq Q$. Also, if a function f is given as the set of ordered pairs, then the set of all first elements of ordered pairs is domain of f and the set of all second element of ordered pairs is range of f.

Thus, domain of $f = \{x: (x, y) \in f\}$, range of $f = \{y: (x, y) \in f\}$.

Example 4.4

Let the function f is defined as

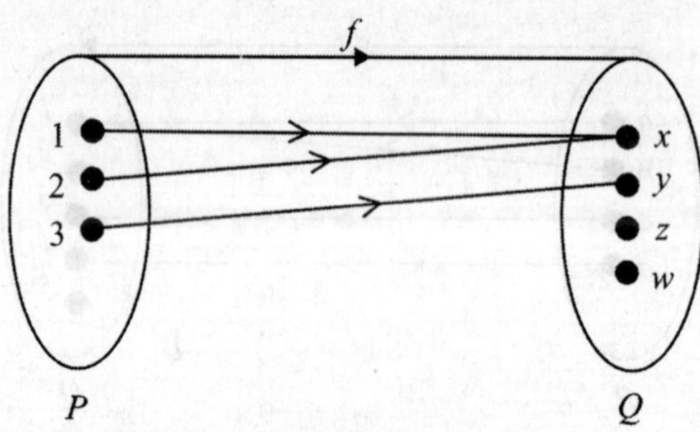

Fig. 4.4

Then domain of $f = \{1, 2, 3\}$, co-domain of $f = \{x, y, z, w\}$, range of $f = \{x, y\}$.

Example 4.5

Let the function f is defined as $f = \{(u, a), (v, b), (w, c)\}$ then domain of $f = \{u, v, w\}$ and range of $f = \{a, b, c\}$.

Example 4.6

Let $f: Z \to Z$ such that

$$f(x) = 3x \ \forall \ x \in Z$$

Then $$f(x) = \{3x: x \in Z\} = \{0, \pm 3, \pm 6, \ldots\}$$

Thus range of f = the set of integer multiple 3.

4.5 Difference between Relation and Function

Suppose A and B are two sets. Let f be a function from A to B. Then from definition of function, f is a subset of $A \times B$ in which each $a \in A$ appears in one and only one ordered pair belonging to f. In other words f is a subset of $A \times B$ satisfying the following two conditions:

i. For each $a \in A$, $(a, b) \in f$ for some $b \in B$.

ii. If $(a, b) \in f$ and $(a, b') \in f$ then $b = b'$

On the other hand every subset of $A \times B$ is a relation from A to B. Thus every function is a relation but every relation is not a function. If R is a relation from A to B then domain of R may be a subset of A. But if f is a function from A to B, then domain of f is equal to A. In a relation from A to B an element of A may be related to more than one element in B. Also there may be some elements of A which may not be related to any element in B. But in a function from A to B each element of A must be associated to one and only one element of B.

Example 4.7

Let $A = \{x, y, z\}$ and $B = \{2, 3, 5, 7\}$ and $f = \{(x, 2), (y, 3), (z, 2)\}$. Then f is a function from A to B. Obviously f is also a relation from A to B. But consider the subset S of $A \times B$ given by $S = \{(x, 2), (x, 3), (y, 5), (z, 7)\}$. Here S is a relation from A to B. But S is not a function from A to B. Because $x \in A$ is associated to two different elements 2 and 3 in B.

4.6 Classification of Functions

Functions can be classified mainly into two groups.

1. Algebraic function

A function which consists of finite number of terms involving powers and roots of independent variable x and four fundamental operations of addition, subtraction, multiplication and division is called algebraic function. The following are the three particular cases of algebraic functions.

(i) Polynomial functions

A function f defined as

$$f(x) = a_0 + a_1 x + a_2 x^2 + ... + a_n x^n \quad ...(1)$$

Where n is any non-negative integer, a_0, a_1, a_n are real constants and $a_0 \neq 0$. Here (1) is a polynomial of x in degree n. For example $f(x) = 3x^3 + 2x^2 + 5x + 7$ is a polynomial of degree 3.

(ii) Rational functions

A function f define as $f(x) = \dfrac{g(x)}{h(x)}$, where $g(x)$ and $h(x)$ are polynomial functions and $h(x) \neq 0$ is called a **rational function**. For example $f(x) = \dfrac{x^2 + 5x + 7}{5x + 9}$ is a rational function.

(iii) Irrational functions

Functions involving radicals are called **irrational functions**. For example $f(x) = \sqrt[5]{x^2 + 7}$ is an irrational function.

2. Transcendental functions

A function which is not algebraic is called a **transcendental function**. In other words a function involving trigonometric, inverse trigonometric, exponential, logarithmic functions of independent variable is called a transcendental function.

(i) Trigonometric functions

A function which consists of trigonometric terms like sin x, cos x, tan x, cosec x, sec x, cot x is called **trigonometric function**. Similarly inverse trigonometric function consists of terms like $\sin^{-1}x$, $\cos^{-1}x$, $\tan^{-1}x$, $\cos^{-1}x$, $\sec^{-1}x$, $\cot^{-1}x$.

(ii) Exponential function

A function f is called **exponential function** if

$f(x) = a^x$, $a > 0$ and satisfies the law $a^1 = a$, $a^x \cdot a^y = a^{x+y}$.

(iii) Logarithmic function

The inverse of exponential function is called logarithmic function. For $y = f(x) = a^x$, $a > 0$ $x \in R$ and $y > 0$, $x = \log y$ is logarithmic function. For example for $0 < x < \infty$, $y = \log x$ if and only if $x = e^y$.

4.7 Transformations or Operators

If f is a mapping from set A to itself then f is said to be a **transformation** or **an operator** on A.

If A is a finite set then domain and range of f is called **permutation** of A.

Example 4.8

If $A = \{1, 2, 3\}$ and $f_1 = \{(1, 2), (2, 3), (3, 1)\}, f_2 = \{(1, 3), (2, 3), (3, 1)\}$ then f_1 and f_2 are two transformations of A and f_1 is a **permutation** of A because in f_1 domain set is $\{1, 2, 3\}$ and range set is $\{1, 2, 3\}$.

4.8 Equal functions

Let f and g be two functions from A to B. Then these functions are said to be **equal** if and only if

i. domain of f is equal to domain of g
ii. co-domain of f is equal to co-domain of g
iii. for all $x \in A, f(x) = g(x)$

If f and g are equal functions then we can write $f = g$.

Example 4.9

Let $A = \{1, 2\}, B = \{3, 6\}$ and $f: A \to B$ is defined as $f(x) = x^2 + 2$ and $g: A \to B$ is defined as $g(x) = 3x$

Then
$$f(1) = 3 = g(1)$$
$$f(2) = 6 = g(2)$$

Hence $f = g$.

Example 4.10

Write the following relations in sets of ordered pairs and find which of them are functions

i. $f_1 = \{(x, y): y = 3x, x \in \{1, 2, 3\}, y \in \{3, 6, 9, 12\}\}$
ii. $f_2 = \{(x, y): y > x + 1, x = 1, 2$ and $y = 2, 4, 6\}$
iii. $f_3 = \{(x, y): x + y = 3, x, y \in \{0, 1, 2, 3\}\}$
iv. $f_4 = \{(x, y): y = 3, x \in N\}$.

Solution

If we express f_1, f_2, f_3 and f_4 as sets of ordered pairs, then we have

i. $f_1 = \{(1, 3), (2, 6), (3, 9)\}$

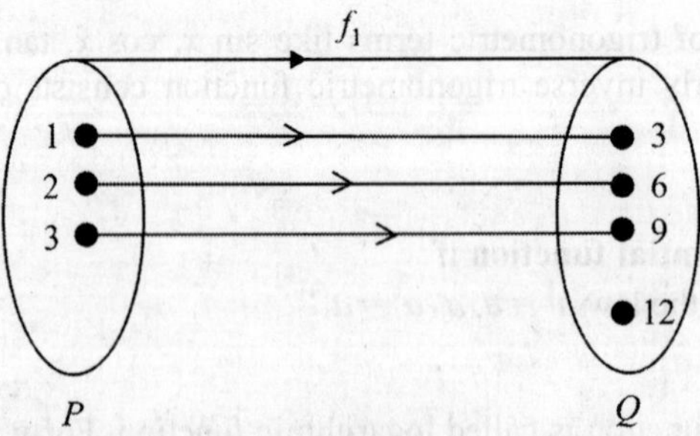

Fig. 4.5

From the diagram it is clear that f_1 is a function.

ii. $f_2 = \{(1, 4), (1, 6), (2, 4), (2, 6)\}$

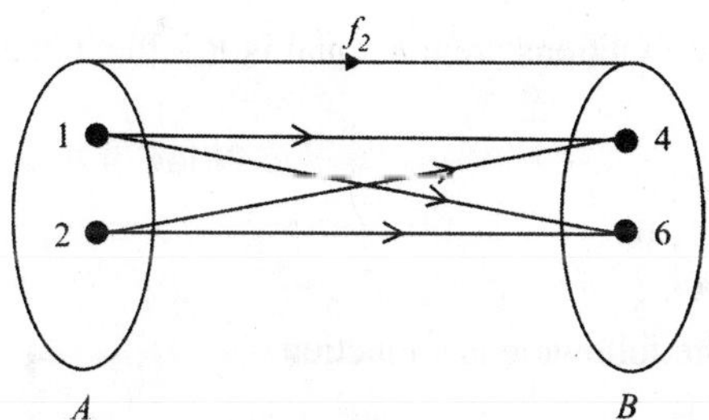

Fig. 4.6

Clearly f_2 is not a function since both 1 and 2 of A are related to two different elements 4 and 6 of B.

iii. The graphical representation of f_1 is

iv. The graphical representation is given by

do yourself.

Example 4.11

Let $A = \{-2, -1, 0, 1, 2\}$ and $f : A \to Z$ such that $f(x) = x^2 - 2x - 3$,
Find (i) the image set of A under f (ii) pre-images of 0, –3 and 5.

Solution

i. Here
$f(-2) = (-2)^2 - 2(-2) - 3 = 4 + 4 - 3 = 5$
$f(-1) = (-1)^2 - 2(-1) - 3 = 1 + 2 - 3 = 0$
$f(0) = (0)^2 - 2(0) - 3 = -3$
$f(1) = (1)^2 - 2(1) - 3 = 1 - 2 - 3 = -4$
$f(2) = (2)^2 - 2(2) - 3 = -3$
Hence the image set of A under f is $\{0, -3, -4, 5\}$

ii. The pre-image of 0 is –1.
The pre-images of –3 are 0 and 2.
The pre-image of 5 is –2.

Example 4.12

Find domain and range of $\{(1, a), (2, a), (3, b), (4, c)\}$.

Solution

The domain set = $\{1, 2, 3, 4\}$ and the range set = $\{a, b, c\}$.

Example 4.13

Find the domain for which the functions $f(x) = 2x^2 - 1$ and $g(x) = 1 - 3x$ are equal. Also find a domain for which the functions are not equal.

Solution

f and g are equal if $\quad f = g \ \forall x$

$$\Rightarrow 2x^2 - 1 = 1 - 3x$$

$$\Rightarrow 2x^2 + 3x - 2 = 0 \Rightarrow x = -2, \frac{1}{2}$$

Therefore, the domain set is $\{-2, \frac{1}{2}\}$

Also the domain for which the functions are not equal is R – the domain for which the functions are equal $\Rightarrow R - \{-2, \frac{1}{2}\}$.

Exercise 4.1

1. State whether or not each of the following is a function
   i. $\{(a, 0), (b, 0), (c, 1), (d, 1)\}$.
   ii. $\{(1, a), (1, b), (2, a), (3, b)\}$.
   iii. $\{(x, y): x, y \in R \text{ and } y^2 = x\}$, where R is the set of real numbers.
   iv. $\{(x, y): x, y \in R \text{ and } y = x^3\}$, where R is the set of real numbers.
2. Find domain and range of each of the following functions
   i. $\{(1, 2), (3, 1), (2, 3)\}$.
   ii. $\{(a, 1), (b, 1), (c, 2)\}$.
   iii. $\{(x, y): x \in Z \text{ and } y = x^2\}$.
3. Let $A = \{-2, -1, 0, 1, 2\}$. Let the function $f: A \to R$ be defined as $f(x) = x^2 + 1$. Find the range of f.
4. Let $f: R \to R$ and $f(x) = x^2$ where R is the set of real numbers.
   Find i. the image set of R under f
   ii. $\{x: f(x) = 4\}$
   iii. $\{y: f(y) = -1\}$.
5. Let $f: R \to R$ and $g: C \to C$ be two functions defined as $f(x) = x^2$ and $g(x) = x^2$, where R and C are set of real and complex numbers respectively. Are they equal functions?
6. Find the domain for which the functions $f(x) = 3x^2 - 1$ and $g(x) = 1 - 5x$ are equal. Also find a domain for which the functions are not equal.
7. Find $f(2), f(8), f(-1), f(-5), f(-2), f(3)$, when $f: R \to R$ be defined by

$$f(x) = \begin{cases} 3x - 1, & x > -3 \\ x^2 - 2, & -2 \le x \le 3 \\ 2x + 3, & x > 2 \end{cases}$$

Answers

1. i. yes ii. No
   iii. No iv. Yes
2. i. $\{1, 2, 3\}, \{1, 2, 3\}$ ii. $\{a, b, c\}, \{1, 2\}$
   iii. $Z, \{0, 1, 4\ldots\}$
3. $\{5, 2, 1\}$
4. i. $\{x^2: x \in R\}$ ii. $\{2, -2\}$
   iii. $\varnothing$
5. No
6. $\{-2, \frac{1}{3}\}, R - \{-2, \frac{1}{3}\}$.
7. $f(2) = 2, f(8) = 19, f(-1) = -1, f(-5) = -16, f(-2) = 2, f(3) = 9$.

4.9 Types of Functions

4.9-1 Into function or Mappings

A mapping $f: A \to B$ is called an **into** mapping or function if there is at least one element in B which have no pre-image in A or if $f(A)$ is a proper subset of B.

The following graphical representation is an example of into mapping.

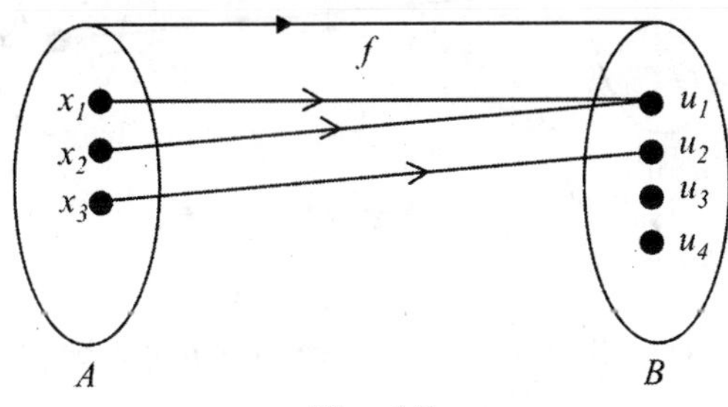

Fig. 4.7

Example 4.14

Let $A = \{-1, -2, 1, 2\}$, $B = \{1, 4, 5\}$ and $f: A \to B$ defined by $f(x) = x^2$. Then f is into mapping from A to B because $f(A) \subset B$.

4.9-2 Onto (or surjective) Mapping (or function)

A mapping $f: A \to B$ is called **onto** (or surjective) if $f(A) = B$, i.e. if each element of B is f-image of some element of A.

The following is a graphical representation of onto mapping.

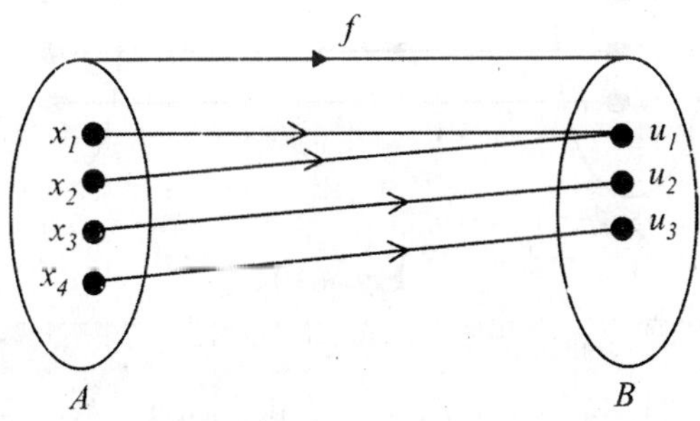

Fig. 4.8

Example 4.15

Let $A = \{-1, -2, 1, 2\}$ and $B = \{1, 4\}$ then $f: A \to B$ defined by $f(x) = x^2$ is onto mapping from A to B because $f(A) = B$.

4.9-3 One-One (Injective) Mapping (or function)

A mapping $f: A \to B$ is called **one-one** mapping or **injective** mapping if different elements of A have different f-images in B.

Thus $f: A \to B$ is one-one if and only if $f(x_1) = f(x_2) \Rightarrow x_1 = x_2 \ \forall \ x_1, x_2 \in A$

or $x_1 \neq x_2 \Rightarrow f(x_1) \neq f(x_2)$.

Following is graphical representation of one-one mapping

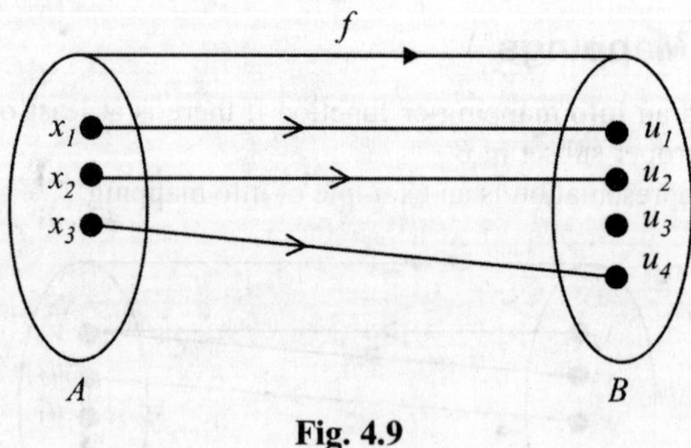

Fig. 4.9

Example 4.16

Let $f : R \to R$ be defined by $f(x) = 2x + 1 \ \forall\ x \in R$ then f is one-one because for $x_1, x_2 \in R, f(x_1) = f(x_2) \Rightarrow 2x_1 + 1 = 2x_2 + 1 \Rightarrow x_1 = x_2$.

4.9-4. Many-One Mapping (or function)

A mapping $f: A \to B$ is called **many-one** if two or more elements in A have the same f-images in B. In other words if $f(x_1) = f(x_2)$ then $x_1 \neq x_2 \ \forall\ x_1, x_2 \in A$.

Following is graphical representation of many-one mapping.

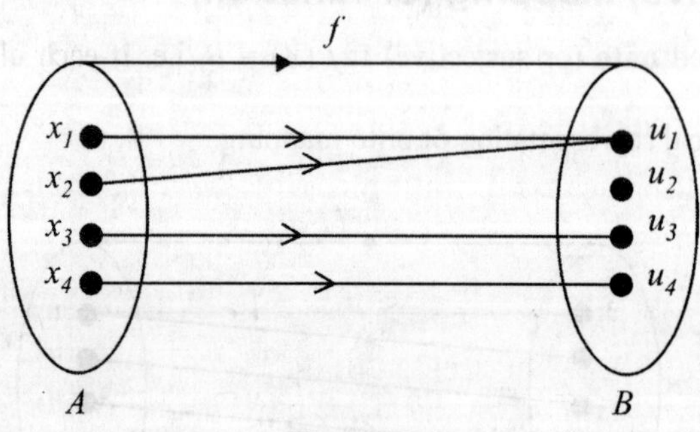

Fig. 4.10

Example 4.17

Let $A = \{1, -2, 2\}$ and $B = \{1, 4, 5\}$ then $f : A \to B$ defined as $f(x) = x^2$ is many–one mapping because $f(-2) = f(2) = 4$ but $-2 \neq 2$.

4.9-5 One-One Onto (or Bijective) Mapping (or function)

A mapping $f : A \to B$ is called **one-one onto** (or **bijective**) if f is both one-one (injective) and onto (surjective). The following is a graphical representation of one-one onto mapping.

If f is injective, each element of B has at most one pre-image. If f is surjective, each element of B has at least one pre-image. If f is bijective, each element of B has exactly one pre-image.

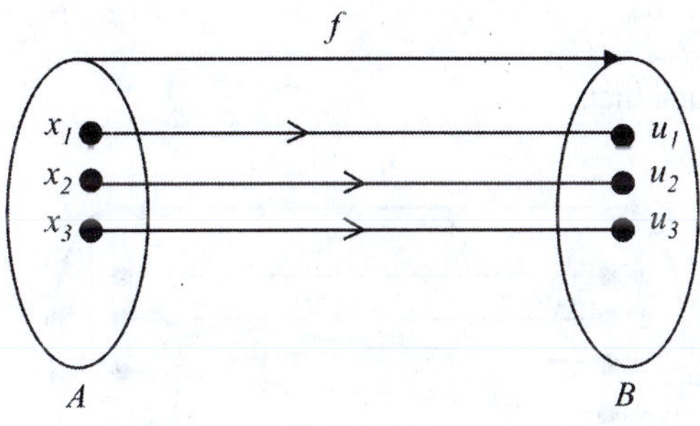

Fig. 4.11

Example 4.18

Let $A = \{1, 2, 3\}$ and $f: A \rightarrow A$ such that $f(x) = 4 - x$. Then $f(x)$ is one–one and onto because $f = \{(1, 3), (2, 2), (3, 1)\}$.

4.9-6 How to prove that a given function *f* is one-one

A function f can be proved to be one to one by showing if $f(x_1) = f(x_2)$ then $x_1 = x_2$ or if $x_1 \neq x_2$ then $f(x_1) \neq f(x_2)$, where x_1 and x_2 are arbitrary elements of the domain f.

4.9-7 How to prove that a given function *f* : A → B is onto

A function $f: A \rightarrow B$ can be proved to be onto by showing if $b \in B$ there exists $a \in A$ such that $b = f(a)$. And $b \in B \Rightarrow b \in f(A)$ for arbitrary chosen element b. Thus every element of B is an element of $f(A)$, hence $B \subseteq f(A)$. But $f(A) \subseteq B \Rightarrow B = f(A)$ and the function is onto.

4.9-8 Identity Mapping (or function)

Let A be a non-empty set. The mapping $f: A \rightarrow A$ is called an **identity** mapping if $f(x) = x \; \forall \; x \in A$.

Example 4.19

Let $A = \{1, 2, 3, 4\}$, then $f = \{ (1, 1), (2, 2), (3, 3), (4, 4)\}$ is an identity mapping of the set A.

An identity mapping is always one-one and onto.

4.9-9 Constant Mapping (or function)

A mapping $f: A \rightarrow B$ is called a **constant** mapping if the range of f consists of only one element. In other words $f: A \rightarrow B$ is constant if every element of A has the same f-image in B. Therefore $f: A \rightarrow B$ such that $f(x) = c \; \forall \; x \in A$ and $c \in B$ where c is a constant.

Example 4.20

Let $f: R \rightarrow R$ be a mapping such that $f(x) = 50 \; \forall \; x \in R$ then f is a constant mapping.

4.9-10 Inverse Image of an Element

Let $f: A \rightarrow B$ be a mapping and $b \in B$. Then the inverse image of the element b under f is the set of all elements of A whose f-image is b and denoted by $f^{-1}(b)$ i.e. if $f: A \rightarrow B$, then $f^{-1}(b) = \{x : x \in A \text{ and } f(x) = b\}$.

Example 4.21

Let $f: A \to B$ be a mapping such that

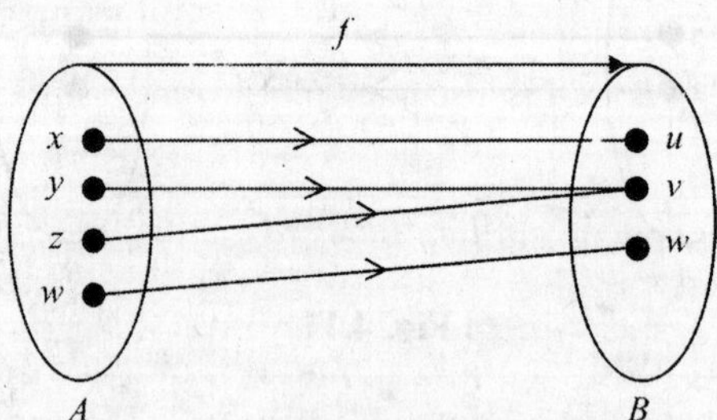

Fig. 4.12

$f(x) = u, f(y) = f(z) = v$ and $f(w) = w$. Then the inverse image of the element $u \in B$ under f is denoted by $f^{-1}(u)$ and $f^{-1}(u) = x$, similarly $f^{-1}(v) = \{y, z\}$ and $f^{-1}(w) = \{w\}$.

4.9-11 Inverse Mapping

Let $f: A \to B$ be a one–one and onto mapping. Then the mapping $f^{-1}: B \to A$ which associates each element $b \in B$ to the element $a \in A$, such that $f(a) = b$, is said to be the **inverse** mapping of the mapping $f: A \to B$.

The graphical representation of inverse mapping is given in Fig. 4.14.

Here we have,

$$f(x) = u, f(y) = v, f(z) = w,$$

and

$$f^{-1}(u) = x, f^{-1}(v) = y, f^{-1}(w) = z.$$

Fig. 4.13: $f: A \to B$

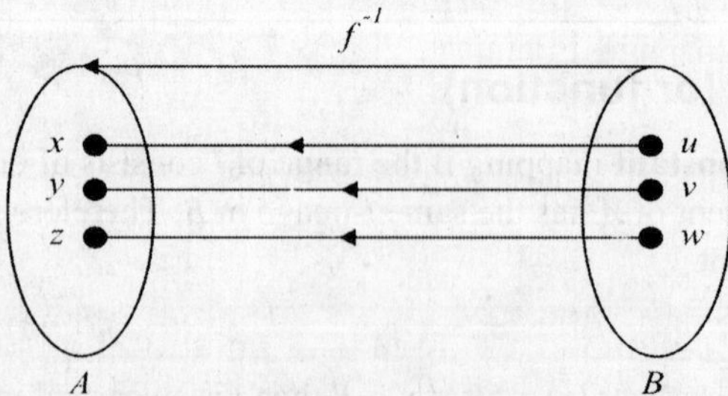

Fig. 4.14: $f^{-1}: B \to A$

Example 4.22

If $A = \{1, 2, 3, 4\}$ and $B = \{3, 6, 9, 12\}$ and $f: A \to B$ defined by $f(x) = 3x$ then we find that $f(1) = 3, f(2) = 6, f(3) = 9, f(4) = 12$

So $f = \{(1, 3), (2, 6), (3, 9), (4, 12)\}$. Clearly, f is one-one and onto.

Thus $f^{-1}: B \to A$ exists and $f^{-1} = \{(3, 1), (6, 2), (9, 3), (12, 4)\}$.

4.10 Properties of Inverse Mapping

Theorem 4.1

If $f: A \to B$ is one-one onto, then $f^{-1}: B \to A$ is also one-one and onto.

Proof

Let $f : A \to B$ is a one-one onto mapping and $f^{-1}: B \to A$ be the inverse of f. Now we have to show that $f^{-1}(b_2)$ is one-one and onto.

1. To show that f^{-1} is one-one.

Let $f^{-1}(b_1) = a_1$ and $f^{-1}(b_2) = a_2$ where $a_1, a_2 \in A$ and $b_1, b_2 \in B$.

Now $f^{-1}(b_1) = f^{-1}(b_2) \Rightarrow a_1 = a_2$

$\Rightarrow f(a_1) = f(a_2)$ [$\because$ f is one-one]

$\Rightarrow b_1 = b_2$ [$\because f(a_1) = b_1, f(a_2) = b_2$]

$\therefore f^{-1}$ is one-one.

2. To show that f^{-1} is onto.

Now if $a \in A$, then by definition of f, there exists $b \in B$ such that $f(a) = b \Rightarrow$ there exists $b \in B$ such that $a = f^{-1}(b)$. Hence f^{-1} is onto.

Thus f^{-1} is one–one and onto mapping.

Theorem 4.2

The inverse of one-one onto mapping is unique.

Proof

Let $f : A \to B$ is one-one and onto. If possible let $f_1: B \to A$ and $f_2: B \to A$ be two inverse mappings of f. Then we have to show that $f_1 = f_2$. Let $b \in B$, then there exists a_1 and a_2 in A such that

$$f_1(b) = a_1 \text{ and } f_2(b) = a_2$$

But f_1 is an inverse mapping of f, thus

$$f_1(b) = a_1 \Rightarrow f(a_1) = b$$

Also $f_2(b) = a_2 \Rightarrow f(a_2) = b$

Since f is one-one mapping

Therefore $f(a_1) = b = f(a_2) \Rightarrow a_1 = a_2 \Rightarrow f_1(b) = f_2(b) \Rightarrow f_1 = f_2$.

Hence inverse of one-one onto mapping is unique.

Example 4.23

Let I be the set of integers then which of the following functions from I to I are one-one, onto or both?

1. $f(x) = x^3$
2. $f(x) = 4x$
3. $f(x) = 3x - 5$
4. $f(x) = x + 7$
5. $f(x) = x^2 + x$.

Solution

1. Let $x_1, x_2 \in I$ such that

$f(x_1) = f(x_2) \Rightarrow (x_1)^3 = (x_2)^3$ [$\because f(x) = x^3$]

Taking cube root of both sides

$$\Rightarrow (x_1^3)^{1/3} = (x_1^3)^{1/3} \Rightarrow x_1 = x_2$$

Thus f is one-one.

Further let $$y = f(x) = x^3 \Rightarrow x = (y)^{1/3}$$

If y is an integer then it is not necessarily true that $(y)^{1/3}$ will also be an integer. For example $3 \in I$ but $3^{\frac{1}{3}} \notin I$. Thus f is not onto. Therefore it is an into mapping.

2. Let $x_1, x_2 \in I$ such that $f(x_1) = f(x_2) \Rightarrow 4x_1 = 4x_2 \Rightarrow x_1 = x_2$
   Hence f is one- one.

   Further, let $$y = 4x \Rightarrow x = \frac{y}{4}.$$

   If y is an integer then it is not necessarily true that $\frac{y}{4}$ will also be an integer. For example $2 \in I$, but $\frac{2}{4} = \frac{1}{2} \notin I$. Thus f is not onto. It is an into mapping.

3. Let $x_1, x_2 \in I$ such that

$$f(x_1) = f(x_2) \Rightarrow 3x_1 - 5 = 3x_2 - 5$$
$$\Rightarrow 3x_1 = 3x_2 \Rightarrow x_1 = x_2$$

   Hence f is one-one.
   Further, let $y = 3x - 5$

$$\Rightarrow x = \frac{y+5}{3}$$

   If y is an integer, then it is not necessarily true that $\frac{y+5}{3}$ will also be an integer. For example, $2 \in I$ but $\frac{2+5}{3} = \frac{7}{3} \notin I$. Thus f is not onto.

4. Let $x_1, x_2 \in I$, such that

$$f(x_1) = f(x_2) \Rightarrow x_1 + 7 = x_2 + 7$$
$$\Rightarrow x_1 = x_2$$

   Hence f is one-one.

   Further, let $\quad y = x + 7 \Rightarrow x = y - 7$

   Now, $\quad y \in I \Rightarrow y - 7 \in I$

   Thus $\quad f(y - 7) = (y - 7) + 7 = y$

   Hence f is onto.

5. Let $x_1, x_2 \in I$, such that

$$f(x_1) = f(x_2) \Rightarrow x_1^2 + x_1 = x_2^2 + x_2$$
$$\Rightarrow x_1(x_1 + 1) = x_2(x_2 + 1) \quad \ldots(1)$$

From equation (1), we observe that $x_1 \neq x_2$. For example if $x_1 = -3$ and $x_2 = 2$, then $f(x_1) = 6 = f(x_2)$ but $x_1 \neq x_2$. Hence f is not one-one. Further, it is also not onto.

Example 4.24

Classify the following functions from R to R as surjective (onto), injective (one-one) or bijective (one-one, onto).

1. $f(x) = |x|$
2. $f(x) = e^x$
3. $f(x) = \log x$
4. $f(x) = \tan x$.

Solution

1. We know that

$$f(x) = |x| = \begin{cases} x, & \text{if } x \geq 0 \\ -x, & \text{if } x < 0 \end{cases}$$

Now let $x_1, x_2 \in R$, such that

$$f(x_1) = f(x_2) \Rightarrow |x_1| = |x_2|$$

But $|x_1| = |x_2|$ does not implies that $x_1 = x_2$. For example, $-3, 3 \in R$ and $|-3| = 3 = |3|$ but $-3 \neq 3$. Hence f is not one–one. Further, f is not onto, because there are many elements in the co-domain which do not have their pre-images in the domain. For example if -2 is in the co-domain then there is no element in the domain whose image is -2. Hence f is not onto.

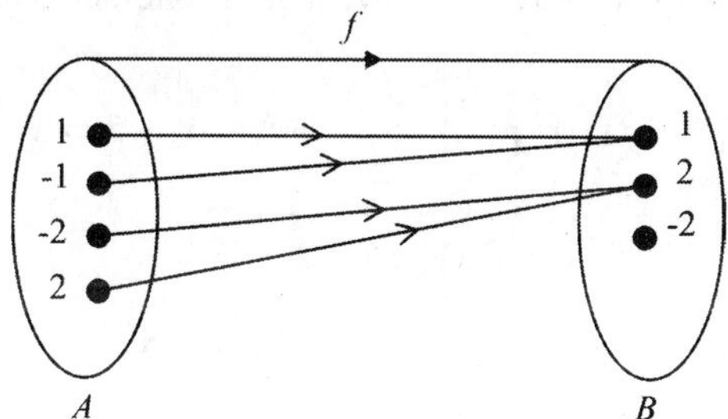

Fig. 4.15

2. Let $x_1, x_2 \in R$, such that

$$f(x_1) = f(x_2) \Rightarrow e^{x_1} = e^{x_2}$$

Taking log of both sides, we have

$$\log_e e^{x_1} = \log_e e^{x_2} \Rightarrow x_1 \log_e e = x_2 \log_e e \Rightarrow x_1 = x_2$$

Hence f is one-one.

Further let $y = e^x \Rightarrow x = \log y$. f is not onto, because logarithm of a negative real number is not a real number. Thus there are many elements in the co-domain which do not have their pre-images in the domain.

3. f is not a mapping from R to R, since the logarithm of a negative real number is not a real number.

4. f is not a mapping from R to R since $\frac{\pi}{2} \in R$ and $\tan\frac{\pi}{2} = \infty$ which is not a real number.

Example 4.25

Let $f: R \to R$ is a mapping defined by $f(x) = \sin x$. Find

i. the image set of R under f and hence conclude whether f is a surjective.

ii. $\{x: x \in R, f(x) = \frac{1}{2}\}$ and hence conclude whether f is an injective.

Solution

i. Since $-1 \le \sin x \le 1 \ \forall\ x \in R$ and $\sin x$ takes all values between -1 and 1 for some $x \in R$. Hence the image set of R under

$$f = f(R) = \{x \in R: -1 \le x \le 1\}$$

Since $f(R)$ is not equal to the co-domain R of f, hence f is not onto.

ii. Since $f(x) = \frac{1}{2}$

Thus $\Rightarrow \sin x = \frac{1}{2} \Rightarrow x = n\pi + (-1)^n \frac{\pi}{6}; n \in I$

$$\left\{x : x \in R, f(x) = \frac{1}{2}\right\} = \left\{n\pi + (-1)^n \frac{\pi}{6} : n \in I\right\}$$

$\Rightarrow \frac{1}{2}$ is the f-image of many elements of R. Hence is f not one-one.

Example 4.26

If $f: R \to R$ is a function such that $f(x) = 3x + 5$. Prove that f is one-one onto. Also find the inverse of f.

Solution

Let $x_1, x_2 \in R$ such that $f(x_1) = f(x_2) \Rightarrow 3x_1 + 5 = 3x_2 + 5 \Rightarrow x_1 = x_2$

$\therefore f$ is one-one.

Again if $y \in R$, then $y = f(x) \Rightarrow y = 3x + 5$

$$\Rightarrow x = \frac{y-5}{3}$$

But $y \in R \Rightarrow \frac{y-5}{3} \in R$.

Thus $y \in R \Rightarrow \exists x = \frac{1}{3}(y-5) \in R$

Such that $f(x) = f\left(\frac{1}{3}(y-5)\right) = 3\left(\frac{1}{3}(y-5)\right) + 5$

$$= y - 5 + 5 = y$$

Therefore f is onto.

Hence f is one-one onto.

Now to find the inverse, let y be the f-image of x, then $f^{-1}: R \to R$ such that

$$f^{-1}(y) = x \Leftrightarrow y = f(x)$$

But $f(x) = 3x + 5$

$\therefore$ $y = 3x + 5$

or $x = \frac{1}{3}(y-5)$

$$f^{-1}(y) = x = \frac{1}{3}(y-5) \quad \forall y \in R$$

Hence $f^{-1}: R \to R$ is defined by

$$f^{-1}(x) = \frac{1}{3}(x-5)$$

Example 4.27

If A and B are non-empty sets, prove that there exists a one-to-one correspondence between $A \times B$ and $B \times A$.

Solution

If $a \in A$ and $b \in B$ then the ordered pair $(a, b) \in A \times B$ and $(b, a) \in B \times A$. Let $f: A \times B \to B \times A$ be a function defined by the rule

$$f(a, b) = (b, a) \ \forall \ (a, b) \in A \times B$$

Here $f(a, b)$ is the image of $(a, b) \in A \times B$ under the function f. Clearly, f is well-defined.

i. f is one-one

Let $(a, b), (c, d) \in A \times B$ such that $f(a, b) = f(c, d)$

$\Rightarrow (b, a) = (d, c)$

$\Rightarrow b = d, a = c$

$\Rightarrow (a, b) = (c, d)$

$\Rightarrow f$ is one-one.

ii. f is onto

Let $(x, y) \in B \times A$ such that $x \in B$ and $y \in A$. Then $(y, x) \in A \times B$.

Now $f(y, x) = (x, y)$

Thus $(x, y) \in B \times A \Rightarrow$ there exists $(y, x) \in A \times B$ such that $f(y, x) = (x, y)$.

Hence f is onto.

Therefore f is a one-to-one correspondence between $A \times B$ to $B \times A$.

Exercise 4.2

1. Classify the following functions as one-one, onto or both.
   i. $f: C \to R; f(z) = |z|$, where C and R denote set of complex and real numbers respectively.
   ii. $f: R \to R; f(x) = ax + b, a, b \subset R, a \neq 0$.
   iii. $f: R \to R; f(x) = \cos x$
   iv. $f: R \to [-1, 1]; f(x) = \sin x$
   v. $f: [0, \pi] \to R; f(x) = \cos x$.
2. Which of the following functions from A to B are one-one and onto?
   i. $\{(1, 3), (2, 5), (3, 7)\}$;
   $A = \{1, 2, 3\}$; $B = \{3, 5, 7\}$.
   ii. $\{(2, a), (3, b), (4, c)\}$;
   $A = \{2, 3, 4\}$; $B = \{a, b, c\}$.
3. Find the maximum domain and the range of the following functions.
   i. $f(x) = 1 - |x|$
   ii. $f(x) = \dfrac{1}{1 - x^2}$
   iii. $f(x) = x^2$ where $x \in R$.

4. Let $A = \{x : x \in N \text{ and } x \text{ is odd}\}$ and $B = \{x : x \in N \text{ and } x \text{ is even}\}$
   A function $f : A \to B$ is defined as $f(x) = x + 1$. Prove that f is one-one onto.
5. Let R_0 denote the set of all non-zero real numbers. Prove that the mapping $f = R_0 \to R_0$ given by $f(x) = \frac{1}{x}\ \forall x \in R_0$ is one-one onto.
6. Find the range of the following function on R.
   i. $f(x) = x^3$ ii. $f(x) = \sin x$
   iii. $f(x) = x^2 + 1$ iv. $f(x) = 1 - \sin x$
   v. $f(x) = 1 - |\sin x|$ vi. $f(x) = |1 - \sin x|$.
7. If $A = \{0, 1\}$, show that the function $f : N \to A$ defined by $f(2x) = 0$ and $f(2x - 1) = 1$; $\forall x \in N$, is many one, onto function.
8. Classify the following functions as surjective, injective, bijective or otherwise; give reasons:
   i. $f : R \to R; f(x) = x^3 - x$ ii. $f : Q - \{3\} \to Q;\ f(x) = \frac{2x+3}{x-3}$
   iii. $f : R_0 \to R^+; f(x) = x^2$ iv. $f : R \to R; f(x) = 5x^3 - 1$
   v. $f : R \to R; f(x) = e^x + e^{-x}$ vi. $f : R \to R; f(x) = \sin^2 x + \cos^2 x$.
9. Given that $A = R - \{3\}$ and $B = R - \{1\}$, where R is the set of real numbers and $f : A \to B$ is defined by $\frac{x-2}{x-3}\ \forall x \in A$. Prove that f is one-one and onto.
10. Is the mapping $f : R \to R - \{0\}$ defined by $\begin{bmatrix} a & b \\ c & d \end{bmatrix} = ad - bc$ bijective?
11. Let $f : R \to R$ be defined by $f(x) = 3x + 4\ \forall x \in R$. Show that f is one- one onto. Also give formula for f^{-1}.
12. Find $f^{-1}(13), f^{-1}(-5)$ when $f : R \to R$ such that $f(x) = x^2 - 3$.
13. Find f^{-1} if $f = \{(a, 2), (b, 4), (c, 6)\}$.
14. If $f : R \to R$ such that $f(x) = x^3 - 3$ find f^{-1}.
15. If $f : R \to R$ is a function such that $f(x) = x^3 + 1$. Is it a bijection or not? Prove your statement. In case it is a bijection, find $f^{-1}(0)$.
16. Let $A = \{-1, 1\}$ and f and g be two functions defined in A as follows $f(x) = x^2$; $g(x) = \sin(\frac{\pi}{2}x)$. Show that g^{-1} exists but f^{-1} does not exist.

Answers

1. i. neither one-one nor onto ii. one-one onto
   iii. neither one-one nor onto iv. not one-one, onto.
   v. one-one, not onto.
2. i. One-one onto ii. One-one onto.
6. i. R ii. $[-1, 1]$
   iii. $[1, \infty]$ iv. $[0, 2]$
   v. $[0, 1]$ vi. $[0, 2]$.
3. i. domain = R, Range = $]-\infty, 1[$ ii. domain = $R - [-1, 1]$, Range = $]-\infty, 0[\cup [1, \infty[$
   iii. domain = R, range = $[0, \infty[$.

8. i neither surjection nor injection
   ii. Injection
   iii. Surjection
   iv. Bijection
   v. Injection
   vi. Neither surjection nor injection.

12. 166, 22

13. $f^{-1} = \{(2, a), (4, b), (6, c)\}$.

14. $f^{-1} = (y + 3)^{1/3}$.

4.11 Product or Composite of Functions

Let X, Y, Z be any three non-empty sets. Let $f : X \to Y$ and $g: Y \to Z$ be two functions. Here the domain of function g is the co-domain of function f. Then the function $g \circ f : X \to Z$ defined by $(g \circ f)(x) = g(f(x))\ \forall x \in X$ is said to be the **composite** (or **product**) of the two functions f and g.

The graphical representation of product of two functions f and g is given by

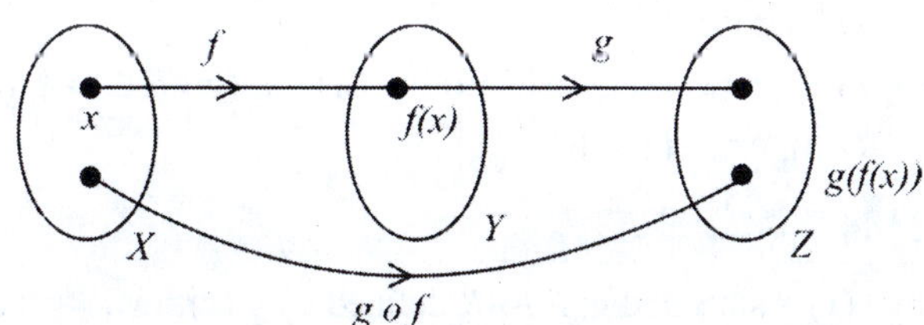

Fig. 4.16

Clearly the image of every $x \in X$ under $g \circ f$ is the g-image of the f-image of x.

The reader should note that it is traditional to write the composition of functions from right to left. Therefore, in the above definition, the first function performed in computing $g \circ f$, which is also written as gf is f. On the other hand, for relations, the composition rs is read from left to right so that first relation is r.

Example 4.28

Let $X = \{1, 2, 3, 4\}$, $Y = \{a, b, c, d, e\}$, $Z = \{u, v, w\}$ and $f : X \to Y$ and $g : Y \to Z$ defined by $f = \{(1, b), (2, a), (3, d), (4, e)\}$ and $g = \{(a, u), (b, u), (c, v), (d, w), (e, v)\}$.

Then $g \circ f: X \to Z$ becomes

$$(g \circ f)(1) = g(f(1)) = g(b) = u$$

$$(g \circ f)(2) = g(f(2)) = g(a) = u$$

$$(g \circ f)(3) = g(f(3)) = g(d) = w$$

$$(g \circ f)(4) = g(f(4)) = g(e) = v$$

Thus $g \circ f : X \to Z$ is given by $g \circ f = \{(1, u), (2, u), (3, w), (4, v)\}$.

Example 4.29

Let $f : R \to R$ defined by $f(x) = 3x + 4$ and $g: R \to R$ defined by $g(x) = x^2$ then $g \circ f: R \to R$ is defined as $(g \circ f)(x) = g(f(x)) = g(3x + 4) = (3x + 4)^2$

Also $f \circ g: R \to R$ is defined as $(f \circ g)(x) = f(g(x)) = f(x^2) = 3x^2 + 4$.

Here $f \circ g$ and $g \circ f$ both are defined but $f \circ g \neq g \circ f$.

Example 4.30

Let $f: X \to Y$ and $g: Y \to Z$ be represented by the following diagram

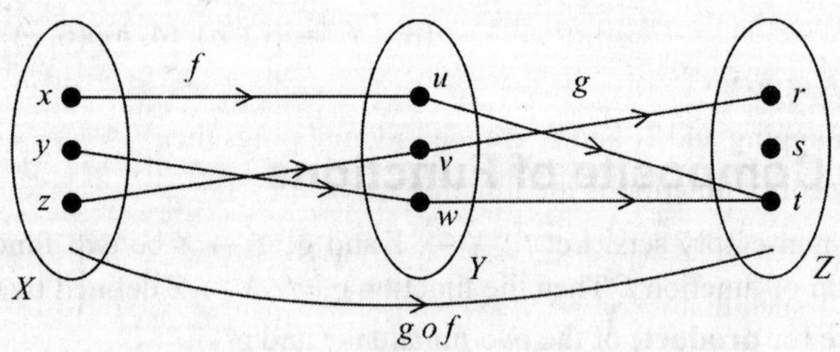

Fig. 4.17

By the definition of $g \circ f$, we find that

$$(g \circ f)(x) = g(f(x)) = g(u) = t, \; (g \circ f)(y) = g(f(y)) = g(w) = t, \; (g \circ f)(z) = g(f(z)) = g(v) = r$$

Thus $\qquad g \circ f = \{(x, t), (y, t), (z, r)\}$.

Example 4.31

Let $f: R \to R$ defined by $f(x) = \sin x$ and $g: R \to R$ defined by $g(x) = x^3$.

Then $\qquad g \circ f: R \to R$ is given by $(g \circ f)(x) = g(f(x)) = g(\sin x) = \sin^3 x$

Also $\qquad f \circ g: R \to R$ is given by $(f \circ g)(x) = f(g(x)) = f(x^3) = \sin x^3$.

Example 4.32

Find $g \circ f$ and $f \circ g$ when $f: R \to R$ and $g: R \to R$ defined as:

$$f(x) = 2x + 3 \text{ and } g(x) = x^2 + 5.$$

Solution

Since $f: R \to R$ and $g: R \to R$, then obviously $g \circ f: R \to R$ and $f \circ g: R \to R$.

also $(g \circ f)(x) = g(f(x)) = g(2x + 3) = (2x + 3)^2 + 5 = 4x^2 + 12x + 14$

and $(f \circ g)(x) = f(g(x)) = f(x^2 + 5) = 2(x^2 + 5) + 3 = 2x^2 + 13$

Clearly $(g \circ f)(x) \neq (f \circ g)(x)$.

4.11-1 Some Properties of Composite of Functions

Theorem 4.3

If $f: X \to Y$ is a one-one and onto mapping and $g: Y \to X$ is the inverse of f, then $f \circ g = I_Y$ and $g \circ f = I_X$, where I_X and I_Y are the identity functions on the sets X and Y respectively.

Proof

We shall first prove that $g \circ f = I_X$.

Since $f: X \to Y$ and $g: Y \to X$, therefore $(g \circ f): X \to X$. Let $x \in X$ and $y \in Y$ such that $f(x) = y$, then by definition of inverse mapping $g(y) = x$. Now $(g \circ f)(x) = g(f(x)) = g(y) = x$.

Thus the mapping $g \circ f$ maps every element $x \in X$ onto itself. Hence $g \circ f = I_X$.

Now we shall show that $f \circ g = I_Y$.

Since $g: Y \to X$ and $f: X \to Y$ then $f \circ g : Y \to Y$. Let $y \in Y$ and let $g(y) = x\ \forall x \in X$, then $f(x) = y$. Now, we have $(f \circ g)(y) = f(g(y)) = f(x) = y$. Therefore the mapping $f \circ g$ maps every element $y \in Y$ onto itself. Thus $f \circ g = I_Y$

Theorem 4.4

If $f: X \to Y$ be a mapping and I_X and I_Y are identity mappings then $I_Y \circ f = f$ and $f \circ I_X = f$ i.e. the product of any function with the identity function is the function itself.

Proof

Let $x \in X$ and $y \in Y$ such that $f(x) = y$.

Now since $f: X \to Y$ and $I_Y = Y \to Y \Rightarrow f(I_Y \circ f) : X \to Y$

We have $(I_Y \circ f)(x) = I_Y(f(x)) = I_Y(y) = y$

Thus for all $x \in X$, $(I_Y \circ f)(x) = f(x)$. Therefore $I_Y \circ f = f$.

Also since $I_X = X \to X$ and $f: X \to Y \Rightarrow f \circ I_X : X \to Y$

We have $(f \circ I_X)x = f(I_X(x)) = f(x) = y$

Thus for all $x \in X$, $(f \circ I_X)(x) = f(x) \Rightarrow f \circ I_X = f$.

Theorem 4.5

If $f: A \to B$ and $g: B \to C$ are two one-one and onto mappings, then $g \circ f: A \to C$ is one-one onto and $(g \circ f)^{-1} = f^{-1} \circ g^{-1}$.

Proof

To prove the theorem, we shall prove that

i. $g \circ f$ is one-one

ii. $g \circ f$ is onto

iii. $(g \circ f)^{-1} = f^{-1} \circ g^{-1}$.

(i) $g \circ f$ is one-one

Let $x_1, x_2 \in A$ such that $(g \circ f)(x_1) = (g \circ f)(x_2) \Rightarrow g(f(x_1)) = g(f(x_2))$.
But g is one-one, therefore $f(x_1) = f(x_2)$. Also f is one-one, thus $x_1 = x_2$
$\therefore$ The mapping $g \circ f$ is one-one.

(ii) $g \circ f$ is onto

Let $z \in C$. Since g is an onto mapping of B to C, therefore there exists $y \in B$, such that $g(y) = z$. Again, since f is an onto mapping of A to B, therefore there exists $x \in A$ such that $f(x) = y$. Hence for every $z \in C$ there exist $x \in A$ such that $(g \circ f)(x) = g(f(x)) = g(y) = z$.

Thus $g \circ f: A \to C$ is onto.

(iii) $(g \circ f)^{-1} = f^{-1} \circ g^{-1}$

Since $g \circ f$ is one-one onto mapping, hence $g \circ f$ is invertible. Let $z \in C$ such that $z = g(y)$ for all $y \in B$, then $y = g^{-1}(z)$. Also let $y = f(x) \Rightarrow x = f^{-1}(y)$

Now $(g \circ f)(x) = g(f(x)) = g(y) = z$

$\therefore$ $(g \circ f)^{-1}(z) = x$ ...(1)

Also $(f^{-1} \circ g^{-1})(z) = f^{-1}(g^{-1}(z)) = f^{-1}(y) = x$

$\therefore \qquad (f^{-1} \circ g^{-1})(z) = x \qquad \ldots(2)$

From (1) and (2), we observe that $(g \circ f)^{-1} = f^{-1} \circ g^{-1}$

Theorem 4.6

The product of functions, i.e. $g \circ f$ is not commutative.

Proof

Let $f: X \to Y$ and $g: Y \to Z$ be any two functions. Here since the co–domain of f is equal to the domain of g, therefore $g \circ f$ exists. But $f \circ g$ does not exist. But if $X = Z$, then $f: X \to Y$ and $g: Y \to X$. In this case both $g \circ f: X \to X$ and $f \circ g: Y \to Y$ exists.

When $X \neq Y$, then $g \circ f$ and $f \circ g$ are not equal. However if $X = Y = Z$, then both $g \circ f$ and $f \circ g$ exist and both are from X to X, even then they may not be equal. Hence in general the product of functions is not commutative.

Theorem 4.7

If f, g and h are three functions such that $f \circ (g \circ h)$ and $(g \circ f) \circ h$ are defined, then $f \circ (g \circ h) = (f \circ g) \circ h$, i.e. the composite of functions is associative.

Proof

Let f, g and h be three functions such that $h : X \to Y$, $g : Y \to Z$, $f: Z \to W$, where X, Y, Z and W are any four non-empty sets. It can be easily seen that both $(f \circ g) \circ h$ and $f \circ (g \circ h)$ are mappings from X to W, i.e.

$$f \circ (g \circ h) : X \to W \text{ and } (f \circ g) \circ h : X \to W$$

To show that $[f \circ (g \circ h)]\,(x) = [(f \circ g) \circ h]\,(x)\ \forall x \in X$.

Let $x \in X$, $y \in Y$, $z \in Z$ such that $h(x) = y$ and $g(y) = z$.

Then
$$\begin{aligned}[f \circ (g \circ h)]\,(x) &= f\,[(g \circ h)(x)]\\ &= f\,[g(h\,(x))]\\ &= f\,[g(y)] = f\,(z)\end{aligned}$$

$\therefore \qquad [f \circ (g \circ h)]\,(x) = f\,(z) \qquad \ldots(1)$

Again
$$\begin{aligned}[(f \circ g) \circ h]\,(x) &= (f \circ g)\,[h\,(x)]\\ &= (f \circ g)\,(y)\\ &= f\,[g\,(y)] = f\,(z)\end{aligned}$$

$\therefore \qquad [(f \circ g)] \circ h]\,(x) = f\,(z) \qquad \ldots(2)$

From (1) and (2), we observe that

$$[f \circ (g \circ h)]\,(x) = [(f \circ g) \circ h\,]\,(x)\ \forall x \in X$$

$$\Rightarrow f \circ (g \circ h) = (f \circ g) \circ h.$$

Hence the composite of functions is associative.

Example 4.33

Verify associativity for the following three mappings:

$f: R \to R$ such that $f(x) = 2x$, $g: R \to R$ such that $g(x) = 1/(2 - x)$,

$h: R \to R$ such that $h(x) = x^2$.

Solution

Obviously $f \circ (g \circ h)$: $R \to R$ and $(f \circ g) \circ h$: $R \to R$ are defined. Now for the verification of associativity we have to show that $[f \circ (g \circ h)]\,(x) = [(f \circ g) \circ h\,]\,(x)\ \forall\ x \in R$.

L.H.S.= $[f \circ (g \circ h)](x) = f[(g \circ h)(x)]$

$= f[g(h(x))] = f[g(x^2)]$

$f(1/(2 \quad x^2)) = 2(1/(2 \quad x^2)) = 2/(2 \quad x^2)$ ...(1)

Also $[(f \circ g) \circ h]x = (f \circ g)[h(x)]$

$= (f \circ g)(x^2) = f[g(x^2)]$

$= f(1/(2 - x^2)) = 2/(2 - x^2)$ ...(2)

From (1) and (2), we observe that

$$[f \circ (g \circ h)](x) = [(f \circ g) \circ h](x) \ \forall\, x \in R$$

Hence composite of mapping is associative.

Example 4.34

If $f: Q \to Q$ such that $f(x) = 2x$ and $g: Q \to Q$ such that $g(x) = x + 2$ be two mappings, then verify that $(g \circ f)^{-1} = f^{-1} \circ g^{-1}$, where Q is a set of rational numbers.

Solution

Since f and g are one-one and onto, then f^{-1} and g^{-1} exist and defined as $f^{-1}: Q \to Q$ such that $f^{-1}(x) = \frac{x}{2}$ and $g^{-1}: Q \to Q$ such that $g^{-1}(x) = x - 2$. As we know that the product of two one-one onto mappings is one-one onto, therefore $g \circ f: Q \to Q$ is one-one onto and

$(g \circ f)(x) = g(f(x)) = g(2x) = 2x + 2$

Since gof is one-one onto therefore $(g \circ f)^{-1}$ exists and defined as

$(g \circ f)^{-1}: Q \to Q$ such that $(g \circ f)^{-1}(x) = (x - 2)/2$ ...(1)

Now $(f^{-1} \circ g^{-1}): Q \to Q$, where

$(f^{-1} \circ g^{-1})(x) = f^{-1}(g^{-1}(x)) = f^{-1}(x - 2) = (x - 2)/2$ ...(2)

From (1) and (2), we observe that

$$(f^{-1} \circ g^{-1})(x) = (g \circ f)^{-1}(x) \ \forall x \in Q$$

$$\therefore \quad f^{-1} \circ g^{-1} = (g \circ f)^{-1}$$

Example 4.35

Let f, g and $h: R \to R$ be defined by (R is the set of real numbers)

$f(x) = x + 2$, $g(x) = 1/(x^2 + 1)$, $h(x) = 3$. Compute

i. $f^{-1} \circ g(x)$

ii. $h \circ f \circ (g \circ f^{-1}) \circ (h \circ f(x))$.

Solution

i. $f: R \to R$ is defined as $f(x) = x + 2 \Rightarrow f^{-1}(x) = x - 2$

$$\text{Now } (f^{-1} \circ g)x = f^{-1}\left[\frac{1}{x^2+1}\right] = \frac{1}{x^2+1} - 2 = \frac{1 - 2x^2 - 2}{x^2+1} = \frac{-2x^2 - 1}{x^2+1}$$

ii. $h \circ f \circ (g \circ f^{-1}) \circ (h \circ f(x))$

$= h \circ f \circ (g \circ f^{-1})(h(x + 2))$

$$= h \circ f \circ (g \circ f^{-1})3$$
$$= h \circ f \circ (g \circ f^{-1}(3))$$
$$= h \circ f \circ g(1)$$
$$= h \circ f(1/2)$$
$$= h(5/2)$$
$$= 3.$$

Example 4.36

Consider $X = \{a, b, c\}$, and define $f: X \to X$ such that $f = \{(a, b), (b, a), (c, c)\}$.
Find (i) f^{-1} (ii) f^2 (iii) f^3 (iv) f^4.

Solution

i. $f^{-1} = \{(b, a), (a, b), (c, c)\} = f$

ii. $f^2 = f \circ f = \{(b, a), (a, b), (c, c)\} \circ (b, a), (a, b), (c, c)$
$= \{(b, b), (a, a), (c, c), \}$

iii. $f^3 = f^2 \circ f = \{(b, b), (a, a), (c, c)\} \circ \{(b, a), (a, b), (c, c)\}$
$= \{(b, a), (a, b), (c, c)\} = f$

iv. $f^4 = f^3 \circ f = \{(b, a), (a, b), (c, c)\} \circ \{(b, a), (a, b), (c, c)\}$
$= \{(b, b), (a, a), (c, c)\} = f^2$.

Example 4.37

Find $g \circ f$ and $f \circ g$ when $f: R \to R$ and $g: R \to R$ defined as : $f(x) = 2x + 3$ and $g(x) = x^3 + 5$.

Solution

Since $f: R \to R$ and $g: R \to R$, then obviously $g \circ f: R \to R$ and $f \circ g: R \to R$.

Also $(g \circ f)(x) = g(f(x)) = g(2x+3) = (2x+3)^3 + 5$

and $(f \circ g)(x) = f(g(x)) = f(x^3+5) = 2(x^3+5) + 5$

thus $(g \circ f)(x) \neq (f \circ g)(x)$.

EXERCISE 4.3

1. Find $g \circ f$ and $f \circ g$ when $f: R \to R$ and $g: R \to R$ defined as
   i. $f(x) = \sin x$ and $g(x) = x^2$ ii. $f(x) = x^2$ and $g(x) = x^3$
   and check whether $g \circ f$ and $f \circ g$ are equal or not.
2. If $f: R^+ \to R^+$ and $g: R^+ \to R^+$ defined as $f(x) = x^2$ and $g(x) = \sqrt{x}$.
   Find $g \circ f$ and $f \circ g$. Are they equal mappings?
3. Find $(f \circ g)(3)$ and $(g \circ f)(2)$ when
   $f: R \to R; f(x) = x^2 + 5$ and $g: R \to R; g(x) = 3x^4 + 2$.
4. Let $A = \{1, 2, 3\}$; $B = \{a, b, c\}$ and let f and g be two mappings from A to B and from B to A defined as $f(1) = b, f(2) = a, f(3) = c, g(a) = 2, g(b) = 1, g(c) = 3$.
   Show that f and g both are one-one and onto and find $f \circ g$ and $g \circ f$.

5. Given f and g as follows, determine $(g \circ f)(x)$. State the domain in each case:

   i. $f: R \to R$ such that $f(x) = 2x + \dfrac{1}{x^2}$

   $g: R \to R$ such that $g(y) = y^4 + 2y + 4$

   ii. $f: R \to R$ such that $f(x) = \dfrac{1}{2-x}$

   $g: R \to R$ such that $g(y) = \dfrac{1}{1+y}$.

6. Find $g \circ f$ and $f \circ g$ when $f: R \to R$ and $g: R \to R$ defined by

   i. $f(x) = 2x + 1$ and $g(x) = x^2 - 2$

   ii. $f(x) = \cos x$ and $g(x) = x^3$.

7. If f and g are two mappings from R to R given by
   $f(x) = x^2 + 3x + 1$, $g(x) = 2x - 3$, then obtain $f \circ g$ and $g \circ f$.

8. If $f: R \to R$ be defined by $f(x) = |x|$, then prove that $f \circ f = f$.

9. Let $A = \{1, 2, 3, 4\}$; $B = \{3, 5, 7, 9\}$; $C = \{7, 23, 47, 79\}$ and $f: A \to B$; $f(x) = 2x + 1$, $g: B \to C$; $g(x) = x^2 - 2$. Express $(g \circ f)^{-1}$ and $f^{-1} \circ g^{-1}$ as the sets of ordered pairs and verify that $(g \circ f)^{-1} = f^{-1} \circ g^{-1}$.

Answers

1. i. $g \circ f = \sin^2 x, f \circ g = \sin x^2$ ii. $(f \circ g)x = (g \circ f)x = x$.

2. $g \circ f = x, f \circ g = x$, yes. 3. $(f \circ g)(3) = 60030, (g \circ f)(2) = 19685$.

4. $f^\circ g = \{(a, a), (b, b), (c, c)\}, g \circ f = \{(1, 1), (2, 2), (3, 3)\}$.

5. i. $(g \circ f)(x) = (2x + x^{-2})^4 + 2(2x + x^{-2}) + 4$, domain $= R$

   ii. $(g \circ f)(x) = \dfrac{2-x}{3-x}$, domain $= R$.

6. i. $(g \circ f)(x) = 4x^2 + 4x - 1, (f \circ g)(x) = 2x^2 - 3$

   ii. $(g \circ f)(x) = \cos^3 x, (f \circ g)(x) = \cos x^3$.

7. $(f \circ g)(x) = 4x^2 - 6x + 1, (g \circ f)(x) = 2x^2 + 6x - 1$.

   $(g \circ f)^{-1} = f^{-1} \circ g^{-1} = \{(7, 1), (23, 2), (47, 3), (79, 4)\}$

4.12 Recursive Definitions

There are many occasions where an object can't be defined explicitly. However, it may be easy to define the object in terms of itself. This process is called **recursion**. We can use recursion to define sequences, functions and sets. For instance, the sequence of powers of 5 is given by $a_n = 5^n$ for $n = 0, 1, 2 \ldots$. But this sequence can also be defined by: $a_{n+1} = 5a_n$, where $a_0 = 1$ and $n = 0, 1, 2, \ldots$.

Let us take some other functions, which play an important role in recursion.

4.12-1 Partial Function

A **partial function** is a relation that associates each element of a set (domain) with at most one element of a set (codomain). This means some elements of the domain may not be associated with any element of the codomain.

4.12-2 Total Function

If a partial function $f: X \rightarrow Y$ associates to every element in X, an element of Y, then f is called a total function. For example see the following diagram.

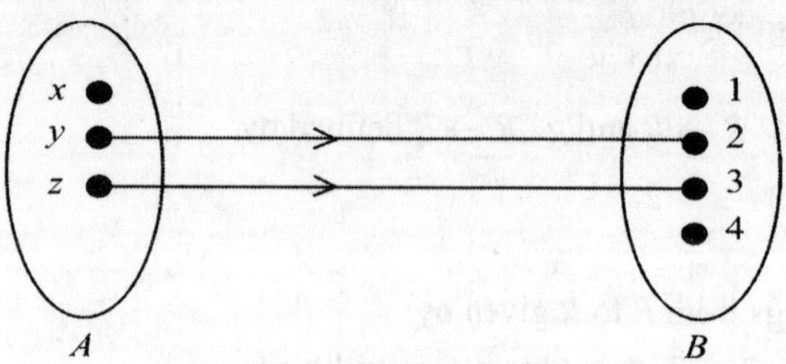

Fig. 4.18

This represents a partial function that is not total since the element x in A is not associated with any element of B.

Let us take another case and consider only those functions whose arguments and values are natural numbers, known as **number- theoretic function**. Consider number theoretic functions of n-variables denoted by $f(x_1, x_2, \ldots, x_n)$.

Any function $f: N^n \rightarrow N$ is called **total function** because it is defined for every n- tuple in N^n.

If $f: D \rightarrow N$, where $D \subseteq N^n$, then f is called **partial function**. For example,

1. $f(x, y) = x + y$, which is defined $\forall\, x, y \in N$ and hence is a total function.
2. $g(x, y) = x - y$, which is defined for only those $x, y \in N$ which satisfy $x \geq y$. Hence $g(x, y)$ is a partial function.

4.12-3 Recursively Defined Functions

To define a function with the set of nonnegative integers as its domain,

1. Mention the value of the function at zero.
2. Give a rule for finding its value as an integer from its values at smaller integers.

Such definition is called a **recursive** or **inductive definition.**

Example 4.38

Let the function f is defined by $f(0) = 5, f(n + 1) = 2f(n) + 5$. Find $f(1), f(2), f(3)$, and $f(4)$.

Solution

From the recursive definition

$$f(1) = 2f(0) + 5 = 2.5 + 5 = 15$$
$$f(2) = 2f(1) + 5 = 2.15 + 5 = 35$$
$$f(3) = 2f(2) + 5 = 2.35 + 5 = 75$$
$$f(4) = 2f(3) + 5 = 2.75 + 5 = 155.$$

Example 4.39

Give recursive definition of a^n, where a is non-zero real number and n is nonnegative integer.

Solution

The recursive definition contains two parts. First specify a^0, i.e. $a^0=1$. Then $a^{n+1}= a.a^n$ for $n = 0, 1, 2, 3, \ldots$. These two equations uniquely defines a^n for all non-negative integer n.

Example 4.40

The sequence is defined by $F_n = F_{n-1} + F_{n-2}$, where $F_0 = 0$, $F_1 = 1$, for $n \geq 2$. Find $F_2, F_3, F_4, F_5, \ldots$.

Solution

From the second part of the definition:

$$F_2 = F_1 + F_0 = 1 + 0 = 1$$
$$F_3 = F_2 + F_1 = 1 + 1 = 2$$
$$F_4 = F_3 + F_2 = 2 + 1 = 3$$
$$F_5 = F_4 + F_3 = 3 + 2 = 5.$$

Example 4.41

Give an inductive definition of the factorial notation $f(n) = n!$.

Solution

Factorial notation can be defined by specifying the initial value namely $f(0) = 1$. And giving a rule for finding $f(n + 1)$ from $f(n)$ we can compute this from the fact that $(n+1)!$ is computed from $n!$ by multiplying $n + 1$. Therefore $f(n + 1) = (n + 1) f(n)$.

4.13 Recursive Functions

The recursive functions, which form a class of computable functions, take their name form process of 'recurrence' or 'recursion'. The recursion is a process of defining the value of a function by using the function itself. There are many occasions where an object can't be defined explicitly but can be defined recursively. We can use recursion to define sequences, functions and sets. For instance, the sequence of powers of 5 is given by $a_n = 5^n$ for $n = 0, 1, 2, \ldots$,. But this sequence can also be defined by $a_{n+1} = 5a_n$, where $a_0 = 1$ and $n = 0, 1, 2, \ldots$.

In this section, we provide an account of the class of recursive functions, with particular emphasis on three basic kinds of recursion:

1. The initial function
2. Iteration
3. Primitive Recursion.

4.14 Types of Recursion

4.14-1 The Initial Function

The initial functions are a certain class of functions (need to be designed) whose values do not in turn depend on the values for smaller arguments. The initial functions are all functions of the natural numbers. Hence, they may take zero as input, but not a negative number and not any rational or irrational number. The class of the initial functions consists of:

Zero Function

The constant function Z, which returns 0 for any argument n, i.e. $Z(n) = 0$ is called a **zero function**.

Successor Function

A function which when given an argument n returns its immediate successor is called a **successor function**. i.e. $S(n) = n + 1 =$ Successor of n

The zero and successor functions take only one argument each at a time. One should note here that an identity function is designed to take any number of arguments. When it takes one argument it returns its argument as its value. When it takes more than one argument, it returns one of them.

Predecessor function

The predecessor function is defined by

$$P(n) = \begin{cases} 0, \text{ if } n = 0 \text{ and } n = 1 \\ P(n-1) + 1, \text{ otherwise} \end{cases}$$

Immediately it can not be derived that the predecessor function can be reduced to an iteration. Basically, $p(n)$ is the second component of the nth iteration of the function of pairs defined as

$$f(x, y) = (x + 1, x)$$

started on (0, 0).

Projection Function

The projection function is also called the generalized identity function, is defined as

$$U_i^n : U_i^n(x_1, x_2, ..., x_n) = x_i,$$

where subscript i denotes that result is ith argument and superscript shows the total number of arguments. e.g. $U_1^2(x, y) = x$, here subscript 1 denotes that result is first argument and superscript shows the total number of arguments.

Similarly, $U_2^3(2,4,6) = 4$

4.14-2 The Building Operations

We can generate other functions by combining functions by a process referred to as **composition**. Let $f_1(x, y)$, $f_2(x, y)$ and $g(x, y)$ be any three functions. The composition of g with f_1 and f_2 is a function h, given by

$$h(x, y) = g(f_1(x, y), f_2(x, y)).$$

Naturally, for h to be nonempty, it is necessary that the **domain of g** include $R_{f_1} \times R_{f_2}$ where R_{f_1} is the range of f_1 and R_{f_2} is the range of f_2. Also the domain of h is $D_{f_1} \cap D_{f_2}$, where D_{f_1} and D_{f_2} are the domains of f_1 and f_2 respectively. In general domain of g is given by $\prod_{i \in I_n} R_{f_i}$, $I_n = \{1, 2,..., n\}$, where R_{f_i} denotes the range of f_i. Domain of h is given by $\bigcap_{i \in I_n} D_{f_i}$, where D_{f_i} denotes the domain of f_i.

1. *If f_1, f_2 and g are total function, then h is also total function.*
2. *In general the function h will be total function iff $f_1, f_2, ..., f_n$ and g are total functions.*
3. *In general, let $f_1, f_2, ..., f_n$ each be partial function of m variables, and let g be a partial function of n variables. Then the composition of g with $f_1, f_2, ..., f_n$ produces a partial function h given by $h(x_1, x_2, ..., x_m) = g(f_1(x_1, x_2, ..., x_m)), ... f_n(x_1, x_2, ... x_m))$.*

Example 4.42

Let $f_1(x, y) = x + y$, $f_2(x, y) = xy + y^2$ and $g(x, y) = xy$.

Solution

Let

$$\begin{aligned} h(x, y) &= g(f_1(x, y), f_2(x, y)) \\ &= g(x + y, xy + y^2) \\ &= (x + y)(xy + y^2). \end{aligned}$$

Here h is a total function, because f_1, f_2 and g are total functions.

Example 4.43

Let $f_1(x, y) = x + y, f_2(x, y) = xy + y^2$ and $g(x, y) = xy$ be functions defined over N. Find the composition of g with f_1 and f_1.

Solution

If $\quad f_1(x, y) = x + y; f_2(x, y) = xy + y^2$ and $g(x, y) = xy$

Then $\quad g(f_1(x, y), f_2(x, y)) = g(x + y, xy + y^2) = (x + y)(xy + y^2)$

is composition of $f_1(x, y)$ and $f_2(x, y)$.

Example 4.44

Let $f_1(x, y) = x - y, f_2(x, y) = y - x$ and $g(x, y) = x + y$ be functions defined over N. f_1 is defined only when $x \geq y$ and f_2 is defined only when $y \geq x$. Find the composition of g with f_1 and f_2.

Solution

If $\quad f_1(x, y) = x - y, f_2(x, y) = y - x$ and $g(x, y) = x + y$.

Then $\quad g(f_1(x, y), f_2(x, y)) = g(x - y, y - x)$

$$= x - y + y - x$$

is the composition of g with f_1 and f_2.

4.14-3 Iteration

Iteration is the simplest form of recursion. Symbolically, the nth iteration of a function f is defined as follows:

$$f^{(0)}(x) = x$$
$$f^{(n+1)}(x) = f(f^{(n)}(x))$$

The first clause is needed to obtain $f^{(1)}(x) = f(x)$ from the second clause.
We also have $f^{(2)}(x) = f(f(x)$ etc.

1. *$f^{(n)}(x)$ is a function of two arguments, n and x, It is better to "fix" the first argument, for obtaining a fixed iteration.*
2. *For example: $S^{(2)}(x)$ = Successor of the successor of x. Such a function enumerates all the even numbers since its values are 0, 2, 4...*

Example 4.45

In each of 7 houses there are 7 cats, each cat kills 7 mice, each mouse would have eaten 7 ears of spelt (wheat), each of spelt would have produced 7 hekat of grain. How much grain is saved by the 7 house cats?

Solution

$$f(x) = 7x$$
$$f^{(6)}(7) = f(f^{(5)}(x))$$
$$= f(f(f^4(x)))$$
$$= f(f(f(f^2(x)))$$
$$= f(f(f(f(f^2(x)))$$
$$= f(f(f(f(f(f(x)))$$
$$= f(f(f(f(f(7x))))$$
$$= f(f(f(f(7(7x))))$$

$$= f(f(f(f(7\,(7x)))))$$
$$= f(f(f(7\,(7\,(7x)))))$$
$$= f(7\,(7\,(7\,(7x))))$$
$$= (7\,(7\,(7\,(7\,(7x)))))$$
$$= 16807. \qquad [\because x = 1]$$

It can also be written in terms of sixth term of G. P. with first term 1 and multiplier 7 i.e. 7^5 *and the sum of last five terms of the progression 19607 (i.e.* $7 + 7^2 + 7^3 + 7^4 + 7^5$*).*

Consider following example:

To compute the sum of first 64 terms of a G. P. with first term 1 and multiplier 2, i.e.

$$1 + 2 + 2^2 + \ldots + 2^{63} = 2^{64} - 1$$
$$= 18, 446, 744, 073, 709, 551, 615.$$

4.14-4 Primitive Recursion

A function f is called **Primitive Recursive** iff it can be obtained from the initial functions by a finite number of operations of composition and recursion.

The operations which defines a function $f(x_1, x_2, \ldots, x_n, y)$ of $n + 1$ variables by using two other functions $g(x_1, x_2, \ldots, x_n)$ and $h(x_1, x_2, \ldots, x_n, y, z)$ of n and $n + 2$ variables respectively, is called **recursion**. A function f of $n + 1$ variables is defined by recursion if there are two functions g and h of n and $n + 2$ variables such that

$$f(x_1, x_2, \ldots, x_n, 0) = g\,(x_1, x_2, \ldots, x_n),$$
$$f(x_1, x_2, \ldots, x_n, y + 1) = h\,(x_1, x_2, \ldots, x_n, y, f(x_1, x_2, \ldots, x_n, y)),$$

i.e. here f at $y + 1$ is expressed in terms of the value of f at y. The variables $x_1, x_2, \ldots, x_n$ are treated as parameters.

Generally, A function f of $n + 1$ variables is defined by recursion if there are two functions g and h of n and $n + 2$ variables such that

$$f\,(x_1, x_2, \ldots, x_n, 0) = g\,(x_1, x_2, \ldots, x_n)$$
$$f\,(x_1, x_2, \ldots, x_n, S(y)) = h\,(x_1, x_2, \ldots, x_n, y, f\,(x_1, x_2, \ldots, x_n, y)$$

Here, the variable y is assumed to be the inductive variable, which means the value of f at $S(y)$ is expressed in terms of the value of f at y. The variables $x_1, x_2, \ldots, x_n$ are treated as parameters and are assumed to remain constant throughout the definition.

Corollary: If $f(0, x) = g\,(x)$ then the primitive recursion for $f(0 + 1, x) = h\,(0, x, f\,(0, x))$.

Example 4.46

Show that the function $f(x, y) = x + y$ is primitive recursive.

Solution

We know that
$$x + (y + 1) = (x + y) + 1$$
$$f(x, y + 1) = f(x, y) + 1 = S(f(x, y))$$

also
$$f(x, 0) = x$$

Now
$$f(x, y) \text{ as } f(x, 0) = x = U_1^1(x)$$
$$f(x, y{+}1) = S(U_3^3(x, y, (x, y)))$$

Here the Base function $g(x) = U_1^1(x)$, and Inductive-step function $h(x, y, z) = S(U_3^3(x, y))$;

Example 4.47

Find the value of (2, 4) when $f(2, 0) = 2$ is given.

Solution

$$\begin{aligned} f(2, 4) &= S\,(f\,(2, 3)) \\ &= S\,(S(f(2, 2))) \\ &= S\,(S(S(f(2, 1)))) \\ &= S\,(S(S(S(f(2, 0))))) \\ &= S\,(S(S(S(2)))) \\ &= S\,(S(S(3))) \\ &= S\,(S(4)) \\ &= S\,(5) \\ &= 6. \end{aligned}$$

Example 4.48

Show that $f(x) = x^2$ are primitive recursive functions. Hence compute the value of $f\,(16)$.

Solution

Here $x^0 = 1$ for $\neq 0$.

And $x^0 = 0$ for $x = 0$

$$x^{1+1} = x * x$$

Hence $f(x) = x^2$ can be defined as

$$\begin{aligned} f(x, 0) &= 1 \\ f(x, 1 + 1) &= x * f(x_1\, 1) \\ &= U_1\,(x, y, f(x, 1) * U_3\,(x, y, f(x, 1)) \end{aligned}$$

$\because$ $f(x, y)$ is a primitive Recursive function.

Here $g(x) = 1 = f(x, 0)$

And $h\,(x, y, z) = x * f(x, 1)$

$$= f(x, 1 + 1)$$

Here

$$\begin{aligned} f(16, 0) &= 1 \\ f\,(16, 1 + 1) &= 16 * f(16, 1) \\ &= 16 * (16 * f(16,0)) \\ &= 16 * 16 * 1 \\ &= 256. \end{aligned}$$

Example 4.49

Show that function $f(x, y) = x + y$ is primitive recursive function. Hence compute the value $f(3, 4)$.

Solution

Here $f(x)$ is a function of two variables x and y which can be defined recursively as follows:

$$\begin{aligned} f(x, y + 1) &= x + (y + 1) \\ &= (x + y) + 1 = f(x, y) + 1 \end{aligned}$$

Also $f(x +0) = x$

Now we define $f(x, y)$ as

$$f(x, 0) = x$$

$$= U_1 (x)$$

$$f(x, y + 1) = f(x, y) + 1$$
$$= S (f (x, y))$$
$$= S (U_3 (x, y, f (x, y)))$$

Here $g(x) = x = f(x, 0)$

And $h(x, y, z) = S (U_3 (x, y, z))$
$$= f(x, y + 1)$$

Here, $g = U_1$ is an initial function. h is obtained from the initial function U_3 and S by composition and f by recursion using g and h. Thus f is evaluated by using composition and finite number of recursion to initial functions U_1, U_3 and S. So f is primitive recursion.

Now $f (3, 0) = 3$
$$f (3, 4) = S(f (3, 3))$$
$$= S(S (f (3, 2)))$$
$$= S(S (S (f (3, 1))))$$
$$= S(S (S (S (f (3, 0)))))$$
$$= S(S (S (S (3)))))$$
$$= S(S (S (4))))$$
$$= S(S (5))$$
$$= S(6)$$
$$= 7$$

There is no negative natural numbers hence subtraction is not defined in case the minuend is smaller than the subtrahend. So we define proper subtraction which is denoted by $\dot{-}$ and defined as follows:

$$x \dot{-} y = x - y \text{ if } x \geq y$$
$$= 0 \text{ if } x < y$$

Example 4.50

Show that the proper subtraction is primitive recursive and hence prove that 6 – 4 = 2.

Solution

Using predecessor function, which is a recursive function, we can define subtraction function as follows:

$$f (x, 0) = x = U_1(x)$$
$$f (x, y + 1) = p (f (x, y)$$
$$= g (f (x, y, f (x, y)$$

where $g (x, y + z) = p(z) = p_3 (x, y, z)$

Hence the proper subtraction function is a recursive function.

Now $f (6, 4) = p (f (6, 3))$
$$= p (p (f (6, 2)))$$
$$= p (p (f (6, 2)))$$
$$= p (p (p (f (6, 1))))$$
$$= p (p (p (p (f (6,0))))$$

$$= p(p(p(p(f(6)))))$$
$$= p(p(p(f(5))))$$
$$= p(p(4)) = p(3) = 2$$

Since $p(5) = p(4) + 1$
$$= [p(3) + 1] + 1$$
$$= [[p(2) + 1] + 1] + 1$$
$$= [[[p(1) + 1] + 1] + 1] + 1$$
$$= [[[0 + 1] + 1] + 1] + 1$$
$$= [[1 + 1] + 1] + 1$$
$$= [2 + 1] + 1$$
$$= 3 + 1 = 4$$

Similarly $p(4) = 3, p(2) = 1.$

4.15 Some More Recursive Functions

Some Definitions:

1. Sign function or non-zero test function (sg):
   sg(0) = 0, sg(y + 1) = 1
   or sg(0) = Z(0), sg(y + 1) = $S(Z(U_2^2(y, \text{sg}(y))))$.
2. **Zero test function** $(\overline{sg})$: $\overline{sg}(0) = 1$, $\overline{sg}(y + 1) = 0$.
3. Predecessor function (P):
   $$P(0) = 0, P(y + 1) = y = U_1^2(y, P(y)).$$
   where $P(0) = 0$; $P(0+1) = P(1) = 0$; $P(1 + 1) = P(2) = 1$;
   $$P(2 + 1) = P(3) = 2 \text{ and so on.}$$
4. Odd and Even Parity function (P_r):
   $$P_r(0) = 0, P_r(y + 1) = \overline{sg}(U_2^2(y, (P_r(y)))$$
   $$P_r(0) = 0; P_r(1) = 1; P_r(1+1) = P_r(2) = \overline{sg}(P_r(1)) = \overline{sg}(1) = 0$$
   $$P_r(3) = P_r(2+1) = \overline{sg}(P_r(2)) = \overline{sg}(0) = 1 \text{ and so on.}$$

4.16 Recursive Function That is Not Primitive Recursive

1. $A(0, y) = y + 1 \; \forall \; y > 0$
2. $A(x + 1, 0) = A(x, 1) \; \forall$ positive integers x
3. $A(x + 1, y + 1) = A(x, A(x + 1, y)) \; \forall$ positive integers x and y.

Example 4.51

Compute $A(1, 2)$.

Solution

$$A(1, 2) = A(0 + 1, 1 + 1)$$
$$= A(0, A(1, 1)) \quad \text{by [3]}$$
$$= A(0, A(0, A(1, 0))) \quad \text{by [3]}$$
$$= A(0, A(0, A(0, 1))) \quad \text{by [2]}$$
$$= A(0, A(0, 2)) \quad \text{by [1]}$$

$= A(0, 3)$ by [1]

$= 4.$ by [1]

Example 4.52

Compute $A(3, 1)$.

Solution

$$
\begin{aligned}
A(3, 1) &= A(2 + 1, 0 + 1) \\
&= A(2, A(3, 0)) && \text{by [3]} \\
&= A(2, A(2, 1)) && \text{by [2]} \\
&= A(2, A(1, A(2, 0))) && \text{by [3]} \\
&= A(2, A(1, A(1, 1))) && \text{by [2]} \\
&= A(2, A(1, A(0, A(1, 0)))) && \text{by [3]} \\
&= A(2, A(1, A(0, A(0, 1)))) && \text{by [2]} \\
&= A(2, A(1, A(0, 2))) && \text{by [1]} \\
&= A(2, A(1, 3)) && \ldots(1) \text{ by [1]} \\
&= A(2, A(0, A(1, 2))) && \text{by [3]} \\
&= A(2, A(0, A(0, A(0, A(1, 1))))) && \text{by [3]} \\
&= A(2, A(0, A(0, 3))) && [\text{using } A(1, 1) = 3 \text{ by [1]} \\
&= A(2, A(0, 4)) && \text{by [1]} \\
&= A(2, 5) && \text{by [1]} \\
&= A(1, A(2, 4)) && \text{by [3]} \\
&= A(1, A(1, A(2, 3))) && \text{by [3]} \\
&= A(1, A(1, A(1, A(2, 2))) \\
&= A(1, A(1, A(1, 7))) \\
&= A(1, A(1, A(0, A(1, 6)))
\end{aligned}
$$

But

$$
\begin{aligned}
A(1, 6) &= A(0, A(1, 5))) && \text{by [3]} \\
&= A(0, A(0, A(1, 4))) && \text{by [3]} \\
&= A(0, A(0, A(0, A(1, 3))) && \text{by [3]} \\
&= A(0, A(0, A(0, A(0, A(1, 2))) && \text{by [3]} \\
&= A(0, A(0, A(0, A(0, A(0, A(1, 1))) && \text{by [3]} \\
&= A(0, A(0, A(0, A(0, A(0, 3))))) && \text{by [3]} \\
&= A(0, A(0, A(0, A(0, 4))))) && \text{by [2]} \\
&= A(0, A(0, A(0, 5)))) && \text{by [2]} \\
&= A(0, A(0, 6)) \\
&= A(0, 7) \\
&= 8.
\end{aligned}
$$

Now

$$
\begin{aligned}
A(3, 1) &= A(1, A(1, A(0, 8))) \\
&= A(1, A(1, 9)) && \text{by [1]} \\
&= A(1, A(0, A(1, 8))) && \text{by [3]} \\
&= A(1, A(0, (10)) && \text{by [2]} \\
&= A(1, 11) \\
&= 13.
\end{aligned}
$$

4.17 Some Special functions

(1) Floor and Ceiling Function

Let x be a real number. Then x lies between two integers called floor and ceiling of x. The **floor function** assigns x the largest integer that does not exceed x. It is denoted by $\lfloor x \rfloor$.

Thus
$$\lfloor x \rfloor = \begin{cases} x, & \text{if } x \text{ is an integer} \\ n, & \text{if } x \text{ is an integer and } n \le x < n+1 \end{cases}$$

Floor function is also known as **Greatest integer function.**

The ceiling function assigns x the smallest integer that is not less than x. It is denoted by $\lceil x \rceil$.

Thus
$$\lceil x \rceil = \begin{cases} x, & \text{if } x \text{ is an integer} \\ n, & \text{if } n \text{ is an integer and } n-1 < x \le n \end{cases}$$

1. *If x is an integer then* $\lceil x \rceil = \lfloor x \rfloor$
2. *The floor function* $\lfloor x \rfloor$ *should not be confused with truncation function. The truncation function only truncates any fraction. For positive real number x, both these functions are same but for the negative x they are different. For example* $\lfloor -4{\cdot}3 \rfloor = -5$, *whereas truncation of* $-4{\cdot}3 = -4$.

Example 4.53

Compute $\lfloor x \rfloor$ and $\lceil x \rceil$ for the value of x.

i,. 3.4 ii. –5·9

iii. 3/4 iv. –1/4

Solution

i. $\lfloor 3{\cdot}4 \rfloor = 3$ and $\lceil 3{\cdot}4 \rceil = 4$

ii. $\lfloor -5{\cdot}9 \rfloor = -6$ and $\lceil -5{\cdot}9 \rceil = -5$

iii. $\lfloor 3/4 \rfloor = 1$ and $\lceil 3/4 \rceil = 1$

iv. $\lfloor -1/4 \rfloor = -1$ and $\lceil -1/4 \rceil = 0$

4.17-1 Properties of floor and ceiling functions

1. $\lfloor x \rfloor = n$ iff $n \le x < n+1$, where n is an integer.
2. $\lceil x \rceil = n$ iff $n-1 < x \le n$, where n is an integer.
3. $\lfloor x \rfloor = n$ iff $x-1 < n \le x$, where n is an integer.
4. $\lceil x \rceil = n$ iff $x \le n < x+1$, where n is an integer.
5. $x-1 < \lfloor x \rfloor \le x \le \lceil x \rceil < x+1$
6. $\lfloor -x \rfloor = -\lceil x \rceil$
7. $\lceil -x \rceil = -\lfloor x \rfloor$
8. $\lfloor x+n \rfloor = \lfloor x \rfloor + n$, when n is an integer.
9. $\lfloor x+n \rfloor = \lceil x \rceil + n$, when n is an integer.

4.18 Countably Infinite Set, Countable set, Uncountable set

A set A is called **countable infinite** or **denumerable** if it can be put into one–one correspondence with Z^+. A set is called **countable** if it is finite or countably infinite. A set which is not countable is called **uncountable**.

The set N and Z are all countably infinite, and therefore countable. A set is countable if it is finite, or it has the same cardinality as Z^+. Since a composition of one-one correspondences is also a one-one correspondence, any set that can be put into one-one correspondence with a countable set is itself countable. That is, to prove that a set of countable it is enough to show that it can be put into one-one correspondence with any set that is known to be countable (so sets like N and Z can be used in place of Z^+, so long as the above provision about compositions is quoted.

Theorem 4.8

Any subset of Z^+ is countable.

Proof

Let $X \subseteq Z^+$. If X is finite, then it is countable. Suppose that X is infinite. Define $f: Z^+ \rightarrow X$ by $f(n)$ is the smallest element is $X - \{f(1), f(2) \dots f(n-1)\}$ (i.e. $f(n)$ is the nth–smallest element is X). To see that f is one–one, observe that if $i<j$ then $f(j)$ is the smallest element in a set that does not contain $f(i)$; thus $f(i) \neq f(j)$. To see that f is onto, note that the positive integer k is larger than at most $k-1$ elements of X hence if $k \in X$ then one of $f(1)$, $f(2), \dots, f(k)$ equals k. The function is therefore a one-one correspondence, and X is countable.

4.18-1 Equipotent

Two sets A and B are said to be **equipotent** (or equivalent or to have the same cardinality or similar) and written as $A \sim B$ if and only if there is one-to-one correspondence between the elements of A and those of B.

One-to-one correspondence can be established by showing a mapping $f: A \rightarrow B$ which is one-to-one and onto.

Example 4.54

Let $N = \{0, 1, 2, \dots\}$ and $N_1 = \{0, 2, 4, \dots\}$. Show that $N \sim N_1$.

Solution

Let $f: N \rightarrow N_1$ be defined by $f(n) = 2n$, $n \in N$. We can see that the function f established a one-to-one correspondence between N and N_1.

Hence $N \sim N_1$. [Observe that $N_1 \subset N$].

Theorem 4.9

An infinite subset of a denumerable set is also denumerable.

Proof

Let S be an infinite subset of a given denumerable set A. Obviously $A \sim N$, there exist a one-to-one correspondence between the elements of A and the set N of natural numbers. Consequently, we have a function $f: N \rightarrow A$ given by $f(n) = x \ \forall \ x \in A$, which is one-to-one and onto.

Thus, the elements of A can be arranged as $f(1), f(2), \dots$. Now delete from this list those elements which are not present in S. The number of the remaining elements is still infinite because S is infinite. Let us denote those elements by $f(i_1), f(i_2), \dots$ Define a function $g : N \rightarrow S$, s.t. $g(n) = f(i_n)$; then g is one-to-one and onto S. Hence S is also denumerable.

4.19 Sequence

An infinite sequence of element of a set X is a function $f: Z^+ \rightarrow X$. The images of the positive integers are usually displayed in list format: $f(1), f(2), f(3), \dots$ Sometimes we write f_i instead of $f(i)$, so that the list becomes $f_1, f_2, f_3 \dots$

Notice that, conversely, if there is an infinite sequence $x_1, x_2, x_3 \dots$. of elements of the set X, then a function $f: Z^+ \rightarrow A$ is obtained by defining $f(i) = x_i$, $i = 1, 2 \ \dots$. Therefore, infinite sequences and functions whose domain is Z^+ are really the same thing. Any subsequence of an infinite sequence $f: Z^+ \to X$ of element of the set X is a restriction of f to a subset of Z^+.

A countable infinite set is one that can be put into one-one correspondence with Z^+. By the definitions of infinite sequence and one-one correspondence, a set X is countably infinite if there is an infinite sequence (function $f: Z^+ \to X$) in which every element of X appears exactly once (one-one and onto function).

Example 4.55

The infinite sequence 1, –1, 3, –3, 5, –5,... contains every odd integer exactly once (the positive odd integer $2k + 1$ appears in position $2k + 1$ and the negative odd integer $-(2k + 1)$ appears in position $2k$). Thus, the set of odd integers is countable.

Theorem 4.10

If A and B are countable sets, then $A \cup B$ is a countable set.

Proof

If A and B are both finite, then so is $A \cup B$, and any finite set is countable. Suppose A is finite and B is countably infinite. Then there are sequences $a_1, a_2, \ldots, a_n$, and $b_1, b_2, \ldots$ that contain every element of these sets, respectively. Thus, $a_1, a_2, \ldots, a_n$, and $b_1, b_2, \ldots$ is an infinite sequence that contains every element of $A \cup B$, so $A \cup B$ is countable.

Similarly, if B is finite and A is countably infinite then $A \cup B$ is countable.

Finally, suppose that A and B are both countably infinite. Then there are infinite sequences $a_1, a_2, \ldots,$ and $b_1, b_2, \ldots$ that contain every element of these sets, respectively. Thus, $a_1, b_1, a_2, b_2, \ldots$ is an infinite sequence that contains every element of $A \cup B$, so $A \cup B$ is countable.

Exercise 4.4

1. Find $F(1), F(2), F(3), F(4)$, and $F(5)$ if $F(n)$ is defined recursively by $F(0) = 3$ and for $n = 0, 1, 2, 3, \ldots$.
   i. $F(n+1) = 3F(n) + 7$ ii. $F(n+1) = 3F\{n\}^2 - 2F(n) - 2$
   iii. $F(n+1) = 3^{F(n)/3}$
2. Find $F(1), F(2), F(3)$, and $F(4)$, if $F(n)$ is defined recursively by $F(0) = 1$ and for $n = 0, 1, 2, \ldots$.
   i. $F(n+1) = F\{n\}^2 + F(n) + 1$ ii. $F(n+1) = F(n) + 2$
3. Find $F(2), F(3), F(4)$ and $F(5)$ if F is defined recursively for $n = 1, 2, \ldots$
   i. $F(1) = 3, F(2) = 13, F(3) = 183, F(4) = 33, 673$
   ii. $F(1) = 3, F(2) = 5, F(3) = 7, F(4) = G$.
4. Show that the function $f\langle x, y\rangle = x - y$ is partial recursive.
5. Let $f_1(x, y) = x + y$, $f_2(x, y) = 3x$, $f_3(x, y) = xy$ and $g(x + y + z)$ be function over N. Find the composition of g with $= 3x f_1, f_2$ and f_3.
6. Let $f(x, y) = 3x^2 + y$, $g_1(x, y, z) = x + y + z$ and $g_2(x, y, z) = x - y - z)$ be defined over N. Find the composition of f with g_1 and g_2.
7. Show that the following function are primitive Recursive:
   i. $f(x, y) = |x - y|$ ii. $f(x, y) = x^y$
   iii. $f(n) = 2^n$
8. Compute $A(1, 1), A(2, 1), A(3, 2), A(3, 3)$.
9. Use the recursive definition of addition & multiplication of two natural numbers, prove that $4 + 5 = 9$ and $4 \times 2 = 8$.

10. Using recursion, define the multiplication function given by $f(x, y) = x * y$.
    [**Hint:** Multiplication of two natural numbers is simply repeated addition].
11. Prove that the set of multiples of three (positive, negative and zero) is countable.
12. Prove that any subset of a countable set is countable.
13. Prove that any Subset of Z is countable.
14. Prove that set R of real numbers is uncountable.
15. Is the set $Q - N$ countable or uncountable? Why? What about $R - R$?
16. Let X be an uncountable set. Prove that if there is a one-one correspondence $f: A \rightarrow X$, then A is uncountable.

Answers

1. i. $F(1) = 16, F(2) = 55.$
2. i. $F(1) = 3, F(2) = 13, F(3) = 183, F(4) = 33{,}673$
   ii. $F(1) = 3, F(2) = 5, F(3) = 7, F(4) = 9.$
3. $x + y + 3x + xy.$
6. $3(x + y + z) + x - y - z.$
8. 3, 5, 17, 37.

Miscellaneous Exercise

1. Let $f: R \rightarrow R$ and $g: R \rightarrow R$ defined by $f(x) = 4x - 1$, $g(x) = \cos x$. find $g \circ f$ and $f \circ g$.
2. Find $g \circ f$, $f \circ g$ when $f: R \rightarrow R$ and $g: R \rightarrow R$ defined by $f(x) = 2x - 1$, $g(x) = x^2 - 2$.
3. Show that the function $f: R \rightarrow R$ defined by $f(x) = 3x - 1$, $x \in R$ is a bijection.
4. Show that the function $f: R \rightarrow R$ given by $f(x) = \sin x$ is neither one-one nor onto.
5. If $f: R \rightarrow R$, then find the range of the following function:
   i. $f(x) = x^2$ ii. $f(x) = \sin x$
   iii. $f(x) = |x|$.
6. Let $f: R \rightarrow R$ defined by $f(x) = x^2$. Find (i) f^{-1} (25) (ii) f^{-1} (–9).
7. Let $A = [-1, 1]$. Let f, g, h be any three functions defined on A by the formula $f(x) = x^2$, $g(x) = x^2$, $h(x) = \sin x$. Find which one is onto.
8. State whether or not each of the following function is one–one :
   i. To each person on earth assign the number, which corresponds to his age.
   ii. To each state in India assign the number of people which live in the state.
9. Let $f: R \rightarrow R$ and $g: R \rightarrow R$ defined by $f(x) = x^2 + 2x - 3$, $g(x) = 3x - 4$. Find the product $g \circ f$ and $f \circ g$.
10. Let $A = [-1, 11]$ and the functions, f_1, f_2, f_3 on A into A defined by $f_1(x) = x^2$, $f_2(x) = x^5$, $f_3(x) = \sin x$. State whether these functions has an inverse function.
11. Let $A = \{1, 2, 3, 4\}$ and $B = \{a, b, c, d\}$. Determine which of the following are functions. Explain:
    i. $f_1 = \{(1, a), (2, b), (3, c), (4, d)\}$ ii. $f_2 = \{(1, a), (2, a), (3, b), (4, d)\}$
    iii. $f_3 = \{(1, a), (2, b), (3, c)\}$ iv. $f_4 = \{(1, a), (2, b), (2, c), (3, a), (4, a)\}$.

12. Find the ranges of the following functions on I.
    i. $f_1 = \{(x, 4x+1): x \in I\}$
    ii. $f_2 = x + 10$.
13. Which of the following functions are one-one, onto or both?
    i. $f_1 : R \to R$ s.t. $f_1(x) = x^3 - x$
    ii. $f_2 : I \to I$ s.t. $f_2(x) = -x + 2$
    iii. $f_3 : N \to N$ s.t. $f_3(x) = x^2 + x$
    iv. $f_4 : R \to R$ s.t. $f_4(x) = -2x$
    v. $f_5 : R \to R$ s.t. $f_5(x) = x^2 - 1$
    vi. $f_6 : R \to R$ s.t. $f_6(x) = x$ if $x < 0$ and x^2 if $x \geq 0$.
14. Let $X = \{1, 2, 3\}$, define $f: X \to X$ by $f(1) = 2, f(2) = 1$ and $f(3) = 3$. Find f^2, f^3, f^4, f^{-1}.
15. Define f, g and h from I to I such that $f(x) = x^2$, $g(x) = x + 1$, $h(x) = x - 1$. Determine
    i. gfh
    ii. fgh
    iii. hfg.

Chapter Objective Type Questions

A. Multiple choice questions.

1. The number of functions from an m element set to an n element set is:
   (A) m^n
   (B) $m + n$
   (C) n^m
   (D) $m \times n$
2. The function $f : [-\pi/2, \pi/2] \to [-1, 1]$, $f(x) = \sin x$ is:
   (A) one-one, into
   (B) one-one, onto
   (C) many-one, into
   (D) many-one, onto
3. Let $A = R - \{3\}$ and $B = R - (1)$. Then, $f : A \to B : f(x) = (x-2)/(x-3)$ is:
   (A) many-one, into
   (B) many-one, onto
   (C) one-one, into
   (D) one-one, onto
4. The domain of $f(x) = \sqrt{9 - x^2}$ is:
   (A) $]-\infty, -3]$
   (B) $[3, -\infty[$
   (C) $[-3, 3]$
   (D) $]-\infty, -3] \cup [3, \infty[$
5. The range of $f(x) - 1 - |x|$ is:
   (A) $]-\infty, 1]$
   (B) $[-1, 1]$
   (C) $[1, \infty[$
   (D) $]-\infty, -1]$
6. Let R^+ be the set of all positive real numbers. Then, the function $f : R^+ \to R^+ : f(x) = e^x$ is:
   (A) many–one, onto
   (B) many–one, into
   (C) one–one, onto
   (D) one–one, into
7. The function $f : C \to R : f(z) = |z|$ is:
   (A) many–one, into
   (B) many–one, onto
   (C) one–one, into
   (D) one-one, onto
8. If $f(x) = \cos x$ and $g(x) = 3x$, then $(f \circ g)(x)$ is:
   (A) $(\cos x)^3$
   (B) $3x^{(\cos x)}$
   (C) $\cos(3x)$
   (D) $\cos x^3$

9. The function $f : R \to R : f(x) = x^3$ is:

(A) one-one, into (B) one-one, onto

(C) many-one, onto (D) many-one, into

B. Fill in the blanks.

1. If f is a function from P to Q, then the set P is said to be the ______ of f and the set Q is said to be the ______ of the function f.
2. If f is a mapping from set A to itself then f is said to be a ______ on A.
3. A mapping $f : A \to B$ is called ________ if $f(A) = B$.
4. If f is surjective, each element of B has ______ one pre-image.
5. Let A be a non-empty set. The mapping $f : A \to A$ is called an _____ mapping if $f(x) = x \ \forall \ x \in A$.
6. A partial function is a relation that associates each element of a set (domain) with _____ element of a set (co-domain).
7. A set A is called ________ if it can be put into one-one correspondence with Z^+.
8. An infinite sequence of element of a set X is a function________.
9. If A and B are countable sets, then $A \cup B$ is a__________.

C. Which of the following statements is True (T)/ False (F)?

1. Every function is a relation but every relation is not a function.
2. If f is a function from A to B, then co-domain of f is equal to A.
3. Let f and g be two functions from A to B. Then these functions are said to be equal iff domain of f is equal to domain of g.
4. A mapping $f : A \to B$ is injective mapping if different elements of A have different f-images in B.
5. If $f : A \to B$, then $f^{-1}(b) = \{x : x \in A \text{ and } f(x) = b\}$.
6. If $f : A \to B$ and $g : B \to C$ are two one-one and onto mappings, then $(g \circ f) : A \to C$ need not be one-one onto.
7. The product of functions, i.e. $g \circ f$ is commutative.
8. The composite of functions is associative.
9. A set is called countable if it is finite or countably infinite.

Answers Objective Type

A.

1. B	2. D	3. C	4. A
5. C	6. A	7. B	8. D

B.

1. domain, co-domain	2. transformation	3. onto	4. at least	5. identity
6. at most one	7. denumerable	8. $f: Z+ \to X$	9. countable set	

C.

1. T	2. F	3. F	4. T
5. T	6. F	7. F	8. T
9. T			

CHAPTER

5 Groups

5.0 Introduction

In chapter 3 and 4 we studied the concept of relations and functions. In this section, first we discuss about an important function called binary composition and then we shall confine ourselves to the study of an algebraic structure known as a group and its properties with examples. This algebraic structure serves as one of the fundamental building blocks for the subject today called abstract algebra.

5.1 Binary Composition

Let S be a non-empty set. Then $S \times S = \{(a, b): a \in S, b \in S\}$. If $f: S \times S \to S$, then f is called a binary composition (operation) in a set S. In other words a binary composition in a set S is a rule which assigns each ordered pair (a, b) $\forall\, a \in S, b \in S$ to a unique element of S.

Under this composition, the image of the ordered pair (a, b) is denoted by $a f b$.

*In this section we shall be concerned only with binary compositions and as such we shall often use the word **composition** instead of 'binary composition'.*

Example 5.1

As the sum of two natural numbers is again a natural number therefore addition is a binary composition on the set of natural numbers N. i.e.

$$a + b \in N \;\forall\, a, b \in N.$$

Similarly, multiplication is also a binary composition over N.

Example 5.2

Subtraction is not a binary composition on the set N. We have $5 - 7 = -2 \notin N$, whereas $5, 7 \in N$. But subtraction is a binary composition on the set of integers I. We have $a - b \in I \;\forall\, a, b \in I$. Similarly, division is not a binary composition on N.

Example 5.3

Let $P(S)$ be a power set of S. We know that, if $A, B \in P(S)$, then

$$A \cup B \in P(S) \text{ and } A \cap B \in P(S)$$

Therefore union and intersection of sets are binary compositions in $P(S)$.

5.2 Closed Composition

A composition $*$ on a set S is said to be **closed** if $a * b \in S \; \forall \; a, b \in S$.

5.3 Commutativity

A composition $*$ in a sets is said to be **commutative composition** (or commutativity) if

$$a*b = b * a \; \forall \; a, b \in S.$$

Example 5.4

Addition and multiplication in I (set of integers) are commutative compositions while subtraction is not so. Because for

$$a, b \in I, a + b = b + a \text{ and } ab = ba \text{ but } a - b \neq b - a \text{ unless } a = b.$$

5.4 Associative Composition

A composition $*$ in a set S is called **associative composition** (associativity) if

$$(a * b) * c = a * (b * c) \; \forall \; a, b, c \in S.$$

Example 5.5

Addition and multiplication in N are associative while subtraction is not so, because for $a, b, c \in N$

$$(a + b) + c = a + (b + c)$$

and $$(ab)\, c = a(bc)$$

but $$(a - b) - c \neq a - (b - c).$$

5.5 Identity Element for a Composition

Let S be a non-empty set and $*$ be a composition in S. Then an element $e_1 \in S$ is called **left identity** for the composition $*$ in S, if

$$e_1 * a = a \; \forall \; a \in S.$$

Similarly, an element $e_2 \in S$ is called **right identity** for the composition $*$ in S, if

$$a * e_2 = a \; \forall \; a \in S.$$

In general, an element $e \in S$ is called an identity element for a composition in S if it is both left identity as well as right identity. Therefore $e \in S$ is called **identity element** in S for the composition $*$ if and only if

$$e * a = a = a * e \; \forall \; a \in S.$$

Example 5.6

In the set of real numbers, 0 is additive identity while 1 is multiplicative identity because

$$0 + a = a = a + 0$$

and $$1.a = a = a.1 \; \forall \; a \in R.$$

Example 5.7

For the power set $P(S)$. Since $A \cap S = A = S \cap A$ and

$$A \cup \varnothing = A = \varnothing \cup A \; \forall \; A \in P(S).$$

Then S is the identity for intersection and $\varnothing$ is the identity element for union.

5.6 Invertible Element for a Composition

Let S be a non-empty set and $*$ be a composition in S and e is its identity. An element $s_1 \in S$ (if it exists) is called **left inverse** of an element

$$a \in S \text{ if } s_1 * a = e$$

Similarly, an element $s_2 \in S$ (if it exists) is called **right inverse** of an element

$$a \in S \text{ if } a * s_2 = e$$

In general, an element $s \in S$ is both a left inverse and a right inverse of $a \in S$ if $s * a = e = a * s$.

It is simply called an **inverse** of a. Now, an element $a \in S$ is called invertible if and only if it has an inverse in S, denoted by a^{-1}.

If the composition is addition, then the inverse of a is –a.

5.7 Algebraic Structure

A non-empty set S equipped with one or more binary compositions is called an **algebraic structure**. Let $*$ be a binary composition in S. Then $(S, *)$ is called an **algebraic structure**. When two binary compositions say + and· are defined on S, then it is denoted by $(S, +, \cdot)$.

Example 5.8

$(N, +)$, $(I, +)$, $(I, -$, $(R, +, \cdot)$ are all algebraic structures.

5.8 Semi-Group

Let G be a non-empty set and $*$ be a composition on G. Then the algebraic structure $(G, *)$ is called a **semigroup** if the composition $*$ is associative in G

i.e. if $(a * b) * c = a * (b * c)\ \forall\ a, b, c \in G$, then $(G, *)$ is a semi-group.

Example 5.9

The structures $(N, +)$, $(Z, +)$, $[P(S), \cup]$, $(N, \cdot)$, $(I,+)$ and $(R, +)$ are all semi groups but $(Z, -)$ is not semi group because subtraction is not associative.

5.9 Monoid

If in a semi-group $(G, *)$, there exists an identity element for its composition, then it is called a **monoid**.

Example 5.10

The semi-group $(N, \cdot)$ is a monoid because $1 \in N$ is the identity for its composition. But $(N, +)$ is not because $0 \notin N$.

5.10 Group

Let G be a non-empty set equipped with a binary composition $*$. The algebraic structure $(G, *)$ is a **group** if the binary composition $*$ satisfies the following postulates:

1. Closure Property

The set G must be closed with respect to the composition $*$.

i.e. $$a * b \in G \; \forall \; a, b \in G.$$

2. Associativity

The composition $*$ must be associative in G.

i.e. $$(a * b) * c = a * (b * c) \; \forall \; a, b, c \in G.$$

3. Existence of identity

The identity element 'e' must be present in the set G.

i.e. $$a * e = a = e * a \; \forall \; a \in G.$$

4. Existence of Inverse

The inverse of each element of G must be in G. In other words the element $x \in G$ is the inverse of $a \in G$ if $x * a = e = a * x$ where e is the identity element for G.

5.11 Abelian Group or Commutative Group

Let $(G, *)$ be a group. If the composition $*$ is commutative in G i.e. $a * b = b * a \; \forall \; a, b \in G$. Then the group $(G, *)$ is said to be **abelian group** or **commutative group**.

A group is called non-abelian or non-commutative if it is not abelian group.

In the definition of group, the composition is denoted by $*$. However we can use any symbol like $\circ, +, \cdot$ etc to denote the composition.

If the composition is addition '+', then identity element will be zero and the inverse of any element $a \in G$ will be the symbol $-a$.

If the composition is multiplication '$\cdot$', then identity element will be one and the inverse of any non-zero element $a \in G$ will be the symbol $1/a$.

5.12 Finite and Infinite group

A group $(G, *)$ is called a **finite group** if the underlying set G consists of a finite number of distinct elements. If a group is not finite then it is said to be an **infinite group**.

5.13 Order of A group

The number of distinct elements in the underling set G of a group $(G,*)$ is called **order of the group**.

We shall denote the order of a group G by the notation $O(G)$.

If the group $(G,*)$ is infinite, then the order of the group will be infinite.

Example 5.11

Prove that the set of integers $I = \{\ldots. -3, -2, -1, 0, 1, 2, 3, \ldots.\}$ forms a group with respect to addition composition.

Solution

Closure Property

We know that the sum of two integers is also an integer. Let $a, b \in I \Rightarrow a + b \in I$. Therefore I is closed for addition.

Associativity

We know that the addition of integers is an associative composition.

Existence of an Identity

We know that the identity element for addition is 0. Since $0 \in I$, therefore identity element exists in I.

Existence of Inverse

For every $a \in I$, there exists $-a \in I$ such that

$$a + (-a) = 0 = (-a) + a.$$

Therefore, every integer has unique additive inverse. Hence, the given set of integers forms a group with respect to the addition composition.

Example 5.12

The set Q of rational numbers is an infinite group with respect to addition composition. But it is not a group for multiplication composition.

Solution

Do yourself.

Example 5.13

Show that the set Q^+ of positive rational numbers does not form a group for the composition $*$ defined by $a * b = a/b \ \forall \ a, b \in Q^+$.

Solution

Closure property

Since $a * b = a/b \in Q^+$, so Q^+ is a closed for $*$. Given that

$$a, b \in Q^+ \Rightarrow \frac{a}{b} \in Q^+ \Rightarrow a * b \in Q^+$$

So, Q^+ is closed for the composition $*$.

Associativity

Let $a, b, c \in Q^+$

Now, $$(a * b) * c = \left(\frac{a}{b}\right) * c = \frac{\left(\frac{a}{b}\right)}{c} = \frac{a}{bc}$$

Also, $$a * (b * c) = a * \left(\frac{b}{c}\right) = \frac{a}{\left(\frac{b}{c}\right)} = \frac{ac}{b}$$

Thus $(a * b) * c \neq a * (b * c)$

Therefore, the composition $*$ is not associative. Hence, Q^+ is not forming a group with respect to the composition $*$.

Example 5.14

Show that the set of all positive rational numbers forms an abelian group under the composition * defined as

$$a * b = \frac{ab}{4} \qquad \forall a, b \in Q^+$$

Solution

Closure Property

Since $a * b = \frac{ab}{4} \in Q^+ \quad \forall\, a, b \in Q^+$, so Q^+ is closed for the composition *.

Associativity

Let $a, b, c \in Q^+$, then

$$(a * b) * c = \left(\frac{ab}{4}\right) * c = \frac{1}{4}\left(\frac{ab}{4}.c\right) = \frac{ab}{4}\frac{c}{4}$$

We know that multiplication is associative in Q^+.

Therefore

$$(a * b) * c = \frac{a}{4}.\frac{bc}{4} = a * \left(\frac{bc}{4}\right) = a * (b * c)$$

$\therefore$ The composition * is associative in Q^+.

Existence of Identity

Let $e \in Q^+$ be an identity element and a be any element of Q^+, then

$$e*a = a \qquad \text{[by def.]}$$

$$\Rightarrow \frac{ea}{4} = a \Rightarrow e = a \cdot \frac{4}{a}$$

$$\Rightarrow e = 4$$

$\therefore$ 4 is the identity for the composition in Q^+.

Existence of inverse

Let a be any element of Q^+ whose inverse is b, then by definition of inverse $a * b = e$ where e is identity element

$$\therefore \frac{ab}{4} = 4 \Rightarrow b = \frac{16}{a}$$

$\therefore \frac{16}{a}$ is the inverse of a in Q^+. Since a is an arbitrary element, so every element of Q^+ is invertible.

Commutativity

Let a, b, $\in Q^+$, then

$$a * b = \frac{ab}{4}$$

because multiplication is commutative in Q^+

$$\Rightarrow a * b = \frac{ba}{4} = b * a$$

So, composition * is commutative. Hence $(Q^+, *)$ is an abelian group.

Example 5.15

Show that the set $Q - \{1\}$ of rational numbers other than 1 is an abelian group under the composition $*$ defined as $x * y = x + y - xy$.

Solution

Let G be a given set, then $G = \{s \in Q : s \neq 1\}$.

Closure Property

Let $x, y \in G$. Then $x + y - xy$ is a rational number.

Assume that

$$x * y = 1$$
$$x + y - xy = 1$$
$$x - xy + y - 1 = 0$$
$$(x - 1)(1 - y) = 0$$
$$x = 1 \text{ or } y = 1$$

But it is a contradiction because $x, y \in G$, therefore $x * y \neq 1$. Hence G is closed for $*$.

Associativity

Let $x, y, z \in G$, then

$$(x * y) * z = (x + y - xy) * z$$
$$= x + y - xy + z - (x + y - xy)z$$
$$= x + y + z - xy - xz - yz + xyz \quad \text{...(1)}$$

Again

$$x * (y * z) = x * (y + z - yz)$$
$$= x + y + z - yz - x(y + z - yz)$$
$$= x + y + z - yz - xy - xz + xyz \quad \text{...(2)}$$

From (1) and (2), we get $(x * y) * z = x * (y * z)$

Therefore, the composition $*$ is associative in G.

Existence of identity

Let $e \in G$ is the identity for its composition $*$ and $x \in G$ is an arbitrary element such that $x \neq 1$. By definition

$$e * x = x \Rightarrow e + x - ex = x \Rightarrow e(1 - x) = x - x \Rightarrow e = 0$$

Therefore $0 \in G$ is the identity element for $*$.

Existence of Inverse

Let $x \in G$ is an arbitrary element such that $x \neq 1$, then $y \in G$ is the inverse of x if $x * y = e$, where e is the identity element of G.

$$x + y - xy = 0 \quad [\because e = 0]$$
$$y(1 - x) = -x$$
$$y = x/x - 1$$

Since $x \neq 1$, thus $\dfrac{x}{x-1} \in G$. The inverse of x is $x/x-1$ in $(G, *)$. Since x is an arbitrary element of G, therefore every element of G is invertible.

Commutativity

Let $x, y \in G$ such that $x \neq 1, y \neq 1$, then

$$x * y = x + y - xy$$

because addition and multiplication are commutative compositions in Q

Therefore $$x * y = y + x - yx$$
$$= y * x$$

Thus $*$ is commutative in G. Hence $(G, *)$ is an abelian group.

Example 5.16

Show that the set C' of all non-zero complex numbers is a group for multiplication of complex numbers.

Solution

We have $C' = \{x + i\,y : x \in R, y \in R \text{ and } x \neq 0, y \neq 0\}$. The number $0 + i\,0$ is the zero complex number.

Closure Property

As the product of two non-zero complex numbers is a non-zero complex number. Therefore C' is closed for multiplication.

Associativity

We know that the multiplication of complex numbers is associative.

Existence of identity

We know that the multiplicative identity is 1. It can be written as $1 + i\,0$.

Thus $$1 = 1 + \mathrm{i}\,0 \in \mathrm{C}'.$$

Therefore $1 + i\,0$ is the identity for multiplication of complex numbers.

Existence of inverse

Let $x + iy \in C'$ where $x \in R, y \in R$ and $x \neq 0, y \neq 0$.

Let $a + ib$ is the inverse of $x + iy$

Then $$(a + ib)(x + iy) = 1 + i\,0$$
$$(ax - by) + i\,(bx + ay) = 1 + i\,0$$
or $$ax - by = 1,\ bx + ay = 0$$

Solving these equations, we get

$$a = \frac{x}{x^2 + y^2},\ b = \frac{-y}{x^2 + y^2}.$$

Since $x^2 + y^2 \neq 0$ and $x \neq 0, y \neq 0$, therefore

$$\frac{x}{x^2 + y^2} + i\frac{(-y)}{x^2 + y^2} \in C'$$

and it is the multiplicative inverse of $x + iy$.

Commutativity

Let $x_1 + iy_1$ and $x_2 + iy_2$ be any two elements of C' with the conditions that x_1, x_2, y_1 and y_2 are non zero real numbers. Then

$$(x_1 + iy_1).(x_2 + iy_2) = (x_1 x_2 - y_1 y_2) + i\,(x_1 y_2 + x_2 y_1) \quad \ldots (1)$$

Also $$(x_2 + iy_2).(x_1 + iy_1) = (x_2 x_1 - y_2 y_1) + i\,(y_2 x_1 + y_1 x_2)$$
$$= (x_1 x_2 - y_1 y_2) + i\,(x_1 y_2 + x_2 y_1) \quad \ldots (2)$$

From (1) and (2), we get

$$(x_1 + iy_1) \cdot (x_2 + iy_2) = (x_2 + iy_2)\,(x_1 + iy_1)$$

Therefore, multiplication of non-zero complex numbers is commutative. Hence C' is an abelian group.

Example 5.17

Let G be the set of all 2×2 matrices $\begin{bmatrix} a & b \\ c & d \end{bmatrix}\begin{bmatrix} a & b \\ c & d \end{bmatrix}$ where a, b, c, d are real numbers, such that $ad - bc \neq 0$. Show that G is a non-abelian group for multiplication of matrices defined as

$$\begin{bmatrix} a & b \\ c & d \end{bmatrix}\begin{bmatrix} w & x \\ y & z \end{bmatrix} = \begin{bmatrix} aw + by & ax + bz \\ cw + dy & cx + dz \end{bmatrix}$$

Solution

Closure Property

Since a, b, c, d, w, x, y, z are all real numbers therefore, $aw + by$, $cw + dy$, $ax + bz$, $cx + dz$ are all real numbers. Now we have to show that this matrix is in G. For this we will show that

$$(aw + by)\,(cx + dz) - (cw + dy)\,(ax + bz) \neq 0$$

Now,
$$\begin{aligned} \text{LHS} &= (aw + by)\,(cx + dz) - (cw + dy)\,(ax + bz) \\ &= (ad - bc)\,(wz - xy) \\ &\neq 0 \qquad [\because ad - bc \neq 0 \text{ and } wz - xy \neq 0] \end{aligned}$$

Associativity

We know that the multiplication of matrices is associative.

Existence of identity

Since $1.1 - 0.0 = 1 \neq 0$, then $\begin{bmatrix} 1 & 0 \\ 0 & 1 \end{bmatrix} \in G$ is the identity for the composition because for every $\begin{bmatrix} a & b \\ c & d \end{bmatrix} \in G$, we have $\begin{bmatrix} 1 & 0 \\ 0 & 1 \end{bmatrix}\begin{bmatrix} a & b \\ c & d \end{bmatrix} = \begin{bmatrix} a & b \\ c & d \end{bmatrix}$

Existence of inverse

Let $\begin{bmatrix} a & b \\ c & d \end{bmatrix} \in G$ and $ad - bc \neq 0$, then the matrix $\begin{bmatrix} \dfrac{d}{ad - bc} & \dfrac{-b}{ad - bc} \\ \dfrac{-c}{ad - bc} & \dfrac{a}{ad - bc} \end{bmatrix}$ exist.

Now $\left(\dfrac{d}{ad - bc}\right)\left(\dfrac{a}{ad - bc}\right) - \left(\dfrac{-b}{ad - bc}\right)\left(\dfrac{-c}{ad - bc}\right) = \dfrac{ad - bc}{(ad - bc)^2} = \dfrac{1}{ad - bc} \neq 0$

Hence the matrix $\begin{bmatrix} \dfrac{d}{ad - bc} & \dfrac{-b}{ad - bc} \\ \dfrac{-c}{ad - bc} & \dfrac{a}{ad - bc} \end{bmatrix} \in \text{G}.$

Also
$$\begin{bmatrix} a & b \\ c & d \end{bmatrix} = \begin{bmatrix} \dfrac{d}{ad - bc} & \dfrac{-b}{ad - bc} \\ \dfrac{-c}{ad - bc} & \dfrac{a}{ad - bc} \end{bmatrix} = \begin{bmatrix} 1 & 0 \\ 0 & 1 \end{bmatrix}.$$

Therefore, this element of G is the inverse of $\begin{bmatrix} a & b \\ c & d \end{bmatrix}$. Hence G is a group.

The matrix multiplication is not always commutative. Therefore G is a non-abelian group.

5.14 Composition Table for Finite Sets

A composition table is a tabular representation of a binary composition in a finite set. Let $S = \{x_1, x_2, \ldots, x_n\}$ is a finite set with n elements and suppose that the composition in S is the multiplication. Construct a composition table by writing the elements of S in a horizontal row as well as in a vertical column. The element $x_i x_j$ related to the ordered pair (x_i, x_j) is placed at the intersection of the row headed by x_i and the column headed by x_j. The composition table for a finite group contains each element exactly once in each of its rows and columns.

Example 5.18

Prove that the set $\{1, -1, i, -i\}$, where $i = \sqrt{(-1)}$, is a finite abelian group with respect to multiplication of complex numbers.

Solution

Let $G = \{1, -1, i, -i\}$. We construct the composition table for $(G, \cdot)$ as follows:

$\cdot$	1	-1	i	$-i$
1	1	-1	i	$-i$
-1	-1	1	$-i$	i
i	i	$-i$	-1	1
$-i$	$-i$	i	1	-1

From the composition table, we observe that

Closure Property

All the entries in the table are elements of the set G. Therefore G is closed for multiplication.

Associativity

Since all the elements of G are complex numbers and multiplication of complex numbers is associative, therefore associativity hold.

Existence of identity

The number 1 is the identity element for the composition table.

Existence of inverse

From the composition table we see that the inverses of $1, -1, i, -i$ are $1, -1, -i$ and i respectively.

Commutativity

The multiplication of complex numbers is commutative.

Since the number of elements in the set G is 4. Therefore $(G, \cdot)$ is a finite abelian group of order 4.

Example 5.19

Show that the set of six transformations $f_1, f_2, f_3, f_4, f_5, f_6$ on the set of complex numbers defined by

$f_1(z) = z, f_2(z) = 1/z, f_3(z) = 1 - z,$

$$f_4(z) = \frac{z}{z-1}, \quad f_5(z) = \frac{1}{1-z}, \quad f_6(z) = \frac{z-1}{z},$$

forms a finite non-abelian group for the composite of functions.

Solution

Let $G = \{f_1, f_2, f_3, f_4, f_5, f_6\}$ and $\circ$ denotes composite of functions. After calculating all the possible products we have,

$$(f_1 \circ f_1)(z) = f_1(f_1(z)) = f_1(2)$$

$$(f_1 \circ f_2)(z) = f_1(f_2(z)) = f_1\left(\frac{1}{z}\right) = \frac{1}{z} = f_2(z)$$

$$(f_2 \circ f_3)(z) = f_2(f_3(z)) = f_2(1 - z) = \frac{1}{1 - z} = f_5(z)$$

$$(f_3 \circ f_4)(z) = f_3(f_4(z)) = f_3\left(\frac{z}{z - 1}\right) = 1 - \frac{z}{z - 1} = f_5(z)$$

$$(f_4 \circ f_6)(z) = f_4(f_6(z)) = f_4\left(\frac{z - 1}{z}\right) = 1 - z = f_3(z)$$

Similarly calculating the other products we get the following composition table

$\circ$	f_1	f_2	f_3	f_4	f_5	f_6
f_1	f_1	f_2	f_3	f_4	f_5	f_6
f_2	f_2	f_1	f_5	f_6	f_3	f_4
f_3	f_3	f_6	f_1	f_5	f_4	f_2
f_4	f_4	f_5	f_6	f_1	f_2	f_3
f_5	f_5	f_4	f_2	f_3	f_6	f_1
f_6	f_6	f_3	f_4	f_2	f_1	f_5

From the table one can easily observe (as in example 5.18) that $(G, \circ)$ is a finite group of order 6. Since the table is not symmetrical about the main diagonal so it is a non-commutative (or non-abelian) finite group.

Example 5.20

Let G be a set consisting of all the ordered pairs (x, y) such that $x, y \in R$ and $x \neq 0$. A composition $*$ is defined in G as follows:

$$(x, y) * (z, w) = (xz, yz + w)$$

Then show that $(G, *)$ is a non-abelian group.

Solution

Closure property

Let (x, y) and (z, w) be any two ordered pairs such that $x, y, z, w \in R$ and $x \neq 0$, $z \neq 0$. Since x and z are non-zero therefore $x.z \neq 0$. Also $yz + w \in R$. Hence $(xz, yz + w) \in G$. G is closed for $*$.

Associativity

Let (x, y), (z, w), (u, v) be elements of G. Then

$$\begin{aligned}[(x, y) * (z, w)] * (u, v) &= (xz, yz+w) * (u, v) \\ &= [\,(xz)\,u, (yz + w)u + v\,] \\ &= (xzu, yzu + wu + v\,] \qquad \ldots(1)\end{aligned}$$

[by associativity and distributivity in R]

Also $(x, y) * [(z, w) * (u, v)] = (x, y) * (zu, wu + v)$

$= [x(zu), yzu + wu + v]$

$= (xzu, yzu + wu + v]$ ...(2)

From (1) and (2), we obtain associativity.

Existence of identity

Because $(x, y) * (1, 0) = (x.1\ y.1+0) = (x, y)$

and $(1, 0) * (x, y) = (1.x, 0.x + y) = (x, y)$

Therefore (1, 0) is the identity and $(1, 0) \in G$.

Existence of inverse

Let $(x, y) \in G$. Then

$\therefore$ $x \neq 0$, $1/x \in R \Rightarrow -y/x \in R$ so $\left(\frac{1}{x}, \frac{-y}{x}\right) \in G$.

Also $(x, y) * \left(\frac{1}{x}, \frac{-y}{x}\right) = \left(x \cdot \frac{1}{x}, y \cdot \frac{1}{x} - \frac{y}{x}\right) = (1, 0)$

Therefore, $\left(\frac{1}{x}, \frac{-y}{x}\right)$ is the inverse of (x, y) in G.

Commutativity

Let (x, y) and (z, w) are in G then

$(x, y) * (z, w) = (xz, yz + w)$ ... (1)

$(z, w) * (x, y) = (zx, wx + y)$ ... (2)

From (1) and (2), we get

$$(x, y) * (z, w) \neq (z, w) * (x, y)$$

therefore, commutativity does not hold. Hence $(G, *)$ is a non-abelian group.

5.15 Addition Modulo *n*

Let a and b are any integers and n is a fixed positive integer. The **addition modulo *n*** is a binary composition written as $a +_n b$ and defined as the least non-negative remainder '*r*' when $a + b$ is divided by n. Therefore, $a +_n b$ is a non-negative remainder after $a + b$ is divided by n.

Example 5.21

$10 +_3 4 = 2$. Since $10 + 4 = 14 = 3(4) + 2$ i.e. 2 is the least non-negative remainder when $10 + 4$ is divided by 3.

5. 16 Multiplication Modulo *m*

The multiplication modulo m written as $a \times_m b$ where a and b are any integers and m is a fixed positive integer. By definition, we have

$$a \times_m b = s, 0 \leq s < m$$

where, s is the least non-negative remainder when $a \cdot b$ is divided by m.

Example 5.22

$5 \times_4 3 = 3$. Since $5 \times 3 = 15 = 4(3) + 3$ i.e. 3 is the least non-negative remainder when 5×3 is divided by 4.

5.17 Additive Group of Integers Modulo *n*

The set $G = \{0, 1, 2, \ldots, n\text{-}1\}$ of first n non-negative integers is a group, the composition being ordinary addition reduced modulo n.

Example 5.23

Show that the set $G = \{0, 1, 2, 3, 4\}$ is a finite abelian group of order 5 under addition modulo 5 as composition.

Solution

We construct the composition table for $(G, +_5)$ as follows.

$+_5$	0	1	2	3	4
0	0	1	2	3	4
1	1	2	3	4	0
2	2	3	4	0	1
3	3	4	0	1	2
4	4	0	1	2	3

Closure Property

Since all the entries in the composition table are elements of the set G. Therefore G is closed for addition modulo 5.

Associativity

It can be verified from the table that for any three elements a, b, c of G, we have

$$(a +_5 b) +_5 c = a +_5 (b +_5 c)$$

for example, let $a = 2, b = 3, c = 4$

then $(2 +_5 3) +_5 4 = 0 +_5 4 = 4$

also $2 +_5 (3 +_5 4) = 2 +_5 2 = 4.$

Existence of identity

From the composition table we see that for any element $a \in G$

$$0 +_5 a = a = a +_5 0$$

Therefore, 0 is the identity element.

Existence of Inverse

From table we observe that the inverses of 0, 1, 2, 3, 4 are 0, 4, 3, 2, 1 respectively. So every element of G is invertible.

Commutativity

The composition $+_5$ is commutative as the corresponding rows and columns in the composition table are identical.

Since the number of elements in the set G is 5, therefore $(G, +_5)$ is a finite abelian group of order 5.

5.18 Multiplicative Group of Integers Modulo *m* where *m* is A Prime

The set G of $(m-1)$ integers 1, 2, 3, …, $m-1$, m being prime is a finite abelian group of order m-1, the composition being multiplication modulo m.

Example 5.24

Prove that the set $G = \{1, 2, 3, 4\}$ under multiplication modulo 5 is an abelian group.

Solution

The composition table for $(G, \times_5)$ is

$\times_5$	1	2	3	4
1	1	2	3	4
2	2	4	1	3
3	3	1	4	2
4	4	3	2	1

Closure property

Since all the entries in the composition table are elements of the set G, therefore G is closed for multiplication modulo 5.

Associativity

It can be verified from the table that for any three elements a, b, c of G,

$$(a \times_5 b) \times_5 c = a \times_5 (b \times_5 c)$$

For example, let $a = 2$, $b = 3$, $c = 4$

$$(a \times_5 b) \times_5 c = (2 \times_5 3) \times_5 4 = 1 \times_5 4 = 4$$

also $$a \times_5 (b \times_5 c) = 2 \times_5 (3 \times_5 4) = 2 \times_5 2 = 4$$

Existence of identity

We have $1 \in G$. If $a \in G$, then from the composition table we see that

$$1 \times_5 a = a = a \times_5 1$$

Therefore, 1 is the identity element.

Existence of Inverse

From table we observe that the inverses of 1, 2, 3, 4 are 1, 3, 2, 4 respectively. So every element of G is invertible.

Commutativity

The composition $\times_5$ is commutative as the corresponding rows and columns in the composition table are identical.

Since the number of elements in the set G is 4, therefore, $(G, \times_5)$ is a finite abelian group of order 4.

Exercise 5.1

1. In the following, determine whether the systems mentioned are groups. In case not, point out which of the group postulates fail to hold:

   i. $(N, +)$ ii. $(N, \cdot)$

   iii. $(R, +)$ iv. $(Q^+, \cdot)$

   v. $(P(S), \cup)$ vi. $(P(S), \cap)$

2. Decide whether (Z, $*$) is a group where the composition $*$ is defined as follows:
   i. $a * b = a - b$ ii. $a * b = ab$
   iii. $a * b = a + b + ab$ iv. $a * b = |ab|$
3. Define the order of a group. Show that the set of all even integers with zero is an abelian group with respect to addition.
4. Does the set of all odd integers form a group with respect to addition?
5. Show that the set of all rational numbers of the form $2^a 3^b$ (a, b integers) is a group for multiplication of rationals.
6. Prove that the set $\{1, w, w^2\}$ where $w^3 = 1$ is a group for multiplication composition.
7. Show that the set $G = \{a + b\sqrt{2}: a, b \in Q\}$ is a group with respect to addition.
8. Prove that the set of all $m \times n$ matrices having their elements as integers (rational or real or complex numbers) is an infinite abelian group with respect to addition of matrices.
9. Show that the set of matrices.

$$A_0 = \begin{bmatrix} \cos\theta & -\sin\theta \\ \sin\theta & \cos\theta \end{bmatrix}$$

   where θ is a real number, forms a group under matrix multiplication
10. Prove that the set of all n, nth roots of unity is a finite abelian group of order n for the multiplication of complex numbers.
11. Show that the set of all matrices of the form $\begin{bmatrix} 0 & a \\ 0 & b \end{bmatrix}, a, b \in R$, is a group for matrix addition.
12. Verify that the set of all matrices of the form $\begin{bmatrix} a & 0 \\ b & 1 \end{bmatrix}, a, b \in R,\ a \neq 0$ is a group under matrix multiplication.
13. Find whether the set of all matrices of the form $\begin{bmatrix} a & a \\ a & a \end{bmatrix}, a \in R;\ a \neq 0$ forms a group under matrix multiplication.
14. Show that the four matrices.

$$\begin{bmatrix} 1 & 0 \\ 0 & 1 \end{bmatrix}, \begin{bmatrix} -1 & 0 \\ 0 & 1 \end{bmatrix}, \begin{bmatrix} 1 & 0 \\ 0 & -1 \end{bmatrix}, \begin{bmatrix} -1 & 0 \\ 0 & -1 \end{bmatrix}$$

   form a multiplicative group.
15. Show that the set of four transformations f_1, f_2, f_3, f_4 on the set of complex numbers defined by $f_1(z) = z, f_2(z) = -z, f_3(z) = 1/z, f_4(z) = -1/z$ forms a finite abelian group with respect to the composite of functions.
16. Show that the set $\{1, 3, 4, 5, 9\}$ is an abelian group under multiplication modulo 11 as composition.
17. Prove that the set $G = \{1, 2, 3, 4, 5, 6\}$ forms a group under multiplication modulo 7.
18. Prove that the set $G = \{0, 1, 2, 3, 4, 5, 6\}$ forms a group under addition modulo 7.
19. Prove that $(\{e\}, *)$ is an abelian group of order one, where $e * e = e$.
20. Give an example of an abelian group of order 2.

21. $*$ is a binary operation defined on G, the set of all positive rational numbers, such that $a * b = \frac{ab}{3} \forall a, b \in G$. Show that $(G, *)$ is a group.

22. Given $G = \{3k : k \in I\}$. Show that $(G, *)$ is a group.

23. Let G be the set of all nonzero real numbers and let $a * b = \frac{ab}{2}$. Show that $(G, *)$ is an abelian group.

Answers

4. No 20. $(\{1, -1\}, \cdot)$

5.19 Properties of Groups

In this section we study some properties of groups. Through out this section we use the symbol '.' for the group composition $*$.

Theorem 5.1 (Uniqueness of Identity)

The identity element in a group is unique.

Proof

Let $(G, \cdot)$ be a group and e e' be its two identities. Then by definition

$$e, e' = e' \text{ if } e \text{ is identity} \quad \ldots(1)$$

and

$$e, e' = e \text{ if } e' \text{ is identity} \quad \ldots(2)$$

But e, e' is unique element in G, from (1) and (2), we get

$$e, e' = e = e'$$

Hence, the identity element is unique.

Theorem 5.2 (Uniqueness of Inverse)

The inverse of each element of a group is unique.

Proof

Suppose x is any element of a group $(G, \cdot)$ and y, z be its two inverses. Then by definition

$$xy = e = yx$$
$$xz = e = zx$$

where e is the identity of G.

Now

$$y(xz) = y\,e \quad [\because xz = e]$$
$$= y \quad \ldots(1)\ [e \text{ is identity}]$$

Also

$$(yx)\,z = e\,z \quad [\because yx = e]$$
$$= z \quad \ldots(2)\ [e \text{ is identity}]$$

But in a group G the composition is associative. Therefore

$$y(xz) = (yx)z \Rightarrow y = z$$

Hence the inverse of an element in a group is unique.

Theorem 5.3

If x^{-1} is the inverse of x of a group G, then $(x^{-1})^{-1} = x$.

Proof

Let e be the identity of G. By definition of inverse element, we have

$$x^{-1} x = e \quad \text{...(1)}$$

because $x^{-1} \in G$ and G is a group therefore $(x^{-1})^{-1} \in G$.

Now, multiply both sides on the left of (1) by $(x^{-1})^{-1}$, we get

$$(x^{-1})^{-1} [x^{-1} x] = (x^{-1})^{-1} e$$

$$= (x^{-1})^{-1} \quad [e \text{ is identity}]$$

since the composition is associative in G, then

$$[(x^{-1})^{-1} x^{-1}] x = (x^{-1})^{-1}$$

$$ex = (x^{-1})^{-1} \quad [(x^{-1})^{-1} \text{ is the inverse of } x^{-1}]$$

$$x = (x^{-1})^{-1} \quad [e \text{ is identity}]$$

$$(x^{-1})^{-1} = x \quad \text{Hence the result.}$$

Theorem 5.4 (Reversal Rule)

If x and y are any elements of a group G, then

$$(xy)^{-1} = y^{-1} x^{-1}$$

Proof

Let x^{-1} and y^{-1} are the inverses of x and y respectively in the group G.

Then

$$x x^{-1} = e = x^{-1} x \quad \text{...(1)}$$

$$y y^{-1} = e = y^{-1} y \quad \text{...(2)}$$

where e is identity element of G.

Now,

$$(xy)(y^{-1} x^{-1}) = x[y (y^{-1} x^{-1})] \quad [\text{by associativity in } G]$$

$$= x [(yy^{-1})x^{-1}]$$

$$= x [ex^{-1}] \quad [\text{From 2}]$$

$$= x x^{-1} \quad [\text{C is identity}]$$

$$= e \quad [\text{From 1}]$$

Therefore, $(xy)(y^{-1} x^{-1}) = e$ ...(3)

Also, $(y^{-1} x^{-1})(xy) = y^{-1} [x^{-1} (xy)]$ [by associativity in G]

$$= y^{-1} [(x^{-1} x)y]$$

$$= y^{-1} [e y] \quad [\text{From 1}]$$

$$= y^{-1} y \quad [\text{c is identity}]$$

$$= e \quad [\text{From 2}]$$

Therefore, $(y^{-1}x^{-1})(xy) = e$ ...(4)

Hence from (3) and (4), we have

$$(xy)(y^{-1}x^{-1}) = e = (y^{-1}x^{-1})(xy)$$

But by the closure property both xy and $y^{-1}x^{-1}$ are the elements of G. Then we conclude that $y^{-1}x^{-1}$ is the inverse of xy in G.

Therefore

$$y^{-1}x^{-1} = (xy)^{-1} \text{ or } (xy)^{-1} = y^{-1}x^{-1}$$

Corollary 5.1

The generalized form of the above theorem by induction is given as follows:

If $a, b, c, \ldots, s, t$ are in G, then

$$(abc \ldots st)^{-1} = t^{-1}s^{-1} \ldots c^{-1}b^{-1}a^{-1}$$

If the composition is addition, then the statement of this theorem will be $-(x + y) = (-y) + (-x)$

Theorem 5.5 (Cancellation Laws)

If x, y, z are any elements of G, then

i. $xy = xz \Rightarrow y = z$ [Left cancellation law]

ii. $yx = zx \Rightarrow y = z$ [Right cancellation law]

Proof

Let e be the identity and x^{-1} be the inverse of x in G. Then

i. $xy = xz$

Multiplying both sides on the left by x^{-1}, we get

$$x^{-1}(xy) = x^{-1}(xz)$$

By associativity

$$(x^{-1}x)y = (x^{-1}x)z$$

Because $x^{-1}x = e$, therefore

$$ey = ez$$

$$\Rightarrow y = z$$ [e is identity]

ii. Similarly, we can prove that

$$yx = zx \Rightarrow y = z.$$

In additive notation these results can be written as

$$x + y = x + z \Rightarrow y = z$$

and $y + x = z + x \Rightarrow y = z$

Theorem 5.6

If x, y are any two elements of a group G, then the equations $xt = y$ and $sx = y$ have unique solutions in G.

Proof

$$x \in G \Rightarrow x^{-1} \in G$$

$\therefore$ $x^{-1} \in G, y \in G \Rightarrow x^{-1} y \in G$ [by closure property]

$\therefore$ $x(x^{-1} y) = (x x^{-1}) y$ [by associativity]

$= ey$ [e is identity]

$= y$

$\therefore t = x^{-1} y$ is a solution of $x t = y$ in G.

Further to show that this solution is unique, let us suppose that at $t = t_1$ and $t = t_2$ are two solutions of the equation $xt = y$

Then $x t_1 = y$ and $x t_2 = y$

$\Rightarrow$ $x t_1 = x t_2$,

$\Rightarrow$ $t_1 = t_2$ [by left cancellation law in G]

Therefore, the solution is unique. Similarly, we can show that $sx = y$ has $y x^{-1}$ as its unique solution in G.

5.19-1 Alternative Definition of a Group

Theorem 5.7

A semigroup G is a group if for all elements $x, y \in G$, the equations $xt = y$ and $sx = y$ have unique solutions in G.

Proof

Let G be a non-empty set with an associative binary composition i.e.

$$x(yz) = (xy)z \;\forall\, x, y, z \in G.$$

To show that G is a group we only need to show that G contains an identity element and each element of G has an inverse.

Existence of Identity

Since G is a non-empty set, so let $x \in G$. Then equations $xt = x$ and $sx = x$ have unique solutions in G. Let these solutions be denoted by e_1 and e_2 respectively.

So $x e_1 = x$ and $e_2 x = x$ ...(1)

Again let y be any other element of G. Then as given there are unique t and s in G such that

$xt = y$ and $sx = y$ ...(2)

Now $sx = y \Rightarrow (sx) e_1 = y e_1$

$\Rightarrow s(x e_1) = y e_1$ [by associativity in G]

$\Rightarrow s x = y e_1$ [by 1]

$\Rightarrow y = y e_1$ [by 2]

therefore e_1 is right identity in G.

Also $xt = y \Rightarrow e_2 (xt) = e_2 y$

$\Rightarrow (e_2 x) t = e_2 y$ [by associativity in G]

$\Rightarrow xt = e_2 y$ [by 1]

$$\Rightarrow y = e_2 y \qquad \text{[by 2]}$$

therefore, e_2 is left identity in G. Since e_1 is right identity in G and $e_2 \in G$.

therefore $\qquad e_2 e_1 = e_2 \qquad$ ...(3)

Also since e_2 is left identity in G and $e_1 \in G$

therefore $\qquad e_2 e_1 = e_1 \qquad$ ...(4)

From (3) and (4) $\qquad e_1 = e_2$

Hence there exists an identity in G.

Existence of Inverse

Again using the given property for the elements $x, e \in G$. The equations $xt = e$ and $sx = e$ have unique solution in G denoted by t_1 and s_1 respectively. Then

$$x t_1 = \text{e and } s_1 x = \text{e} \qquad \text{...(5)}$$

Thus $\qquad x t_1 = e \Rightarrow s_1 (x t_1) = s_1 e$

$$\Rightarrow (s_1 x) t_1 = s_1 \qquad \text{[by associativity in } G\text{]}$$

$$\Rightarrow e t_1 = s_1 \qquad \text{[by 5]}$$

$$\Rightarrow t_1 = s_1$$

Therefore $x \in G$ has an inverse in G. Hence G is a group.

Theorem 5.8

A semigroup G is a group iff the equations $xt = y$ and $sx = y$ have unique solutions in G.

Proof

Combining theorems 5.6 and 5.7 we get theorem 5.8.

Theorem 5.9

If G is a finite semigroup such that for any $x, y, z \in G$.

i. $xy = xz \Rightarrow y = z$ and
ii. $yx = zx \Rightarrow y = z$

Then G is a group.

Proof

Since G is a non-empty finite set, let $G = \{x_1, x_2, \dots x_n\}$.

Consider $x \in G$. Since G is a semigroup, then the n products $xx_1, xx_2, \dots, xx_n$ are all elements of G.

We claim that these elements of G are all distinct, since for any i and j

$$x x_i = x x_j$$

$$\Rightarrow x_i = x_j \qquad \text{[by given property]}$$

but $x_i \neq x_j$, thus $x x_i \neq x x_j$

Therefore, $x x_1, xx_2, \dots, xx_n$ are n distinct elements of G placed in possibly some other order i.e.

$$G = \{x x_1, xx_2, \dots, xx_n\}$$

Now if $y \in G$, then y is one of the above n products. So for every $x, y \in G$, there exists an element $z \in G$ such that $xz = y$

Therefore, for every $x, y \in G$, the equation $xt = y$ has a unique solution in G. Similarly by taking n products

$$x_1 x, x_2 x, \ldots, x_n x.$$

It can be shown that for every $x, y \in G$; the equation $sx - y$ has a unique solution in G.

Hence by Theorem 5.7, G is a group.

Theorem 5.9 is not necessarily valid when G is an infinite group.

Theorem 5.10

A finite semigroup is a group if and only if for $x, y, z \in G$

i. $xy = xz \Rightarrow y = z$ ii. $yx = zx \Rightarrow y = z$

Proof

On combining Theorem 5.5 and Theorem 5.9 we get Theorem 5.10.

5.20 Integral Powers of an Element of a Group

Let $\{G, \cdot\}$ be a group and $a \in G$. Then by closure property a, aa, aaa, $aaaa$, ... are all elements of G, written as $a, a^2, a^3, a^4, \ldots$ respectively.

If n is a positive integer, then

$$a^n = aaa \ldots a \text{ (} n \text{ times)}$$

If n is negative integer, say $n = -p$ where p is a positive integer,

Then $$a^n = a^{-p} = (a^{-1})^p$$

where a^{-1} is the inverse of a in G.

Also, $$a^0 = e \; \forall \; a \in G$$

where e is the identity in G. Thus we have defined a^n for all integral values of n (positive, zero or negative).

5.21 Integral Multiples of an Element of a Group

If in a group G the composition is denoted by addition notation, then we write $n\,a$ in place of a^n where n is a positive integer and we define

$$na = a + a + \ldots. + a \text{ (} n \text{ terms)}$$

If p is a positive integer, then in place of a^{-p} we write $(-p)\,a$

Therefore $$(-p)\,a = p(-a)$$

where $-a$ is the additive inverse of a in G.

Also in place of a^0 we write $0a$, therefore $0a = e$, where e is the identity element in G.

5.22 Laws of Indices

If $a \in G$ then

i. For multiplication composition the laws of indices are

$$a^m a^n = a^{m+n}$$

$$(a^m)^n = a^{mn} \; \forall \; m, n \in I$$

ii. For addition composition the laws of indices are

$$ma + na = (m + n)\, a$$

$$n\,(m\,a) = (n\,m)a \;\forall\; m, n \in I$$

5.23 Order of an Element of a Group

Let $(G, \cdot)$ is a group and e is the identity element in G. If there exists a positive integer n such that

$$a^n = e$$

then the **order** of an element a is finite. If there exists no such positive integer, then a is of **infinite** or **zero** order.

Now, the order of an element $a \in G$, denoted by $O(a)$ is the least positive integer n (if exists), such that $a^n = e$

1. *For addition composition we write $na = e$ in place of $a^n = e$.*
2. *If $a \in G$ and n be a positive integer such that $a^n = e$, then $O(a) \leq n$*

Example 5.25

Let $(G, \cdot)$ be a group and $e \in G$ is identity element in G.

Then, $\quad e^1 = e \Rightarrow O(e) = 1$

Thus in any group the identity element e is always of order one and it is the only element of order one.

Example 5.26

Consider a multiplicative group $\{1, -1, i, -i\}$

Since $\quad 1^1 = 1$ therefore $O(1) = 1$

Now $\quad (-1)^1 = -1$

$(-1)^2 = 1$ [identity element]

therefore $\quad O(-1) = 2$

Again $\quad (i)^1 = i$

$(i)^2 = -1$

$(i)^3 = -i$

$(i)^4 = 1$ [identity element]

thus $\quad O(i) = 4$

Finally $\quad (-i)^1 = -i$

$(-i)^2 = -1$

$(-i)^3 = i$

$(-i)^4 = 1$ [identity element]

Hence $\quad O(-i) = 4$

Example 5.27

Consider the group $[\{0, 1, 2, 3, 4\}, +_5]$. Since 0 is the identity element, therefore

$$O(0) = 1$$

Now $(1)^1 = 1$

$(1)^2 = 1 +_5 1 = 2$

$(1)^3 = 1 +_5 1 +_5 1 = 2 +_5 1 = 3$

$(1)^4 = 1 +_5 1 +_5 1 +_5 1 = 3 +_5 1 = 4$

$(1)^5 = 1 +_5 1 +_5 1 +_5 1 +_5 1 = 4 +_5 1 = 0$ [the identity element]

therefore $O(1) = 5$

Again $2^1 = 2$

$2^2 = 2 +_5 2 = 4$

$2^3 = 2 +_5 2 +_5 2 = 4 +_5 2 = 1$

$2^4 = 2 +_5 2 +_5 2 +_5 2 = 1 +_5 2 = 3$

$2^5 = 2 +_5 2 +_5 2 +_5 2 +_5 2 = 3 +_5 2 = 0$

The identity element, therefore O (2) = 5.

Further by similar discussion O (3) = 5 and O (4) = 5.

Example 5.28

In the group $(Z, +)$, the order of 0 is 1 and the order of every non-zero integer is infinite, because for any $x \in Z$, $x \neq 0$ there exist no positive integer n such that $nx = 0$.

5.24 Properties of The Order of an Element of a Group

In an infinite group elements may be of finite as well as of infinite order. Now we are giving some important theorems on the order of an element of a group.

Theorem 5.11

The order of every element of a finite group is finite and less than or equal to the order of the group.

i.e. $$O(a) \leq O(G) \ \forall \ a \in G.$$

Proof

Let $(G, .)$ be a finite group of order n and $a \in G$. Now consider the elements $a^0 = e, a^1, a^2, \ldots, a^n$

By closure property, all these elements belong to G. These are $n+1$ in number. Since G has a finite number of elements therefore, all these elements are not distinct. Suppose that

$$a^j = a^k \ (0 \leq j < k \leq n)$$

Now $$a^j = a^k$$

$$\Rightarrow a^j . a^{-j} = a^k a^{-j} \qquad [a^{-j} \in G]$$

$$\Rightarrow a^0 = a^{k-j}$$

$$e = a^r$$

where $r = k - j$. Since $k > j$, therefore r is a positive integer less than equal n. Thus there exists a positive integer r such that

$$a^r = e$$

Hence O (a) is finite and less than or equal to O (G).

Theorem 5.12

The order of an element a of group G is the same as that of its inverse a^{-1}.

Proof

Let $O(a) = n$ and $O(a^{-1}) = m$

Now

$$O(a) = n$$

$$\Rightarrow a^n = e \quad \text{[identity element]}$$

$$\Rightarrow (a^n)^{-1} = e^{-1}$$

$$(a^{-1})^n = e$$

$$O(a^{-1}) \leq n$$

$$m \leq n \quad \ldots(1)$$

Also

$$O(a^{-1}) = m$$

$$(a^{-1})^m = e \quad \text{[identity element]}$$

$$(a^m)^{-1} = e$$

$$a^m = e^{-1} = e$$

$$O(a) \leq m$$

$$n \leq m \quad \ldots(2)$$

From (1) and (2) $m = n$

Thus it follows that $O(a^{-1}) = O(a)$

Theorem 5.13

The order of any integral power of an element a cannot exceed the power of a.

Proof

Let $a \in G$ and $O(a) = n$.

Let a^m be any integral power of a.

Since $O(a) = n$, then $a^n = e$

Where e is the identity element in G.

Now

$$(a^n)^m = e^m$$

$$a^{nm} = e$$

$$(a^m)^n = e$$

$$O(a^m) \leq n.$$

Theorem 5.14

If the element a of group G is of order n, then $a^k = e$ iff n is a divisor of k.

Proof

Let $a^k = e$

Since k is an integer and n is a positive integer, then by division algorithm there exists integers p and q, such that

$$k = np + q, 0 \leq q < n$$

Now, $a^k = a^{np+q}$

$= a^{np} . a^q$ [by indices]

$= (a^n)^p . a^q = e^p . a^q$ [$\because a^n = e$]

$= a^q$

Therefore $a^k = a^q = e$

Since $0 \leq q < n$ and $a^n = e$

Then $a^q \neq e$ unless $q = 0$

$\therefore$ $k = np$ or n is a divisor of k.

Conversely, let n is a divisor of k, then there exists an integer P such that

$$k = np$$

Now $a^k = a^{np}$

$= (a^n)^p$

$= e^p$ [$\because a^n = e$]

$= e$

Hence $a^k = e$ iff n is a divisor of k.

Corollary 5.2

The order of any power of an element a of group G is a divisor of the order of a.

Proof

Let O $(a) = n$ and O $(a^r) = m$, where r is any integer

Now $(a^r)^n = (a^n)^r = e^r = e$

Then by theorem 5.14, O (a^r) is a divisor of the order of a.

Theorem 5.15

The orders of the elements a and $x^{-1} a x$ are the same where a, x are any two elements of a group G.

Proof

Let $a \in G$ and $x \in G$ then

$(x^{-1} a x)^2 = (x^{-1} a x) (x^{-1} a x)$

$= x^{-1} a (x x^{-1}) a x$ [by associativity]

$= x^{-1} a e a x$ [$\because x x^{-1} = e$]

$= x^{-1} a^2 x$

Assuming $(x^{-1} a x)^{m-1} = x^{-1} a^{m-1} x$ for any positive integer m, then

$(x^{-1} a x)^m = (x^{-1} a x)^{m-1} (x^{-1} a x)$

$= (x^{-1} a^{m-1} x) (x^{-1} a x)$

$= x^{-1} a^{m-1} (x x^{-1}) a x$ [by associativity]

$= x^{-1} a^{m-1} e a x$ [$\because x x^{-1} = e$]

$= x^{-1} a^{m-1} a x$

$$= x^{-1} a^m x$$

By principle of mathematical induction

$$(x^{-1} a x)^m = x^{-1} a^m x \ \forall\, m \in N.$$

Let O $(a) = m \Rightarrow a^m = e$ where e is the identity element in G.

Now $$(x^{-1} a x)^m = x^{-1} a^m x$$
$$= x^{-1} e x \qquad [\because a^m = e]$$
$$= x^{-1} x$$
$$= e$$

Therefore, $x^{-1} a x$ is also of finite order and

$$\text{O}(x^{-1} a x) \leq m = \text{O}(a) \qquad \dots (1)$$

In the similar way

$$\text{O}(x^{-1} a x) = n$$
$$(x^{-1} a x)^n = e$$
$$x^{-1} a^n x = e$$
$$\Rightarrow a^n = x e x^{-1}$$
$$= x x^{-1}$$
$$= e$$
$$O(a) \leq n = O(x^{-1} a x) \qquad \dots (2)$$

Therefore, if a is of finite order, then $x^{-1} a x$ is also of finite order and

$$\text{O}(x^{-1} a x) \leq \text{O}(a) \leq \text{O}(x^{-1} a x)$$
$$\Rightarrow \text{O}(x^{-1} a x) = \text{O}(a)$$

Corollary 5.3

The orders of ab and ba are equal, where a and b are any elements of a group.

Proof

Consider $$a^{-1}(ab)a = (a^{-1}a)(ba) \text{ [by associativity]}$$
$$= e(ba)$$
$$= ba$$

Thus $$ba = a^{-1}(ab)a$$

Therefore, order of $$ba = \text{order of } a^{-1}(ab)a \qquad \dots (1)$$

As we know that $$\text{O}(x^{-1} a x) = \text{O}(a)$$

From (1), order of ba is equal to order of ab.

Example 5.29

If for every element a of a group G, $a^2 = e$; then prove that G is an abelian group.

Solution

Let a and b be any two elements of the group G. Then from the given property $a^2 = e$

$$\Rightarrow aa = e$$

$$a^{-1} = a \qquad \text{...(1) [by definition of inverse element]}$$

Similarly $$b^2 = e$$

$$\Rightarrow b^{-1} = b \qquad \text{...(2)}$$

Now since $a \in G$, $b \in G$, then by closure property $a\, b \in G$

Therefore by given property

$$(ab)^2 = e$$

$$(ab)\,(ab) = e$$

$$(ab)^{-1} = a\, b \qquad \text{[by definition of inverse element]}$$

$$b^{-1}\, a^{-1} = ab \qquad \text{[by reversal rule]}$$

By using (1) and (2), we get

$$ba = ab$$

Hence G is an abelian group.

Example 5.30

If a, b are any two elements of a group G, then show that

$(ab)^2 = a^2\, b^2$ if and only if G is abelian.

Solution

Suppose G is abelian then we have to show that

$$(ab)^2 = a^2\, b^2$$

Now
$$\begin{aligned}(ab)^2 &= (ab)\,(ab) \\ &= a\,(ba)\,b && \text{[by associativity]} \\ &= a\,(ab)\,b && \text{[}G\text{ is abelian]} \\ &= (aa)\,(bb) && \text{[by associativity]} \\ &= a^2\, b^2\end{aligned}$$

Conversely, let $(ab)^2 = a^2\, b^2$, $\forall\ a, b \in G$, then we have to show that G is abelian.

Now,
$$\begin{aligned}(ab)^2 &= a^2\, b^2 \\ (ab)\,(ab) &= (aa)\,(bb) \\ a\,(ba)\,b &= a(a\, b) && \text{[by associativity]} \\ (ba)\,b &= (a\, b)\,b && \text{[by left cancellation law]} \\ \Rightarrow ba &= ab && \text{[by right cancellation law]}\end{aligned}$$

G is an abelian group.

Example 5.31

Find the solution of the equation $a\,b\,x\,a\,x = c\,b\,x$ in a group G, where a, b and c are given elements of G.

Solution

We have $a, b, c \in G$. Then $a^{-1}, b^{-1}, c^{-1} \in G$.

Now
$$a\,b\,x\,a\,x = c\,b\,x$$

$$\Rightarrow \quad a\,b\,x\,a = c\,b \qquad \text{[by right cancellation law]}$$

$$a^{-1}\,a\,b\,x\,a = a^{-1}c\,b$$

$$ebxa = a^{-1}cb$$

$$b\,x\,a = a^{-1}c\,b \qquad [\because aa^{-1}=e]$$

$$b^{-1}\,b\,x\,a = b^{-1}a^{-1}c\,b$$

$$e\,x\,a = b^{-1}a^{-1}c\,b$$

$$x\,a = b^{-1}a^{-1}c\,b$$

$$x\,a\,a^{-1} = b^{-1}a^{-1}c\,b\,a^{-1}$$

$$x\,e = b^{-1}a^{-1}c\,b\,a^{-1}$$

$$x = b^{-1}a^{-1}c\,b\,a^{-1}$$

Therefore, $x = b^{-1}a^{-1}c\,b\,a^{-1}$ is the solution of $a\,b\,x\,a\,x = c\,b\,x$ in a group G.

Example 5.32

Show that in a group G, we have $ab = e \Rightarrow a = b^{-1}$ and $b = a^{-1}$.

Solution

Given,

$$ab = e$$

$$a^{-1}a\,b = a^{-1}e$$

$$e\,b = a^{-1}$$

$$b = a^{-1}$$

Also

$$ab = e$$

$$a\,bb^{-1} = e\,b^{-1}$$

$$a\,e = b^{-1}$$

$$a = b^{-1}$$

Therefore, if $ab = e$ then $a = b^{-1}$ and $b = a^{-1}$.

Example 5.33

Prove that a group G is abelian if and only if

$$(ab)^{-1} = a^{-1}b^{-1}\ \forall\ a, b \in G.$$

Solution

Suppose G is abelian, then

$$a\,b = b\,a\ \forall\ a, b \in G. \qquad \dots (1)$$

Now

$$(a\,b)\,(a^{-1}\,b^{-1}) = (b\,a)\,(a^{-1}\,b^{-1}) \qquad \text{[from 1]}$$

$$= b\,(a\,a^{-1})\,b^{-1}.$$

$$= b\,e\,b^{-1}$$

$$= bb^{-1} = e$$

Also

$$(a^{-1}\,b^{-1})\,(a\,b) = (a^{-1}\,b^{-1})\,(b\,a) \qquad \text{[from 1]}$$

$$= a^{-1}\,(b^{-1}b)\,a$$

$$= a^{-1}\,e\,a$$

$$= a^{-1} a = e$$

$$(a\, b)^{-1} = a^{-1} b^{-1}$$

Conversely, let $(a\, b)^{-1} = a^{-1} b^{-1} \;\forall\, a, b \in G.$

$$b^{-1} a^{-1} = a^{-1} b^{-1} \qquad \text{[by reversal rule]}$$

$\Rightarrow$ $bb^{-1} a^{-1} = ba^{-1} b^{-1}$

$\Rightarrow$ $e\, a^{-1} = b\, a^{-1} b^{-1}.$

$\Rightarrow$ $a^{-1} = b\, a^{-1} b^{-1}.$

$\Rightarrow$ $a^{-1} b = b\, a^{-1} b^{-1} b$

$\Rightarrow$ $a^{-1} b = b\, a^{-1} e$

$\Rightarrow$ $a^{-1} b = b\, a^{-1}.$

$\Rightarrow$ $a\, a^{-1} b = a\, b\, a^{-1}.$

$\Rightarrow$ $e\, b = a\, b\, a^{-1}.$

$\Rightarrow$ $b = a\, b\, a^{-1}.$

$\Rightarrow$ $b\, a = a\, b\, a^{-1} a$

$\Rightarrow$ $b\, a = a\, b\, e$

$\Rightarrow$ $b\, a = a\, b$

Hence G is abelian.

Example 5.34

If a and b are any elements of a group G, then

$$(b\, a\, b^{-1})^n = b\, a^n\, b^{-1} \text{ for any integer } n.$$

Solution

i. If $n = 0$.

Then $(b\, a\, b^{-1})^0 = e$ [by definition]

Also

$$b\, a^0\, b^{-1} = b\, e\, b^{-1}$$

$$= b\, b^{-1} = e$$

$$(b\, a\, b^{-1})^0 = b\, a^0 b^{-1}$$

ii. If $n > 0$.

Then

$$(b\, a\, b^{-1})^1 = b\, a\, b^{-1}$$

$$= b\, a^1\, b^{-1}$$

Therefore, the result is true for $n = 1$

Suppose that the result is true for $n = m$, then

$$(b\, a\, b^{-1})^m = b\, a^m\, b^{-1}.$$

Consider

$$(b\, a\, b^{-1})^{m+1} = (b\, a b^{-1})^m (b\, a\, b^{-1})$$

$$= (b\, a^m\, b^{-1})(b\, a^1\, b^{-1})$$

$$= b\, a^m\, (b^{-1} b)\, a^1\, b^{-1}$$

$$= b\,a^m\,e\,a^1\,b^{-1}$$
$$= b\,a^m\,a^1\,b^{-1}$$
$$= b\,a^{m+1}\,b^{-1}$$

Therefore the result is true for $n = m + 1$ if it is true for $n = m$. But we have seen that the result is true for $n = 1$. By mathematical induction the result is true for all $n > 0$.

ii. If $n < 0$. Let $n = -k\ \forall\ k > 0$

Now
$$(b\,a\,b^{-1})^n = (b\,a\,b^{-1})^{-k}$$
$$= [\,(b\,a\,b^{-1})^k\,]^{-1}$$
$$= (b\,a^k\,b^{-1})^{-1}$$
$$= (b^{-1})^{-1}\,(a^k)^{-1}\,b^{-1} \qquad \text{[by reversal rule]}$$
$$= b\,a^{-k}\,b^{-1}$$

Hence, from (i), (ii) and (iii) it is clear that $(b\,a\,b^{-1})^n = b\,a^n\,b^{-1}$ for any integer n.

Example 5.35

If G is a group such that $(ab)^n = a^n\,b^n$ for three consecutive integers n for all $a, b \in G$. Show that G is abelian.

Solution

Let $a, b \in G$ and $n, n+1, n+2$ are three consecutive integers such that

$$(ab)^n = a^n\,b^n \qquad \ldots(1)$$
$$(ab)^{n+1} = a^{n+1}\,b^{n+1} \qquad \ldots(2)$$
$$(ab)^{n+2} = a^{n+2}\,b^{n+2} \qquad \ldots(3)$$

Now, we have

$$(ab)^{n+2} = (ab)^{n+1}\,(ab)$$
$$\Rightarrow \quad a^{n+2}\,b^{n+2} = a^{n+1}\,b^{n+1}\,(ab)$$
$$\Rightarrow \quad a\,a^{n+1}\,b^{n+1}\,b = a\,a^n\,b^n\,b\,a\,b$$

By left and right cancellation laws

$$a^{n+1}\,b^{n+1} = a^n\,b^n\,b\,a$$
$$(ab)^{n+1} = (a\,b)^n\,ba$$
$$(ab)^n\,(a\,b) = (ab)^n\,(ba)$$
$$ab = ba \qquad \text{[by left cancellation law]}$$

G is abelian.

Exercise 5.2

1. If a group G has four elements. Show that it must be abelian.
2. Prove that a group G is abelian if $b^{-1}\,a^{-1}\,b\,a = e\ \forall\ a, b \in G$.
3. If G is a group of even order, prove that it has an element $a \neq e$ satisfying $a^2 = e$.
4. If in a group G, $a^5 = e$, $aba^{-1} = b^2$ for $a, b \in G$. Find $O(b)$.

5. If G is a semi-group and for all $a, b \in G, a^2 b = b = b a^2$. Prove that G is an abelian group.
6. If every element of a group G is its own inverse then show that G is abelian.
7. If G is a finite group, show that there exists a positive integer n such that $a^n = e \; \forall \; a \in G$.
8. Give an example of an infinite group each element of which has a finite order.
9. Find the order of each element of the groups.
   i. $(G, +_7)$
   ii. $G = \{1, w, w^2\}$ where $w^3 = 1$ and composition is multiplication
   iii. $G = \{a, a^2, a^3, a^4, a^5, a^6 = e\}$.
10. Give an example of each of the following:
    a. A semi-group without identity, in which cancellation law hold.
    b. A semi-group with identity, in which cancellation law does not hold.
11. If the elements a, b of a group commute and O $(a) = m$, O $(b) = n$, where m and n are relatively prime, prove that O $(ab) = mn$.
12. Prove that the left inverse of an element is also a right inverse of the element

5.25 Permutation Groups

We have already discussed certain examples of group structure in previous section. Most of the examples taken are of abelian group. Here, in this section we shall take some more examples, whose elements are functions called permutations. This is an example of a non-abelian group. Its importance can be judged from the fact established later on as Cayley's theorem that every group is structurally same as some permutation group.

5.25-1 Permutation

Let X be a finite set having n distinct elements. Then a function (mapping) $f: X \to X$ is said to be a **permutation** of degree n if

i. f is one-one
ii. f is onto

The number of distinct elements in the finite set X is called the **degree** of permutation.

5.25-2 Notation for a Permutation

Let X be a finite set having n distinct elements say $\{a_1, a_2, \ldots, a_n\}$. If $f: X \to X$ and f is one-one onto then f is a permutation of degree n. We know that any function on X is well defined iff $f(a_1), f(a_2), \ldots, f(a_n)$ are given. Let $f(a_1) = b_1, f(a_2) = b_2, \ldots, f(a_n) = b_n$, where $b_1, b_2, \ldots, b_n$ are some arrangement of n elements of X. We shall use the two row standard notation denoting such a function f as below

$$f = \begin{pmatrix} a_1 & a_2 & a_3 & \ldots & a_n \\ b_1 & b_2 & b_3 & \ldots & b_n \end{pmatrix}.$$

Example 5.36

If $X = \{1, 2, 3, 4, 5, 6\}$ is a finite set having six elements,

Then $f = \begin{pmatrix} 1 & 2 & 3 & 4 & 5 & 6 \\ 2 & 5 & 4 & 3 & 6 & 1 \end{pmatrix}$ is a permutation of degree 6 such that $f(1) = 2, f(2) = 5, f(3) = 4, f(4) = 3, f(5) = 6$ and $f(6) = 1$.

Example 5.37

If $X = \{1, 2, 3, 4\}$ then $f = \begin{pmatrix} 1 & 2 & 3 & 4 \\ 2 & 2 & 4 & 1 \end{pmatrix}$ is a function on X to itself but is not a permutation. Here $f(1) = 2 = f(2)$ i.e. f is not one-one.

5.25-3 Equality of two Permutations

Let f and g are permutations of a finite set X. The two permutations are equal if

i. degree of f = degree of g

ii. $f(a) = g(a)\ \forall\ a \in X$.

Example 5.38

If $f = \begin{pmatrix} 1 & 2 & 3 \\ 2 & 3 & 1 \end{pmatrix}$ and $g = \begin{pmatrix} 2 & 3 & 1 \\ 3 & 1 & 2 \end{pmatrix}$ are two permutations of a finite set $X = \{1, 2, 3\}$. Here, both the permutations are of degree 3 and $f(a) = g(a)$ for all $a \in X$ i.e. $f(1) = g(1) = 2$

$$f(2) = g(2) = 3$$
$$f(3) = g(3) = 1$$

Hence, f and g are equal.

5.25-4 Total Number of Distinct Permutations of Degree *n*

Consider a finite set X having n distinct elements, then there exists $n!$ distinct combinations of the elements of X. Hence, there will be $n!$ distinct permutations of degree n. If P_n be the set having all such permutations of degree n, then the set P_n will have $n!$ distinct elements.

This set P_n is said to be a **symmetric set of permutations** of degree n.

Example 5.39

The set P_3 of all permutations of degree 3 have $3! = 6$ elements If $X = \{a, b, c\}$, then

$$P_3 = \left\{\begin{pmatrix} a & b & c \\ a & b & c \end{pmatrix}, \begin{pmatrix} a & b & c \\ b & a & c \end{pmatrix}, \begin{pmatrix} a & b & c \\ a & c & b \end{pmatrix}, \begin{pmatrix} a & b & c \\ c & b & a \end{pmatrix}, \begin{pmatrix} a & b & c \\ b & c & a \end{pmatrix}, \begin{pmatrix} a & b & c \\ c & a & b \end{pmatrix}\right\}$$

5.25-5 Identity permutation

Let X be a finite set having n distinct elements. A permutation I is called **identity permutation** if

$$I(a) = a\ \forall\ a \in X.$$

Thus, if $X = \{1, 2, \ldots, n\}$

Then
$$I = \begin{pmatrix} 1 & 2 & 3 & \ldots & n \\ 1 & 2 & 3 & \ldots & n \end{pmatrix}$$

is the identity permutation of degree n.

5.25-6 Product or Composite of two Permutations

If f and g be any two permutations of degree n, then the **product** or **composite** of two permutations is denoted by fg or $f \circ g$ and is obtained by first carrying out the operation defined by g and then by f. The degree of fg is also n.

Example 5.40

Let $f = \begin{pmatrix} 1 & 2 & 3 & 4 & 5 \\ 2 & 4 & 5 & 1 & 3 \end{pmatrix}$ and $g = \begin{pmatrix} 1 & 2 & 3 & 4 & 5 \\ 1 & 3 & 5 & 4 & 2 \end{pmatrix}$ be permutations of the set $X = \{1, 2, 3, 4, 5\}$.

Then
$$fg(1) = f[g(1)] = f(1) = 2$$
$$fg(2) = f[g(2)] = f(3) = 5$$
$$fg(3) = f[g(3)] = f(5) = 3$$
$$fg(4) = f[g(4)] = f(4) = 1$$
$$fg(5) = f[g(5)] = f(2) = 4$$

Hence

$$fg = \begin{pmatrix} 1 & 2 & 3 & 4 & 5 \\ 2 & 4 & 5 & 1 & 3 \end{pmatrix}\begin{pmatrix} 1 & 2 & 3 & 4 & 5 \\ 1 & 3 & 5 & 4 & 2 \end{pmatrix}$$
$$= \begin{pmatrix} 1 & 2 & 3 & 4 & 5 \\ 2 & 5 & 3 & 1 & 4 \end{pmatrix}$$

Example 5.41

Show that the multiplication of permutations is not in general commutative.

Solution

To show this consider

$$f = \begin{pmatrix} 1 & 2 & 3 & 4 \\ 2 & 4 & 3 & 1 \end{pmatrix} \text{ and } g = \begin{pmatrix} 1 & 2 & 3 & 4 \\ 4 & 2 & 1 & 3 \end{pmatrix}$$

Now $fg = \begin{pmatrix} 1 & 2 & 3 & 4 \\ 2 & 4 & 3 & 1 \end{pmatrix}\begin{pmatrix} 1 & 2 & 3 & 4 \\ 4 & 2 & 1 & 3 \end{pmatrix} = \begin{pmatrix} 1 & 2 & 3 & 4 \\ 1 & 4 & 2 & 3 \end{pmatrix}$

$gf = \begin{pmatrix} 1 & 2 & 3 & 4 \\ 4 & 2 & 1 & 3 \end{pmatrix}\begin{pmatrix} 1 & 2 & 3 & 4 \\ 2 & 4 & 3 & 1 \end{pmatrix} = \begin{pmatrix} 1 & 2 & 3 & 4 \\ 2 & 3 & 1 & 4 \end{pmatrix}$

Thus $fg \neq gf$

Again let $f = \begin{pmatrix} 1 & 2 & 3 \\ 2 & 3 & 1 \end{pmatrix}$ and $g = \begin{pmatrix} 1 & 2 & 3 \\ 3 & 1 & 2 \end{pmatrix}$

Then $fg = \begin{pmatrix} 1 & 2 & 3 \\ 1 & 2 & 3 \end{pmatrix}$ and $gf = \begin{pmatrix} 1 & 2 & 3 \\ 1 & 2 & 3 \end{pmatrix}$

Thus $fg = gf$.

5.26 Group of Permutations

Theorem 5.16

The set P_n of all permutations on n symbols is a finite group of order $n!$ with respect to product of mappings as the operation. For $n \leq 2$, P_n is abelian and for $n > 2$ it is always non-abelian.

Proof

Let $X = \{a_1, a_2, \ldots, a_n\}$ is a finite set with n distinct elements.

Let $f = \begin{pmatrix} a_1 & a_2 & \cdots & a_n \\ b_1 & b_2 & \cdots & b_n \end{pmatrix}$ be a permutation of degree n, where $b_1 = f(a_1)$, $b_2 = f(a_2)$, ... , $b_n = f(a_n)$ are n distinct elements of X.

Since the different arrangements of the elements of X are $n!$, then the number of distinct permutations of degree n will be $n!$. If P_n is the set of all such permutations, then P_n has $n!$ distinct elements. Now, P_n will be a group if the following properties hold

i. Closure Property

Let

$$f = \begin{pmatrix} b_1 & b_2 & \cdots & b_n \\ c_1 & c_2 & \cdots & c_n \end{pmatrix} \text{ and}$$

$$g = \begin{pmatrix} a_1 & a_2 & \cdots & a_n \\ b_1 & b_2 & \cdots & b_n \end{pmatrix}$$

be any two permutations of degree n. Then

$$fg \; \begin{matrix} b_1 & b_2 & \ldots & b_n \\ c_1 & c_2 & \ldots & c_n \end{matrix} \quad \begin{matrix} a_1 & a_2 & \ldots & a_n \\ b_1 & b_2 & \ldots & b_n \end{matrix} \quad \begin{matrix} a_1 & a_2 & \ldots & a_n \\ c_1 & c_2 & \ldots & c_n \end{matrix}$$

Since, $c_1, c_2, \ldots, c_n$ are also an arrangement of n elements $a_1, a_2, \ldots, a_n$ of X, then fg is a permutation of degree n.

Thus $fg \in P_n \; \forall f, g \in P_n$

Hence P_n is closed for the composition known as product of two permutations.

ii. Associativity

Let $\quad f = \begin{pmatrix} c_1 & c_2 & \cdots & c_n \\ d_1 & d_2 & \ldots & d_n \end{pmatrix}, g = \begin{pmatrix} b_1 & b_2 & \cdots & b_n \\ c_1 & c_2 & \cdots & c_n \end{pmatrix}$ and $h = \begin{pmatrix} a_1 & a_2 & \cdots & a_n \\ b_1 & b_2 & \cdots & b_n \end{pmatrix}$

be any three permutations of degree n, then

$$fg = \begin{pmatrix} c_1 & c_2 & \cdots & c_n \\ d_1 & d_2 & \ldots & d_n \end{pmatrix}\begin{pmatrix} b_1 & b_2 & \cdots & b_n \\ c_1 & c_2 & \cdots & c_n \end{pmatrix} = \begin{pmatrix} b_1 & b_2 & \cdots & b_n \\ d_1 & d_2 & \cdots & d_n \end{pmatrix}$$

$$(fg)h = \begin{pmatrix} b_1 & b_2 & \cdots & b_n \\ d_1 & d_2 & \cdots & d_n \end{pmatrix}\begin{pmatrix} a_1 & a_2 & \cdots & a_n \\ b_1 & b_2 & \cdots & b_n \end{pmatrix} = \begin{pmatrix} a_1 & a_2 & \cdots & a_n \\ d_1 & d_2 & \ldots & d_n \end{pmatrix} \quad \ldots (1)$$

Also

$$gh = \begin{pmatrix} b_1 & b_2 & \cdots & b_n \\ c_1 & c_2 & \cdots & c_n \end{pmatrix}\begin{pmatrix} a_1 & a_2 & \cdots & a_n \\ b_1 & b_2 & \cdots & b_n \end{pmatrix} = \begin{pmatrix} a_1 & a_2 & \cdots & a_n \\ c_1 & c_2 & \cdots & c_n \end{pmatrix}$$

$$f(gh) = \begin{pmatrix} c_1 & c_2 & \cdots & c_n \\ d_1 & d_2 & \cdots & d_n \end{pmatrix}\begin{pmatrix} a_1 & a_2 & \cdots & a_n \\ c_1 & c_2 & \cdots & c_n \end{pmatrix} = \begin{pmatrix} a_1 & a_2 & \cdots & a_n \\ d_1 & d_2 & \cdots & d_n \end{pmatrix} \quad \ldots (2)$$

Now from (1) and (2), we get $(fg)h = f(gh)$. Hence the composition is associative in P_n.

iii. Existence of Identity

The identity permutation of degree n is the identity element of P_n.

Let $\quad f = \begin{pmatrix} a_1 & a_2 & \cdots & a_n \\ b_1 & b_2 & \cdots & b_n \end{pmatrix}$ and $I = \begin{pmatrix} a_1 & a_2 & \cdots & a_n \\ a_1 & a_2 & \cdots & a_n \end{pmatrix}$

Then $\quad f\,I = \begin{pmatrix} a_1 & a_2 & \cdots & a_n \\ b_1 & b_2 & \cdots & b_n \end{pmatrix}\begin{pmatrix} a_1 & a_2 & \cdots & a_n \\ a_1 & a_2 & \cdots & a_n \end{pmatrix} = \begin{pmatrix} a_1 & a_2 & \cdots & a_n \\ b_1 & b_2 & \cdots & b_n \end{pmatrix} = f$

Also $\quad If = \begin{pmatrix} b_1 & b_2 & \dots & b_n \\ b_1 & b_2 & \dots & b_n \end{pmatrix}\begin{pmatrix} a_1 & a_2 & \dots & a_n \\ b_1 & b_2 & \dots & b_n \end{pmatrix} = \begin{pmatrix} a_1 & a_2 & \dots & a_n \\ b_1 & b_2 & \dots & b_n \end{pmatrix} = f$

Thus $\quad fI = If = f \ \forall f, I \in P_n$

iv. Existence of Inverse

Let $\quad f = \begin{pmatrix} a_1 & a_2 & \dots & a_n \\ b_1 & b_2 & \dots & b_n \end{pmatrix}$ be a permutation of degree n then the permutation

$f^{-1} = \begin{pmatrix} b_1 & b_2 & \dots & b_n \\ a_1 & a_2 & \dots & a_n \end{pmatrix}$ is also a permutation of degree n.

Now, $\quad ff^{-1} = \begin{pmatrix} a_1 & a_2 & \dots & a_n \\ b_1 & b_2 & \dots & b_n \end{pmatrix}\begin{pmatrix} b_1 & b_2 & \dots & b_n \\ a_1 & a_2 & \dots & a_n \end{pmatrix} = \begin{pmatrix} b_1 & b_2 & \dots & b_n \\ b_1 & b_2 & \dots & b_n \end{pmatrix} = I$

Also $\quad f^{-1}f = \begin{pmatrix} b_1 & b_2 & \dots & b_n \\ a_1 & a_2 & \dots & a_n \end{pmatrix}\begin{pmatrix} a_1 & a_2 & \dots & a_n \\ b_1 & b_2 & \dots & b_n \end{pmatrix} = \begin{pmatrix} a_1 & a_2 & \dots & a_n \\ a_1 & a_2 & \dots & a_n \end{pmatrix} = I$

Therefore, f^{-1} is the inverse of f.

Hence P_n is a group of order $n!$ for the given composition. Now, for $n = 1$, the set P_n has only one element and we know that every group of order one is abelian. Thus P_n is abelian for $n = 1$.

For $n = 2$, the number of elements in the set P_n will be $2! = 2$ and every group of order 2 is abelian therefore P_n is abelian for $n = 2$. Now for $n > 2$ we shall show that P_n is non-abelian.

Let $\quad f = \begin{pmatrix} 1 & 2 & 3 \\ 2 & 1 & 3 \end{pmatrix}$ and $g = \begin{pmatrix} 1 & 2 & 3 \\ 3 & 1 & 2 \end{pmatrix}$ then

$$fg = \begin{pmatrix} 1 & 2 & 3 \\ 2 & 1 & 3 \end{pmatrix}\begin{pmatrix} 1 & 2 & 3 \\ 3 & 1 & 2 \end{pmatrix} = \begin{pmatrix} 1 & 2 & 3 \\ 3 & 2 & 1 \end{pmatrix} \qquad \dots(1)$$

$$gf = \begin{pmatrix} 1 & 2 & 3 \\ 3 & 1 & 2 \end{pmatrix}\begin{pmatrix} 1 & 2 & 3 \\ 2 & 1 & 3 \end{pmatrix} = \begin{pmatrix} 1 & 2 & 3 \\ 1 & 3 & 2 \end{pmatrix} \qquad \dots(2)$$

From (1) and (2), we get $fg \neq gf$. Hence, P_n is non abelian for $n > 2$.

*The group P_n of all permutations of degree n is said to be the **symmetric group** of degree n.*

5.27 Cyclic Permutations or Cycles

Let X be a finite set. A permutation of X is said to be a **cyclic permutation** or a **cycle** if there exists elements $a_1, a_2, \dots, a_n$ in X such that $f(a_1) = a_2, f(a_2) = a_3, \dots, f(a_{n-1}) = a_n, f(a_n) = a_1$ and for any $x \in X$ different from $a_1, a_2, \dots, a_n$, $f(x) = x$. We denote f by **one-row notation** $(a_1, a_2 \dots a_n)$. Further n is called the **length of the cycle** f.

For example, suppose $A = (a, b, c, d)$ is a set of 4 elements. A permutation on the elements denoted by α_1 is represented as $\alpha_1 = \begin{pmatrix} a & b & c & d \\ \downarrow & \downarrow & \downarrow & \downarrow \\ b & d & c & a \end{pmatrix}$. This permutation can also be represented with the help of a digraph as shown in Fig.5.1

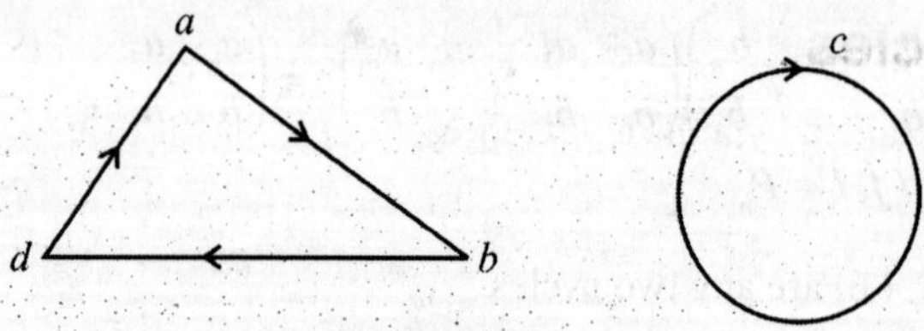

Fig. 5.1

This permutation can also be represented as ((*abd*) (*c*)), where (*abd*) and (*c*) are called **cycles of permutation**. The number of edges in a cycle is called **length of the cycle**. In (*abd*), the length is 3 and in (*c*) it is one. The number of elements in the object set A is called the **degree** of the permutation. The degree of the permutation (*abd*) (*c*) is 4. A permutation of degree k is said to be of type $(c_1, c_2, \ldots, c_k)$ if it has c_i cycles of length i, $i = 1, 2, \ldots, k$. The permutation ((*abd*)(*c*)) is of the type (1, 0, 1, 0), since we have one cycle (*c*) of length 1, 0 cycle of length 2, 1 cycle of length 3 and 0 cycles of length 4.

Further for a permutation group of order n, if we add cycle structures of all n permutations and divide the sum by its order n, we get the cycle index.

Example 5.42

Let $X = \{1, 2, 3, 4, 5\}$, then $f = \begin{pmatrix} 1 & 2 & 3 & 4 & 5 \\ 2 & 3 & 4 & 5 & 1 \end{pmatrix}$ is a cyclic permutation. Here $f(1) = 2, f(2) = 3, f(3) = 4, f(4) = 5, f(5) = 1$. Thus $f = (1\ 2\ 3\ 4\ 5)$ is a cyclic permutation of length 5.

Example 5.43

Let $X = \{1, 2, 3, 4\}$, then $f = (1\ 3\ 2) = \begin{pmatrix} 1 & 2 & 3 & 4 \\ 3 & 1 & 2 & 4 \end{pmatrix}$ is a cyclic permutation of length 3.

Here, $f(1) = 3, f(2) = 1, f(3) = 2, f(4) = 4$.

A cycle remains unchanged by changing the place of its elements provided their cyclic order is not changed. Therefore

$$(1\ 2\ 3\ 4) = (2\ 3\ 4\ 1) = (3\ 4\ 1\ 2)$$

5.28 Transposition

A cyclic permutation of length two is called a **transposition**.

Example 5.44

$f = (4\ 5)$ is a transposition such that $f(4) = 5, f(5) = 4$ and $f(x) = x$ if $x \neq 4$ or 5.

Example 5.45

If the transposition (*a b*) is a permutation of degree 4 on four symbols *a*, *b*, *c*, *d* then the corresponding permutation will be

$$f = \begin{pmatrix} a & b & c & d \\ b & a & c & d \end{pmatrix}$$

Example 5.46

In a cycle of length one the image of the element involved is the element itself and the missing elements remain unchanged. Hence every cycle of length one is an **identity permutation**.

5.29 Product of Cycles

The product of two cycles is obtained by multiplying the permutations represented by them.

Example 5.47

Let $f = (1\ 3\ 5)$ and $g = (1\ 2\ 4\ 6)$ are any two cycles.
The permutation of order 7 on seven symbols 1, 2, 3, 4, 5, 6, 7 is

$$(1\ 3\ 5)\ (1\ 2\ 4\ 6)$$

$$f = \begin{pmatrix} 1 & 2 & 3 & 4 & 5 & 6 & 7 \\ 3 & 2 & 5 & 4 & 1 & 6 & 7 \end{pmatrix} \begin{pmatrix} 1 & 2 & 3 & 4 & 5 & 6 & 7 \\ 2 & 4 & 3 & 6 & 5 & 1 & 7 \end{pmatrix}$$

$$= \begin{pmatrix} 1 & 2 & 3 & 4 & 5 & 6 & 7 \\ 2 & 4 & 5 & 6 & 1 & 3 & 7 \end{pmatrix}$$

$$= (1\ 2\ 4\ 6\ 3\ 5)\ (7)$$

Since a cycle of length one is an identity permutation, then

$$(1\ 3\ 5)\ (1\ 2\ 4\ 6) = (1\ 2\ 4\ 6\ 3\ 5).$$

5.30 Disjoint Cycles

Let f and g be any two cycles. The two cycles are said to be **disjoint** if the set of numbers that represent the cycles have no common element.

Example 5.48

Consider two cycles (1 2 5) and (4 6). Clearly, these are disjoint because they have no common element whereas the cycles (1 2 3 4) and (3 4 5 6) are not disjoint cycles as 3 and 4 are common in (1 2 3 4) and (3 4 5 6).

Theorem 5.17

The product of disjoint cycles is commutative.

Proof

By definition, two cycles f and g of same degree are disjoint cycles iff

$$f(x) \neq x \Rightarrow g(x) = x$$

and $\quad g(x) = x \Rightarrow f(x) \neq x \quad$ for all $x \in X$.

i. Let $f(x) \neq x$, then $g(x) = x$. Now if $y = f(x)$.

then $\quad f(y) = f[f(x)] \neq y \quad [\because f(x) \neq x]$

Thus $\quad g(y) = y$

Now
$$gf(x) = g[f(x)]$$
$$= g[y] \quad [\because y = f(x)]$$
$$= y \quad \ldots(1)$$

Also
$$fg(x) = f[g(x)]$$
$$= f(x) \quad [g(x) = x]$$
$$= y \quad \ldots(2)\ [y = f(x)]$$

From (1) and (2), we get $gf = fg$.

ii. Let $f(x) = x$, then $g(x) \neq x$.

By similar discussion given above we get $fg = gf$.

Hence the product of two disjoint cycles of same degree is commutative.

Example 5.49

Let $f = (1\ 6\ 7)$ and $g = (2\ 5\ 3\ 4)$ be any two distinct cycles represent permutations on 7 symbols 1, 2, ..., 7.

Then
$$fg = \begin{pmatrix} 1 & 2 & 3 & 4 & 5 & 6 & 7 \\ 6 & 2 & 3 & 4 & 5 & 7 & 1 \end{pmatrix}\begin{pmatrix} 1 & 2 & 3 & 4 & 5 & 6 & 7 \\ 1 & 5 & 4 & 2 & 3 & 6 & 7 \end{pmatrix}$$
$$= \begin{pmatrix} 1 & 2 & 3 & 4 & 5 & 6 & 7 \\ 6 & 5 & 4 & 2 & 3 & 7 & 1 \end{pmatrix} \quad \ldots (1)$$

Also
$$gf = \begin{pmatrix} 1 & 2 & 3 & 4 & 5 & 6 & 7 \\ 1 & 5 & 4 & 2 & 3 & 6 & 7 \end{pmatrix}\begin{pmatrix} 1 & 2 & 3 & 4 & 5 & 6 & 7 \\ 6 & 2 & 3 & 4 & 5 & 7 & 1 \end{pmatrix}$$
$$= \begin{pmatrix} 1 & 2 & 3 & 4 & 5 & 6 & 7 \\ 6 & 5 & 4 & 2 & 3 & 7 & 1 \end{pmatrix} \quad \ldots (2)$$

From (1) and (2), we get $fg = gf$

Hence product of two disjoint cycles is commutative.

Theorem 5.18

Every permutation of a finite set is a product of transpositions.

5.31 Inverse of a cyclic permutation

Inverse of a cyclic permutation is obtained by reversing the order of its elements.

Thus, if $f = (1\ \ 2\ \ 3\ \ \ldots\ \ n-1\ \ n)$ is a cyclic permutation of degree n, then

$$f^{-1} = (n\ \ n-1\ \ \ldots\ \ 3\ \ 2\ \ 1)$$ is **inverse** of f.

As we know that the product of any element and its inverse gives an identity element, therefore

$$ff^{-1} = I = f^{-1}f.$$

Every transposition is its own inverse. If (1 2) is a transposition, then

$$(1\ 2)^{-1} = (2\ 1) = (1\ 2)$$

Theorem 5.19

Every permutation can be expressed as a product of disjoint cycles.

Proof

Consider $f = \begin{pmatrix} 1 & 2 & 3 & 4 & 5 & 6 \\ 2 & 1 & 4 & 5 & 3 & 6 \end{pmatrix}$ a permutation of degree 6 on the set {1, 2, 3, 4, 5, 6}.

First we write down all the cycles of length one i.e. the cycles whose f-image remain same. Then we take the cycles of length greater than one. Thus we start with 1. After 1 we write the image of 1 which is 2. After 2 we do not write any element because the image of 2 is 1 which is in the bracket. Now we start a new bracket. This bracket will contain an element, which has not been taken earlier. Thus we write 3. After 3 we write the image of 3 which is 4. After 4 we write the image of 4 which is 5. After 5 we close the bracket since the image of 5 is the first element 3.

Since each element has been included in one or the other bracket, therefore our task is over.

Thus, f can be written as $f = (6)\ (1\ 2)\ (3\ 4\ 5)$

Since cycles of length one represent identity permutation, therefore

$$f = (1\ 2)\ (3\ 4\ 5).$$

Theorem 5.20

Every cycle of length n can be expressed as a product of n-1 transpositions.

Proof

If $f = (a_1\ a_2 \ldots a_{n-1}\ a_n)$ be a cycle of length n, then we have

$$f = (a_1\ a_n)\ (a_1\ a_{n-1}) \ldots (a_1\ a_2)$$

is the product of $(n - 1)$ transpositions.

Example 5.50

Express $f = \begin{pmatrix} 1 & 2 & 3 & 4 & 5 & 6 & 7 \\ 2 & 4 & 5 & 1 & 3 & 7 & 6 \end{pmatrix}$ as a product of transpositions.

Solution

The permutation f can be expressed as a product of disjoint cycles.

Thus $f = (1\ 2\ 4)\ (3\ 5)\ (6\ 7)$

By Theorem 5.20 $(1\ 2\ 4) = (1\ 4)\ (1\ 2)$

Thus $f = (1\ 4)\ (1\ 2)\ (3\ 5)\ (6\ 7).$

The representation of a permutation as a product of transpositions is not unique. But the number of transpositions in each representation is same. Hence every permutation can be expressed as product of transpositions in infinitely many ways.

5.32 Even and Odd Permutations

A permutation f of a finite non-empty set X is said to be **even** or **odd** permutation if it can be expressed as a product of an even or odd number of transpositions respectively.

Example 5. 51

Let $f = \begin{pmatrix} 1 & 2 & 3 & 4 & 5 & 6 & 7 & 8 \\ 2 & 3 & 1 & 5 & 6 & 4 & 8 & 7 \end{pmatrix}$, find whether f is even or odd permutation.

Solution

The permutation can be expressed as a product of disjoint cycles.

Thus $f = (1\ 2\ 3)\ (4\ 5\ 6)\ (7\ 8)$

$= (1\ 3)\ (1\ 2)\ (4\ 6)\ (4\ 5)\ (7\ 8)$

which is the product of 5 (i.e. odd) number of transpositions.

Hence, the permutation is an odd permutation.

Example 5.52

Let $f = \begin{pmatrix} 1 & 2 & 3 & 4 & 5 & 6 \\ 6 & 5 & 4 & 3 & 1 & 2 \end{pmatrix}$, find whether f is even or odd permutation.

Solution

We have $f = (1\,6\,2\,5)\,(3\,4)$

$f = (1\,6)\,(1\,2)\,(1\,5)\,(3\,4)$

which is the product of 4 (i.e. even number) of transpositions.

Hence, the permutation is an even permutation.

SOME IMPORTANT RESULTS

1. *A permutation can not be both even and odd.*
2. *A cycle of length n can be expressed as the product of (n-1) transpositions. Therefore, it is to be noted that a cycle of length n will be an even permutation if n is odd and it will be an odd permutation if n is even.*
3. *Every transposition is an odd permutation.*
4. *Identity permutation is always an even permutation.*
5. *The product of two even permutations is an even permutation.*
6. *The product of two odd permutations is an even permutation.*
7. *The product of an even permutation and an odd permutation is an odd permutation and vice-versa.*
8. *The inverse of an even permutation is an even permutation and inverse of an odd permutation is an odd permutation.*

Theorem 5.21

Out of $n!$ permutations on n symbols, $n!/2$ are even permutations and $n!/2$ are odd perm- utations.

Proof

If P_n be the set of all permutations of degree n, then the set P_n will have $n!$ distinct elements. Since a permutation is either an even permutation or an odd permutation but not both, therefore

$n!$ = number of even permutations + number of odd permutations.

If $e_1, e_2, \ldots, e_s$ be the even permutations and $o_1, o_2, \ldots, o_t$ be the odd permutations, then

$$n! = s + t$$

And hence $P_n = \{e_1, e_2, \ldots e_s, o_1, o_2, \ldots, o_t\}$

Now let $l \in P_n$ and l is a transposition then by closure property $le_1, le_2, \ldots, le_s, lo_1, lo_2, \ldots, lo_t$ are all elements of P_n and $le_1, le_2, \ldots, le_s$ are all odd permutations and $lo_1, lo_2, \ldots, lo_t$ are all even permutations.

Now we claim that the permutations $le_1, le_2, \ldots, le_s$ are all distinct.

For it, let $le_i = le_j \Rightarrow e_i = e_j$ [left cancellation law]

But $e_i \neq e_j \therefore le_i \neq le_j$

Thus $le_1, le_2, \ldots, le_s$ are all odd permutations. But we have t odd permutations,

Therefore $s \leq t$ ...(1)

Similarly, we can show that the even permutations $lo_1, lo_2, \ldots, lo_t$ are distinct elements of P_n.

Therefore $t \leq s$ ...(2)

From (1) and (2), we get $s = t = \frac{n!}{2}$. Therefore, the total number of even permutations of degree n are $\frac{n!}{2}$ and the total number of odd permutations of degree n are $\frac{n!}{2}$.

Theorem 5.22

If A_n is the set of all even permutations of degree n, then A_n is a finite group of order $\frac{n!}{2}$ with respect to product of permutations.

Proof

Closure Property

As we know that the product of two even permutations is an even permutation, therefore the set A_n is closed for product of permutations.

Associativity

We know that multiplication of permutations is an associative composition.

Existence of identity

We know that an identity permutation is an even permutation. If I is an identity permutation then $I \in A_n$.

Now, we have $Ig = g = gI \ \forall \ g \in A_n$

$\therefore$ I is an identity element.

Existence of Inverse

Let $g \in A_n$. If g^{-1} is the inverse of g in the group of all permutations of degree n, then g is an even permutation because $g\,g^{-1} = I$ and I is an even permutation

Therefore $\qquad g \in A_n \Rightarrow g^{-1} \in A_n$ such that $g\,g^{-1} = I = g^{-1}\,g$

Hence, every element of A_n have multiplicative inverse. Hence A_n is a group of all even permutations of degree $\frac{n!}{2}$ because the total number of even permutations of degree n is $\frac{n!}{2}$.

This group A_n is known as **alternating group**.

Since the product of two odd permutations is an even permutation, therefore the set of all odd permutations is not a group.

Example 5.53

Show that the set X of four permutations I, (1 2) (3 4), (1 3) (2 4) and (1 4) (2 3) on four symbols 1, 2, 3, 4 is an abelian group with respect to the permutation multiplication.

Solution

Let $\qquad I = f_1$

$(1\ 2)\ (3\ 4) = f_2$

$(1\ 3)\ (2\ 4) = f_3$

$(1\ 4)\ (2\ 3) = f_4.$

Now we will prepare the composition table for multiplication composition.

Since $\qquad f_1 = I,$

then $\qquad f_1 f_1 = f_1,$

$$f_1 f_2 = f_2 = f_2 f_1$$

$$f_1 f_3 = f_3 = f_3 f_1$$

$$f_4 f_1 = f_4 = f_4 f_1$$

Also $$f_2 f_2 = f_3 f_3 = f_4 f_4 = I$$

Further $$f_2 f_3 = (1\ 2)\ (3\ 4)\ (1\ 3)\ (2\ 4)$$

$$= (1\ 4)\ (2\ 3)$$

$$= f_4$$

$$= f_3 f_2$$

Similarly, $f_3 f_4 = f_2 = f_4 f_3$ and so on.

Hence the composition table is

Product of Permutations	f_1	f_2	f_3	f_4
f_1	f_1	f_2	f_3	f_4
f_2	f_2	f_1	f_4	f_3
f_3	f_3	f_4	f_1	f_2
f_4	f_4	f_3	f_2	f_1

From the table, we can see that

i. the product of any two permutations of the given set is again in the set therefore the set is closed for product of permutations.

ii. f_1 is identity element .

iii. each element have its inverse in the given set.

In fact

$$f_1^{-1} = f_1$$

$$f_2^{-1} = f_2$$

$$f_3^{-1} = f_3$$

$$f_4^{-1} = f_4$$

iv. The multiplication of permutations is associative and commutative.

∴ the given set is a finite abelian group of order 4 with respect to the multiplication of permutations.

Exercise 5.3

1. Let $f = \begin{pmatrix} 1 & 2 & 3 & 4 \\ 2 & 3 & 4 & 1 \end{pmatrix}$ and $g = \begin{pmatrix} 1 & 2 & 3 & 4 \\ 3 & 2 & 4 & 1 \end{pmatrix}$ be two permutations of degree 4. Find *fg* and *gf*.

2. Find the inverse of the permutation $f = \begin{pmatrix} 1 & 2 & 3 & 4 \\ 1 & 3 & 4 & 2 \end{pmatrix}$.

3. Express the following permutations as the product of disjoint cycles.

   a. $\begin{pmatrix} 1 & 2 & 3 & 4 & 5 & 6 \\ 3 & 6 & 5 & 4 & 2 & 1 \end{pmatrix}$

   b. $\begin{pmatrix} 1 & 2 & 3 & 4 & 5 & 6 & 7 & 8 & 9 \\ 2 & 8 & 3 & 9 & 5 & 7 & 6 & 1 & 4 \end{pmatrix}$

   c. (1 2 3 4) (5 6 7) (1 6 7) (1 5)

4. Decompose the permutation $\begin{pmatrix} 1 & 2 & 3 & 4 & 5 & 6 & 7 \\ 6 & 5 & 2 & 4 & 3 & 1 & 7 \end{pmatrix}$ into transpositions.

5. Examine whether the following permutation is even or odd.

$$f = \begin{pmatrix} 1 & 2 & 3 & 4 & 5 & 6 & 7 & 8 & 9 \\ 2 & 5 & 4 & 3 & 6 & 1 & 7 & 9 & 8 \end{pmatrix}$$

6. Prove that a cycle containing an odd number of symbols is an even permutation whereas a cycle containing an even number of symbols is an odd permutation.

7. Show that the four permutations *I*, (*ab*), (*cd*), (*ab*) (*cd*) on four symbols *a*, *b*, *c*, *d* form a finite abelian group with respect to the permutation multiplication.

8. Show that the set P_3 of all permutations of three symbols 1, 2, 3 is a finite non-abelian group of order 3! with respect to permutation multiplication as composition.

9. Compute $a^{-1}ba$, where $a = (1\ 3\ 5)\ (1\ 2)$ and $b = (1\ 5\ 7\ 9)$.

10. Prove that there is no a such that $a^{-1}\ (1\ 3\ 5)\ a = (1\ 3)\ (5\ 7\ 8)$.

11. Determine which of the following are even permutations :

    a. (1 2 3) (1 2)

    b. (1 2 3 4 5) (1 2 3) (4 5)

    c. (1 2) (1 3) (1 4) (2 5).

12. Prove that $(1\ 2\ 3\ \ldots\ n-1\ n)^{-1} = (n\ n-1 \ldots 3\ 2\ 1)$.

Answers

1. $fg = \begin{pmatrix} 1 & 2 & 3 & 4 \\ 4 & 3 & 1 & 2 \end{pmatrix}$ and $gf = \begin{pmatrix} 1 & 2 & 3 & 4 \\ 2 & 4 & 1 & 3 \end{pmatrix}$

2. $\begin{pmatrix} 1 & 2 & 3 & 4 \\ 1 & 4 & 2 & 3 \end{pmatrix}$

3. a. (1 3 5 2 6) (4)

   b. (1 2 8) (3) (4 9) (5) (6 7)

   c. (1 2 3 4 6 5 7).

4. (16) (25) (13).

5. odd.

11. a. odd

    b. odd

    c. even.

5.33 Subgroups

Let (*G*, *) be a group and let *H* be a non empty subset of *G* called a **complex of the group** *G*. If *H* itself forms a group with respect to the same operation *, then the group (*H*, *) is said to be a subgroup of the group (*G*, *).

As we know that every set is a subset of itself, therefore if *G* is a group, then *G* itself is a subgroup of *G*. Also let $e \in G$, then $\{e\}$ is a subset of *G* and hence a subgroup of *G*.

Thus *G* and $\{e\}$ are subgroups of any group *G*. These subgroups are called **trivial** or **improper** sub groups of the group *G*.

A subgroup H of a group G, is called a **proper subgroup** if it is different from G as well as from $\{e\}$.

Example 5.54

The set E of all even integers is a subgroup of additive group of all integers.

Example 5.55

Let $G = \{1, -1, i, -i\}$, $H = \{1, -1\}$, where $i^2 = -1$.

Then G is a group under usual multiplication of complex numbers and H is a subgroup of G.

Example 5.56

The additive group of integers is a subgroup of the additive group of all rational numbers.

Example 5.57

Let G be the multiplicative group of all non-singular 2×2 matrices over complex numbers. Let H be the set of the following eight matrices.

$$\pm\begin{bmatrix}1 & 0\\0 & 1\end{bmatrix}, \pm\begin{bmatrix}i & 0\\0 & -i\end{bmatrix}, \pm\begin{bmatrix}0 & i\\i & 0\end{bmatrix}, \pm\begin{bmatrix}0 & 1\\-1 & 0\end{bmatrix},$$

Clearly $H \subseteq G$. Since H is a group under matrix multiplication, then H is a subgroup of G.

Example 5.58

The **alternating group** A_n of all even permutations of degree n is a subgroup of the symm- etric group P_n of all permutations of degree n.

Example 5.59

Let G be the group of all real 2×2 matrices $\begin{bmatrix}a & b\\c & d\end{bmatrix}$ with $ad - bc \neq 0$ under matrix multiplication.

Let $H = \left\{\begin{bmatrix}a & b\\0 & d\end{bmatrix} \in G : ad \neq 0\right\}$ then one can easily verify that H is a subgroup of G.

Example 5.60

Let G be the group of all non-zero complex numbers $a + bi$ (a, b real, not both 0) under multiplication, and let

$$H = \{a + bi \in G : a^2 + b^2 = 1\}$$

Then H is a subgroup of G.]

Theorem 5.23

Let H be a subgroup of G. Then prove that

i. the identity of H is the same as that of G.

ii. the inverse of any element of H is the same as the inverse of the same regarded as an element of G.

Proof

i. Let e_1 and e_2 be the identities of G and H respectively.

Now if $\quad a \in H \Rightarrow a \in G \Rightarrow e_1 a = a \quad \dots (1)$

Also $\quad a \in H \Rightarrow e_2 a = a \quad \dots (2)$

From (1) and (2), we get

$$e_1 a = e_2 a \Rightarrow e_1 = e_2$$ [by right cancellation law in G]

Hence, the identity elements are same.

ii. Let e be the identity of G. So by (i) e is the identity of H. Let $a \in H$ and b and c are inverses of a in H and G respectively. Then

$$ab = ba = e \quad \text{...(1)}$$

and
$$ac = ca = e \quad \text{...(2)}$$

From (1) and (2) we have

$$ab = ac$$

$$\Rightarrow b = c. \quad \text{[by left cancellation law in } G\text{]}$$

5.34 Complex of a Group

Any non-empty subset H of a group G is called a **complex** of G.

5.35 Algebra of Complexes

5.35-1 Multiplication of Complexes

Let H and K be any two complexes of *a* group G, then their multiplication denoted by HK is the set defined by

$$HK = \{h\,k : h \in H, k \in K\}$$

Clearly $HK \subset G$. Thus $H\,K$ is a complex of G having elements of G obtained on multiplying each member of H with each member of K.

The multiplication of complexes is associative i.e. for any three complexes H_1, H_2, H_3 of G

$$H_1\,(H_2\,H_3) = (H_1\,H_2)\,H_3.$$

5.35-2 Inverse of a Complex

The inverse of a complex H of a group G is defined by

$$H^{-1} = \{h^{-1} : h \in H\}.$$

THEOREM 5.24 (Reversal Rule)

If H and K are any complexes of a group G, then show that

$$(HK)^{-1} = K^{-1}\,H^{-1}.$$

Proof

First we shall show that

$$(HK)^{-1} \subseteq K^{-1}\,H^{-1}$$

Let $t \in (HK)^{-1}$ and is arbitrary.

Then
$$t = (h\,k)^{-1}, \text{ where } h \in H, k \in K.$$

$$= k^{-1}\,h^{-1} \quad \text{[by reversal rule]}$$

$$\in K^{-1}\,H^{-1} \quad [\because k^{-1} \in K^{-1}, h^{-1} \in H^{-1}]$$

$$\Rightarrow t \in K^{-1}\,H^{-1}$$

Now since $t \in (H\,K)^{-1} \Rightarrow t \in K^{-1}\,H^{-1}$

$$\Rightarrow (HK)^{-1} \subseteq K^{-1}\,H^{-1} \quad \text{...(1)}$$

Now we shall show that $K^{-1}\,H^{-1} \subseteq (HK)^{-1}$

Again, let $x \in K^{-1}\,H^{-1}$ and is arbitrary.

Then
$$x = k^{-1}\,h^{-1}.$$

$$= (hk)^{-1} \quad \text{[by reversal rule]}$$

$$\in (HK)^{-1} \quad [\because hk \in HK]$$

$$\Rightarrow x \in (HK)^{-1}$$

Now $\quad x \in K^{-1} H^{-1} \Rightarrow x \in (HK)^{-1} \Rightarrow K^{-1} H^{-1} \subseteq (HK)^{-1} \quad \ldots(2)$

From (1) and (2), we get $(HK)^{-1} = K^{-1} H^{-1}$.

Theorem 5.25

If H is any subgroup of the group G, then prove that $H^{-1} = H$.

Proof

Let $h \in H$

$$\Rightarrow h^{-1} \in H \quad [\because H \text{ is a group}]$$

$$\Rightarrow (h^{-1})^{-1} \in H^{-1} \quad [\text{by def. of } H^{-1}]$$

$$\Rightarrow h \in H^{-1}$$

$$\therefore H \subseteq H^{-1} \quad \ldots (1)$$

Again, let $\quad h^{-1} \in H^{-1}$

$$\Rightarrow h \in H$$

$$\Rightarrow h^{-1} \in H \quad [\because H \text{ is a group}]$$

$$H^{-1} \subseteq H \quad \ldots (2)$$

From (1) and (2), we have

$$H^{-1} = H.$$

The converse of Theorem 5.25 is not true. i.e. if H is a complex of a group G and $H^{-1} = H$, then it is not necessarily true that H is a sub group of G.

Example 5.61

Consider a multiplicative group $G = \{-1, 1, i, -i\}$, where $i^2 = -1$. Let $H = \{-1\}$, then H is a complex of G. Since the inverse of -1 in G is -1, then $H^{-1} = \{-1\} = H$. But H is not a subgroup of G because $(-1)(-1) = 1 \notin H$.

Thus H is not closed with respect to multiplication.

Theorem 5.26

If H is a subgroup of G, then prove that

$$HH = H.$$

Proof

First, show that $HH \subseteq H$. Let $h_1 h_2$ be any element of HH where $h_1 \in H$, $h_2 \in H$. Since H is a subgroup, then by closure property

$$h_1 \in H, h_2 \in H \Rightarrow h_1 h_2 \in H$$

Now $\quad h_1 h_2 \in HH \Rightarrow h_1 h_2 \in H \Rightarrow HH \subseteq H \quad \ldots(1)$

Again let h is any arbitrary element of H.

$$\therefore\ h = he \qquad \text{[e is identity element of } G\text{]}$$

$$\in HH \qquad [\because\ h \in H \text{ and } e \in H]$$

Thus $\qquad H \subseteq HH \qquad \ldots(2)$

From (1) and (2), we get

$$HH = H.$$

5.36 Criterion for a Complex to be a Subgroup

The following theorems give us a criterion for a complex to be a subgroup.

Theorem 5.27

The necessary and sufficient conditions for a non-empty subset H of a group G to be a subgroup is that

$$a \in H, b \in H \Rightarrow ab^{-1} \in H.$$

Proof

The condition is necessary

Let H be a subgroup of G; and let $a, b \in H$. Since H is a group, then

$$b \in H \Rightarrow b^{-1} \in H$$

Further, by closure property in H.

$$a \in H, b^{-1} \in H \Rightarrow ab^{-1} \in H$$

The condition is necessary.

The condition is sufficient

Now suppose that

$$a \in H, b \in H \Rightarrow ab^{-1} \in H$$

Then we have to show that H is a subgroup of G.

Since H is non-empty, then there exists $a \in H$ such that $a \neq 0$. Now by given condition.

$$a \in H, a \in H \Rightarrow a\, a^{-1} \in H \Rightarrow e \in H$$

$\therefore$ H has the identity element.

Again by the given condition

$$e \in H, a \in H \Rightarrow e\, a^{-1} \in H \Rightarrow a^{-1} \in H$$

$\therefore$ every element of H is invertible. Again from the same condition

$$a \in H, b \in H \Rightarrow a \in H, b^{-1} \in H$$

$$\Rightarrow a\,(b^{-1})^{-1} \in H$$

$$\Rightarrow a\, b \in H \qquad [\because (b^{-1})^{-1} = b]$$

$\therefore$ H is closed for the composition of G.

Since H is subset of G, therefore associative law will hold in H as in G.

Hence H is subgroup of G. Thus the condition is sufficient.

For additive composition, the above condition becomes

$a \in H, b \in H \Rightarrow a - b \in H.$

REMARK

Above theorem can be used whenever it is required to show that a non empty subset H of a group G is a subgroup of G. For this, we take any arbitrary elements $a, b \in H$ and show that $ab^{-1} \in H$.

Theorem 5.28

The necessary and sufficient condition for a non-empty subset H of a finite group G to be a subgroup is that

$$a \in H, b \in H \Rightarrow a\,b \in H.$$

Proof

The condition is necessary

Let H be a subgroup of G. Then H must be closed with respect to multiplication composition in G. Thus we have $a \in H, b \in H \Rightarrow a\,b \in H$.

The condition is necessary.

The condition is sufficient

Now we have

$$a \in H, b \in H \Rightarrow a\,b \in H$$

Then we have to show that H is a subgroup of G.

Since H is non-empty subset of G, then there exists $a \in H$.

By given property $a \in H \Rightarrow a^2 = aa \in H \Rightarrow a^3 = a^2\,a \in H \Rightarrow a^n = a^{n-1}\,a \in H$

where n is a positive integer. Since G is a finite group, therefore the order of each element in G is finite.

Let $\qquad O(a) = m \ \forall\ a \in G \Rightarrow a^m = e$

But $\qquad a^m \in H \Rightarrow e \in H \Rightarrow$ identity exists in H

Also $\qquad a^{-1} = e\,a^{-1} = a^m\,a^{-1} = a^{m-1} \in H$

Thus each element of H possesses inverse.

Since all the elements of H are in G, then the composition in H must be associative.

Hence H is a subgroup of G.

The criterion given in the Theorem 5.28 is necessary for all finite or infinite groups but it is not sufficient for infinite groups.

Example 5.62

$(N, +)$ is not a subgroup of $(I, +)$ although it satisfies the condition of Theorem 5.28.

5.37 Algebra of Subgroups

5.37-1 Intersection of Subgroups

Theorem 5.29

Prove that the intersection of any two subgroups of a group G is again a subgroup of G.

Proof

Suppose H_1 and H_2 be any two subgroups of a group G, then we have to prove that $H_1 \cap H_2$ is also a subgroup of G.

Since the identity element is common to H_1 and H_2, then $H_1 \cap H_2$ is non-empty. In order to prove that $H_1 \cap H_2$ is a subgroup, it is sufficient to prove that $a \in H_1 \cap H_2, b \in H_1 \cap H_2 \Rightarrow a\,b^{-1} \in H_1 \cap H_2$

Now $\quad a \in H_1 \cap H_2 \Rightarrow a \in H_1$ and $a \in H_2$

$b \in H_1 \cap H_2 \Rightarrow b \in H_1$ and $b \in H_2$

But H_1 and H_2 are subgroups, so by Theorem 5.27

$$a \in H_1, b \in H_1 \Rightarrow a\,b^{-1} \in H_1$$

and $\quad a \in H_2, b \in H_2 \Rightarrow a\,b^{-1} \in H_2$

Now, $\quad a\,b^{-1} \in H_1$ and $a\,b^{-1} \in H_2 \Rightarrow a\,b^{-1} \in H_1 \cap H_2$

Thus $\quad a \in H_1 \cap H_2, b \in H_1 \cap H_2 \Rightarrow a\,b^{-1} \in H_1 \cap H_2$

Therefore, $H_1 \cap H_2$ is also a subgroup of the group G.

Corollary

If $H_1, H_2, ..., H_n$ be a finite family of subgroups of G then $H_1 \cap H_2 \cap ... \cap H_n$ is also a subgroup of G.

$H_1 \cap H_2$ is the largest subgroup of G contained in both H_1 and H_2 because $H_1 \cap H_2$ is the largest subset of G which is contained in H_1 as well as in H_2. By largest we mean that it is contained in H_1 and in H_2 and contains every subgroup contained in both H_1 and H_2.

5.37-2 Union of Subgroups

Generally, the union of two subgroups of a group may not be a subgroup.

Example 5.63

Let $(I, +)$ be an additive group of integers.

Now let $H_1 = \{ ...-6, -4, -2, 0, 2, 4, 6 ...\}$

and $H_2 = \{ ...-15, -10, -5, 0, 5, 10, 15 ...\}$ are two subgroups of $(I, +)$

then $H_1 \cup H_2 = \{... -10, -6, -5, -4, -2, 0, 2, 4, 5, 6, 10, ...\}$

$H_1 \cup H_2$ is not a group for addition composition because $H_1 \cup H_2$ is not closed for addition as $2 \in H_1 \cup H_2$, $5 \in H_1 \cup H_2$ but $2 + 5 = 7 \notin H_1 \cup H_2$.

Therefore $H_1 \cup H_2$ is not a subgroup of G. Now if we take one more subgroup H_3 of G such that $H_3 = \{ ... -8, -4, 0, 4, 8 ...\}$.

Then $H_1 \cup H_3 = H_1$ and H_1 is a subgroup of G. Hence in this case $H_1 \cup H_3$ is a subgroup of G.

The following Theorem 5.30 gives us some additional condition for $H_1 \cup H_2$ to become a subgroup.

Theorem 5.30

The union of two subgroups of a group G is a subgroup if and only if one is contained in the other.

Proof

Consider two subgroups H_1 and H_2 of a group G and let

$$H_1 \subset H_2 \text{ or } H_2 \subset H_1.$$

If $H_1 \subset H_2$, then $H_1 \cup H_2 = H_2$

But H_2 is a subgroup, therefore $H_1 \cup H_2$ is a subgroup of G.

Similarly, if $H_2 \subset H_1$, then $H_1 \cup H_2 = H_1$ and hence $H_1 \cup H_2$ is subgroup. Conversely, suppose that $H_1 \cup H_2$ is a subgroup, then we have to show that either $H_1 \subset H_2$ or $H_2 \subset H_1$.

Let $H_1 \not\subset H_2$. Then there exists an element a such that $a \in H_1$ but $a \notin H_2$. Let $b \in H_2$ therefore $a \in H_1$ and $b \in H_2 \Rightarrow a \in H_1 \cup H_2$ and $b \in H_1 \cup H_2$

But $H_1 \cup H_2$ is a group $\Rightarrow a\,b \in H_1 \cup H_2 \Rightarrow a\,b \in H_1$ or $a\,b \in H_2$

Now $a\,b \in H_2$, $b \in H_2$

$$\Rightarrow (a\,b)b^{-1} \in H_2$$
$$\Rightarrow a(b\,b^{-1}) \in H_2$$
$$\Rightarrow a\,e \in H_2$$
$$\Rightarrow a \in H_2$$

But $a \notin H_2$

Hence $a\,b \notin H_2$

$$\Rightarrow a\,b \in H_1$$

But $a\,b \in H_1$, $a \in H_1$

$$\Rightarrow a^{-1}(ab) \in H_1$$
$$\Rightarrow (a^{-1}a)\,b \in H_1$$
$$\Rightarrow e\,b \in H_1$$
$$\Rightarrow b \in H_1$$
$$\Rightarrow H_2 \subset H_1$$

Hence the union of two subgroups is a subgroup iff one is contained in the other.

5.37-3 Multiplication of Two Subgroups

In general the multiplication of two subgroups of a group may not be a subgroup. For example, let

$$H_1 = \{(1), (1\ 2)\} \text{ and } H_2 = \{(1), (2\ 3)\}$$

are two subgroups of the permutation group S_3. Then

$$H_1 H_2 = \{(1), (1)(2\ 3), (1\ 2)(1), (1\ 2)(2\ 3)\}$$
$$= \{(1), (2\ 3), (1\ 2), (1\ 2\ 3)\}$$

$H_1\,H_2$ is not a subgroup of S_3 because $(2\ 3)(1\ 2) = (1\ 3\ 2) \notin H_1\,H_2$.

The following Theorem 5.31 gives us some additional condition for the product of two subgroups to become a subgroup.

Theorem 5.31

If H and K are two subgroups of a group G, then HK is a subgroup of G iff $HK = KH$.

Proof

First suppose that $HK = KH$ then we have to show that HK is a subgroup of G. For this, it is sufficient to show that $(HK)(HK)^{-1} = (HK)$

Now $(HK)(HK)^{-1} = (HK)(K^{-1}H^{-1})$ [by reversal rule]

$= H(KK^{-1})H^{-1}$ [by associativity]

We know that if K is a subgroup then $KK^{-1} = K$.

$= (HK)H^{-1}$,

$= (KH)H^{-1}$, [$\because HK = KH$]

$= K(HH^{-1})$

$= KH$ [$\because$ H is a subgroup $\Rightarrow HH^{-1} = H$]

$\therefore$ HK is a subgroup.

Conversely, let HK is a subgroup,

Then we have

$(HK)^{-1} = HK$ [by $H^{-1} = H$]

$K^{-1}H^{-1} = HK$

But H, K are subgroups [Reversal Rule]

$\Rightarrow H^{-1} = H, K^{-1} = K$

$\therefore$ $KH = HK$.

Hence, HK is a subgroup iff $HK = KH$.

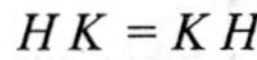

If H and K are subgroups of a commutative group G, then HK is a subgroup of G.

Example 5.64

Let G be a group, $a \in G$ and $N(a) = \{x \in G : ax = xa\}$. Show that $N(a)$ is a subgroup of G.

Solution

Let G be a group and $e \in G$ be an identity element, then

$$ea = ae \;\forall\; a \in G$$

$\Rightarrow e \in N(a)$ [by def. of $N(a)$]

$\Rightarrow N(a)$ is not empty.

Let $x_1, x_2 \in N(a)$, then $a x_1 = x_1 a$ and $a x_2 = x_2 a \;\forall\; a \in G$

Also $a x_2 = x_2 a$

$$\Rightarrow x_2^{-1}(ax_2)x_2^{-1} = x_2^{-1}(x_2a)x_2^{-1}$$

$$\Rightarrow (x_2^{-1}a)(x_2x_2^{-1}) = (x_2^{-1}x_2)(ax_2^{-1})$$

$$\Rightarrow x_2^{-1}a = ax_2^{-1}$$

$$x_2^{-1} \in N(a)$$

Thus $x_2 \in N(a) \Rightarrow x_2^{-1} \in N(a)$

Again

$$a\left(x_1x_2^{-1}\right) = \left(ax_1\right)x_2^{-1} = \left(x_1a\right)x_2^{-1}$$
$$= x_1\left(ax_2^{-1}\right) = x_1\left(x_2^{-1}a\right)$$
$$= \left(x_1x_2^{-1}\right)a$$
$$\Rightarrow x_1x_2^{-1} \in N(a)$$

Therefore, we have $x_1 \in N(a)$, $x_2 \in N(a) \Rightarrow x_1x_2^{-1} \in N(a)$

Hence $N(a)$ is a subgroup of G.

*N(a) is called **normalizer** of a in the group G.*

Exercise 5.4

1. Let G be a group and $H = \{x \in G : x\,g = g\,x \;\forall\; g \in G\}$, then show that H (centre of the group) is a subgroup of G.
2. Let G be a group and $b \in G$. Then prove that the set
$$H = \{x \in G : x\,b^2 = b^2\,x\}$$
is a subgroup of G.
3. Show that a nonempty subset H of a group G is a subgroup iff
   i. $a \in H, b \in H \Rightarrow a\,b \in H$ ii. $a \in H \Rightarrow a^{-1} \in H$.
4. Let S_A denotes the group of all permutations of the set $A = \{a, b, c, d\}$. Prove that $H = \{(a), (ab), (cd), (ab)(cd)\}$ is a finite abelian subgroup of S_A.
5. If H_1 is a subgroup of H_2 and H_2 is a subgroup of G, then H_1 is also a subgroup of G.
6. Let H be a subgroup of G and define
$$T = \{x \in G : xH = Hx\}.$$
Prove that T is a subgroup of G.
7. Let G be a group, H is a subgroup of G. Let for $x \in G$
$$x\,H\,x^{-1} = \{x\,h\,x^{-1} : h \in H\}.$$

Prove that $x\,H\,x^{-1}$ is a subgroup of G.]

5.38 Cyclic Groups

In this section we will study special type of groups which are generated by a single element called cyclic groups.

A group G is said to be **cyclic group** if there exists an element $a \in G$, such that every element of G is of the form a^n, where n is some integer. In such a case the element a is called the **generator of G** and G is called as **cyclic group generated by** a denoted by $G = [a]$. Thus if the composition of a cyclic group G is multiplication and a is its generator, then

$$G = [a] = \{\ldots a^{-3}, a^{-2}, a^{-1}, a^{0} = e, a^{1}, a^{2}, a^{3}, \ldots\}$$

If the composition of a cyclic group G is addition and a is its generator then

$$G = [a] = \{\ldots -2a, -a, 0\,a = e, a, 2a, \ldots\}.$$

There may be more than one generators of a cyclic group.

Example 5.65

The multiplicative group $G = \{1, -1, i, -i\}$ is a cyclic group.

The elements of G can be written as $G = \{i, i^2, i^3, i^4\}$.

Thus, G is a cyclic group with generator i i.e $G = [i]$.

Also, $G = \{(-i), (-i)^2, (-i)^3, (-i)^4\}$.

Therefore $G = [-i\,]$.

Example 5.66

The additive group Z of integers is an infinite cyclic group. The generators of Z are 1 and –1.

Example 5.67

The multiplicative group $G = \{1, w, w^2\}$ is a cyclic group with generators w and w^2.

Example 5.68

$A_3 = \{(1), (1\ 2\ 3), (1\ 3\ 2)\}$ is a cyclic subgroup of the symmetric group P_3 with generators (1 2 3) and (1 3 2).

Example 5.69

The group $G = (\{0, 1, 2, 3, 4\}, +_5)$ is a cyclic group with generators 1 and 4.

We see that

$$1^1 = 1$$
$$1^2 = 1 +_5 1 = 2$$
$$1^3 = 1 +_5 1 +_5 1 = 3$$
$$1^4 = 1 +_5 1 +_5 1 +_5 1 = 4$$
$$1^5 = 1 +_5 1 +_5 1 +_5 1 +_5 1 = 0$$

$\therefore$ $G = \{1, 1^2, 1^3, 1^4, 1^5 = 0\} = [1]$.

5.39 Some Properties of Cyclic Groups

Theorem 5.32

Every cyclic group is abelian.

Proof

Suppose that $G = [b]$ is a cyclic group generated by b. Let $x_1, y_1 \in G$. Then by definition of cyclic group, there exists integers m and n such that

$$x_1 = b^m \text{ and } y_1 = b^n$$

Now

$$\begin{aligned} x_1 y_1 &= b^m b^n \\ &= b^{m+n} && \text{[by indices]} \\ &= b^{n+m}. && \text{[addition is commutative in } Z\text{]} \\ &= b^n b^m. && \text{[by indices]} \\ &= y_1 x_1 \end{aligned}$$

Thus, we have $x_1 y_1 = y_1 x_1\ \forall\ x_1, y_1 \in G$.

Therefore, G is an abelian group.

Theorem 5.33

If a is an element of a group G such that the order of a is n, then order of the cyclic subgroup generated by a is also n.

Proof

Let H be a cyclic subgroup of a group G generated by $a \in G$.

Thus $$H = [a]$$

Now consider a group $K = \{a, a^2, a^3, \ldots, a^n = e\}$ of n elements. Then by definition of H and K, we have

$$K \subset H \quad \ldots(1)$$

Now, we will show that $H \subset K$. For it, let $a^m \in H$

$$\Rightarrow a^m \in K \text{ whenever } 1 \le m \le n.$$

If $m > n$, then by division algorithm there exist integers p and q such that

$$m = np + q, 0 \le q < n.$$

Now
$$\begin{aligned} a^m &= a^{np+q} \\ &= a^{np}.a^q && \text{[by indices]} \\ &= (a^n)^p a^q \\ &= (e)^p a^q && [\because a^n = e] \\ &= a^q \end{aligned}$$

But $q < n$, then $a^q \in K \Rightarrow a^m \in K \; \forall \; m \ge n$

Thus $$H \subset K \quad \ldots(2)$$

From (1) and (2), we get $H = K$

Now, we claim that the elements of K are all distinct.

Let $$a^i = a^j.$$

$\Rightarrow$ $$a^i a^{-j} = a^j a^{-j}$$

$\Rightarrow$ $$a^{i-j} = e$$

$\Rightarrow$ $$O(a) = i - j$$

But $i \ne j$ and $0 \le j < i \le n$

Therefore $$O(a) < n$$

Which is a contradiction because $O(a) = n$

Hence the elements of K are all distinct.

$\Rightarrow$ $$O(K) = n$$

and hence $$O(H) = n$$

Thus, order of a cyclic subgroup generated by a is n.

Corollary

The order of a finite cyclic group is same as the order of its generator.

Proof

Let G be a finite cyclic group generated by a. Now $a \in G$ and G is finite therefore $O(a)$ must be finite. Thus by Theorem 5.33, order of cyclic group is equal to the order of a. Hence $O(G) = O(a)$.

Theorem 5.34

A finite group of order n is cyclic if and only if it contains an element of order n.

Proof

Let G be a finite group of order n and $a \in G$ such that order of a is n. i.e. $O(a) = n$. If H be the subgroup of G generated by a, then by corollary of Theorem 5.33

$$O(a) = n \Rightarrow O(H) = n$$

$$\Rightarrow O(H) = O(G)$$

Thus $H \subset G$, G is finite and $O(H) = O(G)$

$$\Rightarrow H = G$$

$$\Rightarrow G \text{ is a cyclic group.}$$

Conversely, if $G = [a]$ is a finite cyclic group of order n, then by corollary if Theorem 5.33, $a \in G$ such that $O(a) = n$.

Theorem 5.35

If a is a generator of a cyclic group, then a^{-1} is also its generator.

Proof

Let G be a cyclic group generated by a and $x \in G$ then $x = a^n$.

$$\Rightarrow x = (a^{-1})^{-n}$$

Thus x can be expressed as some integral power of a^{-1}.

Also $$a \in G \Rightarrow a^{-1} \in G \qquad \text{[G is a group]}$$

$\therefore$ $$G = [a^{-1}]$$

Therefore a^{-1} is also a generator of G.

Theorem 5.36

Every infinite cyclic group has two and only two generators.

Proof

Let G be an infinite cyclic group generated by a i.e. $G = [a]$. Since a is the generator of G, then a^{-1} is also a generator of G.

If two generators are equal i.e. $a = a^{-1}$.

$\Rightarrow$ $$a^2 = e$$

$\Rightarrow$ $$O(a) = 2$$

$\Rightarrow$ $$O(G) = 2$$

$\Rightarrow G$ is finite cyclic group.

But G is infinite cyclic group, therefore $a \neq a^{-1}$.

Now, we have to show that a^{-1} and a are only generators of G. Let, if possible, a^m be a generator of G where $m \neq \pm 1$.

Now since $a \in G$ and a^m is a generator of G, then by definition there exists an integer n such that

$$a = (a^m)^n = a^{mn} \qquad \text{[by indices]}$$

or $$a\,a^{-1} = a^{mn}\,a^{-1}$$

$$\Rightarrow \qquad e = a^{mn-1}$$

$$\Rightarrow \qquad O(a) = mn-1$$

$\Rightarrow$ O (a) is finite

$\Rightarrow$ G is finite

which is a contradiction.

Hence a^m cannot be a generator of G unless $m = 1$ or $m = -1$.

Thus G has exactly two generators namely a and a^{-1}.

Theorem 5.37

Every subgroup of a cyclic group is cyclic.

Proof

Let G be a cyclic group generated by a i.e. $G = [a]$ and H be a subgroup of G. If H is improper subgroup of G, then $H = \{e\}$ or $H = G$. where $e \in G$ is an identity element .

In this case the theorem is obviously proved. Let H is a proper subgroup of G, then the elements of H are integral powers of a i.e.

$$H = \{a^t \neq e \text{ for some } t \in Z\}$$

Now if $\qquad a^t \in H \Rightarrow a^{-t} \in H \qquad$ [H is a subgroup]

Therefore H contains elements which are positive as well as negative integral powers of a. Suppose that t be the least positive integer such that $a^t \in H$. Now we have to show that H is a cyclic group generated by a^t.

Let $a^r \in H$. Then by division algorithm, there exist two integers p and q such that $r = tp + q \quad 0 \le q < t$

or $\qquad r - tp = q$

Now $\qquad a^t \in H \Rightarrow (a^t)^p = a^{tp} \in H$

$\qquad (a^{tp})^{-1} = a^{-tp} \in H \qquad$ [H is a subgroup]

Now $\qquad a^r \in H, a^{-tp} \in H \Rightarrow a^r . a^{-tp} = a^{r-tp} = a^q \in H$

But t is the least positive integer such that $a \in H$ and $0 \le q < t$,

Therefore q must be zero. Hence $r = tp$

$\therefore \qquad a^r = a^{tp} = (a^t)^p.$

Thus every element of H can be expressed as some integral power of a^t. Hence H is cyclic.

Theorem 5.38

Every proper subgroup of an infinite cyclic group is infinite.

Proof

Let G be an infinite cyclic group generated by a i.e. $G = [a]$. Let H be a proper subgroup of G. As we know that a subgroup of a cyclic group is cyclic, thus H is cyclic and $H = [a^t]$ where t is the least positive integer such that $a^t \in H$.

Now we have to show that H is of infinite order.

Let us suppose that H be a finite subgroup of order m. Since a^t is a generator of H,

Therefore, $\qquad (a^t)^m = e$ where $tm > 0$

$\Rightarrow a^{tm} = e$

$\Rightarrow$ O(a) is finite $\Rightarrow$ G is finite

which is a contradiction

Hence, H must be an infinite cyclic subgroup of G.

Example 5.70

Show that the group $(\{1, 2, 3, 4\}, \times_5)$ is cyclic.

Solution

Let us denote the given group by G and assume that G is a cyclic group. Let $a \in G$. If all the elements of G can be expressed as some integral power of a, then a is called generator of G denoted by $G = [a]$.

We see that $\quad O(G) = 4$ and

$$1^1 = 1$$

$$\Rightarrow \quad O(1) = 1$$

$$2^1 = 2, 2^2 = 2 \times_5 2 = 4, 2^3 = 2^2 \times_5 2 = 4 \times_5 2 = 3$$

$$2^4 = 2^3 \times_5 2 = 3 \times_5 2 = 1$$

$$\Rightarrow \quad O(2) = 4$$

$$3^1 = 3, 3^2 = 3 \times_5 3 = 4, 3^3 = 3^2 \times_5 3 = 4 \times_5 3 = 2$$

$$3^4 = 3^3 \times_5 3 = 2 \times_5 3 = 1$$

$$\Rightarrow \quad O(3) = 4$$

$$4^1 = 4, 4^2 = 4 \times_5 4 = 1$$

$$\Rightarrow \quad O(4) = 2.$$

We find that 2 and 3 in G are such that

$$O(2) = 4 = O(G)$$

and $\quad O(3) = 4 = O(G)$

Hence, the generators of G are 2 and 3. Thus $G = [2] = [3]$.

Example 5.71

Find all the generators of the cyclic group of order 8.

Solution

Let $G = [a]$, then $O(a) = 8$

$$\therefore \quad G = \{a, a^2, a^3, a^4, a^5, a^6, a^7, a^8 = e\}$$

Now we have $O(a) = O(G) = 8$

Now the other generators of G are elements of the form a^n, where n is relatively prime to 8. Since 3 is prime to 8.

$$\Rightarrow a^3 \text{ is also a generator of } G.$$

Also $\quad$ 5 is prime to 8 $\Rightarrow a^5$ is a generator of G.

Again $\quad$ 7 is prime to 8 $\Rightarrow a^7$ is a generator of G.

Thus G has four generators a, a^3, a^5, a^7.

Example 5.72

Prove that the set of an nth roots of unity is a finite cyclic group of order n.

Solution

Let G be the set of n n^{th} roots of unity. Then $G = \{a : a^n = 1\}$.

Now we have to show that G is a group for multiplication composition.

Let $$a, b \in G \Rightarrow a^n = 1 \text{ and } b^n = 1$$
$$\Rightarrow a^n b^n = 1.1 = 1$$
$$\Rightarrow (ab)^n = 1 \qquad \text{[by indices]}$$
$$a\,b \in G$$

Thus G is closed for multiplication.

Also $1^n = 1 \Rightarrow 1 \in G$ then, identity element for multiplication exists in G.

Further, for each $a \in G$

$$\Rightarrow a^n = 1 \Rightarrow 1 = \frac{1}{a^n}$$

or $$\frac{1}{a^n} = 1 \Rightarrow \left(\frac{1}{a}\right)^n = 1$$
$$\Rightarrow \frac{1}{a} \in G$$

which is the inverse of a. Thus each element in G is inevitable.

As the multiplication of numbers is associative, so it is also associative in G. Hence $(G, \times)$ is a group. Since G is the set of roots of equation $a^n = 1$ which being of nth degree has exactly n roots.

Then $O(G) = n$

Now, we have to show that G is a cyclic group.

As $$1 = \cos 0 + i \sin 0 = \cos 2p\pi + i \sin 2p\pi \quad \forall\ p \in Z$$

$$\therefore\ 1^{\frac{1}{n}} = (\cos 2p\pi + i \sin 2p\pi)^{\frac{1}{n}}$$
$$= \cos\left(\frac{2p}{n}\pi\right) + i \sin\left(\frac{2p}{n}\pi\right), \text{ where } p = 1, 2, 3, \ldots, n \qquad \text{[By De Moivre's theorem]}$$
$$= e^{\frac{2p\pi}{n} i}, p = 1, 2, 3, \ldots, n$$

$$\therefore\ G = \left\{e^{\frac{2p\pi}{n} i} : p = 1, 2, 3, \ldots, n\right\}$$
$$= \left\{e^{\frac{2\pi}{n} i}, e^{\frac{4\pi}{n} i}, e^{\frac{6\pi}{n} i}, \ldots, e^{\frac{2\pi n}{n} i}\right\}$$

Therefore, G is a cyclic group of order n generated by $e^{\frac{2\pi}{n} i}$.

Exercise 5.5

1. Give an example of a finite abelian group which is not cyclic.

   [Hint: Let G be the set of four matrices

   $$I = \begin{bmatrix} 1 & 0 \\ 0 & 1 \end{bmatrix}, A = \begin{bmatrix} -1 & 0 \\ 0 & 1 \end{bmatrix}, B = \begin{bmatrix} 1 & 0 \\ 0 & -1 \end{bmatrix}, C = \begin{bmatrix} 1 & 0 \\ 0 & 1 \end{bmatrix},$$

   and the composition is multiplication of matrices.

2. Show that every finite group of order less than six must be abelian.
3. How many generators are there of the cyclic group G of order

   i. 10   ii. 12

   iii. 60
4. Show that the group $(\{1, 2, 3, 4, 5, 6\}, \times_7)$ is cyclic. How many generators are there?
5. Find the order of each element in the cyclic group.

$$G = \{a, a^2, a^3, a^4, a^5, a^6 = e\}$$

e being identity element.

6. Let G be a group, $G \neq \{e\}$. Then G has no proper subgroups if and only if, G is a cyclic group of prime order.

Answers

3. i. three   ii. Four

   iii. fifteen.

5. $O(a) = 6$, $O(a^2) = 3$, $O(a^3) = 2$, $O(a^4) = 3$, $O(a^5) = 6$, $O(a^6) = 1$.

5.40 Cosets

Let G be a multiplicative group and H be a subgroup of G. Let $a \in G$, then the set $Ha = \{ha : h \in H\}$ is said to be a **right coset** of H in G generated by a.

Similarly, the set $aH = \{ah : h \in H\}$ is said to be a **left coset** of H in G generated by a.

Thus for every element of G, we have a left coset and a right coset of H in G generated by that element

Obviously $\quad aH \subset G$ and $Ha \subset G \ \forall\ a \in G$.

If e is the identity element of G, then the left and the right coset of H corresponding to identity e coincide with H.

In other words $\quad eH = H = He$.

Hence, H itself is a left as well as a right coset of H in G. Further, no left coset can be empty. As H is a subgroup of G and

$$e \in G \Rightarrow e \in H$$

So, if aH is a left coset of H in G generated by a, than

$$ae \in aH$$

$$\Rightarrow a \in aH$$

$$\Rightarrow aH \text{ is not empty.}$$

Similarly, we can show that Ha is not empty.

The right coset Ha will be equal to the corresponding left coset aH if G is abelian group. In case of non–abelian group G, either $aH = Ha$ or $aH \neq Ha$.

If the composition of group G is addition, then aH and Ha is defined as $a + H = \{a + h : h \in H\}$ *and* $H + a = \{h + a : h \in H\}$.

Example 5.73

Find all the cosets of $4Z$ in the additive group $(Z, +)$ of integers.

Solution

Let $H = 4Z$ then $H = \{\ldots -8, -4, 0, 4, 8 \ldots\}$

The different cosets of H in Z are as follows:

$$H = H + 0 = \{ \ldots -8, -4, 0, 4, 8, \ldots\}$$
$$H + 1 = \{ \ldots -7, -3, 1, 5, 9, \ldots\}$$
$$H + 2 = \{ \ldots -6, -2, 2, 6, 10, \ldots\}$$
$$H + 3 = \{ \ldots -9, -5, -1, 3, 7, \ldots\}$$
$$H + 4 = \{ \ldots -8, -4, 0, 8, 4, \ldots\}$$
$$H + 5 = \{ \ldots -7, -3, 1, 5, 9, \ldots\}$$

We see that the right cosets H, $H + 1$, $H + 2$ and $H + 3$ are all distinct. After these four cosets we observe that

$$H + 0 = H + 4 = H + 8 = H + 12 = \ldots$$
$$H + 1 = H + 5 = H + 9 = H + 13 = \ldots$$
$$H + 2 = H + 6 = H + 10 = H + 14 = \ldots$$
$$H + 3 = H + 7 = H + 11 = H + 15 = \ldots$$

Thus H has only four distinct cosets in Z.

Example 5.74

Let G be the group of all permutations of degree three on three symbols 1, 2, 3. Find all the cosets of $H = \{(1), (1\ 2)\}$.

Solution

Here the elements of G are the permutations

$$f_1 = (1), f_2 = (12), f_3 = (23), f_4 = (31), f_5 = (123), f_6 = (132)$$

We find left cosets of H in G as follows :

$$f_1 H = \{ (1)(1), (1)(12)\}$$
$$= \{ (1), (12)\} = H$$
$$f_2 H = \{ (12)(1), (12)(12)\}$$
$$= \{ (12), (1)\} = H$$
$$f_3 H = \{ (23)(1), (23)(12)\}$$
$$= \{ (23), (132)\}$$
$$f_4 H = \{ (31)(1), (31)(12)\}$$
$$= \{ (31), (123)\}$$
$$f_5 H = \{ (123)(1), (123)(12)\}$$
$$= \{ (123), (13)\}$$
$$= f_4 H$$
$$f_6 H = \{ (132)(1), (132)(12)\}$$
$$= \{ (132), (23)\}$$
$$= f_3 H$$

Thus, we get only three distinct left cosets i.e. $H, f_3 H, f_4 H$.

Therefore $\qquad G = H \cup f_3 H \cup f_4 H.$

Also the right cosets of H in G are

$$H f_1 = \{ (1)(1), (12)(1)\}$$

$$= \{ (1), (12)\} = H$$

$$Hf_2 = \{ (1)(12), (12)(12)\}$$

$$= \{ (12), (1)\} = H$$

$$Hf_3 = \{ (1)(23), (12)(23)\}$$

$$= \{ (23), (123)\}$$

$$Hf_4 = \{ (1)(31), (12)(31)\}$$

$$= \{ (31), (132)\}$$

$$Hf_5 = \{ (1)(123), (12)(123)\}$$

$$= \{ (123), (23)\} = Hf_3$$

$$Hf_6 = \{ (1)(132), (12)(132)\}$$

$$= \{ (132), (13)\} - Hf_4$$

Thus, we get only three distinct right cosets of H in G.

5.41 Properties of Cosets

Theorem 5.39

If H is any subgroup of G and $h \in H$, then $Hh = H = hH$.

Proof

Let $h \in H$ and h' is any arbitrary element of H. Then $h'h$ is an arbitr-ary element of Hh.
But H is a subgroup of G. Then by closure property

$$h' \in H, h \in H \Rightarrow h' h \in H$$

$\Rightarrow$ every element of Hh is also an element of H.

$\therefore$ $$Hh \subseteq H \qquad ...(1)$$

Again
$$h' = h'e$$
$$= h'(h^{-1} h) \qquad [e = h^{-1} h]$$
$$= (h'h^{-1})h \qquad \text{[by associativity]}$$
$$\in Hh \qquad \text{[by inverse and closure property]}$$

$\therefore$ every element of H is also an element of Hh.

Hence $$H \subseteq Hh \qquad ...(2)$$

From (1) and (2), we get $H = Hh$

Similarly, we can show that $Hh = H$

Theorem 5.40

Let H be a subgroup of group G and let $a \in G$. Then $a \in Ha$ and $a \in aH$.

Proof

Let e is an identity element. Then $a \in H, e \in H$

$$\Rightarrow ae \in aH \; \forall \; a \in G$$

$$\Rightarrow a \in aH$$

Also
$$e \in H$$
$$\Rightarrow ea \in Ha \; \forall \; a \in G$$

$$\Rightarrow a \in Ha$$

This implies that aH and Ha are not empty.

Theorem 5.41

Let a and b are any two elements of a group G and H be any subgroup of G, then $Ha = Hb \Leftrightarrow ab^{-1} \in H$ and $aH = bH \Leftrightarrow a^{-1}b \in H$.

Proof

Let $a \in Ha$

Since, Ha = Hb

$$\Rightarrow a \in Hb$$

$$\Rightarrow ab^{-1} \in H bb^{-1}.$$

$$\Rightarrow ab^{-1} \in He \qquad [bb^{-1} = e]$$

$$\Rightarrow ab^{-1} \in H \qquad \ldots(1)$$

Again, let $ab^{-1} \in H$

$$\Rightarrow Ha\, b^{-1} = H \qquad [h \in H \Rightarrow Hh = H]$$

$$\Rightarrow Ha\, b^{-1} b = Hb$$

$$\Rightarrow Ha\, e = Hb$$

$$\Rightarrow Ha = Hb$$

Thus $Ha = Hb \Leftrightarrow ab^{-1} \in H$

Similarly, we can show that

$$aH = bH \Leftrightarrow a^{-1}b \in H.$$

Theorem 5.42

Any two right (left) cosets of a subgroup are either identical or disjoint.

Proof

Suppose H be a subgroup of a group G and let Ha and Hb are two right cosets of H in G which are not disjoint. Then there exists at least one element say y, such that $y \in Ha$ and $y \in Hb$.

Let $y = h_1 a$ and $y = h_2 b$ for some $h_1, h_2 \in H$

Now $$h_1 a = h_2 b$$

$$\Rightarrow \quad h^{-1}_1 h_1 a = h^{-1}_1 h_2 a$$

$$\Rightarrow \quad e\, a = h^{-1}_1 h_2 b$$

$$\Rightarrow \quad a = h^{-1}_1 h_2 b$$

Since H is a subgroup, then $h_1, h_2 \in H \Rightarrow h^{-1}_1 h_2 \in H$

So for any $h \in H$, we have

$$ha = h(h^{-1}_1 h_2 b)$$

$$= (h h^{-1}_1 h_2)b$$

$$\in Hb \qquad [\because hh^{-1}_1 h_2 \in H]$$

$$\Rightarrow \quad ha \in Hb$$

$\Rightarrow$ $Ha \subset Hb$ ...(1)

Similarly, we can show that $Hb \subset Ha$ ...(2)

From (1) and (2), we get $Ha = Hb$

Thus any two right cosets are identical if they are not disjoint.

A similar proof can be given for any two left cosets.

5.42 Lagrange's Theorem

Theorem 5.43

The order of each subgroup of a finite group is a divisor of the order of the group.

Proof

Let H be a subgroup of a finite group G such that

$$O(H) = m \text{ and } O(G) = n$$

Let $h_1, h_2, \ldots, h_m$ are m distinct elements of H.

Let $a \in G$, then Ha is a right coset of H in G and we have

$$Ha = \{h_1a, h_2a, \ldots, h_ma\}$$

Clearly Ha has m distinct elements.

As we know that any two right cosets of H in G are disjoint and G is a finite group, then the number of distinct right cosets of H in G will be finite. Let these are p in numbers. Therefore

$$G = Ha_1 \cup Ha_2 \cup \ldots \cup Ha_p$$

Since all the cosets are pairwise disjoint, so

$$O(G) = O(Ha_1) + O(Ha_2) + \ldots + O(Ha_p)$$

$$\Rightarrow \quad O(G) = m + m + \ldots + m \ (p \text{ times})$$

$$\Rightarrow \quad n = pm$$

$$\Rightarrow \quad p = \frac{n}{m}$$

$\Rightarrow$ m is a divisor of n

$\Rightarrow$ $O(H)$ is a divisor of $O(G)$.

Remark

Converse of Lagrange's theorem is not necessarily true.

Theorem 5.44

The order of every element of a finite group is a divisor of the order of the group.

Proof

Suppose G is a finite group of order m. Let $a \in G$ such that $O(a) = n$. Now, we have to prove that n is a divisor of m.

Let H be a cyclic subgroup of G generated by a i.e. $H = [a]$. But we know that, if a be an element of a group G such that $O(a) = n$, then order of the cyclic subgroup $H = [a]$, generated by a, is also n. Therefore $O(H) = n$. By Lagrange's theorem $O(H)$divides $O(G)$, so n divides m.

Theorem 5.45

If G is a finite group of order n and $a \in G$, then $a^n = e$.

Proof

Since the order of every element of a finite group is finite, so let $O(a) = m$. Then by Theorem 5.44, m is a divisor of n so let $n = mt$, $t \in N$.

Now
$$a^n = a^{mt}$$
$$= (a^m)^t$$
$$= e^t \qquad [\because O(a) = m \Rightarrow a^m = e]$$
$$= e.$$

Theorem 5.46

Every group of prime order is cyclic.

Proof

Suppose G is a finite group of order p, where p is a prime (i.e. a number other than one having factors ± 1 and $\pm p$). Since $p > 1$, then there exists $a \neq e$ such that $a \in G$. Let H be a cyclic subgroup of G generated by a. Then order of H is a divisor of order of G.

$\Rightarrow$ O(H) is a divisor of p. Since p is prime, then O(H) = 1

But $H \neq [e]$, therefore $O(H) \neq 1$

Hence $O[H] = p$

Thus $H \subset G$ and $O(H) = O(G)$

$\Rightarrow H = G$

But H is cyclic, hence G is cyclic.

Theorem 5.47

A group of prime order can have no proper subgroups.

Proof

Suppose G is a group of prime order p and H is a subgroup of G of order n. By Lagrange's theorem n is a divisor of p. Since p is a prime then $n = 1$ or $p \Rightarrow H = [e]$ or G. Thus G has no proper subgroups.

5.43 Index of a Subgroup

Let G be a group and H be a subgroup of G. Then the number of distinct left (right) cosets of H in G is said to be the index of H in G and denoted by $[G : H]$ or by $i_G(H)$.

If H has m different cosets in a finite group G, then by Lagrange's theorem

$$O(G) = m[O(H)]$$

$$\Rightarrow \qquad m = \frac{O(G)}{O(H)}$$

Index of H in $G = \dfrac{O(G)}{O(H)}$.

Thus $$[G{:}H] = \frac{O(G)}{O(H)}$$

Example 5.75

If $H = 4Z$ and $G = (Z, +)$, where $Z = \{\ldots -3, -2, -1, 0, 1, 2, 3, \ldots\}$

Then $[G : H] = 4$

Because there are only four different right cosets of H in G.

Theorem 5.48

If H and K be finite subgroups of a group G. Then show that

$$\Rightarrow O(HK) = \frac{O(H)O(K)}{O(H \cap K)}$$

Proof

Let $R = H \cap K$. Then R is a subgroup of G and $R \subseteq H$ and $R \subseteq K$. Thus R is a subgroup of H. Since H is finite, then the number of distinct right cosets in right coset decomposition of H with respect to R is finite.

Suppose it is t, then by Lagrange's theorem, we have

$$t = \frac{O(H)}{O(R)}$$

If $h_1R, h_2R, \ldots, h_tR$ are distinct left cosets of R in H, then

$$H - h_1R \cup h_2R \cup \ldots \cup h_tR$$

$$= \bigcup_{i=1}^{t} h_iR$$

Now
$$HK = \left(\bigcup_{i=1}^{t} h_iR\right)K = \bigcup_{i=1}^{t} h_i(RK) = \bigcup_{i=1}^{t} h_iK$$

$[\because R \subseteq K \Rightarrow RK = K]$

$$= h_1K \cup h_2K \cup \ldots \cup h_tK$$

Further, we claim that the cosets $h_1K, h_2K, \ldots, h_tK$ are pairwise distinct. We have

$$h_iK = h_jK \Rightarrow h_j^{-1}h_i \in K$$

Since H is a subgroup, then

$$h_i, h_j, \in H \Rightarrow h_j^{-1}h_i \in H$$

$$\Rightarrow h_j^{-1}h_i \in H \cap K$$

$$\Rightarrow h_j^{-1}h_i \in R$$

$$\Rightarrow h_iR = h_jR$$

which is a contradiction, because the cosets of R are all distinct. Thus $h_1K, h_2K, \ldots, h_tK$ are all distinct left cosets and they are pairwise disjoint also. The number of elements in each of h_iK is equal to the number of elements in K, therefore

$$O(HK) = t.O(K)$$

$$= \frac{O(H)}{O(R)}O(K) = \frac{O(H)O(K)}{O(H \cap K)}$$

Theorem 5.49

If $H \subset K$ be two subgroups of a finite group G, then show that

$$[G: H] = [G : K]\,[K : H].$$

Proof

Let G be a finite group and H and K are two subgroups of G such that $H \subset K$. Since $H \subset K$, therefore H is a subgroup of K also.

By Lagrange's theorem, we have

$$[G:H] = \frac{O(G)}{O(H)} \quad ...(1)$$

$$[G:K] = \frac{O(G)}{O(K)} \quad ...(2)$$

and $$[K:H] = \frac{O(K)}{O(H)} \quad ...(3)$$

Now from (2) and (3), we get

$$[G:K][K:H] = \frac{O(G)\,O(K)}{O(K)\,O(H)}$$

$$= \frac{O(G)}{O(H)}$$

$$= [G:H]. \quad \text{[From 1]}$$

Exercise 5.6

1. Let H be a subgroup of a group G. If $[G: H] = 2$, then show that
$$aH = Ha \ \forall \ a \in G.$$
2. Write down the cosets of H in G where
   i. $H = \{ (1), (123), (132) \} : G = P_3$.
   ii. $H = (3Z, +) : G = (Z, +)$
   iii. $H = \{1, a^2\}$ of the cyclic group $G = [a]$ of order 4.
   iv. $H = [\{0, 4\}, +_8] : G = \{(0, 1, 2, 3, 4, 5, 6, 7), +_8\}$.
3. Give an example of a group G having a subgroup H and two elements a, b such that $Ha = Hb$ but $aH \neq bH$.
4. Show that the set of inverses of the elements of a right coset is a left coset i.e. $(Ha)^{-1} = a^{-1}H$.
5. Consider two subgroups $H = \{I, (12)\}$ and $K = \{I, (13)\}$ of P_3. Determine HK. Using Lagrange's theorem or otherwise prove that HK is not a subgroup of P_3.
6. Prove that the order of a finite group is a multiple of the order of each one of its subgroups.

5.44 Homomorphism and Isomorphism of Groups

In this section we discuss the very important concept of homomorphism and isomorphism of groups.

5.44-1 Homomorphism

Suppose G and G' be any two groups and let $\circ$ and $*$ denote their respective binary compositions. Then a mapping $f: G \to G'$ is called a **homomorphism** if $f(a \circ b) = f(a) * f(b) \ \forall a, b \in G$.

Thus if f is a homomorphism from G to G' then f preserves the compositions in G and G' i.e. image of the composite = composite of images

A group G' is said to be a **homomorphic image** of a group G, if there exists a homomorphism f of G onto G'.

5.44-2 Endomorphism

A homomorphism of a group G into G itself, is called **endomorphism** of G.

5.44-3 Monomorphism

A homomorphism f from a group G to a group G' is said to be a **monomorphism** if f is one-one.

5.44-4 Epimorphism

If a homomorphic mapping f from a group G to a group G' is onto then f is said to be **epimorphism.**

5.44-5 Isomorphism

A homomorphism f of a group $(G,\circ)$ into $(G', *)$ is an isomorphism from G to G' if

i. f is one-one i.e.

$$f(a) = f(b) \Rightarrow a = b \;\forall\; a, b \in G.$$

ii. f is onto means for every $b \in G'$ there exists $a \in G$ such that $f(a) = b$.

5.44-6 Automorphism

An isomorphism of a group G onto itself is called an **automorphism** of G.

5.44-7 Isomorphic Groups

A group G is said to be isomorphic to a group G' if there exists an isomorphism of G onto G' denoted by $G \cong G'$.

Example 5.76

Let $(I, +)$ be a group. Prove that the mapping $f: I \to I$ defined by

$f(x) = 5x \;\forall\; x \in I$ is homomorphism from I to itself.

Solution

The mapping $f: I \to I$ will be homomorphism if it preserves the composition in I. Now

$$\begin{aligned} f(x_1+x_2) &= 5(x_1+x_2) \\ &= 5x_1+5x_2 \\ &= f(x_1)+f(x_2) \end{aligned}$$

Thus f preserves compositions in I and hence homomorphism.

Example 5.77

Show that the mapping $f: R \to R_0$ given by $f(x) = 2^x \;\forall\; x \in R$ is a homomorphism of R into R_0 where R is the additive group of real numbers and R_0 is the multiplicative group of non-zero real numbers.

Solution

The mapping $f: R \to R_0$ will be homomorphism if it preserves the compositions in R and R_0.

Let $x_1, x_2 \in R$, then

$$f(x_1 + x_2) = 2^{x_1+x_2} = 2^{x_1} . 2^{x_2} = f(x_1).f(x_2)$$

Thus f preserves the compositions in R and R_0. Hence $f: R \to R_0$ is homomorphism.

Example 5.78

Let P_n be the symmetric group of degree n and $G = [\{1, -1\}; .]$

The mapping $f: P_n \rightarrow G$ given by

$$f(r) = \begin{cases} 1, & \text{if } r \text{ is even permutation} \\ -1, & \text{if } r \text{ is odd permutation} \end{cases}$$

is a homomorphism of P_n onto G.

Proof

Let r, s $\in P_n$. Then there exists following cases:

Case I : When both *r* and *s* are even permutations

We know that the product of two even permutations is even permutation, then $r\,s$ is even permutation. Thus

$$f(r\,s) = 1 = 1.1 = f(r).f(s)$$

Case II : When both *r* and *s* are odd permutations

We also know that the product of two odd permutations is an even permutation, then $r\ s$ is even permutation. Thus

$$f(r\,s) = 1 = (-1).(-1) = f(r).f(s)$$

Case III : When *r* is even and *s* is odd permutation

Since the product of even and odd permutation is an odd permutation, then $r\,s$ is an odd permutation. Thus

$$f(r\,s) = -1 = (1).(-1) = f(r).f(\text{s})$$

Case IV : When *r* is odd and s is even permutation

Then r s is an odd permutation. Thus

$$f(r\,s) = -1 = (-1).(1) = f(r).f(s)$$

Hence, we conclude that in all cases

$$f(r\,s) = f(r).f(s)\ \forall\ r, s \in P_n.\text{ and } f(P_n) = G$$

Thus f is a homomorphism of P_n onto G.

Example 5.79

Show that the mapping $f: G \rightarrow G'$ given by $f(x) = 2x\ \forall\ x \in G$ is an isomorphism of G onto G' where G is the additive group of integers and G' is the additive group of even integers including zero.

Solution

The mapping $f: G \rightarrow G'$ will be an isomorphism if it is

1. one-one
2. onto
3. preserves compositions in G and G'.

***f* is one-one**

Let $x_1, x_2 \in$ G. Then $f(x_1) = f(x_2) \Rightarrow 2x_1 = 2\,x_2 \Rightarrow x_1 = x_2$. [by left cancellation law]

Thus two elements in G have the same f-image in G' only if they are equal.

Therefore, distinct elements in G have distinct f-images in G'. Hence, f is one-one.

***f* is onto**

Let y is any element of G' i.e. y is an even integer, then $\frac{y}{2} \in G$

Now $$f\left(\frac{y}{2}\right) = 2\left(\frac{y}{2}\right) = y$$

i.e. $y \in G' \Rightarrow$ there exists $\frac{y}{2} \in G$ such that $f\left(\frac{y}{2}\right) = y$. Thus each element of G' is the f-image of some elements of G. Therefore f is onto.

***f* preserves compositions in G and G′**

Let $x_1, x_2 \in G$. Then

$$f(x_1+x_2) = 2(x_1+x_2) = 2x_1 + 2x_2 = f(x_1)+f(x_2)$$

where, $f(x_1) = 2x_1$ and $f(x_2) = 2x_2$.

Thus f preserves compositions in G and G'. Therefore f is an isomorphism of G onto G'.

Example 5.80

Let a be an element of a group G. Then the mapping f_a: $G \rightarrow G$ given by

$$f_a(x) = a\,x\,a^{-1},\ \forall\, x \in G \text{ is an automorphism of } G.$$

Proof

Let $a \in G$ and $x \in G$. Since G is a group, then

$$a\,x\,a^{-1} \in G$$

$\therefore f_a$ is a mapping form G to G

***f*$_a$ is one-one**

Let $x_1, x_2 \in G$, then $f_a(x_1) = f_a(x_2)$

$$\Rightarrow a\,x_1\,a^{-1} = a\,x_2\,a^{-1}$$

By left and right cancellation laws $x_1 = x_2$

$\therefore f_a$ is one-one

***f*$_a$ is onto**

For every $y \in G$, there exists $a^{-1}\,y\,a \in G$ such that

$$\begin{aligned} f_a(a^{-1}\,y\,a) &= a\,(a^{-1}\,y\,a)\,a^{-1}. \\ &= (a\,a^{-1})\,y\,(a\,a^{-1}) \quad \text{[Associativity]} \\ &= e\,y\,e = y \\ &\Rightarrow f_a \text{ is onto.} \end{aligned}$$

***f*$_a$ Preserves composition in G.**

Let $x_1, x_2 \in G$, then

$$\begin{aligned} f_a(x_1\,x_2) &= a(x_1\,x_2)\,a^{-1}. \\ &= a\,x_1\,e\,x_2\,a^{-1}. \end{aligned}$$

$$= a x_1 a^{-1} \; a x_2 a^{-1} \qquad [\because a^{-1} a = e]$$

$$= (a x_1 a^{-1})(a x_2 a^{-1})$$

$$= f_a(x_1) . f_a(x_2)$$

$\therefore f_a$ preserves composition in G. Hence f_a is an isomorphism.

Since f_a is onto itself, therefore it is an automorphism.

5.45 Some Important Properties of Homomorphism and Isomorphism

Theorem 5.50

Let f be a homomorphic (or an isomorphic) mapping of a group G into a group G'. Then

i. If e is the identity of G, then $f(e)$ is the identity of G'.

Proof

Let $a \in G \Rightarrow f(a) \in G'$. Suppose e is the identity of G and e' be the identity of G'.

Now $f(a) \in G', e' \in G'$

$$\Rightarrow e' f(a) = f(a)$$

$$= f(ea) \qquad [e \text{ is identity of } G]$$

Now f is isomorphism

$$\therefore e' f(a) = f(e) f(a)$$

Since G' is a group, therefore by right cancellation law

$$e' = f(e)$$

Thus $f(e)$ is the identity of G'.

ii. If $a \in G$, then $f(a^{-1}) = [f(a)]^{-1}$.

Proof

Let $\quad a \in G \Rightarrow a^{-1} \in G$

$$\Rightarrow a^{-1} a = a a^{-1} = e$$

Since

$$e' = f(e) = f(a a^{-1}) \qquad [\text{from (i)}]$$

$$= f(a) f(a^{-1}) \qquad [f \text{ is isomorphism}]$$

$$\Rightarrow f(a^{-1}) = [f(a)]^{-1}.$$

Theorem 5.51

The relation of isomorphism in the set of all groups is an equivalence relation.

Proof

As we know that a relation is said to be an equivalence relation if it is reflexive, symmetric and transitive. Therefore we shall prove that the relation of isomorphism denoted by $\cong$ in the set of all groups is reflexive, symmetric and transitive.

Reflexive

Suppose G be a group. Then we have to show that G is isomorphic to G. Let $f : G \to G$ be the identity mapping on G.

$$\Rightarrow f(x) = x \;\forall\; x \in G.$$

We know that an identity mapping is one–one onto, then f is one-one and onto. Also if a and b be any elements of G. then

$$f(a) = a \text{ and } f(b) = b$$

Again $\qquad f(ab) = ab = f(a) f(b) \qquad$ [by def.]

Therefore f preserves composition in G. Hence f is isomorphism of G onto G.

Thus f is reflexive.

Symmetric

Let G and G_1 be two groups such that G is isomorphic to G_1, i.e. $G \cong G_1$ and $f : G \to G_1$ be the corresponding isomorphism. Then f is one-one onto and preserves compositions in G and G_1. Since f is bijective, so f^{-1} exists. i.e. $f^{-1} : G_1 \to G$ exists and it is also a bijection (one-one and onto).

Now we have to show that $f : G_1 \to G$ is also composition preserving.

If $a, b \in G$ and $a_1, b_1 \in G_1$ such that

$$f(a) = a_1, f(b) = b_1$$

or $\qquad a = f^{-1}(a_1)$ and $b = f^{-1}(b_1)$

Now

$$\begin{aligned} f^{-1}(a_1 b_1) &= f^{-1}[f(a) f(b)] \\ &= f^{-1}[f(a\,b)] \qquad [f \text{ is isomorphism}] \\ &= a\,b \\ &= f^{-1}(a_1) f^{-1}(b_1) \end{aligned}$$

Thus f^{-1} is composition preserving.

Hence $G_1 \cong G$ and relation $\cong$ is symmetric.

Transitive

Let G, G_1 and G_2 be three groups such that $G \cong G_1$, $G_1 \cong G_2$.

Further let $f : G \to G_1$ and $g : G_1 \to G_2$ are respective isomorphism.

Since by definition f and g are bijection, thus $g \circ f : G \to G_2$ is also a bijection.

Again if x_1, y_1 are any elements of G, then

$$\begin{aligned} (g \circ f)(x_1 y_1) &= g[f(x_1 y_1)] \\ &= g[f(x_1) f(y_1)] \qquad [\because f \text{ is isomorphism}] \end{aligned}$$

Also g is isomorphism

$$\begin{aligned} &= g[f(x_1)]\, g[f(y_1)] \\ &= (g \circ f)(x_1)\, (g \circ f)(y_1) \end{aligned}$$

$\Rightarrow g \circ f$ is composition preserving.

Hence $g \circ f$ is an isomorphism of G onto G_2.

Thus the relation $\cong$ is transitive.

Therefore the relation of isomorphism in the set of all groups is an equivalence relation.

5.46 Cayley's Theorem

Theorem 5.52

Every finite group is isomorphic to some permutation group.

Proof

Suppose G be a finite group. Then $a \in G, x \in G \Rightarrow ax \in G$. Consider a mapping $f_a: G \to G$ such that

$$f_a(x) = ax \ \forall \ x \in G.$$

f_a is one-one

Let $x, y \in G$, then $f_a(x) = f_a(y)$

$$\Rightarrow ax = ay$$

$$\Rightarrow x = y \qquad \text{[By left cancellation law]}$$

Thus f_a is one-one.

f_a is onto

For every $y \in G$, there exists an element $a^{-1}y$ in G such that

$$f_a(a^{-1}y) = a(a^{-1}y)$$

$$= (a a^{-1})y \qquad \text{[by associativity]}$$

$$= ey \qquad [\because a a^{-1} = e]$$

$$= y$$

Therefore f_a is onto.

Since $f_a: G \to G$ is one-one onto, thus f_a is a permutation of G.

Let $G_1 = \{f_a : a \in G\}$

Now we claim that G_1 is a group with respect to the product of two functions.

Closure Property

Let $f_a, f_b \in G_1$ for $a, b \in G$.

Now,

$$(f_a f_b)\, x = f_a\, [f_b(x)]$$

$$= f_a\, [bx]$$

$$= a\,(bx)$$

$$= (ab)\, x \qquad \forall\, x \in G$$

$$= f_{ab}(x)$$

$$\Rightarrow \quad f_a f_b = f_{ab} \qquad \ldots(1)$$

Because $a\,b \in G$, therefore $f_{ab} \in G_1$. Hence G_1 is closed with respect to the product of functions.

Associativity

Let $f_a, f_b, f_c \in G_1$, where $a, b, c \in G$. Then

$$f_a(f_b f_c) = f_a f_{bc} \qquad \text{[from 1]}$$

$$= f_{a(bc)} \qquad \text{[from 1]}$$

$$= f_{(ab)\,c} \qquad \text{[Associativity in } G\text{]}$$

$$= f_{ab} f_c \quad \text{[from 1]}$$

$$= (f_a f_b) f_c \quad \text{[from 1]}$$

Therefore, product of functions is associative in G_1.

Existence of Identity

If $e \in G$ be the identity element, then by definition f_e is the identity element of G_1. Now for each $f_a \in G_1$, we have

$$f_e f_a = f_{ea} = f_a$$

Also $$f_a f_e = f_{ae} = f_a$$

Hence f_e is identity element of G_1.

Existence of Inverse

As G is a group, therefore for every $a \in G \Rightarrow a^{-1} \in G$. Also for $a \in G \Rightarrow f_a \in G_1$ then $a^{-1} \in G \Rightarrow f_{a^{-1}} \in G_1$

We claim that $f_{a^{-1}}$ is the inverse of f_a in G_1.

Consider $$f_a f_{a^{-1}} = f_{aa^{-1}} \quad \text{[From 1]}$$

$$= f_e \quad [a\,a^{-1} = e]$$

Also $$f_{a^{-1}} f_a = f_{a^{-1}a} \quad \text{[From 1]}$$

$$= f_e \quad [a\,a^{-1} = e]$$

Hence $f_{a^{-1}}$ is the inverse of f_a in G_1.

Therefore G_1 is a group. Now we have to show that G is isomorphic to G_1 i.e. $G \cong G_1$. Consider the mapping ψ from G into G_1 given by.

$$\psi(x) = f_x \quad \forall\, x \in G.$$

ψ is one-one

For any $x, y \in G$, $\psi(x) = \psi(y)$

$$\Rightarrow \quad f_x = f_y$$

$$\Rightarrow \quad f_x(x_1) = f_y(x_1) \ \forall\, x_1 \in G$$

$$\Rightarrow \quad x_1 x = x_1 y \quad \text{[by def.]}$$

$$\Rightarrow \quad x = y \quad \text{[left cancellation law]}$$

Hence ψ is one-one.

ψ is onto

Let $f_x \in G_1$, then $x \in G$

$$\Rightarrow \psi(x) = f_x$$

Therefore ψ is onto.

ψ Preserves Compositions in G and G_1

For any $x, y \in G$

$$\psi(xy) = f_{xy}$$
$$= f_x f_y \quad \text{[From 1]}$$
$$= \psi(x)\,\psi(y) \quad \text{[by def.]}$$

Thus ψ preserves compositions in G and G_1.

Hence ψ is an isomorphism and consequently $G \cong G_1$.

Example 5.81

Find the permutation group isomorphic to the multiplicative group $\{1, -1, i, -i\}$.

Solution

Suppose $G = \{1, -1, i, -i\}$ and $f_a : G \to G$ be a mapping defined by

$$f_a(x) = ax \;\forall\; x \in G.$$

Then, we have

$$f_1 = \begin{pmatrix} 1 & -1 & i & -i \\ 1.1 & 1(-1) & 1.i & 1(-i) \end{pmatrix}$$
$$= \begin{pmatrix} 1 & -1 & i & -i \\ 1 & -1 & i & -i \end{pmatrix} = I$$
$$f_{-1} = \begin{pmatrix} 1 & -1 & i & -i \\ -1.1 & -1(-1) & -1.i & -1(-i) \end{pmatrix}$$
$$= \begin{pmatrix} 1 & -1 & i & -i \\ -1 & 1 & -i & i \end{pmatrix} = (1\;-1)(i\;-i)$$
$$f_i = \begin{pmatrix} 1 & -1 & i & -i \\ i.1 & i(-1) & i.i & i(-i) \end{pmatrix}$$
$$= \begin{pmatrix} 1 & -1 & i & -i \\ i & -i & -1 & 1 \end{pmatrix}$$
$$= (1\;\; i\;\; -1\;\; -i)$$
$$f_{-i} = \begin{pmatrix} 1 & -1 & i & -i \\ -i.1 & -i(-1) & -i(i) & -i.(-i) \end{pmatrix}$$
$$= \begin{pmatrix} 1 & -1 & i & -i \\ -i & i & 1 & -1 \end{pmatrix}$$
$$= (1\;\; -i\;\; -1\;\; i)$$

Hence by Cayley's theorem, the permutation group is isomorphic to G, is given by

$$G_1 = \{f_1, f_{-1}, f_i, f_{-i}\}$$

and the corresponding isomorphism is defined by

$$\psi(x) = f_x \;\forall\; x \in G.$$

*The group G_1 is known as the **regular permutation group** isomorphic to the group G.*

Exercise 5.7

1. Prove that following groups are isomorphic groups.
   a. The multiplicative group $\{1, w, w^2\}$ and the permutation group $\{e, (abc), (acb)\}$.
   b. The multiplicative group $\{1, -1, i, -i\}$ and $(\{0, 1, 2, 3\}, +_4)$.
2. Show that in a group G, the mapping $f: G_1 \to G$, defined by
$$f(x) = x^{-1}\ \forall\, x \in G \text{ is an automorphism of G iff G is abelian.}$$
3. Find the permutation group which is isomorphic to the group.
   i. $[\{1, w, w^2\}, .]$ ii. $[\{0, 1, 2, 3\}, +_4]$
   iii. $[\{1, 2\}, \times_3]$.
4. Let G, G_1 and G_2 be groups, and let $f: G \to G_1$ and $g: G_1 \to G_2$. Prove that if f and g are isomorphism, then $g \circ f$ is an isomorphism.
5. Show that any two cyclic groups of the same order are isomorphic.
6. Show that every homomorphic image of an abelian group is abelian.
7. Show that each of the following mappings is a homomorphism.
   i. $f_1 : (C, +) \to (R, +)$ where $f_1\,(x + iy) = x$
   ii. $f_2 : (C, +) \to (C, +)$ where $f_2\,(x + iy) = iy$
   iii. $f_3 : (R_0, \cdot) \to (R_0, \cdot)$ where $f_3\,(x) = x^2$.
8. Prove that every group of order 3 is isomorphic to every other group of order 3.

Answers

3. i. $\{e, (abc), (acb)\}$ ii. $\{1, -1, i, -i\}$
   iii. $\{1, -1\}$.

5.47 Normal Subgroups

Let G be a group and H is a subgroup of G. Subgroup H is called **normal subgroup** (or **self-conjugate subgroup** or **invariant subgroup**) of G if and only if for every $x \in G$ and all $h \in H$, $x\,h\,x^{-1} \in H$.

By definition, one can observe that H is a normal subgroup of G if and only if for every $x \in G$, $x\,H\,x^{-1} \subset H$. For every group G there are at least two normal subgroups namely, the group G itself and the group having identity element alone; i.e. G and $\{e\}$. These two subgroups are said to be **improper normal subgroups** of G. A normal subgroup of a group G is said to be a **proper normal subgroup** if it is different from G as well as from $\{e\}$.

5.47-1 Simple Group

A group having improper normal subgroup is said to be **simple group**. For example, a group of prime order is simple.

Example 5.82

The alternating group A_n of all even permutations of degree n is a normal subgroup of the symmetric group P_n of all permutations of degree n.

Solution

Let $f \in P_n$ and $g \in A_n$. If f is an even permutation, then f^{-1} is an even permutation and hence $f g f^{-1}$ is an even permutation.

Thus $$f g f^{-1} \in A_n$$

If f is an odd permutation, then f^{-1} is an odd permutation. Since the product of two odd permutations is even, therefore

$$f g f^{-1} \in A_n$$

hence in both cases, we get

$$f \in P_n, g \in A_n \Rightarrow f g f^{-1} \in A_n$$

Thus A_n is a normal subgroup of P_n.

5.48 Properties of Normal Subgroups

Theorem 5.53

Every subgroup of an abelian group is normal.

Proof

Let G be an abelian group and H be a subgroup of G. Let $h \in H$ and $x \in G$ then $xh = hx$

$$\Rightarrow \quad xhx^{-1} = (hx)\,x^{-1} \quad [G \text{ is abelian}]$$
$$= h(x\,x^{-1}) \quad [\text{by associativity in } G]$$
$$= h\,e \quad [\because x\,x^{-1} = e]$$
$$= h$$

Since $h \in H \Rightarrow xhx^{-1} \in H.$

Hence H is a normal subgroup of G.

Corollary

Every subgroup of a cyclic group is normal.

Proof

Since every cyclic group is abelian then by Theorem 5.53 every subgroup of a cyclic group is normal.

Theorem 5.54

A subgroup H of a group G is a normal subgroup if and only if for every x in G, $x H x^{-1} = H$.

Proof

Let H be a normal subgroup of G. Then we have to show that

$$xHx^{-1} = H \;\forall\, x \in G.$$

Now $x \in G$ and $h \in H \Rightarrow xhx^{-1} \in H$

$$\Rightarrow xHx^{-1} \subset H \quad \ldots(1)$$

Again $$xHx^{-1} \subset H \;\forall\, x \in G.$$

$$\Rightarrow \quad x^{-1} H (x^{-1})^{-1} \subset H \quad [\because x^{-1} \in G]$$

$$x^{-1} H x \subset H$$

$$x(x^{-1} Hx)\,x^{-1} \subset x\,Hx^{-1}$$

$$(x x^{-1}) H (x x^{-1}) \subset x H x^{-1}$$

$$e H e \subset x H x^{-1}$$

$$\Rightarrow \quad H \subset x H x^{-1} \qquad \ldots(2)$$

From (1) and (2), we get $x H x^{-1} = H$.

Conversely, suppose that $x H x^{-1} = H \; \forall \; x \in G$. Then we have to show that H is a normal subgroup of G.

Now $\quad x H x^{-1} = H$

$$\Rightarrow x H x^{-1} \subset H$$

$$\Rightarrow x h x^{-1} \in H \; \forall \; h \in H \text{ and } \forall \; x \in G.$$

By definition of normal subgroup, H is a normal subgroup of G.

Thus a subgroup H is normal iff it is self conjugate.

Theorem 5.55

The intersection of any two normal subgroups of a group is a normal subgroup.

Proof

Let G be a group and H_1, H_2 be any two normal subgroups of a group G. Then obviously $H_1 \cap H_2$ is a subgroup of G.

Now let $x \in G$ and $h \in H_1 \cap H_2$. Then $h \in H_1 \cap H_2 \Rightarrow h \in H_1$ and $h \in H_2$.

But H_1 is a normal subgroup of G, then by definition

$$x h x^{-1} \in H_1 \; \forall \; x \in G \text{ and } h \in H_1 \qquad \ldots(1)$$

Also H_2 is a normal subgroup of G, then

$$x h x^{-1} \in H_2 \; \forall \; x \in G \text{ and } h \in H_2 \qquad \ldots(2)$$

Now from (1) and (2), we get

$$\forall \; x \in G, \; h \in H_1 \cap H_2 \Rightarrow x h x^{-1} \in H_1 \cap H_2.$$

Thus $H_1 \cap H_2$ is a normal subgroup of G.

Theorem 5.56

A subgroup H of a group G is a normal subgroup of G iff the product of two left (right) cosets of H in G is again a left (right) coset of H in G.

Proof

Let G be a group and H is a normal subgroup of G. Suppose that aH and bH are two left cosets of H in G for all $a, b \in G$.

Now by definition, if $h_1 \in H$ and $h_2 \in H \Rightarrow a h_1 \in a H$ and $b h_2 \in bH$

Then $\quad a h_1 \, bh_2 \in aH \, bH$

Also $\quad a h_1 \, bh_2 = a (h_1 \, b) \, h_2$

But we know that for some $h_1 \in H$

if $bH = Hb$. then $\quad h_1 \, b = bh_1$

$$\therefore \quad a (h_1 \, b) h_2 = a (bh_1) \, h_2$$

$$= (ab) \, h_1 \, h_2 \in ab \, H$$

Thus $aHbH \subset abH$ ... (1)

Now for $h \in H$

$$(ab)h = a(bh)$$
$$= (ae)(bh) \in a H b H \qquad [\because e \in H]$$
$$\Rightarrow \quad abH \subset aH\, bH \qquad ...(2)$$

From (1) and (2), we get $aH\,bH = ab\,H$

Thus H is a normal subgroup of a group G if $aH\,bH = ab\,H$.

Conversely

Let the product of two left cosets of H in G is again a left coset of H in G. Let $y \in G$. Since G is a group therefore $y \in G \Rightarrow y^{-1} \in G$. Thus $y\,H$ and $y^{-1}\,H$ are two left cosets of H in G, then by assumption $y\,H\,y^{-1}\,H$ is a left coset of H in G.

Now $$y \in y\,H \Rightarrow y^{-1} \in y^{-1}\,H$$
$$\Rightarrow y\,y^{-1} \in y\,H\,y^{-1}\,H$$
$$\Rightarrow e \in y\,H\,y^{-1}\,H$$

But H itself is a left coset of H containing e and any two left cosets are either identical or disjoint, so $yHy^{-1}\,H = H\,\forall\, y \in G.$

$$\Rightarrow y\,h_1 y^{-1}\,h_2 \in H\,\forall\, y \in G \text{ and } h_1, h_2 \in H$$
$$\Rightarrow (yh_1y^{-1}h_2)h_2^{-1} \in Hh_2^{-1}$$

but $h_2 \in H \Rightarrow h_2^{-1} \in H \Rightarrow Hh_2^{-1} = H$

$$\Rightarrow (yh_1y^{-1})(h_2h_2^{-1}) \in H$$
$$\Rightarrow yh_1y^{-1} \in H \;\; \forall\, x \in G \text{ and } h_1 \in H$$

$\Rightarrow$ H is a normal subgroup of G.

Theorem 5.57

A subgroup H of a group G is a normal subgroup if and only if every left coset of H is a right coset of H.

Proof

Let G is a group and H be the normal subgroup of G. Then

$$x\,H\,x^{-1} = H\,\forall\, x \in G.$$
$$\Rightarrow \quad (x\,H\,x^{-1})\,x = Hx$$
$$\Rightarrow \quad x\,H\,(x^{-1}\,x) = Hx \qquad [\text{ by associativity}]$$
$$\Rightarrow \quad x\,H\,e = Hx$$
$$\Rightarrow \quad x\,H = Hx$$

Thus, every left coset of H in G is a right coset of H in G.

Conversely

Let every left coset of H in G is a right coset of H in G.

i.e. $$x\,H = H\,x\,\forall\, x \in G.$$
$$\Rightarrow \quad x\,H\,x^{-1} = H\,x\,x^{-1}$$

$\Rightarrow \quad x\,H\,x^{-1} = H\,e \quad [\because x\,x^{-1} = e]$

$\Rightarrow \quad x\,H\,x^{-1} = H$

$\Rightarrow$ H is a normal subgroup of G.

Hence, H is normal subgroup of G iff $x\,H = H\,x\ \forall\ x \in G$.

Example 5.83

Let G be a group and $H = \{x \in G : x\,g = g\,x\ \forall\ g \in G\}$. Show that H is normal subgroup of G.

Solution

First, we will show that H is a subgroup of G. Clearly H is non–empty because

$$e \in G \Rightarrow e\,g = g\,e\ \forall\ g \in G.$$

Let $x_1, x_2 \in H$

$\Rightarrow \quad x_1\,g = g\,x_1$ and $x_2\,g = g\,x_2\ \forall\ g \in G \quad \ldots(1)$

Now
$$x_1\,x_2\,g = x_1\,(x_2\,g)$$
$$= x_1\,(g\,x_2) \quad \text{[From (1)]}$$
$$= (x_1\,g)\,x_2 \quad \text{[Associativity]}$$
$$= (g\,x_1)\,x_2 \quad \text{[From (1)]}$$
$$= g\,x_1\,x_2$$

$\Rightarrow x_1\,x_2 \in H$. Hence H is closed for composition.

Again $x_2 \in H \Rightarrow x_2\,g = g\,x_2\ \forall\ g \in G$.

$\Rightarrow x_2^{-1}(x_2g) = x_2^{-1}(gx_2)$

$\Rightarrow (x_2^{-1}x_2)g = (x_2^{-1}g)x_2$

$\Rightarrow g = (x_2^{-1}g)x_2$

$\Rightarrow gx_2^{-1} = (x_2^{-1}g)x_2x_2^{-1}$

$\Rightarrow g\,x_2^{-1} = x_2^{-1}g$

$\Rightarrow \quad x_2^{-1} \in H$

Hence, the inverse of each element of H are in H.

Now $\quad x_1 \in H, x_2 \in H \Rightarrow x_1 \in H, x^{-1}{}_2 \in H$

Then
$$\left(x_1x_2^{-1}\right)g = x_1\left(x_2^{-1}g\right)$$
$$= x_1\left(gx_2^{-1}\right) \quad \text{[by def.]}$$
$$= \left(x_1g\right)x_2^{-1} \quad \text{[Associativity]}$$
$$= \left(gx_1\right)x_2^{-1} \quad \text{[by def.]}$$
$$= gx_1\,x_2^{-1}$$
$$\Rightarrow x_1\,x_2^{-1} \in H$$

Hence, H is a subgroup of G. Now we claim that H is a normal subgroup of G. Let $x \in G, h \in H$

Then
$$
\begin{aligned}
xHx^{-1} &= (xh)x^{-1} \\
&= (hx)x^{-1} && \text{[by def.]} \\
&= hxx^{-1} \\
&= he \\
&= h \in H
\end{aligned}
$$

Thus $xhx^{-1} \in H \; \forall \, x \in G$ and $h \in H$. Hence H is a normal subgroup of G.

Example 5.84

If H is a subgroup of G such that $x^2 \in H \; \forall \, x \in G$, then prove that H is a normal subgroup of G.

Solution

Let G be a group and H is its subgroup then $g \in G, h \in H \Rightarrow gh \in G$.

By definition $\quad gh \in G \Rightarrow (gh)^2 \in H$

Again $\quad g^{-1} \in G \Rightarrow (g^{-1})^2 \in H$ $\quad$ [by def.]

and $\quad h \in G \Rightarrow h^{-1} \in H$ $\quad$ [H is a subgroup]

Now $h^{-1} \in H, (g^{-1})^2 \in H$ and H is a subgroup, then $h^{-1}g^{-2} \in H$

Further
$$
\begin{aligned}
(gh)^2 &\in H, h^{-1}g^{-2} \in H \\
&\Rightarrow (gh)^2 h^{-1}g^{-2} \in H \\
&\Rightarrow gh\,gh\,h^{-1}g^{-2} \in H \\
&\Rightarrow gh\,g\,g^{-1}g^{-1} \in H \\
&\Rightarrow gh\,g^{-1} \in H \; \forall \, g \in G.
\end{aligned}
$$

Hence H is a normal subgroup of G.

Exercise 5.8

1. If H and K are two normal subgroups of a group G, then HK is also a normal subgroup of G.
2. If H is a subgroup of G and N is a normal subgroup of G, then $H \cap N$ is a normal subgroup of H.
3. Every subgroup of a group with index 2, is a normal subgroup.
   [**Hint:** Group with index 2 is a group having two distinct left (right) cosets.]
4. Let H be a subgroup of a group G and
   $$N_H = \{ x \in G : xH = Hx \}$$
   Prove that N_H is a subgroup of G and H is a normal subgroup of N_H.
5. Show that $H = \{(1), (12)(34), (13)(24), (14)(23)\}$ is a normal subgroup of A_4.
6. If H is a subgroup of G and N is a normal subgroup of G then show that
   i. HN is a subgroup of G.
   ii. N is a normal subgroup of HN.
7. Let $G = \{(a, b): a, b \in Q, a \neq 0\}$
   Define * in G as follows
   $$(a, b) * (c, d) = (ac, ad + b)$$
   Show that $(G, *)$ is a non-abelian group. Consider $H = \{(1, b): b \in Q\}$. Prove that H is a normal subgroup of G.

8. If N is a normal subgroup of G, and H is a subgroup of G then show that $NH = HN$.

5.49 Factor Group (or Quotient Group)

Let G be a group and H be a normal subgroup of G. As we know that the right and left cosets of H in G are equal, so there is no need to distinguish between left and right cosets of a normal subgroup H. We say that Hx is a coset of G $\forall\, x \in G$. Let G/H denote the set of all cosets of H in G. Now we shall see that it is a group. This group is said to be **quotient group** (or **factor group**) of G by H.

Theorem 5.58

Let G be a group and H be a normal subgroup of G. The set G/H of all cosets of H in G, is a group under the binary composition defined as follows: For any Ha, $Hb \in G/H$,

$$Ha\, Hb = Hab \ \forall\ a, b \in G.$$

Proof

Let G/H is the family of all cosets of H in G. Then $G/H = \{Ha: a \in G\}$

Now we have to show that the composition of product of cosets is well defined.

Let $\quad Ha = Hx$ and $Hb = Hy$

Then $\quad a\,x^{-1} \in H$ and $by^{-1} \in H$

Let $\quad a\,x^{-1} = h_1$ and $by^{-1} = h_2$

Then $\quad a = h_1 x$ and $b = h_2 y$

Now
$$\begin{aligned} ab &= h_1 x\, h_2 y \\ &= h_1 h_3 xy && [\because H \text{ is a normal}] \\ &= h_4\, xy && [\because H \text{ is a normal}] \end{aligned}$$

$$\Rightarrow \quad ab(xy)^{-1} = h_4\, xy\, (xy)^{-1} = h_4 \in H$$

By definition $Hab = Hxy$

Thus $Ha = Hx$, $Hb = Hy \Rightarrow Hab = Hxy$.

Hence the composition is well defined. Now we claim that G/H form a group with respect to this composition.

Closure Property

Obviously G/H is closed for the composition.

Associativity

Let $a, b, c \in G$, then

$$Ha, Hb, Hc \in G/H$$

Now
$$\begin{aligned} (Ha\, Hb)\, Hc &= (Hab)\, Hc && [\text{by def.}] \\ &= H(ab)c \\ &= Ha\,(bc) && [\text{since } G \text{ is associative}] \\ &= Ha\,(Hbc) && [\text{by def.}] \\ &= Ha\,(Hb\, Hc) \end{aligned}$$

Hence the composition is associative in G/H.

Existence of Identity

$\because H = He$ where $e \in G$ is identity $\Rightarrow He \in G/H$.

Now $\qquad Ha\,He = Hae = Ha \;\forall\, Ha \in G/H$

and $\qquad He\,Ha = Hea = Ha$

Thus He is identity element for G/H.

Existence of Inverse

Since G is a group therefore $\forall\, a \in G, \exists\, a^{-1} \in G$

Thus if $\qquad Ha \in G/H \Rightarrow Ha^{-1} \in G/H$

Now
$$\begin{aligned} Ha\,Ha^{-1} &= Ha\,a^{-1} && \text{[by def.]} \\ &= He. && [a\,a^{-1} = e] \\ &= H \end{aligned}$$

and
$$\begin{aligned} Ha^{-1}\,Ha &= Ha^{-1}a && \text{[by def.]} \\ &= He && [a^{-1}a = e] \\ &= H \end{aligned}$$

Hence Ha^{-1} is the inverse of Ha in G/H. Therefore every element of G/H is invertible. Thus G/H is a group for the composition of product of cosets.

Quotient Group (Factor Group)

Let G be a group and H is a normal subgroup of G. Then the set G/H of all cosets of H in G form a group with respect to the composition defined by

$$Ha\,Hb = Hab \;\forall\, Ha, Hb \in G/H.$$

This group is said to be **quotient group** (or **factor group**) of G by H.

If the binary composition in G is addition, then each coset of H in G is denoted by $H + (a + b) = (H + a) + (H + b)$.

Theorem 5.59

For a finite group G and for each normal subgroup H of G,

$$O(G/H) = \frac{O(G)}{O(H)}$$

Proof

Let t be the number of distinct right cosets of H in G, then by Lagrange's theorem

$$O(G) = O(H)\,t$$

$$\Rightarrow \qquad t = \frac{O(G)}{O(H)}$$

Thus
$$O(G/H) = \frac{O(G)}{O(H)}$$

Theorem 5.60

Every quotient group of an abelian group is abelian.

Proof

Let G be an abelian group and H is the normal subgroup of G, then G/H is a quotient group.

Let $a \in G \Rightarrow Ha \in G/H$ and $b \in G \Rightarrow Hb \in G/H$

Now let G be an abelian group $\Rightarrow ab = ba \ \forall \ a, b \in G$. We have to show that G/H is abelian.

Now
$$\begin{aligned} Ha\,Hb &= Hab && \text{[by def. of quotient group]} \\ &= Hba && \text{[G is abelian]} \\ &= Hb\,Ha && \text{[by def.]} \end{aligned}$$

Hence, G/H is abelian.

The converse of Theorem 5.60 is not necessarily true because P_3 is not abelian but P_3/A_3 is abelian group of order 2.

Theorem 5.61

Every quotient group of a cyclic group is cyclic.

Proof

Let G be a cyclic group generated by a and H be a normal subgroup of G. As each element of G is of the form a^n, then $H a^n \in G/H$.

Now
$$\begin{aligned} H a^n &= H(aaa \ldots n \text{ times}) \\ &= Ha\,Ha \ldots n \text{ times} && [Hab = Ha\,Hb] \\ &= (Ha)^n \end{aligned}$$

$$\Rightarrow \qquad G/H = [Ha]$$

Hence, G/H is cyclic group generated by Ha.

The converse of the Theorem 5.61 is not necessarily true because P_3 is not cyclic but P_3/A_3 is cyclic.

Theorem 5.62

A group is homomorphic to its quotient group.

Proof

Suppose G be a group and H is the normal subgroup of G.

Consider a mapping $f: G \rightarrow G/H$ given by $f(x) = Hx \ \forall \ x \in G$.

By definition of the mapping it is clear that the mapping is onto, because for each $Ha \in G/H$, there exists $a \in G$ such that

$$f(a) = Ha$$

Also if $a, b \in G$, then

$$f(ab) = Hab = Ha\,Hb = f(a)\,f(b)$$

Therefore f preserves the composition. Hence f is homomorphism of G onto G/H.

Example 5.85

Find the quotient group G/H when $G = [\ \{1, -1, i, -i\ \}, \cdot\ ]$ and $H = [\ \{1, -1\}, .]$. Also prepare the composition table for G/H.

Solution

As G is an abelian group and H is a subgroup of G, then H is a normal subgroup of G and hence G/H is a group. Now the right cosets of H in G are as follows:

$$H{\cdot}1 = \{1{\cdot}1, -1{\cdot}1\} = \{1, -1\} = H$$

$$H{\cdot}(-1) = \{1{\cdot}(-1), -1{\cdot}(-1)\} = \{-1, 1\} = H$$

$$H{\cdot}i. = \{1{\cdot}i, -1{\cdot}i\} = \{i, -i\}$$

$$H{\cdot}(-i) = \{1{\cdot}(-i), -1{\cdot}(-i)\} = \{-i, i\} = Hi$$

Therefore $G/H = \{H, Hi\}$.

The composition table of G/H is given as

.	H	Hi
H	H	Hi
Hi	Hi	H

Example 5.86

Find the quotient group G/H where $G = (I, +)$, $H = (3I, +)$.

Solution

Since $(I, +)$ is an abelian group, then $(3I, +)$ is a normal subgroup of $(I, +)$. Thus G/H is a group and the different cosets of H in G are as follows:

$$H + 0 = \{\ \ldots.. -6, -3, 0, 3, 6, \ldots\} = H$$

$$H + 1 = \{\ \ldots.. -5, -2, 1, 4, 7, \ldots\}$$

$$H + 2 = \{\ \ldots.. -4, -1, 2, 5, 8 \ldots\}$$

$$H + 3 = \{\ \ldots.. -6, -3, 0, 3, 6 \ldots\}$$

We observe that

$$H = H + 3 = H + 6 = H + 9 = \ldots$$

$$H + 1 = H + 4 = H + 7 = H + 10 = \ldots$$

$$H + 2 = H + 5 = H + 8 = H + 11 = \ldots$$

Thus H has three distinct cosets and then $G/H = \{H, H + 1, H + 2\}$

The composition table of G/H is as follows:

+	H	$H + 1$	$H + 2$
H	H	$H + 1$	$H + 2$
$H + 1$	$H + 1$	$H + 2$	H
$H + 2$	$H + 2$	H	$H + 1$

Example 5.87

Find the quotient group *G/H* and give its composition table when $G = P_3$ and $H = A_3$.

Solution

We have $P_3 = \{(1), (123), (132), (12), (23), (13)\}$

and $A_3 = \{(1), (123), (132)\}$

Since A_3 is a normal subgroup of P_3 and $O\left(\frac{P_3}{A_3}\right) = \frac{O(P_3)}{O(A_3)} = \frac{6}{3} = 2$

Also $a \in H \Rightarrow aH = H$ and $a \notin H \Rightarrow aH \neq H$

and there are two different cosets of A_3, then

$$(1)\,A_3 = A_3$$

$$(123)\,A_3 = A_3$$

$$(132)\,A_3 = A_3$$

Also

$$(12)\,A_3 \neq A_3$$

$$(23)\,A_3 \neq A_3$$

$$(13)\,A_3 \neq A_3$$

$$(12)\,A_3 = (23)\,A_3 = (13)\,A_3$$

Hence $P_3 / A_3 = \{A_3, (12)\,A_3\}$.

The composition table for P_3 / A_3 is given below.

·	A_3	$(12)\,A_3$
A_3	A_3	$(12)\,A_3$
$(12)\,A_3$	$(12)\,A_3$	A_3

Exercise 5.9

1. Find the quotient group *G/H* in each and also give the composition table for it.
   i. $G = (\{0, 1, 2, 3, 4, 5\}, +_5); H = [\{0, 4\}, +_5]$
   ii. $G = (I, +), H = (4I, +)$.
2. Give an example of a non-abelian group *G* and a proper normal subgroup *H* of *G*, such that *G/H* is abelian. [**Hint:** Consider the group P_4 and its subgroup A_4.]
3. Let *G* be a group and *H* be a normal subgroup of *G* such that *G/H* is abelian. Show that

$$a\,b\,a^{-1}\,b^{-1} \in H\ \forall\ a, b \in$$

Chapter Objective Type Questions

A. Multiple choice questions.

1. The set of integers *Z* with respect to multiplication is -

   (A) Quasi group (B) Semi-group

   (C) Monoid (D) Group

2. The set of natural numbers N under binary operation $a \cdot b = a - b$ is:
   (A) Quasi group (B) Semi-group
   (C) Monoid (D) Group
3. The set S of square matrices of same order with respect to matrix addition, is a:
   (A) Quasi group (B) Semi-group
   (C) Group (D) Abelian group
4. The set of square matrices of order 2, with respect to matrix multiplication, is a
   (A) Quasi group (B) Semi-group
   (C) Monoid (D) Group
5. The set of all non singular square matrices order with respect to matrix multiplication, is a
   (A) Quasi group (B) Abelian group
   (C) Monoid (D) Group
6. If G is a group, for $a \in G$, $N(a)$ is the normalizer of a, then $\forall x \in N(a)$ –
   (A) $xa = ax$ (B) $xa = e$
   (C) $ax = e$ (D) $xa \neq ax$
7. If G is a group, then for all $a, b \in G$ –
   (A) $(ab)^{-1} = a^{-1} b^{-1}$ (B) $(ab)^{-1} = b^{-1}a^{-1}$
   (C) $(ab)^{-1} = ab$ (D) $(ab)^{-1} = ba$
8. If G is a set of integers and $a \cdot b = a - b$, then G is –
   (A) Quasi group (B) Semi-group
   (C) Monoid (D) Group
9. In a group G, for each element $a \in G$, there is –
   (A) No inverse (B) A unique inverse $a^{-1} \in G$
   (C) More than one inverse (D) none of these
10. If $a, b \in G$, a group then b is conjugate to a if there exist $c \in G$ -
    (A) $b = c^{-1}\, a\, c$ (B) $a = c\, b$
    (C) $b = ac^{-1}$ (D) $b = c\, c^{-1} a$
11. If G is a group of order n, then order of identity element is -
    (A) 1 (B) greater than 1
    (C) n (D) none of these
12. If $a \in G$ is of order n and p is prime to n, then the order of a^p is
    (A) n (B) 1
    (C) less than n (D) greater than n
13. If the orders of elements $a, a^{-1} \in G$ are m and n respectively, then –
    (A) $m = n$ (B) $m \neq n$
    (C) $m = n = 0$ (D) none of these
14. If $G = \{1, -1, i, -i\}$ is multiplicative group, then order of $-i$ is –
    (A) 1 (B) 2
    (C) 3 (D) 4

15. If $G = \{(0, 1, 2, 3, 4), +5\}$, then the order of 2 is –

(A) 1 (B) 2

(C) 4 (D) 5

16. If G is a group of even order, $\forall\ a \neq e$ if $a^2 = e$ then G is –

(A) Abelian group (B) Sub group

(C) Normal group (D) none of these

17. Every group of prime order is –

(A) Cyclic (B) Abelian

(C) Sub-group (D) Normal group

18. If G is a group and $a \in G$ such that $a^2 = a$, then a is equal to -

(A) Identity element (B) Inverse

(C) zero element (D) none of these

19. If in a group G, $a \in G$ the order of a is n and order of a^p is m, then -

(A) $m \leq n$ (B) $m \geq n$

(C) $m = 0$ (D) none of these

20. The generators of a group $G = \{ a, a^2, a^3, a^4, a^5, a^6 = e \}$ are –

(A) a and a^5 (B) a^2 and a^4

(C) a^3 and a^5 (D) a^2 and a^4

21. If G is a finite group and order of group is p then $\forall\ a \in G$ –

(A) $a^p = e$ (B) $a^p \neq e$

(C) $a^p = a$ (D) $a^p = a^{-1}$

22. The number of elements in a group is –

(A) Identity of group (B) Order of group

(C) Inverse of group (D) none of these

23. The identity permutation is

(A) Even permutation (B) Odd permutation

(C) neither even nor odd (D) none of these

24. The product of even permutation is

(A) Odd permutation (B) Even permutation

(C) Even or odd permutation (D) none of these

25. The inverse of an even permutation is –

(A) Odd permutation (B) Even permutation

(C) Even or odd permutation (D) none of these

26. The inverse of an odd permutation is –

(A) Odd permutation (B) Even permutation

(C) Even or odd permutation (D) none of these

27. The product of (1 2 4 5) (3 2 1 5 4) is –

(A) (2 3) (B) (1 5)

(C) (3 4 1) (D) (1 5 3 1)

28. If $f = (2\ 3)$ and $g = (4\ 5)$ be two permutation on five symbols 1, 2, 3, 4, 5 then gf is –

(A) $\begin{pmatrix} 1 & 2 & 3 & 4 & 5 \\ 1 & 3 & 5 & 4 & 2 \end{pmatrix}$ (B) $\begin{pmatrix} 1 & 2 & 3 & 5 & 6 \\ 1 & 4 & 6 & 5 & 4 \end{pmatrix}$

(C) $\begin{pmatrix} 1 & 2 & 3 & 5 & 7 \\ 1 & 4 & 6 & 4 & 1 \end{pmatrix}$ (D) $\begin{pmatrix} 1 & 2 & 3 & 4 & 5 \\ 1 & 3 & 2 & 5 & 4 \end{pmatrix}$

29. If b and c are the inverse of some element $a \in G$ then –

(A) $b = c$ (B) $b \neq c$

(C) $b = kc$ for some k (D) none of these

30. HK is a sub-group of G iff

(A) $HK = HK$ (B) $HK \subset HK$

(C) $HK \supset KH$ (D) $HK \neq KH$

31. If number of left cosets of H in G are n and number of right cosets of H in G are m then –

(A) $m = n$ (B) $m \geq n$

(C) $m \leq n$ (D) none of these

32. If H_1 and H_2 are two right coset sets of sub-group H then –

(A) $H_1 \cap H_2 = \varnothing$ or $H_1 = H_2$ (B) $H_1 \cap H_2 \neq \varnothing$

(C) $H_1 \cup H_2 = \varnothing$ (D) $H_1 \neq H_2$ and $H_1 \cap H_2 \neq \varnothing$

33. A one-one mapping of a finite group onto itself is-

(A) Isomorphism (B) Homomorphism

(C) Automorphism (D) none of these

34. If H_1 and H_2 are two subgroups of g then following is also a subgroup of G –

(A) $H_1 \cap H_2$ (B) $H_1 \cup H_2$

(C) $H_1\ H_2$ (D) none of these

35. Statement A: All cyclic groups are Abelian

Statement B: The order of cyclic group is same as the order of its generator.

(A) A and B are false (B) A and B are true

(C) A is true, B is false (D) A is false, B is true

36. If G is a finite group of order n, $a \in G$ and order of a is m, if G is cyclic then –

(A) $m = n$ (B) $m > n$

(C) $m < n$ (D) none of these

37. Statement A: Every isomorphic image of a cyclic group is cyclic

Statement B: Every homomorphic image of a cyclic group is cyclic

(A) A and B are false (B) A and B are true

(C) A is true only (D) A is false only

B. Fill in the blanks.

1. The set M of square matrices (of same order), w.r.t. matrix multiplication is ________.
2. If G is a finite group of order n, then for every $a \in G$, $a^n =$ ______.
3. The order of identity element in a group G is ______.

4. If $a, b \in G$, a group of order m then order of ab and ba are _______.
5. If $G = \{1, -1\}$ is a group, then order of 1 is ______.
6. If permutation $A = \begin{pmatrix} 1 & 2 & 3 & 4 & 5 \\ 2 & 3 & 1 & 5 & 4 \end{pmatrix}$ and $B = \begin{pmatrix} 1 & 2 & 3 & 4 & 5 \\ 1 & 3 & 4 & 5 & 2 \end{pmatrix}$ then $BA =$ ______.
7. Permutation $\begin{pmatrix} 1 & 2 & 3 & 4 & 5 & 6 \\ 6 & 1 & 2 & 5 & 4 & 3 \end{pmatrix}$ is equivalent to ______.
8. The product of permutation (1 2 3) (2 4 3) (1 3 4) is equal to _______.
9. If permutation $C =$ (1 2 3 4 5 6 7) then C^3 is ______.
10. In the additive group G of integers, the order of inverse element a^{-1}, $\forall\, a \in G$ is _______.
11. The group $G = (\{0, 1, 2, 3, 4\}, +_5)$ is a cyclic group with generators ________.
12. The intersection of any two normal subgroups of a group is a _________ subgroup.
13 Every quotient group of a cyclic group is ______.
14. If H is a subgroup of G then _______.
15. Out of $n!$ permutations on n symbols, even and odd permutations are ______.
16. Every finite group is isomorphic to some _________ group.

C. Which of the following is True (T) / False (F)?

1. If $(G, *)$ is a group and $\forall\, a, b \in G$, $b^{-1} * a^{-1} * b * a = e$ then G is an Abelian group.
2. If $b, a^{-1} \in G$, a group and order of a and a^{-1} are m and n respectively, then $m > n$.
3. The permutation $\begin{pmatrix} 1 & 2 & 5 & 3 & 4 \\ 3 & 4 & 1 & 5 & 2 \end{pmatrix}$ is equal to (1 5) (1 3) (2 4).
4. If $C =$ (1 2 3 4) then $C^2 =$ (2 3) (3 1).
5. The idempotent element in a group is an identity element of a group.
6. If e_1 and e_2 are two identity elements of a group G then $e_1 = ke_2$, for some k.
7. Let $G = \{1, -1\}$ then under ordinary multiplication $(G, \cdot)$ is a Semi-group.
8. Let Q be a set of rational numbers then under ordinary addition $(Q, +)$ is Quasi group.
9. Let G be a group of square matrices of same order w.r.t. matrix multiplication then it is not an abelian group.
10. If G is a finite group, then for every $a \in G$, the order of a is zero.
11. In the additive group of integers the order of every element $a \neq 0$ is infinity.
12 Every infinite cyclic group has two and only two generators.
13. Every subgroup of an abelian group is normal.
14. Every finite group of order less than six must be abelian.
15. The order of each subgroup of a finite group is a divisor of the order of the group.

Answers Objective Type

A.

1.	C	2.	A	3.	D	4.	B
5.	D	6.	A	7.	B	8.	A
9.	B	10.	A	11.	A	12.	A

13. A	14. D	15. D	16. A
17. A	18. A	19. A	20. A
21. A	22. B	23. A	24. B
25. B	26. A	27. A	28. D
29. A	30. A	31. A	32. A
33. C	34. A	35. B	36. A
37. B			

B.

1. Monoid	2. E	3. One	4. Equal
5. 1	6. $\begin{pmatrix} 1 & 2 & 3 & 4 & 5 \\ 2 & 1 & 5 & 4 & 3 \end{pmatrix}$	7. (1 6 3 2)(4 5)	8. I
9. (1 4 7 3 6 2 5)	10 infinity	11. 1 and 4	12. Normal
13. Cyclic	14. $HH = H$ or $H^{-1} = H$	15. $n\,!/2$	16 permutation

C.

1. T	2. F	3. T	4. F
5. T	6. F	7. F	8. F
9. T	10. F	11. T	12. T
13. T	14. F	15. T	16. T

CHAPTER

6 Rings and Fields

6.0 Introduction

In previous chapter we have studied an algebraic structure with only one binary composition. Such types of algebraic structures are called groups. But there are non-empty sets equipped with more than one binary compositions, namely the set of integers, set of rational numbers etc. In this chapter we are going to introduce such algebraic structures which are equipped with two binary compositions.

6.1 Ring

An algebraic structure $(R, +, \cdot)$, where R is a non-empty set and $+$ and $\cdot$ are two binary operations known as addition and multiplication respectively, is called a **Ring** if the following postulates are satisfied:

i. $(R, +)$ is an abelian group.

ii. $(R, \cdot)$ is a semi-group.

iii. Multiplication is distributive with respect to addition.

Thus a non-empty set R with two well defined binary compositions addition and multiplication is a ring if:

1. Addition is associative i.e. $\forall\ a, b, c \in R$

$$(a+b)+c=a+(b+c)$$

2. There exists an element $0 \in R$ such that

$$a+0=a\ \forall\ a \in R$$

3. For each $a \in R$, there exists $-a \in R$ such that

$$a+(-a)=0$$

4. Addition is commutative

$$a+b=b+a\ \forall a, b \in R$$

5. Multiplication is associative

$$(a \cdot b) \cdot c=a \cdot (b \cdot c)\ \forall\ a, b, c \in R$$

6. Multiplication is distributive with respect to addition. For all $a, b, c \in R$.

   i. Left distributive law

$$a \cdot (b+c)=a \cdot b+a \cdot c$$

   ii. Right distributive law

$$(b+c) \cdot a=b \cdot a+c \cdot a.$$

*We shall use the symbol 0 to denote the additive identity which is called **zero element** of the ring. Similarly, the additive inverse of an element a of ring R is denoted by – a which is called the negative of a.*

6.1-1 Commutative Ring

A ring $(R, +, \cdot)$ is called a **commutative ring** if $a \cdot b = b \cdot a \ \forall \ a, b \in R$

6.1-2 Ring with Unity

A ring $(R, +, \cdot)$ is said to be **ring with unity** if R has the identity element for its multiplicative composition. If there exists $e \in R$ such that

$$a \cdot e = a = e \cdot a \ \forall a \in R.$$

Example 6.1

The set R consisting of a single element 0 with two compositions + and · defined as $0 + 0 = 0$ and $0.0 = 0$ is a ring. This ring is called **zero ring** or **null ring**.

Example 6.2

The set $R = \{a, b\}$ with addition and multiplication defined by the following table

+	a	b
a	a	b
b	b	a

·	a	b
a	a	a
b	a	b

is a commutative ring with unity, a is its zero element and b is its unit element .

Example 6.3

The set I of integers is a ring under addition and multiplication of integers. This ring is said to be the **ring of integers.**

It is a commutative ring with unity because multiplication of integers is commutative and 1 is the unity element in I.

Similarly $(Q, +, \cdot)$, $(R, +, \cdot)$, $(C, +, \cdot)$, $(kI, +, \cdot)$ where k is a scalar are commutative rings.

Example 6.4

The set M of all 2×2 matrices over integers form a ring under matrix addition and multiplication. M is a non-commutative ring with unity. The zero matrix is the additive identity and the matrix $\begin{bmatrix} 1 & 0 \\ 0 & 1 \end{bmatrix}$ is unity.

For $\quad A = \begin{bmatrix} a_{11} & a_{12} \\ a_{21} & a_{22} \end{bmatrix}, B = \begin{bmatrix} -a_{11} & -a_{12} \\ -a_{21} & -a_{22} \end{bmatrix}$ is the negative of A.

To show that M is non-commutative it is sufficient to find two matrices $A, B \in M$ such that

$$AB \neq BA$$

$$A = \begin{bmatrix} 2 & 3 \\ 5 & 4 \end{bmatrix} \text{ and } B = \begin{bmatrix} 2 & 0 \\ 1 & 5 \end{bmatrix}$$

Then

$$AB = \begin{bmatrix} 7 & 15 \\ 14 & 20 \end{bmatrix} \text{ and } BA = \begin{bmatrix} 4 & 6 \\ 27 & 23 \end{bmatrix}$$

Clearly $AB \neq BA$

Hence M is not commutative.

Example 6.5

The set $R = \{0, 1, 2, 3, 4\}$ is a commutative ring with unity with respect to compositions $+_5$ and $\times_5$.

Solution

The composition tables for the two compositions are as follows:

Table 6.1

$+_5$	0	1	2	3	4
0	0	1	2	3	4
1	1	2	3	4	0
2	2	3	4	0	1
3	3	4	0	1	2
4	4	0	1	2	3

Table 6.2

$\times_5$	0	1	2	3	4
0	0	0	0	0	0
1	0	1	2	3	4
2	0	2	4	1	3
3	0	3	1	4	2
4	0	4	3	2	1

From Table 6.1, one can observe easily that $(R, +_5)$ is an abelian group. And from Table 6.2, $(R, \times_5)$ is a semi-group.

For any three elements say 1, 3, 4 of R.

$$1 \times_5 (3 +_5 4) = 1 \times_5 2 = 2$$

and

$$(1 \times_5 3) +_5 (1 \times_5 4) = 3 +_5 4 = 2$$

$\therefore$

$$1 \times_5 (3 +_5 4) = (1 \times_5 3) +_5 (1 \times_5 4)$$

Hence $\times_5$ is distributive over $+_5$. Since $1 \in R$ and $\times_5$ is commutative in R. Therefore $(R, +_5, \times_5)$ is a commutative ring with unity.

Example 6.6

Let R be a set of real numbers and $f: R \to R$ be a mapping which is continuous on R. Let F is the set of all such mappings. The sum and product of the mappings for $f, g \in F$ are defined as

$$(f + g)(x) = f(x) + g(x) \; \forall \, x \in R$$

$$(f \cdot g)(x) = f(x)\, g(x) \; \forall \, x \in R$$

Then $(F, +, \cdot)$ is a commutative ring with unity.

Example 6.7

Let R be a system satisfying all the postulates required for a ring with the possible exception of $a + b = b + a$. If there exists one element $c \in R$ such that $ac = bc \Rightarrow a = b \; \forall \, a, b \in R$, prove that R is a ring.

Solution

$$(a + b)(c + c) = a(c + c) + b(c + c)$$

$$= ac + (ac + bc) + bc \quad \text{...(1)}$$

Also

$$(a + b)(c + c) = (a + b)c + (a + b)c$$

$$= ac + (bc + ac) + bc \quad \text{...(2)}$$

From (1) and (2), we get

$$ac + bc = bc + ac$$

$$\Rightarrow \quad (a + b)c = (b + a)c \quad \text{[by right distributive law]}$$

$$\Rightarrow \quad a + b = b + a \quad [\because ac = bc \Rightarrow a = b]$$

Hence R is a ring.

6.2 Field

A ring $(F, +, .)$ is said to be a field if

i. it is commutative
ii. it has unity
iv. it is such that every non-zero element has multiplicative inverse.

Example 6.8

The rings $(Q, +, .)$, $(R, +, .)$, $(C, +, .)$ are all fields.

Example 6.9

The ring $(\{0, 1, 2, 3, 4\}, +_5, \times_5)$ is an example of finite field.

It may be observed from the Table 6.1 and Table 6.2 that

i. $\{0, 1, 2, 3, 4\}$ is an abelian group for $+_5$. 0 is the zero element and each element is invertible in the set.
ii. $\{1, 2, 3, 4\}$ is an abelian group for $\times_5$. 1 is unity and each element has its multiplicative inverse.

Example 6.10

Prove that the set of real numbers of the form $a + b\sqrt{2}$ where a and b are integers with ordinary addition and multiplication forms a ring. Is it a field?

Solution

Let $G = \{a + b\sqrt{2} : a, b \in I\}$, obviously G is a subset of the set of real numbers R.

Let $x, y \in G$, where $x = a_1 + b_1\sqrt{2}$ and $y = a_2 + b_2\sqrt{2}$ $\forall a_1, a_2, b_1, b_2 \in I$

Now $$x + y = (a_1 + b_1\sqrt{2}) + (a_2 + b_2\sqrt{2}) = (a_1 + a_2) + (b_1 + b_2)\sqrt{2}$$

Since $a_1 + a_2$ and $b_1 + b_2$ are integers, hence $x + y \in G$.

Also $$xy = (a_1 + b_1\sqrt{2}).(a_2 + b_2\sqrt{2}) = (a_1 a_2 + 2 b_1 b_2) + (a_1 b_2 + a_2 b_1)\sqrt{2}$$

Since $a_1 a_2 + 2 b_1 b_2$ and $a_1 b_2 + a_2 b_1$ are integers, then $x.y \in G$.

Thus G is closed for addition and multiplication of real numbers.

Again $$x - y = (a_1 + b_1\sqrt{2}) - (a_2 + b_2\sqrt{2}) = (a_1 - a_2) + (b_1 - b_2)\sqrt{2}$$

Since $a_1 - a_2$ and $b_1 - b_2$ are integers, hence $x - y \in G$.

Hence $(G, +)$ is a subgroup of the abelian group $(R, +)$.

Consequently, $(G, +)$ is an abelian group. Since multiplication of real numbers is associative, commutative and distributive over addition. Hence $(G, +, \cdot)$ is a commutative ring.

It is not a field because the multiplicative inverse of a non-zero number $a + b\sqrt{2} \in G$ in R is $\left(\frac{a}{a^2 - 2b^2}\right) + \left(\frac{-b}{a^2 - 2b^2}\right)\sqrt{2}$ which is not necessarily in G because $\left(\frac{a}{a^2 - 2b^2}\right)$ and $\left(\frac{-b}{a^2 - 2b^2}\right)$ need not be integers.

For example, for the number $5 + 6\sqrt{2} \in G$, the multiplicative inverse in R is $\frac{-5}{47} + \frac{6}{47}\sqrt{2}$ which is not in G.

Example 6.11

Show that the set G of all real valued functions defined on the closed interval [0, 1] is a commutative ring with unity with respect to pointwise addition and pointwise multiplication of functions, defined as

$$(f+g)(x) = f(x) + g(x)$$
$$(fg)(x) = f(x)\, g(x) \ \forall f, g \in G$$

Solution

Let $f, g \in G$, then $f(x)$ and $g(x)$ are real numbers for all $x \in [0, 1]$. Since the sum and product of two real numbers are also real numbers, therefore $f + g \in G$ and $fg \in G \ \forall f, g \in G$.

Hence, the compositions defined on G are binary compositions.

Consider an algebraic structure $(G, +, \cdot)$ and observe the following:

As we know that addition and multiplication of real numbers are commutative and associative, therefore

$$nnf + g = g + f, fg = gf$$

and $$n(f+g) + h = f + (g+h), (fg)h = f(gh) \ \forall f, g, h \in G$$

Existence of zero and unity elements

Consider the mappings 0 and I such that

$$0(x) = 0 \text{ and } I(x) = 1 \ \forall x \in [0, 1]$$

Then $$(0+f)x = 0(x) + f(x) = 0 + f(x) = f(x) \ \forall x \in [0, 1]$$

Thus $$0 + f = f \ \forall f \in G$$

Also, $$(If)(x) = I(x) f(x)$$
$$= 1 f(x) = f(x) \ \forall x \in [0, 1]$$

Thus $$If = f \ \forall f \in G.$$

Hence 0 and I are zero and unity elements in G.

Existence of additive Inverse

Let $f \in G$. Consider a function $-f$ as follows:

$$(-f)x = -f(x) \ \forall x \in [0, 1]$$

$\Rightarrow$ $$-f \in G \ \forall f \in G.$$

such that $$(-f+f)x = (-f)x + f(x) = -f(x) + f(x)$$
$$= 0 = 0(x)$$

$\Rightarrow$ $$-f + f = 0$$

Thus $-f$ is the additive inverse of f in G.

Distributive Law

Suppose $f, g, h \in G$, then

$$[f(g+h)](x) = f(x)(g+h)(x)$$
$$= f(x)[g(x) + h(x)]$$
$$= f(x) g(x) + f(x) h(x)$$
$$= (fg)(x) + (fh)(x) \quad \text{[by def.]}$$
$$= (fg + fh)(x)$$

Similarly, $[(g+h)f](x) = (gf)(x) + (hf)(x)$

Hence, multiplication is distributive over addition.

Thus $(G, +, \cdot)$ is a commutative ring with unity.

Example 6.12

Prove that if $a, b \in R$, then $(a+b)^2 = a^2 + ab + ba + b^2$ where $x^2 = x\,x$.

Solution

We have

$$
\begin{aligned}
(a+b)^2 &= (a+b)(a+b) && [x^2 = xx] \\
&= a(a+b) + b(a+b) \\
&= aa + ab + ba + bb && \text{[distributive law]} \\
&= a^2 + ab + ba + b^2 && [\because x^2 = xx]
\end{aligned}
$$

Example 6.13

If R is a ring such that $a^2 = a \; \forall \; a \in R$ then prove that

i. $a + a = 0 \; \forall \; a \in R$

ii. $a + b = 0 \Rightarrow a = b$

iii. R is a commutative ring.

Solution

i. $a \in R \Rightarrow a + a \in R$

Now $(a+a)^2 = a + a$ [given]

$\Rightarrow (a+a)(a+a) = a + a$

$\Rightarrow (a+a)a + (a+a)a = a + a$ [left distributive law]

$\Rightarrow (a^2 + a^2) + (a^2 + a^2) = a + a$

$\Rightarrow (a+a) + (a+a) = a + a$ $[\because a^2 = a]$

$\Rightarrow (a+a) + (a+a) = (a+a) + 0$

by left cancellation law for addition in R

$$a + a = 0$$

ii. Since $a + a = 0$ ...(1)

then consider $a + b = 0$ ...(2)

From (1) and (2), we get

$a + a = a + b \Rightarrow a = b$ [left cancellation Law]

iii. We have $(a+b)^2 = (a+b)$

$\Rightarrow (a+b)(a+b) = (a+b)$ [By left distributive law]

$\Rightarrow (a+b)a + (a+b)b = (a+b)$ [By right distributive law]

$\Rightarrow a^2 + ba + ab + b^2 = a + b$

But $a^2 = a, b^2 = b$

$\therefore a + ba + ab + b = a + b$

$\Rightarrow (a+b) + (ab+ba) = (a+b) + 0$ [By cancellation law]

$$ab + ba = 0$$

$ab = ba$ [by 2]

R is commutative

Exercise 6.1

1. Prove that the set of rational numbers is a field with respect to addition and multiplication.
2. Prove that the set of integers $R = \{0, 1, 2, 3, 4\}$ forms a field under addition modulo five $(+_5)$ and multiplication modulo five $(\times_5)$.
3. If operations $*$ and $\circ$ on the set of integers are defined as: $a * b = a + b - 1$, $a \circ b = a + b - ab$.
4. Prove that the algebraic structure $(R, *, \circ)$ is a commutative ring with unity.
5 If R is a ring with unity, then show that $(-1)\, a = -a = a\,(-1)\ \forall\ a \in R$ and $(-1)(-1) = 1$.
6. Prove that a ring R is commutative iff $(a + b)^2 = a^2 + 2ab + b^2\ \forall\ a, b \in R$.
7. Prove that the set C of all ordered pairs (a, b) of real numbers is a field under addition and multiplication defined as,

$$(a, b) + (c, d) = (a + c, b + d)$$
$$(a, b) \cdot (c, d) = (ac - bd, bc + ad).$$

8. Show that the following are rings:

   i. $[C; +, \cdot]$ ii. $[M_{2\times2}(R); +, \cdot]$

   iii. $[I_6; +, \cdot]$ iv. $[I_8; +_8, \times_8]$.
9. Which of the above rings in Ex. 7 are commutative? Are they rings with unity? Determine the unity of the above rings.
10. Prove that, if R is a ring with unity, then this unity is unique.
11. Let $Q(\sqrt{2}) = \{a + b\sqrt{2} : a, b \in Q\}$. Prove that $[Q\sqrt{2}; +, \cdot]$ is a field where $+, \cdot$ stand for regular addition and multiplication.
12. Show that $[I_3; +_3, \times_3]$ and $[I_7; +_7, \times_7]$ are fields.
13. Consider a ring $(R, +, \cdot)$ defined by $a \cdot a = a$. Determine whether the ring is commutative.

Chapter Objective Type Questions

1. The number of unit element(s) of the ring (Z, +, .) is
   (A) 2 (B) 3
   (C) 1 (D) infinite
2. The number of binary operation(s) involved in a ring is–
   (A) 1 (B) 2
   (C) 3 (D) infinite
3. The number of unit element(s) of the ring Z is-
   (A) 1 (B) 2
   (C) finite but more than 2 (D) infinite
4. If $(F, +, .)$ is a field, then
   (A) $(F, +)$ is an abelian group (B) $(F, .)$ is an abelian group
   (C) F is an integral domain (D) none of these
5. In a ring $(Z_5, +, .)$ the additive inverse of [3] is
   (A) –[3] (B) 0
   (C) [2] (D) [–3]

6. The number of unit element(s) of the ring Z_{10} is

(A) 0 (B) 4

(C) 5 (D) 9

7. If order of group G is p^2, where p is prime then–

(A) G is Abelian (B) G is not Abelian

(C) G is Ring (D) None of these

8. A ring $(F, +, .)$ is said to be a field if

(A) it is commutative (B) it has unity

(C) it is such that every non–zero (D) All of these element has multiplicative inverse

B. Fill in the blanks.

1. The singleton {0} with binary operations addition and multiplication is ring and it is called ______.
2. If E is the set of even integers under ordinary addition and multiplication, then E is a ______ ring.
3. A ring $(R, +, \cdot)$ is called a ________ if $a \cdot b = b \cdot a \; \forall \; a, b \in R$.
4. The set I of integers is a ring under addition and multiplication of integers. This ring is known as the ______.

C. Which of the following is True (T) / False (F) in a ring?

1. $1.a = a = a.1 \forall a \in R$.
2. $a.b = 0 \Rightarrow$ either $a = 0$ or $b = 0 \; \forall a, b \in R$.
3. $-(-a) = a, \forall a \in R$.
4. Every ring is a field.
5. $a.b = b.a \; \forall a, b \in R$.
6. Every field is a ring.

Answers Objective Type

A.

1. A 2. B 3. B 4. A
5. C 6. B 7. A 8. D

B.

1. zero ring 2. commutative 3. commutative
4. ring of integers.

C.

1. F 2. F 3. T 4. F
5. F 6. T

CHAPTER

7 Logic

7.0 Introduction

The notion of *logic* is basic to all kinds of reasoning, so we should begin with a careful analysis of this concept.

The word "logic" itself is quite interesting; more than hundred definitions have been given by philosophers and mathematicians according to its practical applications in subjects as diverse as computer science, natural and physical sciences, social sciences and in our daily lives to solve a multitude problems.

As per oxford English Dictionary logic, "is science of reasoning, thinking or inference, particular scheme of or treatise on this etc".

Logic is specifically concerned with whether the reasoning is correct or not for determining the validity of arguments, it provides rules and techniques. Mathematical statements accurately express the rules of logic, which are used to distinguish between valid and invalid inferences. Logic concentrates on the forms of arguments and focuses on the relationship among statements rather than with any particular argument or statement. For example, consider the following statements:

All mathematicians are intelligent.

Anyone who is intelligent is a genius.

Therefore, all mathematicians are genius.

In virtue of its form, logic is concerned only with principles of valid inferencing and not concerned with reasoning. In above example, first two statements are true, which conclude that the third statement 'All mathematician are genius' is also true. The truthness of third statement depends on truthness of first two statements but it is not logic. Hence, truthness or validity of argument should be decided by following rules and techniques of reasoning individually on our own feelings about the argument.

But for expressing the rules and theories we need a formal language, called the object language, which is defined by some syntax and semantics. Syntax and semantics of any language consists of certain well-defined terms and well-specified uses of these terms. To avoid ambiguity, we can use natural language, programming language, symbols, which are easy to write and manipulate. All these things discussed above will be dealt in detail in this chapter.

7.1 Propositions or Statements

Whatever we communicate in our natural language are either sentences or statements. Sentences like:

Please wait.	(a request)
May God bless you!	(a wish)
Do you speak English?	(an inquiry)
Wish you many happy returns of the day!	(a wish)

and statements like:

Mumbai is a big city.	(True)
41 is not a prime number.	(False)
$11 + 1 = 100$. (Binary addition)	(True)
India is a country.	(True)
The earth is round.	(True)

The declarative sentences which can have only one possible value are called **statements** and the two possible values are called truth-values. These two truth-values are 'true' and 'false' denoted by 'T' and 'F' or sometimes by '1' and '0' respectively.

These "declarative sentences with truth values either true or false but not both are called **propositions**[*]". Propositions are the basic building blocks of any theory of logic.

Example 7.1

Which of the following sentences are statements?

a. $2 + 3 = 5$
b. $4x + 5 = 7$
c. $110 + 11 = 1001$ (Binary addition)
d. Learn it.
e. Do your work.
f. Moon is the satellite of the Earth.
g. Snow is white.
h. Who are you?
i. Joe is a programmer.
j. Jupiter is the biggest planet in the universe.

Solution

(a), (f), (g) and (j) are statements because they happen to be true.
(b) is not a proposition because it is neither true nor false. This is because no value of variable x is given.
(c) is true in case of binary numbers.
(d) and (e) are imperative sentences. Hence they will not be having either true or false values. This implies that they are not statements.
(h) is an interrogative sentence, so it is not a statement.
(i) is either true or false (not both), but the actual value is not known at this time.

7.1-1 Representation of Proposition

To represent the proposition, we shall use upper case letters such as $A, B, C, P, Q, R, \ldots$ etc, in declarative sentences. These letters are known as **logical variables** and can be replaced by statements. For example:

P: Laksdweep is an Island.

Q: $111 + 1 = 1000$

These propositions are simple propositions also known as **primitive** or **atomic propositions.**

---

[*] In logic, propositions and statements are not different but they are one and the same.

George Boole in 1854, gave a method to produce new propositions by combining one or more propositions using certain symbols or punctuation marks called **connectives**. These new propositions obtained after combining propositions are known as **compound** or **molecular propositions**. The propositions are combined by connectives such as "and" and "or". For example:

P: The sun is not shining and it is raining today.

The truth value of statement *P* is completely depends upon the truth values of two statements P_1 and P_2 which are

P_1: The sun is not shining

P_2: It is raining today.

Hence, we conclude that the truth value of compound statement can be completely determined by the truth values of its sub statements and by the types of connectives being used between them.

The representation of propositions using logical variables/symbols is studied under **symbolic logic**.

Another most important connective is "not" or "negation". It is defined as "it is not the case that *P*." For example:

P: It is raining.

Not *P*: It is not raining.

"Not" is a unary operation on primitive or compound statements. It does not join two statements. If the statement *P* is true then "not *P*" truth value is false. We shall discuss these connectives in the next section in detail.

Exercise 7.1

1. Which of the followings are statements?
   a. What time is it?
   b. $x + y = z + y$
   c. Are you O.K.?
   d. Paris is in France.
   e. India is a country.
   f. The set of integers is infinite.
   g. $1 + 1 = 2$
   h. $9 < 5$.
   i. $10001 + 11 = 10100$
   j. Do your homework.
2. Which of the following sentences are propositions? What are the truth values of those that are propositions?
   a. $a + b = 5$ if $a = 3$ and $b = 2$.
   b. Madhya Pradesh is a big state.
   c. Today is Thursday.
   d. $x + y = y + z$ if $x = z$.
   e. $2x + 3 < 10\ \forall\ x \in [0, 3]$.
3. What is the negative of each of the following propositions?
   a. It is raining today.
   b. You miss the final examination.
   c. $2 + 3 = 5$
   d. All mathematicians are intelligent.
   e. New Delhi is the capital of England.
4. Express each of the following in logical form and write down its negation.
   a. Indira Gandhi was first lady Prime Minister of India.
   b. Mars is not the only planet in the universe that has life.
   c. All people like cricket.
   d. $101+1 = 110$.
   e. All humans are mortal.

5. Which of the following statements are the negation of the statement "8 is even and –3 is negative"
   a. 8 is odd and –3 is not negative.
   b. 8 is even and –3 is not positive.
   c. 8 is odd and –3 is negative.
   d. 8 is even and –3 is not negative.

Answers

1. a. F
   b. F
   c. F
   d. T
   e. T
   f. T
   g. T
   h. F
   i. T (in case of binary digits)
   j. F
2. a. T
   b. T
   c. Neither true nor false, it depends upon the day
   d. T
   e. T
4. a. *P*: Indira Gandhi was first lady Prime Minister of India.

   not *P*: It is not the case that *P*.

   d. *P*: 101 + 1 = 110

   not *P*: 101 + 1 ≠ 110

   e. *P*: All humans are mortal

   not *P*: All humans are not mortal.

7.2 Truth Table and Connectives

We have defined earlier that the two valued logic "true" and "false" can be represented by 'T' or '1' and 'F' or '0' respectively. In digital electronics, used in computer hardware, engineers use these two-valued logic for 'high' voltage and 'low' voltage representation. A value '1' corresponds to a 'true', 'T' or 'high' voltage and a value '0' to a 'false', 'F' or 'low' voltage.

Once we know the truth values of each logic variables, a computational table, called **truth table,** can be formed by which we can determine the truth value of a proposition (compound/simple).

As in linguistic, logical connectives such as and, or, not, if... then..., if and only if... etc. are used to combine two or more sentences, similarly here in the symbolic logic, we use connectives to combine the two or more statements which are shown in following table with their symbols.

Connective English word	Name of connective	Symbol
Not	negation	$\rceil$ or ~ or $\bar{\ }$ or ′
And	conjunction	$\wedge$
Or	disjunction	$\vee$
if ... then	implics (conditional)	$\rightarrow$
if and only if	biconditional	$\leftrightarrow$

Truth table contains all possible values of the input variable and the corresponding values of the output variables. Hence it is advantageous to represent any declarative statement in proposition form and then in symbolic innovations. Hence, truth table is a special technique to define the operational behavior of logical connectives. Here, in this section, we discuss in detail about the logical connectives with the help of truth table.

7.2-1 Negation

As defined earlier, **negation** of statement P is stated as

"It is not the case that P"

The negation of any statement P may be represented by $\urcorner P$ or $\sim P$ or $\bar{P}$ and read as "not P". The truth value of $\urcorner P$ depends on the truth value of P. If truth value of P is true then $\urcorner P$ is false and vice-versa. Since, negation, is an unary operator it constructs a new proposition from a given single proposition. For example,

P: Null set is subset of every set.

Then the negation of P, i.e. $\urcorner P$ can be represented by

$\urcorner P$: It is not the case that null set is subset of every set.

or $\urcorner P$: Null set is not the subset of every set.

Let us draw the truth table for negation of a statement. For example, P is a statement then the first column of the table is for the variable P and it contains all the truth values of P. The second column is for $\urcorner P$ which contains the corresponding truth values of input values. The Table 7.1 summaries the negation of P.

Table 7.1: Truth table for negation

P	$\urcorner P$
T	F
F	T

The Table 7.1 can also be represented with the help of symbol '0' and '1' as shown in the following table.

P	$\urcorner P$
1	0
0	1

Negation of negation of proposition is proposition itself

i.e. $\urcorner\urcorner P = P$.

Example 7.2

Find the negation of the proposition

"Health is more precious than wealth"

and express it in simple English.

Solution

Suppose

P: Health is more precious than wealth.

Then negation is

$\urcorner P$: It is not the case that health is more precious than wealth.

The expression in simple English form is

$\neg P$: Health is not more precious than wealth.

Example 7.3

Consider the following statements

P: 2 + 3 = 7

Q: It is not the case that 2 + 3 = 7

R: 2 + 3 ≠ 7

Q and R are the negation of statement P. Since P is false; Q and R are true.

Example 7.4

What is the value of $\neg\neg P$ in example 7.2?

Solution

P: Health is more precious than wealth

then negation is

$\neg P$: Health is not more precious than wealth.

$\neg\neg P$: Health is more precious than wealth.

Exercise 7.2

Write the negation of the following propositions:

1. Mars is the biggest planet in the universe.
2. Mathematicians are not intelligent.
3. The sets of prime numbers are infinite.
4. 1+1 = 4
5. It is cold.
6. All students are laborious.
7. Square root of 8 is less than 2.
8. 45 is divisible by 7.
9. $x^2 + 9 = 0$.
10. Port Blair is the capital of Andaman and Nicobar islands.
11. $2 + 7 \leq 11$.
12. There will be snow in January.

Answers

1. Mars is not the biggest planet in the universe.
2. Mathematicians are intelligent.
3. The set of Prime numbers is not infinite.
4. 1+1 ≠ 4.
5. It is not cold.
6. All students are not laborious.
7. Square root of 8 is not less than 2.

8. 45 is not divisible by 7.
9. $x^2 + 9 \neq 0$.
10. Port Blair is not the capital of Andaman and Nicobar islands.
11. $2 + 7$ is not less than or equal to 11.
12. There will not be snow in January.

7.2-2 Conjunction

In spoken language we use the word "and" for connecting two or more sentences, words and clauses, that are to be taken jointly.

Similarly, in logic, logical connective **conjunction** of two statements P and Q is the statement having truth value *true* when both P and Q are true else it has the truth value *false*. The conjunction of P and Q is represented by $P \wedge Q$ and read as "P and Q". For example

P: Today is Sunday

Q: It is holiday

$P \wedge Q$: Today is Sunday and it is holiday.

Since conjunction, is a binary operator, operates on two logical statements (simple or compound) and constructs a new proposition from these two statements. $P \wedge Q$ is also called a *truth function* of P and Q and its value is dependent on the value of P and Q.

For drawing the truth table for conjunction of two statements, let us consider statements P and Q. The first and second columns of the table consist of all the possible combinations of truth values of P and Q. The third column represents the truth values of $P \wedge Q$. The values under $P \wedge Q$ is determined by the truth values of P and Q. If P and Q both have truth value 'T' or '1' then the $P \wedge Q$ has truth value 'T' or '1' otherwise it is 'F' or '0'. The Table 7.2 gives the truth table for $P \wedge Q$.

Table 7.2: Truth table for conjunction

P	Q	$P \wedge Q$
F	F	F
F	T	F
T	F	F
T	T	T

or

P	Q	$P \wedge Q$
0	0	0
0	1	0
1	0	0
1	1	1

Truth tables contain all possible values of resultant variable corresponding to the input variables. The information given by truth table can also be represented in tree form:

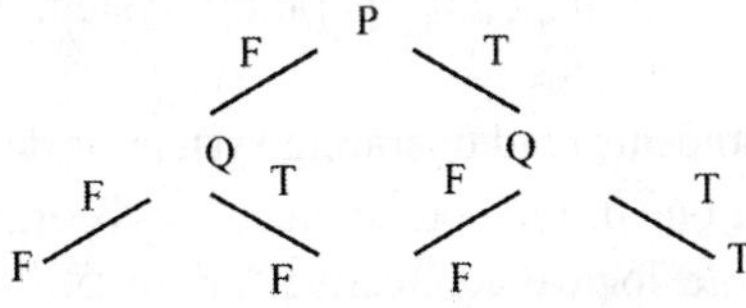

Nodes (pendant vertices) represent the input variables and labelled branches contain the possible truth values. Leaves show the resultant value of conjunction of two variables P and Q.

Example 7.5

Form the conjunction of P and Q where

P: Algebraist are intelligent.

Q: George Boole was algebraist.

Solution

$P \wedge Q$: Algebraist are intelligent and George Boole was algebraist.

Example 7.6

Form the conjunction of

P: $2 + 3 < 7$

Q: Today is cold.

Solution

$P \wedge Q$: $2 + 3 < 7$ and today is cold.

1. *In logic we may combine two totally unrelated statements by logical "and" which is not the case of our formal English language.*
2. *Conjunction of proposition with itself gives the same proposition, i.e.*

$$P \wedge P \equiv P$$

and represented with the help of truth table as

P	P	$P \wedge P = P$
F	F	F
T	T	T

7.2-3 Disjunction

The word 'or' in English language is used for uniting two alternatives, commonly used in two distinct sense i.e. either of the two alternatives, one occurs but not both (exclusively) and at least one of the two alternatives occurs (inclusively). e.g.

"South Korea or Germany will be moving to the final of football world cup."

In this compound statement we have connective "or" in exclusive manner i.e. exactly one of the two possibilities will occur not both.

In symbolic logic, logical connective **disjunction** of two statements P and Q is the statement having truth value *false* when both P and Q are false else it has the truth value *true*. The disjunction of P and Q may be represented by $P \vee Q$ and read as "P or Q". For example:

P : Computer Science students read operating system.

Q : Computer Science students read discrete mathematics.

$P \vee Q$: Computer Science students read operating system or Computer Science students read discrete mathematics. or

$P \vee Q$: Computer Science students read operating system or discrete mathematics.

It is clear from above example that one or the both subjects i.e. operating system and discrete mathematics are read by computer Science students. i.e. logical connective "or" or "$\vee$ *" is used in inclusive manner.

Since disjunction is a binary operator, therefore to draw the truth table we need at least two propositions, let they are P and Q. Like conjunction table in disjunction truth table also, first two columns contains all possible combinations of truth values of P and Q and third column has the corresponding values determined by the truth

* The latin word 'vel' means the inclusive sense of 'or' and denoted by symbol '$\vee$', called wedge (or a vee), is a truth functional connective.

values of P and Q. In disjunction table, the value under $P \vee Q$ column is 'F' when both P and Q have truth values 'F' otherwise it is 'T'. Table 7.3 summarises the truth table of disjunction of two statements P and Q.

Table 7.3: Truth table for disjunction

P	Q	$P \vee Q$
F	F	F
F	T	T
T	F	T
T	T	T

or

P	Q	$P \vee Q$
0	0	0
0	1	1
1	0	1
1	1	1

The combination of truth values P and Q in disjunction table can be represented in the form of tree as

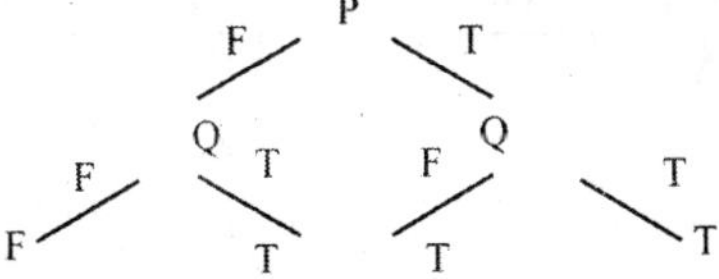

Like conjunction, disjunction of propositions with itself gives the same proposition.

i.e. $P \vee P \equiv P$

and can be represented by truth table as

P	P	$P \vee P \equiv P$
F	F	F
T	T	T

Example 7.7

Form the disjunction of

P: $2 + 3 < 7$

Q: Tomorrow is a holiday.

Solution

$P \vee Q$: $2 + 3 < 7$ or tomorrow is a holiday.

So far, we have learnt that compound statements consist of two proposition and joined with one connective, but it is not necessary that these propositions are always primitive i.e. propositions used for construction of compound statements may again be a compound statement or primitive. These compound statements may have many connectives collectively joining them. This is denoted by an expression $P(P_1, P_2 \ldots)$, where $P_1, P_2 \ldots$ are logical variables and expression $P(P_1, P_2 \ldots)$ is called **proposition.** For example, to construct the truth table for compound statement $\urcorner(P \wedge Q) \vee \urcorner Q$ we can have this type of situation. As shown in Table 7.4, the first two columns of the table are for all the possible combinations of truth values of propositional variables i.e. P and Q. Then there are columns for all intermediate elementary stages i.e. $(P \wedge Q)$, $\urcorner Q$ and $\urcorner(P \wedge Q)$ and the last column contains the truth values of proposition $\urcorner(P \wedge Q) \vee \urcorner Q$.

Table 7.4

P	Q	$P \wedge Q$	$\neg Q$	$\neg(P \wedge Q)$	$\neg(P \wedge Q) \vee \neg Q$
0	0	0	1	1	1
0	1	0	0	1	1
1	0	0	1	1	1
1	1	1	0	0	0

We may construct the truth table in another way as shown in following tables.

Table 7.5

P	Q	$\neg$	$(P$	$\wedge$	$Q)$	$\vee$	$\neg$	Q
0	0							
0	1							
1	0							
1	1							
Step								

Firstly we construct the truth table as shown in Table7.5 in which first two columns contain all the possible combinations of truth values of input variables. The other columns will be corresponding to different connectives and propositions contained in the compound statements separately. Last row of the table contains the step numbers which show the order in which operations are performed according to the precedence* of connectives.

The final table may be evaluated in systematic way as follows:

Step 1:

Enter the truth values of variable under the corresponding columns of propositions as shown in Table 7.6 and indicate this step number as 1.

Table 7.6

P	Q	$\neg$	$(P$	$\wedge$	$Q)$	$\vee$	$\neg$	Q
0	0		0		0			0
0	1		0		1			1
1	0		1		0			0
1	1		1		1			1
Step			1		1			1

Step 2:

Enter the truth values under each logical connective according to the precedence and indicate the step numbers in last row corresponding to the order in which operations are being performed.

• Similar to the precedence of operators in mathematics there is the precedence of connectives in logic. "$\neg$ has more precedence than $\wedge$ which has precedence over $\vee$".

Table 7.7

P	Q	$\neg$	$(P$	$\wedge$	$Q)$	$\vee$	$\neg$	Q
0	0	1	0	0	0	1	1	0
0	1	1	0	0	1	1	0	1
1	0	1	1	0	0	1	1	0
1	1	0	1	1	1	0	0	1
Step		3	1	2	1	4	2	1

The last (highest) value of step number, i.e. 4, contains the truth values of given compound statement.

We can observe that if there are two input variables then the total number of possible combinations are 4 i.e. 2^2. If we have three input variables then the possible number of all the combinations will be 2^3 i.e. 8 and the truth values of compound statement containing three input variables depend on the truth values of all the three input statements. In general, if compound statement contains n propositions then there will be 2^n possible combinations of truth values (i.e. there are 2^n rows). The final truth values of compound statement depend on the corresponding truth values of input variable. e.g. consider the compound proposition $(P \wedge Q) \vee (\neg R \wedge Q)$ where P, Q and R are propositions. In this case truth table is having eight different combinations of truth values of input variables. The result is shown in Table 7.8 which depends on the corresponding truth values of input variables.

Table 7.8

P	Q	R	$P \wedge Q$	$\neg R$	$\neg R \wedge Q$	$(P \wedge Q) \vee (\neg R \wedge Q)$
0	0	0	0	1	0	0
0	0	1	0	0	0	0
0	1	0	0	1	1	1
0	1	1	0	0	0	0
1	0	0	0	1	0	0
1	0	1	0	0	0	0
1	1	0	1	1	1	1
1	1	1	1	0	0	1

Example 7.8

Let

P: Saturn is the second biggest planet in the Universe.

Q: George Boole was born in Lincoln.

Write down the meaning of the following propositions:

a. $P \wedge \neg Q$

b. $\neg(P \vee Q)$ (Try yourself).

Solution

(a) $\neg Q$: George Boole was not born in Lincoln.

$P \wedge \neg Q$: Saturn is the second biggest planet in the Universe and George Boole was not born in Lincoln.

$P \wedge \neg Q$: is false, if either of the statement is false.

Example 7.9

Construct the truth table for $S: (P \wedge Q) \vee (Q \wedge R) \vee (R \wedge P)$.

Solution

The given compound statement can be represented with the help of X, Y, Z and W propositions where X is $P \wedge Q$, Y is $Q \wedge R$, Z is $R \wedge P$ and W is $X \vee Y$ then the resultant expression S is $W \vee Z$. In general, we can write the compound statements in this way to reduce the complexities of the propositions. The resultant truth table is given below.

Table 7.9

P	Q	R	$X: P \wedge Q$	$Y: Q \wedge R$	$Z: R \wedge P$	$W: X \vee Y$	$S: W \vee Z$
0	0	0	0	0	0	0	0
0	0	1	0	0	0	0	0
0	1	0	0	0	0	0	0
0	1	1	0	1	0	1	1
1	0	0	0	0	0	0	0
1	0	1	0	0	1	0	1
1	1	0	1	0	0	1	1
1	1.	1	1	1	1	1	1

Example 7.10

Construct the truth table of $S: (\urcorner P \vee Q) \wedge (\urcorner R \vee P)$.

Table 7.10

P	Q	R	$\urcorner P$	$X: \urcorner P \vee Q$	$\urcorner R$	$zY: \urcorner R \vee P$	$S: X \wedge Y$
0	0	0	1	1	1	1	1
0	0	1	1	1	0	0	0
0	1	0	1	1	1	1	1
0	1	1	1	1	0	0	0
1	0	0	0	0	1	1	0
1	0	1	0	0	0	1	0
1	1	0	0	1	1	1	1
1	1	1	0	1	0	1	1

Exercise 7.3

1. Prepare the truth tables for the following statements:

a. $(\urcorner P \vee Q)$

b. $(P \wedge \urcorner Q)$

c. $(P \wedge Q) \vee \urcorner(\urcorner P \wedge \urcorner Q)$

d. $\urcorner(\urcorner P \wedge Q) \wedge (\urcorner P \wedge Q) \wedge (P \wedge Q)$

e. $P \wedge \urcorner(Q \wedge P) \wedge (P \wedge \urcorner Q) \vee (Q \wedge \urcorner P)$.

2. Let P: Lazy persons do not work properly.

   Q: Population of Senegal is 80 lakhs.

   Write down the meaning of following propositions:

   a. $\neg(P \vee Q)$

   b. $(P \vee Q) \wedge (\neg P \vee \neg Q)$

   c. $(\neg P \wedge \neg Q) \vee (P \wedge Q)$

   d. $\neg P \wedge \neg Q$

   e. $(P \wedge Q) \vee (\neg Q \wedge R)$.

3. For arbitrary propositions P, Q and R draw the truth table for the following statements:

   a. $P \vee (Q \wedge R)$

   b. $P \wedge \neg(Q \vee R)$

   c. $(\neg P \wedge Q) \vee (R \wedge \neg Q)$

   d. $(P \wedge \neg Q) \vee (P \wedge \neg R)$

   e. $(\neg P \vee R) \wedge (P \vee \neg Q) \wedge (Q \vee \neg R)$.

4. Form the conjunction and the disjunction of P and Q of each of the following:

   a. P: I am poor Q: I am not happy

   b. P: $3 + 4 < 12$ Q: $8 = 4 \times 3$

   c. P: $x > 10$ Q: $x \leq 10$

   d. P: Today is Sunday Q: Grass is wet.

5. Write each of the following in terms of P, Q and logical connectives.

   a. Today is Sunday and the fish cannot fly today.

   b. Today is Sunday and we will go to Delhi.

   c. It will rain tomorrow or it will snow tomorrow.

   d. 2 is an integer and it belongs to the set of natural number.

Answers

1. c.

P	Q	$\neg P$	$\neg Q$	$X: \neg P \vee \neg Q$	$Y: P \vee Q$	$\neg Y$	$\neg Y \vee X$
0	0	1	1	1	1	1	1
0	1	1	0	1	0	0	1
1	0	0	1	1	1	0	1
1	1	0	0	0	0	0	0

e.

P	Q	$X: P \wedge Q$	$\neg X$	$\neg Q$	$Y: P \wedge \neg Q$	$\neg P$	$Z: Q \wedge \neg P$	$(P \wedge \neg X \wedge Y) \vee Z$
0	0	0	1	1	0	1	0	0
0	1	0	1	0	0	1	1	1
1	0	0	1	1	1	0	0	1
1	1	1	0	0	0	0	0	0

2. a. It is not the case that lazy persons do not work properly or population of Senegal is 80 lakhs.

   b. Lazy persons do not work properly or population of Senegal is 80 lakhs and lazy persons work properly or population of Senegal is not 80 lakhs.

c. Lazy persons do not work properly and population of Senegal is 80 lakhs or population of Senegal is not 80 lakhs and lazy persons do not work properly.

4. a. $P \wedge Q$:I am poor and I am not happy.

   $P \vee Q$: I am poor or I am not happy.

   Both statements are true.

   b. $P \wedge Q$:$3 + 4 < 12$ and $8 = 4 \times 3$ is false since $8 \neq 4 \times 3$

   $P \vee Q$:$3 + 4 < 12$ or $8 = 4 \times 3$ is true since $3 + 4 < 12$.

   c. $P \wedge Q$: $x > 10$ and $x \leq 10$ is false since whatever be the value of x one of the statement will be false

   $P \vee Q$: $x > 10$ or $x \leq 10$ is true

   d. $P \wedge Q$: Today is Sunday and Grass is wet.

   $P \vee Q$: Today is Sunday or Grass is wet.

Its truthiness depends upon the situation.

5. a. Let P: Today is Sunday.

   Q: Fish can fly today

   The statement is $P \wedge \rceil Q$.

   b. Let P: Today is Sunday.

   Q: We will go to Delhi

   The statement is $P \wedge Q$.

   c. Let P: It will rain tomorrow

   Q: It will snow tomorrow

   The statement is $P \vee Q$.

   d. Let P: 2 is an integer

   Q: 2 belongs to the set of natural no.

   The statement is $P \wedge Q$.

7.2-4 Conditional and Biconditional

In our daily life conditional statements play an important role. As we can see the conditional statements such as 'if ..., then ...', 'whenever', 'if and only if' etc. are frequently used in daily life. Above conditional statements show some kind of relationship or implication that means fulfillment of a thing depends upon another. These type of statements are called implication statement. For example, the statement "If you work hard then you will get good marks" shows the dependency of second statement "you will get good marks" upon the first part of the statement.

In Mathematical reasoning, the meaning of implications and relationships are different from philosophical conditions. In symbolic logic, the proposition 'If P then Q' or 'P implies Q' or 'P is sufficient for Q' or 'Q if P' or 'Q is necessary for P' is denoted as $P \rightarrow Q$ (or $P \Rightarrow Q$) and is called **implication** or **conditional.** The truth table of statement $P \rightarrow Q$ is shown in Table 7.11. The truth value of statement is false (or '0') only for the '0' truth value of Q and for '1' of P. In $P \rightarrow Q$ proposition P is called **antecedent** (or **hypothesis** or **premise**) and Q is called the **consequence** (or **conclusion**).

Table 7.11: Truth table for implication

P	Q	$P \to Q$
0	0	1
0	1	1
1	0	0

or

P	Q	$P \to Q$
F	F	T
F	T	T
T	F	F

For example:

P: Today is Sunday

Q: $3 + 5 = 8$

$P \to Q$: If today is Sunday then $3 + 5 = 8$.

In natural language, there is no relation between these two given statements P and Q. But the mathematical reasoning is independent of relationship between hypothesis and conclusion; the result of implication is specified by its truth table which is not dependent on linguistic concept.

The truth table for implication can be represented in the form of tree as:

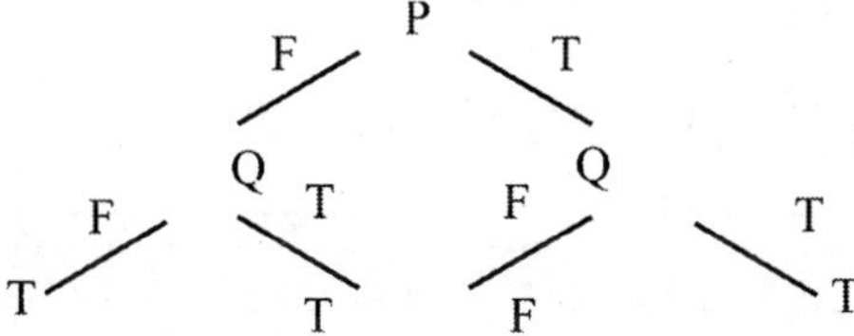

Implication of statement on itself always give the truth value 'T' i.e.

P	P	$P \to P$
F	F	T
T	T	T

Example 7.11

Form the implication $P \to Q$ of

P: Ram is a student of computer Science and Engineering.

Q: Ram is a good programmer.

Solution

$P \to Q$: If Ram is a student of Computer Science and Engineering then Ram is a good programmer.

Example 7.12

Write the following statement in symbolic form.

"If Govind buys a book of computer Science and English then he is a student of Science".

Solution

Let P: Govind buys a book of Computer Science.
Q: Govind buys a book of English.
R: Govind is a student of Science,

then the given statement symbolized as $P \wedge Q) \rightarrow R$.

Example 7.13

Construct the truth table for $(P \rightarrow Q) \wedge (Q \rightarrow R)$.

Solution

P	Q	R	$P \rightarrow Q$	$Q \rightarrow R$	$(P \rightarrow Q) \wedge (Q \rightarrow R)$
0	0	0	1	1	1
0	0	1	1	1	1
0	1	0	1	0	0
0	1	1	1	1	1
1	0	0	0	1	0
1	0	1	0	1	0
1	1	0	1	0	0
1	1	1	1	1	1

Example 7.14

Assuming that P is true, Q is false and R is true, find the truth value of each proposition.

a. $P \rightarrow (Q \rightarrow R)$

b. $P \rightarrow (Q \vee \urcorner R)$

c. $(P \vee R) \rightarrow \urcorner Q$.

Solution

For finding the truth value of each proposition, we replace each symbol P, Q and R by its truth value T, F and T:

a. $\mathrm{T} \rightarrow (\mathrm{F} \rightarrow \mathrm{T}) = \mathrm{T} \rightarrow \mathrm{T} = \text{true}$

b. $\mathrm{T} \rightarrow (\mathrm{F} \vee \urcorner \mathrm{T}) = \mathrm{T} \rightarrow (\mathrm{F} \vee \mathrm{F}) = \mathrm{T} \rightarrow \mathrm{F} = \text{false}$

c. $(\mathrm{T} \vee \mathrm{T}) \rightarrow \urcorner \mathrm{F} = \mathrm{T} \rightarrow \mathrm{T} = \text{true}$.

Example 7.15

Consider P: $3 > 5$, Q: $2 < 8$, find $P \rightarrow Q$ and $Q \rightarrow P$.

Solution

Here P is false and Q is true, hence

$P \rightarrow Q$: If $3 > 5$ then $2 < 8$ is true

$Q \rightarrow P$: If $2 < 8$ then $3 > 5$ is false

Hence $P \rightarrow Q$ is true, while $Q \rightarrow P$ is false.

Example 7.16

Compute the truth table for

a. $\sim P \rightarrow \sim Q$ and $Q \rightarrow P$

b. $\sim Q \rightarrow \sim P$ and $P \rightarrow Q$

c. $\sim P \vee Q$ and $P \rightarrow Q$, where '~' denotes negation.

Solution

a.

P	Q	$\sim P$	$\sim Q$	$\sim P \rightarrow \sim Q$	$Q \rightarrow P$
0	0	1	1	1	1
0	1	1	0	0	0
1	0	0	1	1	1
1	1	0	0	1	1

b.

P	Q	$\sim Q$	$\sim P$	$\sim Q \rightarrow \sim P$	$P \rightarrow Q$
0	0	1	1	1	1
0	1	0	1	1	1
1	0	1	0	0	0
1	1	0	0	1	1

c.

P	Q	$\sim P$	$\sim P \vee Q$	$P \rightarrow Q$
0	0	1	1	1
0	1	1	1	1
1	0	0	0	0
1	1	0	1	1

7.2-5 Converse, inverse and contrapositive implications

1. Example 7.16 (b) shows that the truth values of implication $P \rightarrow Q$ and $\sim Q \rightarrow \sim P$ are same, then $\sim Q \rightarrow \sim P$ is called the **contrapositive** of $P \rightarrow Q$. Similarly $\sim P \rightarrow \sim Q$ is the contrapositive of $Q \rightarrow P$ (as shown in example 7.16(a)).
2. The implication $\sim P \rightarrow \sim Q$ is called the **inverse** of $P \rightarrow Q$.
3. If $P \rightarrow Q$ is an implication proposition then $Q \rightarrow P$ is called the **converse** of $P \rightarrow Q$.

The truth table for conditional, converse, inverse and contrapositive is as follows:

Table 7.12

P	Q	**Conditional** $P \rightarrow Q$	**Converse** $Q \rightarrow P$	**Inverse** $\neg P \rightarrow \neg Q$	**Contrapositive** $\neg Q \rightarrow \neg P$
T	T	T	T	T	T
T	F	F	T	T	F
F	T	T	F	F	T
F	F	T	T	T	T

Example 7.17

Give the converse and contrapositive of the statement

"If you are mathematician then you are algebraist".

Solution

Let us consider

P: You are mathematician

Q: You are algebraist.

The converse is

$Q \rightarrow P$: If you are algebraist then you are mathematician.

The contrapositive is

$\sim Q \rightarrow \sim P$: If you are not algebraist then you are not mathematician.

Example 7.18

Construct the truth table for

$$(P \rightarrow Q) \wedge (Q \rightarrow P).$$

Solution

P	Q	$X: P \rightarrow Q$	$Y: Q \rightarrow P$	$X \wedge Y$
0	0	1	1	1
0	1	1	0	0
1	0	0	1	0
1	1	1	1	1

Example 7.19

If $5x + 3 = 13$ then $x = 2$. Find converse, inverse and contrapositive.

Solution

Let $P(x) \equiv 5x + 3 = 13$

and $Q(x) \equiv x = 2$ then

i. Converse: If $x = 2$ then $5x + 3 = 13$.

ii. Inverse: If $5x + 3 \neq 13$ then $x \neq 2$.

iii. Contrapositive: If $x \neq 2$ then $5x + 3 \neq 13$.

Example 7.20

Consider the following conditional statement.

"If the floods destroy my house or the fires destroy my house, then my insurance company will pay me". Write the converse, inverse, and contrapositive of the statement.

Solution

Let P = The floods destroy my house

Q = The fires destroy my house

R = My insurance company will pay me.

The symbolic forms of given statement is

$$(P \vee Q) \rightarrow R$$

i. Converse is $R \Rightarrow (P \vee Q)$

ii. Inverse is $\rceil(P \vee Q) \Rightarrow \rceil R$

iii. Contrapositive is $\rceil R \Rightarrow \rceil(P \vee Q)$.

7.2-6 Biconditional

In the English language and in mathematics, necessary and sufficient conditions are applicable on many statements at the same time. In logic these statements are known as **biconditional statements** i.e. if there are two statements *P* and *Q* then the statement $P \leftrightarrow Q$ or $P \Leftrightarrow Q$, read as "*P* if and only if *Q*" or "*P* iff *Q*" or "*P* is necessary and sufficient for *Q*", is called an **equivalence** or **biconditional** statement. The truth values of $P \leftrightarrow Q$ is true if *P* and *Q* both have identical truth values i.e. either both are true or false otherwise it is false. Table 7.13 shows the truth values of biconditional statement $P \leftrightarrow Q$.

Table 7.13: Truth table for equivalence or biconditional

P	*Q*	$P \leftrightarrow Q$
0	0	1
0	1	0
1	0	0
1	1	1

The resultant truth values in table of example 7.18 are similar to the Table 7.13. This imply $P \Leftrightarrow Q$ is identical to $(P \rightarrow Q) \wedge (Q \rightarrow P)$.

Truth table for biconditional statement is represented in tree from as:

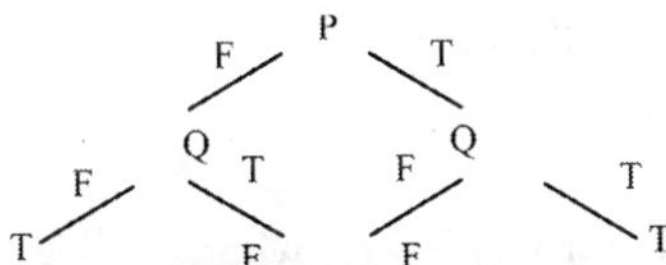

Example 7.21

Construct the truth table for

a. $(P \rightarrow Q) \leftrightarrow (Q \rightarrow P)$

b. $((P \rightarrow Q) \wedge (Q \rightarrow R)) \leftrightarrow (P \leftrightarrow R)$.

Solution

a.

P	*Q*	$X: P \rightarrow Q$	$Y: Q \rightarrow P$	$X \leftrightarrow Y$
0	0	1	1	1
0	1	1	0	0
1	0	0	1	0
1	1	1	1	1

b.

P	Q	R	$X: P \to Q$	$Y: Q \to R$	$Z: P \leftrightarrow R$	$U: X \wedge Y$	$U \leftrightarrow Z$
0	0	0	1	1	1	1	1
0	0	1	1	1	0	1	0
0	1	0	1	0	1	0	0
0	1	1	1	1	0	1	0
1	0	0	0	1	0	0	1
1	0	1	0	1	1	0	0
1	1	0	1	0	0	0	1
1	1	1	1	1	1	1	1

Example 7.22

Write the following statements in symbolic form:

a. "If the determinant of a system of linear equation is zero then either the system has no solution or has an infinite number of solutions."

b. Triangle ABC is an equilateral if and only if all the angles A, B and C are of 60°.

Solution

a. Let

P: The determinant of a system of linear equation is zero.
Q: The system has no solution.
R: The system has infinite number of solutions.
Then symbolic form is

$$P \to (Q \vee R).$$

b. Given statement can be written in form $(P \leftrightarrow Q)$ where:
P: Triangle ABC is an equilateral
Q: All angles A, B and C are of 60°.

Example 7.23

Express the statement $(\neg(P \vee Q) \vee (\neg P) \wedge Q)$ in simplest possible form.

Solution

We have $(\neg(P \vee Q) \vee (\neg P) \wedge Q)$

$$= ((\neg P) \wedge (\neg Q) \vee (\neg P) \wedge Q) \qquad [\because \neg(p \vee q) = \neg p \wedge \neg q]$$
$$= ((\neg P) \wedge (\neg Q) \vee Q)$$
$$= (\neg P) \wedge T$$
$$= \neg P. \qquad [\because (\neg q) \vee q = t]$$

Exercise 7.4

1. Write each conditional proposition symbolically. Write the converse of each statement symbolically and in words. Also, find the truth value of each conditional proposition and its converse.
   a. If sun set in west then Mohan learn the lesson.
   b. If $1 < 3$ then $x^2 = 4$.
   c. $3 > 2$ iff $0 < 5^2 - 6$.
   d. There is snow falling iff temperature is at or below 0° Celsius.
   f. John will pass in calculus if he work hard.
2. Write down the converse, contrapositive and negation of each of the following sentences.
   a. If Kolkata and Mumbai are old cities then they are popular.
   b. 21 is divisible by 3 iff it is divisible by 7.
   c. If an integer is a multiple of 10, then it is always divisible by 5 and 2.
   d. A statement is biconditional iff both the statements have the same truth values.
   e. If today is Sunday then we will go either to Delhi or Meerut.
3. Construct truth tables for the following formulas.
   a. $(P \vee (Q \wedge R)) \leftrightarrow [(P \vee Q) \wedge (P \vee R)]$
   b. $\rceil[(P \wedge \rceil Q)] \rightarrow (\rceil P \wedge Q)$
   c. $[P \wedge (Q \leftrightarrow R)] \vee (R \leftrightarrow P)$
   d. $(P \rightarrow \rceil R) \vee (Q \rightarrow \rceil R)$
   e. $(\rceil P \leftrightarrow \rceil Q) \leftrightarrow (P \leftrightarrow Q)$.
4. Let P, Q and R are the propositions

   P: Sabnam is not feeling well.

   Q: Sabnam miss the final examination.

   R: Sabnam passes the course.

   Express each of the following propositions as an English sentence.
   a. $P \rightarrow Q$
   b. $P \vee Q \vee R$
   c. $(P \rightarrow \rceil R) \vee (Q \rightarrow \rceil R)$
   d. $Q \rightarrow \rceil R$
   e. $\rceil Q \leftrightarrow R$.
5. Let P, Q and R are the propositions,

   P: Rahim does every exercise of this book.

   Q: Rahim gets highest marks in Discrete mathematics.

   R: Rahim gets highest marks in final examination.

   Write the following propositions using P, Q and R and logical connectives.
   a. If Rahim does every exercise in this book then he will get highest marks in Discrete mathematics.
   b. Rahim gets highest marks in discrete mathematics iff he gets height marks in final examination.
   c. Rahim gets the highest marks in final examination iff either he gets the highest marks in Discrete mathematics or do every exercise this book.
   d. If Rahim does not do every exercise of this book then he will not get the highest marks in discrete mathematics.
   e. If Rahim gets the highest marks in discrete mathematics and in final examination then Ram does every exercise in this book.

6. Assuming that P is 'true', Q is 'true' and R is 'false', find the truth value of each proposition.

   a. $\neg P \rightarrow (\neg Q \rightarrow \neg R)$
   b. $P \rightarrow (\neg Q \rightarrow R)$
   c. $(P \rightarrow Q) \wedge (Q \rightarrow R) \wedge (Q \rightarrow \neg P)$
   d. $(P \leftrightarrow Q) \wedge (Q \leftrightarrow R)$
   e. $(P \vee Q) \rightarrow R$
   f. $P \vee (\neg Q \rightarrow R)$.

7. Write down the converse, inverse and contrapositive of the following implications:

   i. If $ABCD$ is a square then $ABCD$ is a rectangle.
   ii. $x + 4 = 10 \rightarrow x = 3$.
   iii. If it is raining then he will fall down.

8. There are two restaurants next to each other. One has a sign that says, "Good food is not cheap" and the other has a sign that says, "Cheap food is not good". Are the signs saying the same thing.

Answers

1. a. P: Sun set in west;

   Q: Mohan learn the lesson

   Symbolic form: $P \rightarrow Q$;

   Converse: $Q \rightarrow P$

   Converse (in words): If Mohan learn the lesson then sun set in west.

   b. P: There is snow falling;

   Q: Temperature is at 0°C;

   R: Temperature is below 0°C.

   Symbolic form: $P \leftrightarrow (Q \vee R)$;

   Converse: $(Q \vee R) \leftrightarrow P$

   Converse (in words): Temperature is at or below 0°C iff there is snow falling.

2. a. Converse: If Kolkata and Mumbai are popular then they are old cities

   Contrapositive: If Kolkata and Mumbai are not popular they are not old cities.

   Negation: If Kolkata and Mumbai are not old cities then they are not popular.

   c. Converse: If an integer is divisible by 5 and 2 then it is a multiple of 10.

   Contrapositive: If an integer is not divisible by 5 and 2 then it is not a multiple of 10.

   Negation: If an integer is not multiple of 10 then it is not always divisible by 5 and 2.

   f. Converse: If we will go either Delhi or Meerut then today is Sunday.

   Contrapositive: If we will not go to Delhi or Meerut then today is not Sunday.

   Negation: If today is not Sunday then we will not go either to Delhi or Meerut.

3. a. Truth table

P	Q	R	$X: Q \wedge R$	$Y: P \vee X$	$Z: P \wedge Q$	$W: P \vee R$	$T: Z \wedge W$	$Y \leftrightarrow T$
0	0	0	0	0	0	0	0	1
0	0	1	0	0	0	1	0	1
0	1	0	0	0	1	0	0	1
0	1	1	1	1	1	1	1	1
1	0	0	0	1	1	1	1	1
1	0	1	0	1	1	1	1	1
1	1	0	0	1	1	1	1	1
1	1	1	1	1	1	1	1	1

c. Truth table

P	Q	R	$X: Q \leftrightarrow R$	$Y: P \wedge X$	$Z: R \leftrightarrow P$	$Y \vee Z$
0	0	0	1	0	1	1
0	0	1	0	0	0	0
0	1	0	0	0	1	1
0	1	1	1	0	0	0
1	0	0	1	1	0	1
1	0	1	0	0	1	1
1	1	0	0	0	0	0
1	1	1	1	1	1	1

d. Truth table

Table 7.14

P	Q	R	$\neg R$	$X: P \to \neg R$	$Y: Q \to \neg R$	$X \vee Y$
0	0	0	1	1	1	1
0	0	1	0	1	1	1
0	1	0	1	1	1	1
0	1	1	0	1	0	1
1	0	0	1	1	1	1
1	0	1	0	0	1	1
1	1	0	1	1	1	1
1	1	1	0	0	0	0

4. a. If Sabnam is not feeling well then she misses the final examination.

   c. If Sabnam is not feeling well then she does not pass the course or if Sabnam misses the final examination then she does not pass the course.

   e. Sabnam does not miss the final examination iff Sabnam passes the course.

5. a. $P \to Q$ b. $Q \leftrightarrow R$

   c. $R \leftrightarrow (Q \vee P)$ d. $\rceil P \to \rceil Q$

   e. $(Q \wedge R) \to P$.

6. a. true   b. true

   c. false   d. false

   e. false.

7. i. Converse: $Q \to P$, inverse $\rceil P \to \rceil Q$, Contrapositive $\rceil Q \to \rceil P$, where P = $ABCD$ is a square, Q= $ABCD$ is a rectangle.

   ii. Converse: $Q \to P$, Inverse $\rceil P \to \rceil Q$, Contrapositive $\rceil Q \to \rceil P$, where $P \equiv (x + 4 = 10)$, $Q \equiv (x = 6)$.

7.2-7 Tautologies

Till now, we have learnt many logical connectives and construction of the truth tables of molecular statements. We saw that the truth table represents the resultant truth values of statement which depend upon the truth values of input variables. These resultant values may either be true or false depending on the values of its propositional variables. The statements which contain both truth values (i.e. 0 and 1) are called **contigent form**. But it is perfectly natural to feel, that the statement "Taj Mahal is the most beautiful monument" (symbolized as B) and "Either Taj Mahal is the most beautiful monument or its not" (i.e. $B \vee \sim B$) both may be "true". Similarly, the statement "Rajiv Gandhi was assassinated" (Symbolized as W) and "Rajeev Gandhi was both assassinated and not assassinated" (i.e. $W \wedge \sim W$) are both 'false.' Both statements are historical facts and truth value depends upon the truth of history.

Logically, any statement $B \vee \sim B$ always true (as shown in Table 7.15) and its truth value is independent to the truth values of input variable B. Hence, A statement that has only 'true' value for all possible values of its propositional variables is called a tautology or tautologous statement form or Logical truth or universally valid formula.

Table 7.15

B	$\sim B$	$B \vee \sim B$
0	1	1
1	0	1

7.2-8 Contradiction

A statement that has only 'false' value for all possible values of its propositional variable is said to be **self-contradictory** or **contradiction** or a **absurdity,** e.g. the statement $W \wedge \sim W$ is a contradictory statement.

W	$\sim W$	$W \wedge \sim W$
0	1	0
1	0	0

Negation of contradiction is tautology and vice-versa.

Example 7.24

Verify that the proposition $P \vee \rceil (P \wedge Q)$ is tautology.

Solution

To construct the truth table, we have

P	Q	$P \wedge Q$	$\neg(P \wedge Q)$	$P \vee \neg(P \wedge Q)$
0	0	0	1	1
0	1	0	1	1
1	0	0	1	1
1	1	1	0	1

Since the last column has all values '1' for all possible values of P and Q hence given proposition is a tautology.

Example 7.25

Use the truth tables to characterize that following statement forms as tautologous, contradictory or contigent.

a. $[P \to (P \to Q)] \to Q$

b. $[P \to (Q \to R)] \to ((P \to Q) \to (P \to R))$

c. $(P \to P) \to (Q \wedge \sim Q)$

d. $P \to (P \to (Q \wedge \sim Q))$.

Solution

a.

P	Q	$X: P \to Q$	$Y: P \to X$	$Y \to Q$
0	0	1	1	0
0	1	1	1	1
1	0	0	0	1
1	1	1	1	1

Hence the statement is contigent.

b.

P	Q	R	$U: P \to Q$	$V: P \to R$	$X: Q \to R$	$Y: P \to X$	$Z: U \to V$	$Y \to Z$
0	0	0	1	1	1	1	1	1
0	0	1	1	1	1	1	1	1
0	1	0	1	1	0	1	1	1
0	1	1	1	1	1	1	1	1
1	0	0	0	0	1	1	1	1
1	0	1	0	1	1	1	1	1
1	1	0	1	0	0	0	0	1
1	1	1	1	1	1	1	1	1

Hence the statement is tautologous.

c.

P	Q	Q	$X: P \to P$	$Y: Q \wedge \sim Q$	$X \to Y$
0	0	1	1	0	0
0	1	0	1	0	0
1	0	1	1	0	0
1	1	0	1	0	0

The statement is Contradictory.

d.

P	Q	$\sim Q$	$X: Q \wedge \sim Q$	$Y: P \to X$	$P \to Y$
0	0	1	0	1	1
0	1	0	0	1	1
1	0	1	0	0	0
1	1	0	0	0	0

The last column of truth table has both values '1' and '0', hence the given statement is contigent.

7.2-9 Principle of Substitution

As we know, on increasing the number of variables n to a set number of rows in truth table increases simultaneously in the ratio of n: 2^n. The process to prove the statement is tautologous using truth tables become tedious for large values of n. An another method namely, **Principle of Substitution** can be used to prove the tautology of the statements. According to it,

"If P is a tautology and let P_1, P_2 ... are any propositions such that on substituting P_1, P_2... in the P in place of independent input variables, we still have a tautology".

1. *Whenever we use principle of substitution, the same formula is substituted for the variable throughout the statement.*
2. *Substitutions are made for the atomic formula, not for molecular formula.*
3. *Formula obtained after substitution is called a **substitution instance**.*

   e.g. Consider $(P \to (Q \to P))$, *substitute* $R \leftrightarrow S$ *for* P *and* $R \to S$ *for* Q *then its substitution instance is*

$$((R \leftrightarrow S) \to ((R \to S) \to (R \leftrightarrow S))).$$

Exercise 7.5

1. Determine which one is tautology, contradiction or contigent.

   a. $(P \wedge Q) \leftrightarrow P$
   b. $((\urcorner P \to Q) \to (Q \to P))$
   c. $((P \to (Q \to R)) \to ((\urcorner P \to \urcorner Q) \to \urcorner R))$
   d. $\urcorner(P \vee (Q \wedge R)) \leftrightarrow ((P \vee Q) \wedge (P \vee R))$
   e. $((\urcorner Q \wedge P) \wedge Q)$

2. Construct the truth tables to determine whether the given statement is a tautology, a contigent or an absurdity.

   a. $P \wedge \urcorner P$
   b. $Q \vee (\urcorner Q \wedge P)$
   c. $P \wedge Q \to P$
   d. $Q \to (Q \to P)$
   e. $(P \wedge Q) \vee (Q \wedge \urcorner P)$
   f. $P \to (Q \wedge P)$
   g. $Q \vee (\urcorner P \wedge Q)$

3. If P and Q are true and R, S and T are false then find the truth value of each statement.

   a. $\urcorner(P \to Q)$
   b. $\urcorner P \to R$
   c. $(R \wedge S \wedge T) \to (P \vee Q)$
   d. $(Q \to (R \to S)) \wedge ((P \to S) \to (\urcorner T))$
   e. $P \to ((R \to Q) \to S)$.

4. Produce the substitution instances of the following formulas for the given substitution.
   a. $(P \rightarrow Q) \rightarrow P$
   Substitute $(P \rightarrow Q)$ for P and $((P \wedge Q) \rightarrow R)$ for Q.
   b. $(P \wedge Q) \leftrightarrow P$
   Substitute $(P \wedge \urcorner Q)$ for P and P for Q.

Answers

1. a. Contigent

P	Q	$P \wedge Q$	$(P \wedge Q) \leftrightarrow P$
0	0	0	0
0	1	0	0
1	0	0	0
1	1	1	1

   b. Contigent
   c. Contigent
   d. Contigent
   e. Contradiction

3. a. false
   b. true
   c. true
   d. true
   e. false.

4. a. $((P \rightarrow Q) \rightarrow ((P \wedge Q) \rightarrow R)) \rightarrow (P \rightarrow Q)$
   b. $((P \wedge \urcorner Q) \wedge P) \leftrightarrow (P \wedge \urcorner Q)$.

7.3 Equivalence of Formulas or Logical Equivalence

In mathematical/ metalanguage argument or expression, we can replace the expression or variable with another formula or value without loss of integrity and still the result is equivalent to the original expression or argument. Similarly in logic the two statements said to be **equivalent**, if the resulting truth values of both statements are same for all possible combinations of input variables i.e. the final column in the truth tables of both statements are identical (provided the order of truth values of input variables are same in both statements). These two statements are known as **logical equivalent** statements. If A and B are two logical equivalent statements, then they are denoted as $A \equiv B$, or $A \cong B$ and read as: "A is equivalent to B." For example, let us consider two propositions:

$$A: P \rightarrow Q$$

and
$$B: \sim P \vee Q$$

From the truth Table 7.16, we can see that the two statements are equivalent as both have same resultant truth values.

Table 7.16

P	Q	$A: P \rightarrow Q$	$\sim P$	$B: \sim P \vee Q$	$A \leftrightarrow B$
0	0	1	1	1	1
0	1	1	1	1	1
1	0	0	0	0	1
1	1	1	0	1	1
Step		1	2	3	4

The column numbered 1 and 3 in Table 7.16 shows the logical equivalence of $P \rightarrow Q$ and $B \sim P \vee Q$.

We also know that in biconditional statement, the final truth values are true when both the statements have same truth values. Hence the statement $A \equiv B$ iff $A \leftrightarrow B$ is tautology (as shown in column no. 4 in table 7.16).

Since, logically equivalent statements have same truth values therefore one value may be substituted for another in any formula without leading to ambiguity and loss of originality.

Some important logical equivalent relations and properties are summarized in Table 7.17. Consider P, Q and R are propositions and 't' and 'f' as special variables representing the tautology and contradiction respectively.

Table 7.17: Important Logical Equivalent Identities

1. Commutative Laws
   a. $P \vee Q \equiv Q \vee P$
   b. $P \wedge Q \equiv Q \wedge P$
2. Associative Laws
   a. $P \vee (Q \vee R) \equiv (P \vee Q) \vee R$
   b. $P \wedge (Q \wedge R) \equiv (P \wedge Q) \wedge R$
3. Distributive Laws
   a. $P \wedge (Q \vee R) \equiv (P \wedge Q) \vee (P \wedge R)$
   b. $P \vee (Q \wedge R) \equiv (P \vee Q) \wedge (P \vee R)$
4. Idempotent Laws
   a. $(P \vee P) \equiv P$
   b. $(P \wedge P) \equiv P$
5. Absorption Laws
   a. $P \vee (P \wedge Q) \equiv P$
   b. $P \wedge (P \vee Q) \equiv P$
6. De Morgan's Laws
   a. $\neg(P \vee Q) \equiv (\neg P) \wedge (\neg Q)$
   b. $\neg(P \wedge Q) \equiv (\neg P) \vee (\neg Q)$
7. Detachment
   $((P \rightarrow Q) \wedge P) \rightarrow Q$
8. Chain rule or Transitive Property
   $((P \rightarrow Q) \wedge (Q \rightarrow R)) \rightarrow P \rightarrow$
9. Conditional Rules
   a. $(P \rightarrow Q) \equiv (\neg P) \vee Q$
   b. $(P \rightarrow Q) \equiv (\neg Q) \rightarrow (\neg P)$
10. Equivalence Rules
   a. $P \leftrightarrow Q \equiv (P \rightarrow Q) \wedge (Q \rightarrow P)$
   b. $\neg(P \leftrightarrow Q) \equiv (P \leftrightarrow \neg Q)$
   c. $P \leftrightarrow Q \equiv (P \wedge Q) \vee (\neg P) \wedge (\neg Q)$

11. Implication Rules
    a. $(P \to Q) \to ((P \wedge R) \to (Q \wedge R))$
    b. $(P \to Q) \to ((P \vee R) \to (Q \vee R))$
12. Contrapositive Inference
    $$(P \to Q) \wedge (\neg Q) \to (\neg P)$$
13. Identities
    i. a. $(P \vee \mathbf{f}) \equiv P$
       b. $(P \wedge \mathbf{t}) \equiv P$
    ii. a. $P \vee \mathbf{t} \equiv \mathbf{t}$
        b. $P \wedge \mathbf{f} \equiv \mathbf{f}$
    iii. a. $P \vee \neg P \equiv \mathbf{t}$
         b $P \wedge \neg P \equiv \mathbf{f}$
14 Involution Law
    $$\neg(\neg P) \equiv P$$

All the formulas given in Table 7.17 can be proved using truth table and substitution principle.

Example 7.26

State and prove the commutative law using truth table.

Solution

Commutative Laws: Let P and Q be two propositions then

a. $P \vee Q \equiv Q \vee P$ and

b. $P \wedge Q \equiv Q \wedge P$.

Truth Table

a.

P	Q	$X: P \vee Q$	$Y: Q \vee P$	$X \leftrightarrow Y$
0	0	0	0	1
0	1	1	1	1
1	0	1	1	1
1	1	1	1	1
Step		1	2	3

The column marked 1 and 2 are identical and so the biconditional shown in column number 3, have the truth values '1'. Hence the statement $P \vee Q$ is equivalent to $Q \vee P$.

b.

P	Q	$X: P \wedge Q$	$Y: Q \wedge P$	$X \leftrightarrow Y$
0	0	0	0	1
0	1	0	0	1
1	0	0	0	1
1	1	1	1	1
Step		1	2	3

Again column 1 and 2 are identical and column 3 shows that it is a tautology hence the statement $P \wedge Q \equiv Q \wedge P$.

Example 7.27

State and prove the De Morgan's laws using truth table.

Solution

De Morgan's Laws: If P and Q are two propositions then the two statements

a. $\neg(P \vee Q) \equiv (\neg P) \wedge (\neg Q)$ b. $\neg(P \wedge Q) \equiv (\neg P) \vee (\neg Q)$

are called **De Morgan's Laws**. It can be defined as

"The negation of the disjunction (or conjunction) of two statements is logically equivalent to the conjunction (or disjunction) of the negations of the two statements respectively".

Let us prove these statements using truth table:

Truth table:

a.

P	Q	$\neg P$	$\neg Q$	$X: P \vee Q$	$\neg X$	$(\neg P) \wedge (\neg Q)$
0	0	1	1	0	1	1
0	1	1	0	1	0	0
1	0	0	1	1	0	0
1	1	0	0	1	0	0

b.

P	Q	$\neg P$	$\neg Q$	$X: P \wedge Q$	$\neg X$	$\neg P \vee \neg Q$
0	0	1	1	0	1	1
0	1	1	0	0	1	1
1	0	0	1	0	1	1
1	1	0	0	1	0	0

Last two columns of both tables are identical. Hence the theorem.

Example 7.28

Show that $\neg(P \vee (\neg P \wedge Q)) \equiv \neg P \wedge \neg Q$ using logical equivalent formulas.

Solution

$\neg(P \vee (\neg P \wedge Q) \equiv (\neg P) \wedge \neg(\neg P \wedge Q)$ (De Morgan's Law)

$\equiv \neg P \wedge (\neg(\neg P) \vee \neg Q)$ (De Morgan's Law)

$\equiv \neg P \wedge (P \vee \neg Q)$ (Involution Law)

$\equiv (\neg P \wedge P) \vee (\neg P \wedge \neg Q)$ (Distribution Law)

$\equiv \mathrm{f} \vee (\neg P \wedge \neg Q)$ (Identity (iii))

$\equiv \neg P \wedge \neg Q$ (Identity (i))

Hence $\neg(P \vee (\neg P \wedge Q)) \equiv \neg P \wedge \neg Q$.

Example 7.29

Show that $(P \wedge Q) \rightarrow (P \vee Q)$ is a tautology.

Solution

To prove the above statement we will use logical equivalence (we may also prove it using truth table).

$$(P \wedge Q) \rightarrow (P \vee Q) \equiv \urcorner(P \wedge Q) \vee (P \vee Q)$$
$$\equiv (\urcorner P \vee \urcorner Q) \vee (P \vee Q) \qquad \text{(De Morgan's Law)}$$
$$\equiv \urcorner P \vee (\urcorner Q \vee P) \vee Q$$
$$\equiv \urcorner P \vee (P \vee \urcorner Q) \vee Q \qquad \text{(Commutative Law)}$$
$$\equiv (\urcorner P \vee P) \vee (\urcorner Q \vee Q) \qquad \text{(Associative Law)}$$
$$\equiv t \vee t \qquad \text{(Identity (iii))}$$
$$\equiv t.$$

Example 7.30

Show that contrapositives are logically equivalent, i.e. $\urcorner q \rightarrow \urcorner P \equiv P \rightarrow Q$.

Solution

Let us form the truth table for this

P	Q	$\urcorner P$	$\urcorner Q$	$\urcorner Q \rightarrow \urcorner P$	$P \rightarrow Q$
T	T	F	F	T	T
T	F	F	T	F	F
F	T	T	F	T	T
F	F	T	T	T	T

Observe that column 5 and 6 are identical.

Example 7.31

Show the implication:

i. $((P \vee \urcorner P) \rightarrow Q) \rightarrow ((P \vee \urcorner P) \rightarrow R) \Rightarrow (Q \rightarrow R)$

ii. $(P \rightarrow Q) \rightarrow Q \Rightarrow P \vee Q.$

Solution

i. $((P \vee \urcorner P) \rightarrow Q) \rightarrow ((P \vee \urcorner P) \rightarrow R) \Rightarrow (Q \rightarrow R)$

L.H.S.: $((P \vee \urcorner P) \rightarrow Q) \rightarrow ((P \vee \urcorner P) \rightarrow R$

$$\Rightarrow (\urcorner(P \vee \urcorner P) \vee Q) \rightarrow (\urcorner(P \vee \urcorner P) \vee R) \qquad [\because \text{P} \rightarrow \text{Q} = (\urcorner \text{P}) \vee Q]$$
$$\Rightarrow (\urcorner(t) \vee Q) \rightarrow (\urcorner f) \vee R) \qquad [\because \text{P} \vee \urcorner P = t]$$
$$\Rightarrow (f \vee Q) \rightarrow (t \vee R) \qquad [\because \urcorner t = f]$$
$$\Rightarrow Q \rightarrow R \qquad [\because t \vee Q = Q]$$
$$\Rightarrow \text{RHS}$$

ii. $(P \rightarrow Q) \rightarrow Q \Rightarrow P \vee Q$

L.H.S.: $(P \rightarrow Q) \rightarrow \text{Q} = (\urcorner(P) \vee Q) \rightarrow Q$

$$= \urcorner((\urcorner P) \vee Q) \vee Q$$
$$= (\urcorner(\urcorner P) \wedge \urcorner Q) \vee Q$$

$= (P \wedge \urcorner Q) \vee Q$

$= (P \vee Q) \wedge (\urcorner Q) \vee Q)$ [distributivity]

$= (P \vee Q) \wedge t)$ $[\because (\urcorner Q) \vee Q = t)]$

$= P \vee Q$

= RHS.

Example 7.32

Show that

i. $\urcorner(P \Leftrightarrow Q) \Leftrightarrow (P \vee Q) \wedge \urcorner(P \wedge Q)$

ii. $\urcorner(P \vee Q) \Leftrightarrow \urcorner P \wedge \urcorner Q$.

Solution

i. $\urcorner(P \leftrightarrow Q) \leftrightarrow (P \vee Q) \wedge \urcorner(P \wedge Q)$

LHS $= \urcorner(P \Leftrightarrow Q) \leftrightarrow \urcorner((P \wedge Q) \vee ((\urcorner P) \wedge (\urcorner Q)))$

$\leftrightarrow \urcorner(P \wedge Q) \wedge \urcorner((\urcorner(P) \wedge (\urcorner Q))$ $[\because \urcorner(P \vee Q) = \urcorner P \wedge \urcorner Q]$

$\leftrightarrow \urcorner(P \wedge Q) \wedge (\urcorner P) \wedge \urcorner(\urcorner Q))$

$\leftrightarrow \urcorner(P \wedge Q) \wedge (P \vee Q)$

$\leftrightarrow$ RHS.

ii. $\urcorner(P \vee Q) \leftrightarrow \urcorner P \wedge \urcorner Q$

P	Q	$P \vee Q$	$\urcorner(P \vee Q)$	$\urcorner P$	$\urcorner Q$	$\urcorner P \wedge \urcorner Q$
0	0	0	1	1	1	1
0	1	1	0	1	0	0
1	0	1	0	0	1	0
1	1	1	0	0	0	0

From table $\urcorner(P \vee Q) \leftrightarrow \urcorner P \wedge \urcorner Q$.

7.4 Argument, Valid Argument and Fallacy Argument

An **argument** is an assertion that a set of given propositions called premises yield another proposition called the **conclusion**. If $P_1, P_2, P_3 \ldots, P_n$ is the set of proposition that yields the conclusion Q then it is denoted by

$$P_1, P_2, P_3 \ldots, P_n \vdash Q.$$

An argument is called **Valid argument** if the conclusion is true whenever all the premises are true. The argument is valid if the 'And' of the group of proposition implies conclusion is a tautology, i.e.

$$P_1 \wedge P_2 \wedge P_3 \ldots \wedge P_n \to Q \text{ is a tautology}$$

where, $P_1, P_2, P_3 \ldots, P_n$ are proposition and Q is the conclusion.

An argument is called **fallacy** or an **invalid argument** if it is not a valid argument.

Example 7.33

If P, Q and R are three propositions then $(P \to Q) \wedge (Q \to R) \to (P \to R)$ is a tautology, and is therefore a valid argument. This is symbolised as

$$P \to Q \text{ (premise)}$$

$$Q \rightarrow R \text{ (premise)}$$
$$\therefore\ P \rightarrow R \text{ (conclusion)}$$

This is known as transitive rule or law of syllogism.

Example 7.34

If P and Q are two propositions then $(P \wedge (P \rightarrow Q) \rightarrow Q$ is a tautology, and is therefore a valid argument. This is symbolised as:

$$P \text{ (premise)}$$
$$P \rightarrow Q \text{ (premise)}$$
$$\therefore\ Q \text{ (conclusion)}$$

This is known as law of **detachment**.

Example 7.35

Prove the validity of the following argument using truth tables.

P: If there is a storm it rain

Q: If it rains then sea is rough.

Solution

Let P: There is a storm

Q: It rains

R: Sea is rough

The above argument in symbolic form is

$$P \rightarrow Q, Q \rightarrow R, \vdash P \rightarrow R$$

This argument is verified by constructing the truth table for

$$(P \rightarrow Q) \wedge (Q \rightarrow R) \rightarrow (P \rightarrow R)$$

P	Q	R	$P \rightarrow Q$	$Q \rightarrow R$	$(P \rightarrow Q) \wedge (P \rightarrow Q)$	$(P \rightarrow Q) \wedge (Q \rightarrow R) \wedge P$	$(P \rightarrow Q) \wedge (Q \rightarrow R) \rightarrow (P \rightarrow R)$
T	T	T	T	T	T	T	T
T	T	F	T	F	F	F	T
T	F	F	F	T	F	F	T
T	F	T	F	T	F	F	T
F	T	T	T	T	T	F	T
F	T	F	T	F	F	F	T
F	F	F	T	T	T	F	T
F	F	F	T	T	T	F	T

Since last column contains all T's, hence the argument is valid.

Example 7.36

Test the validity of following argument if we have a square, then diagonals bisect each other. If we don't have a square then diagonals don't bisect each other.

Solution

Let P: We have a square.

Q: Diagonals bisect each other.

The arguments symbolically expressed as:

Symbolically:

$$P \rightarrow Q \text{ (premise)}$$
$$\neg P \text{ (premise)}$$
$$\therefore \neg Q \text{ (conclusion)}$$

Let us form the truth table for the proposition

$$((P \rightarrow Q) \wedge \neg P) \rightarrow \neg Q$$

P	Q	$P \rightarrow Q$	$(P \rightarrow Q) \wedge \neg P$	$(P \rightarrow Q) \wedge \neg P] \rightarrow \neg Q$
F	F	T	T	T
F	T	T	T	F
T	F	F	F	T
T	T	T	F	T

All the truth values are not 'T in the last column hence argument is not valid, i.e. it is a fallacy.

7.4-1 Logical Implications

A proposition P is said to logically imply a proposition Q if implication $P \rightarrow Q$ is a tautology. Logical implication is denoted by $P \Rightarrow Q$. (Some author represent this also by the symbolic $\rightarrow$).

Caution! Care must be taken to distinguish implication (or conditional) with logical implication.

Implication ($\rightarrow$)	Logical implication ($\Rightarrow$)
1. Implication is only a way to connect two propositions P and Q. 2. All implications are not logical implication.	1. If P logically implies Q then they are related to the extent that whenever P has truth value 'T so does Q. 2. All logical implications are implications.

Following three statements are equivalent:

1. *P logically implies Q.*
2. *The proposition $P \rightarrow Q$ is a tautology.*
3. *The argument $P \vdash Q$ is valid.*
4. *Q is true whenever P is true.*

Readers are advised to prove these.

Example 7.37

Show that $(P \vee Q) \wedge \neg Q \rightarrow P$ is a logical implication.

Solution

Construct the truth table

P	Q	$\rceil P$	$P\vee Q$	$(P\vee Q)\wedge\rceil Q$	$(P\vee Q)\wedge\rceil Q\to P$
T	T	F	T	F	T
T	F	T	T	T	T
F	T	F	T	F	T
F	F	T	F	F	T

All truth vales of last column is T, hence, it is a logical implication. This is read as $(P \vee Q) \wedge \rceil Q$ logically implies P.

7.5 Rules of Inferences

To evaluate the validity of arguments in the propositional calculus, we already have a reliable method i.e. construction of truth tables. We know argument is nothing but a sequence of statements. Some statements appear as grounds, some as warrants and some are known as part of the frame of reference. All the statements except the final one are called **premises** or **assumption** or **hypothesis**, assumed to be true in the argument. To construct proofs, we want to draw the conclusion and the final statement is called **conclusion**. This can also be done by using **rules of inference**. We can always tabulate the truth-values of premises and conclusion, checking for each horizontal line on which the premises are true, while the conclusion is false. This method is not always convenient since for n number of statements, there must be 2^n lines. Therefore for checking the validity of an argument, it is sufficient to check only those horizontal lines (i.e. rows) of the truth table for which all the premises are true and correspondingly their conclusion is true. This whole process is known as **truth table technique** (and we call it **method I**).

Method I

Step wise procedure for checking the validity using truth table technique

Step 1: In the given argument, identify the premises and conclusion.

Step 2: Construct the truth table for premises and conclusion.

Step 3: Mark the rows in which all the premises are true called **critical rows**.

Step 4: (i) If in each critical row the conclusion is also true, then argument is valid.

(ii) If there is at least one critical row in which conclusion is false then argument is invalid.

Method II

When an argument is valid and its premises are true, the truth conclusion is said to be **inferred** or **deduced** form the **truth of premises**. Each inference can be written as an implication form as follows:

(Conjunction of premises) $\to$ (Conclusion).

We can say that this inference is valid if the implication is a tautology].

There is another shorter way to proof the validity of an argument as given below.

Method III

As we know in an argument there is a sequence of statements (or arguments). A chain of interrelated arguments is valid so long as each of its links (statements) is valid. We can demonstrate the validity of an argument by starting with its premises, taking one tiny valid step at a time, and finally reaching its conclusion. A single limitation of this procedure is that each of the tiny steps must be a substitution instance of some valid argument from which we can discover transparent path leading from premises to conclusion.

In spite of the fact that any valid pattern of inferences could be used in this **proof procedure**, we will make things easier by relying on a very short list of valid argument forms, which are shown in Table 7.18. Each new tiny step that we take in constructing a proof must then be a substitution instance of one of these rules of inference. We have already seen these rules in Table 7.15 (tautological implications). Rules of inference named

Modus Pones (M.P) and Hypothetical Syllogism (H.S.) are also known as **fundamental Rules**. Since some of the other rules of inferences in the Table 7.19 are nothing more than a reinterpretation of two fundamental rules.

Fundamental Rule I

If argument P is assumed as true and also the argument $P \rightarrow Q$ is accepted as true then Q must be true. Symbolically, it can be written in following pattern:

$$\begin{array}{c} P \quad Q \\ P \\ \hline Q \end{array}$$

In this tabular presentation, all the hypothesis or premises are above the horizontal line and conclusion is below the horizontal line. This rule is named as **modus pones** or **the rule of detachment**. The validity of the argument can also be shown using truth table. To do so, construct the truth table for premises and conclusion as shown in Table 7.18. From method I, only fourth row is critical row in which all premises are true. But in this case the conclusion is also true hence the argument is valid.

Table 7.18

		1st Premise	2st Premise	Conclusion
P	Q	$P \rightarrow Q$	P	Q
F	F	T	F	F
F	T	T	F	T
T	F	F	T	F
T	T	T	T	T

Another way to check the validity of the statement is by method II, In this method stating that the above argument is valid when $((P \rightarrow Q) \wedge P) \rightarrow Q$ is a tautology, see Table 7.19 (b).

Rule of Inference

Table 7.19 (a)

Tautological form	Name	Rule of inference
$P \wedge Q \rightarrow P$ $P \wedge Q \rightarrow Q$	Simplification	$P \quad P$ $Q \quad Q$ $P \quad Q$
$P \rightarrow P \vee Q$ $Q \rightarrow P \vee Q$	Addition	P $\overline{P \quad Q}$ Q $\overline{P \quad Q}$
$(P) \wedge (Q) \rightarrow (P \wedge Q)$	Conjunction	P Q $\therefore P \wedge Q$
$(P \rightarrow Q) \wedge P \rightarrow Q$	Modus Ponens	$P \rightarrow Q$ P $\therefore Q$

$((P \to Q) \wedge \neg Q) \to \neg P$	Modus Tollens	$P \to Q$ $\neg Q$ $\therefore \neg P$
$(P \to Q) \wedge (Q \to R) \to (P \to R)$	Hypothetical Syllogism	$P \to Q$ $Q \to R$ $\therefore\ P \to R$
$(P \vee Q) \wedge \neg P \to Q$	Disjunctive Syllogism	$P \vee Q$ $\neg P$ $\therefore\ Q$
$((P \to Q) \wedge (R \to S)) \wedge (P \vee R)$ $\to (Q \to S)$	Constructive Dilemma	$(P \to Q) \wedge (R \to S)$ $P \vee R$ $Q \vee S$
$((P \to Q) \wedge (R \to S)) \wedge (\neg Q \vee \neg S)$ $\to (\neg P \vee R)$	Destructive Dilemma	$(P \to Q) \wedge (R \to S)$ $\neg Q \vee \neg S$ $\therefore\ \neg P \vee \neg S$

Table 7.19 (b)

P	Q	$X: P \to Q$	$Y: X \wedge P$	$Y \to Q$
F	F	T	F	T
F	T	T	F	T
T	F	F	F	T
T	T	T	T	T

Example 7.38

Represent the following argument symbolically and determine whether the argument is valid.

If today is David's birthday, then today is Oct. 12.

Today is David's birthday.

$\therefore$ Today is Oct 12.

Solution

Let us assume:

P: Today is David's birthday

Q: Today is Oct. 12

Symbolically, we can write the argument form as

$$\begin{array}{l} P \to Q \\ P \\ \hline \therefore\ Q \end{array}$$

Hence by modus ponens the argument is valid.

Example 7.39

Suppose the implication "If the number is divisible by 6 then it is divisible by 3" and its hypothesis the number is divisible by 6 are true. Then by modus ponens, it follows that the conclusion of the implication", number is divisible by 3".

Fundamental Rule 2

Whenever the two implications $P \to Q$ and $Q \to R$ are accepted as true, we must accept the implication $P \to R$ as true.

Symbolically

$$\begin{array}{c} P \to Q \\ \underline{Q \to R} \\ \therefore\ P \to R \end{array}$$

This rule of inference commonly called the **law** of **hypothetical syllogism** or **the transitive rule**.

Now, construct the truth table for premises and conclusion. ←

			1st Premise	2nd Premise	Conclusion	
P	Q	R	$P \to Q$	$Q \to R$	$P \to R$	
F	F	F	T	T	T	←
F	F	T	T	T	T	←
F	T	F	T	F	T	
F	T	T	T	T	T	←
T	F	F	F	T	F	
T	F	T	F	T	T	
T	T	F	T	F	F	
T	T	T	T	T	T	←

We can observe form truth table, first, second, fourth and eighth rows have both premises true marked by ←. Since in each case the conclusion is also true hence the argument is valid.

The above argument is valid because $((P \to Q) \wedge (Q \to R)) \to (P \to R)$ is a tautology.

The transitive rule can be extended to a larger number of implications, e.g.

$$\begin{array}{c} P \to Q \\ Q \to R \\ \underline{R \to S} \\ \therefore\ P \to S \end{array}$$

Example 7.40

Represent the following argument symbolically and determine whether the argument is valid.

If today is children's day then today is Pt. Jawaharlal Nehru's birthday.

It today is Pt. Jawaharlal Nehru's birthday then today is 14th Nov.

Hence if today is children's day then today is 14th Nov.

Solution

Let us assume:

P: Today is children's day

Q: Today is Pt. Jawaharlal Nehru's birthday

R: Today is 14th Nov.

The argument is of the form

$$\begin{array}{c} P \rightarrow Q \\ \underline{Q \rightarrow R} \\ \therefore P \rightarrow R \end{array}$$

Hence, by hypothetical syllogism the argument is valid. There are some other valid rules of inference, which are listed in Table 7.19 (a). Here we discuss them.

1. Simplification

The simplification (Simple.) rule permits us to infer the truth of a conjunct from that of a

$$\frac{P \wedge Q}{p}$$

Table 7.20

		Premise	Conclusion
P	Q	$P \wedge Q$	P
F	F	F	F
F	T	F	F
T	F	F	T
T	T	T	T

Notice that simplification warrants only an inference to the last of the two conjuncts, even though the truth of the third conjunct could also be derived. In the truth table fourth row is critical row and the conclusion is also true for this row hence the argument is valid. And $(P \wedge Q) \rightarrow P$ is a tautology.

2. Addition

The argument of the form

$$\frac{P}{p \quad Q}$$

This rule warrants the inference from any true statement to its disjunction with anything what so over. As seen in truth table last two rows are critical rows for that conclusion is also true hence the argument is valid. And $(P \rightarrow (P \vee Q))$ is tautology.

		Premise	Conclusion
P	Q	P	$P \vee Q$
F	F	F	F
F	T	F	T
T	F	T	T
T	T	T	T

3. Conjunction

Conjunction (Conj.) permits the derivation of a conjunction from the truth value of both the conjuncts.

$$\begin{array}{c} P \\ Q \\ \hline \therefore\ p \wedge Q \end{array}$$

Table 7.21

1ST Premise	2nd Premise	Conclusion
P	Q	$P \wedge Q$
F	F	F
F	T	F
T	F	F
T	T	T

As this truth table illustrates, this is a natural inference from our definition of the connective.

4. Modes Tollens

The modes Tollens (M.T.) has the simpler form:

$$\begin{array}{c} P \to Q \\ \neg Q \\ \hline \therefore\ \neg P \end{array}$$

		1ST Premise	2ST Premise	Conclusion
P	Q	$P \to Q$	$\neg Q$	$\neg P$
F	F	T	T	T
F	T	T	F	T
T	F	F	T	F
T	T	T	F	F

The above truth table shows the validity of all substitution-instance. Whenever its premise is true, the conclusion is true as well. **Modus tollens (M.T.)** is a latin word meaning "method of denying".

The validity of modus tollens can be shown to follow from modus ponens together that a conditional statement is logically equivalent to its contrapositive. By forming the truth table, we can also establish the validity of argument by method II. i.e.

$((P \to Q) \wedge \neg Q) \to \neg P$ is a tautology.

Example 7.41

Represent the argument:

The sum of digits of a number is divisible by 3 then the number is divisible by 3.
The number is not divisible by 3.
∴ The sum of digits of a number is not divisible by 3.

Symbolically determine whether the argument is valid.

Solution

Let us assume:

P: The sum of digits of a number is divisible by 3.

Q: The number is divisible by 3.

$\urcorner Q$: The number is not divisible by 3.

The argument can be written as

$$\begin{array}{c} P \rightarrow Q \\ \urcorner Q \\ \hline \therefore \urcorner P \end{array}$$

Thus, by modus tollens the argument is valid.

5. Disjunctive Syllogism

The argument form of a Disjunctive Syllogism (D. S.) is

$$\begin{array}{c} P \vee Q \\ \urcorner P \\ \hline \therefore Q \end{array}$$

Table 7.22

		1ST Premise	2nd Premise	Conclusion
P	***Q***	$P \vee Q$	$\urcorner P$	***Q***
F	*F*	*F*	*T*	*F*
F	*T*	*T*	*T*	*T*
T	*F*	*T*	*F*	*F*
T	*T*	*T*	*F*	*T*

The premises are true on line 2 and on this line the conclusion is also true. Thus the inference is valid. This argument states that when there are possibilities and one can rule one out, the other must be the case. This can be established by the truth table of

$$((P \vee Q) \wedge \urcorner P) \rightarrow \urcorner Q \text{ is a tautology.}$$

6. Constructive Dilemma

The most complex of rules of inference is constructive Dilemma (C. D.). It involves four statement variables, hence the truth-table that shows its validity must take into account 16 different combinations of truth values. The argument form of a constructive dilemma is

$$\begin{array}{c} (P \rightarrow Q) \wedge (R \rightarrow S) \\ P \vee R \\ \hline \therefore Q \vee S \end{array}$$

The premises are true on rows 4, 8 13, 14 and 16 of the following table. We con observe that on each of these lines the conclusion is also true. Thus the inference is valid and we can be sure that every argument that is a substitution instance of this argument form must be valid.

						1st premise	2nd premise	Conclusion
P	***Q***	***R***	***S***	$P \rightarrow Q$	$R \rightarrow S$	$(P \rightarrow Q) \wedge (R \rightarrow S)$	$P \vee R$	$Q \vee$
F	*F*	*F*	*F*	*T*	*T*	*T*	*F*	*F*
F	*F*	*F*	*T*	*T*	*T*	*T*	*F*	*T*
F	*F*	*T*	*F*	*T*	*F*	*F*	*T*	*F*
F	*F*	*T*	*T*	*T*	*T*	*T*	*T*	*T*

						1st premise	2nd premise	Conclusion
P	Q	R	S	$P \to Q$	$R \to S$	$(P \to Q) \wedge (R \to S)$	$P \vee R$	$Q \vee$
F	T	F	F	T	T	T	F	T
F	T	F	T	T	T	T	F	T
F	T	T	F	T	F	F	T	T
F	T	T	T	T	T	T	T	T
T	F	F	F	F	T	F	T	F
T	F	F	T	F	T	F	T	T
T	F	T	F	F	F	F	T	F
T	F	T	T	F	T	F	T	T
T	T	F	F	T	T	T	T	T
T	T	F	T	T	T	T	T	T
T	T	T	F	T	F	F	T	T
T	T	T	T	T	T	T	T	T

7. Destructive Dilemma

As Constructive Dilemma, **Destructive Dilemma** also involves four statement variables and the truth table of it consists of 2^4 i.e. 16 different combinations of truth values. The argument form of the Destructive Dilemma is

$$\begin{array}{l} (P \to Q) \wedge (R \to S) \\ \quad\quad\quad\; \urcorner Q \vee \urcorner S \\ \hline \therefore\ \urcorner P \vee \urcorner R \end{array}$$

From the truth table we observe that the premises are true on lines 1, 2, 4, 5 and 13 and on each of these lines the conclusion is also true. Hence the argument is valid. Also

$(((P \to Q) \wedge (R \to S) \wedge (\urcorner Q \vee \urcorner S)) \to (\urcorner P \vee \urcorner R)$ is a tautology.

						1st premise	2nd premise	Conclusion
P	Q	R	S	$X: P \to Q$	$Y: P \to S$	$X \wedge Y$	$\urcorner Q \vee \urcorner S$	$\urcorner P \vee \urcorner R$
F	F	F	F	T	T	T	T	T
F	F	F	T	T	T	T	T	T
F	F	T	F	T	F	F	T	T
F	F	T	T	T	T	T	T	T
F	T	F	F	T	T	T	T	T
F	T	F	T	T	T	T	F	T
F	T	T	F	T	F	F	T	T
F	T	T	T	T	T	T	F	T
T	F	F	F	F	T	F	T	T
T	F	F	T	F	T	F	T	T
T	F	T	F	F	F	F	T	F
T	F	T	T	F	T	F	T	F

						1st premise	2nd premise	Conclusion
P	Q	R	S	$X: P \to Q$	$Y: P \to S$	$X \wedge Y$	$\rceil Q \vee \rceil S$	$\rceil P \vee \rceil R$
T	T	F	F	T	T	T	T	T
T	T	F	T	T	T	T	F	T
T	T	T	F	T	F	F	T	F
T	T	T	T	T	T	T	F	F

7.6 Tautological Implications

Two statements A and B are said to be tautologically implies iff $A \to B$ is a tautology. It is denoted by $A \Rightarrow B$ (symbol "$\Rightarrow$" is not a connective) and read as "A is tautologically implies B".

Some important implication relations are summarized in Table 7.23. Let P, Q and R are statements, then

Table 7.23: Tautological Implications

$P \wedge Q \Rightarrow P$
$P \wedge Q \Rightarrow Q$
$P \Rightarrow P \vee Q$
$\rceil P \Rightarrow P \to Q$
$Q \Rightarrow P \to Q$
$\rceil(P \to Q) \Rightarrow P$
$\rceil(P \to Q) \Rightarrow \rceil Q$
$P \wedge (P \to Q) \Rightarrow Q$
$\rceil Q \wedge (P \to Q) \Rightarrow \rceil P$
$\rceil P \wedge (P \vee Q) \Rightarrow Q$
$(P \to Q) \wedge (Q \to R) \Rightarrow P \to R$
$(P \vee Q) \wedge (P \to R) \wedge (Q \to R) \Rightarrow R$

Example 7.42

Prove the following implications:

a. $Q \Rightarrow (P \to Q)$

b. $P \wedge (P \to Q) \Rightarrow Q$.

Solution

a. Draw the truth table for $Q \to (P \to Q)$.

P	Q	$P \to Q$	$Q \to (P \to Q)$
0	0	1	1
0	1	1	1
1	0	0	1
1	1	1	1
Step		1	2

The column numbered 2 shows the $Q \to (P \to Q)$ is a tautology. Hence Q is tautologically implies $(P \to Q)$.

b. Draw the truth table for $P \wedge (P \to Q) \Rightarrow Q$.

P	Q	$P \to Q$	$P \wedge (P \to Q)$	$P \wedge (P \to Q) \to Q$
0	0	1	0	1
0	1	1	0	1
1	0	0	0	1
1	1	1	1	1
Step		1	2	3

Column 3 shows the tautology. Hence $P \wedge (P \to Q) \Rightarrow Q$.

Theorem 7.1

If $A_1, A_2, \ldots, A_n, P, Q$ are propositions and $(A_1 \wedge A_2 \wedge A_3 \wedge \ldots \wedge A_n \wedge P) \Rightarrow Q$ then prove that

$$(A_1 \wedge A_2 \wedge A_3 \wedge \ldots \wedge A_n \wedge P) \to (P \to Q).$$

Proof

We know that if P, Q and R are propositions then

$$P \to (Q \to R) \equiv (P \wedge Q) \to R^{\#} \qquad \ldots (1)$$

Hence from our assumption

$(A_1 \wedge A_2 \wedge A_3 \wedge \ldots \wedge A_n \wedge P) \Rightarrow Q$ means $(A_1 \wedge A_2 \wedge A_3 \wedge \ldots \wedge A_n \wedge P) \to Q$ is a tautology. Then from (1) we say that $(A_1 \wedge A_2 \wedge A_3 \wedge \ldots \wedge A_n) \to (P \to Q)$ is a tautology.

Exercise 7.6

1. Use truth tables to verify the following equivalencies:
   a. $\neg(P \leftrightarrow Q) \equiv (P \vee Q) \wedge \neg(P \wedge Q)$
   b. $[\neg Q \wedge (P \to Q)] \equiv P$
   c. $((P \vee Q) \to R) \equiv (P \to R) \wedge (Q \to R)$
   d. $\neg P \wedge (\neg Q \wedge R) \vee (Q \wedge R) \vee (R \wedge P) \equiv R$
   e. $\neg(P \leftrightarrow Q) \equiv (P \wedge \neg Q) \vee (\neg P \wedge Q)$.
2. Show that $\neg(\neg P)$ and P are logically equivalent.
3. Show that each of the following implications is tautology by using truth tables.
   a. $[(P \vee Q) \wedge (P \to R) \wedge (Q \to R)) \to R$
   b. $(P \to (Q \to R)) \to ((P \to Q) \to (P \to R))$
   c. $((P \to Q) \to Q) \to (P \vee Q)$
   d. $(\neg P \to \neg Q) \to (Q \to P)$
   e. $(\neg P) \to (\neg(P \to \neg Q))$.

$^{\#}$ $\therefore P \to (Q \to R) \equiv P \to (\neg Q \vee R)$
$\equiv \neg P \vee (\neg Q \vee R) \equiv (\neg P \vee \neg Q) \vee R$
$\equiv \neg(P \wedge Q) \vee R \equiv (P \wedge Q) \to R$

4. Show the following equivalencies without constructing the truth tables.

   a. $P \to (Q \to P) \equiv \rceil P \to (P \to Q)$
   b. $((P \vee \rceil P) \to Q) \to ((P \vee \rceil P) \to R \equiv (Q \to R)$
   c. $(P \wedge Q) \vee (\rceil P \vee \rceil (P \vee Q) \equiv \rceil P \vee Q$
   d. $(P \leftrightarrow Q) \equiv (\rceil P \vee Q) \vee (P \wedge Q)$
   e. $P \to (Q \vee R) \equiv (P \to Q) \vee (P \to R)$.

5. Let P: Amit fails in Discrete Mathematics

   Q: Amit is sad

   R: Amit has not got the job.

   Write out the following propositions.

   a. $\rceil (P \to Q)$
   b. $\rceil (P \to \rceil (Q \wedge R))$
   c. $(R \vee P) \leftrightarrow Q$
   d. $Q \to (P \wedge R)$
   e. $R \vee (P \to Q)$.

6. Give the negation of each of the following statements:

   a. If you are rich, then I will walk.
   b. If I do not win the game then I will not enter the contest.
   c. If 2 is odd then Madhya Pradesh has small population.
   d. If $3 \leq 4$ then $2 > 2$.
   e. If I am not President then I will go to Dehradun.
   f. If 2 is even then England has large population.

7. Write the following statements in terms of P, Q, R... and logical connectives.

   a. If I am not in a good mood, then I will go to see a movie.
   b. If $2 + 2 = 5$ then fish can fly.
   c. If $1 + 1 = 3$ then there is snowfall in Delhi.
   d. If I am not in humour then I will not study discrete structures.
   e. I go to market only if I do not study discrete structures.
   f. Working hard is sufficient for me to be successful.

8. State the contrapositive of each implication in exercise 6.

9 Let P: $4 > 1$; Q: $4 < 5$; R: $3 \leq 3$; S: $2 > 2$

   Write English sentences corresponding to the following statements

   a. $(P \wedge S) \to Q$
   b. $(Q \wedge P) \vee (Q \wedge \rceil P)$
   c. $(Q \wedge P) \to P$
   d. $\rceil R \to P$
   e. $(P \to S) \wedge (S \to T)$
   f. $(R \wedge S \wedge P) \to (P \vee Q)$.

Answers

5. a. It is not the matter that if Amit fails in Discrete Mathematics then he is sad.

b. It is not a matter that if Amit fails in Discrete Mathematics then there is no case he is sad and not get the job.

c. Amit has not got the job and fails in Discrete Mathematics iff he is sad.

7.7 Duality Law

We saw that two propositions or formula may be equivalent and can be replaced by another without loss of originality. Similarly, we can obtain new formula in set theory by replacing the $\cap$ by $\cup$ and $\cup$ by $\cap$ and the originality of formula is maintained. In logic, connectives $\wedge$ and $\vee$ and variables **t** and **f** are dual to each other. On replacing $\wedge$ by $\vee$ and $\vee$ by $\wedge$ in a formula we obtain the **dual** of given formula. Similarly, if the formula contains the variable **t** and **f** then replace **t** by **f** and **f** by **t** respectively to obtain the dual.

e.g. Dual of $P \wedge R \vee Q$ is $P \vee R \wedge Q$.

Example 7.43

Write the duals of

a. $\neg(P \vee Q) \wedge (P \vee \neg Q \wedge \neg R)$

b. $\neg P \wedge (\neg Q \wedge R) \vee (Q \wedge R) \vee (P \wedge R)$.

Solution

Duals are

a. $\neg(P \wedge Q) \vee (P \wedge \neg Q \vee \neg R)$

b. $\neg(P \vee (\neg Q \vee R) \wedge (Q \vee R) \wedge (P \vee R)$.

7.8 Well Formed formula

Mathematical expressions are represented in well-defined form using parenthesis, braces and square brackets to avoid the ambiguity. For solving the expressions we use the precedence of operators and follow the rules of BODMAS recursively. Similarly, in logic a statement, which is the collection of variables or statements, represented in well-defined form using parentheses according to priority of operation. These statements are known as **well-formed formula**. We can generate well-formed formula recursively using following rules:

1. An atomic-statement P is a well formed formula.
2. If P is a well-formed formula then $\neg P$ is also a well formed formula.
3. If P and Q are well formed formula then $(P \wedge Q)$, $(P \vee Q)$, $(P \rightarrow Q)$ and $(P \leftrightarrow Q)$ are also well formed formula.
4. A statement consist of variables, parentheses and connectives is recursively a well formed formula iff it can be obtained by applying the above three rules.
5. Nothing else is a well formed formula.

According to rule 4, following are also well formed formula:

$\neg(P \vee Q)$, $\neg(P \wedge Q)$, $(P \rightarrow (P \wedge Q))$, $(P \rightarrow (P \vee Q))$, $((P \rightarrow Q) \leftrightarrow R)$ *etc.*

But, now a question comes into mind, when a formula is not well formed? The answer is

1. *Any of the parenthesis is missing (either in beginning or at the end of the statement).*
2. *Binary operator* $(\wedge, \vee, \rightarrow, \leftrightarrow)$ *contains single operand in place of double.*

e.g. Following are not well formed formula:

$(P \wedge Q) \rightarrow (\wedge P)$, $(\neg P \wedge Q$,

$(P \vee Q) \rightarrow (Q \rightarrow R) \rightarrow (R)$ *etc.*

7.9 Some more Connectives (Exclusive OR, NAND, NOR)

As discussed earlier that the word 'or' in linguistic is used in exclusive sense of oring. Logically in many situations we have to use the word 'or', exclusively in place of inclusive 'or'. The "**exclusive OR**" is denoted by symbol '$\oplus$' and defined as "if P and Q are two statements then exclusive oring of P and Q is true when either P or Q is true otherwise it is false". Table 7.24 shows the truth values of exclusive OR.

Table 7.24 Truth table for ex-OR

P	Q	$P \oplus Q$
0	0	0
0	1	1
1	0	1
1	1	0

Properties of Exclusive OR

1. $P \oplus Q \equiv Q \oplus P$ (Symmetric)
2. $(P \oplus Q) \oplus R \equiv P \oplus (Q \oplus R)$ (Associative)
3. $P \wedge (Q \oplus R) \equiv (P \wedge Q) \oplus (P \wedge R)$ (Distributive)

Above statements can be proved easily either using truth table or by principle of substitution. Readers are advised to try themselves.

Example 7.44

Prove that $P \oplus Q \equiv (P \wedge \urcorner Q) \vee (\urcorner P \wedge Q)$.

Solution

To prove this let us make a truth table with the sequences shown below:

P	Q	$P \oplus Q$	$\urcorner P$	$\urcorner Q$	$X: P \wedge \urcorner Q$	$Y: \urcorner P \wedge Q$	$X \vee Y$
0	0	0	1	1	0	0	0
0	1	1	1	0	0	1	1
1	0	1	0	1	1	0	1
1	1	0	0	0	0	0	0
Step		1	2	3	4	5	6

Column number 1 and 6 shows the similar truth values for same combinations of input values, hence the statement.

Example 7.45

Prove that $P \oplus Q \equiv \urcorner (P \leftrightarrow Q)$.

Solution

P	Q	$X: P \oplus Q$	$P \leftrightarrow Q$	$Y: \urcorner (P \leftrightarrow Q)$	$X \leftrightarrow Y$
0	0	0	1	0	1
0	1	1	0	1	1
1	0	1	0	1	1
1	1	0	1	0	1
Step		1	2	3	4

Column number 4 shows the tautology which signifies the statement is true.

7.9-1 NAND and NOR

Example 7.44 shows that exclusive OR can be represented using connectives $\urcorner$, $\vee$, and $\wedge$ only.

Some times in real life we have to take decision true when either or both alternatives have truth values 'false'. This can happen due to some type of psychological pressure. If we think logically, this situation is useful in the design of computers (specially in case of architecture and robotics).

The connectives used for these purposes are known as NAND and NOR. "NAND" is nothing but a combination of "NOT" and "AND", i.e. "negation of conjunction".

Let us assume that if P and Q be two statements then NAN Ding of P and Q is denoted by $P \uparrow Q$ which is equivalent to $\urcorner(P \wedge Q)$. The truth table of $P \uparrow Q$ is shown in Table 7.25.

Table 7.25: Truth table for NAND

P	Q	$P \uparrow Q$
0	0	1
0	1	1
1	0	1
1	1	0

Again, "NOR" is a combination of "NOT" and "OR" i.e. negation of disjunction". NORing of the statements P and Q is denoted by $P \downarrow Q$ and equivalent to $\urcorner(P \vee Q)$. Table 7.26 shows the truth values of NOR.

Table 7.26: Truth table for NOR

P	Q	$P \downarrow Q$
0	0	1
0	1	0
1	0	0
1	1	0

1. *Similar to $\wedge$ and $\vee$ connectives $\uparrow$ and $\downarrow$ are dual to each other.*
2. *Connectives $\uparrow$ and $\downarrow$ can be expressed in terms of $\{\wedge, \urcorner\}$ and $\{\vee, \urcorner\}$ respectively.*

7.9-2 Properties of NAND and NOR

1. Commutatively
   a. $P \uparrow Q \equiv Q \uparrow P$ b. $P \downarrow Q \equiv Q \downarrow P$
2. a. $P \uparrow P = \urcorner P$ b. $P \downarrow P = \urcorner P$

Readers are advised to prove this.

Example 7.46

Prove or disprove that $P \uparrow (Q \uparrow R) \equiv (P \uparrow Q) \uparrow R$.

Solution

From L.H.S.

$$P \uparrow (Q \uparrow R) \equiv P \uparrow \neg(Q \wedge R) \quad \text{(According to definition of NAND)}$$
$$\equiv \neg(P \wedge \neg(Q \wedge R) \quad \text{(According to definition of NAND)}$$
$$\equiv \neg P \vee \neg\neg(Q \wedge R) \quad \text{(De Morgan's Law)}$$
$$\equiv \neg P \vee (Q \wedge R) \quad (\neg\neg P \equiv P) \ldots (1)$$

R.H.S.:

$$(P \uparrow Q) \uparrow R \equiv (\neg(P \wedge Q)) \uparrow R \quad \text{(According to definition)}$$
$$\equiv \neg((\neg(P \wedge Q)) \wedge R) \quad \text{(According to definition)}$$
$$\equiv \neg\neg(P \wedge Q) \vee \neg R) \quad \text{(De Morgans' Law)}$$
$$\equiv (P \wedge Q) \vee \neg R \quad \ldots (2)$$

and (2) are not identical. Hence the given statement is not satisfied.

Example 7.46 shows that connective NAND is not associative. Similarly, due to the property of duality connective NOR is also not associative.

Example 7.47

Express $P \to (\neg P \to Q)$ in terms of $\uparrow$ only.

Solution

$$P \to (\neg P \to Q) \equiv P \to (\neg(\neg P) \vee Q) \quad \text{(Conditional Rule)}$$
$$\equiv P \to (P \vee Q) \quad (\because \neg\neg P \equiv P)$$
$$\equiv P \to (\neg(\neg P \wedge \neg Q)) \quad \text{(De Morgan's Law)}$$
$$\equiv P \to (\neg P \uparrow \neg Q) \quad \text{(By definition of NAND)}$$
$$\equiv (\neg P) \vee ((\neg P) \uparrow \neg Q)) \quad \text{(Conditional Rule)}$$
$$\equiv \neg((\neg\neg P) \wedge \neg(\neg P \uparrow \neg Q)) \quad \text{(De Morgan's Law)}$$
$$\equiv \neg(P \uparrow \neg(\neg P \uparrow \neg Q)) \quad \text{(By definition of NAND)}$$

$$P \to (\neg P \to Q) \equiv P \uparrow \neg(\neg P \uparrow \neg Q).$$

Example 7.48

The peirce arrow $\downarrow$ (*NOR*) is a logical binary operation which is defined as follows:

$$P \downarrow Q = (\neg P \vee Q).$$

i. Prove that $P \wedge Q = (P \downarrow P) \downarrow (Q \downarrow Q)$.

ii. Write $P \to Q$ using pierce arrows only.

Solution

i. R.H.S.

$$(P \downarrow P) \downarrow (Q \downarrow Q) = (\neg(P \vee P) \downarrow (\neg(Q \vee Q))$$
$$= (\neg P) \downarrow (\neg Q)$$
$$= \neg((\neg P) \vee (\neg Q))$$

$$= \neg(\neg P) \wedge \neg(Q))$$

$$= P \wedge Q$$

$$= \text{LHS.}$$

ii. $P \to Q = (\neg P) \vee Q = \neg(P \wedge (\neg Q)) = P \downarrow (\neg Q)$.

7.10 Functionally Complete Sets of Connectives

We have already discussed that all the statements or formula has an equivalent formula which can be replaced by it. We have also seen that the connectives $\uparrow$, $\downarrow$, $\oplus$ can be expressed in an equivalent form using connectives $\wedge, \vee, \neg, \to$ or $\leftrightarrow$. Similarly, connective $\to$ and $\leftrightarrow$ can also be expressed using only $\wedge$, $\vee$, and $\neg$. This implies a set of connectives can be used to express the formula in equivalent form. The set of connectives in which every formula can be expressed in terms of an equivalent formula containing the connectives from this set is known as **functionally complete set**. We have seen previously that

$$P \to Q \equiv \neg P \vee Q \qquad \ldots (1)$$

$$P \leftrightarrow Q \equiv (P \to Q) \wedge (Q \to P) \qquad \ldots (2)$$

From (1) and (2) one can conclude that the conditional operator can be replaced by $\neg$ and $\vee$ while biconditional by $\to$ and $\wedge$.

1. *The functionally complete set for* $\uparrow$ *is* $\{\neg, \wedge\}$.
2. *The functionally complete set for* $\downarrow$ *is* $\{\neg, \vee\}$.
3. *Singleton set* $\{\neg\}$, $\{\vee\}$ *or* $\{\wedge\}$ *are not functionally complete set.*

Example 7.49

Write an equivalent formula for $((P \vee Q) \wedge R) \to (P \vee R)$ which is free from conditional operator $\wedge$.

Solution

$$((P \vee Q) \wedge R) \to (P \vee R) \equiv \neg((P \vee Q) \wedge R) \vee (P \vee R) \qquad \text{(Conditional Rule)}$$

$$\equiv (\neg(P \vee Q) \vee \neg R) \vee (P \vee R). \qquad \text{(De Morgan's Law)}$$

Example 7.50

Express $(P \wedge \neg Q) \vee (\neg P \vee \neg Q)$ in terms of $\{\neg, \wedge\}$ only.

Solution

$$(P \wedge \neg Q) \vee (\neg P \wedge \neg Q) \equiv \neg\neg((P \wedge \neg Q) \vee (\neg P \wedge \neg Q)) \qquad (\because \neg\neg P \equiv P)$$

$$\equiv \neg(\neg(P \wedge \neg Q) \wedge (\neg P \wedge \neg Q)). \qquad \text{(De Morgan's Law)}$$

Example 7.51

Show that $\{\neg, \to\}$ is functionally complete set.

Solution

We can prove it by assuming any formula.

We know that

$$P \to Q \equiv \neg P \vee Q,$$

We have

$$P \vee Q \equiv \rceil P \vee Q$$
$$\equiv P \rightarrow Q.$$

Hence $\{\rceil, \rightarrow\}$ is a functionally complete set as $\{\rceil, \vee\}$.

Exercise 7.7

1. Prove that
   a. $(P \uparrow Q) \oplus (P \uparrow Q) \equiv (P \vee Q) \wedge (P \downarrow Q)$.
   b. $(P \oplus Q) \equiv (P \wedge \sim Q) \vee (\sim P \wedge Q)$.
2. Form truth table for following
   a. $P \oplus Q \oplus R$ b. $P \uparrow Q \uparrow R$.
3. Establish the equivalence and write its dual equivalence.
   a. $(P \wedge Q) \vee (\rceil P \vee (\rceil P \vee Q)) \equiv (\rceil P \vee Q)$ b. $\rceil P \wedge (\rceil Q \wedge R) \vee (Q \wedge R) \vee (P \wedge R) \equiv R$
   c. $P \leftrightarrow Q \equiv \rceil (P \vee Q) \vee (P \wedge Q)$ d. $P \rightarrow (Q \vee R) \equiv (P \rightarrow Q) \vee (P \rightarrow R)$
   e. $\rceil (P \leftrightarrow Q) \equiv \rceil (P \vee \rceil Q) \vee (\rceil P \wedge Q)$ e. $(P \rightarrow Q) \rightarrow (P \wedge Q) \equiv (\rceil P \rightarrow Q) \wedge (Q \rightarrow P)$
   g. $P \rightarrow (Q \rightarrow R) \equiv (P \wedge Q) \rightarrow R$.
4. Express the given formulas in terms of $\{\wedge, \rceil\}$, $\{\rceil, \vee\}$ and $\{\rceil, \vee, \wedge\}$.
   a. $\rceil (P \leftrightarrow (Q \rightarrow (R \vee P)))$ b. $P \leftrightarrow Q$
   c. $P \uparrow Q$.
5. Show that $\{\vee, \wedge\}$ is not functionally complete.

7.11 Predicate Calculus

We have seen earlier in this chapter that the propositional logic deals with complete statements, (though it cannot examine the internal structure of a statement). Since the statements are also known as propositions hence the "Statement calculus" is also called "**Propositional Calculus.**" Further in propositional calculus two atomic statements are not compared for their common features. For example,

Rita goes to college.

Anita goes to college.

These two statements have same predicate* "goes to college" but not for same person. Statement calculus express these statements with two different symbols which does not show the same property of statements.

Now, if we recall our mathematical knowledge of functions i.e. relationship of two or more variables such that each value of one variable is related to at least one value of second variable. For example

$$y = f(x) \text{ where } f(x) = 2x + 3.$$

In this function we can evaluate the value of y for different values of x. But one thing is common for all values of x that it is multiplied by 2 and added in 3.

Hence, from a mathematical point of view the statement can be represented in the form of predicates (verb + object) and subject. As in functions one variable depends upon another variable and also be represented by assumed function name, common statements can also be represented by assumed predicate name on which

* **Predicate** (in linguistic) is the part of a sentence or clause containing a verb and stating something about the subject.

subject depends. This type of logic is known as **predicate** and represented by **propositional function** or **predicate logic**. In this section, we discuss predicate logic and predicate calculus.

Readers may think about the applications area of predicate logic. If one analyses the psychology of human beings, then he will reach on conclusion that human beings take most of the decision on the basis of his previous experiences. Similarly, Robots work on some predefined algorithm's where programs can take new values and give decision (output) according to previous fed programs. These algorithms/programs are written in the languages using predicate logic. Predicate logic is widely applicable in robotics engineering, artificial intelligence, neural networks etc. The important role of predicate logic (PL) is in building intelligent systems to perform complex robotics task and develop a knowledge base system.

7.11-1 Predicate

Predicate logic is used mostly in computer related fields. For example, propositions logic are used in computer architecture and electronic parts of computers because output is obtained in terms of truth values (i.e. either 'true' or 'false'). But most of the statements in mathematics and computer Science are neither true nor false. For example $y < 5$, $x + 7 = 9$. $x^3 - y^3 = (x - y)(x^2 + y^2 - xy)$ etc. The truth values of these statements depend upon the values of variable x and y. Before learning how to convert these statements into proposition we must learn how to make predicates. Now consider the example

1. Sunita goes to college.
2. Rita goes to college.

The predicate in linguistic "goes to college" is known as predicate in logic also. Symbolically, it can be represented by any capital letter. For our learning convenience we will prefer a letter (alphabet) near to the meaning of predicate. In the predicate goes to college "college" is main word (or object in linguistic) then we may take "*C*" as a predicate symbol. Statement 1, i.e. "Sunita goes to college", "Sunita", is the object (Subject in linguistic) about which an action or feeling is directed, denoted by small alphabetical letter such as "*s*". Then the statement 1 can be written as $C(s)$ and 2 as $C(r)$. '*s*' and '*r*' in this representation are objects and '*C*' is the predicate symbol.

Example 7.52

Represent each of the following in symbolic form.

a. Anand is a Computer Scientist.
b. Anita is a bachelor.
c. Karan is a president.
d. X is an odd number.
e. Copper conducts electricity.

Solution

Predicates and objects in each part can be represented as

C : is a computer Scientist

B : is a bachelor

P : is a president

O : is an odd number

E : conducts electricity

a : Anand

an : Anita

ab : Karan

x : X

c : copper.

Then the statements (a) to (d) can be written in symbolic form as:

a. $C(a)$ b. $B(an)$

c. $P(ab)$ d. $O(x)$

e. $E(c)$.

We saw that predicates claim about the objects, and have at least one name of an object which shows the relationship of object and predicate. If predicate symbol contains only one object or argument then it is known as **'first order predicate logic'** or **'one place predicate'** and denoted by P^1. 1 at the superscript of P shows the **arity** or order of predicate. That means 'n' number of small alphabet attached the predicate symbol, where $n > 0$, represent the relationship between predicates and one or more that one objects, known as ***n*-place predicate** where n denotes the order or arity of predicate and written in superscript of predicate symbol, i.e. P^n.

All the names, which depend upon predicate, are enclosed in parentheses and separated by commas. For example

Sangita and Rita went to college with Anita and Beena.

The predicate 'went to college' and objects Sangita, Rita, Anita and Beena can be denoted by symbol C, s, r, a and b respectively then the statement symbolized as $C(s, r, a, b)$. This predicate is of 4-place predicate. It can also be represented as $C^4(s, r, a, b)$.

Here we can conclude that predicate is a sentence that contains a finite number of variables and becomes a proposition when specific values are substituted for the variable.

$n = 0$ i.e. 0-place predicate is nothing but a statement, has no letter attached with it.

Example 7.53

Represent each of the following in symbolic form.

a. Lucknow is the capital of Uttar Pradesh.

b. Ramesh is younger than Saksham.

c. Sunil, Sudhir and Suresh are scientist.

Solution

All the parts involves following predicates and objects:

C : is the capital
Y : is younger
S : are scientist
l : Lucknow
up : Uttar Pradesh
r : Ramesh
s : Saksham
s_1 : Sunil
s_2 : Sudhir
s_3 : Suresh

The statements (a) to (d) can be represented in symbolic form as

a. C (l, up)

b. Y (r, s)

c. S (s_1, s_2, s_3)

REMARK

1. (a) and (b) are the examples of 2-place predicate whereas (c) is of 3-place predicate.
2. Subscript attached to argument names are used to distinguish the names represented by same letter. Similarly we may use subscript number in predicate symbols to distinguish two predicates names represented by same letter i.e. P_1, P_2 etc.

In this situation complete representation of predicate name is P_i^n where n shows the **order** or **arity** and i distinguish the predicates of same arity.

7.12 Functions, Variables and Quantifiers

We saw objects used in predicate names has constant value, e.g.

Saksham is brilliant

B represents the predicate "is brilliant" and *r* for "Saksham" then predicate represented in symbolized form by *B*(*r*). Similarly, another objects such as

Gita is brilliant,

Ramesh is brilliant,

Amit is brilliant etc.

may be symbolized by B (s), B (r_1), B (a) etc. We can write B (x) for "x is brilliant". And the value of "x" is replaced by 'r', 's', 'r_1' and 'a' accordingly. 'x' is not a constant, its values can be changed according to the requirement of statement, hence 'x' is a **variable** also called **predicate variable** and B (x) is a **simple statement function** or **propositional function** of one variable. The proposition function will have truth value either 'true or 'false'. One can observe here that when can create values are substituted in place of predicate variable 'x' we get a statement.

Similarly the statement function of two or more variables may be expressed. See the following examples.

Example 7.54

Represent the following in symbolic form:

a. x is greater than 5

b. x, y, z are odd numbers

c. x is greater than y.

Solution

Predicates and objects in all parts may be represented as

G: is greater than

O: are odd number

Symbolic forms of statement (a) to (c) are given by

a. G (x, 5)   b. O (x, y, z)

c. G (x, y).

Example 7.55

Express the following in symbolized form

i. Everybody loves God.

ii. For all x, x is greater than 5.

iii. Every natural number is either even or odd.

iv. Every student of Computer Science and Engineering read mathematics.

Solution

i. Predicates and objects are

L: Loves

g: God

x: Everybody

Symbolic form is

$$(\forall x)\, L(x, g)$$

ii. Predicates:

G: is greater than

Symbolic form:

$$(\forall x)\, G(x, 5)$$

iii. Objects and Predicates

x : number

$E(x)$: x is an Even

$O(x)$: x is an odd

$N(x)$: x is natural number.

Symbolic form:

$$(\forall x)\, N(x) \rightarrow E(x) \vee O(x).$$

iv. Objects are predicates

x : student

$C(x)$: x is a student of Computer Science and Engineering Student.

$M(x)$: x reads mathematics

Symbolic form:

$$(\forall x)\, C(x) \rightarrow M(x).$$

7.13 Quantifiers

All the above discussed statements are simple statement functions. Now we discuss compound statement functions. Compound statement functions are nothing but the combination of simple statement functions using connectives ($\wedge$, $\vee$, $\rightarrow$, $\leftrightarrow$, $\neg$). For example:

1. x is an integer and odd. If $O(x)$ be "x is odd" and $I(x)$ be "x is an integer", then the statement can be written in symbolic form as

$$I(x) \wedge O(x)$$

2. We may take another example

"Persons are mortal"

M stands for "are mortal" and P for "persons". We may symbolize it as

$$P(x) \rightarrow M(x)$$

i.e. If x is a person then he is mortal. But it is incomplete. It tells only about one x not for all persons."

3. Similarly equations of mathematics such as

$$(x^3 - 1) = (x - 1)((x^2 - x + 1)$$
$$(x^2 - y^2) = (x - y)(x + y)$$
$$x + 3 = 7 \text{ etc.}$$

The variable x and y are replaced by numbers, either real, integer, complex or rational etc. Similarly, we can take one more example:

4. Theorem: for every triangle, the sum of the angles is equal to 180^0.
5. Let us consider the following arguments:

All mathematicians are intelligent.
Anyone who is intelligent is an algebraist.
Therefore, all mathematicians are algebraist.

Thus if I stands for the predicate "are intelligent"

M_2 stands for the predicate "mathematicians"
A stands for the predicate "is an algebraist"
x stands for the "Anyone".

In symbolize form, the above argument may be written as

$$\forall x\ (M_2(x) \rightarrow I(x)) \rightarrow \text{A}(x)$$

Again consider the following propositions:

1. All rings are circle.
2. Some computers are Pentium -1.
3. Any one who is intelligent is Scientist.
4. For every integer x, x^2 is a positive integer.

Again consider example (2) which can be translated as

" There are some x such that x is a computer and x is pentium – 1". We symbolize it as $C(x) \wedge P(x)$,

where $C(x) \quad \equiv x$ is a computer
$P(x) \quad = x$ is pentium -1

And translated as

"There are some x, $[C(x) \wedge P(x)]$"

Similarly consider example (5) which can be translated as

For every x, $I(x) \rightarrow I^+(x)$

Where $I(x) \equiv x$ is an integer
$I^+(x) \equiv x$ is a positive integer

In the above propositions the words "for every", "all", "some", "anyone" are associated with the idea of quantity. The phrases like "there are some x", "for every x" etc. which specifies the extent to which predicate holds over a range of things are called **quantifiers**.

1. Universal Quantifier
2. Existential Quantifiers

In other words we can say: In logic quantification is a construct that specifies the extent to which a predicate holds over a range of things. A language element which generates quantification is called quantifiers.

7.13-1 Universal Quantifier

The phrases "for every or "for all" is called **universal quantifier**. In other words in predicate logic, universal quantification is an attempt to formalize the notion that something (a logical predicate) is true for everything (or every relevant thing). Universal quantifier is symbolized by '$\forall$' and expressed in any of the following way.

$$\forall x \in A, P(x) \text{ or } P(x), \forall x \in A$$

or

$$\forall x, P(x) \text{ or } P(x), \text{ is true for all } x \in A.$$

7.13-2 Existential Quantifiers

The phrase "there exists" or "for some" or "for at least one" is called an **existential quantifier**. In other words in predicate logic, an existential quantification is the property or relation to at least one member of domain. Existential quantifier is symbolized by '$\exists$' and read as "there exist", "for some", "at least one". It is expressed by any of the following way:

$$(\exists x \in A)\, P(x) \text{ or } \exists x \in A \text{ such that } P(x)$$

or

$$(\exists x)\, P(x) \text{ or } P(x) \text{ is true for some } x \in A.$$

7.14 Range of quantification

Every quantification involves one specific variable and a **domain of discourse** or **universe of discourse** or **universe** or **range of quantification** of that variable. The range of quantification specifies the set of values that variable takes. In the example 4, above the range of quantification is the set of integer.

Now in terms of universe U, universal quantifier is written as:

If $P(x)$ is the proposition over the universe U by $\forall x \in U, P(x)$ and read as "For every $x \in U$, $P(x)$ is true and existential quantifiers as: "If $P(x)$ is proposition over universe U, denoted by $(\exists x \in U)\, P(x)$ and read as 'There exist at least one value in the universe of variable x such that $P(x)$ is true".

Example 7.56

Express the following in symbolic form using quantifiers.

i. Every student of computer science reads Discrete structure.
ii. All judges are lawyers.
iii. All beds have four legs.

Solution

i. Objects are predicates

x : student
$C(x)$: x is a student of Computer science.
$D(x)$: x reads discrete structure.
Symbolic form

$$(\forall x)\, C(x) \rightarrow D(x).$$

ii. $J(x)$: x is a judges
$L(x)$: x is a lawyer.

Symbolic form:

$$(\forall x)\, J(x) \rightarrow L(x).$$

iii. $B(x)$: x is a bed
$L(x)$: x is a leg.
Symbolic form:

$$(\forall x)\, B(x) \rightarrow L(x).$$

7.15 Negation of Quantified propositions

Negation of quantified statement is done in the following way:

i. When a unified quantified proposition is negated we obtain an existentially quantified proposition.

ii. When a existentially quantified proposition is negated we obtain an universally quantified proposition.

a. $\rceil \{\exists x\, P(x)\} \equiv \forall x, \{\rceil P(x)\}$

b. $\rceil \{\forall x\, P(x)\} \equiv \exists x \{\rceil P(x)\}$.

(a) and (b) are also called **De Morgan's law**.

Here, it is to be noted that to obtain the negation of a statement which contains universal ($\forall$) and existential ($\exists$) quantifiers, interchange the $\forall$ by $\exists$ and $\exists$ by $\forall$ and apply the De Morgan's laws on the statements.

Example 7.57

i. Negation of $\exists x\, (x^2 < 0)$ is $\forall x \rceil (x^2 < 0)$ or $\forall x\, (x^2 < 0)$

ii. Negation of is $\forall x\, (x + 1 < 0)$

$$\rceil x \rceil (x + 1 < x) \text{ or } \exists x\, (x + 1 \geq x).$$

7.16 Propositions with Multiple Quantifiers

If we have more than one variable in a proposition then it is quantified with multiple quantifiers. The multiple universal quantifiers as well as multiple existential quantifier can be arranged in any order without changing the meaning of the resulting proposition. But the proposition which contains both universal and existential quantifiers cannot be exchanged without changing the meaning of the proposition.

For example, the proposition $\exists x\, \forall y\, P(x, y)$ translates "there exists some x such that $P(x, y)$ is true for every y".

Example 7.58

Express the following in symbolic form and write down its negation.

i. Some integers are even and some are odd.

ii. Some apples are sweet.

iii. For at least one x, $x < 1$.

iv. Some students of chemical engineering are good in mathematics.

Solution

Writing predicates and objects as

$I(x)$: x is an integer

$O(x)$: x is an odd

$E(x)$: x is an even]

$A(x)$: x is an apple

$S(x)$: x is sweet

$L(x, a)$: x is less than a

$C(x)$: x is a student of chemical Engineering.

$M(x)$: x is good in mathematics

And using the existential quantifier, we can write the statements (i) to (iv) as

i. $(\exists x) I(x) \rightarrow E(x) \wedge O(x))$

ii. $(\exists x) A(x) \wedge S(x)$

iii. $(\exists x) L(x, 1)$

iv. $(\exists x) C(x) \wedge M(x)$.

Negation of the statements is:

i. $\sim(\exists x)(I(x)) \rightarrow E(x) \wedge O(x)) \equiv (\forall x) \sim (I(x) \wedge \sim (E(x) \wedge O(x)))$

$\equiv (\forall x) \sim I(x) \vee \sim\sim (E(x) \wedge O(x)))$

$\equiv (\forall x) \sim I(x) \vee (E(x) \wedge O(x)))$

ii. $\sim(\exists x) A(x) \wedge S(x) \equiv (\forall x) \sim A(x) \vee \sim S(x)$

iii. $\sim(\exists x) L(x, 1) \equiv (\forall x) \sim L(x, 1)$

iv. $\sim(\exists x) C(x) \rightarrow M(x) \equiv (\forall x) \sim (C(x) \wedge \sim M(x))$

$\equiv (\forall x) \sim C(x) \vee M(x)$.

Example 7.59

Negate each of the statements in Example 7.55.

i. $\sim(\forall x) L(x, g) \equiv (\exists x) \sim L(x, g)$

ii. $\sim(\forall x) G(x, 5) \equiv (\exists x) \sim G(x, 5)$

iii. $\sim(\forall x) N(x) \rightarrow E(x) \vee O(x) \equiv (\exists x) \sim (N(x) \wedge \sim (E(x) \vee O(x)))$

$\equiv (\exists x) \sim N(x) \vee (E(x) \vee O(x)))$

iv. $\sim(\forall x) C(x) \rightarrow M(x) \equiv (\exists x) \sim (C(x) \wedge \sim M(x))$

$\equiv (\exists x) \sim C(x) \vee M(x)$.

Example 7.60

Negate the each of the following propositions:

i. $\exists x P(x) \vee \exists y q(y)$

ii. $\forall x P(x) \wedge \forall y q(y)$

iii. $\forall x P(x) \wedge \exists y q(y)$

iv. $\exists x P(x) \vee \forall y q(y)$

v. $\forall x \exists y (P(x, y) \rightarrow q(x, y))$.

Solution

i. $\urcorner(\exists x P(x) \vee \exists y q(y))$

$\equiv \urcorner \exists x P(x) \wedge \urcorner \exists y q(y)$ $[\because \urcorner(p \vee q) = \urcorner(p \wedge \urcorner q)]$

$\forall x \urcorner P(x) \wedge \forall y \urcorner q(y)$

ii. $\urcorner(\forall x P(x) \wedge \forall y q(y))$

$= \urcorner(\forall x P(x) \vee \urcorner \forall y q(y)$ $[\because \urcorner(p \wedge q) = \urcorner(p \wedge \urcorner q]$

$\equiv \exists x \urcorner P(x) \vee \exists y \urcorner q(y)$

iii. $\rceil(\forall x\, P(x) \wedge \exists y\, q(y))$

$\equiv \rceil(\forall x\, P(x) \vee \rceil\exists y\, q(y)$

$= \exists x\, \rceil P(x) \vee \forall y \rceil q(y)$

iv. $\rceil(\exists x\, P(x) \vee \forall y\, q(y))$

$\equiv \rceil\exists x\, P(x) \wedge \rceil\forall y\, q(y)$

$\equiv \forall x \rceil P(x) \wedge \exists y \rceil q(y)$

v. $\rceil(\forall x\, \exists y\, (P(x, y) \rightarrow q(x, y))$

$\equiv \rceil\exists x \forall y \rceil (P(x, y) \rightarrow q(x, y)$.

REMARK

All the examples and statements discussed above have no truth values, as these statements contain variable x and value of x is not substituted in either of quantifiers. This means the second problem discussed above is as it is.

7.17 Free and Bound Variables

We have already define the predicate in the beginning of this topic but the same can also be defined in terms of universe of discourse as "If $P(x)$ be a propositional function involving the variable x and consider a set U, then for every values of x in U. $P(x)$ is a proposition where U is the universe of discourse".

The variables involved in propositional function may be **free** or **bound**. The quantified variables, $\forall$ or $\exists$, are bound by values of universe of discourse on which that function is defined i.e. in all the strings or sentences contains $(\forall x)$ or $(\exists x)$, x is a **bounded variable** otherwise x is a **free variable**. For example:

1. $(\forall x)\, P(x)$
2. $(\forall x)\, P(x, y)$
3. $(\exists x)\, P(x) \wedge Q(y)$
4. $(\forall x)\, (P(x) \rightarrow (\exists y)\, R(x, y))$.

In all the above examples all the occurrences of x are bound where as in (2) and (3) occurrences of y are free but in (4) y is also bound.

1. *Variables must be bound in statements and predicates while they may be free in statement function.*
2. *The specific symbol used to denote the occurrence of bound variable is not importani. Any letter can be changed throughout the string without affecting the string, provided that new letter is not used previously in the string. i.e.*

$$\forall x\, P(x, y) \equiv \forall z\, P(z, y).$$

3. *A constant value can be substituted in place of free variable, i.e. $(\forall x)\, P(x, a)$ is a substitution instance of $(\forall x)\, P(x, y)$.*

Example 7.61

Let $P(x)$ be the statement x is greater than 5. Find the truth value of $\exists x\, P(x)$ where the domain of x is the set of integers.

Solution

The truth value of $\exists x\, P(x)$ is true for $x \geq 6$ and false for $x \leq 5$.

Example 7.62

Find the universe of discourse of the statement x is less than 3 when x is a natural number.

Solution

x is less than 3. If domain of x is {1, 2} when $x \in N$, then sentence is true and for domain {4, 5 ...} the sentence returns false.

Example 7.63

Translate the following sentence in symbolic form and write the universe of discourse for truth values of sentence.

a. For all n belong to set of natural numbers, s.t. $n + 1 > 8$.

b. Every natural number is either even or odd.

c. Some researchers from artificial intelligence and psychology form a separate discipline of cognitive science.

d. All real numbers hold the property of associativity.

e. Some rational numbers are not integers.

Solution

a. Let N : Set of natural numbers

then $\forall x \in N : n + 1 > 8$ and domain is { 8, 9, 10, ...}.

b. Let N : Set of natural numbers

$O(x)$: x is odd

$E(x)$: x is even

In symbolic form

$$\forall x \in N: O(x) \vee E(x)$$

or

$$(\forall x) N(x) \rightarrow O(x) \vee E(x)$$

Domain Set: N.

c. Let $A(x)$: x belongs to AI streams

$P(x)$: x belongs to Psychology streams

$C(x)$: x belongs to cognitive science stream

In symbolic form

$$(\exists x) A(x) \wedge P(x) \rightarrow C(x)$$

Domain set: {researchers of AI and Psychology}.

d. Let R : Set of Real number

$A(x)$: x hold the property of Associativity.

In symbolic form

$$\forall x \in R, A(x) \text{ or } (\forall x) R(x) \rightarrow A(x)$$

where $R(x)$ denotes that x is a real number.

e. Let $Q(x)$: x is a rational number

$I(x)$: x is an integer

In symbolic form

$$(\forall x) Q(x) \rightarrow (\exists x) \sim I(x)$$

Domain set: Q (set of rational numbers).

Example 7.64

i. Write the negation for the statement:

$$\forall x \in R, x > 3 \Rightarrow x^2 > 9.$$

Solution

Let $P(x)$ and $Q(x)$ denote "$x > 3$" and $x^2 > 9$. Then the given statement is

$$\forall x (P(x) \Rightarrow Q(x)) \text{ equivalent to } \forall x (\neg P(x) \vee Q(x))$$

The negation of the statement is

$$\exists x (P(x) \wedge \neg Q(x))$$

There exists a real number x such that $x > 3$ and $x^2 \leq 9$.

ii. Write the negation for the statement:

$$\forall \text{ sets } A, \text{ if } A \subseteq R \text{ then } A \subseteq Z.$$

State which one is true - the negation or statement? Explain your reasoning, where R is the set of real numbers and Z is the set of integers.

Solution

Try yourself.

Exercise 7.8

1. Represent each of the following in symbolic form, assuming a suitable domain of interpretation in each case:
   i. Allahabad is a holy city.
   ii. Ramanujan was a great Mathematician.
   iii. U.P. Tech. Univ. is situated at Lucknow.
   iv. Kanpur is an industrial city.
   v. Mahatma Gandhi was assassinated.
   vi. Jupiter is the biggest planet of the universe.
   vii. Moon is the satellite of Earth.
   viii. If one is poor in mathematics, he cannot be a good engineer.
   ix. Work hard or face the consequences.
   x. If air conditioner is on then environment of room is cool.
2. Represent each of the following in symbolic form:
   i. Anyone who does not study or submit assignment in time or attend tutorials will loose marks.
   ii. If it is raining then there is a traffic jam.
   iii. Sum of any two natural numbers is again a natural number.
   iv. All students score very less marks in tough question paper.
   v. Neither Rosen nor Rahim achieved their goal.
   vi. He is intelligent and smart.
   vii. If one likes discrete mathematics then he will be expert in computing.
   viii. Every natural number is either even or odd.

ix. Anyone who learns mathematics, computer Science and psychology will become a good robotics engineer.

x. All the students of computer science are intelligent.

3. Let $P(x)$ be a propositional function on X then prove that

i. $\sim(\forall x)P(x) \equiv (\exists x)\sim P(x)$ ii. $\sim(\exists x)P(x) \equiv (\forall x)\sim P(x)$

4. Determine the truth value of each of the following statements where $U = R$.

i. $\exists x, x^2 = x$ ii. $\exists x, x - 3 < x$

iii. $\forall x, x + 2 > x$ iv. $\forall x\ \exists y, x^2 + y^2 < 12$

v. $\exists x\ \exists y\ \forall z, x^2 + y^2 < 2z^2$

5. Negate each of the following statements

i. $(\exists x)\ x + 3 = 10$ ii. $(\forall x)\ x + 3 < 10$

iii. $\exists x)\ x + 2 = x$ iv. $\forall x, x - 3 < x$

v. $\forall x\ P(x) \wedge \exists y\ Q(y)$ vi. $\exists x\ \forall y\ P(x) \vee Q(y)$

vii. $\exists x\ \forall y\ P(x, y)$ viii. $\exists y\ \exists x\ \forall z\ P(x, y, z)$

ix. $\exists x\ \forall y\ P(x, y) \rightarrow Q(x, y)$ x. $\exists x\ \exists y\ \forall z, P(x) \wedge Q(y) \wedge P(x, z)$.

Answers

4. a. True b. True

c. True d. True

e. False

5. i. $(\forall x)\ x + 3 \neq 10$ ii. $(\exists x)\ x + 3 > 10$

iii. $(\forall x)\ x + 2 \neq x$ iv. $(\exists x)\ x - 3 > 10$

v. $\exists x \sim P(x) \vee \forall y \sim Q(y)$ vi. $\forall x\ \exists y \sim P(x) \wedge \sim Q(y)$

vii. $\forall x\ \exists y \sim P(x, y)$ viii. $\forall x\ \forall y\ \exists z \sim P(x, y, z)$

ix. $\forall x\ \exists y \sim P(x, y) \vee Q(x, y)$ x. $\forall x\ \forall y\ \exists z, \sim P(x) \vee \sim Q(y) \vee \sim P(x, z)$.

Chapter Objective Type Questions

A. Multiple choice questions.

1. Which of the following statement is the negation of the statement, "4 is even and –5 is negative"?

(A) 4 is odd or –5 is not negative (B) 4 is even or –5 is not negative

(C) 4 is odd and –5 is not negative (D) 4 is even and –5 is not negative

2. If P and Q stands for the statement

P: It is cold

Q: It is chilly,

then $(P \wedge \sim Q)$ means

(A) It is not cold and it is chilly (B) It is cold and it is not chilly

(C) It is cold or it is not chilly (D) It is not cold or it is not chilly

3. Consider the statement, "Either $-4 \leq x \leq -2$ or $2 \leq x \leq 4$." The negation of this statement is

(A) $x < -4$ or $4 < x$ or $-2 < x < 2$ (B) $-2 < x < 2$

(C) $x \leq -4$ or $4 \leq x$ or $-2 < x < 2$ (D) $x < -4$ or $4 < x$

4. Which of the following is the negation of the statement, "For all odd primes $x < y$ there exists positive non-primes $s < t$ such that $x^2 + y^2 = s^2 + t^2$".

(A) There exists odd primes $x < y$ and there exists positive non-primes $s < t$ such that $x^2 + y^2 = s^2 + t^2$.

(B) For all odd primes $x < y$ and for all positive non-primes $s < t$, $x^2 + y^2 = s^2 + t^2$.

(C) There exists odd primes $x < y$ such that for all positive non-primes $s < t$, $x^2 + y^2 = s^2 + t^2$.

(D) There exists odd primes $x < y$ such that for all positive non-primes $s < t$, $x^2 + y^2 = s^2 + t^2$.

5. Let P be "he is short" and let Q "he is brilliant". Then the statement "It is false that he is tall or brilliant" is:

(A) $P \wedge Q$ (B) $\sim P \wedge Q$

(C) $\sim(\sim P \vee Q)$ (D) $P \vee \sim Q$

6. $P \rightarrow Q$ is logically equivalent to:

(A) $\sim P \vee Q$ (B) $\sim P \wedge Q$

(C) $\sim P \rightarrow Q$ (D) $\sim Q \rightarrow P$

7. Which of the following statement is the negation of the statement "6 is even or –7 is negative"?

(A) 6 is even and -7 is negative (B) 6 is odd and -7 is not negative

(C) 6 is odd or –7 is not negative (D) 6 is even or –7 is not negative

8. The truth table for $(P \vee Q) \vee (P \wedge R)$ is the same as the truth table for-

(A) $(P \vee Q) \wedge (P \vee R)$ (B) $(P \wedge Q) \vee P$

(C) $(P \vee Q) \wedge (P \wedge R)$ (D) $P \vee Q$

9. $(P \wedge Q) \wedge \sim (P \vee Q)$ is a ____

(A) tautology (B) fallacy

(C) Invalid (D) neither fallacy nor tautology

10. Which of the following proposition is a tautology?

(A) $P \vee (P \rightarrow Q)$ (B) $P \vee (Q \rightarrow P)$

(C) $(P \vee Q) \rightarrow P$ (D) $P \rightarrow (P \rightarrow Q)$

11. What is the converse of the following assertion?

I stay only if you come.

(A) If I stay then you come (B) If you do not come then I do not stay

(C) I stay if you come (D) If you do not stay then you come

12. Which one is the contrapositive of $Q \rightarrow P$?

(A) $\neg P \rightarrow \neg Q$ (B) $P \rightarrow Q$

(C) $\neg Q \rightarrow \neg P$ (D) None of these

13. Which of the following is quantifier used in predicate calculus?

(A) Universal (B) existential

(C) Universal and existential (D) None of these

14. A sufficient condition that a triangle Δ be a right triangle is that $x^2+y^2=t^2$. An equivalent statement is:
   (A) If $x^2+y^2=t^2$ then Δ is a right triangle
   (B) Δ is a right triangle only if $x^2+y^2=t^2$
   (C) Δ is a right triangle unless $x^2+y^2=t^2$
   (D) If $x^2+y^2=t^2$ then Δ is not a right triangle

Consider a propositional language with two propositional constants "red", and "dangerous" - each indicating the property suggested by its spelling. Using these propositional constants, encode the following English sentences as Propositional Logic sentences.

15. An object is dangerous if it is red.
   (A) dangerous ⇐ red
   (B) dangerous ⇒ red
   (C) dangerous ⇔ red
   (D) dangerous ⇐ ¬red
16. An object is dangerous only if it is red.
   (A) dangerous ⇐ red
   (B) dangerous ⇒ red
   (C) dangerous ⇔ red
   (D) dangerous ⇐ ¬red
17. An object is dangerous if and only if it is red.
   (A) dangerous ⇐ red
   (B) dangerous ⇒ red
   (C) dangerous ⇔ red
   (D) dangerous ⇐ ¬red
18. An object is dangerous unless it is red.
   (A) dangerous ⇐ red
   (B) dangerous ⇒ red
   (C) dangerous ⇔ red
   (D) dangerous ⇐ ¬red
19. Let

   M = "Rohan is a Math major,"

   C = "Rohan is a Commerce major,"

   G = "Rohan's grandfather is a literature major,"

   D = "Rohan's grandfather has read Discovery of India," and

   K = "Rohan's grandfather has read Kamayani."

   Which of the following expresses the statement "Rohan is Commerce major and a Math major, but his grandfather is a literature major who hasn't read both Discovery of India and Kamayani?"
   (A) $C \wedge M \wedge (G \vee (\sim D \vee \sim K))$
   (B) $C \wedge M \wedge G \wedge (\sim D \wedge \sim K)$
   (C) $C \wedge M \wedge G \wedge (\sim D \vee \sim K)$
   (D) $C \wedge M \wedge G \wedge (D \vee K)$
20. "If the sky is cloudy then it will rain and it will not rain", is:
   (A) a contradiction
   (B) not a contradiction
   (C) fallacy
   (D) None of these
21. The statement $(X \wedge Y) \Rightarrow X$ is a/an:
   (A) Absurdity
   (B) Tautology
   (C) Contingency
   (D) None of the above
22. $[\sim Q \wedge (P \rightarrow Q)] \rightarrow \sim P$ is:
   (A) Tautology
   (B) Contradiction
   (C) Invalid
   (D) None of these

23. Consider the statement form $P \Rightarrow Q$, where P = "If Abhinash is Kiran's father then Kiran is Vipin's daughter" and Q = "Vipin is Abhinash's brother." Which of the following statements is equivalent to this statement?
   (A) If Vipin is not Abhinash's brother, then Abhinash is not Kiran's father and Kiran is Vipin's daughter.
   (B) If Vipin is not Abhinash's brother, then Abhinash is Kiran's father and Kiran is not Vipin's daughter.
   (C) If Vipin is not Abhinash's brother, then Abhinash is Kiran's father or Kiran is Vipin's daughter.
   (D) If Vipin is Abhinash's brother, then Abhinash is Kiran's father and Kiran is Vipin's daughter.
24. Consider the statement, "If n is divisible by 60 then n is divisible by 2 and by 3 and by 5." Which of the following statements is equivalent to this statement?
   (A) If n is divisible by 2 and divisible by 3 and divisible by 5 then n is divisible by 60.
   (B) If n is divisible by 2 or divisible by 3 or divisible by 5 then n is divisible by 60.
   (C) If n is not divisible by 2 or not divisible by 3 or not divisible by 5 then n is not divisible by 60.
   (D) If n is not divisible by 60 then n is not divisible by 2 or not divisible by 3 or not divisible by 5.
25. The converse of a statement is: If an iron rod is stretched, then it has been heated. Write the inverse of the statement.
   (A) If a iron rod has not been heated then it is not stretched
   (B) If a iron rod is not stretched then it has not been heated
   (C) If a iron rod has been heated then it is not stretched
   (D) None of these
26. Which of the following statements is the contrapositive of the statement, "You win the race if you know the tricks but are not over anxious".
   (A) If you know the tricks and are over anxious then you win the race.
   (B) If you lose the race then you don't know the tricks or you are over anxious.
   (C) A necessary condition that you know the tricks or you are not over anxious is that you win the race.
   (D) If you don't know the tricks or are over anxious you lose the race.
27. $((P\vee(R\vee Q))\wedge\sim(\sim Q\wedge\sim R)$ is logically equivalent to:
   (A) $Q\vee R$
   (B) $(P\wedge R)\vee(P\wedge Q)$
   (C) $(P\vee Q)\wedge\sim(P\vee R)$
   (D) $((P\vee R)\vee Q))\wedge(P\vee R)$

B. Fill in the blanks.

1. These declarative sentences with truth values either true or false but not both are called ______.
2. The representation of propositions using logical variables/symbols is studied under ______.
3. In logic, logical connective ______ of two statements P and Q is the statement having truth value true when both P and Q are true else it has the truth value false.
4. Disjunction of propositions with itself gives the ______ proposition.
5. In the truth table of statement $P \rightarrow Q$, the truth value of statement is '0' only for the __ truth value of Q and for ___ of P.
6. In $P \rightarrow Q$ proposition P is called _____ and Q is called the _____.
7. The implication $\sim P \rightarrow \sim Q$ is called the _____ of $P \rightarrow Q$.
8. If $P \rightarrow Q$ is an implication proposition then $Q \rightarrow P$ is called the _____ of $P \rightarrow Q$.
9. A statement that has only "true" value for all possible values of its propositional variables is called a_____.

10. In logic the two statements said to be _________, if the resulting truth values of both statements are same for all possible combinations of input variables.
11. The phrases "for every" or "for all" is called _________ quantifier.
12. The variables involved in propositional function may be ___ or ____.

C. Which of the following statements is True (T)/ False (F)?

1. $(P \wedge Q) \vee (\sim P \wedge Q) \vee (P \wedge \sim Q)$ is equal to $\sim Q \wedge \sim P$.
2. $(P \wedge Q) \vee (\sim P \wedge Q) \vee (P \wedge \sim Q)$ is equal to $[(P \vee \sim P) \wedge Q] \vee (P \wedge \sim Q)$.
3. $(P \wedge Q) \vee (\sim P \wedge Q) \vee (P \wedge \sim Q)$ is equal to $Q \vee (P \wedge \sim Q)$.
4. $(P \wedge Q) \vee (\sim P \wedge Q) \vee (P \wedge \sim Q)$ is equal to $P \vee (Q \wedge \sim P)$.
5. $(P \wedge Q) \vee (\sim P \wedge Q) \vee (P \wedge \sim Q)$ is equal to $Q \vee P$.
6. In symbolic logic, logical connective disjunction of two statements P and Q is the statement having truth value false when both P and Q are false else it has the truth value true.
7. Implication of statement on itself always gives the false value 'F'.
8. The statements which contain both truth values (i.e. 0 and 1) are called contingent form.
9. A statement that has only 'false' value for all possible values of its propositional variable is said to be self tautology.
10. If P and Q are well formed formula then $(P \leftrightarrow Q)$ may not be well formed formula.
11. Singleton set $\{\rceil\}$, $\{\vee\}$ or $\{\wedge\}$ are not functionally complete set.

Answers Objective Type

A.

1. A	2. B	3. A	4. C
5. C	6. A	7. B	8. D
9. B	10. A	11. B	12. A
13. C	14. A	15. A	16. B
17. C	18. D	19. C	20. B
21. B	22. A	23. B	24. C
25. B	26. B	27. A	

B.

1. Propositions	2. symbolic logic	3. conjunction	4. Same
5. 0, 1	6. antecedent, consequence	7. inverse	8. Converse
9. Tautology	10. equivalent	11. universal quantifier	12. free or bound

C.

1. F	2. T	3. T	4. T
5. T	6. F	7. F	8. T
9. F	10. F	11. T	

CHAPTER

8 Lattices

8.0 Introduction

We have seen how sets, relations and logic play an important role in representation modeling and processing of information. Here in this chapter we will focus our attention on a special type of structure: lattices, that has applications in mathematics (algebra), sorting, searching, construction of computer circuits and finite state machines etc. According to the knowledge of molecular science the "lattices" is a regular arrangement of atoms or molecules in a crystal. If we take the elements of sets in place of crystals then lattices can be used in mathematics and computer science. Before we discuss lattices in detail, readers are advised to recall the terms related to posets, ordered set and Hasse diagram studied earlier.

8.1 Lattice

Consider a poset L under ordering relation $\preceq$ such that every subset $\{a, b\}$ of two elements has unique greatest lower bound or infimum (glb or inf) and unique least upper bound or supremum (lub or sup). Then L is known as **lattice**. We denote

glb $\{a, b\} = a \wedge b$ (call it the **meet** of a and b)

and lub $\{a, b\} = a \vee b$ (call it the **join** of a and b)

Other notations of meet are '', '·' and '∩' and that of join are '⊕', '+' and '∪'. In other words, lattice is a poset consisting of a greatest lower bound and a least upper bound with two binary operations, meet and join.*

Example 8.1

Let D be a set of $\{1, 3, 5, 15\}$. Prove that partially order set D under the relation 'divides' is a lattice.

Solution

Let $D = \{1, 3, 5, 15\}$ is a poset under the relation 'divides' written as $\langle D, |\rangle$. To determine the lub of any two elements such as 3 and 5, we see for all $d_l \in D$, 3 divides d_l and 5 divides d_l. Then the lub of 3 and 5 is 15. Similarly to find the glb of 5 and 15, we consider for all $d_g \in D$ such that d_g divides 5 and d_g divides 15. Here, $d_g = 1$ and $d_g = 5$ satisfy the conditions. But 1 divides 5 hence $d_g = 5$ is the glb of 5 and 15. In this manner, we can calculate glb and lub for each pair of elements in the set D. glb and lub is given in Table 8.1 and Table 8.2 respectively.

Table 8.1: (glb)

∧	1	3	5	15
1	1	1	1	1
3	1	3	1	3
5	1	1	5	5
15	1	3	5	15

Table 8.2: (lub)

∨	1	3	5	15
1	1	3	5	15
3	3	3	15	15
5	5	15	5	15
15	15	15	15	15

We can see from the tables that D, has both a meet and a join, hence is a lattice under 'divides'. Lattice $\langle D, | \rangle$ can be represented with the help of Hasse diagram as shown in Fig. 8.1.

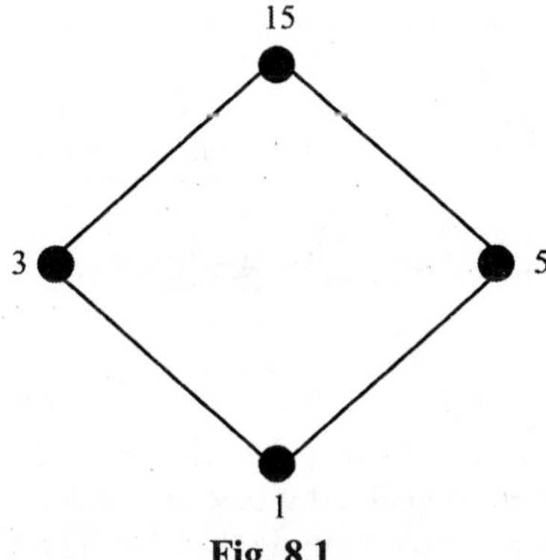

Fig. 8.1

Example 8.2

Determine whether each of the posets {1, 2, 3, 4, 5} and {1, 2, 4, 8, 16} is a lattice under the relation 'divides'.

Solution

Let the poset

$$D_1 = \{1, 2, 3, 4,5\}$$

and

$$D_2 = \{1, 2, 4, 8, 16\}.$$

Table 8.3

∧	1	2	4	8	16
1	1	1	1	1	1
2	1	2	2	2	2
4	1	2	4	4	4
8	1	2	4	8	8
16	1	2	4	8	16

Table 8.4

∨	1	2	4	8	16
1	1	2	4	8	16
2	2	2	4	8	16
4	4	4	4	8	16
8	8	8	8	8	16
16	16	16	16	16	16

To determine the glb and lub of D_1 we take all the elements $d_{11} \in D$ such that for any pair of elements a and b, a divides d_{11} and b divides d_{11}. But for the set D_1 elements 2 and 3 have no upper bound, hence certainly does not have a least upper bound. Therefore, the poset D_1 is not a lattice.

Now consider the poset D_2 and calculate glb and lub for all the pair of elements in D_2, which is shown in Table 8.3 and Table 8.4.

One can observe from the table that D_2, has both a meet and join, is a lattice under the relation 'divides'. The lattice is represented with the help of Hasse diagram as shown in Fig. 8.2:

Fig. 8.2

In Example 8.2, greatest lower bound of two elements is the smaller of the elements and least upper bound is the greater of the elements i.e. glb and lub of two elements in this case is nothing but the greatest common divisor and least common multiple of these two numbers respectively.

Example 8.3

Show that $\langle Z^+, |\rangle$ is a lattice.

Solution

Let Z^+ be the set of positive integers and $|$ denotes the relation "division" in Z^+. Let $a, b \in Z^+$. Any upper bound of $\{a, b\}$ is nothing but an element divisible by both a and b i.e. common multiple of a and b. Hence least upper bound of a and b is the least common multiple (l cm) of a and b.

i.e. $$\text{lub } \{a, b\} = \text{lcm } (a, b).$$

Similarly, greatest lower bound is the greatest common divisor (gcd) of a and b

i.e. $$\text{glb } \{a, b\} = \text{gcd } (a, b)$$

Because elements of poset $\langle Z^+, |\rangle$, has both a lub and glb. Hence $\langle Z^+, |\rangle$, is a lattice.

Example 8.4

Consider a positive integer z and set of all divisors of z, D_z. Then prove that $\langle D_z, |\rangle$ is a lattice for $z = 6, 24, 30$ with the help of Hasse diagram.

Solution

Hasse diagrams of D_6, D_{24}, D_{30} are shown in Fig. 8.3(a), 8.3(b), 8.3(c) respectively. It can be seen from the Fig. 8.3 that all the elements have a glb and lub. So, $\langle D_z, |\rangle$ is a lattice.

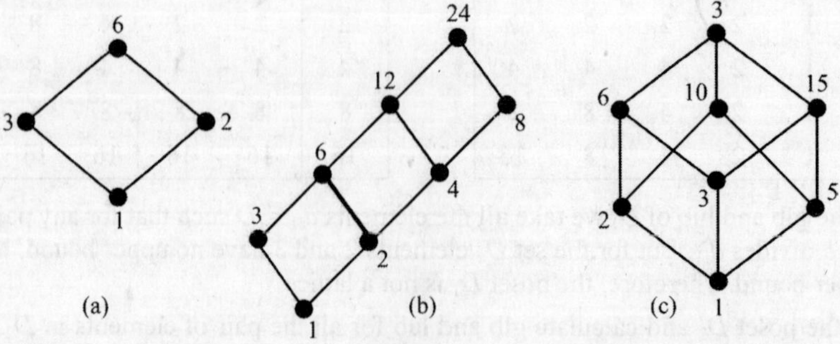

Fig. 8.3

Example 8.5

Which of the partially ordered sets shown in Fig. 8.4 are lattices?

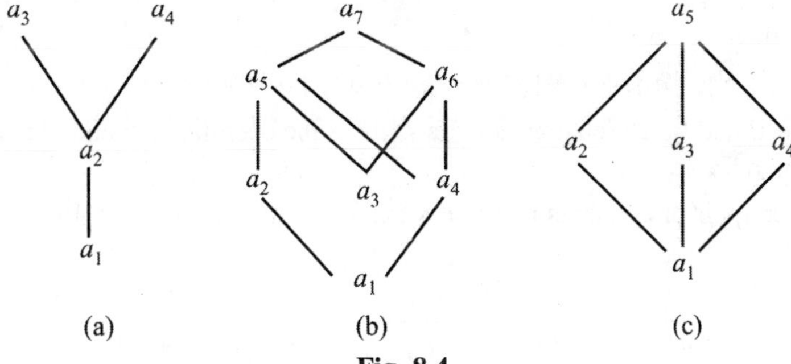

Fig. 8.4

Solution

Poset shown in Fig. 8.4(a) is not a lattice since the elements a_3 and a_4 have no upper bound and hence sup (a_3, a_4) does not exists.

Poset (b) is also not a lattice since the elements a_5 and a_6 have two lower bounds a_3 and a_4, but neither precedes the other, hence inf (a_5, a_6) does not exist.

Poset shown in Fig. 8.4(c) is a lattice. All the elements in poset (c) have unique lub and glb.

Example 8.6

Determine whether or not each of the following sets is a lattice with respect to divisibility:

$$A = \{2, 3, 4, 12\} \text{ and } B = \{1, 2, 3, 9, 18\}.$$

Solution

Set $A = \{2, 3, 4, 12\}$ is not a lattice. Here, every pair of elements do not contain the glb. If we take the pair of elements 2 and 3, the glb(2, 3) = 1 but $1 \notin A$. So inf (2, 3) does not exist in the set.

Similarly, examining the set B sup (2, 3) = 6 but $6 \notin B$. Hence B is also not a lattice with respect to divisibility.

Example 8.7

Explain why any linearly ordered set is a lattice.

Solution

We know that linearly ordered set is also known as 'chain' can be represented generally of the type, as shown in Fig. 8.5. Consider any pair of elements a and b of a linearly ordered set where a relates to b. Thus inf $(a, b) = a$ and sup $(a, b) = b$.

Since a glb and a lub exist for any pair of elements, hence ordered set is a lattice.

Fig. 8.5

Example 8.8

Let $X = \{a, b, c\}$. Then prove that poset $P(X)$ of all subsets of X is a lattice under the ordering relation $\subseteq$.

Solution

$X = \{a, b, c\}$ then

$P(X) = \{\varnothing, \{a\}, \{b\}, \{c\}, \{a, b\}, \{a, c\}, \{b, c\}, \{a, b, c\}\}$.

In this case, the 'meet' and 'join' relations are the same as the operations $\cap$ and $\cup$ respectively with respect to relation $\subseteq$. It is shown in Table 8.5.

We observe that every pair of elements in Table 8.5 have unique glb and lub. Hence poset $\langle P(X), \subseteq \rangle$ is a lattice.

Table 8.5

$\wedge$	$\varnothing$	$\{a\}$	$\{b\}$	$\{c\}$	$\{a, b\}$	$\{a, c\}$	$\{b, c\}$	$\{a, b, c\}$
$\varnothing$	$\varnothing$	$\varnothing$	$\varnothing$	$\varnothing$	$\varnothing$	$\varnothing$	$\varnothing$	$\varnothing$
$\{a\}$	$\varnothing$	$\{a\}$	$\varnothing$	$\varnothing$	$\{a\}$	$\{a\}$	$\varnothing$	$\{a\}$
$\{b\}$	$\varnothing$	$\varnothing$	$\{b\}$	$\varnothing$	$\{b\}$	$\varnothing$	$\{b\}$	$\{b\}$
$\{c\}$	$\varnothing$	$\varnothing$	$\varnothing$	$\{c\}$	$\varnothing$	$\{c\}$	$\{c\}$	$\{c\}$
$\{a, b\}$	$\varnothing$	$\{a\}$	$\{b\}$	$\varnothing$	$\{a, b\}$	$\{a\}$	$\{b\}$	$\{a, b\}$
$\{a, c\}$	$\varnothing$	$\{a\}$	$\varnothing$	$\{c\}$	$\{a\}$	$\{a, c\}$	$\{c\}$	$\{a, c\}$
$\{b, c\}$	$\varnothing$	$\varnothing$	$\{b\}$	$\{c\}$	$\{b\}$	$\{c\}$	$\{b, c\}$	$\{b, c\}$
$\{a, b, c\}$	$\varnothing$	$\{a\}$	$\{b\}$	$\{c\}$	$\{a, b\}$	$\{a, c\}$	$\{b, c\}$	$\{a, b, c\}$

$\vee$	$\varnothing$	$\{a\}$	$\{b\}$	$\{c\}$	$\{a,b\}$	$\{a, c\}$	$\{b, c\}$	$\{a,b,c\}$
$\varnothing$	$\varnothing$	$\{a\}$	$\{b\}$	$\{c\}$	$\{a, b\}$	$\{a, c\}$	$\{b, c\}$	$\{a,b,c\}$
$\{a\}$	$\{a\}$	$\{a\}$	$\{a, b\}$	$\{a, c\}$	$\{a, b\}$	$\{a, c\}$	$\{a,b,c\}$	$\{a,b,c\}$
$\{b\}$	$\{b\}$	$\{a, b\}$	$\{b\}$	$\{b, c\}$	$\{a, b\}$	$\{a,b,c\}$	$\{b, c\}$	$\{a,b,c\}$
$\{c\}$	$\{c\}$	$\{a, c\}$	$\{b, c\}$	$\{c\}$	$\{a,b,c\}$	$\{a, c\}$	$\{b,c\}$	$\{a,b,c\}$
$\{a,b\}$	$\{a, b\}$	$\{a, b\}$	$\{a, b\}$	$\{a, b, c\}$	$\{a, b\}$	$\{a,b,c\}$	$\{a,b,c\}$	$\{a,b,c\}$
$\{a, c\}$	$\{a, c\}$	$\{a, c\}$	$\{a,b,c\}$	$\{a, b, c\}$	$\{a,b,c\}$	$\{a,c\}$	$\{a,b,c\}$	$\{a,b,c\}$
$\{b, c\}$	$\{b, c\}$	$\{a,b,c\}$	$\{b, c\}$	$\{b, c\}$	$\{a,b,c\}$	$\{a,b,c\}$	$\{b,c\}$	$\{a,b,c\}$
$\{a,b,c\}$	$\{a,b,c\}$	$\{a,b,c\}$	$\{a,b,c\}$	$\{a,b,c\}$	$\{a,b,c\}$	$\{a,b,c\}$	$\{a,b,c\}$	$\{a,b,c\}$

Example 8.9

Find two subsets X_1 and X_2 of poset $X = \{a, b, c, d, e\}$ shown in Fig. 8.6 such that X_1 and X_2 are lattices.

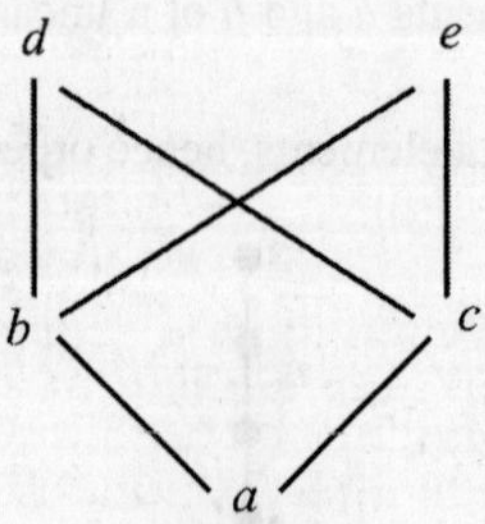

Fig. 8.6

Solution

We observe from Fig. 8.6, that poset $X = \{a, b, c, d, e\}$ is not a lattice. Because glb (d, e) is not unique and sup (d, e) does not exist. But we know that any linearly ordered subset is a lattice. Hence the two subsets which are lattices are given by $\{a, b, c, d\}$ and $\{a, b, c, e\}$.

Exercise 8.1

1. Explain why the following partially ordered sets given in Fig.8.7 are not lattices?

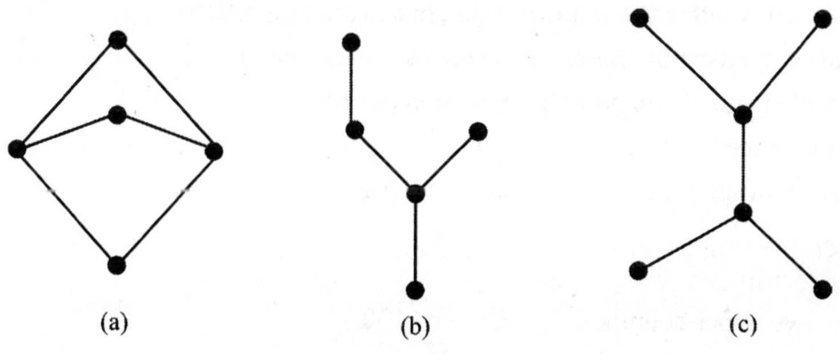

Fig. 8.7

2. Define the term lattices.
3. Prove that the poset $\langle P(X), \subseteq \rangle$ is a lattice where $X = \{1, 2, 3\}$ and $P(X)$ is the power set of X.
4. Prove that the set $\{1, 2, 3, 4, 6, 12\}$ of factors of 12 under divisibility form a lattice.
5. Draw the diagram of $\langle P(X), \subseteq \rangle$ when $X = \{a, b, c\}$ and $P(X)$ the power set of X.
6. Identify which of the posets shown in Fig. 8.8 are lattices.

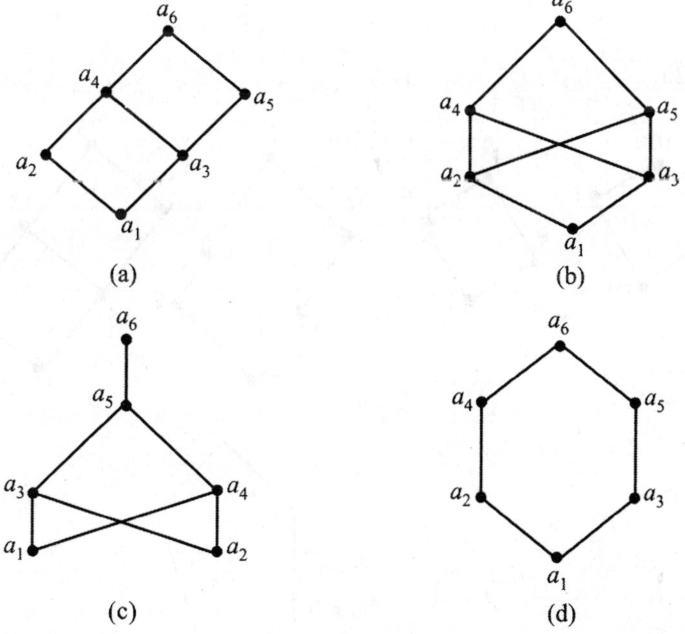

Fig. 8.8

7. Determine whether or not each of the following sets is a lattice with respect to divisibility $A = \{1, 3, 4, 9\}$ and $B = \{1, 2, 3, 9, 18\}$.
8. Let A be any subset of the natural number N with the usual order. Prove that A is a lattice.
9. Draw the diagram of lattices $\langle S_n, | \rangle$ for $n = 10$, 15, 20 and 75. For what values of n do you expect that $\langle S_n, | \rangle$, to be a chain? Where symbol "|" denotes the relation of 'divisibility'.
10. Draw the Hasse diagram of two lattices having five elements.
11. Consider the partially ordered set < {2, 4, 6, 8}, 1>, where "2/4" means 2 divides 4.
    Show with reason that whether the following statements are true or false:
    i. Every pair of elements in the poset has a greatest lower bound.
    ii. Every pair of elements in the poset has a least upper bound.
    iii. This poset is a lattice.
12. Draw Hasse diagram of set {1, 2, 3, 6, 12} under the partial ordering relation, 'divides'.

Answers

1. a. No, there are two upper bounds. b. Yes
   c. No, two upper and lower bounds.
6. a. Yes b. Yes
   c. No d. Yes
7. a.

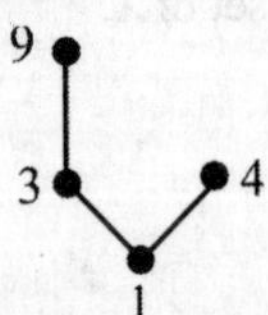

b.

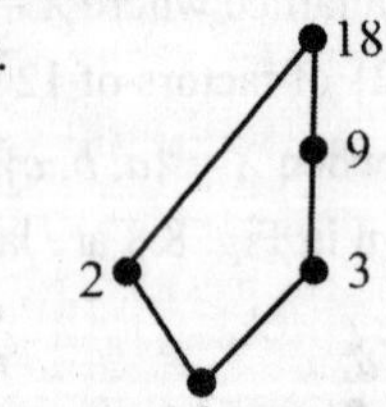

9. $n = 10$ $n = 15$ $n = 20$

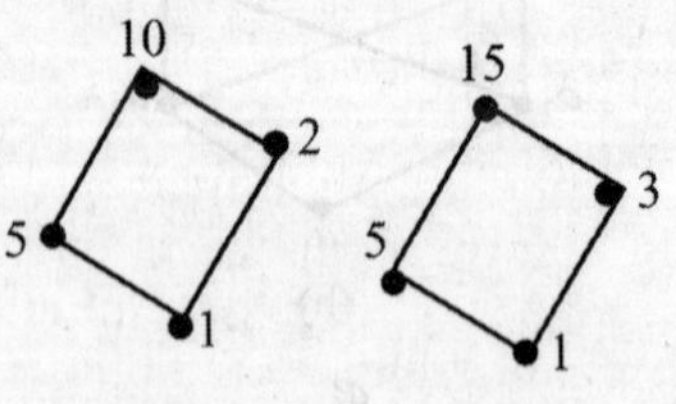

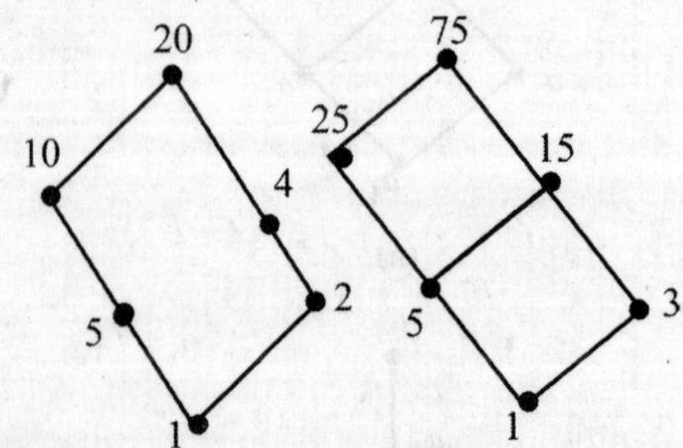

10. i. ii.

12.

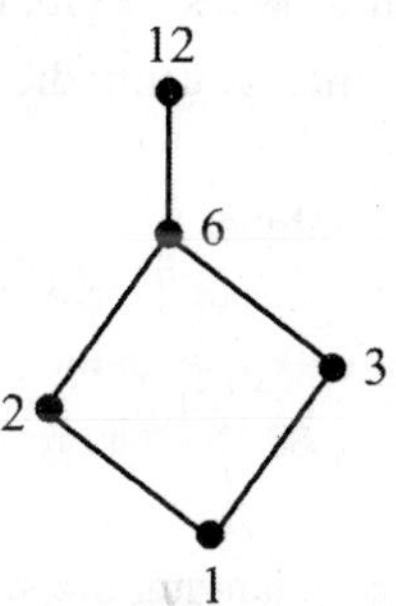

Theorem 8.1

A poset $\langle L, \preceq \rangle$ is a lattice if and only if every non empty finite subset of L has glb and lub.

Proof

Consider a Lattice $\langle L, \preceq \rangle$ and a non-empty finite subset S of L. Three cases arise:

I. If S has only one element x then glb (S) = lub $(S) = x$.

II. If S has two elements such as x, y then according to the definition of lattice lub S and glb S exist.

III. S has three elements i.e. $S = \{x, y, z\}$.

Again by the definition of lattice any two elements have glb and lub.

Let us consider $u = \text{glb } \{x, y\}$ ...(1)

and $v = \text{glb } \{z, u\}$ ...(2)

then by definition of u and v, $u \preceq x$, $u \preceq y$, $v \preceq z$, $v \preceq u$

Thus $v \preceq x$, $v \preceq y$, $v \preceq z$.

$\Rightarrow v$ is lower bound of $\{x, y, z\}$.

Now, we wish to prove that v is the greatest lower bound of $S = \{x, y, z\}$. Consider any other lower bound, w, of $\{x, y, z\}$ then.

$w \preceq x$ ...(3)

$w \preceq y$ ...(4)

$w \preceq z$ ...(5)

i.e. relations (3), (4) and (1) gives $w \preceq u$ ...(6)

and relations (5), (6) and (2) gives $w \preceq v$ ...(7)

Hence $v = \text{glb } \{x, y, z\} = \text{glb } S$.

Similarly, we can prove lub of S also exist.

Conversely, if glb and lub of every non empty finite subset of L exist, it implies subset of pair of elements has glb and lub hence L is a lattice under relation $\preceq$.

Theorem 8.2 (Uniqueness of glb and lub)

The elements a and b of a poset $\langle L, \preceq \rangle$ has unique glb and lub.

Proof

Let us assume that g_1 and g_2 are two different glb of a and b. By definition of glb, we have

$$a \preceq g_1,\ b \preceq g_1, \text{ and } a \preceq g_2,\ b \preceq g_2,$$

Since g_1 is a glb of set L then $g_1 \preceq g_2$. But g_2 is also a glb, hence $g_2 \preceq g_1$ and according to the property of antisymmetry of relation $\preceq$, we have $g_1 = g_2$ which is contradiction of our assumption. Thus poset $\langle L, \preceq \rangle$ has unique glb.

Similarly, we can prove that lub of poset is unique.

8.2 Properties of Lattices

8.2-1 Principle of Duality

In the set theory and logic duality is obtained by changing the statement by interchanging the notation $\cap$ and $\wedge$ by $\cup$ and $\vee$ respectively and vice versa. In the same way here the **dual** of any statement in a lattice $\langle L, \wedge, \vee \rangle$ or $\langle L, \oplus, * \rangle$ or $\langle L, \cap, \cup \rangle$ is defined by the statement which is obtained by interchanging $\wedge$, $\oplus$ and $\cap$ by $\vee$, $*$ and $\cup$ respectively and vice versa, i.e. if $\preceq$ is a partial ordering relation on any set then its inverse relation is $\succeq$ (also known as 'converse' relation).

Example 8.10

Write the dual of each statement

1. $(a \wedge b) \vee c = (b \vee c) \wedge (c \vee a)$
2. $(a \wedge b) \vee a = a \wedge (b \vee a)$.

Solution

On replacing $\wedge$ by $\vee$ and $\vee$ by $\wedge$ in each statement we obtain the dual of statements:

1. $(a \vee b) \wedge c = (b \wedge c) \vee (c \wedge a)$
2. $(a \vee b) \wedge a = a \vee (b \wedge a)$.

Theorem 8.3

Consider a lattice $\langle L, \preceq \rangle$ where $a, b, c \in L$, then the following are true:

Idempotent Law

(P1) $a \wedge a = a$ (P1)′ $a \vee a = a$

Commutative law

(P2) $a \wedge b = b \wedge a$ (P2)′ $a \vee b = b \vee a$

Associative Law

(P3) $a \wedge (b \wedge c) = (a \wedge b) \wedge c$ (P3)′ $a \vee (b \vee c) = (a \vee b) \vee c$

Absorption Law

(P4) $a \wedge (a \vee b) = a$ (P4)′ $a \vee (a \wedge b) = a$.

Proof

Properties (P1) to (P4) can be proved easily using the definitions of supremum and infimum. Let us consider the axiom

(P1) (Idempotent law): We have already observed that the sup and inf of a singleton element is again that element and its dual, i.e. (P1)′ also holds good.

(P2) commutative law: As we know that inf and sup of a pair of elements in a lattice must be unique. Hence

$$\inf(a, b) = \inf(b, a) \Rightarrow a \wedge b = b \wedge a$$

It is also applicable for its dual. Hence the (P2)′.

(P3) associative law: Consider $x = a \vee (b \vee c)$ and $y = (a \vee b) \vee c$

Now,

$$x = a \vee (b \vee c) \Rightarrow a \preceq x, (b \vee c) \preceq x$$

$$\Rightarrow a \preceq x, b \preceq x, c \preceq x$$

$\therefore$

$$a \preceq x, b \preceq x \Rightarrow a \vee b \preceq x$$

$$\Rightarrow a \vee b \preceq x, c \preceq x \Rightarrow (a \vee b) \vee c \preceq x$$

$$\Rightarrow y \preceq x$$

Similarly, we can show that $x \preceq y$

$$\Rightarrow a \vee (b \vee c) = (a \vee b) \vee c$$

In a similar manner we can prove the axiom (P3) of associative law.

(P4) Absorption law: By definition of $\vee$, for any $a \in L$, $a \preceq a$, and $a \preceq a \vee b$, hence $a \preceq a \wedge (a \vee b)$.

By the antisymmetry, we have $a \wedge (a \vee b) \preceq a$. So $a = a \wedge (a \vee b)$.

8.2-2 Stability

Theorem 8.4

For any $a, b, c \in L$, we have.

i. $a \preceq b, a \preceq c \Rightarrow a \preceq (b \vee c), a \preceq (b \wedge c)$,

ii. $a \succeq b, a \succeq c \Rightarrow a \succeq (b \wedge c), a \succeq (b \vee c)$.

Proof

The proof is left as an exercise.

Theorem 8.5

Let L be a lattice. Then for every a and b in L, if $\vee$ and $\wedge$ denote join and meet, then the following are equivalent.

1. $a \preceq b$
2. $a \wedge b = a$
3. $a \vee b = b$.

Proof

Let L be a lattice and $a, b \subset L$. Firstly, we prove

$$a \preceq b \Leftrightarrow a \wedge b = a \qquad \text{...(A)}$$

For this, consider $a \preceq b$

We know that, $\quad a \preceq a$ and $a \preceq b \Rightarrow a \preceq a \wedge b$

On the other hand, by definition of $a \wedge b$ we have $a \wedge b \preceq a$

hence, $\quad a \preceq b \Rightarrow a \wedge b = a \qquad$ ...(1)

Now, if we assume that $a \wedge b = a$, (which is only possible when $a \preceq b$)

i.e. $\quad a \wedge b = a \Rightarrow a \preceq b \qquad$ ...(2)

On combining (1) and (2), we obtain the required equivalence (A)

In a similar manner, it can also be shown that.

$$a \preceq b \Leftrightarrow a \vee b = b \qquad \text{...(B)}$$

Again, let us assume that $a \preceq b$. Also, we have $b \preceq b$,

$$\Rightarrow a \vee b \leq b \quad \ldots(1)$$

By the definition of supremum $\quad a \vee b \geq b \quad \ldots(2)$

From (1) and (2), one can conclude that $a \vee b = b$

i.e. $\quad a \preceq b \Rightarrow a \vee b = b \quad \ldots(3)$

On the other hand, we assume that $a \vee b = b$ but it is only possible if $a \preceq b$

Hence $\quad a \vee b = b \Rightarrow a \preceq b \quad \ldots(4)$

So $\quad a \preceq b \Leftrightarrow a \vee b = b$

Hence all the above three axioms are equivalent.

8.2-3 Isotonicity

Theorem 8.6

Let L be a lattice. Then, for every a, b and c in L, if $a \preceq b$ then

i. $a \vee c \preceq b \vee c$

ii. $a \wedge c \preceq b \wedge c$

i.e. join and meet taken with fixed element, preserve the inequality.

Proof

Since $b \preceq b \vee c$, $c \preceq b \vee c$, hence if $a \preceq b$ then $b \vee c$ is an upper bound of $\{a, b\}$.

$\therefore$ $a \vee c \preceq b \vee c$, $a \vee c$ is the least among the upper bounds of $\{a, c\}$.

Similarly, $a \wedge c \preceq a$ and $a \wedge c \preceq c$

But if $\quad a \preceq b$ then $a \wedge c$ is a lower bound of $\{b, c\}$

$\Rightarrow \quad a \wedge c \preceq b \wedge c$ i.e. $a \wedge c$ is the greatest lower bound of $\{a, c\}$.

8.2-4 Distributive Inequalities

Theorem 8.7

Consider a lattice $\langle L, \preceq \rangle$ and $a, b, c \in L$, the distributive in equalities.

i. $a \wedge (b \vee c) \geq (a \wedge b) \vee (a \wedge c)$

ii. $a \vee (b \wedge c) \leq (a \vee b) \wedge (a \vee c)$ hold good.

Proof

i. We have $a \succeq (a \wedge b)$, $a \succeq (a \wedge c)$

By theorem 8.7 (ii)

$$a \succeq (a \wedge b) \vee (a \wedge c) \quad \ldots(1)$$

Similarly $\quad (b \vee c) \succeq b \succeq (a \wedge b)$

and $\quad (b \vee c) \succeq c \succeq (a \wedge c)$

Thus, Again using Theorem

$$(b \wedge c) \succeq (a \wedge b) \vee (a \wedge c) \quad \ldots(2)$$

from (1) and (2) and by using theorem 8.5, we obtain

$$a \wedge (b \vee c) \succeq (a \wedge b) \vee (a \wedge c) \text{ which is (1).}$$

ii. may be proved as above or by the principle of duality.

REMARK

In many lattices distributive inequalities behave like equalities when the operations union '$\cup$' and intersection '$\cap$' are distributive over each other.

8.2-5 Modular Inequality

Theorem 8.8

If $\langle L, \preceq \rangle$ is any lattice, then for any $a, b, c \in L$, we have

$$a \preceq c \Leftrightarrow a \vee (b \wedge c) \preceq (a \vee b) \wedge c.$$

Proof

Firstly, suppose $a \preceq c$ holds

then $\quad a \preceq a \vee b, a \preceq c \Rightarrow a \preceq (a \vee b) \wedge c \quad$ [by definition of $\wedge$]

also $\quad b \wedge c \preceq b \preceq a \vee b, b \wedge c \preceq c \Rightarrow b \wedge c \preceq (a \vee b) \wedge c$

this implies that $(a \vee b) \wedge c$ is an upper bound of a and $b \wedge c$.

Hence $\quad a \vee (b \wedge c) \preceq (a \vee b) \wedge c \; \forall\, a, b, c \in L$

On the other hand if the given inequality

$$a \vee (b \wedge c) \preceq (a \vee b) \wedge c$$

holds for every $a, b, c \in L$ then for $b = a$, we get

$$a \vee (a \wedge c) \preceq (a \vee a) \wedge c$$

or $\quad a \preceq a \wedge c \quad$ [by absorption and Idempotent]

but we have $a \wedge c \preceq c$

$\therefore\ a \preceq c$. Hence the theorem.

The result of modular inequality is self dual (on interchanging the roles of a and c in the statement)

Exercise 8.2

1. If L be a lattice and $a, b, c \in L$ then prove that
$$(a \wedge b) \vee (b \wedge c) \vee (c \wedge a) \preceq (a \vee b) \wedge (b \vee c) \wedge (c \vee a).$$
2. Show that $(a \wedge b) \vee (c \wedge d) \preceq (a \vee c) \vee (b \vee d)$ holds if a, b, c, d are in a lattices L.
3. If $a \preceq b$, $c \preceq d$, then show that $(a \vee c) \preceq (b \vee d)$. What would be its dual?
4. Let a lattice L and $a \preceq b \preceq c$ then show that
$$a \vee b = b \wedge c \text{ and } (a \wedge b) \vee (b \wedge c) = b = (a \vee b) \wedge (a \vee c).$$
5. Prove the following version of modular inequality
$$a \vee [b \wedge (a \vee c)] \preceq (a \vee b) \wedge (a \vee c) \; \forall\, a, b, c \in L, \text{ where } L \text{ is a lattice.}$$
6. Consider $\oplus$ and $*$ as the operator join and meet respectively and show the commutative, associative and absorption identity hold good.

7. Prove all the properties of lattice for the relation $\subseteq$ in a lattice $\langle L, \cup, \cap \rangle$.
8. In a lattice L if $a, b, c \in L$ then prove that
$$a \preceq b \Leftrightarrow a \wedge b = b \Leftrightarrow a \vee b = a.$$
9. Show that in a Lattice L if $a, b, c, d \in L$ then
$$(a \wedge b) \vee (c \wedge d) \preceq (a \vee c) \wedge (b \vee d)$$
$$(a \wedge b) \vee (b \wedge c) \vee (c \wedge a) \preceq (a \vee b) \wedge (b \vee c) \wedge (c \vee a).$$
10. Prove that every finite subset of a lattice has unique lub and glb.
[**Hint**: Using mathematical Induction].
11. Prove that $\forall\ a, b, c \in L$ when $\langle L, \preceq \rangle$ be a lattice
$$(a \wedge b) \vee [\,(a \wedge c) \wedge [\,(a \wedge b) \wedge (b \wedge c)\,]\,] = a \wedge b.$$
[**Hint**: we have
$$(a \wedge b) \preceq [\,(a \wedge b) \vee [\,(a \wedge c)\,]$$
and $$(a \wedge b) \preceq [\,(a \wedge b) \vee (b \wedge c)\,]$$
$\therefore$ $$(a \wedge b) \preceq [\,(a \wedge b) \vee (a \wedge c)\,] \wedge [\,(a \wedge b) \vee (b \wedge c)\,] \quad \text{[by stability]}$$
On the other hand,
$$a \preceq (a \wedge b) \text{ and } a \preceq (a \wedge c)$$
$\therefore$ $$a \preceq [(a \wedge b) \vee (a \wedge c)\,] \quad \ldots(1)$$
Similarly $$b \preceq [\,(a \wedge b) \vee (a \wedge c)\,] \quad \ldots(2)$$
from (1) and (2)
$$(a \wedge b) \preceq a \wedge [\,(a \wedge b) \vee (b \wedge c)\,] \quad \text{[by Isotonicity]}$$
$$\prec [\,(a \wedge b) \vee (a \wedge c)\,] \wedge [\,(a \wedge b) \vee (b \wedge c)\,]$$
Now use antisymmetry.
12. Prove that dual of a lattice is a lattice
[**Hint**: Let $\langle L, R \rangle$ be a lattice and $\langle L', R' \rangle$ be its dual. Then $L = L'$ and R' is converse of R.
Let us consider $x, y \in L' \Rightarrow x, y \in L$
$\Rightarrow$ sup $\{x, y\}$ exists in L and let $x \vee y = \sup\{x, y\}$.
Then $x\,R\,(x \vee y)$ and $y\,R\,(x \vee y)$
$\Rightarrow (x \vee y)\,R'\,x, (x \vee y)\,R'y,$
$\Rightarrow x \vee y$ is lower bound of $\{x, y\}$ in L'.
Let another element z such that z is any lower bound (x, y) in L' then
$z\,R'x, z\,R'y$
$\Rightarrow x\,R\,z, y\,R\,z$
$\Rightarrow z$ is an upper bound of $\{x, y\}$ in L.
$\Rightarrow (x \vee y.\ R\,z$ as $x \vee y = \sup\{x, y\}$ in L
$\Rightarrow z\,R'\,(x \vee y)$
i.e. $(x \vee y)$ is greatest lower bound of $\{x, y\}$ in L'. Similarly $(x \wedge y)$ is least upper bound of $\{x, y\}$ in L'].

8.3 Lattices as an Algebraic Systems

As we know the algebraic system is a set together with some rules for combining elements of the set to form other elements of the set. In this section, we define a **lattice as an algebraic system**. Later with the help of this we take some special type of lattices such as sub lattices, complete, bounded, semi-lattices etc. and operations like direct product, homomorphism etc.

Theorem 8.9

Consider a lattice as an algebraic system $\langle L, \oplus, *\rangle$ together with two binary operations $\oplus$ and $*$ such that

1. $a * b = b * a \,;\, a \oplus b = b \oplus a$ (Commutative Law)
2. $a * (b * c) = (a * b) * c \,;\, a \oplus (b \oplus c) = (a \oplus b) \oplus c$ (Associative Law)
3. $a * (a \oplus b) = a;\; a \oplus (a * b) = a$ (Absorption Law)

hold for each $a, b, c \in L$. Then we can define a partial order relation R on L such that (L, R) become a lattice with $a \vee b = a \oplus b$ and $a \wedge b = a * b \;\forall\; a, b \in L$.

Proof

Firstly, we prove the idempotent law for $*$ and $\oplus$ operations.

i.e. $$a * a = a \quad \ldots (1)$$

and $$a \oplus a = a \quad \ldots (2)$$

Due to the fact of property (3), if we replace second a in equation (1) by $a \oplus (a * a)$ then it becomes

$$a * a = a * (a \oplus (a * a)) = a \;\forall\; a \in L$$

Similarly $$a \oplus a = a \oplus (a * (a \oplus a)) = a \;\forall\; a \in L \quad \text{[by property (3)]}$$

Now, establish the relation R as a partial order relation.

I. We have $a\,R\,a \;\forall\; a$

$\because$ $$a * a = a \;\forall\; a \in L \quad \text{[by idempotent Law]}$$

Hence R is reflexive.

II. Consider $a\,R\,b$ and $b\,R\,a$ for some $a, b \in L$ such that $a * b = a$ and $b * a = b$. But by commutative law, we have

$$a * b = b * a \Rightarrow a = b \text{ i.e. } a\,R\,b \text{ and } b\,R\,a \Rightarrow a = b.$$

Hence the relation R is ant symmetric.

III. Finally, let us consider that for $a, b, c \in L$, $a\,R\,b$ and $b\,R\,c$.

i.e. $$a * b = a \quad \ldots (3)$$

and $$b * c = b \quad \ldots (4)$$

$\therefore$ $$\begin{aligned} a * c &= (a * b) * c = a * (b * c) && \text{[by associativity]} \\ &= a * b && \text{[by (4)]} \\ &= a && \text{[by (3)]} \\ &\Rightarrow a\,R\,c \end{aligned}$$

i.e. the relation R is transitive.

Thus we conclude that relation R is a partial ordering relation. i.e. $\langle L, R\rangle$ is a poset.

Now to show that it is indeed a lattice, we observe that

(1) $$\begin{aligned} (a * b) * a &= a * (b * a) && \text{[by associativity]} \\ &= a * (a * b) && \text{[by commutativity]} \end{aligned}$$

$$= (a * a) * b \quad \text{[by associativity]}$$
$$= a * b \quad \text{[by idempotent]}$$
$$\Rightarrow (a * b)\ R\ a$$

Similarly,

$$(a * b) * b = a * (b * b) \quad \text{[by associativity]}$$
$$= a * b \quad \text{[by idempotent]}$$
$$\Rightarrow (a * b)\ R\ b$$

Thus $(a * b)$ is a lower bound of $\{ a, b \}$.

Now let us assume that $\exists\, c \in L$ such that $c\ R\ a$ and $c\ R\ b$ then we have

$$c * a = c \quad \text{...(5)}$$

and
$$c * b = c \quad \text{...(6)}$$

This means that
$$c * (a * b) = (c * a) * b \quad \text{[by associativity]}$$
$$= c * b \quad \text{[by (5)]}$$
$$= c \quad \text{[by (6)]}$$
$$\Rightarrow c\ R\ (a * b)$$

i.e. the last step shows that $a * b$ is the greatest lower bound of $\{a, b\}$.

Hence,
$$a * b = \inf \{ a, b \} = a \wedge b.$$

Similarly, we can show that $a \oplus b$ is the least upper bound of $\{a, b\}$ w.r.t partial ordering relation R. i.e.

$$a \oplus b = \sup \{a, b\} = a \vee b.$$

Hence the theorem.

Example 8.11

Show that in a lattice if $a \le b \le c$, then

i. $a \oplus b = b * c$ ii. $(a * b) \oplus (b * c) = b = (a \oplus b) * (a \oplus c)$.

Solution

i
$$a \oplus b = \sup (a, b) = b \quad \text{… (1)}$$
$$b * c = \inf (b, c) = b \quad \text{… (2)}$$

From (1) and (2) $a \oplus b = b * c$

ii.
$$(a * b) \oplus (b * c) = \sup ((a * b), (b * c))$$
$$= \sup (\inf (a, b), \inf(b, c))$$
$$= \sup (a, b) \quad [\because a \le b, b \le c]$$
$$= b [\because a \le b \quad \text{… (3)}$$

Again
$$(a \oplus b) * (a \oplus c) = \inf ((a \oplus b), (a \oplus c))$$
$$= \inf (\sup (a, b), \sup(a, c))$$
$$= \inf (b, c)\ [\because a \le b, a \le c]$$
$$= b\ [\because b \le c] \quad \text{...(4)}$$

From (3) and (4), we get

$$(a * b) \oplus (b * c) = b = (a \oplus b) * (a \oplus c).$$

8.3-1 Lattice homomorphism, isomorphism and endomorphism

If $\langle L, \wedge, \vee\rangle$ and $\langle L', *, \oplus\rangle$ are two lattices, then a **lattice homomorphism** is a function $f: L \to L'$ such that for any $a, b \in L$

$$f(a \wedge b) = f(a) * f(b) \text{ and } f(a \vee b) = f(a) \oplus f(b)$$

We observe that both the operations of meet and join are preserved.

On the other hand, let us consider the two lattices $\langle L, \wedge, \vee\rangle$ and $\langle L', *, \oplus\rangle$. Let the partial ordering relations on L and L' are R and R', corresponding to the operations meet and join respectively.

Now, if homomorphism $f: L \to L'$ of two lattices $\langle L, \wedge, \vee\rangle$ and $\langle L', *, \oplus\rangle$ is also one- one onto (i.e. bijective), then f is called an **isomorphism**. And the lattices are called **isomorphic lattices**.

If a homomorphism $f: L \to L$, where $\langle L, *, \oplus\rangle$ is a lattice, i.e. mapping on itself is known as **endomorphism.** If this mapping is isomorphic, i.e. $f: L \to L$ is an isomorphism, then f is called **automorphism**.

This theory of homomorphism and isomorphism is also applicable on posets.

8.3-2 Order Preserving

If we consider two partially ordered sets $\langle P, R\rangle$ and $\langle P', R'\rangle$ then the mapping $f: P \to P'$ is known to be **order preserving** iff a $R\, b\ \forall\ a, b \in P, f(a)\, R' f(b)$ in P' with respect to the ordering R in P and R' in P'.

Example 8.12

Show that if $f: L \to L$ is homeomorphic then f preserves the ordering relation.

Solution

If the function $f: L \to L'$ is a homomorphism then we can easily show that f preserves the ordering relations also. From $a\, R\, b \Leftrightarrow a \wedge b = a$

We have $\quad f(a \wedge b) = f(a) * f(b) = f(a) \Leftrightarrow f(a)\, R' f(b)$

i.e. $\quad a\, R\, b \Rightarrow f(a)\, R' f(b)$ if f is a homomorphism.

8.3-3 Order Isomorphic

For two posets $\langle P, R\rangle$ and $\langle P', R'\rangle$ if there exist a mapping $f: P \to P'$ which is bijective and if both f and f^{-1} are order preserving, then f is known as **order isomorphic**.

1. *Two lattices which are isomorphic can be represented by the same diagram.*
2. *If two lattices are order-isomorphic then they are also isomorphic lattice. Two order-isomorphism lattices are an equivalent to lattice isomorphism.*
3. *Hence by 1 and 2 two posets which are ordered isomorphic can be represented by the same diagram.*

Example 8.13

Let $L = \{a, b, c\}$ and $L' = \{1, 2, 3, 5, 6, 10, 15, 30\}$ are set of three alphabets and multiple of 30 respectively. Prove that the lattice $\langle P(L), \cap, \cup\rangle$ and $\langle D(L'), \text{gcd}, \text{lcm}\rangle$ are isomorphic.

Solution

We know that if two lattices are isomorphic then both have same Hasse diagram as shown in Fig. 8.9 and Fig.8.10. From the Hasse diagram it is clear that L and L' are isomorphic.

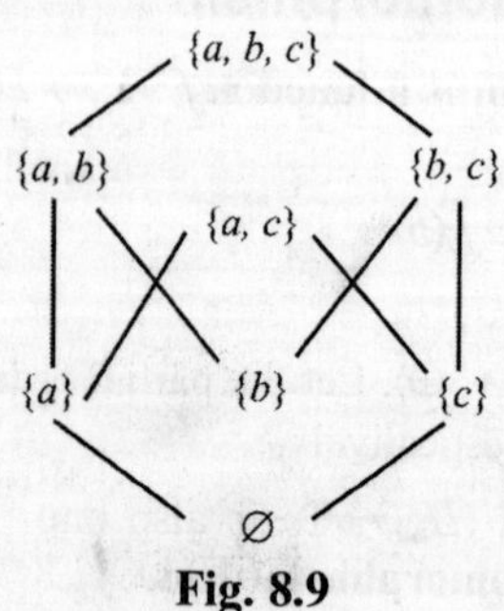

Fig. 8.9

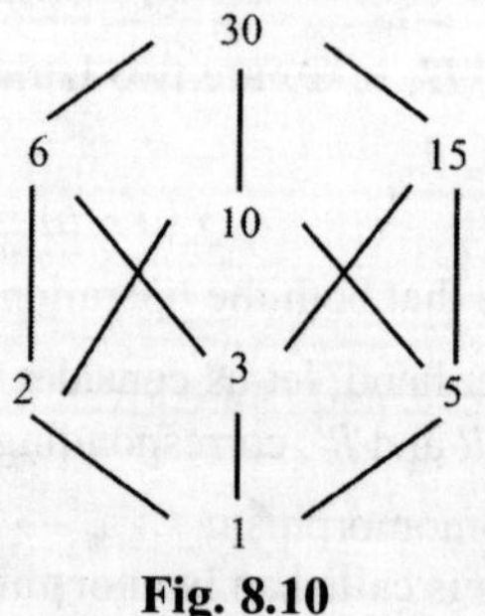

Fig. 8.10

Example 8.14

Are the two posets shown in Fig. 8.11 isomorphic?

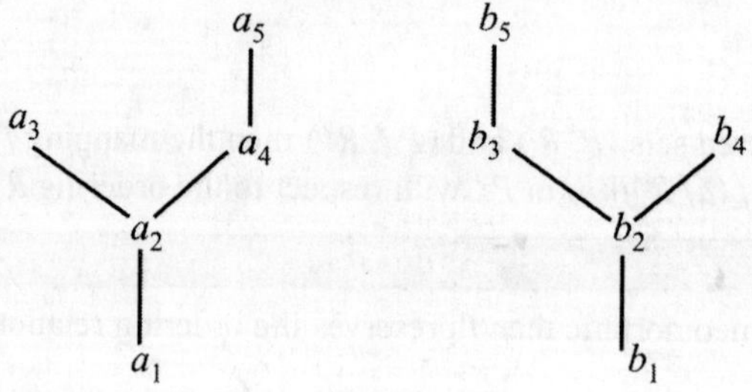

Fig. 8.11

Solution

Consider the mapping

$$f = \{(a_1, b_1), (a_2, b_2), (a_3, b_4), (a_4, b_3), (a_5, b_5)\}.$$

We see that this mapping preserves the structure of the original set. e.g.

$$f(a_3 \wedge a_4) = f(a_2) = b_2$$

and $$f(a_3) \wedge f(a_4) = b_4 \wedge b_3 = b_2.$$

Hence, two lattices shown in Fig. 8.11 are isomorphic.

Example 8.15

Are the lattices L and L' in Fig. 8.12 are isomorphic?

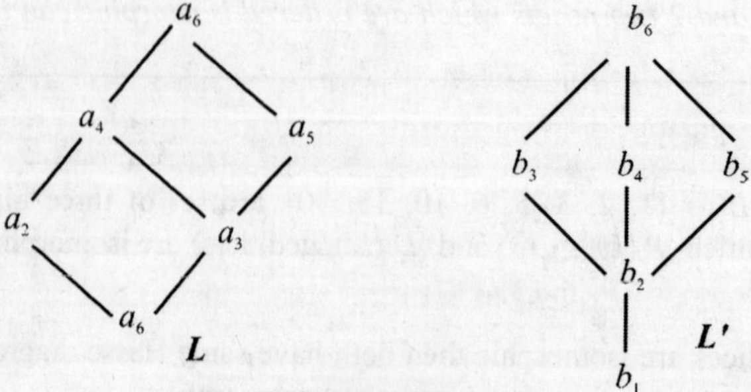

Fig. 8.12

Solution

From the Fig. 8.12 we observe that a_1 maps to b_1 and a_6 maps to b_6. Also, inf $(b_2, x) = b_2$ for $x \neq b_1$ in L'. But this is not true for a_2, a_3, a_4 or a_5 in L. Thus, there is no such $f : L \to L'$ which preserves mapping. Hence L and L' are not isomorphic.

Example 8.16

Two lattices L and L' are given in Fig. 8.13. Are they isomorphic?

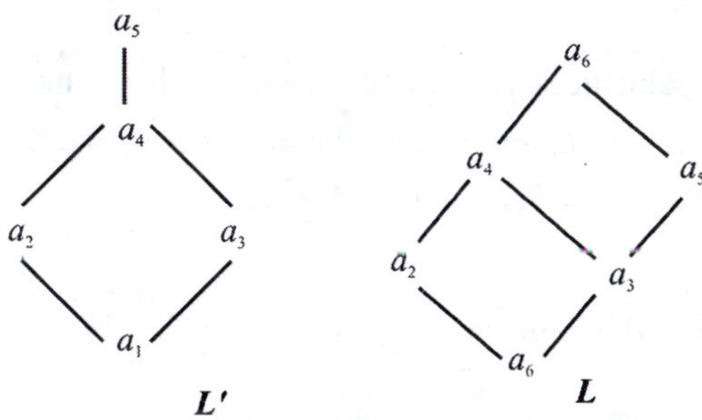

Fig. 8.13

Solution

From the Fig. 8.13, we notice that two lattices do not have same number of elements. Hence, trivially, no one-to-one and onto mapping exists that satisfies the condition. i.e. L and L' are not isomorphic.

8.3-4 Sublattice

Let $\langle L, \wedge, \vee\rangle$ be a lattice and consider a non-empty subset S of L. Then $\langle S, \wedge, \vee\rangle$ is a **sublattice** of $\langle L, \wedge, \vee\rangle$ iff S is closed under the operations $\wedge$ and $\vee$ i.e.

$$a \wedge b, a \vee b \in S \ \forall\ a, b \in S.$$

REMARK

For any two elements $a, b \in L$ such that $a \leq b$ and if $(L, \wedge, \vee)$ be a lattice and for a closed interval $[a, b]$; $x \in [a, b]$ such that $a \leq x \leq b$, $[a, b]$ is a sublattice of L.

8.3-5 Convex Sublattice

A sublattice $\langle S, \wedge, \vee\rangle$ of a lattice $\langle L, \wedge, \vee\rangle$ is called a **convex sublattice** if the closed set of inf and sup is a subset of S i.e.

$$[a \wedge b, a \vee b] \subseteq S \ \forall\ a, b \in S.$$

Example 8.17

Consider a lattice $L = \langle N, \text{gcd}, \text{lcm}\rangle$ and set D_n be all positive divisors of a fixed $n \in N$ (Natural number). Prove that $\langle D_n, \text{gcd}, \text{lcm}\rangle$ is a sublattice of $\langle N, \text{gcd}, \text{lcm}\rangle$.

Solution

Let d_1 and d_2 are two elements of set D_n. We know that lcm$\{d_1, d_2\}$ and gcd$\{d_1, d_2\}$ are again divisors of n i.e. D_n is closed under the operations lcm and gcd. Hence $\langle D_n, \text{gcd}, \text{lcm}\rangle$ is a sublattice of $\langle N, \text{gcd}, \text{lcm}\rangle$.

Example 8.18

Consider the lattice $\langle L, \wedge, \vee\rangle$ in Fig. 8.14. Determine whether or not each of the following is a sublattice of L:

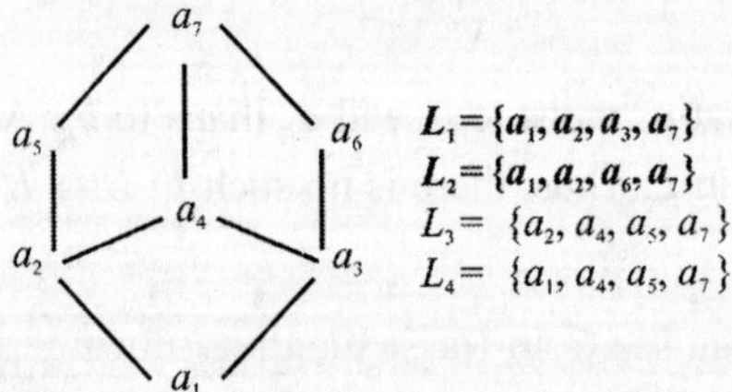

Fig. 8.14

Solution

According to the definition of sublattices, L' is a sublattice if it is closed under the operation of L, i.e. $\wedge$ and $\vee$. Hence from Fig. 8.14 we notice that L_2 and L_3 are sublattices but L_1 and L_4 are not. Because for $a_2, a_3 \in L_1$, $a_2 \vee a_3 = a_4 \notin L_1$, and similarly for $a_4, a_5 \in L_4$, $a_4 \wedge a_5 = a_2 \notin L_4$.

Example 8.19

Consider the lattice $L = \{a, b, c, d, e\}$ shown in Fig. 8.15. Find all sublattices with three or more elements.

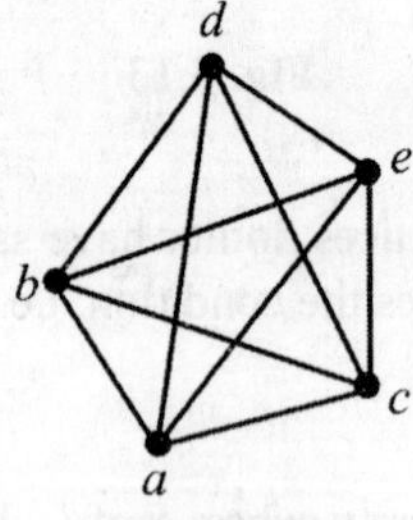

Fig. 8.15

Solution

On examining the figure and recalling the definition of sublattice, we observe that all the subsets with three or more elements where the supremum and infimum exist are sublattice.

Readers are advised to find them.

Example 8.20

Consider a lattice $L = \langle P(S), \cap, \cup \rangle$ where $P(S)$ is the power set of any set S. Then only those sub collections of $P(S)$ are sublattices of L which are closed under $\cap$ and $\cup$. For this, let us consider a set $S = \{a, b, c, d\}$ then $P(S) = \{\varnothing, \{a\}, \{b\}, \{c\}, \{d\}, \{a, b\}, \{a, c\}, \{a, d\}, \{b, c\}, \{b, d\}, \{c, d\}, \{a, b, c\}, \{a, b, d\}, \{b, c, d\}, \{a, c, d\}, \{a, b, c, d\}\}$.

And $$L' \subseteq P(S) = \{\varnothing, \{a\}, \{a, b\}, \{b\}, \{a, b, c\}\}$$

would be a sublattice but another set

$$L'' \subseteq P(S) = \{\varnothing, \{a\}, \{b\}, \{c\}, \{b, c\}\}$$

would not be a sublattice because $\{a\} \cup \{c\} = \{a, c\}$ is not in L''.

The collection of subsets of any set S such that for any two subsets S_1 and S_2 in this family both $S_1 \cap S_2$ and $S_1 \cup S_2$ are in the family is obviously a sublattice of L. Such a family is called a ***ring**** ***of subsets of S*** *or* ***lattice of subsets.***

* *ring of subset* is different from the ring of group theory discussed in Chapter 6.

Example 8.21

Show that it is not necessary that union of two sublattices is again a sublattice.

Solution

Let us consider a lattice $L = \{1, 2, 3, 4, 6, 12, 24\}$ of factors of 24 under divisibility. Then $L' = \{1, 2\}$ and $L'' = \{1, 3\}$ are sublattices of L as we can see from Hasse diagram shown in Fig. 8.16. But the subset $L' \cup L'' = \{1, 2, 3\}$ is not a sublattice.

$$\because\ 2, 3 \in L' \cup L''$$

But $2 \cup 3 = 6 \notin L' \cup L''$.

Hence union of two sublattices L' and L'' may not be sublattice.

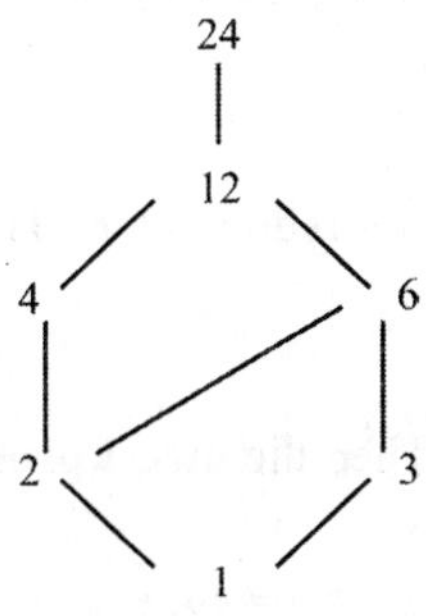

Fig. 8.16

Exercise 8.3

1. Consider the Lattice in Fig. 8.17 under relation $\leq$. Determine whether or not each of the following is a sublattice of L.

   $L_1 = \{a_1, a_2, a_4, a_6\}$, $L_2 = \{a_3, a_5, a_7, a_8\}$, $L_3 = \{a_1, a_2, a_4, a_8\}$.

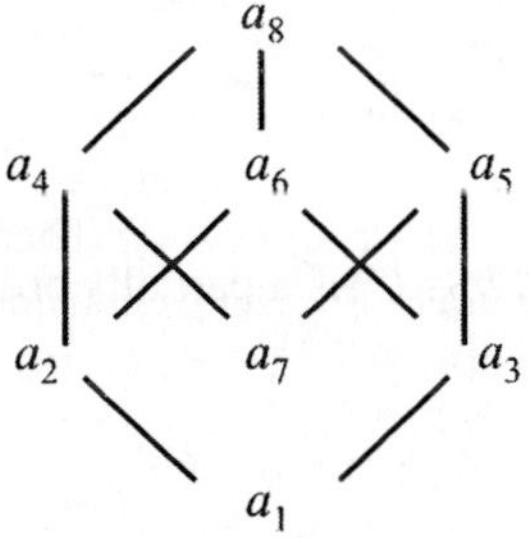

Fig. 8.17

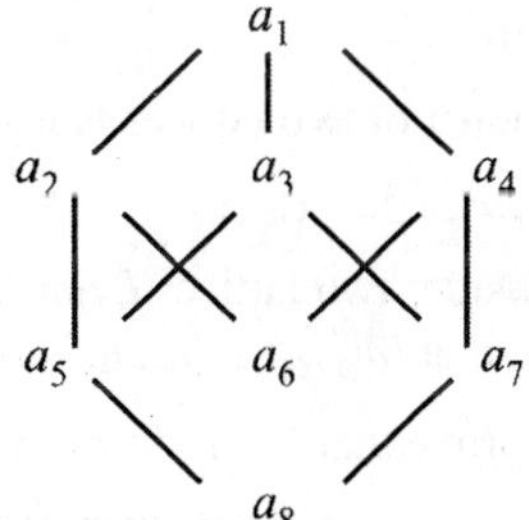

Fig. 8.18

2. Show that the set $L_1 = \{a_1, a_2, a_4, a_6\}$, $L_2 = \{a_3, a_5, a_7, a_8\}$ are ring of subsets of $S = \{p, q, r\}$ and sublattices of $\langle P(S), \cap, \cup\rangle$, where $a_1 = \{p, q, r\}$, $a_2 = \{p, q\}$, $a_3 = \{p, r\}$, $a_4 = \{q, r\}$, $a_5 = \{p\}$, $a_6 = \{q\}$, $a_7 = \{r\}$, $a_8 = \varnothing$ in Fig. 8.18.

3. Find all the sublattices of lattices with following Hasse diagram.

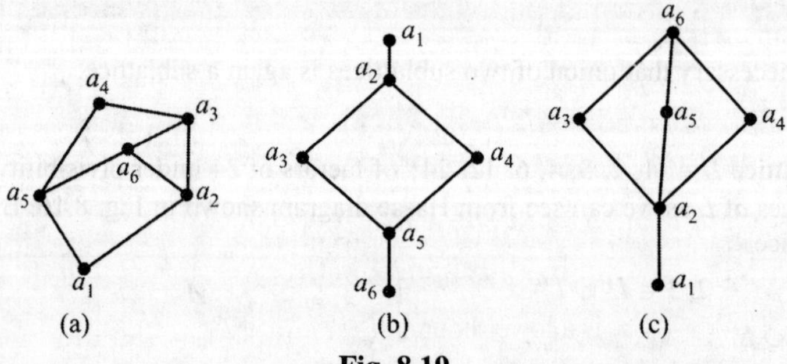

Fig. 8.19

4. Find all the sublattices of lattice D_{24}, where D_{24} is a set of all factors of 24 under divisibility.

Answers

1. L_3 is sublattice as it satisfies all the properties required for a lattice.

8.3-6 Direct Product

Consider two lattices $\langle L \wedge, \vee\rangle$ and $\langle L', *, \oplus\rangle$. Then the **direct product** of the lattices $\langle L \times L', \cdot, +\rangle$ in which the binary operations + and · on $L \times L'$ are given by

$$\langle a_1, b_1\rangle \cdot \langle a_2, b_2\rangle = \langle a_1 * a_2, b_1 \wedge b_2\rangle$$

and $$\langle a_1, b_1\rangle + \langle a_2, b_2\rangle = \langle a_1 \oplus a_2, b_1 \vee b_2\rangle$$

for any $$\langle a_1, b_1\rangle \text{ and } \langle a_2, b_2\rangle \in L \times L'.$$

All the properties of lattices such as commutativity, associativity and absorption also hold on the direct product, L × L′, under operations + and '·'.

Theorem 8.10

Direct product of two lattices is itself a lattice.

Proof

Let us consider two lattices L and L' and $L \times L' = \{(a, b): a \in L, b \in L'\}$is a partially ordered set under the relation ≤ such that $\langle a_1, b_1\rangle \leq \langle a_2, b_2\rangle \Leftrightarrow a_1 \leq a_2$ in L and $b_1 \leq b_2$ in L'.

Now, we prove that $L \times L'$ forms a lattice.

If $\langle a_1, b_1\rangle, \langle a_2, b_2\rangle \in L \times L'$ then $a_1, a_2 \in L$ and $b_1, b_2 \in L'$

∵ L and L' are lattices, hence sup and inf of $\{a_1, a_2\}$ and $\{b_1, b_2\}$ must exist in L and L' respectively.

Again, consider $$a_1 \wedge a_2 = \inf \{ a_1, a_2\}$$

and $$b_1 \wedge b_2 = \inf \{ b_1, b_2\}$$

Then $$a_1 \wedge a_2 \leq a_1, a_1 \wedge a_2 \leq a_2$$

and $$b_1 \wedge b_2 \leq b_1, b_1 \wedge b_2 \leq b_2$$

⇒ $$\langle a_1 \wedge a_2, b_1 \wedge b_2\rangle \leq \langle a_1, b_1\rangle$$

and $$\langle a_1 \wedge a_2, b_1 \wedge b_2\rangle \leq \langle a_2, b_2\rangle$$

$$\Rightarrow \qquad \langle a_1 \wedge a_2, b_1 \wedge b_2 \rangle \leq \langle a_1, b_1 \rangle$$

$$\text{and} \qquad \langle a_1 \wedge a_2, b_1 \wedge b_2 \rangle \leq \langle a_2, b_2 \rangle$$

$$\Rightarrow \qquad \langle a_1 \wedge a_2, b_1 \wedge b_2 \rangle \text{ is lower bound of } \{ \langle a_1, b_1 \rangle, \langle a_2, b_2 \rangle \}.$$

Now, suppose that $\langle a_3, b_3 \rangle$ is any lower bound of $\{ \langle a_1, b_1 \rangle, \langle a_2, b_2 \rangle \}$.

$$\text{Then} \qquad \langle a_3, b_3 \rangle \leq \langle a_1, b_1 \rangle \text{ and } \langle a_3, b_3 \rangle \leq \langle a_2, b_2 \rangle$$

$$\Rightarrow \qquad a_3 \leq a_1 ; a_3 \leq a_2 ; b_3 \leq b_1 ; b_3 \leq b_2$$

$$\Rightarrow \qquad a_3 \text{ is a lower bound of } \langle a_1, a_2 \rangle \text{ in } L$$

$$\text{and} \qquad b_3 \text{ is a lower bound of } \langle b_1, b_2 \rangle \text{ in } L'$$

$$\Rightarrow \qquad a_3 \leq a_1 \wedge a_2 = \inf \{ a_1, a_2 \}$$

$$\text{and} \qquad b_3 \leq b_1 \wedge b_2 = \inf \{ b_1, b_2 \}$$

$$\Rightarrow \qquad \langle a_3, b_3 \rangle \leq \langle a_1 \wedge a_2, b_1 \wedge b_2 \rangle$$

$$\Rightarrow \qquad \langle a_1 \wedge a_2, b_1 \wedge b_2 \rangle \text{ is greatest lower bound } \{ \langle a_1, b_1 \rangle, \langle a_2, b_2 \rangle \}.$$

Similarly, by duality, we have $\langle a_1 \vee a_2, b_1 \vee b_2 \rangle$ is least upper bound of $\{ \langle a_1, a_2 \rangle, \langle b_1, b_2 \rangle \}$. Hence $L \times L'$ is a lattice.

1. *All lattices not necessarily be written as a direct product of others.*
2. *Direct product of lattices can be used to construct large lattices from smaller ones.*
3. *The order of lattice formed by the direct product of two lattices is equal to the product of the orders of the lattices appearing in the direct product.*

8.3-7 Complete Lattice

A lattice L is said to be **complete** if every non-empty subset has a least upper bound and a greatest lower bound.

For example, if S be the class of all subsets of some universal set A and an equivalence relation $\leq$ defined as

$$X \leq Y \Rightarrow X \text{ is a subset of } Y$$

Then since $X \leq Y = X \cap Y$ and $X \vee Y = X \cup Y$, hence it is a lattice. Further it follows that every subset have a glb and lub and hence it is a complete lattice.

8.3-8 Finite Lattice

If every finite subset S of a lattice $\langle L, \wedge, \vee \rangle$ has a least upper bound and a greatest lower bound then S is known as **finite lattice**. The lub and glb of S may be represented by

$$\text{lub of } S \quad \overset{n}{\underset{i\;1}{}} a_i \text{ and glb of } S \quad \overset{n}{\underset{i\;1}{}} a_i \;\; \forall\, a_i \in S, i = 1, 2, \ldots, n.$$

REMARK

1. Every finite lattice is complete.
2. $(I^+, \preceq)$, $(R, \preceq)$, $(I, \preceq)$ etc. are not complete lattice since the set I^+, R and I are not finite.

8.3-9 Bounded Lattice

The least and the greatest element of a lattice are called **bounds** (units, universal bound) of the lattice and are denoted by 0 and 1 respectively. A lattice $\langle L, \wedge, \vee \rangle$ is said to be **bounded** if there exist $0, 1 \in L$ such that $\forall\, a \in L$, $0 \leq a \leq 1$. It is also represented as $\langle L, \wedge, \vee, 0, 1 \rangle$ and

$$\text{lub of } L = \bigvee_{i=1}^{n} a_i = 1 \text{ and glb of } L = \bigwedge_{i=1}^{n} a_i = 0$$

1. *It is to be noted that*

$$a \vee 0 = a; \quad a \wedge 0 = 0$$
$$a \vee 1 = 1; \quad a \wedge 1 = a$$

i.e. 0 and 1 are the identity of the operations $\vee$ and $\wedge$ respectively.

2. *In bounded lattice, 1 and 0 are dual to each other.*
3. *All the properties of lattices such as commutativity, associativity and absorption also hold good in case of bounded lattice.*

Theorem 8.11

Every finite lattice L is bounded.

Proof

Consider a finite lattice $L = \{a_1, a_2, ..., a_n\}$. Then the greatest element of L is $a_1 \vee a_2 \vee a_3 \vee ... \vee a_n$ and least element is $a_1 \wedge a_2 \wedge a_3 \wedge ... \wedge a_n$. Hence if lattice is finite then it is also bounded.

REMARK

1. $\langle P(S), \cap, \cup \rangle$ is bounded lattice with $0 = \varnothing$ and $1 = S$, where $P(S)$ is the power set of any set S.
2. $\langle N, | \rangle$ is not bounded. Since 1 is the lower bound of the lattice but it has no upper bound.
3. Lattice $\langle D_n, \preceq \rangle$ is bounded. Since 1 is the lower bound and n is the upper bound in the set D_n, where D_n is the set of all multiples of n.

Example 8.22

Consider a set $L = \{x \in Q : 0 \le x \le 2\}$ i.e. the set of all rational numbers between 0 and 2. Show that L is not complete and find out the bounds of L.

Solution

L is bounded with 0 and 2 as the lower and upper bounds of L. As we know for completeness, every non-empty subsets of L has a lub and glb hence L is complete. But if x belongs to irrational number then L is not complete. e.g. the subset $A = \{x \in L : x^2 < 2\}$ does not have the lub in L because $\sqrt{2}$ is not a rational number, hence lub of A does not exist. Thus L is not a complete lattice although it is bounded.

8.3-10 Complemented Lattice

Let $\langle L, \wedge, \vee, 0, 1 \rangle$ be bounded lattice and $a, b \in L$ such that $a \wedge b = 0$ and $a \vee b = 1$ then b is known as **complement** of a and a is the complement of b, i.e. complement is symmetric. Complement of a is represented by $\bar{a}$ or a'.

If in a bounded lattice $\langle L, \wedge, \vee, 0, 1 \rangle$ every element has at least one complement then the lattice is known as **complemented lattice**.

0 and 1 are complement of each other.

Example 8.23

Find out whether the lattice $\langle D_{12}$, gcd, lcm$\rangle$ is complemented, where D_{12} is the set of all factors of 12 under divisibility.

Solution

Here in $D_{12} = \{1, 2, 3, 4, 6, 12\}$, 3 and 4 are complements of each other because gcd $\{3, 4\} = 1$ and lcm $\{3, 4\} = 12$.

Since gcd $\{2, 3\} = 1$; lcm $\{2, 3\} = 6 \neq 12 \Rightarrow 3$ is not a complement of 2 and vice versa.

Similarly, gcd $\{2, 4\} = 2 \neq 1$; lcm $\{2, 4\} = 4 \neq 12 \Rightarrow 4$ is not a complement of 2 and vice versa.

Proceeding is this way we find, $\langle D_{12}$, gcd, lcm$\rangle$ is not a complemented lattice.

Example 8.24

Show that whether the lattice $\langle L^n, \leq_n\rangle$ is a complemented lattice whenever if $L = \{0, 1\}$ then the lattice $\langle L, \leq\rangle$, $L = \{\langle 0, 0\rangle, \langle 0, 1\rangle, \langle 1, 0\rangle, \langle 1, 1\rangle\}$ then the lattice is $\langle L^2, \leq_2\rangle$ and so on, where L^n is a lattice of n tuple.

Solution

Let $\langle L^n, \leq_n\rangle$ be the lattice of n tuples of $\langle 0, 1\rangle$. We can prove it by taking a special case $n = 3$. The lattice $\langle L^3, \leq_3\rangle$ can be represented by Hasse diagram as shown in Fig. 8.20. The bounds of lattice are $\langle 0, 0, 0\rangle$ and $\langle 1, 1, 1\rangle$. The complement of $\langle 1, 0, 0\rangle$ is $\langle 0, 0, 1\rangle$ Similarly we can prove it for n tuples. i.e. $\langle L^n, \leq_n\rangle$ is bounded and each element has at least one complement. Hence $\langle L^n, \leq_n\rangle$ is a complemented lattice.

Example 8.25

Consider a set $X = \{a, b, c\}$ then the lattice $\langle P(X), \cap, \cup\rangle$ is bounded lattice with $0 = \varnothing$ and $1 = X$ bounds. Now if we take $S \subseteq X = \{a, b\}$ then $S = \{c\}$ such that $S \cap \bar{S} = \varnothing$ and $S \cup \bar{S} = X$, i.e. every element of $P(X)$ has a complement. Hence $\langle P(X), \cap, \cup, \varnothing, X\rangle$ is a complemented lattice.

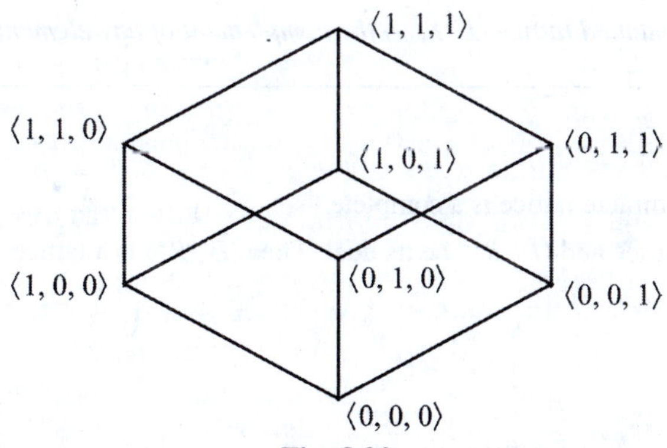

Fig. 8.20

Example 8.26

Consider the following lattices with their Hasse diagrams.

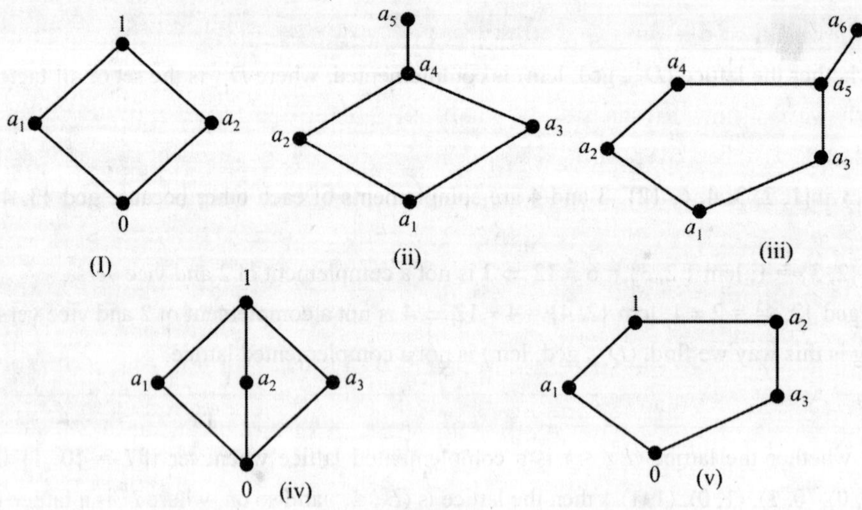

Fig. 8.21

Solution

i. a_1 and a_2 are complement of each other.

ii. Here $0 = a_1$ and $1 = a_5$ but there is no x corresponding to a_3 such that $a_3 \vee x = a_5$ and $a_3 \wedge x = a_1$.

iii. It is complemented lattice with $0 = a_1$ and $1 = a_5$.

iv. The complements of a_1 are a_2 and a_3. The complements of a_2 are a_1 and a_3. The complements of a_3 are a_1 and a_2. Further 0 and 1 are complement of each other. Hence the lattice is complemented lattice.

v. The complements of a_1 are a_2 and a_3. The complements of a_2 are a_1 and a_3. The complements of a_3 are a_1 and a_2. Hence the lattice is complemented lattice.

For any complemented lattice $\langle L, \wedge, \vee\rangle$ the complement of any element of L need not be unique.

Exercise 8.4

1. Prove that the dual of complete lattice is a complete.

[**Hint**: Let $\langle L, R\rangle$ be a lattice and $\langle L', R'\rangle$ be its dual. Then $\langle L', R'\rangle$ is a lattice.

Let $S \subseteq L'$ and $\emptyset \notin S$

Now let $a = \inf$ of $S \in L$

Then $a\, R\, x\ \forall\, x \in L$

$\Rightarrow x\, R'\, a\ \forall\, x \in L'$

$\Rightarrow a$ is an upper bound of S in L'.

Again, let us assume any element b such that b be any upper bound of S in L then

$x\, R'\, a\ \forall\, x \in L'$

$\Rightarrow b\, R\, x\ \forall\, x \in L$

$\Rightarrow b\, R\, a$ as a is inf of S in L

$\Rightarrow a\, R'\, b$

$\Rightarrow a$ is least upper bound of S in L.

Similarly we can show that sup of S in L will be inf of S in L'. Hence $\langle L', R'\rangle$ is complete.]

2. Check whether $\langle L, \vee, \wedge\rangle$ be a complemented lattice under the relation "$|$" i.e. divides where $L = \{1, 2, 3, 5, 30\}$.
3. For the following complemented lattice $\langle L, \vee, \wedge\rangle$ determine the complement of every element in L.

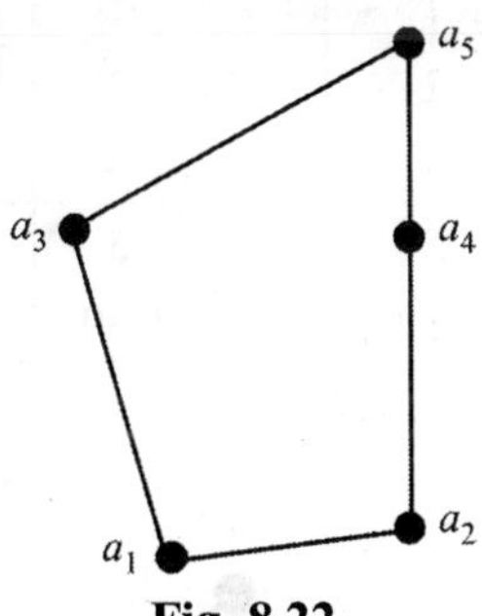

Fig. 8.22

4. Find the complement of each element in D_{105} where D_{105} is divisors of 105.
5. Which of the two lattices $\langle D_n, |\rangle$ for $n = 30$ and $n = 45$ are complemented?
6. Prove that in a lattice with two or more elements, no element is its own complement.
7. Consider the lattice $D_{12} = \{1, 2, 3, 4, 6, 12\}$ the divisors of 12 ordered by divisibility. Find
   a. The lower and upper bound of D_{12}.
   b. The complement (s) of 4 and 6.
   c. Is D_{12} a complemented lattice?
8. Any chain more than two elements [other than the bounds 0 and 1] is not a complemented lattice. Prove it.

Answers

7. a. Lower bound = 1, upper bound = 12.
   b. Complement of 4 = 3, 6 has no complement.
   c. D_{12} is not a complemented lattice.

8.3-11 Distributive lattice

Consider a lattice $\langle L, \vee, \wedge\rangle$ and let $a, b, c \in L$ then L is known as **distributive lattice** if following distributive properties holds:

$$a \wedge (b \vee c) = (a \wedge b) \vee (a \wedge c)\ \forall\ a, b, c \in L$$

and

$$a \vee (b \wedge c) = (a \vee b) \wedge (a \vee c)\ \forall\ a, b, c \in L$$

otherwise L is said to be non-distributive.

1. *Both the above equalities are equivalent, therefore to check whether the lattice is distributive or not, it is sufficient to verify one of them.*
2. *Distributive Lattice follows the duality principle.*

Example 8.27

Consider a poset $L = \{1, 2, 3, 5, 30\}$ under the relation 'divides'. Show that $\langle L, \vee, \wedge\rangle$ is a distributive lattice.

Solution

For the poset $L = \{1, 2, 3, 5, 30\}$ the operation tables for meet and 'join' are shown in Table 8.6 (a) and Table 8.6(b) respectively.

Table 8.6(a)

$\wedge$	1	2	3	5	30
1	1	1	1	1	1
2	1	2	1	1	2
3	1	1	3	1	3
5	1	1	1	5	5
30	1	2	3	5	30

Table 8.6(b)

$\vee$	1	2	3	5	30
1	1	2	3	5	30
2	2	2	30	30	30
3	3	30	3	30	30
5	5	30	30	5	30
30	30	30	30	30	30

Hasse diagram of given poset may be represented as

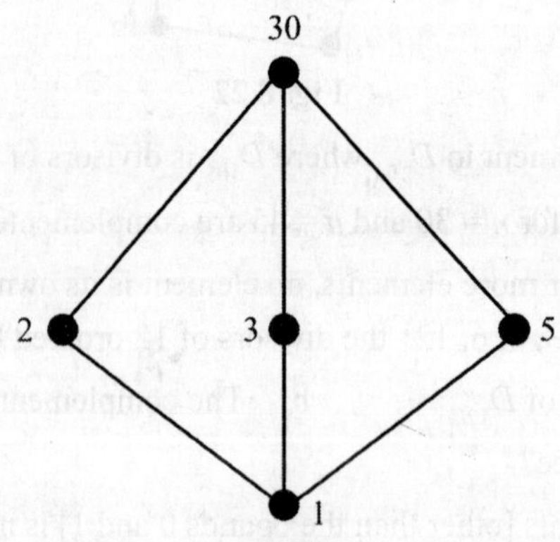

Fig. 8.23

As, every pair of elements in L has both a glb and a lub, therefore $\langle L, \vee, \wedge\rangle$ is a lattice under the relation 'divides'.

Now to show that it is distributive, let us consider three elements 2, 5 and 30 of poset L then

$$2 \vee (5 \wedge 30) = 2 \vee (5) = 30$$

and
$$(2 \vee 5) \wedge (2 \vee 30) = 30 \wedge (30) = 30$$

it shows that $\langle L, \wedge, \vee\rangle$ is distributive lattice. But if we take the elements 2, 5 and 3 of poset L then

$$2 \vee (5 \wedge 3) = 2 \vee 1 = 2 \quad \ldots (1)$$

and
$$(2 \vee 5) \wedge (2 \vee 3) = 30 \wedge 30 = 30 \quad \ldots (2)$$

(1) and (2) are not equal i.e. it is not a distributive lattice. Hence, we can conclude that $\langle L, \vee, \wedge\rangle$ is not a distributive lattice.

Theorem 8.12

Let $\langle L, \vee, \wedge\rangle$ is a complemented and distributive lattice, and any element $a \in L$, then the complement $\bar{a}$ is unique.

Proof

To prove the theorem, let us consider an element $a \in L$ having two complements say a_1 and a_2, then by definition of complement.

$$a \wedge a_1 = 0 \quad \ldots (1); \qquad a \vee a_1 = 1 \quad \ldots (1')$$

Similarly $$a \wedge a_2 = 0 \quad \ldots (2); \qquad a \vee a_2 = 1 \quad \ldots (2')$$

Now,

$$\begin{aligned}
a_1 &= a_1 \wedge 1 = a_1 \wedge (a \vee a_2) && \text{[by (2')]} \\
&= (a_1 \wedge a) \vee (a_1 \wedge a_2) && \text{[by distributivity]} \\
&= 0 \vee (a_1 \wedge a_2) && \text{[by(1)]} \\
&= (a_2 \wedge a) \vee (a_2 \wedge a_1) && \text{[by(2)]} \\
&= a_2 \wedge (a \vee a_1) && \text{[by distributivity]} \\
&= a_2 \wedge 1 = a_2 && \text{[by (1')]}
\end{aligned}$$

Hence $a_1 = a_2$, which contradicts our assumption that $a \in L$ has two different complements, i.e. complement of any element of L is unique.

Example 8.28

Prove that if a and b are elements in a bounded distributive lattice and if a has a complement a', then

i. $a \vee (a' \wedge b) = a \vee b$ ii. $a \wedge (a' \vee b) = a \wedge b$.

Solution

i. $$\begin{aligned}
\text{LHS} = a \vee (a' \wedge b) &= (a' \vee a') \wedge (a \wedge b) \\
&= 1 \wedge (a \vee b) && [\because a \vee a' = 1] \\
&= a \vee b && [1 \wedge a = a] \\
&= \text{RHS}
\end{aligned}$$

ii. $$\begin{aligned}
\text{LHS} = a \wedge (a' \vee b) &= (a' \wedge a') \vee (a \wedge b) && \text{[distributivity]} \\
&= 0 \vee (a \wedge b) && [\because a \wedge a' = 0] \\
&= a \wedge b && [\because 0 \vee a = a] \\
&= \text{RHS.}
\end{aligned}$$

Theorem 8.13

Let $\langle L, \vee, \wedge \rangle$ be a complemented, distributive lattice then for any element $a \in L$, $\bar{\bar{a}} = a$.

Proof

Readers can easily prove it from the definitions and uniqueness of complement of a.

The statement of Theorem 8.13 *is also known as involution law.*

8.3-12 Properties of Distributive lattices

1. De morgan's Laws

Theorem 8.14

If $\langle L, \wedge, \vee \rangle$ be a complemented and distributive lattice then for any $a, b \in L$

$$\overline{a \vee b} = \bar{a} \wedge \bar{b} \quad \ldots (1)$$

$$\overline{a \wedge b} = \bar{a} \vee \bar{b} \quad \ldots (2)$$

Proof

Let us consider

$$(a\vee b)\wedge(\bar{a}\wedge\bar{b}) = (a\wedge(\bar{a}\wedge\bar{b}))\vee(b\wedge(\bar{a}\wedge\bar{b})) \quad \text{(by distributivity)}$$

$$= ((a\wedge\bar{a})\wedge\bar{b})\vee(b\wedge(\bar{b}\wedge\bar{a})) \quad \text{(by associativity)}$$

$$=(0\wedge\bar{b})\vee((b\wedge\bar{b})\wedge\bar{a}) = 0\vee(0\wedge\bar{a}) = 0\vee 0 = 0 \quad \text{...(3)}$$

also, $$(a\vee b)\vee(\bar{a}\wedge\bar{b}) = (a\vee(\bar{a}\wedge\bar{b}))\vee(b\vee(\bar{a}\wedge\bar{b})) \quad \text{(by distributivity)}$$

$$= ((a\vee\bar{a})\wedge(a\vee\bar{b}))\vee((b\vee\bar{b})\wedge(b\vee\bar{a})) \quad \text{(by distributivity)}$$

$$= (1\wedge(a\vee\bar{b}))\vee(1\wedge(b\vee\bar{a})) = (a\vee\bar{b})\vee(b\vee\bar{a})$$

$$= (a\vee\bar{b}\vee b)\vee\bar{a} \quad \text{(by associativity)}$$

$$= (a\vee 1)\vee\bar{a} = a\vee\bar{a} = 1 \quad \text{...(4)}$$

Thus from (3) and (4), we observe that $(a \vee b)$ and $\bar{a} \wedge \bar{b}$ are complement to each other. Hence $\overline{a\vee b} = \bar{a}\wedge\bar{b}$. (2) can be proved easily by using the principle of duality.

2. Joint Cancellation Law

Theorem 8.15

For any distributive lattice $\langle L, \vee, \wedge\rangle$ the joint cancellation law i.e.

$$(a \vee b = a \vee c) \wedge (a \wedge b = a \wedge c) \Rightarrow b = c$$

holds for any $a, b, c \in L$.

Proof

For any $a, b, c \in L$, suppose

$$(a \vee b) = (a \vee c) \text{ ...(1) and } (a \wedge b = a \wedge c) \text{ ...(2)}$$

Then

$$\begin{aligned}
B &= b \vee (a \wedge b) && \text{[by absorption]}\\
&= b \vee (a \wedge c) && \text{[by hypothesis (2)]}\\
&= (b \vee a) \wedge (b \vee c) && \text{[by distributivity]}\\
&= (a \vee b) \wedge (b \vee c) && \text{[by commutativity]}\\
&= (a \vee c) \wedge (b \vee c) && \text{[by hypothesis (1)]}\\
&= (a \wedge b) \vee c && \text{[by distributivity]}\\
&= (a \wedge c) \vee c && \text{[by hypothesis (2)]}\\
&= c. && \text{[by absorption]}
\end{aligned}$$

Theorem 8.16

In a distributive lattice $\langle L, \vee, \wedge\rangle$ for any $a, b, c \in L$

$$(a \wedge b) \vee (b \wedge c) \vee (c \wedge a) = (a \vee b) \wedge (b \vee c) \wedge (c \vee a) \text{ holds.}$$

Proof

Let L be a distributive lattice, then we have

$$(a \wedge b) \vee (b \wedge c) \vee (c \wedge a) = \{ ((a \wedge b) \vee b) \wedge ((a \wedge b) \vee c) \} \vee (c \wedge a) \quad \text{[by distributivity]}$$

$$= \{b \wedge ((a \vee c) \wedge (b \vee c))\} \vee (c \wedge a) \quad \text{[by absorption]}$$

$$= \{(a \vee c) \wedge (b \wedge (b \wedge c))\} \vee (c \wedge a)$$

$$= (a \vee c) \wedge b)\} \vee (c \wedge a) \quad \text{[by absorption]}$$

$$= ((a \vee c) \vee (c \wedge a)) \wedge (b \vee (c \wedge a)) \quad \text{[by distributivity]}$$

$$= \{((a \vee c) \vee c) \wedge ((a \vee c)) \vee a)\} \wedge \{ ((b \vee c) \wedge (b \vee a))\} \quad \text{[by distributivity]}$$

$$= \{(a \vee c) \wedge (a \vee c)\} \wedge \{(b \vee c) \wedge (a \vee b)\}$$

$$= (a \vee c) \wedge (b \vee c) \wedge (a \vee b)$$

$$= (a \vee b) \wedge (b \vee c) \wedge (c \vee a)$$

Hence the theorem.

Example 8.29

Show that the pentagonal lattice shown in Fig. 8.24 is not a distributive lattice.

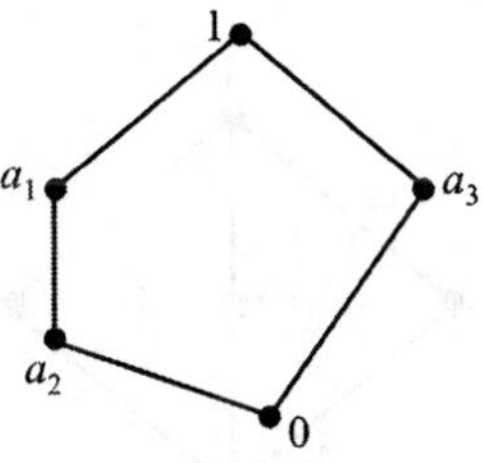

Fig. 8.24

Solution

From the figure one can observe that:

The complements of a_1 are a_2 and a_3.

The complements of a_2 are a_1 and a_3.

The complements of a_3 are a_2 and a_3.

Hence, $\quad a_1 \wedge (a_2 \vee a_{3)} = a_1 \wedge 1 = a_1 \quad$ [by $a \vee \bar{a} = 1$]

while $\quad (a_1 \wedge a_2) \vee (a_1 \wedge a_{3)} = a_2 \vee 0 = a_2$

Thus, $\quad a_1 \neq a_2$.

But for the combination

$$a_3 \wedge (a_1 \vee a_{2)} = 0 = (a_3 \wedge a_{1)} \vee (a_3 \wedge a_{2)}$$

Hence Lattice L is not distributive for first combination, but it is distributive for the second combination.

REMARK

A lattice is distributive if and only if it does not contain five elements pentagonal.

Modular Lattice

Let us consider a lattice $\langle L, \vee, \wedge\rangle$ and let $a, b, c \in L$ then L is known as **modular lattice** if it holds following:

$$a \leq c \Rightarrow a \vee (b \wedge c) = (a \vee b) \wedge c.$$

Theorem 8.17

Every distributive lattice is modular but converse is not true.

Proof

Let $\langle L, \wedge, \vee\rangle$ be a lattice under the relation $\leq$. For any $a, b, c \in L$

if $a \leq c$ then $a \vee c = c$ ...(1)

If L is distributive, then we have

$$a \vee (b \wedge c) = (a \vee b) \wedge (a \vee c)$$

$$= (a \vee b) \wedge c \quad \text{[by (1)]}$$

Hence, distributive lattice is modular.

But there are some modular lattices (see example 8.30) which are not distributive.

Example 8.30

Show that the diamond lattice (which is not distributive) is modular.

Solution

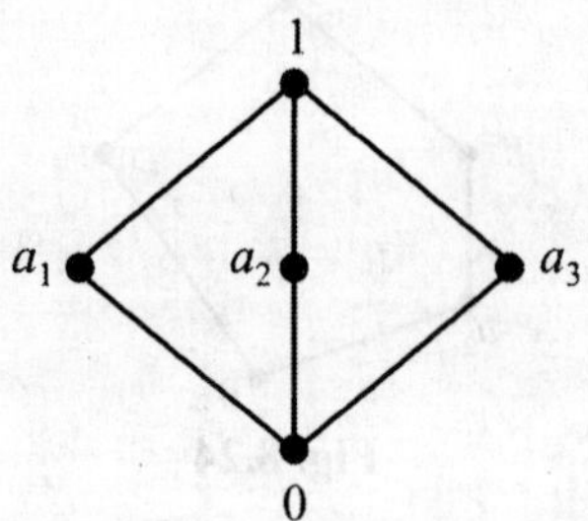

Since the diamond lattice $\langle L, \vee, \wedge\rangle$ is symmetric w.r.t. a_1, a_2, a_3 whenever

$a < c$ are $a_1 < 1$ and $0 < a_1$.

Thus if we take $a = a_1$ and $c = 1$

Then $a \vee (b \wedge c) = a_1 \vee (b \wedge 1) = a_1 \vee b$

and $(a \vee b) \wedge c = (a_1 \vee b) \wedge 1 = b_1 \vee b$ for any $b \in L$.

Similarly, if $a = 0, c = a_1$.

$a \vee (b \wedge c) = 0 \vee (b \wedge a_{1)} = b \wedge b_1$

while $(a \vee b) \wedge c = (0 \vee b) \wedge b_1 = b \wedge b_1$

Hence the modularity holds even if the lattice is non-distributive.

Example 8.31

Prove that the pentagonal lattice is not modular.

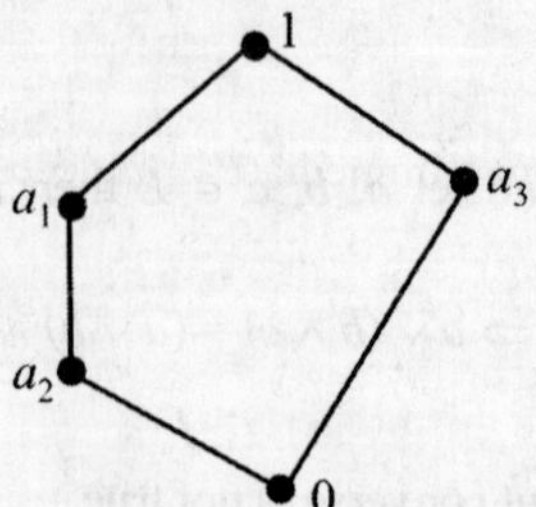

Solution

Let us consider a lattice $\langle L, \vee, \wedge\rangle$ and let $a, b, c \in L$.

If we take $a = a_2, b = a_3, c = a_1$ then $a < c$.

But $a \vee (b \wedge c) = a_2 \vee (a_3 \wedge a_1) = a_2 \vee 0 = a_2$

whereas $(a \vee b) \wedge c = (a_2 \vee a_3) \wedge a_1 = 1 \wedge a_1 = a_1$

i.e. $a_2 \neq a_1$. Hence pentagonal lattice is not modular.

REMARK

A modular lattice is distributive if and only if it does not contain a diamond lattice as one of its sublattices.

Example 8.32

Every lattice $\langle L, \vee, \wedge\rangle$ is modular for all $a, b, c \in L$ if

$$(a \wedge b) \vee (b \wedge c) \vee (c \wedge a) = (a \vee b) \wedge (b \vee c) \wedge (c \vee a).$$

Solution

Let us consider

$$(a \wedge b) \vee (b \wedge c) \vee (c \wedge a) = (a \vee b) \wedge (b \vee c) \wedge (c \vee a)$$

If $a \leq c \Rightarrow a \wedge c = a$ and $a \vee c = c$

Thus for $a \leq c$, we get

$$(a \wedge b) \vee (b \wedge c) \vee a = (a \vee b) \wedge (b \vee c) \wedge c$$

or $$\{a \vee (a \wedge b)\} \vee (b \wedge c) = (a \vee b) \wedge (b \vee c) \wedge c$$

or $$a \vee (b \wedge c) = (a \vee b) \wedge c$$ [by absorption]

Hence, L is a modular lattice.

8.4 Join Irreducible Elements and Atoms

In a lattice $\langle L, \wedge, \vee\rangle$ with lower bound 0, an element $a \in L$ is said to be **join irreducible** if

$$a = x_1 \vee x_2 \Rightarrow a = x_1 \text{ or } a = x_2{}^*.$$

And those elements which immediately succeed 0, are called **atoms**. These elements are also join irreducible.

It is not necessary that all join irreducible elements are atoms.

Theorem 8.18

In a lattice L, if $a \in L$ such that $a \neq 0$ then a is join irreducible if and only if it has a unique immediate predecessor.

Proof

Let us consider an element $a \in L$ has two immediately predecessors say a_1 and a_2 as shown in Fig. 8.25(a).

---

* Prime numbers under multiplication have this property

i.e. $P = x_1 x_2 \Rightarrow P = x_1$ or $P = x_2$.

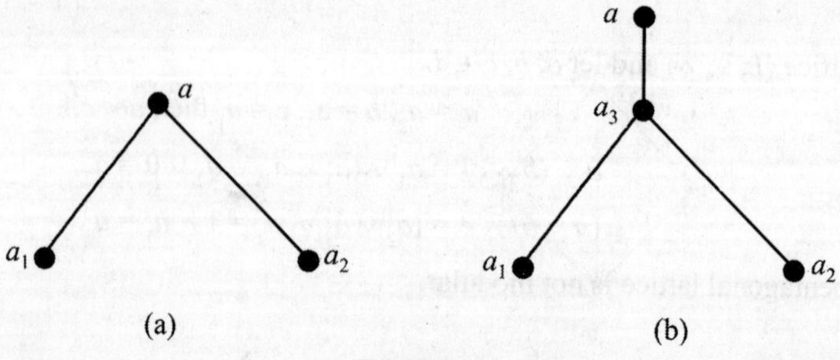

Fig 8.25

Then $a = a_1 \vee a_2$ i.e. $a \neq a_1$ or $a \neq a_2$.

Thus a is not join-irreducible.

Now if we take another example in which a has a unique predecessor as shown in Fig. 8.25(b) then for a_1 and a_2

$a \neq \sup \{a_1, a_2\}$ since a_3 lie between them. Hence the theorem.

Example 8.33

Find the join-irreducible elements of lattice L shown in Fig. 8.26.

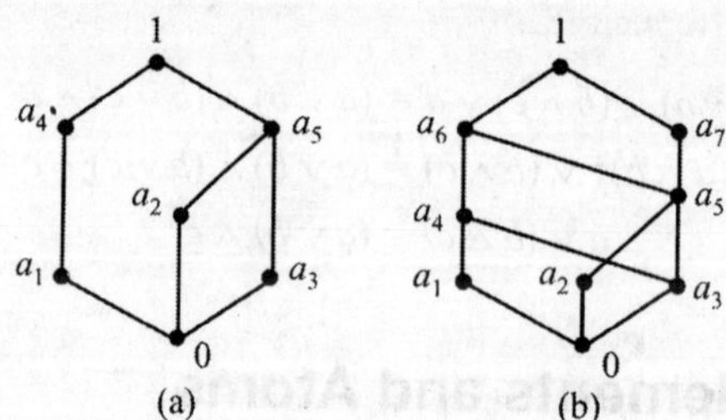

Fig. 8.26

Solution

Since the elements with unique immediate predecessor are join irreducible. Hence in Fig. 2.26(a) a_1, a_2, a_3, a_4 are join irreducible. And in Fig. 8.26(b) a_1, a_2, a_3 and a_7 are join irreducible elements.

8.5 Meet Irreducible

If L is a lattice with an upper bound 1 then an element $a \in L$ is said to be **meet irreducible** if $a = a_1 \wedge a_2 \Rightarrow a = a_1$ or $a = a_2$. $a \neq 1$ is meet irreducible iff a has only one immediate successor.

Example 8.34

Find the meet-irreducible elements of lattice L shown in Fig. 8.26.

Solution

In Fig. 8.26(a), the meet irreducible elements are a_1, a_2, a_3, a_4, a_5.

In Fig. 8.26(b), the meet irreducible elements are a_1, a_2, a_4, a_6, a_7.

Example 8.35

Find the atoms for the lattices L shown in Fig. 8.27 where a_1 is the lower bound.

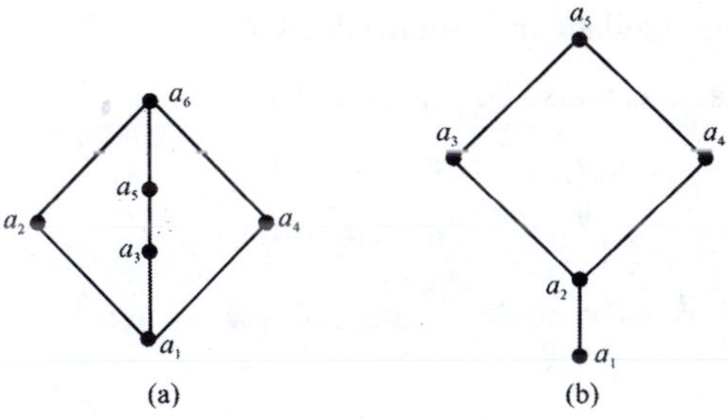

Fig. 8.27

Solution

By the definition of atoms, we can see in Fig. 8.27(a) a_2, a_3, and a_4 are atoms and in Fig. 8.27(b) a_2 is an atom.

Example 8.36

Let $\langle N, | \rangle$ be a lattice under the relation divisibility, where N is a set of natural number. (Here 1 is the lower bound). Find the elements of N, which are atoms and which are join irreducible?

Solution

All the prime numbers are the atoms because they immediately succeed the 0 element i.e. 1. All powers of prime numbers and 1 itself are join irreducible.

8.6 Irredundant Expression

In a lattice $\langle L, \wedge, \vee \rangle$ an element a can be expressed by $a = a_1 \vee a_2 \vee \ldots \vee a_n$ where a_i *s* are join irreducible and if $a_j \vee a_k = a_k$ then we can delete a_k from the expression a. Thus the expression is irredundant if no a_j precedes a_k.

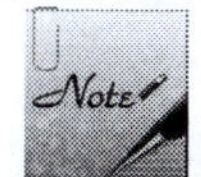

The decomposition of an element $a \in L$ into irredundant and join-irreducible elements is not unique.

Example 8.37

Find an irredundant and join irreducible decomposition for the element a_7 in Fig. 8.28.

Solution

a. Here $a_7 = a_4 \vee a_6$.

But $a_4 = a_1 \vee a_2$ and $a_6 = a_2 \vee a_3 \Rightarrow a_7 = a_1 \vee a_2 \vee a_2 \vee a_3$.

Since $a_2 \vee a_2 = a_2$, by deleting a_2 from our expression, we get $a_7 = a_1 \vee a_2 \vee a_3$

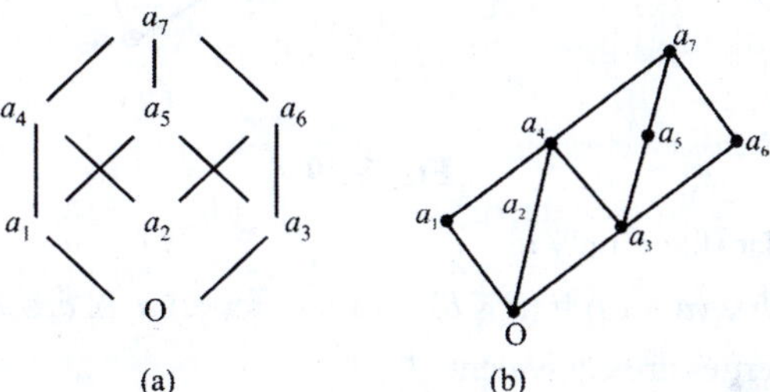

Fig. 8.28

where a_1, a_2 and a_3 are irredundant and join irreducible.

Readers may observe that a_7 can also be expressed by

$$a_7 = a_4 \vee a_5$$

$$a_7 = a_5 \vee a_6.$$

(b) Here $a_7 = a_4 \vee a_6$. But $a_4 = a_1 \vee a_2$.

$\therefore a_7 = a_1 \vee a_2 \vee a_6$.

Since a_1, a_2 and a_6 are irredundant and join irreducible.

But there is also an alternative solution:

$$a_7 = a_5 \vee a_6 \text{ and } a_7 = a_2 \vee a_3 \vee a_5$$

Also $\qquad a_7 = a_2 \vee a_3 \vee a_6$

Exercise 8.5

1. Show that the lattice of the power set X, i.e. $\langle P(X), \cap, \cup \rangle$ is a distributive lattice under the relation $\subseteq$.
2. Prove that in a lattice direct product of any two distributive lattice is a distributive lattice.
3. Check whether the lattices shown in Fig. 8.29 are distributive or modular.

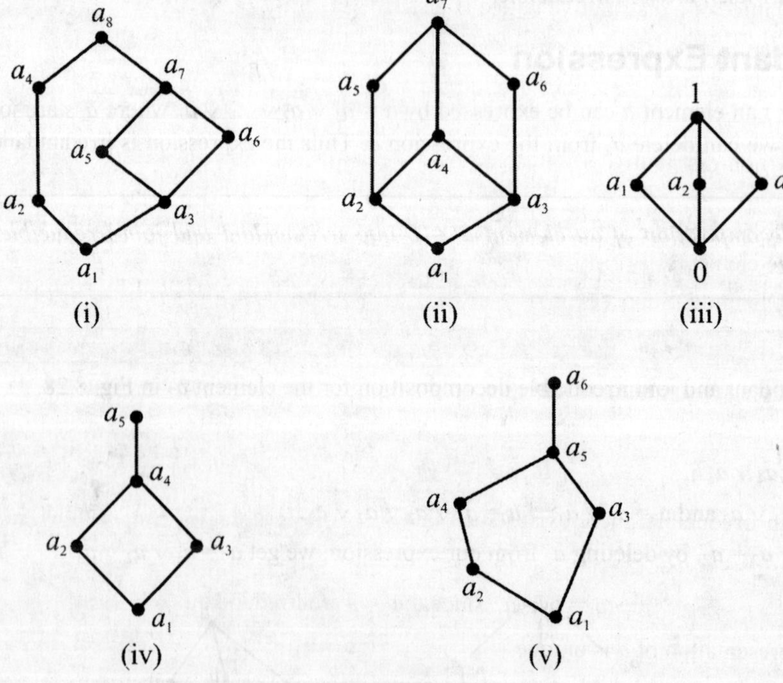

Fig. 8.29

4. Show that a lattice is modular if and only if

$$a \wedge (b \vee (a \wedge c)) = (a \wedge b) \vee (a \wedge c) \text{ for all } a, b, c \in L.$$

5. Show that distributive properties are equivalent.
6. Prove that in a distributive lattice, the distributive laws can be generalized as

$$a \wedge \left(\bigvee_{i=1}^{n} b_i \right) = \bigvee_{i=1}^{n} (a \wedge b_i) \text{ and } a \vee \left(\bigwedge_{i=1}^{n} b_i \right) = \bigvee_{i=1}^{n} (a \vee b_i)$$

7. Show that if $(L, \wedge, \vee)$ is a bounded distributive lattice then every element has unique complement.
8. Show that in a bounded distributive lattice, the elements which have complements form a sublattice.
9. Determine which of the following lattices shown in Fig. 8.30 are non distributive.

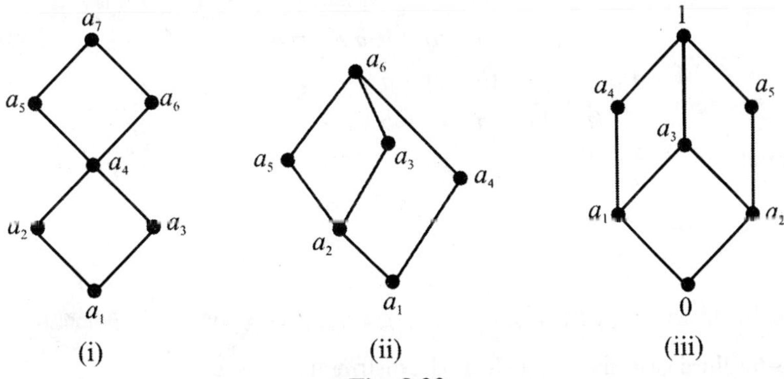

Fig. 8.30

10. Show that sublattice of a distributive lattice is distributive.
11. Consider a lattice $\langle P(X), \text{gcd}, \text{lcm} \rangle$. Find which element of $P(X)$ are atoms and which are join irreducible?
12. Express each number that is not join irreducible and irredundant in the lattice D_{60}, divisors of 60, ordered by divisibility.
13. A lattice L is non-distributive iff it contains a sublattice isomorphic to the lattices of pentagonal and diamond.
14. Let L be a finite distributive lattice. Then every a in L can be written uniquely as the join of irredundant and join-irreducible elements.

    [**Hint:** Since L is finite. Suppose

    $$a = a_1 \vee a_2 \vee \ldots \vee a_p = b_1 \vee b_2 \vee \ldots \vee b_q .$$

    where a's and b's are irredundant and join irreducible. For any i, we have $a_i \preceq (a_1 \vee a_2 \vee \ldots \vee a_{p)} = (b_1 \vee b_2 \vee \ldots \vee b_{q)}$

    Hence $a_i = a_i \wedge (b_1 \vee b_2 \vee \ldots \vee b_{q)} = (a_i \wedge b_1) \vee (a_i \wedge b_2) \vee \ldots \vee (a_i \wedge b_q)$

    Since a_i is join irreducible, there exist j such that $a_i = a_i \wedge b_j$ and so $a_i \preceq b_j$. Similarly for b_j there exist a_k such that $b_j \preceq b_k$ i.e. $a_i \preceq b_j \preceq a_k$

    $$\Rightarrow a_i = b_j = a_k, \text{ since the } a\text{' s are irredundant.}$$

    Hence the representation of a is unique.]
15. Let L be a complemented lattice with unique complements. Then the join irreducible elements of L other than 0 are its atoms.

    [**Hint:** Prove it by using contradiction consider a is a join irreducible and not an atom. Then a has a unique immediate predecessor $\neq 0$].
16. Prove that every chain is distributive lattice.

    [**Hint:** Let $\langle L, \vee, \wedge \rangle$ be a chain and $a, b, c \in L$. Consider the following three cases:

Case I: $a \le b \le c$

so $a \wedge b = a, a \wedge c = a$

$\Rightarrow$ $a \wedge (b \vee c) = a \wedge c = a$

and $(a \wedge b) \vee (a \wedge c) = a \vee a = a$

so the law holds.

Case II: $b \le a \le c$

then $a \wedge b = b, a \wedge c = a$

so $a \wedge (b \vee c) = a \wedge c = a$

and $(a \wedge b) \vee (a \wedge c) = b \vee a = a$ $[\because a \ge b]$

Again the law holds.

Case III: $b \le c \le a$

then $a \wedge b = b, a \wedge c = c$

Now, $a \wedge (a \vee c) = a \wedge c = c$

while $(a \wedge b) \vee (a \wedge c) = b \vee c = c$

hence law holds for third case also i.e. a chain is a distributive lattice

Answers

9. a. Distributive lattice.
   b. Non distributive lattice. Since its sublattice $\{a_1, a_2, a_4, a_6\}$ is a pentagonal.
   c. Non-distributive. Since a_4 has two complements a_5 and a_2.
13. Singleton set are the atoms and join irreducible of $P(X)$.

8.7 Consistent Enumeration

Consider a finite partially ordered set P. There are many occasions where we want to assign positive integers to the elements of P in such a way that order is preserved. In other words we can say that we seek a function $f: P \to N$ where N is a natural number so that if $x \prec y$ then $f(x) < f(y)$. Such a function is called **Consistent Enumeration** of P.

Theorem 8.19

There exists a consistent enumeration for any finite poset X.

Proof

Consider a finite poset P with n elements. Then there exists a consistent enumeration

$$f: P \to \{1, 2, 3, \ldots, n\}.$$

Let us prove this theorem by mathematical induction on number n of elements. Suppose $n = 1$, say $P = \{p\}$. Then $f(p) = 1$ is a consistent enumeration of P. Now suppose $n > 1$ and the theorem holds for posets with fewer that n elements. Since P is a finite set there must exist a minimal element say x. Let $Q = P \mid \{x\}$. Then Q is a finite poset with $n-1$ elements and hence by induction, Q admits a consistent enumeration; say $g : Q \to \{1, 2, 3, \ldots, n-1\}$.

Define $f: P \to \{1, 2, \ldots, n\}$ by

$$f(x)=\begin{cases}1, & \text{if } a = x\\ g(x)+1, & \text{if } a \neq x\end{cases}$$

Then f is required consistent enumeration.

Example 8.38

Show that for the poset shown in Fig. 8.31, consistent enumeration need not be unique.

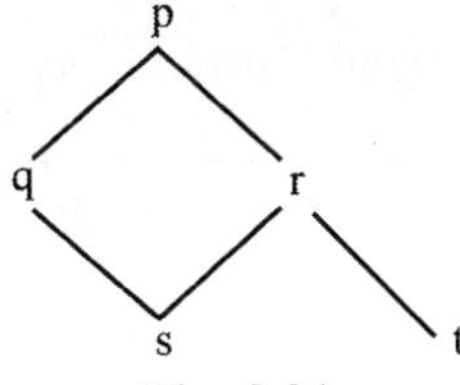

Fig. 8.31

Solution

Here, we can see that there are two different enumeration for the poset.

i. $f(s) = 1, f(t) = 2, f(q) = 3, f(r) = 4, f(p) = 5$

ii. $g(t) = 1$, $g(s) = 2$, $g(r) = 3$, $g(q) = 4$, $g(p) = 5$.

Example 8.39

Show that the chain shown in Fig. 8.32 admits only one consistent enumeration if we map the set into $\{1, 2, 3, 4, 5\}$.

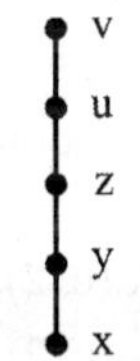

Fig. 8.32

Solution

The chain admits only one consistent enumeration if we map the set into $\{1, 2, 3, 4, 5\}$. Specifically we can say

$$f(x) = 1, f(y) = 2, f(z) = 3, f(u) = 4, f(v) = 5.$$

Example 8.40

Let $P = \{p, q, r, s, t\}$ be ordered as shown in Fig. 8.33. Find all possible consistent enumeration $f: P \rightarrow \{1, 2, 3, 4, 5\}$.

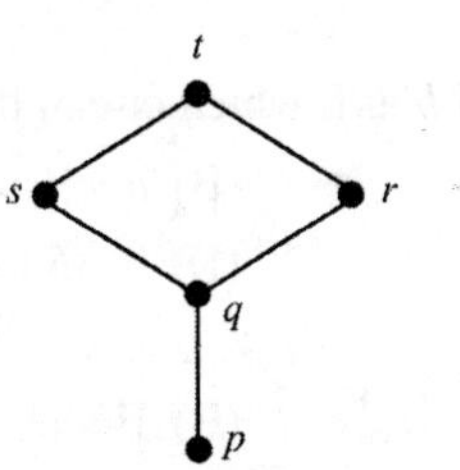

Fig. 8.33

Solution

There is only one minimal element p in P, therefore $f(p) = 1$, and the only maximal element in P is t therefore $f(t) = 5$. Now the only successor of p is q. The choices for s and r are $f(s) = 3$ and $f(d) = 4$ or vice versa. Thus there are two possible enumerations, as follows:

i. $f(p) = 1, f(q) = 2, f(r) = 3, f(s) = 4, f(t) = 5.$
ii. $f(p) = 1, f(q) = 2, f(r) = 4, f(s) = 3, f(t) = 5.$

Usually we cannot recreate the original partial order from a given consistent enumeration.

Exercise 8.6

1. Let $P = \{a, b, c, d, e\}$ be ordered as in Fig. 8.34.

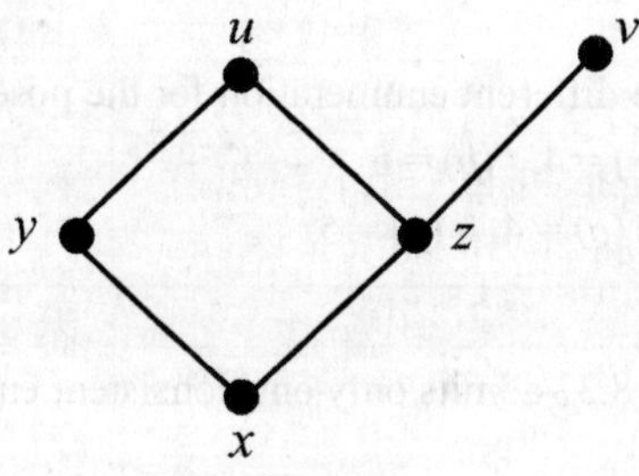

Fig. 8.34

List all consistent enumeration of P into $\{1, 2, 3, 4, 5\}$.

2. Suppose the following are three consistent enumerations of an ordered set $A = \{a, b, c, d\}$. [(*a*, 1), (*b*, 2), (*c*, 3), (*d*, 4)].

[(a, 1), (b, 3), (c, 2), (d, 4)].
[(a, 1), (b, 4), (c, 2), (d, 3)].

Assuming the Hasse diagram of A is connected, draw D.

Answers

1. *xyzuv, xyzvu, xzyuv, xzyvu, xzvyu.*
2. $a \preceq b, a \preceq c, c \preceq d.$

Chapter Objective Type Questions

A. Multiple choice questions.

1. Let L be a lattice. Then for every a and b in L which one of the following is correct?
(A) $a \vee b = a \wedge b$ (B) $a \vee (b \vee c) = (a \vee b) \vee c$
(C) $a \vee (b \wedge c) = a$ (D) $a \vee (b \vee c) = b$
2. Hasse diagram are drawn for:
(A) Partially ordered sets (B) Boolean algebra
(C) Lattices (D) None of these

3. A poset $\langle L, \preceq \rangle$ is a lattice if and only if every non empty finite subset of L has:
   (A) glb (B) lub
   (C) glb and lub (D) glb or lub
4. Let L be a lattice. Then, for every a, b and c in L, if $a \preceq b$ then.
   (A) $a \vee c \preceq b \vee c$ (B) $a \wedge c \preceq b \wedge c$
   (C) either (A) or (B) (D) both (A) and (B)
5. If $\langle L, \wedge, \vee \rangle$ and $\langle L', *, \oplus \rangle$ are two lattices, then a lattice homomorphism is a function $f : L \to L'$ such that for any $a, b \in L$:
   (A) $f(a \wedge b) = f(a) * f(b)$ (B) $f(a \vee b) = f(a) \oplus f(b)$
   (C) both (A) and (B) (D) either (A) or (B)
6. If a homomorphism $f : L \to L$, where $\langle L, *, \oplus \rangle$ is a lattice, i.e. mapping on itself. Then it is known as_____.
   (A) endomorphism (B) automorpism
   (C) endomorphim (D) none of these

B. Fill in the blanks.

1. A lattice $(L, \leq)$ is said to be a complete lattice iff every non-empty subset S of L has a glb and a_______.
2. Let A be set of all real numbers in [1, 5] and $\leq$ is relation of 'less than equal to'. Then, lattice (A, $\leq$) is ____.
3. Consider a lattice $\langle L, \preceq \rangle$ where $a, b \in L$, then $a \wedge (a \vee b) =$ _______ .
4. In two lattices $\langle L, \wedge, \vee \rangle$ and $\langle L', *, \oplus \rangle$ if homomorphism $f : L \to L'$ is also bijective, then f is called______ .
5. Let $a, b, c \in L$ in a lattice $\langle L, \vee, \wedge \rangle$ then L is known as modular lattice if it holds________ .
6. A modular lattice is distributive if and only if it does not contain a _______as one of its sublattices.
7. There exists a consistent enumeration for any_______.

C. Which of the following statements is True (T)/ False (F)?

1. The elements a and b of a poset $\langle L, \preceq \rangle$ has unique glb and lub.
2. A linearly ordered set is a lattice.
3. In f and sup of a pair of elements in a lattice may not be unique.
4. A homomorphism $f : L \to L$, where $\langle L, *, \oplus \rangle$ is a lattice, is known as endomorphism.
5. Every distributive lattice is modular but converse is not true.
6. The pentagonal lattice is modular.
7. It is necessary that all join irreducible elements are atoms.
8. $a = a_2$. $a \neq 1$ is meet irreducible iff a has only one immediate successor.
9. The decomposition of an element $a \in L$ into irredundant and join-irreducible elements is unique.

Answers (Objective Type)

A.

1. B 2. A 3. C 4. D
5. C 6. A

B.

1. least upper bound
2. complete lattice
3. a
4. an isomorphism
5. $a \le c \Rightarrow a \vee (b \wedge c) = (a \vee b) \wedge c$
6. diamond lattice
7. finite poset

C.

1. T
2. T
3. F
4. T
5. T
6. F
7. F
8. T
9. F

CHAPTER 9

Boolean Algebra

9.0 Introduction

George Boole (1815-1864), a great mathematician was the founder of logic. He gave basic rules of logic in his book "*An investigation of the laws of thought*" that was used to design circuits. Circuits are used in computers, electronic and electrical devices which take input in the form of 0 or 1; 'on' or 'off'; 'lit' or 'unlit'. These circuits are used to design the arithmetic logic unit (ALU) of a digital computer. The algebra based on these rules is known as **Boolean algebra**. The elements of Boolean algebra are used in several identities and expressions, known as Boolean functions, widely used to minimize the size of circuits.

We now introduce the notion of Boolean algebra and its properties as a particular kind of lattices.

9.1 Boolean Algebra

If a lattice $\langle B, \wedge, \vee\rangle$ with two binary operations $\wedge$ and $\vee$ bounded by 0 and 1 where 0 is the lower bound and 1 is the upper bound, and each element of B has a unique complement i.e. B is bounded and complemented lattice then it is known as **Boolean Algebra** and the lattice is known as **Boolean lattice**. It is denoted by $\langle B, \wedge, \vee, ', 0, 1\rangle$ or : by $\langle B, \cdot, +, ', 0, 1\rangle$ where $\wedge$ or . and $\vee$ or + are **meet** and **join** respectively and symbol $'$ denotes the complement of any element in B.

Other symbols used for meet and join are represented in Table 9.1.

Table 9.1: Terminology used in Boolean Algebra

Symbols used			Read as
In logic	In Set theory	By Computer Designers	
$\vee$	$\cup$	$+$	"or", "join", "sum"
$\wedge$	$\cap$	.	"and", "meet", "product"
$\urcorner$	c	$\overline{\ }$ or $'$	"complement"
$\preceq$	$\subseteq$	$\leq$	" Inclusion Relation"

9.2 Properties of Boolean Algebra

As, $\langle B, \cdot, +, ', 0, 1\rangle$ is a complemented and bounded lattice, hence all the identities of lattices, bounded lattices and complemented lattices hold good in Boolean algebra also. Some important Boolean expressions are summarized in Table 9.2. Consider that $a, b, c \in B$ and 0 and 1 are lower and upper bounds respectively.

Table 9.2: Properties of Boolean Algebra

1. Idempotent Laws

$$a \vee a = a \quad \text{or} \quad a + a = a$$
$$a \wedge a = a \quad \text{or} \quad a \cdot a = a$$

2. Commutative Laws

$$a \vee b = b \vee a \quad \text{or} \quad a + b = b + a$$
$$a \wedge b = b \wedge a \quad \text{or} \quad a \cdot b = b \cdot a$$

3. Associative Laws

$$(a \cdot b) \cdot c = a \cdot (b \cdot c)$$
$$(a + b) + c = a + (b + c)$$

4. Absorption Laws

$$a \cdot (a + b) = a$$
$$a + (a \cdot b) = a$$

5. Distribution Laws

$$a \cdot (b + c) = (a \cdot b) + (a \cdot c)$$
$$a + (b \cdot c) = (a + b) \cdot (a + c)$$
$$(a \cdot b) + (b \cdot c) + (c \cdot a) = (a + b) \cdot (b + c) \cdot (c + a)$$
$$(a \cdot b) = (a \cdot c) \text{ and } (a + b) = (a + c) \Rightarrow b = c$$

6. Boundedness Laws

$0 \leq a \leq 1$

$$a \cdot 0 = 0 \qquad \{\text{Dominance Law}\}$$
$$a + 1 = 1 \qquad \{\text{Dominance Law}\}$$
$$a \cdot 1 = 1 \qquad \{\text{Identity Law}\}$$
$$a + 0 = 0 \qquad \{\text{Identity Law}\}$$

7. Complement Laws

$$a \cdot a' = 0$$
$$a + a' = 1$$
$$0' = 1$$
$$1' = 0$$
$$\left.\begin{aligned}(a \cdot b)' &= a' + b'\\ (a + b)' &= a' \cdot b'\end{aligned}\right\} \text{De Morgan's Laws}$$

8. If partial ordering relation $\preceq$ exist on B, then

$$(I14)\ a \cdot b = \text{glb}\ \{a, b\}$$
$$(I14)'\ a + b = \text{lub}\ \{a, b\}$$
$$(I15)\ a \preceq b \Leftrightarrow a \cdot b = a \Leftrightarrow a + b = b$$
$$(I16)\ a \preceq b \Leftrightarrow a \cdot b' = 0 \Leftrightarrow b' \preceq a' \Leftrightarrow a' + b = 1$$

9. Involution Law

$$(a')' = a$$

10. Duality

If $\langle B, \cdot, +, ', 0, 1\rangle$ be a Boolean algebra then dual of Boolean expression can easily be obtained by interchanging "join" by "meet" and "0s" by "1s" and vice-versa, as we have done in lattices.

1. *If a and b are two elements of Boolean algebra B, then a is included in B iff*

$$a \cdot b' = 0, \text{ i.e. } a \le b \Leftrightarrow a \cdot b' = 0.$$

2. *If there is no scope for confusion, the symbol '·' can be deleted, that means algebraic product $a \cdot b$ can be written as ab.*

Following are the algebraic proofs of some of the laws taken above.

Theorem 9.1

i. Additive identity is unique.
ii. Multiplicative identity is unique.

Proof

i. Let 0 and 0_1 be two additive identities of Boolean algebra B, then

$$0 + 0_1 = 0 = 0_1 + 0 \qquad [\because 0_1 \text{ is additive identity}]$$

$$0_1 + 0 = 0_1 = 0 + 0_1 \qquad [\because 0 \text{ is additive identity}]$$

But $\quad 0 + 0_1 = 0_1 + 0 \qquad [\because \text{By commutativity}]$

$\Rightarrow \quad 0_1 = 0$

Therefore additive identity is unique in B.

Similarly we can prove (ii).

Theorem 9.2 Idempotent laws

i. $a + a = a$
ii. $a \cdot a = a \quad \forall\, a \in B$

Proof

i.

$$\begin{aligned} a + a &= (a + a) \cdot 1 && [\text{By identity law}] \\ &= (a + a) \cdot (a + a') && [\because a + a' = 1] \\ &= a + a \cdot a' && [\text{By distributive law}] \\ &= a + 0 && [\because a \cdot a' = 0] \\ &= a && [\text{ 0 is identity w.r.t. '+'}] \end{aligned}$$

Hence $\quad a + a = 1.$

ii.

$$\begin{aligned} a \cdot a &= a \cdot a + 0 && [\text{By identity law}] \\ &= a \cdot a + a' && [\because a \cdot a' = 0] \\ &= a \cdot (a \cdot a') && [\text{ By distributive law}] \\ &= a \cdot 1 && [\because a + a' = 1] \\ &= a && [\because \text{1 is multiplicative identity}] \end{aligned}$$

Hence $\quad a \cdot a = a, \ \forall\, a \in B.$

Theorem 9.3

Boundedness Laws (i) $a + 1 = 1$ (ii) $a \cdot 0 = 0, \forall\, a \in B$.

Proof

i. Let a be an arbitrary element of B, then

$$a + 1 = 1 \cdot (a+1) \quad [\because 1 \text{ is multiplicative identity}]$$
$$= (a + a') \cdot (a+1) \quad [\because a + a' = 1]$$
$$= a + (a' \cdot 1) \quad [\text{By distributive law}]$$
$$= a + a' \quad [\because a' \cdot 1 = a']$$
$$= 1 \quad [\because a + a' = 1]$$

Hence $a + 1 = 1.$

ii. $$a \cdot 1 = 0 + a \cdot 0 \quad [\because 0 \text{ is additive identity}]$$
$$= a \cdot a' \cdot a \cdot 0 \quad [\because a \cdot a' = 0]$$
$$= a \cdot (a' \cdot 0) \quad [\text{By distributive law}]$$
$$= a\ a' \quad [\because a' + 0 = a']$$
$$= 0 \quad [\because a \cdot a' = 0]$$

Hence $a \cdot 0 = 0.$

Theorem 9.4 Absorption Law

i. $a + a \cdot b = a$

ii. $a \cdot (a + b) = a\ \forall\ a, b \in B.$

Proof

i. $$a + a \cdot b = a \cdot 1 + a \cdot b \quad [1 \text{ being multiplicative identity}]$$
$$= a \cdot (1 + b) \quad [\text{By distributive law}]$$
$$= a \cdot (b + 1) \quad [\text{By commutative law}]$$
$$= a \cdot 1 \quad [\because b + 1 = 1]$$
$$= a$$

Hence $a + a \cdot b = a,\ \forall\ a, b \in B.$

ii. $$a \cdot (a + b) = (a + 0) \cdot (a + b) \quad [\because 0 \text{ is additive identity}]$$
$$= a + 0 \cdot b \quad [\text{By distributive law}]$$
$$= a + b \cdot 0 \quad [\text{By commutative law}]$$
$$= a + 0 \quad [b \cdot 0 = 0]$$
$$= a$$

Hence $a \cdot (a + b) = a.$

Theorem 9.5

For each $a \in B$ its complement a' is unique.

Proof

Let x and y be two complements of $a \in B$, then

$$x = x + 0 \quad [\because 0 \text{ is additive identity}]$$
$$= x + a \cdot y \quad [\because y \text{ is a complement of } a]$$
$$= (x + a) \cdot (x + y) \quad [\text{By distributive law}]$$
$$= (a + x) \cdot (x + y) \quad [\text{By commutative law}]$$

$$= 1\cdot(x+y) \qquad [\because a' = x \text{ and } a + a' = 1]$$

$$= x + y \qquad \ldots(1)\ [1 \text{ is multiplication identity}]$$

Similarly, $$y = y + 0$$

$$= y + a\cdot x \qquad [\because x \text{ is a complement of } a]$$

$$= (y+a)\cdot(y+x) \qquad [\text{By distributive law}]$$

$$= (a+y)\cdot(x+y) \qquad [\text{By commutative law}]$$

$$= 1\cdot(x+y) \qquad [y = a' \text{ and } a + a' = 1]$$

$$= x + y \qquad \ldots(2)\ [1 \text{ is multiplicative identity}]$$

From (1) and (2), we find that

$$x = y.$$

Therefore, two complements of a are equal. So, every element in B has unique complement in it.

Theorem 9.6 Complements Laws

i. $0' = 1$

ii. $1' = 0.$

Proof

i. $$0' = 0 + 0' \qquad [\because 0 \text{ is an additive identity}]$$

$$= 1 \qquad [\because a + a' = 1]$$

Hence $$0' = 1$$

ii. $$1' = 1\cdot 1' \qquad [\because 1 \text{ is multiplicative identity}]$$

$$= 0 \qquad [\because a\cdot a' = 0]$$

Hence $$1' = 0.$$

Theorem 9.7 Involution Law

$$(a')' = a.$$

Proof

Let $a \in B$, then

$$(a')' = 1\cdot(a')' \qquad [\because 1 \text{ is multiplicative identity}]$$

$$= (a + a')\cdot(a')' \qquad [\because a + a' = 1]$$

$$= a\cdot(a')' + (a') + (a')' \qquad [\text{By distributive law}]$$

$$= a\cdot(a')' + 0 \qquad [\because a\cdot a' = 0]$$

$$= a\cdot(a')' + a\cdot a' \qquad [\because a\cdot a' = 0]$$

$$= a\cdot a' + a\cdot(a')' \qquad [\text{By commutative law}]$$

$$= a\cdot(a' + (a')') \qquad [\text{By distributive law}]$$

$$= a\cdot 1 \qquad [\because a + a' = 1]$$

$$= a \qquad [\because 1 \text{ is multiplicative identity}]$$

Hence $$(a')' = a.$$

Theorem 9.8 De Morgan's Law

i. $(a\cdot b)' = a' + b'$

ii. $(a + b)' = a'\cdot b', \forall\, a, b \in B.$

Proof

i. From (i) we observe that $a' + b'$ is complement of $(a \cdot b)$. Therefore to prove (i) it is sufficient to show that

$$(a \cdot b) + (a' + b') = 1$$

and $$(a \cdot b) + (a' + b') = 0$$

where 1 is multiplicative and 0 is additive identity in B.

Now $(a \cdot b) + (a' + b') = [a + (a' + b')] \cdot [b + (a' + b')]$ [By distributive law]

$= [(a + a') + b')] \cdot [(b + b') + a']$ [By associative and commutative law]

$= (1 + b') \cdot (1 + a')$ [$\because a + a' = 1$ and $b + b' = 1$]

$= (b' + 1) \cdot (a' + 1)$ [By commutative law]

$= 1 \cdot 1$ [$\because a + 1 = a$]

$= 1.$

Therefore $(a \cdot b) + (a' + b') = 1$ ... (1)

and $(a \cdot b)(a' + b') = (a \cdot b) \cdot a' + (a \cdot b) \cdot$ [By distributive law]

$= a \cdot (b \cdot a') + a \cdot (b \cdot b')$ [By associative law]

$= a \cdot (a' \cdot b) + a \cdot (b \cdot b')$ [By commutative law]

$= (a \cdot a') \cdot b + a \cdot (b \cdot b')$ [By associative law]

$= 0 \cdot b + a \cdot 0$ [$\because a\,a' = 0,\ b\,b' = 0$]

$= b \cdot 0 + a \cdot 0$ [By commutative law]

$= 0 + 0$

$= 0$

Therefore $(a \cdot b) \cdot (a' + b') = 0$ ... (2)

Hence from (1) and (2), we conclude that $a' + b'$ is complement of $a \cdot b$, consequently

$$(a \cdot b)' = (a' + b').$$

ii. From (ii) we observe that $a' \cdot b'$ is complement of $a + b$. Therefore to prove (ii), it is sufficient to show that

$$(a + b) + a' \cdot b' = 1$$

and $$(a + b) \cdot (a' \ b') = 0$$

where 1 is multiplicative and 0 is additive identity in B.

Now $(a + b) + (a' \cdot b') = [(a + b) + a'] \cdot [(a + b) + b']$ [By distributive law]

$= [a + (b + a')] \cdot [a + (b + b')]$ [By associative law]

$= [a + (a' + b)] \cdot [a + (b + b')]$ [By commutative law]

$= [(a + a') + b] \cdot [a + (b + b')]$ [By associative law]

$= (1 + b) \cdot (a + 1)$ [$\because a + a' = 1, b + b' = 1$]

$= 1 \cdot 1$ [$\because a + 1 = 1,\ b + 1 = 1$]

$$= 1.$$

Therefore $(a + b) + a' \cdot b' = 1.$ ... (1)

and $(a + b) \cdot (a' \cdot b') = a \cdot (a' \cdot b') + b \cdot (a' \cdot b')$ [By distributive law]

$= (a \cdot a') \cdot b' + (b \cdot b') \cdot a'$

[By commutative law and associative law]

$= 0 \cdot b' + 0 \cdot a'$ [$\because a \cdot a' = 0 = b \cdot b'$]

$= b' \cdot 0 + a' \cdot 0$ [$\because a \cdot 0 = 0$]

$= 0 + 0$

$= 0$ [$\because$ 0 is additive identity]

Therefore $(a + b) \cdot (a' \cdot b') = 0$ ... (2)

Hence from (1) and (2) we conclude that $a' \cdot b'$ is complement of $a + b$, i.e.

$$(a + b)' = a' \cdot b'.$$

Theorem 9.9 Boundedness Laws

For any element a of a Boolean algebra B, show that

i. $a \leq 1$

ii. $0 \leq a, \forall a \in B.$

Hence $0 \leq a \leq 1$.

Proof

i. Since for any $a \in B$, we have

$$a \cdot 0 = 0$$

$\therefore$ $a \cdot 1' = 0$ [$\because 1' = 0$]

Thus, $a \cdot 1' = 0 \Rightarrow a \leq 1$ [$\because a \leq b \Rightarrow a \cdot b = 0$]

ii. If $a \in B$ then $a' \in B$.

$\because$ $a' \cdot 0 = 0$

$\Rightarrow$ $0 \cdot a' = 0$ [By commutative law]

$\Rightarrow$ $0 \leq a,$

Hence from (i) and (ii), we conclude for all $a \in B$,

$$0 \leq a \leq 1.$$

Theorem 9.10

In a Boolean algebra $(B, +, \cdot, ')$ if $a, b \in B$, then

Prove that

i. $a + b = \text{lub } \{a, b\}$

ii. $a \cdot b = \text{glb } \{a, b\}$.

Proof

i. For all $a, b, c \in B$, we know that

$a + (a + b) = (a + a) + b$ [By associative law]

$= a + b$ [$\because a + a = a$]

$\therefore \quad a + (a + b) = a + b$

$\Rightarrow \quad a \le a + b \quad \ldots (1)$

Now $\quad b + (a + b) = \{b + a\} + b$ [By associative law]

$= (a + b) + b$ [By commutative law]

$= a + (b + b)$ [By associative law]

$= a + b$ [$\because b + b = b$]

$\therefore \quad b + (a + b) = a + b$

$\Rightarrow \quad b \le a + b \quad \ldots (2)$

From (1) and (2) we conclude that $(a + b)$ is upper bound of a as well as b. Let c be some other upper bound of a and b then

$$a \le c \text{ and } b \le c.$$

$\Rightarrow \quad a + c = c \text{ and } b + c = c$

But $\quad (a + b) + c = a + (b + c)$ [By associative law]

$= a + c$ [$\because b + c = c$]

$= c$ [$\because a + c = c$]

$\therefore \quad a + b \le c$

Hence $\quad a + b = \text{lub } \{a, b\}$

i.e. $(a + b)$ is least upper bound of a as well as b.

ii. Similar to the proof of (i) here, we have

$a + a \cdot b = a$ and $b + a \cdot b = b$ [By absorption law]

$\Rightarrow \quad a \cdot b + a = a$ and $a \cdot b + b = b$ [By commutative law]

$\therefore \quad a \cdot b \le a$ and $a \cdot b \le b$.

Thus $a \cdot b$ is the lower bound of a as well as b.

Let c be any other lower bound of a as well as b, then.

$c \le a$ and $c \le b$

$\Rightarrow \quad c + a = a$ and $c + b = b$.

But $\quad c + a \cdot b = (c + a) \cdot (c + b)$ [By distributive law]

$= a \cdot b$ [$\because c + a = a, c + b = b$]

$\therefore \quad c \le a \cdot b$

Hence $\quad a \cdot b \le \text{glb } \{a, b\}$

i.e. $a \cdot b$ is the greatest lower bound of a and b.

9.3 Striking Features of Boolean Algebra

i. Boolean algebra deals with a finite set of elements, namely 0 and 1which are usually called false and true, whereas ordinary algebra have an infinite number of elements. In Boolean algebra all the variables and constants can have only two possible values 0 and 1.

ii. Boolean algebra does not have any operations equivalent to subtraction and division.

iii. In Boolean algebra the cancellation laws do not hold, i.e.

a. $a + b = a + c \nRightarrow b = c$

b. $a \cdot b = a \cdot c \nRightarrow b = c$

iv. In ordinary algebra, there is no equivalent of the unary operation over a variable known as complementing.

Example 9.1

Let <B, +, · > be a Boolean algebra and $a, b \in B$, then show that

i. $a + (a + b) = a + b$,   ii. $a \cdot b = a \Rightarrow a \cdot b' = a$

iii. $a + a \cdot b = a' + b$ iv. $a + b = b \Rightarrow a \cdot b = a$

Solution

i.
$$a + (a + b) = (a + a) + b \quad \text{[By associative law]}$$
$$= a + b \quad [\because a + a = a]$$

Hence $a + (a + b) = a + b$.

ii. Given $a \cdot b = a$

Now
$$a \cdot b' = (a \cdot b) \cdot b'$$
$$= a \cdot (b \cdot b') \quad \text{[By associative law]}$$
$$= a \cdot 0 \quad [\because b \cdot b' = 0]$$
$$= 0$$

Hence $a \cdot b = a \Rightarrow a \cdot b' = 0$

iii.
$$a' + a \cdot b = (a' + a) \cdot (a' + b) \quad \text{[By distributive law]}$$
$$= (a + a') \cdot (a' + b) \quad \text{[By commutative law]}$$
$$= 1 \cdot (a' + b) \quad [\because a + a' = 1]$$
$$= a' + b$$

Hence $a' + a \cdot b = a' + b$ [1 being multiplicative identity]

iv. Given $a + b = b$

Now
$$a \cdot b = a \cdot (a + b)$$
$$= a \cdot a + a \cdot b \quad \text{[By distributive law]}$$
$$= a + a \cdot b \quad \text{[By Idempotent law]}$$
$$= a \quad \text{[By absorption law]}$$
$$= a + b$$

Hence $a + b = b \Rightarrow a \cdot b + a$.

Example 9.2

Simplify the following

i. $(a + b)\, a' \cdot b'$ ii. $a \cdot b + [(a + b') \cdot b]'$

iii. $[a + (a' \cdot b')] \cdot [a' + (a \cdot b)]$.

Solution

i.
$$(a + b) = (a + b) \cdot (a' \cdot b')$$
$$= (a + b) \cdot (a + b)' \quad \text{[By De-Morgan's law } (a + b)' = a' \cdot b'\text{]}$$
$$= 0 \quad [\because a \cdot a' = 0]$$

ii. $$a \cdot b + [(a+b') \cdot b]' = a \cdot b + [b \cdot (a+b')]' \quad \text{[By commutative law]}$$
$$= a \cdot b + [b \cdot a + b \cdot b']' \quad \text{[By distributive law]}$$
$$= a \cdot b + [a \cdot b + 0]' \quad [\because\ b \cdot a = a \cdot b \text{ and } b \cdot b' = 0]$$
$$= a \cdot b + (a \cdot b)'$$
$$= 1. \quad [\because\ a + a' = 1]$$

iii. $$[a + (a' \cdot b')] \cdot [a' + (a \cdot b)]$$
$$= [(a+a')(a+b')]\,[(a'+a) \cdot (a'+b)]$$
[By distributive law]
$$= [1 \cdot (a+b')][1 \cdot (a'+b)] \quad [\because\ a + a' = 1]$$
$$= (a+b')(a'+b) \quad \text{[1 being multiplicative identity]}$$
$$= a \cdot a' + a \cdot b + b' \cdot a' + b' \cdot b \quad \text{[By distributive law]}$$
$$= 0 + a \cdot b + b' \cdot a' + 0 \quad [\because\ a \cdot a' = 0]$$
$$= a'.b' + a \cdot b$$
[By commutative law and 0 is additive identity]

Example 9.3

Show that:

i. $a \cdot b + a' \cdot b' = (a'+b) \cdot (a+b')$

ii. $a \cdot b \cdot c + a \cdot b \cdot c' + a \cdot b' \cdot c + a' \cdot b \cdot c = a \cdot b + b \cdot c + c \cdot a, \forall\ a, b, c \in B.$

Solution

i. $$a \cdot b + a' \cdot b' = (a \cdot b + a') \cdot (a \cdot b + b') \quad \text{[By distributive law]}$$
$$= [(a+a') \cdot (b+a')] \cdot [(a+b') \cdot (b+b')]$$
[By distributive law]
$$= [\,1 \cdot (a'+b)] \cdot [(a+b') \cdot 1]$$
$$= (a'+b) \cdot (a+b') \quad \text{[1 being multiplicative identity]}$$

ii. $$a \cdot b \cdot c + a \cdot b \cdot c' + a \cdot b' \cdot c + a' \cdot b \cdot c$$
$$= a \cdot b\,(c \cdot c') + a \cdot b' \cdot c + a' \cdot b \cdot c \quad \text{[By distributive law]}$$
$$= a \cdot b \cdot 1 + a \cdot b' \cdot c + a' \cdot b \cdot c \quad [\because\ c \cdot c' = 1]$$
$$= a \cdot b + a \cdot b' \cdot c + a' \cdot b \cdot c$$
$$= a \cdot [b + b' \cdot c] + a' \cdot b \cdot c \quad \text{[By distributive law]}$$
$$= a \cdot [(b+b'] \cdot [(b+c)] + a' \cdot b \cdot c \quad \text{[By distributive law]}$$
$$= a \cdot [1 \cdot (b+c)] + a' \cdot b \cdot c \quad [\because\ c + c' = 1]$$
$$= a \cdot (b+c) + a' \cdot b \cdot c \quad \text{[1 being multiplicative identity]}$$
$$= a \cdot b + b \cdot c + a' \cdot b \cdot c$$
$$= a \cdot b + (a + a' \cdot b) \cdot c$$
[By distributive and associative law]
$$= a \cdot b + [a + a') \cdot (a+b) \cdot c \quad \text{[By distributive law]}$$
$$= a \cdot b + [1 \cdot (a+b)] \cdot c \quad [\because\ a + a' = 1]$$

$$
\begin{aligned}
&= (a \cdot b) + (a + b) \cdot c \\
&= a \cdot b + a \cdot c + b.\, c && \text{[By distributive law]} \\
&- \text{R.H.S.}
\end{aligned}
$$

Example 9.4

Prove that $a + b = a + c$ and $a \cdot b = a{\cdot}c \Rightarrow b = c$.

Solution

Let $a + b = a + c$ and $a \cdot b = a \cdot c$, then

$$
\begin{aligned}
b &= b + b \cdot a && \text{[By absorption law]} \\
&= b \cdot b + b \cdot a \\
&= b \cdot (b + a) \\
&= b \cdot (a + b) && \text{[By commutative law]} \\
&= b \cdot (a + c) && \text{[Given } a + b = a + c\text{]} \\
&= b \cdot a + b \cdot c \\
&= a \cdot b + b \cdot c \\
&= a \cdot c + b \cdot c \\
&= (a + b) \cdot c \\
&= (a + c) \cdot c && \text{[Given } a + b = a + c\text{]} \\
&= a \cdot c + c \cdot c \\
&= a \cdot c + c \\
&= c + c \cdot a && \text{[By commutative law]} \\
&= c \cdot (1 + a) && [1 + a = 1, c \cdot 1 = c] \\
&= c\,.
\end{aligned}
$$

Example 9.5

In a Boolean algebra $<B, +, \cdot, ' >$, prove that $(a + b).\, (a' + c) = a'\, b + a \cdot c$.

Solution

$$
\begin{aligned}
(a + b).\, (a' + c) &= (a \cdot a' + a \cdot c) + (b \cdot\ a' + b \cdot c) && \text{[By distributive law]} \\
&= a \cdot (a' + c) + b \cdot (a' + c) && \text{[By distributive law]} \\
&= (0 + a \cdot c) + (b \cdot a' + b \cdot c) && [\because\ a + a' = 0] \\
&= a \cdot c + b \cdot a' + b \cdot c \\
&= a \cdot c + b \cdot a' + (b \cdot c) \cdot 1 \\
&= a \cdot c + b \cdot a' + b \cdot c\, (a + a') && [\because\ a + a' = 1] \\
&= a \cdot c + b \cdot a' + (b \cdot c) \cdot a + (b \cdot c) \cdot a' \\
&= a \cdot c\, (\, b \cdot c) \cdot a + a' \cdot b + a' \cdot (b \cdot c) \\
& && \text{[By commutative law]} \\
&= a \cdot c + b\, (c \cdot a) + a' \cdot b + (a' \cdot b) \cdot c \\
& && \text{[By associative law]} \\
&= a \cdot c + b \cdot (a \cdot c) + a' \cdot b + (a' \cdot b) \cdot c \\
&= (a \cdot c) \cdot (1 + b) + (a' \cdot b) \cdot (1 + c) && \text{[By distributive law]}
\end{aligned}
$$

$$= (a \cdot c) \cdot 1 + (a' \cdot b) \cdot 1 \qquad [\because 1 + b = 1, 1 + c = 1]$$

$$= a \cdot c + a' \cdot b$$

Hence $\qquad (a + b).\,(a' + c) = a \cdot\, c + a' \cdot b.$

Example 9.6

If $<B, +, \cdot, '>$ is a Boolean algebra, then prove $(ab' + bc) \cdot (ac + bc') = ac$.

Solution

$$\begin{aligned}
&(ab' + bc) \cdot\ (ac + bc') \\
&= (ab' + b) \cdot (ab' + c) \cdot (ac + b) \cdot (ac + c') \\
&= (a + b) \cdot (b' + b) \cdot (a + c) \cdot (b' + c) \cdot (a + b) \cdot (c + b) \cdot (a + c') \cdot (c + c') \\
&= (a + b) \cdot 1 \cdot (a + c) \cdot (b' + c) \cdot (a + b) \cdot (c + b) \cdot (a + c') \cdot\ 1 \\
&= (a + b) \cdot (a + b) \cdot (a + c) \cdot (a + c') \cdot (c + b) \cdot (c + b') \\
&= (a + b)\,(a + c \cdot c')\,(c + b \cdot\ b') \\
&= (a + b)\,(a + 0)\,(c + 0) \\
&= (a + b) \cdot a \cdot c \\
&= a \cdot\ a \cdot c + a \cdot b\ \cdot c \\
&= a \cdot c + a \cdot b\ \cdot c \\
&= a \cdot c + a \cdot b\ \cdot c \\
&= a \cdot c + b \cdot a\ \cdot c \\
&= a \cdot c \cdot (1 + b) \\
&= a \cdot c \cdot 1 \\
&= a \cdot c.
\end{aligned}$$

9.4 Basic Operations

Boolean algebra basically consists of two truth value elements represented by the symbols '0' and '1', where '0' is the symbol for 'F' (i.e. false) and '1' is the symbol for 'T' (i.e. true). It is to be noted here that '0' and '1' are used as symbols they do not represent actual numbers. In digital logic field several other terms are used for '0' and '1'. The other terms corresponding to these symbols have been given in Table 9.3.

Table 9.3

0	1
False	True
No	Yes
Off	On
Open	Closed
Low	High
Unlit	Lit

Boolean algebra consists of three basic operations, namely logical addition or (Boolean sum), logical multiplication (or Boolean product) and complementation (negation). The terms **sum, OR**, **joint** and **disjunction** are also used for logical addition and the terms **product, AND, meet** and **conjunction** are also used for logical multiplication.

9.4-1 Logical addition (or Boolean Sum)

Logical addition (or **Boolean Sum**) is represented by the symbol '+' or 'v'. If $a, b, c \in B$ then the Boolean sum of a, b is given by

$$a + b = c \text{ and read as } a \text{ or } b \text{ equals } c.$$

Here c can be obtained by listing all possible combinations of a and b. See following Table 9.4

Table 9.4

Input *a*	**Input *b***	**Output $c = a + b$**
0	0	0
0	1	1
1	0	1
1	1	1

By observing the table one can conclude that the logical operation '+' is not exactly matches with ordinary addition. Since in logical addition $1 + 1 = 1$ which is not the case in ordinary addition. However except for this case the logical addition is similar to ordinary addition.

9.4-2 Logical multiplication (or Boolean Product)

Logical multiplication (or **Boolean Product**) is represented by the symbol ' · ' or '∧'. If $a, b, c \in B$ then the Boolean product of a, b is given by,

$$a \cdot b = c \text{ and read as } a \text{ and } b \text{ equals } c.$$

Here c can be obtained by listing all possible combinations of a and b. See following Table 9.5

Table 9.5

INPUT		**OUTPUT**
a	***b***	**c = a – b**
0	0	0
0	1	0
1	0	0
1	1	1

By observing the truth table one can conclude that the logical operation ' · ' is exactly matches with the ordinary multiplication.

9.4-3 Complementation (or negation)

The complement of an element denoted by bar '–' or ' ' ' is defined by $0' = 1$ and $1' = 0$.

Example 9.7

Consider $B^n = B \times B \times B \times \ldots\times$ Operations·, + and ′ are defined component wise using Table 9.4 and 9.5. The elements of B^n are B. n-bit sequences (for convenience, we will use n-bits representation without using commas). Let us consider $x, y \in B^6$, where $x = 110010, y = 101011$.

Then

$$x \cdot y = 100010$$
$$x + y = 111011$$
$$x' = 001101$$
$$y' = 010100$$

Thus B^6 is a Boolean algebra.

Here 0 = 000000 is the **zero element**, and 1 = 111111 is the **unit element** and treated as bounded elements.

Also, B^n has 2^n elements then in general form it can be written as $\forall\, x, y \in B^n$ and

$$x = \langle x_1, x_2, x_3 \ldots x_n \rangle, y = \langle y_1, y_2, y_3 \ldots y_n \rangle, \text{ then}$$

$$x \cdot y = \langle x_1 \cdot y_1, x_2\, y_2, \ldots, x_n \cdot y_n \rangle,$$

$$x + y = \langle x_1 + y_1, x_2 + y_2, \ldots, x_n + y_n \rangle,$$

$$x' = \langle x_1, x_2, \ldots, x_n \rangle \text{ and } 0 = 0^n,\ 1 = 1^n.$$

This algebra is known as a ***switching algebra*** *and used to represent the switching network with n inputs and one output as shown in this Fig. 9.1*

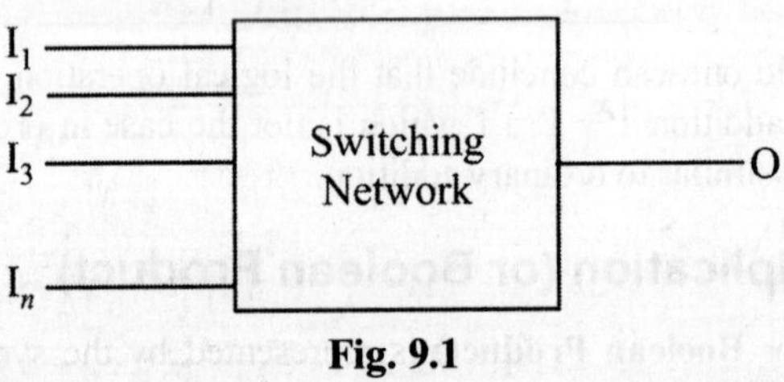

Fig. 9.1

Example 9.8

Consider a set $X = \{a, b, c\}$. Then show that $\langle P(X), \cap, \cup, ', \varnothing, X \rangle$ be a Boolean algebra. where $P(X)$ is a power set of X.

Solution

If set $X = \{a, b, c\}$ then $P(X)$ can be represented by Hasse diagram as shown in Fig. 9.2.

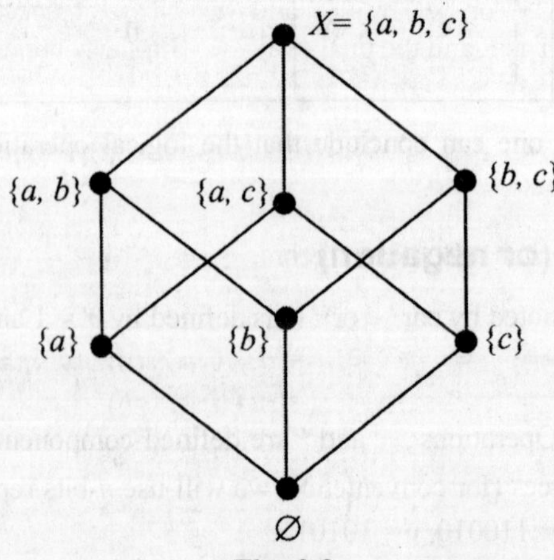

Fig. 9.2

Every element of lattice $\langle P(X), \cap, \cup \rangle$ has complement and bounded by $\varnothing$ and $X = \{a, b, c\}$. Thus, the lattice is bounded and complemented. Hence $\langle P(X), \cap, \cup, ', \varnothing, X \rangle$ is a Boolean algebra.

Example 9.9

Show the following in a Boolean algebra

a. $(x + y)(x' + z) = xz + x'y + yz$ b. $xy' + yz' + z' = x'y + y'z + z'x$.

Solution

Let $\langle B, +, \cdot, ', 0, 1\rangle$ be a Boolean algebra and $x, y, z \in B$, then

a. L.H.S.

$$\begin{aligned}(x + y)(x' + z) &= (x + y)x' + (x + y)z && \text{[By distributive law]}\\ &= xx' + yx' + xz + yz && \text{[By distributive law]}\\ &= 0 + x'y + xz + yz && [\because xx' = 0]\\ &= xz + x'y + yz = \text{R.H.S.}\end{aligned}$$

b. L.H.S.

$$\begin{aligned}xy' + yz' + zx' &= xy'(z + z') + yz'(x + x') + zx'(y + y') && [\because x + x' = 1]\\ &= xy'z + xy'z' + yz'x + yz'x' + zx'y + zx'y\\ & && \text{[By distributive law]}\\ &= xy'z + xy'z' + xyz' + x'yz' + x'yz + x'y'z\\ &= (xy'z + x'y'z) + (xyz' + xy'z') + (x'yz' + x'yz)\\ &= y'z(x + x') + xz'(y + y') + x'y(z + z') && [\because x + x' = 1]\\ &= x'y + y'z + xz' = \text{R.H.S.}\end{aligned}$$

Example 9.10

Define dual of a Boolean expression. Write the dual of the following expression:

$$(x - 1)(0 + x') = 0.$$

Solution

Two Boolean expressions are said to be dual of each other if either one can be obtained from the other by replacing "·" by "+" and "+" by "·" or "∨" by "∧" and "∧" by "∨" respectively. Out of two connectives one is called **dual** of other. The dual of T is F and the dual of F is T. Similarly the dual of 1 is 0 and that of 0 is 1.

The dual of given expression is

$$(x - 0)(1 \cdot x') = 1.$$

Example 9.11

Show that $a \vee (\bar{a} \wedge b) = a \vee b$ in a Boolean algebra.

Solution

$$\begin{aligned}\text{LHS} &= a \vee (\bar{a} \wedge b) = (a \vee \bar{a}) \wedge (a \vee b) && \text{[By distributive]}\\ &= 1 \wedge (a \vee b) && [\because a \vee \bar{a} = 1]\\ &= a \vee b\\ &= \text{RHS.}\end{aligned}$$

Example 9.12

Find the dual of $(x \cdot 1) + (x \cdot y) + (y \cdot z) + (z \cdot 0)$.

Solution

The dual is $(x + 0) \cdot (x + y) \cdot (y + z) \cdot (z + 1)$.

9.5 Sub-Boolean Algebra

Consider a Boolean algebra $\langle B, +, \cdot, ', 0, 1\rangle$ and subset S of B such that $0, 1 \in S$ and S is closed under the binary operations $+, \cdot$ and unary operation $'$, then $\langle S, \wedge, \vee, ', 0, 1\rangle$ is known as **sub Boolean algebra** or **Boolean sub algebra**.

In other words the necessary and sufficient condition for a nonempty subset B_1 of a Boolean algebra B to be a sub-Boolean algebra is that.

i. $a, b \in B_1 \Rightarrow a \cdot b \in B_1$ ii. $a \in B_1 \Rightarrow a' \in B_1$.

REMARK

To prove that S is a sub algebra, it is enough to show the closureness with respect to functionally complete set of operation $\{+, '\}$ and $\{\cdot, '\}$. Because $\forall\, a, b \in B, a + b = (a' \cdot b')'$; $(a \cdot a')' = 1$; $(a \wedge a') = 0$. Therefore, it is not necessary to verify the closureness with respect to all three operations $+$, $\cdot$, and $'$ and $0, 1 \in S$.

Example 9.13

Consider a set $D_{30} = \{1, 2, 3, 5, 6, 10, 15, 30\}$. D_{30} is a Boolean algebra because $30 = 2 \times 3 \times 5$ is a product of distinct primes 2, 3 and 5. Then Boolean subalgebra of D_{30} can be $S_1 = \{1, 2, 15, 30\}$, $S_2 = \{1, 3, 10, 30\}$ and $S_3 = \{1, 5, 6, 30\}$. There are many sublattices that are Boolean algebra but not Boolean subalgebra such as

$$S_4 = \{1, 2, 3, 6\}, \quad S_5 = \{1, 3, 5, 15\},$$
$$S_6 = \{1, 2, 5, 10\}, \quad S_7 = \{2, 6, 10, 30\},$$
$$S_8 = \{5, 10, 15, 30\}, \quad S_9 = \{3, 6, 15, 30\}.$$

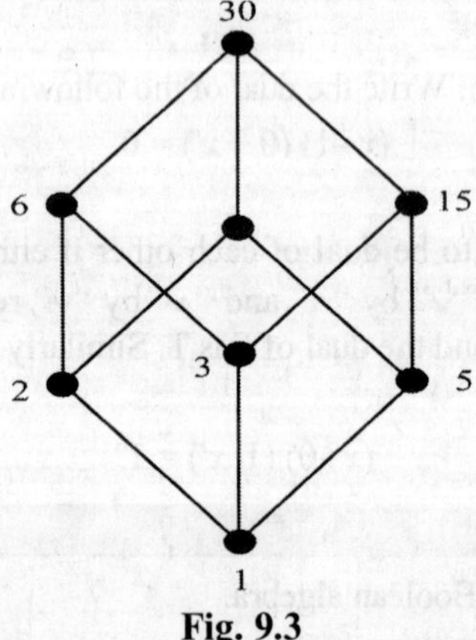

Fig. 9.3

This is because of the reason that these sublattices are not closure with respect to operations $+$, $\cdot$ and $'$. Similarly, the sets $\{1, 3, 6, 15, 30\}$ and $\{1, 2, 3, 6, 15, 30\}$ are sublattices but not Boolean algebra under any operation.

9.6 Boolean Homomorphism

If $\langle B, \wedge, \vee, ', 0, 1\rangle$ and $\langle B', \cdot, +, \bar{\ }, 0', 1'\rangle$ are two Boolean algebra then the function $f: B \to B'$ such that for any $a, b \in B$

$$f(a \wedge b) = f(a) . f(b), \quad f(a \vee b) = f(a) + f(b) \text{ and } f(a') = (\overline{f}(a))$$

is known as **Boolean Homomorphism,** i.e. f preserves both the binary operations of meet and join and unary operation complement. If Boolean Homomorphism f is one-one and onto, i.e. bijective then f is known as **Boolean isomorphism**.

Example 9.14

Lattices $\langle P(S_1), \subseteq\rangle$ and $\langle P(S_2), \subseteq\rangle$ shown by Hasse diagrams in Fig. 9.4 are isomorphic where $S_1 = \{a, b, c\}$ and $S_2 = \{2, 3, 5\}$. Actually we see that one possible isomorphism $f: S_1 \to S_2$ is given by

$f(\{a\}) = \{2\}$, $f(\{b\}) = \{3\}$, $f(\{c\}) = \{5\}$, $f(\{a, b\}) = \{2, 3\}$, $f(\{b, c\}) = \{3, 5\}$,
$f(\{a, c\}) = \{2, 5\}$, $f(\{a, b, c\}) = \{2, 3, 5\}$, $f(\{\varnothing\}) = \varnothing$.

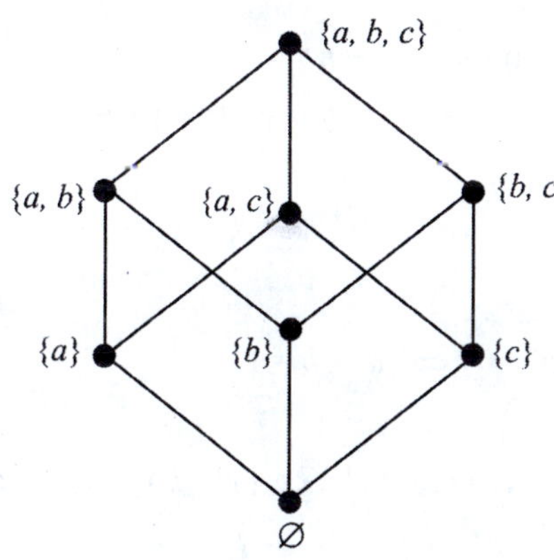

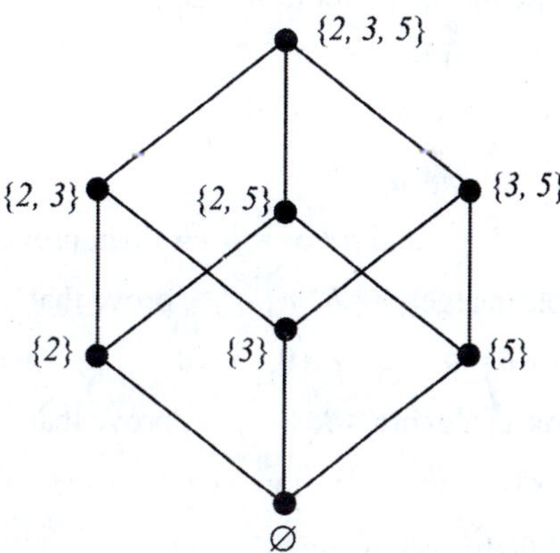

Fig. 9.4

9.7 Direct Product of Boolean Algebras

Let us consider two Boolean algebras $\langle B, \wedge_1, \vee_1, ', 0_1, 1_1\rangle$ and $\langle B', \wedge_2, \vee_2, '', 0_2, 1_2\rangle$. The direct product of two Boolean algebra defined as a Boolean algebra $\langle B \times B', \wedge_3, \vee_3, ''', 0_3, 1_3\rangle$ in which the binary operations $\wedge_3$ and $\vee_3$ on $B \times B'$ are such that

$$\langle a_1, b_1\rangle \wedge_3 \langle a_2, b_2\rangle = \langle a_1 \wedge_1 a_2, b_1 \wedge_2 b_2\rangle$$

and

$$\langle a_1, b_1\rangle \vee_3 \langle a_2, b_2\rangle = \langle a_1 \vee_1 a_2, b_1 \vee_2 b_2\rangle$$

$$\langle a_1, b_1\rangle''' = \langle a'_1, b''_1\rangle$$

for any

$$\langle a_1, b_1\rangle, \langle a_2, b_2\rangle \in B \times B'.$$

Exercise 9.1

1. Verify $B = \{1, 2, 3, 5, 6, 10, 15, 30\}$ be a set of multiple of 30, is a Boolean algebra under the relation $\subseteq$.
2. Find the values of the following expression.
   a. $1 \cdot 0'$
   b. $1 + 1'$
   c. $0' \cdot 0$
   d. $(1 + 0)'$.
3. Verify
   a. Idempotent law
   b. Identity law
   c. Dominance law
   d. Commutative law
   e. Associative law
   f. De Morgan's law

g. Involution law.

4. Find the duals of the following Boolean expressions

a. $xyz + x'y'z'$

b. $xz' + x \cdot 0 + y' \cdot z + x' \cdot 1$

c. $x\,((y'\,x\,) + y\,)' = 1.$

5. Prove the following Boolean Identities

a. $x + (x\,'y\,) = x + y$

b. $(x\,y\,) + (x\,y'\,) = x$

c. $x\,y\,z + xy = xy$

d. $x \le y \Rightarrow x + y\,z = y\,(x + z).$

6. Prove the following for $a, b \in B$.

a. $a \cdot (a \cdot b) = a \cdot b$

b. $ab' = 0 \Rightarrow a + b = b$

c. $(a + b)' + (a + b)' = a'$

d. $a \cdot b + a \cdot b' + a' \cdot b + a' \cdot b' = 1$

e. $a' + a \cdot b = a' + b\cdot$

7. If $a + x = b + x$ and $a + x' = b + x'$ then prove that $a = b$.

8. In Boolean algebra $<B, +, \cdot, '>$, prove that

i. $a + b = 0 \Leftrightarrow a = 0,\ b = 0$

ii. $a \cdot b' = 0 \Leftrightarrow a' + b = 1.$

9. In Boolean algebra $<B, +, \cdot, '>$, prove that

i. $x'(x \quad y)\,' \quad y\ (y \quad x')\,' \quad y(y' \quad x)\,' \quad 1$

ii. $(x \quad yz)\ (y' \quad x)\ (y' \quad z') \quad x(y' \quad z')$.

10. Simplify the following:

i. $(a + b + c)'$

ii. $a \cdot b + b \cdot c + b \cdot c'$

iii. $a \cdot b \cdot c + a' + b' + c'$

iv. $a \cdot b + a\ b' + a' \cdot b \cdot a' \cdot b'$

v. $(a \cdot b') \cdot (a' + b) \cdot (a' + b')$

vi. $(a \cdot b' + c)'$.

11. Show that in any Boolean algebra if

$$a \cdot x = 0 \text{ and } a + x = 1, \text{ then } x = a'.$$

12. Show that following four equations are mutually equivalent:

$$a \cdot b = a,\ a + b = b,\ a' + b = 1,\ a \cdot b' = 0.$$

13. Prove or disprove the following equations

a. $x_1 \quad x_2 \quad (x_1 \quad x_3) \quad x_1 \quad x_2 \quad x_1 \quad x_3$

b. $\left((x_1 + x_2) \cdot (x_3' + x_4)' \cdot (x_3 \cdot x_2')\right) = 0$

c. $(x_1 + x_2) \cdot (x_3 + x_4) = (x_3 \cdot x_1) + (x_3 \cdot x_2) + (x_4 \cdot x_1) + (x_4 \cdot x_2)$

d. $(x_1 \cdot (x_2 \cdot x_3)') = (x_1 \cdot x_2') + (x_1 \cdot x_3')$

e. $x_1 + (x_1 \cdot x_2) = x_1$.

14. Prove that if n is a positive integer and p^2 / n, i.e. p^2 divides n, where p is a prime number, then D_n is not a Boolean algebra.

15. Show that the lattice shown by Hasse diagram in Fig. 9.5 is not a Boolean algebra.

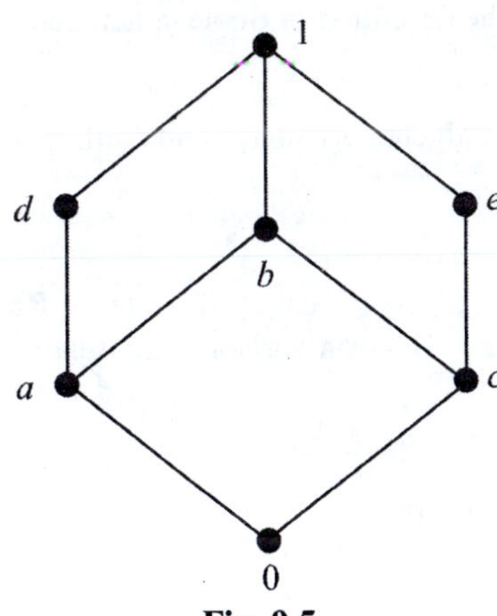

Fig. 9.5

Fig. 9.6

16. Prove that D_n is not a Boolean algebra when $n = 40, 75$.
17. Write all the sub Boolean algebra, subalgebra of given Boolean algebra shown in Fig. 9.6.
18. A Boolean algebra is self dual. Prove it.

    [**Hint:** Consider $f : B \to B$ such that $f(a) = a'$ where b is a Boolean algebra and $a \in B$. Then f is well-defined for each $a \in B$, a' exists and unique.

    Now $\forall\, a, b \in B$

$$f(a) = f(b) \Rightarrow a' = b'$$
$$\Rightarrow (a')' = (b')' \Rightarrow a = b \text{ i.e. one-one}$$

    For any $a \in B$, a' is its pre-image i.e. f is onto.

    Also, for $a, b \in B$

$$f(a \wedge b) = (a \wedge b)' = a' \vee b' = f(a) \vee f(b)$$

and
$$f(a \vee b) = (a \vee b)' = a' \wedge b' = f(a) \wedge f(b)$$

    i.e. f is a dual – homomorphism. Hence f is self-dual].
19. Prove that every finite Boolean algebra B is isomorphic to B_{2^n} where n is the number of atoms in B.

Answers

10. i. $a' \cdot b' \cdot c'$ ii. B

    iii. 1

17. $S_1 = \{a, a', 0, 1\}$,

    $S_2 = \{a' + b, a \cdot b', 0, 1\}$,

    $S_3 = \{a \cdot b', b', a, 1\}$,

    $S_4 = \{b', a \cdot b', a', 0\}$,

    $S_5 = \{a, b', 0, 1\}$.

9.8 Boolean Functions

Here first we define some of the fundamental terms required for the definition of Boolean function.

Constant

Any symbol used to represent a specific element of a Boolean algebra is called a **constant**.
For example 0 and 1 are constants.

Variable

A symbol used to represent any arbitrary element of Boolean algebra is called variable (or **Boolean variable** of Boolean Algebra. Boolean variable assumes values either 0 or 1. Boolean variables are usually represented by small letters $x, y, z,\ldots$ or letters with subscript $x_1, x_2, x_3, \ldots$.

Literal

A Boolean variables x or its complement, x' is referred to as **literal**.

Monomial

In Boolean algebra a product in which every factor is some variable or its complement is called a **monomial**. For example xy, $x'yz$ are monomials but $x(y + z')t$ is not a monomial by itself; it is a sum of two monomials, $xyt + xz't$.

Polynomial

A finite sum of monomials is called a **polynomial** or **Boolean polynomial**. In a polynomial each term is called a term of the polynomial. For example $x + y,\ x' + y, y + xy,\ x'yz' + xy$ etc.

Boolean Expression

An expression built up from the variables $x\ y, z, \ldots$ with the help of connectives '+' , '·',' ' ' a finite number of times is said to be **Boolean expression**.

1. 0 and 1 are Boolean expressions.
2. The literals $x_1, x_2, \ldots.., x_n$ are Boolean expressions.
3. If E_1 and E_2 are Boolean expressions then $E_1 \cdot E_2$ and $E_1 + E_2$ are Boolean expressions.
4. If E is a Boolean expression, then E' is also a Boolean expression.
5. An expression consists of literals, parenthesis and symbols is a Boolean expression iff it can be obtained by applying any of the above four rules.

Boolean expression can be represented by $E_1, E_2, E_3, \ldots.$ and more explicitly defined by $E_1(x_1, x_2, \ldots, x_n)$.

1. *A Boolean expression of n variables may or may not have all the n variables.*
2. *Binary operation '+' and '·' and unary operation ' satisfy all the identities of Boolean algebra. It implies that we can write one Boolean expression in the form of the other.*
3. *If one Boolean expression $E_1(x_1, x_2, \ldots, x_n)$ obtained from the other $E_2(x_1, x_2, \ldots, x_n)$ by a finite number of applications of the identities of Boolean algebra then E_1 and E_2 are known as **equivalent** (**or equal**).*

Atom

Let $<B, +, \cdot, ', 0, 1>$ be a Boolean algebra. An Element $a \in B$ is called an **atom** if $a \neq 0$ and for every $x \in B$,

$$x \cdot a = a \text{ or } x \cdot a = 0.$$

Boolean Function

A Boolean function is defined as an algebraic expression formed with constants and Boolean variables with finite applications of logical operator, '+', ' · ' and ' ′ '.

For example $xy' + z(xy + x'z') + 0$ is a Boolean function.

In other words: Let $(B, +, \cdot, ', 0, 1)$ be a Boolean algebra. Then a Boolean function of n variables defined as a function $f: B^n \to B$ which is associated with a Boolean expression of n variables.

Two Boolean variables are said to be independent if each can assume value independently of the other. For example for two variables x and y any of the following four possibilities can occur:

$$x = 0,\ y = 0;\ x = 0,\ y = 1;\ x = 1,\ y = 0;\ x = 1,\ y = 1.$$

It is to be noted here that variable and its complement can never be independent.

We can conclude from here

1. If we have n independent Boolean variables $x_1, x_2, \ldots, x_n$ then together they can be assigned values in 2^n number of ways.
2. Number of variables in a function is equal to the number of independent variables in the function. For example if a function has variables a and a' then it is a function of one variable. Similarly $a' \cdot b$ and $a \cdot b$ are functions of two variables.
3. Different Boolean expressions may determine the same Boolean functions. In such a case these Boolean expressions are said to be **equivalent**.

Example 9.15

Write the Boolean function $f(x, y, z) = [(x + y')(x + y' \cdot z)']'$ in polynomial.

Solution

$$f(x, y, z) = [(x + y')(x + y' \cdot z)']'$$

$$= (x + y')' + (x \cdot y' \cdot z)']' \quad \text{[By D' Morgan's law]}$$

$$= x' \cdot (y')' + (x \cdot y' z) \quad \text{[By D' Morgan's law } (x')' = x\text{]}$$

$$= x' \cdot y' + x \cdot y' \cdot z \quad \ldots (1)$$

(1) is the required polynomial in two variables.

9.9 Minterms and Maxterms

Let $x_1, x_2, x_3, \ldots, x_n$ are n Boolean variables. Then a Boolean expression formed by these variables which is the product (conjunction or meet) of n distinct literals is called a **minterm**. Minterm is alternatively called as **product term** or **complete product** or **standard product** or **fundamental product** of n variables. Minterms are also represented by min j or m_j.

For example, minterms generated by the two variables, x and y in B^2 are xy, $x'y$, xy' and $x'y'$. Similarly, the sum (disjunction or join) of a set of n distinct variables is known as **maxterm**. Maxterm is also alternatively called **as standard sum** or **sum term** and may be represented by max j or M_j.

For example $x + y$, $x' + y$, $x + y'$, $x' + y'$ are maxterms of two variables x and y.

Example 9.16

Consider the three binary variables x, y and z. Total minterms and maxterms can be written as shown in Table 9.6 for three variables.

Table 9.6 : Minterms and Maxterms of three variables

x	y	z	Minterms Term	Minterms Notation	Maxterms Term	Maxterms Notation
0	0	0	$x'y'z'$	m_0	M_0	$x+y+z$
0	0	1	$x'y'z$	m_1	M_1	$x+y+z'$
0	1	0	$x'yz'$	m_2	M_2	$x+y'+z$
0	1	1	$x'yz$	m_3	M_3	$x+y'+z'$
1	0	0	$xy'z'$	m_4	M_4	$x'+y+z$
1	0	1	$xy'z$	m_5	M_5	$x'+y+z'$
1	1	0	xyz'	m_6	M_6	$x'+y'+z$
1	1	1	xyz	m_7	M_7	$x'+y'+z'$

Here, from Table 9.6, we noticed that $m_5 = x\ y'\ z$ (minterm) $= x' + y + z'$ (maxterm) because 5 has the binary representation 101, where x is treated as 1, y' as 0 and z as 1 in case of minterms and complemented in maxterms.

From the above examples it is clear that when there are two variables there will be $2 \times 2 = 2^2$ number of minterm /maxterms. When there are three variables, there will be 2^3 number of minterms/ maxterms. In general when there are n variables, there will be 2^n minterms/ maxterms.

9.10 Normal Form

A Boolean function of n variables is said to be in **normal form** if it contains all the n variables. For example $x+y, x+y', x+xy', xy$ etc. are normal forms in two variables.

9.10-1 Disjunctive normal Form

When a Boolean function is expressed as a sum of minterms then it is called **disjunctive normal form** (DNF) or **minterm expansion of Boolean function** or **canonical form.** It is also referred to as **sum of products form** (SOP) or **standard sum of products.**

Following are some of the examples of disjunctive normal forms

i. $f(x, y) = xy' + xy$

ii. $f(x, y, z) = xyz + x'yz + x'y'z.$

9.10-2 Complete disjunctive normal form

When a Boolean function is expressed as a sum of all 2^n minterms of n variables then it is called **complete disjunctive normal form** (CDNF) or **complete minterm expansion.** For example complete disjunctive normal form in two variables is

$$xy + xy' + x'y + x'y'.$$

9.10-3 Conjunctive normal form

When a Boolean function is expressed as a product of maxterms then it is called conjunctive normal form (CNF) or maxterm expansion of Boolean function. It is also referred to as product of sums form (POS) or standard product of sums or dual canonical form. For example conjunctive normal form in two variables in x and y is

$$(x+y)\cdot(x'+y)$$

and conjunctive normal form in three variables x, y and z is

$$(x+y+z)\cdot(x'+y'+z).$$

9.10-4 Complete Conjunctive normal form

When a Boolean function is expressed as a product of all 2^n maxterms of n variables then it is called **complete conjunctive normal form (CCNF)** or **complete maxterm expansion.** For example complete conjunctive normal form in two variables is

$$(x+y)\cdot(x'+y)\,(x+y')\cdot(x'+y')$$

and complete conjunctive normal form in three variables is

$$(x+y+z)\cdot(x'+y+z)\,(x+y'+z)\,(x+y+z')\,(x'+y'+z)\,(x+y'+z')\,(x'+y+z')\,(x'+y'+z').$$

9.10-5 Working Rule to find disjunctive and Conjunctive normal form

Conjunctive and disjunctive normal forms can be obtained from Boolean expression by algebraic method.

Disjunctive normal form/ complete disjunctive normal form

Step 1. Write the Boolean expression as a sum of products by using Boolean identities.

Step 2. Multiply those terms of the Boolean expression by as many 1 as the number of missing variables in that term.

Step 3. Replace 1 by sum of variables and its complement. For example if variable x is missing replace $1 = x + x'$. Similar replacement should be made for each 1.

Step 4. Apply distributive law and other identities so that every term is a minterm and no term is repeated (or contained in another product term). The result will be the required DNF.

Step 5. To obtain complete disjunctive normal form CDND.

Since (CDNF) must contain all 2^n minterms of n variables hence if it is short of some minterms add same number of 0s and put $0 = xx'$.

Conjunctive Normal Form/ Complete Conjunctive normal Form

Step 1. Write the Boolean function as product of sums by using Boolean identities.

Step 2. If any sum term of the Boolean function obtained in step 1 missing some variables, add 0 to that term.

Step 3. Replace 0 by product of missing variable and its complement.

Step 4. Apply distributive laws and other identities to obtain the required CNF.

Example 9.17

Express each Boolean expression $E(x, y, z)$ as a disjunctive normal form.

a. $E = x(xy' + x'y + y'z)$.

b. $E = (x' + y)' + x'y$.

Solution

a.

$$\begin{aligned} E &= x(xy' + x'y + y'z) \\ &= xxy' + x'xy + xy'z && \text{[By distributivity]} \\ &= xy' + xy'z && [\text{By } x\cdot x = x \text{ and } x\cdot x' = 0] \\ &= xy'\cdot 1 + xy'z \end{aligned}$$

$$= xy'(z+z') + xy'z \qquad [\because z+z'=1]$$
$$= xy'z + xy'z' + xy'z$$
$$= xy'z + xy'z'. \qquad [\text{By } x+x=x]$$

Example 9.18

Express the Boolean expression $E = xyz' + y'z + xz'$ in a sum of products.

Solution

It is to be observed here that the expression $E = xyz' + y'z + xz'$ is not in sum of product form because the product xz' is contained in the product xyz'. A fundamental product P_1 is said to be contained in another fundamental product if the literals of P_1 are also literals of P_2. Now to convert the expression in sum of product form apply following steps.

$$E = xyz' + y'z + xz'$$
$$= xyz' + xz' + y'z \qquad [\text{By commutative law}]$$
$$= xz' + xyz' + y'z$$
$$= xz' + (xz')y + y'z \qquad [\text{By commutative law}]$$
$$E = xz' + y'z \qquad [\text{By absorption law}]$$

This is the required sum of products form for E.

Example 9.19

Express the following function in disjunctive normal form.

i. $f(x, y, z) = x'z + xy'z$. ii. $f(x, y, z) = (x + y)(z'y')'$.

Solution

i. The given function is in sum of product form $f(x, y, z) = x'z + xy'z$. Since the first product does not contain y, hence multiply it by 1 and then put $1 = y + y'$.

$$f(x, y, z) = x'z \cdot 1 + xy'z$$
$$= x'z(y + y') + xy'z \qquad [\because y + y' = 1]$$
$$= x'zy + x'zy' + xy'z. \qquad [\text{By distributive law}]$$

This is the required disjunctive normal form.

ii. $$f(x, y, z) = (x + y)(z'y')'$$
$$= (x + y)[(z')' + (y')'] \qquad [\text{By De- Morgan's law}]$$
$$= (x + y)(z + y) \qquad [\because (a')' = a]$$
$$= xz + xy + yz + y \cdot y$$
$$= xz + xy + yz + y$$
$$= xz \cdot 1 + xy \cdot 1 + yz \cdot 1 + y \cdot 1 \cdot 1$$

[∵ first product does not contain y and the last term does not contain x and z].

$$= xz(y + y') + xy(z + z') + yz(x + x') + y(x + x')(z + z')$$
$$= xzy + xzy' + xzy + xyz' + xyz + yzx' + y(xz + xz' + x'z + x'z')$$
$$= xzy + xzy' + xyz + xyz' + yzx + yzx' + yxz + yxz' + yx'z + yx'z'$$
$$= (xzy + xyz + yxz) + xzy' + (xyz' + yxz') + (yzx' + yx'z) + yx'z'$$
$$= xyz + xy'z + xyz' + x'yz + x'yz'.$$

Example 9.20

Express the function $f(x, y, z) = x\,(yz')'$ in complete disjunctive normal form.

Solution

$$
\begin{aligned}
f(x, y, z) &= x(yz')' \\
&= x\,[y' + (z')'] \qquad \text{[By De- Morgan's Rule]} \\
&= x\,[y' + z] \\
&= xy' + xz \\
&= xy' \cdot 1 + xz \cdot 1 \\
&= xy'(z + z') + xz\,(y + y') \\
&= xy'z + xy'z' + xzy + xzy'.
\end{aligned}
$$

Example 9.21

Prove that the value of complete disjunctive normal form of a Boolean function is reduced to identically one.

Solution

Let us take the complete disjunctive normal form in two variables

$$
\begin{aligned}
f(x, y) &= xy + xy' + x'y + x'y' \\
&= x(y + y') + x'(y + y') \qquad \text{[By distributive law]} \\
&= x \cdot 1 + x'\,1 \qquad [\because x + x' = 1] \\
&= x + x' \\
&= 1
\end{aligned}
$$

Similarly we can prove for three and higher variables.

Example 9.22

Write the conjunctive normal form of the following functions:

i. $f(x, y, z) = (x + y)\,(x + z')$

ii. $f(x, y, z) = x$

iii. $f(x, y, z) = (xy' + xz)' + x'$

iv. $f(x, y, z) = [x + (x' + y')'] \cdot [x + (y' \cdot z')']$.

Solution

i. Here first sum (or factor) is missing z and second sum (or factor) is missing y. Hence add zeros to both Sums. Replace 0 of first sum by zz' and that of second by yy'. Now, we have

$$
\begin{aligned}
f(x, y, z) &= (x + y)\,(x + z') \\
&= (x + y + 0)\,(x + z' + 0) \\
&= (x + y + zz')\,(x + z' + yy') \\
&= (x + y + z) \cdot [(x + y + z') \cdot (x + z' + y)] \cdot (x + z' + y') \qquad \text{[By distributive law]} \\
&= (x + y + z) \cdot [(x + y + z') \cdot (x + y + z')] \cdot (x + y' + z') \qquad \text{[By commutative law]} \\
&= (x + y + z) \cdot (x + y + z') \cdot (x + y' + z'). \qquad [\because a \cdot a = a]
\end{aligned}
$$

This is the required conjunctive normal form.

ii.

$$
\begin{aligned}
f(x, y, z) &= x \\
&= x + 0 \\
&= x + y \cdot y' \qquad [\because yy' = 0]
\end{aligned}
$$

$$= (x + y) \cdot (x + y') \qquad \text{[By distributive law]}$$
$$= (x + y + 0) \cdot (x + y' + 0)$$
$$= [(x + y) + zz'] \cdot [(x + y') + zz']$$
$$= (x + y + z) \cdot (x + y + z') \cdot (x + y' + z) \cdot (x + y' + z') \qquad \text{[By distributive law]}$$

This is the required conjunctive normal.

iii. $$f(x, y, z) = (xy' + xz')' + x'$$
$$= (xy')' \cdot (xz)' + x' \qquad \text{[By De- Morgan's law]}$$
$$= [x' + (y')'] \cdot (x' + z') + x' \qquad \text{[By De- Morgan's law]}$$
$$= [x' + y) \cdot (x' + z'] + x' \qquad [\because (y')' = y]$$
$$= [x' + y + x') \cdot (x' + z' + x'] \qquad \text{[By distributive law]}$$
$$= [(x' + x') + y] \cdot [(x' + x') + z']$$

[By commutative law and Associative law]

$$= [x' + y) \cdot (x' + z'] + x' \qquad [\because a' + a' = a']$$
$$= [x' + y + 0) \cdot (x' + z' + 0] \qquad [\because \text{first sum missing } z \text{ and second sum missing } y]$$
$$= [(x' + y + zz') \cdot (x' + z' + yy']$$
$$= (x' + y + z) \cdot [(x' + y + z') \cdot (x' + z' + y)] \cdot (x' + z' + y) \qquad \text{[By distributive law]}$$
$$= (x' + y + z) \cdot [(x' + y + z') \cdot (x' + y + z')] \cdot (x' + y' + z'] \text{[By commutative law]}$$
$$= (x' + y + z) \cdot (x' + y + z') \cdot (x' + y' + z'] \qquad [\because a \cdot a = a]$$

iv. $$f(x, y, z) = [x + (x' + y')'] \cdot [x + (y' \cdot z')']$$
$$= [x + \{(x')' y'\}] \cdot [x + \{y'\}' + (z')'\}] \qquad \text{[By De-Morgan's law]}$$
$$= (x + xy') \cdot (x + y + z) \qquad [\because (a')' = a]$$
$$= x \cdot (x + y + z) \qquad \text{[Absorption law } a + a \cdot b = 0]$$
$$= x \cdot (x + y) + x \cdot z \qquad \text{[By distributive law]}$$
$$= x + x \cdot z$$
$$= x \qquad \ldots (1)$$
$$= x + 0 + 0 \qquad [\because y \text{ and } z \text{ variables are missing}]$$
$$= x + yy' + zz'$$
$$= [(x + y) + z \cdot z'] \cdot [(x + y') + z \cdot z'] \qquad \text{[By distributive law]}$$
$$= (x + y + z) \cdot (x + y + z') \cdot (x + y' + z) \cdot (x + y' + z') \cdot \qquad \text{[By distributive law]}$$

This is the required CNF.

REMARK

In the part (i) we have added one zero while in the part (iv) we have $f(x, y, z) = x$ and added two zeros. But CNF of both the exercises are same.

Example 9.23

Prove that the value of the complete conjunctive normal form of a Boolean function is identically reduced to 0.

Proof

Let us prove this for three variables namely x, y and z. Complete conjunctive normal form of three variable is given by

$$f(x, y, z) = (x + y + z)\,(x + y + z')\,(x + y' + z)\,(x' + y + z)$$
$$(x + y' + z')\,(x' + y + z')\,(x' + y' + z)(x' + y' + z')$$
$$= [(x + y + z)\,(x + y + z')]\,[(x + y' + z)\,(x + y' + z')]$$
$$[\,(x' + y + z)\,(x' + y + z')]\,[(x' + y' + z)\,(x' + y' + z')]$$
$$= [(x + y + z \cdot z'\,)\,(x + y' + z \cdot z')\,(x' + y + z \cdot z'\,)\,(x' + y' + z \cdot z')$$

[By distributive law]

$$= (x + y)\,(x + y')\,(x' + y)\,(x' + y') \qquad [\because\ aa' = 0 \text{ and } a + 0 = a]$$
$$= (x + y \cdot y'\,) \cdot (x' + y \cdot y'\,)$$
$$= x\,x' \qquad [\because\ xx' = 0,\ yy' = 0,\ a + 0 = a]$$
$$= 0.$$

9.10-6 Conversion of CNF to DNF and Vice-versa

The following steps will be helpful in conversion of DNF to CNF or vice-versa.

Conversion of DNF to CNF

Let the Boolean function in DNF is f, then

Step 1: Determine f' by finding sum of those terms of the complete DNF in f which are not present in the given function.

Step 2: Determine complement of f' i.e. f''.

Step 3: Simplify f'' obtained in step 2 by using De-Morgan's laws.

Step 4: As $f'' = f$ the simplified f'' is the required CNF.

Conversion of CNF to DNF

Let the Boolean function in CNF is f, then

Step 1: Determine f' by finding product of those terms of the complete CNF in f which are not present in the given function.

Step 2: Determine complement of f' i.e. f''.

Step 3: Simplify f'' obtained in step 2 by using De-Morgan's laws.

Step 4: As $f'' = f$ the simplified f'' is the required DNF.

Example 9.24

Convert the disjunctive normal form $xy + xy' + x'y'$ to conjunctive normal form.

Solution

Given DNF is $f(x, y) = xy + xy' + x'y'$ complete DNF in two variables x and y is given by

$$xy + xy' + x'y + x'y'$$

Therefore, complement of f is given by

$$f(x, y) = xy$$
$$\Rightarrow f'\,(x, y) = (x' \cdot\ y)' = (x')' + y'$$

This is the required CNF of the given function.

Example 9.25

Convert the disjunctive normal form $(x + y)\ \ (x' + y)\,(x + y')$ into disjunctive normal form.

Solution

Given CNF is $f(x, y) = (x + y)(x' + y)(x + y')$

Complete CNF in two variables x and y is given by $(x + y')(x' + y)(x + y')$

Therefore, complement of f is given by

$$f'(x, y) = x' + y'$$

$$\Rightarrow f''(x, y) = (x' + y')' = x \cdot y$$

This is the required DNF of the given function.

Example 9.26

Convert the DNF $xyz + xy'z + x'yz' + xy'z' + x'y'z'$ into CNF.

Solution

Let, $f(x, y, z) = xyz + xy'z + x'yz' + xy'z' + x'y'z'$.

Complete disjunctive normal form in three variables x, y and z is

$$xyz + xyz' + xy'z + x'yz + xy'z' + x'yz' + x'y'z + x'y'z'$$

Hence $$f'(x, y, z) = xyz' + x'yz' + xy'z + x'yz$$

$$= xyz' + x'yz + x'y'z$$

Now taking the complement of f', we have

$$f''(x, y, z) = (xyz' + x'yz + x'y'z)'$$

$$= (xyz')'(x'yz)'(x'y'z)'$$

$$= (x' + y' + z) \cdot (x + y' + z') \cdot (x + y + z')$$

This is the required CNF.

Example 9.27

Convert the CNF into DNF

$$f(x, y, z) = (x + y + z')(x + y' + z)(x' + y + z')(x' + y' + z)(x' + y' + z').$$

Solution

Complete CNF in three variable x, y and z is given by

$$(x + y + z)(x + y + z')(x + y' + z)(x' + y + z)(x' + y' + z)(x' + y + z')(x + y' + z')(x' + y' + z')$$

Hence $$f'(x, y, z) = (x + y + z)(x + y' + z')(x' + y + z')$$

Hence $$f''(x, y, z) = [(x + y + z)(x + y' + z')(x' + y + z')]'$$

$$= [(x + y + z)(x + y' + z')]' + (x' + y + z')'$$

$$= (x + y + z)' + (x + y' + z')]' + xy'z$$

$$= x'y'z' + x'yz + xy'z$$

This is the required disjunctive normal form.

Example 9.28

Express $E(x, y, z) = z(x' + y)' + y'$ in its disjunctive normal form.

Solution

Applying De Morgan's law and involution

$$E = z(x' + y)' + y'$$

$$= z(x''y') + y'$$

$$= z(xy') + y'$$
$$= zxy' + y'$$
$$= xy'z + y'.1.1 \quad \text{[By commutative law]}$$

Multiply $(x + x')$ and $(z + z')$ is second term, we have

$$E = xy'z + y(x + x')(z + z')$$
$$= xy'z + (yx + yx')(z + z')$$
$$= xy'z + yxz + yxz' + yx'z + yx'z'$$
$$= xy'z + xyz + xyz' + x'yz + x'yz'$$

This is a required disjunctive normal form.

1. *Every Boolean expression can be represented in minterms normal form with the help of identities.*
2. *If there are n variables, then possible minterms are 2^n and we can get only 2^{2^n} different sum-of-product canonical forms. They will be the elements of Boolean algebra $\langle B_{2^n}, +, \cdot, ', 0, 1\rangle$.*

9.11 Simplification of Boolean expression by Algebraic method

Boolean expression can be reduced to simpler forms by using Boolean algebraic identities and consensus theorem. Simplification of Boolean expression can be done by eliminating some of literal and terms involved in it. The elimination of literals and terms can be achieved by deliberate introduction of redundant terms like adding aa' multiplying $(a + a')$ or adding ab to a in the original expression.

1. *An equivalent Boolean sum-of-product expression which is simpler than given Boolean expression and no other expression is simpler than that is known as **minimal sum-of-product**.*
2. *If a product term T_p in the Boolean expression E is such that*
   $$T_p + E = E$$
   *and no other term has this property then T_p is known as **prime implicant**.*
3. *Above two notes imply that sum of prime implicants for a Boolean expression are in minimal sum-of-products form.*
4. *Minimal sum-of-product form may also be obtained by following steps:*
   ***Step 1**: Express the Boolean expression E as a complete S-O-P form in which each product can be expressed in the form of prime implicant.*
   ***Step 2**: Now, remove one by one those prime implicants whose summands appear among the summands of the remaining prime implicants.*

Example 9.29

Simplify the following Boolean expressions to minimum number of literals.

a. $x + x'y$.

b. $x'y'z + x'yz + xy'$.

Solution

a. $$x + (x'y) = (x + x')(x + y) \quad \text{[Distributivity]}$$
$$= 1(x + y) = x + y \quad [x + x' = 1]$$

b. $$x'y'z + x'yz + xy' = x'z(y' + y) + xy' \quad \text{[Distributivity]}$$
$$= x'z + xy'.$$

Example 9.30

Express the following in minimal S-O-P form
$$E = xy' + xz' + yz'.$$

Solution

Step 1: Express each prime implicant of E as a complete S-O-P form as
$$xy' = xy'(z + z') = xy'z + xy'z'$$
$$xz' = xz'(y + y') = xyz' + xy'z'$$
$$yz' = yz'(x + x') = xyz' + x'yz'.$$

Step 2: Prime implicants of xz' i.e. xyz' and $xy'z'$ appear in other summands, hence xz' can be deleted.

Thus $E = xy' + yz'$ is minimal sum of products.

The prime implicants can be obtained with the help of another method called ***consensus method****. In this method if there are two product terms T_{P1} and T_{P2} such as exactly one variable x_i exist in one term and their complement in another term then the consensus of T_{P1} and T_{P2} can be obtained by the product of literals of T_{P1} and T_{P2} after removing the x_i without repetition.*

Following steps will be helpful to find the prime implicants

Step 1: Remove any product term T_{Pi} which contains any other product term T_{Pj}.

Step 2: Add consensus of T_{Pi} and T_{Pj}.

Step 3: Continue step 1 and 2 until either of them cannot be applied further.

Example 9.31

Consider $T_{P_1} = xyz'w$ and $T_{P_2} = xz'w'$ then consensus is xyz' which is obtained by deleting w and w'.

Example 9.32

Let, $E = xy'z' + x'z' + xyz' + xy'$, then find

a. Prime implicant.

b. Minimal sum-of-product form.

Solution

a. $$E = xy'z' + x'z' + xyz' + xy'$$
$$= xy'z' + x'z' + xy' \quad \text{[Remove } xy'z \text{ as } xy'z \text{ includes } xy'\text{]}$$
$$= xy'z' + x'z' + xy' + y'z' \quad \text{[Remove } xy'z' \text{ as } xy'z' \text{ includes } y'z'\text{]}$$

This is a sum of prime implicants.

b. **Step 1:** Express each prime implicants obtained in (a) of E as a S-O-P form as
$$x'z' = x'z'(y + y') = x'yz' + x'y'z'$$
$$xy' = xy'(z + z') = xy'z + xy'z'$$
$$y'z' = y'z'(x + x') = xy'z' + x'y'z'.$$

Step 2: Delete $y'z'$ as its prime implicants $xy'z'$ and $x'y'z'$ appear in other product terms. Hence $E = x'z' + xy'$ which is a minimal S-O-P form.

Example 9.33

Write the expression $E = x(y+z)$ in conjunctive normal form.

Solution

$$\begin{aligned}
E &= x(y+z)\\
&= [x + (y y')]\,[(y+z) + (x x')]\\
&= ((x+y)(x+y'))\,[(y+z+x)(y+z+x')]\\
&= [\{(x+y)(x+y')\} + (z z')]\;[(x+y+z)(x'+y+z)]\\
&= (x+y+z)(x+y+z')(x+y'+z)(x+y'+z')(x+y+z)(x'+y+z)\\
&= (x+y+z)(x+y+z')(x+y'+z)(x+y'+z')(x'+y+z).
\end{aligned}$$

Example 9.34

Find the disjunctive normal form of the expression which is represented in conjunctive normal form as $E = (x+y+z)\ (x+y+z')\ (x'+y+z)$.

Solution

Here we find DNF by taking the complement directly.

We know that $E = (E')'$

$$\begin{aligned}
\Rightarrow E &= [\{(x+y+z)\ (x+y+z')\ (x'+y+z)\}']'\\
&= [(x+y+z)' + (x+y+z')' + (x'+y+z)']'\\
&= [x'y'z' + x'y'z + x y'z']'\\
&= x y z + x y z' + x'yz + xy'z + x'yz'.
\end{aligned}$$

Exercise 9.2

1. Express
   a. $E(x, y, z) = z(x'+y) + y'$.
   b. $E(x, y, z) = y(x+y+z)'$.
   c. $E(x, y, z) = x(xy + y' + x'y)$.

   in its complete disjunctive normal form.
2. Expand the following into their canonical disjunction form
   a. $E(x, y, z) = xy + yz$.
   b. $E(x, y, z) = x + y'z$.
   c. $E(x, y, z) = x(y'z' + yz)$.
   d. $E(x, y, z) = yz' + x'z + yz$.
3. Express the Boolean function in disjunctive normal form
   a. $E(x, y, z) = xy + x'z$.
   b. $E(x, y) = xy$.
   c. $E(x, y, z) = y' + [\{z' + x + (yz)'\}']\,(z + x'y)$.
4. Obtain the DNF and CNF form of the following expression
   a. $E(x, y, z, w) = xy' + w$.
   b. $E(x, y, z, w) = ((x+y)(zw)')'$.
5. Consider
   i. $E = xyz + x'z' + xyz' + x'y'z + x'yz'$.
   ii. $E = xy + y't + x'yz + xy'zt$.
   iii. $E = x'z' + xy + x'y' + yz'$.

Find

a. Prime implicants. b. Minimal sum for each E.

6. Express the Boolean expression in three variable x, y and z.

a. $E = (x + y)(x' + y')$. b. $E = xyz + (x + y)(x + z)$.

Conjunctive normal form.

7. Express the following expression in S-O-P and P-O-S forms.

a. $E(x, y, z) = (x' + y)(y' + z)$. b. $E(x, y, z, w) = (yz' + xz)(xy' + zw')$.

c. $E(x, y, z) = y' + (z' + x + (yz)')(z + yx')$.

8. Put the Boolean expression in complete CNF form.

a. $[(xy')' + z'](x'z)'$. b. $x(y + z)$.

9. Convert the function $f(x, y, z) = (x + y + z)(xy + x'z)'$ in CNF and DNF. Also write the complement of CNF.

10. Write down the DNF in minimum possible variables of the following function:

i. $xy + xz + xy'$. ii. $(x + y + z) \cdot (xy + x'z)'$.

11. Find the complement of the following Boolean expression.

i. $x(y'z' + yz)$. ii. $xy' + xz + y'z$.

12. Write the DNF of $f(x, y, z) = (x + y) \cdot (x + z) + y + z'$.

13. Write the CNF of the Boolean function $f(x, y, z) = (x'y + xyz' + xy'z + xy'z't + t')'$.

Answers

5. i. (a) $E = x'z' + xy + x'y' + yz'$ (b) $E = x'z' + xy + x'y'$ [delete yz'].

ii. (a) $E = xy + y't + yz' + xt + xz + z'$ (b) $E = y't + xz + yz'$.

6. (a) $(x + y + z)(x + y + z')(x' + y' + z)(x' + y' + z')$

(b) $(x + y + z)(x + y + z')(x + y' + z)$.

8. (a) $E = (x + y + z)(x' + y + z')(x + y + z')(x + y' + z)(x + y' + z')(x' + y' + z')$

(b) $E = (x + y + z)(x + y + z')(x + y' + z)(x + y' + z')(x' + y + z)$.

9. CNF : $(x + y + z)(x + y + z')(x' + y' + z)(x + y' + z')(x' + y + z')$

DNF : $xy'z + xy'z' + x'yz'$.

Complement of CNF : $(x + y + z')(x' + y + z)(x + y' + z')(x + y' + z')$.

10. (i) $xyz + xy'z + xyz' + xy'z'$. (ii) $xy'z + xy'z' + x'yz'$.

11. (i) $x' + (y + z) \cdot (y' + z')$. (ii) $(x' + y) \cdot (x + z) \cdot (y + z')$.

12. $xyz + xy'z + xyz' + x'yz + x'yz' + xy'z' + x'y'z'$.

9.12 More about Boolean Expressions and Boolean Functions

9.12-1 Valuation Process

Consider a Boolean algebra $\langle B, +, \cdot, ', 0, 1\rangle$ and an expression $E(x_1, x_2, \ldots, x_n)\ \forall\ x_1, x_2, \ldots, x_n \in B$. For any $(\alpha_1, \alpha_2, \ldots, \alpha_n) \in B^n$, where B^n represents the n-tuple in B. we may replace the literals $x_1, x_2, \ldots, x_n$ in Boolean expression by the elements $\alpha_1, \alpha_2, \ldots, \alpha_n$ respectively. As the expression involve join, meet and complements of the variables and their combinations which gives us a well defined element of B, denoted by $E(\alpha_1, \alpha_2, \dots, \alpha_n)$ and known as **values of expression** E at the point $(\alpha_1, \alpha_2, \dots, \alpha_n)$ of B^n. And this process to determine the value of Boolean expression is known as **valuation process**. If this valuation process is over and we are getting only two elements 0 and 1 of Boolean algebra then it is known as **binary valuation process**.

Two expressions E_1 and E_2 generated by $x_1, x_2, \dots, x_n$ are said to be **equivalent** if they can be reduced to each other by identities of Boolean algebras, i.e.

$$E_1(x_1, x_2, \dots, x_n) = E_2(x_1, x_2, \dots, x_n).$$

Example 9.35

Find the value of expression $E = xy\,[xw + y' + zx']$ for $x = \alpha$, $y = 1$, $z = \beta$ and $w = 1$, where $\alpha, \beta, 1 \in B$. The Hasse diagram of this is shown in Fig. 9.7.

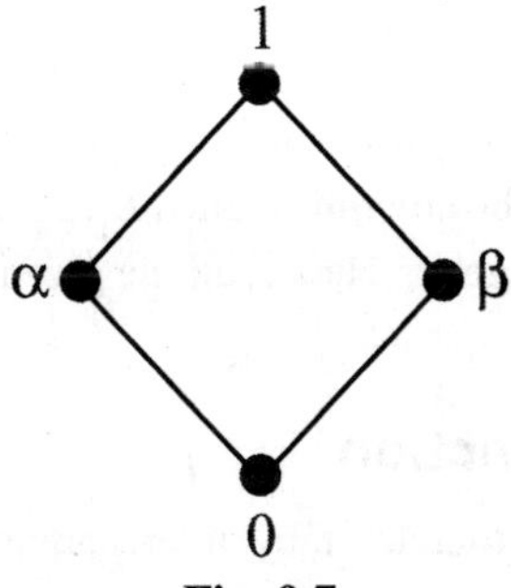

Fig. 9.7

Solution

Substituting the values of x, y, z and w, we get

$$\begin{aligned} E &= \alpha 1\,[\alpha 1 + 0 + \beta\alpha'] \\ &= \alpha\,[\alpha + \beta\beta] = \alpha\,[\alpha + \beta] \\ &= \alpha \cdot 1 = \alpha. \end{aligned}$$

In another way, if we write the expression in equivalent form as

$$\begin{aligned} E &= xy\,[xw + y' + x'z] = xyxw + xyy' + xyx'z \\ &= xyw + 0 + 0 = xyw \end{aligned}$$

Now, putting the given values, we obtain $E = \alpha.1.1 = \alpha$.

Example 9.36

Show that the following Boolean expressions are equivalent to one another $E_1 = xz + x'y$

$$E_2 = (x + y)(x' + z)$$

$$E_3 = (x + y)(x' + z)(y + z)$$

$$E_4 = xz + x'y + yz.$$

Solution

Equivalency of all expressions may also be proved with the help of the following truth table.

Table 9.7

x	y	z	x'	xz	$x'y$	E_1	$x+y$	$x'+z$	E_2	yz	E_4	$y+z$	E_3
0	0	0	1	0	0	0	0	1	0	0	0	0	0
0	0	1	1	0	0	0	0	1	0	0	0	1	0
0	1	0	1	0	1	1	1	1	1	0	1	1	1
0	1	1	1	0	1	1	1	1	1	1	1	1	1
1	0	0	0	0	0	0	1	0	0	0	0	0	0
1	0	1	0	1	0	1	1	1	1	0	1	1	1
1	1	0	0	0	0	0	1	0	0	0	0	1	0
1	1	1	0	1	0	1	1	1	1	1	1	1	1

On examining the truth values under the columns named E_1, E_2, E_3 and E_4, we observe that these are equal for every triplet of the two-element Boolean algebra. Hence, all the given expressions are equivalent.

9.12-2 Symmetric Boolean Function

A Boolean expression is known as **symmetric** if by interchanging any two variables we get an equivalent expression.

Example 9.37

a. $xy' + x'y$ is symmetric to $yx' + y'x$ [by interchanging x to y and vice versa].

b. $x'y' + (x + y)$ is symmetric to $y'x' + (y + x)$.

Boolean function forms a Boolean algebra under the operations meet, join, complement and boundedness (by 0 *and* 1*).*

Theorem 9.11

A necessary and sufficient condition for a Boolean expression in n variable to be symmetric is that there exist a set of numbers $n_1, n_2, \ldots, n_k$ between 0 and n (called its **characteristic numbers**) such that the expression has the value 1 when exactly n_i of the variables are assigned the value 1 and the rest are assigned 0.

Example 9.38

Let us consider the function $f: B_3 \to B$ whose truth table is given in Table 9.8(a), then f is produced by the Boolean expression

$$\begin{aligned} E(x, y, z) &= E(0, 1, 1) + E(1, 1, 1) \\ &= (x' \cdot y \cdot z) + (x \cdot y \cdot z) \\ &= (x' \cdot y \cdot z + x) \cdot (x' \cdot y \cdot z + y) \cdot (x' \cdot y \cdot z + z) \\ &= (x' \cdot y \cdot z + x) \cdot (x' \cdot y \cdot z + y) \\ &= (x' + x) \cdot (y \cdot z) = y \cdot z \end{aligned}$$

Thus $$f = y \cdot z.$$

Example 9.39

Let $f: B_2 \rightarrow B$ be the function whose truth table is shown in Table 9.8(b).

On examining the Table 9.8, we observe that the value 1 is for input (0, 0) and (0, 1) then

$$E(x, y, z) = E(0, 0) + E(0, 1) = (x' \cdot y') \vee (x' \cdot y)$$
$$= x' \cdot (y' + y) = x' \text{ i.e. } f \text{ is produced by } x'.$$

Table 9.8(a)

x	Y	z	$f(x, y, z)$
0	0	0	0
0	0	1	0
0	1	0	0
0	1	1	1
1	0	0	0
1	0	1	0
1	1	0	0
1	1	1	1

Table 9.8(b)

x	y	f
0	0	1
0	1	1
1	0	0
1	1	0

Example 9.40

Simplify Boolean function algebraically and form circuit diagram.

$$f(x_1, x_2, x_3, x_4) = (x_1 \cdot x_2 + x_3) \cdot (x_2 + x_3) + x_3.$$

Solution

$$f(x_1, x_2, x_3, x_4) = (x_1 \cdot x_2 + x_3) \cdot (x_2 + x_3) + x_3$$
$$= x_1 \cdot x_2 \cdot x_2 + x_1 \cdot x_2 \cdot x_3 + x_3 \cdot x_2 \cdot x_2 + x_3 \cdot x_3 + x_3$$
$$= x_1 \cdot x_2 + x_1 \cdot x_2 \cdot x_3 + x_3 \cdot x_2 + (x_3 + x_3)$$
$$= x_1 \cdot x_2 + x_1 \cdot x_2 \cdot x_3 + x_3 \cdot x_2 + x_3$$
$$= x_1 \cdot x_2 + x_1 \cdot x_2 \cdot x_3 + x_3 \cdot (x_2 + 1)$$
$$= x_1 \cdot x_2 + x_1 \cdot x_2 \cdot x_3 + x_3 \qquad [\because 1 + x_2 = 1]$$
$$= x_1 \cdot x_2 + x_3 (x_1 \cdot x_2 + 1)$$
$$= x_1 \cdot x_2 + x_3.$$

Exercise 9.3

1. Prove that:

   a. $x'((y' + z)' + yz) + ((x + y')' z) = x'y$

   b. $x'y'z'w' + x'y'z'w + x'y'zw + x'y'zw' = x'y'$

   c. $(x(y' + z))'(y' + (xz')')' = xyz'$

   d. $(x + y')(y + z')(z + x') = (x' + y)(y' + z)(z' + x)$.

2. Prove that the symmetric functions are symmetric.

3. Determine whether the following functions are symmetric

   a. $xyz' + xy'z + x'yz + xy'z' + x'yz' + x'y'z$

   b. $x'yz' + x'z'w + x'yzw + xyz'w$.

9.13 Logic Gates

Logic operations have already been discussed in chapter 7 and we have also gone through Boolean algebra which is used to model the circuitry of electronic devices earlier in this chapter. As we have seen that Boolean algebraic theorems are used for the manipulation of logical expressions. Here in this section, we will take the behavior of **circuits** (or **gatess**) which used to be in either of the states 0 or 1. These could be states of being open or closed in the case of electrically or mechanically operated switches allowing no current or some specified current to pass through or they could be some other electronic devices like semi-conductors in conducting/ non-conducting states or charged/ uncharged or magnetized/ oppositely magnetized state or some other devices carrying symbols 0 or 1 etc.

We can describe electrical circuits containing switches by logical or Boolean expressions. Thus, we can define **logic gates** (or **circuits**) as an electronic circuit which produces a typical output signal depending on its input signal. The output signal of a gate is a **simple Boolean operation** of its input signal. Gates, blocks of hardware, are the basic logic elements that produce signals of binary 0 or 1. In general we can represent gates in three ways, as given in Table 9.9.

- Graphical symbol
- Algebraic Notation (expressions)
- Truth table.

1. *Above mentioned logic circuits are also known as* ***combinational circuits*** *or* ***gating networks.***
2. *Logic networks are of two types*
   (*a*) *AND-to-OR* (*S-O-P expression*)
   (*b*) *OR-to-AND* (*P-O-S expression*)

i.e. in case (a), some of the inputs or their complements are input in each AND gate and whatever be the outputs of AND gates are treated as input in single OR gate. And the output of OR gate is the resultant output of circuit. Similarly, in case (b) OR gates follow the single AND gate.

Table 9.9 Digital Logic Gates

Operation (read as)	Graphical Symbol: Input Variable	Graphical Symbol: Algebraic Expression	Truth Table	Properties
AND (and)	X, Y	$F = X.Y$	X Y | F 0 0 | 0 0 1 | 0 1 0 | 0 1 1 | 1	(i) Any of the input is zero; output is zero (ii) Both the input is one; output is one
OR (or)	X, Y	$F = X + Y$	X Y | F 0 0 | 0 0 1 | 1 1 0 | 1 1 1 | 1	(i) Either of the input is one; output is one (ii) Both the input is zero; output is zero
NOT (not) also called inverter	X	$F = \bar{X}$	X | F 0 | 1 1 | 0	(i) Input is zero; output is one (ii) Input is one; output is zero, i.e. output is complement of input

NAND (not and) Equivalent to NOT gate in series with an AND gate	X, Y → $F=(X.Y)'$ $= X'+Y'$	X Y | F 0 0 | 1 0 1 | 1 1 0 | 1 1 1 | 0	(i) Either of the input is zero; output is one (ii) Both the input is one; output is zero
NOR (nor or) Equivalent to NOR gate with an OR	X, Y → $F=(X+Y)'$ $= X'.Y'$	X Y | F 0 0 | 1 0 1 | 0 1 0 | 0 1 1 | 0	(i) Either input is one; output is zero (ii) Both input is zero; output is one
Exclusive-OR (ex-or)	X, Y → $F=(X\oplus Y)$ $= XY' + X'Y$	X Y | F 0 0 | 0 0 1 | 1 1 0 | 1 1 1 | 0	(i) Both input is same; output is zero (ii) Dissimilar input gives the output one
Exclusive-NOR (ex-nor or equivalence)	X, Y → $F=(X\oplus Y)'$ $= X'Y' + XY$	X Y | F 0 0 | 1 0 1 | 0 1 0 | 0 1 1 | 1	(i) Same input gives the output one (ii) Dissimilar input gives the output zero
Buffer	X → $F=X$	X | F 0 | 0 1 | 1	(i) Input is zero; output is zero (ii) Input is one; output is one

Example 9.41

Design a circuit using gates to realize the following function

$$F = (X + YZ)\,(Y + Z\,\bar{X}).$$

Solution

Let there are three input logical variables *X*, *Y*, *Z* and *F* is the output. The literal *X* appears two times one in uncomplemented form and other in complemented form. A circuit using gates are simply designed by analyzing the expression (F).

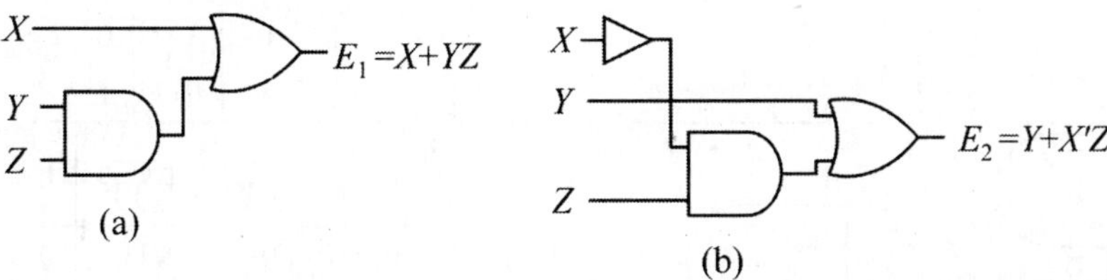

– First term $X + YZ$ consists of two terms, one is *X* and other is *YZ*. *YZ* is recognised as an AND operator between two input *Y* and Z. Then *X* is ORed with *YZ* and get a complete realization of first term. (Fig. 9.8(a))

- Second term $Y + Z\,\overline{X}$ consists of two terms Y and $Z\,\overline{X}$. $Z\,\overline{X}$ is recognized as an AND operator between the two input Z and complemented X. Then Y is ORed with $Z\,\overline{X}$ term and get complete realization. (Fig 9.8(b))
- Now, the complete realization is obtained by using AND operator between the two output which got from previous two realization and this will be the required output F. (Fig. 9.8(c))

The whole circuit may be represented in steps as shown in Fig. 9.8.

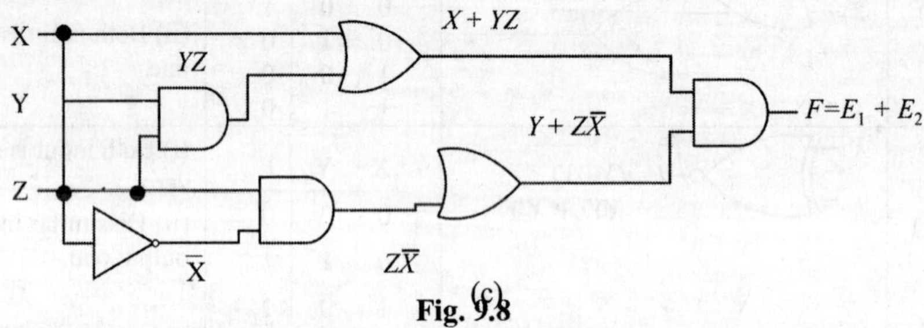

(c)
Fig. 9.8

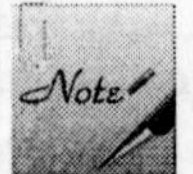

We have already seen the algebraic expression and truth table of NAND and NOR which can be made from NOT-AND and NOT-OR respectively. A circuit can be obtained by using only NAND or only NOR gates to replace the AND, OR and NOT gates.

Example 9.42

Let us consider the NAND gate and try to represent NOT, AND and OR operations through it.

Solution

Following is the graphical symbol, algebraic expression, truth table and the final conclusion drawn.

<table>
<tr><th>Graphical Symbol</th><th>Algebraic Expression</th><th>Truth table</th><th>Result</th></tr>
<tr><td>X — F=X'</td><td></td><td>X F
0 1
1 0</td><td>NOT</td></tr>
<tr><td>X, Y — $\overline{XY}$ — F= X.Y</td><td></td><td>X Y $\overline{XY}$ F
0 0 1 0
0 1 1 0
1 0 1 0
1 1 0 1</td><td>AND</td></tr>
<tr><td>X, Y — F=X+Y</td><td></td><td>X Y $\overline{XX}$ $\overline{YY}$ F
0 0 1 1 0
0 1 1 0 1
1 0 0 1 1
1 1 0 0 1</td><td>OR</td></tr>
</table>

As in logic, a set of gates by which any Boolean function can be implemented is called ***functionally complete set****. Such as {AND, NOT}, {NOR}, {NAND}, {OR, NOT}.*

Example 9.43

Find Boolean function corresponding to the following logic-gates.

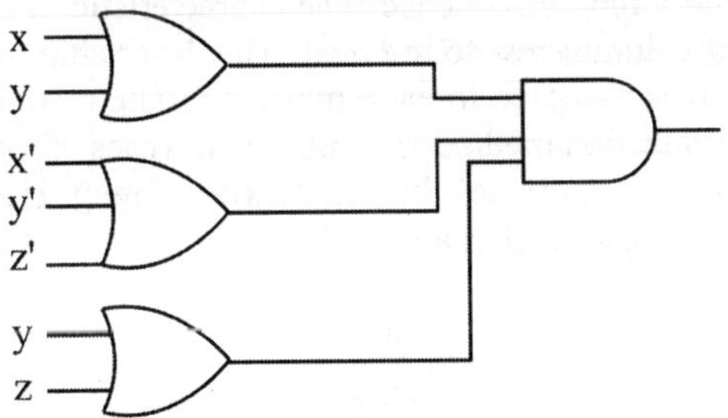

Solution

Boolean function corresponding to the logic gates is

$$f(x, y, z) = (x + y) \cdot (x' + y' + z)' \cdot (y + z).$$

9.14 Representation and Minimization of Boolean Functions

As we have discussed in previous sections, the combinational circuits are interconnected circuits of gates according to certain rules to produce an output depending on its input values. These combinational functions are independent of time. But, actually there is slight delay due to the complexity of circuits. This delay is approximately proportional to the depth or number of levels i.e. number of paths from input to output. We can represent these Boolean functions in the form of:

a. Truth table [discussed in chapter 7]

b. Algebraic Expression [discussed earlier in this chapter]

c. Cubic representation etc.

But the basic problem is how to get an equivalent combinational circuit which is minimal (i.e. circuits use minimum number of gates) by which the depth of the circuits should not exceed a specific level. For minimization of gates, we can use

a. Boolean algebraic simplification [already discussed in previous section]

b. Karnaugh Maps.

9.15 Minimization of Boolean Functions (Karnaugh Maps)

We have seen that the complexity of the logic gates directly depends on the complexity of the algebraic expressions. Boolean expression may be simplified by algebraic identities and used as a technique to minimize the whole circuit. Another method developed by E. W. Veitch in 1952, known as **Veitch diagram**, is a polished form of Venn diagram. In this diagram circle is replaced by square and arranged in the form of two dimensional array (matrix). In 1953, M. Karnaugh improved the idea of veitch who rearranged the assignments on the cells of matrix and gave it a new name **K-map**. The map method, is a pictorial representation of truth table, provides a procedure for minimizing Boolean functions. If the Boolean function contains n Variable then the k-map will be made up of 2^n squares, with each square representing one minterm of the function. In case of one variable i.e. $n = 1$ there are only two cells, as shown in Fig. 9.9(a), labelled by X and X' that are associated to 1 and 0 respectively. In the Boolean function of two variables i.e. $n = 2$, there are four (i.e. 2^2) cells, as shown in Fig. 9.9(b) labelled X, X', Y, Y' where left column is assigned to input the values of Y' and right for Y. The same is applicable for the values of X and X' in rows. The decimal number is written in the upper right corner which represents the binary number corresponding to that cell and the minterms written in the centre of cells. Binary number of each cell is assigned by reading the binary value corresponding to row and then column. When these two numbers are concatenated we will get the binary number corresponding to that cell. The map for 3 variables is represented in the Fig. 9.9(c) which has 8 cells (i.e. can have eight minterms for three variables). It is notable that the minterms are not arranged in a binary sequence. The characteristic of this sequence is that only one bit changes in value from one adjacent column/ row to the next. The 4-variable maps is also shown in Fig. 9.9(d) which has 16 minterms and one cell is assigned to each minterms. The rows and columns are numbered in a Gray code* sequence. The map minimization method is similar in all types of maps. A map for more than that of 6 variables is difficult to draw because it increases the complexity of map. Hence we will try to minimize the Boolean function upto 6 variable using Karnaugh map.

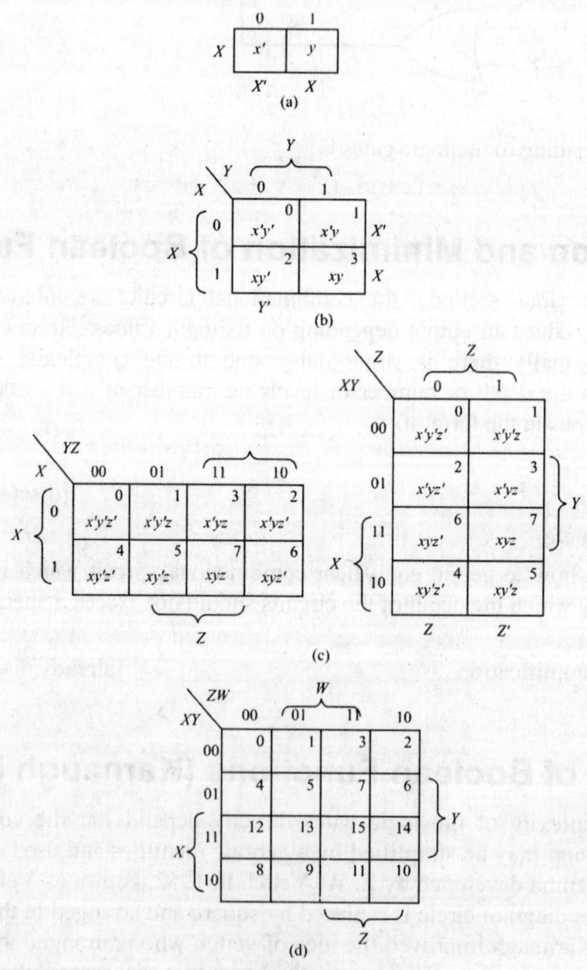

Fig. 9.9

* In Gray code sequence one bit changes at a time in value from one adjacent row/column.

One can observe from the maps shown in Fig. 9.9 (c) and (d), that any two adjacent squares in the map differ by only one variable, which is provided in one square and complemented in the other. From the postulates of Boolean algebra, it follows that the sum of two minterms in adjacent squares can be simplified to a single AND term consisting of only two literals (in case of 3-variable). As we know adjacent squares are defined to be squares next to each other. But in addition to this, the map is considered to lie on a surface on which with the top and bottom edges, as well as the right and left edges touch each other to form adjacent squares. In other words, the map act as lying on the surface of sphere. The combination of adjacent squares is useful during the simplification process of function. We know that one square represents one minterms of n variables. Two adjacent squares represent a term of (n–1) literals, similarly four adjacent squares represent a term of (n–2) literals and so on. Likewise n adjacent squares represent the function equal to 1.

For representing the Boolean function using K-map, first select the structure of K-map appropriate for that function according to their number of literals. Now, to reduce the complexity, using 1 only to those squares for which their minterms produce 1 for function. Then search for adjacent pairs of 2, 2^2, 2^3 and so on. A 1 can appear in more than one adjacent pairs. Type of adjacencies are shown in Fig. 9.10 according to the said rules. The following examples show the procedure how to simplify the Boolean function using K-map. The expressions so obtained are in the form of S-O-P form and can be easily expressed in P-O-S form by the fact $F'' = F'$.

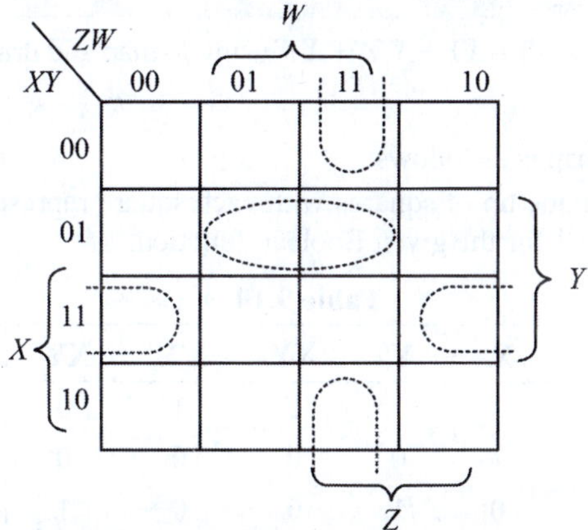

(a) Type of adjacencies for two squares

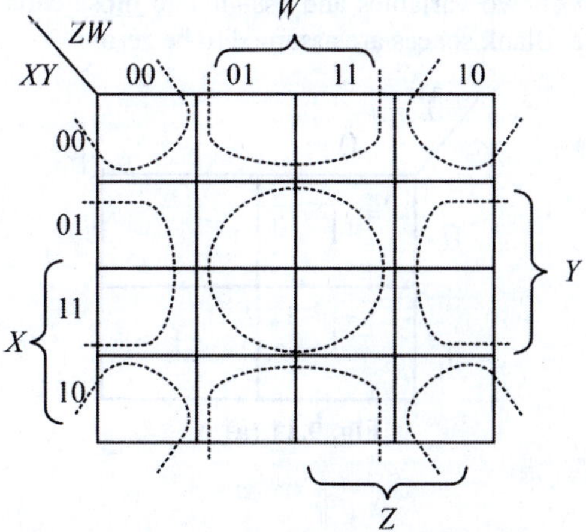

(b) Type of adjacencies for four squares

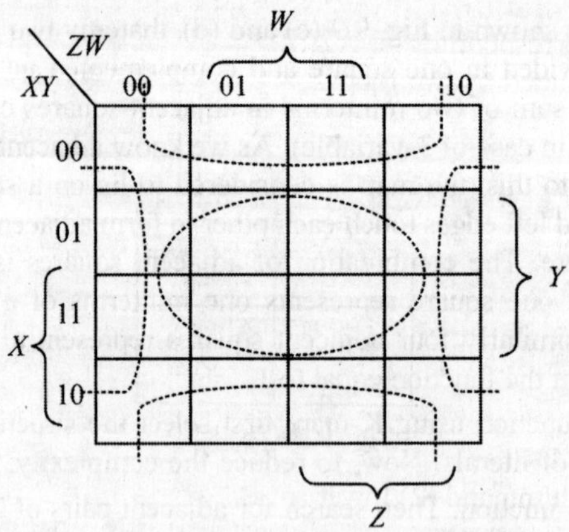

(c) Type of adjacencies for eight squares

Fig. 9.10

Example 9.44

Simplify the Boolean function $F(X, Y, Z) = XY + X'Y' + XY'$ using K-map and draw the circuit diagram for it.

Solution

The stepwise procedure for K-map is as follows:

Step 1: As map is a diagram made up of squares with each square representing one minterm. Firstly draw the truth table for the value 1 for the given Boolean function.

Table 9.10

X	Y	X′	Y′	XY	X′Y′	XY′	F
0	0	1	1	0	1	0	1
0	1	1	0	0	0	0	0
1	0	0	1	0	0	1	1
1	1	0	0	1	0	0	1

Step 2: Create a simple map of two variables and assign 1 to those cells corresponding to minterms for which F is 1 in truth table. Blank spaces are assumed to be zero.

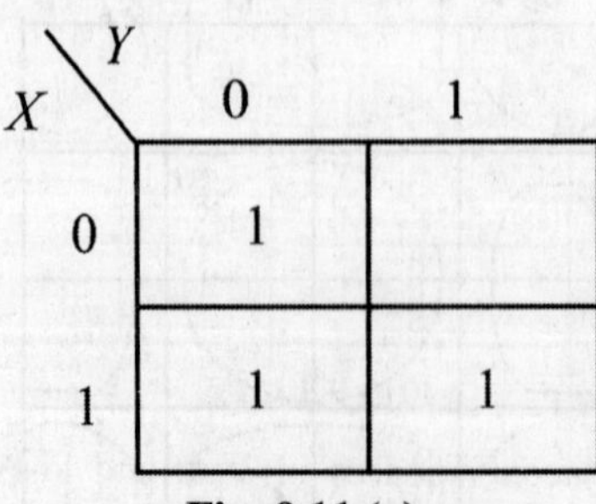

Fig. 9.11 (a)

Step 3: Observe the squares corresponding to XY, $X'Y'$, XY' in Fig. 9.11(a) and pair the adjacent 1's as shown in Fig. 9.11(b) designated by two loops.

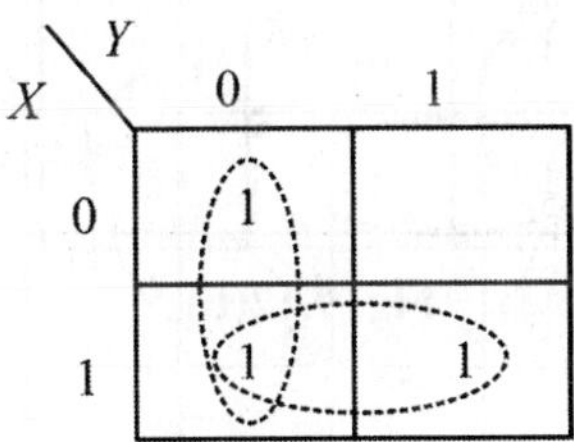

Fig. 9.11 (b)

Step 4: The pairs are: (1) left column and (2) lower row

Left column can be represented as

1. $X'Y' + XY' = Y'(X' + X) = Y'$

Lower row can be represented as

2. $XY' + XY = X\ (Y' + Y) = X$

Thus the Boolean function derived from the Karnaugh map is

$$F = Y' + X$$

The circuit diagram of F is shown in Fig. 9.11(c)

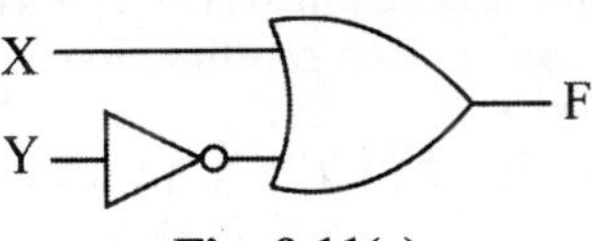

Fig. 9.11(c)

Example 9.45

Simplify the Boolean functions $f(x, y, z) = \Sigma\ (0, 2, 4, 5, 6)$ which indicate the decimal equivalent for x, y, z values for which the output is 1.

Solution

Step 1: Draw the map for three variables and assign 1 to the corresponding cells for given decimal value in expression.

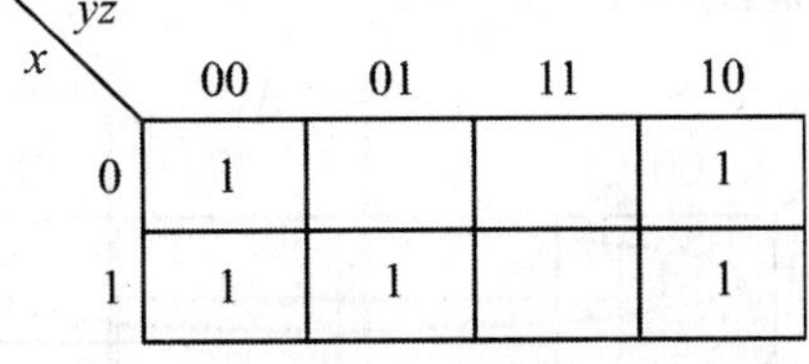

x \ yz	00	01	11	10
0	1			1
1	1	1		1

Fig. 9.12(a)

Step 2: First, combine the four adjacent squares in first and last columns. Then for remaining single minterm, we can combine it with an adjacent square that is already used once, only to reduce the complexity of expression as shown in Fig. 9.12(b) designated by loops.

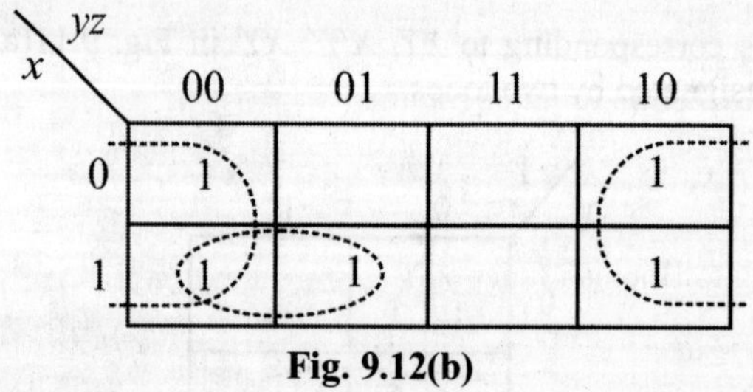

Fig. 9.12(b)

Step 3: The pairs are

1. Four 1's in first and last column.
2. Two 1's in lower left corner.

Expressions for the pairs may be written as.

1. $x'y'z' + xy'z' + x'yz' + xyz'$

$= (x' + x)y'z' + (x' + x)yz'$

$= y'z' + yz'$

$= z'(y' + y) = z'$

2. $xy'z' + xy'z$

$= xy'(z' + z) = xy'$

Thus, the Boolean function derived from Karnaugh map is $f = z' + xy'$.

Example 9.46

Simplify the Boolean function $F(x, y, z, w) = \Sigma(0, 1, 2, 3, 13, 15)$.

Solution

Step 1: Construct the K-map for the given function and assign 1 at appropriate position.

Step 2: Combine the four adjacent 1's in first row and two 1's in third row as shown in Fig. 9.13.

Step 3: The expressions given in pairs are

1. $x'y'z'w' + x'y'z'w + x'y'zw + x'y'zw'$

$= x'y'z' + x'y'z = x'y'$

2. $xyz'w + xyzw = xyw.$

Step 4: Then the Boolean function is $F = x'y' + xyw$.

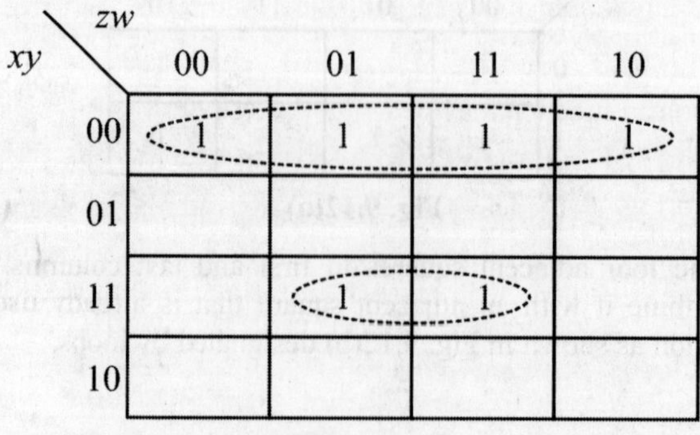

Fig. 9.13

Example 9.47

Use the Karnaugh map representation to find a minimal S-O-P expression of each of the following functions.

$$F(x, y, z, w) = \Sigma\ (0, 2, 5, 7, 8, 10, 12, 14\ ).$$

Solution

Karnaugh map corresponding to Boolean function F is shown in Fig. 9.14

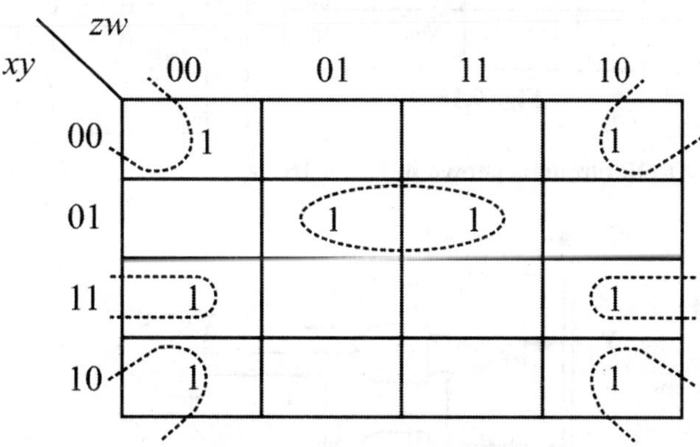

Fig. 9.14

Now, combine the pairs of adjacent 1's as

1. Combine all the four 1's at the corners.
2. Two 1's at first and last column.
3. Two middle 1's.

Thus, the expressions of all the pairs are

i. $x'y'z'w' + xy'z'w' + x'y'zw' + xy'zw'$

$$= (x' + x)\,y'z'w' + (x' + x)\,y'zw'$$
$$= y'z'w' + y'zw'$$
$$= y'w'$$

2. $xyz'w' + xyzw' = xyw'$
3. $x'yz'w + x'yzw = x'yw$

Then, simplified functions is $F = y'w' + xyw' + x'yw$.

Example 9.48

Design a minimal AND-OR circuit which has the following truth table:

Solution

From the truth table, the Boolean function may be written as

$$F = x'\,y'\,z' + x'\,yz' + xy'z' + xyz$$

The K-map of F is shown in Fig. 9.15. There are three combination as shown in Fig. 9.16 indicated by three loops. Hence the function is

$$F = (x'y'z' + xy'z') + (x'y'z' + x'yz') + xyz$$
$$= y'z' + x'z' + xyz.$$

x	y	z	F
0	0	0	1
0	0	1	0
0	1	0	1
0	1	1	0
1	0	0	1
1	0	1	0
1	1	0	0
1	1	1	1

Table 9.11

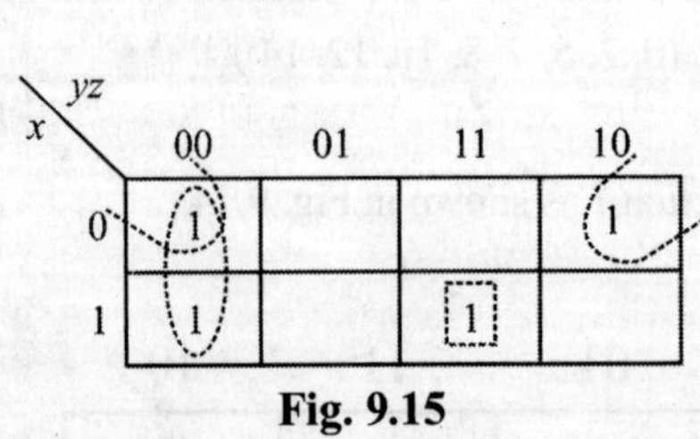

Fig. 9.15

The corresponding AND-OR circuit is shown in Fig. 9.16.

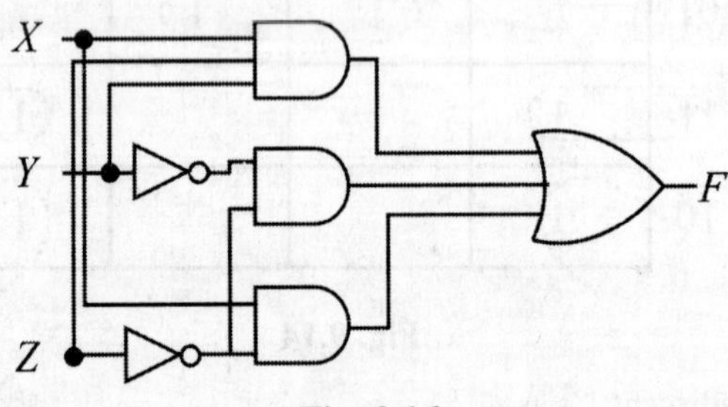

Fig. 9.16

Exercise 9.4

1. Represent NOT, AND and OR operations through NOR operator.
2. For each of the following functions give (a) The circuits diagram representation (b) The truth table representation (c) The Karnaugh map.
   i. $f = xy + yz'$ ii. $f = w + y'z + x'y$
   iii. $f = x'y'z + x'yz' + xy'z'$ iv. $f = xy' + yz' + zx'$.
3. Draw the logic diagram of the function $f = ((X' + Y')' + (X + Y')')'$ and then simplified function.
4. Express the Boolean functions as logic circuit.
   a. $F = X'YZ + XY'Z' + XY'$ b. $F = XYZ + XY'Z + XY$.
5. Find the truth table of the logic circuits expressed by
   a. $F = XYZ + XY' + X'Y$. b. $F = XY'Z + XY + Z'$.
   c. $F = XYZ' + X'YZ$. d. $F = (X'Y)'YZ' + X'(YZ + Z')$.
6. Find the output sequence for the function = X + Y + Z′ where
   a. X = 10000010; Y = 10010011; Z = 11001100.
   b. X = 00111111; Y = 11111100; Z = 11000011.
   c. X = 11000000; Y = 10101010; Z = 00000011.
7. Use a Karnaugh map to find a minimal sum for following functions
   a. $F = y'z' + y'w' + yz'w + x'y'wz$. b. $F = xy' + xyz + x'y'z' + x'yzw'$
   c. $F = x'y'z' + x'y'z + xyz' + xy'z + xyz$ d. $F = x' + y' + z'$.

8. Use the Karnaugh map representation to find a minimal S-O-P expression of each of the following function:
   a. $F(x, y, z) = \Sigma(2, 3, 5, 7)$.
   b. $F(x, y, z, w) = \Sigma(0, 2, 3, 4, 13, 15, 8, 10)$.
   c. $f(x, y, z, w, t) = \Sigma(0, 1, 3, 8, 12, 13, 17, 18, 24, 27, 31)$.
   d. $f(x, y) = \Sigma(0, 1)$.

Answers

2. (i) (a)

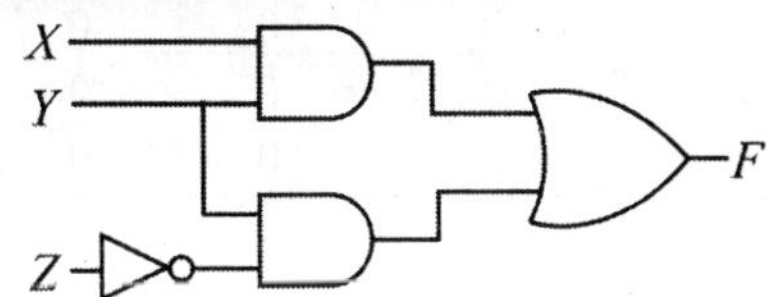

(b)

x	y	z	xy	z'	yz'	$xy + yz'$
0	0	0	0	1	0	0
0	0	1	0	0	0	0
0	1	0	0	1	1	1
0	1	1	0	0	0	0
1	0	0	0	1	0	0
1	0	1	0	0	0	0
1	1	0	1	1	0	1
1	1	1	1	0	0	1

(c)

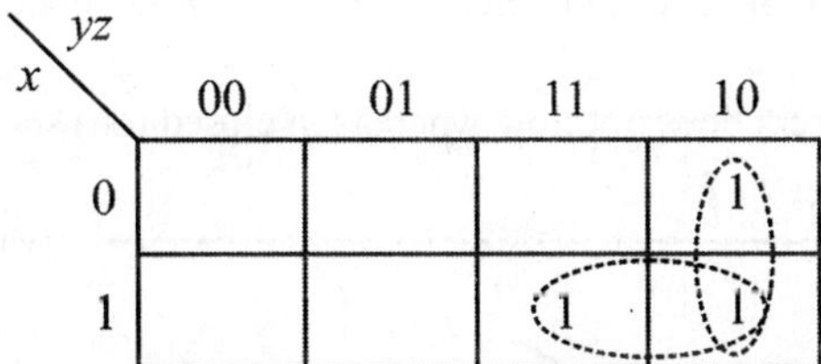

ii. (a)

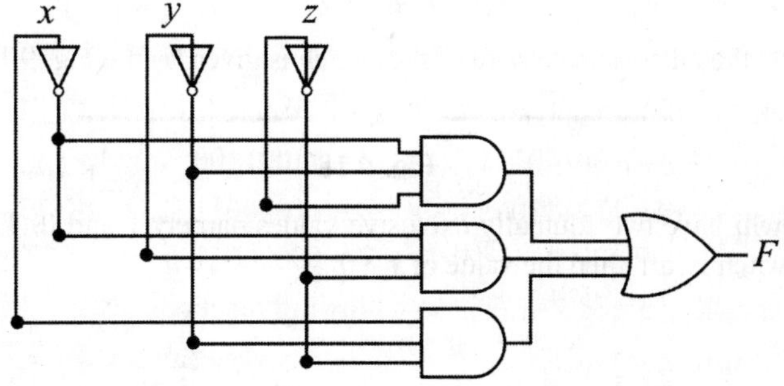

iii. (b)

x	y	z	x'	y'	z'	$p: x'y'z$	$q: x'yz'$	$r: xy'z'$	pqr
0	0	0	1	1	1	0	0	0	0
0	0	1	1	1	0	1	0	0	1
0	1	0	1	0	1	0	1	0	1
0	1	1	1	0	0	0	0	0	0
1	0	0	0	1	1	0	0	1	1
1	0	1	0	1	0	0	0	0	0
1	1	0	0	0	1	0	0	0	0
1	1	1	0	0	0	0	0	0	0

(c)

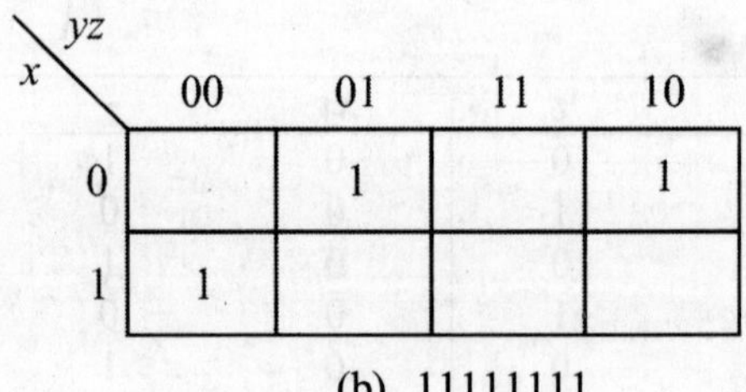

6. (a) 10011011 (b) 11111111
(c) 11111110.

9.16 Switching Circuit and Boolean Algebra

An electric circuit having a network of wires and switches is known as a **switching circuit**. A switch is a tool, which is used to control the flow of the current in electric circuits. The switches can have two stable states, namely, open (off) and closed (on).

In the open (off) state the current does not flow whereas in closed (on) state the current flows.

The two states are given in Fig 9.17.

Fig. 9.17

If x is a switch variable then the variable state of the switch is given by the Fig. 9.18.

x

Fig. 9.18

Here, the variable x will have two mutually exclusive values namely 1 and 0. Thus, if the switch is 'on' the value of $x = 1$ and if the switch is 'off' then the value of $x = 0$.

9.17 Combination of Two Switches in a Circuit

If there are two switches in an electric circuit, then they can be connected in the following two different ways: (i) In Series (ii) In Parallel

i. **In Series:** Consider two switches having variable state x and y respectively, connected in series as shown in fig. 9.19

—— x —— y ——

Fig. 9.19

From Fig.9.20 we can say that the current will flow if both switches are 'on' and the current will not flow if either switch is 'off'. So, the combined state of the two switches is given by xy.

From this discussion, we can obtain the Table 9.12.

Table 9.12

x	Y	Xy
1	1	1
1	0	0
0	1	0
0	0	0

ii. **In parallel:** Consider two switches are connected in parallel as shown in Fig. 9.20.

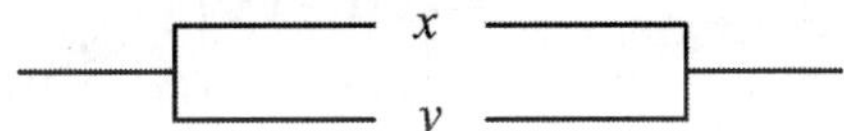

Fig. 9.20

From Fig. 9.20 it is clear that the current will not flow if both switches are 'off' otherwise the current will flow. So, the combined state of the two switches is given by $x + y$.

The truth table for this is given by

Table 9.13

x	y	x + y
1	1	1
1	0	1
0	1	1
0	0	0

Example 9.49

Draw a simple network for the following functions.

i. $f(x, y) = x + xy$ ii. $f(x, y) = xy' + x'y$.

Solution

i. Here, x and xy are in parallel and x and y are in series. The required network is given as

x

x —— y

ii. Here, xy' and $x'y$ are in parallel. x and y' are in series. x' and y are in series. The required network is given as

x — y'

x' — y

Example 9.50

Translate the given diagram into switching function, simplify it and draw the equivalent network.

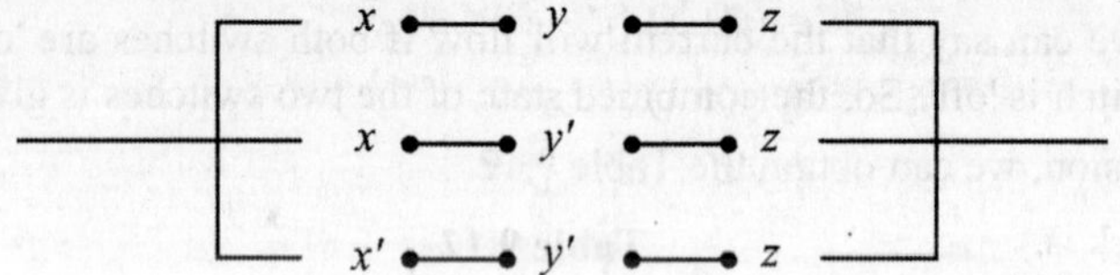

Solution

The required switching function will be

$$\begin{aligned}
f(x, y, z) &= xyz + xy'z + x'y'z \\
&= (xy + xy' + x'y')\, z \\
&= z\,[x\,(y + y') + x'y'] \\
&= z\,[x \cdot z + x'y'] \\
&= z\,[x + x'y'] \\
&= z\,[\,(x + x')\,(x + y')] \\
&= z\,[\,1\,(x + y')] \\
&= z\,(x + y').
\end{aligned}$$

The simplified network is

— z — [x in parallel with y']

Example 9.51

The following table gives the value of the function $f(x, y, z)$. Find the corresponding function. Draw a simplified circuit diagram of the function.

Table 9.14

x	y	Z	f (x, y, z)
0	0	0	1
0	0	1	0
0	1	0	0
1	0	0	1
1	1	0	1
1	0	1	0
0	1	1	1
1	1	1	1

Solution

As we know that if $f(x, y, z)$ have 1 then current will flow. So, the required function is

$$
\begin{aligned}
f(x, y, z) &= x'y'z' + xy'z' + xyz' + x'yz + xyz \\
&= (x' + x)\, y'z' + xyz' + (x' + x)\, yz \\
&= 1 \cdot y'z' + xyz' + 1 \cdot yz \\
&= y'z' + xyz' + yz \\
&= (y' + xy)\, z' + yz \\
&= (y' + x)\,(y' + y)\, z' + yz \\
&= (y' + x)\, z' + yz \\
&= y'z' + xz' + yz
\end{aligned}
$$

The simplified network is.

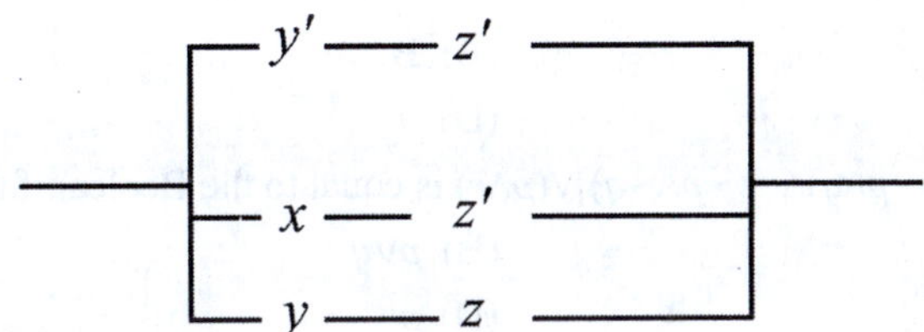

Exercise 9.5

1. Draw a simple network for the following functions:
   i. $f(x, y, z) = xyz + x + yz$
   ii. $f(x, y, z) = (x + yz)\, z + (x + y)\,(x + z)$
2. Find the function corresponding to the following networks:

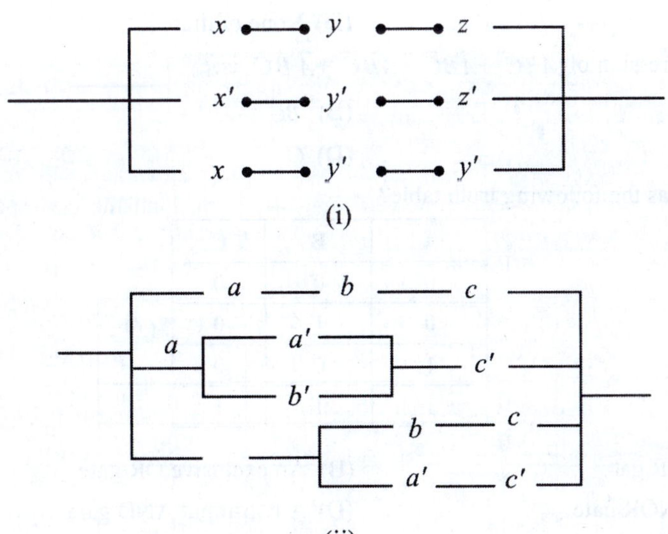

Chapter Objective Type Questions

A. Multiple choice questions

1. If B is a Boolean Algebra, then which of the following is true?
   (A) B is a finite, distributive but not complemented lattice.
   (B) B is a finite but not complemented lattice.
   (C) B is a finite, complemented and distributive lattice.
   (D) B is not distributive lattice.
2. The expression $a+\overline{a}c$ is equivalent to-
   (A) $\overline{a}$ (B) c
   (C) $a \quad c$ (D) 1
3. Let * be a Boolean operation defined by $A * B = AB + \overline{A}\,\overline{B}$, then $A*A$ is:
   (A) 1 (B) B
   (B) A (D) 0
4. The Boolean function $[\sim(\sim p \wedge q) \wedge \sim(\sim p \wedge \sim q)] \vee (p \wedge r)$ is equal to the Boolean function:
   (A) r (B) $p \vee q$
   (C) $p \wedge r$ (D) p
5. The Boolean expression $XY + XY' + X'Z + XZ'$ is independent of the Boolean variable:
   (A) X (B) Y
   (C) Z (D) none of these
6. The Boolean expression $A + AB + A\overline{B}$ is independent to:
   (A) A (B) B
   (C) Both A and B (D) None of these
7. The minimized expression of $AB\overline{C} + \overline{A}B\overline{C} + A\overline{B}\overline{C} + \overline{A}\,\overline{B}\overline{C}$ is:
   (A) $A \quad \overline{C}$ (B) $\overline{B}C$
   (C) $\overline{C}$ (D) C
8. Which logic gate has the following truth table?

A	B	C
0	0	0
0	1	0
1	0	0
1	1	1

   (A) A two-input OR gate (B) An exclusive OR gate
   (C) An exclusive NOR gate (D) A two-input AND gate

9. The necessary and sufficient condition for a nonempty subset B_1 of a Boolean algebra B to be a sub-Boolean algebra is:

(A) $a, b \in B_1 \Rightarrow a \cdot b \in B_1$ (B) $a \in B_1 \Rightarrow a' \in B_1$

(C) both (A) and (B) (C) none of these

10. For any element a of a Boolean algebra B-

(A) $a \leq 1$ (B) $0 \leq a, \forall a \in B$

(C) $0 \leq a \leq 1$ (D) all of these

11. $(x+y) \cdot (x'+y)(x+y') \cdot (x'+y')$ in two variables is a:

(A) CDNF (B) CCNF

(C) CNF (D) DNF

12. In an AND gate:

(A) any of the input is zero then output is zero (B) both the input is one then output is one

(C) either (A) or (B) (D) both (A) and (B)

13. In a NAND gate:

(A) either of the input is zero then output is one (B) both the input is one then output is zero

(C) both (A) or (B) (D) either (A) or (B)

14. A set of gates by which any Boolean function can be implemented is called_______.

(A) functionally complete set (B) complete set

(C) totally complete set (D) none of these

15. If B is a Boolean Algebra, then which of the following is true?

(A) B is a finite, complemented and distributive lattice

(B) B is a finite, distributive but not complemented lattice

(C) B is not distributive lattice

(D) B is a finite but not complemented lattice

16. A self complemented, distributive lattice is called-

(A) Modular lattice (B) Boolean Algebra

(C) Self dual lattice (D) Complete lattice

B. Fill in the blanks.

1. B is bounded and complemented lattice then it is known as ________.
2. If $a \in B$ and 1 is upper bound, then $a + 1 =$________.
3. If partial ordering relation $\preceq$ exist on B, then $a \cdot b =$ _________.
4. If a and b are two elements of Boolean algebra B, then a is included in B iff $a \cdot b' =$_______.
5. In ordinary algebra, there is no equivalent of the unary operation over a variable known as _____.
6. If $\langle B, \wedge, \vee, ', 0, 1\rangle$ and $\langle B', \cdot, +, \bar{\ }, 0', 1'\rangle$ are two Boolean algebra then the function $f: B \rightarrow B'$ such that for any $a, b \in f$ preserves both the binary operations of meet and join and unary operation complement is known as _____.

7. A Boolean variables x or its complement, x' is referred to as_____.
8. A finite sum of monomials is called a _____.
9. If we have n independent Boolean variables $x_1, x_2, \ldots, x_n$ then together they can be assigned values in ___ number of ways.
10. If there are n variables, then possible minterms are___ and we can get only___ different sum-of-product canonical forms.
11. When a Boolean function is expressed as a product of maxterms then it is called _______.
12. When a Boolean function is expressed as a product of all 2^n maxterms of n variables then it is called_______.
13. A Boolean expression is known as ______if by interchanging any two variables we get an equivalent expression.
14. In ______gate if both input is same then output is zero and if input is dissimilar then output is one.

C. Which of the following statements is True (T)/ False (F)?

1. Consider that $a, b, c \in B$ then $a + (b \cdot c) = (a + b) \cdot (a + c)$.
2. If $a \in B$, then $a \cdot a' = 1$.
3. If $\langle B, \cdot, +, ', 0, 1\rangle$ be a Boolean algebra then dual of Boolean expression can easily be obtained by interchanging "join" by "meet".
4. Boolean algebra have operations equivalent to subtraction and division.
5. If Boolean Homomorphism f is bijective then f is known as Boolean isomorphism.
6. If $a, b \in B$, then $(a + b)\, a' \cdot b' = 1$.
7. Logical addition is similar to ordinary addition.
8. Logical multiplication exactly matches with the ordinary multiplication.
9. A Boolean expression of n variables always have all the n variables.
10. In Boolean algebra Boolean variable and its complement can never be independent.
11. Different Boolean expressions may determine the same Boolean functions.
12. In Boolean algebra $(x + y) \cdot (x' + y)$ is a disjunctive normal form in two variables x and y.
13. The prime implicants can be obtained by consensus method.
14. In NOR gate if either input is one then output is zero.

Answers Objective Type

A.

1. C	2. C	3. A	4. D
5. B	6. B	7. C	8. A
9. C	10. D	11. B	12. D
13. C	14. A	15. A	16. B

B.

1. Boolean Algebra
2. 1
3. glb{ a, b}
4. 0
5. complementing
6. Boolean Homomorphism
7. Literal
8. Polynomial
9. 2^n
10. $2^n, 2^{2^n}$
11. conjunctive normal form
12. complete conjunctive normal form
13. Symmetric
14. Exclusive-OR
15. functionally complete set

C.

1. T
2. F
3. T
4. F
5. T
6. F
7. F
8. T
9. F
10. T
11. T
12. F
13. T
14. T

CHAPTER 10

Graph Theory

10.0 Introduction

Consider a number of towns connected by roads. Given a road map and we want to know whether the two towns are connected. Here, let $S = \{a, b, c, d, e, f, \ldots\}$ be the set of towns and R be a Binary Relation on S such that (a, b) is in R if town a is linked to town b by road. Furthermore, we want to know whether for a given a and b in S, there exist $c, d, e, \ldots, h$ in S such that $\{(a, c), (c, d), (d, e) \ldots, (h, b)\} \subseteq R$.

We can see many situations and problems of this type in real life that can be dealt with discrete objects and binary relations. These objects and binary relations can be represented conveniently with the help of graphs. This leads us naturally to a study of the theory of graphs. In this chapter, we will identify and discuss the basic components of a graph and some optional features that many graphs have, the most commonly asked questions related to graphs and how graphs are used to model different situations.

10.1 Graph

A graph $G = (V, E)$ consists of:

i. A set of objects $V = \{v_1, v_2, \ldots\}$ whose elements are called **vertices, points, or nodes** of G.

ii. A set $E = \{e_1, e_2, e_3, \ldots\}$ of unordered pairs of distinct vertices whose elements are called **edges** of G such that each edge e_r is identified with an unordered pair (v_i, v_j) of vertices. Fig.10.1 represents the graph $G(V, E)$.

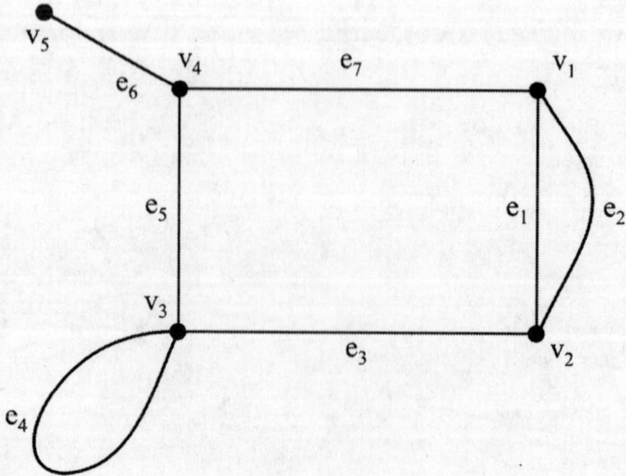

Fig. 10.1

The vertices v_i and v_j are said to be **adjacent** if there is an edge $v_i v_j = e_k$ to which it is associated. In such a case v_i and v_j are called the **end points** of e_k and e_k is said to **connect** or **join** of the vertices v_i and v_j.

The graphs are commonly represented by diagram. Every diagram consists of a set of vertices, which are represented by dots or circles, and in every diagram lines or arcs connect certain pairs of such points. The other details, such as geometry of the edges, their lengths and the positions of the points are of no importance at present. In the graph shown in the Fig.10.1 we can see it has five vertices and seven edges.

10.2 Finite Graph, Infinite Graph and Trivial Graph

A graph is said to be **finite** if it has finite number of vertices as well as finite number of edges, otherwise it is an **infinite graph**. It is to be noted here that if a graph has finite number of vertices it must automatically have a finite number of edges and hence must be finite. The finite graph with one vertex and no edge, i.e. a single point is called the **trivial graph**. Portions of two infinite graphs are shown in Fig. 10.2. The graphs shown in Fig.10.3 (a) and (b) are examples of finite graphs.

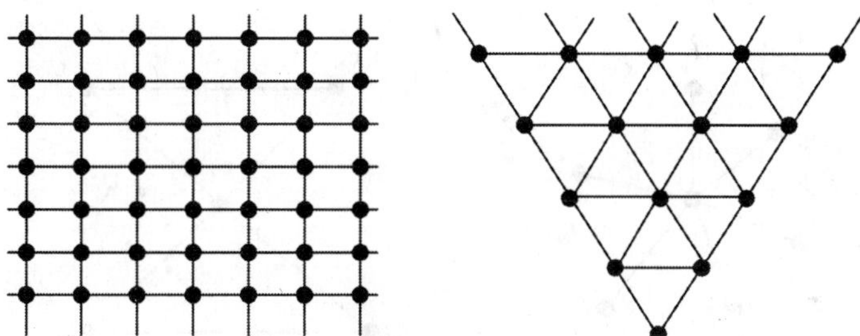

Fig. 10.2: Portions of Infinite graph

10.3 Loop and parallel edges

From the Fig.10.3 one can observe that every edge is associated with two vertices. An edge of a graph which join a vertex to itself is called a **loop** or **self-loop** (not to be confused with loop of electrical network theory or with a loop in a computer program). In Fig.10.3 (b) e_7 is a loop. It is quite possible that two vertices may be connected with more than one edge. Such edges are called **parallel edges.**

In Fig. 10.3(a) there are two parallel edges namely e_2 and e_3 between the vertices v_2 and v_3, and there are two parallel loops e_5 and e_6 at v_1. In Fig. 10.3(b) there is one loop at v_2 and three parallel edges between v_1 and v_2.

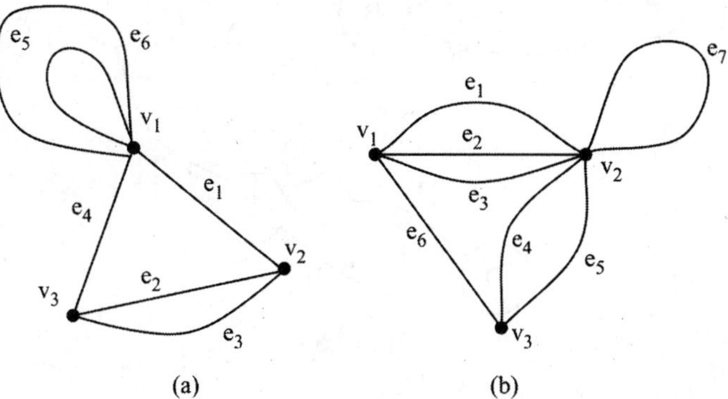

Fig. 10.3

10.4 Simple Graph and Multigraph

A **simple graph** is one for which there is no more than one edge between a pair of vertices. Any graph, which contains some parallel edges, is called a **multigraph**. The graph shown in Fig. 10.3(a) and (b) are multigraph graphs while Fig. 10.4(b) is a simple graph.

A graph, in which loops and parallel edges are allowed, is called a **pseudo graph**. Graph shown in Fig.10.3 (a) is a Pseudo graph.

Example 10.1

An intercom in an institute can be easily described using a graph as shown in Fig.10.4 (a), which describes a network of six-telephone t_1, t_2, t_3, t_4, t_5 and t_6. An edge between two vertices (telephones) indicates that direct two-way communication is possible between the two persons. It is to be noted that edges of this graph is not directed. Similarly, we can take example of network of computers as shown in the Fig. 10.4(b). This describes a network of five-computer c_1, c_2, c_3, c_4, c_5 and c_6, which are directly or indirectly linked to each other. There are several situations for which this graph can serve as model.

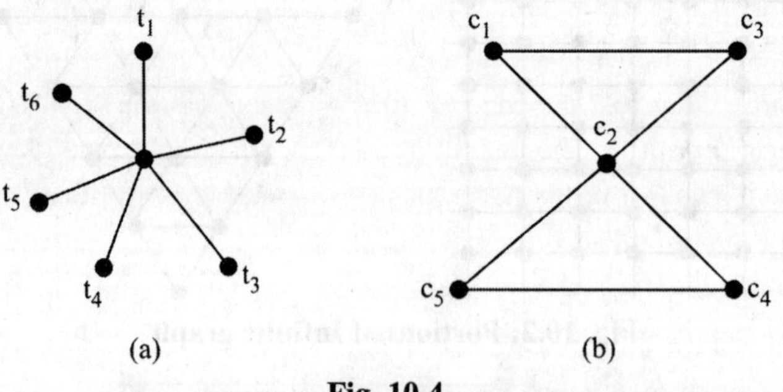

Fig. 10.4

Example 10.2

A road map is a common example of a multigraph. In a road map we can see there are many cities which are connected to other cities in more than one ways. In order to give clear travel directions, we use to give names to the different roads connecting different cities as can be seen in the multigraph shown in Fig. 10.5.

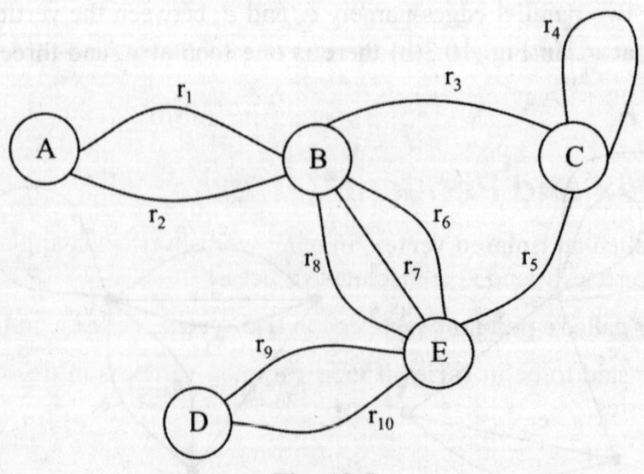

Fig. 10.5

Example 10.3

A single railway track connecting different cities is a very common example of a simple graph.

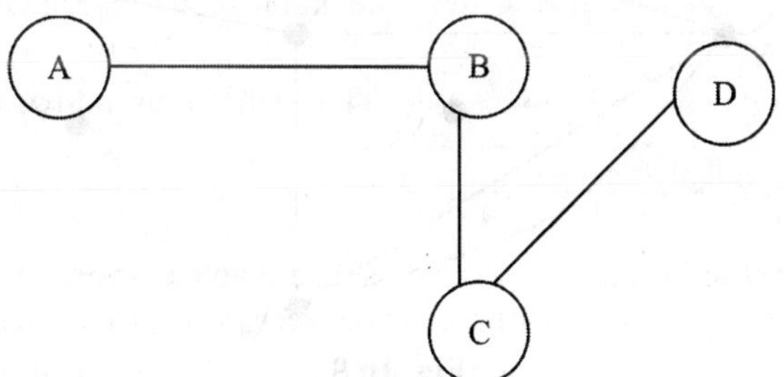

Fig. 10.6: Simple Graph

10.5 Incidence and degrees

Let $G(V, E)$ be a graph and edge $e_k \in E$ which joins the vertices v_i and v_j, then the edge e_k is said to be **incident** to the vertices v_i and v_j. Some times when we wish to be more specific, we say that the edge (v_i, v_j) is incident from v_i and is incident into v_j. Fig.10.7 the edge (a, b) is incident with the vertices a and b. The degree of a vertex v_i is denoted by $\deg(v_i)$ or $d(v_j)$. The **degree** of a vertex in an undirected graph is the number of edges incident with it, except that a loop at a vertex contributes twice to the degree of that vertex.

A vertex is said to be **even** or **odd** according as its degree is an even or odd number. Thus vertex v_1 is even whereas v_2, v_3, v_4 and v_5 are odd in Fig.10.8.

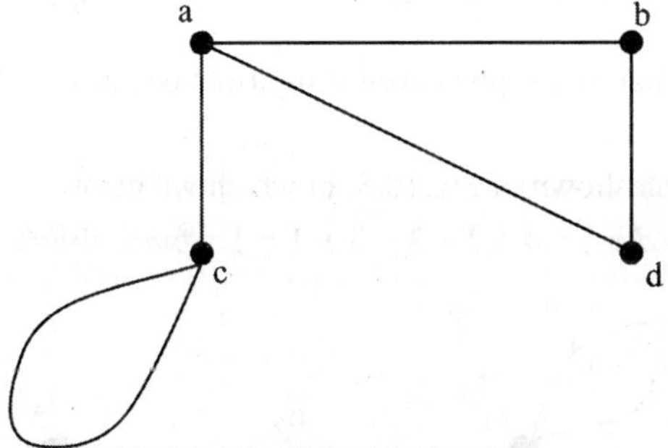

Fig. 10.7

10.6 Isolated Vertex and Pendant Vertex

A vertex of degree zero is called an **isolated vertex**. In other words, vertex having no incident edge is called an isolated vertex. In Fig.10.9 vertices v_5 and v_6 are isolated vertices.

A vertex of degree one is called a **pendant vertex** or an **end vertex**. Vertex v_3 in Fig.10.8 is a pendant vertex.

Two adjacent edges are said to be in **series** if their common vertex is of degree two. In Fig.10.8 the two edges incident on v_7 are in series.

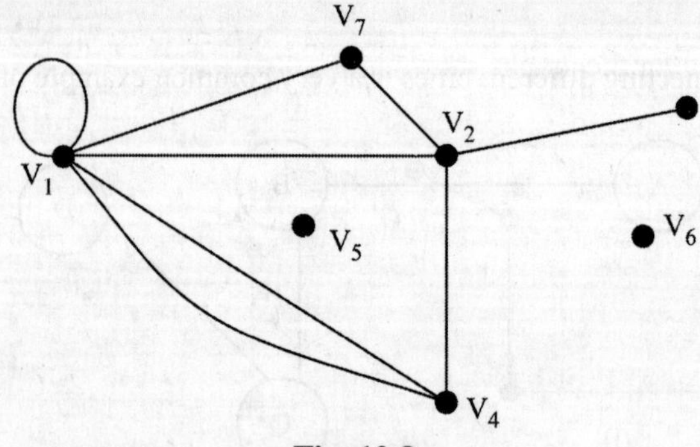

Fig. 10.8

Theorem 10.1 (Handshaking Lemma)

The sum of the degrees of all vertices of a graph *G* is twice the number of edges in *G*, i.e.

$$\sum_{i=1}^{n} d(v_i) = 2e, \text{ where } e \text{ is the number of edges in } G.$$

This result is due to Euler in 1736. This is called handshaking lemma because if several people shake hands, then the total number of hands shaking must be even, since just two hands are involved in each handshake.

Proof

Consider a graph with *m* edges and *n* vertices $v_1, v_2, v_3, \ldots, v_n$. Since each edge contributes two degree, the sum of the degrees of all vertices, in *G* is twice the number of edges in *G*. That is

$$\sum_{i=1}^{n} d(v_i) = 2e.$$ Note that this applies even if multiple edges and loops are present.

For example, consider the graph shown in Fig.10.8, in which we have
$d(v_1) + d(v_2) + d(v_3) + d(v_4) + d(v_5) = 4 + 3 + 3 + 3 + 1 = 1 =$ twice the number of edges which is 7.

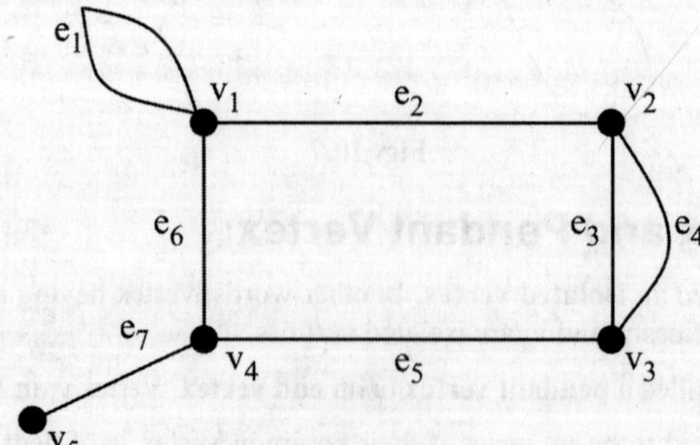

Fig. 10.9

Example 10.4

Verify handshaking lemma for the graphs shown in Fig. 10.10.

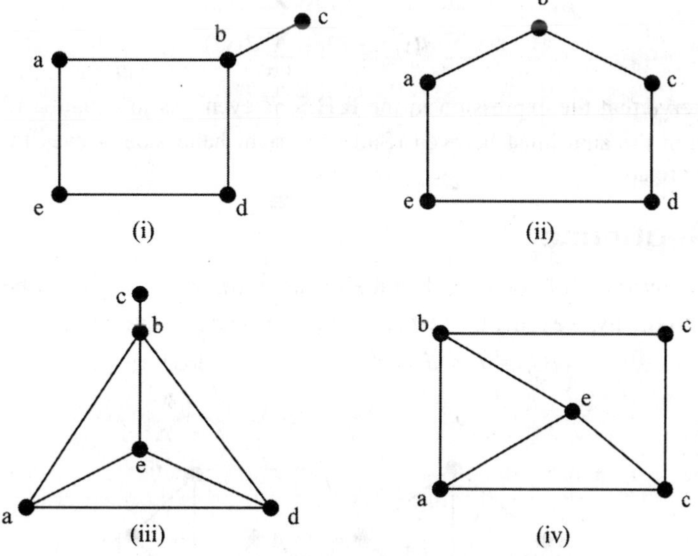

Fig. 10.10

Solution

i. In Fig 10.10 (i)

$$d(a) + d(b) + d(c) + d(d) + d(e) = 2 + 3 + 1 + 2 + 2 = 10 = 2 \times \text{No. of edges}$$

ii. In Fig 10.10 (ii)

$$d(a) + d(b) + d(c) + d(d) + d(e) = 2 + 2 + 2 + 2 + 2 = 10 = 2 \times \text{No. of edges}$$

iii. In Fig 10.10 (iii)

$$d(a) + d(b) + d(c) + d(d) + d(e) = 3 + 4 + 1 + 3 + 3 = 14 = 2 \times 7 = 2 \times \text{No. of edges}$$

iv. In Fig 10.10 (iv)

$$d(a) + d(b) + d(c) + d(d) + d(e) = 3 + 3 + 2 + 3 + 3 = 14 = 2 \times 7 = 2 \times \text{No. of edges.}$$

Hence, in all the graphs handshaking lemma is verified.

Theorem 10.2

The number of vertices of odd degree in a graph *G* is always even.

Proof

Since the sum of the degrees of all vertices in *G* is twice the number of edges in *G*, i.e.

$$\sum_{i=1}^{n} d(v_i) = 2e \qquad \ldots (1)$$

where *e* is the number of edges in *G*.

Let us consider the vertices with odd and even degrees separately. In equation (1) L.H.S. can be expressed as the sum of two sums that of vertices of even and odd degrees, i.e.

$$\sum_{i=1}^{n} d(v_i) = \sum_{\text{even}} d(v_j) + \sum_{\text{odd}} d(v_k)$$

or
$$\sum_{\text{even}} d(v_j) + \sum_{\text{odd}} d(v_k) = 2e$$

or
$$\sum_{\text{odd}} d(v_k) = 2e - \sum_{\text{even}} d(v_j) \qquad \ldots(2)$$

Here, we can observe that the expression in the R.H.S. is even. As in equation (2) each $d(v_k)$ is odd, the total number of terms in the sum must be even (Since the right hand side is even)to make the sum an even number. Hence the theorem.

10.7 Degree Sequence

If $v_1, v_2, v_3, \ldots v_n$ are the vertices of graph G, then the sequence $(d_1, d_2, d_3, \ldots. d_n)$, where d_i = degree(v_i), is the degree sequence of G. Normally, we order the vertices so that the degree sequence is monotone increasing, that is

$$\delta(G) = d_1 \leq d_2 \leq d_3 \leq \ldots \leq d_n = \Delta(G).$$

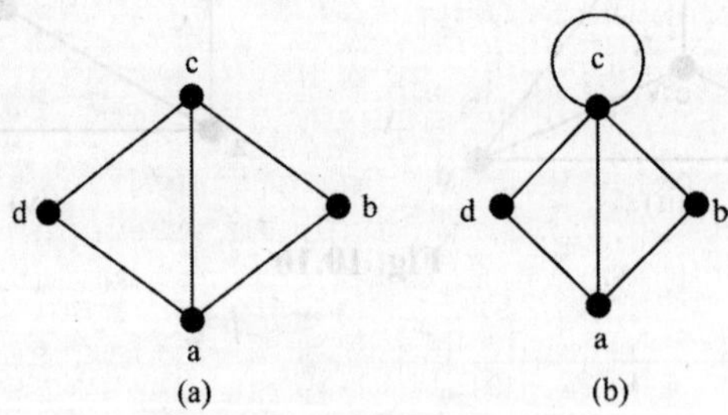

Fig. 10.11

where $\delta(G)$ = minimum of all degrees

and $\Delta(G)$ = maximum of all degrees

Degree sequence of graph shown in the Fig.10.11 (A) and (B) is (2, 2, 3, 3) and (2, 2, 3, 5) respectively.

Example 10.5

Is there exist a non simple graph G with degree sequence (1, 1, 3, 3, 3, 4, 6, 7)?

Solution

Yes, a non-simple graph of degree sequence (1, 1, 3, 3, 3, 4, 6, 7) is possible, which is exhibited in the following graph.

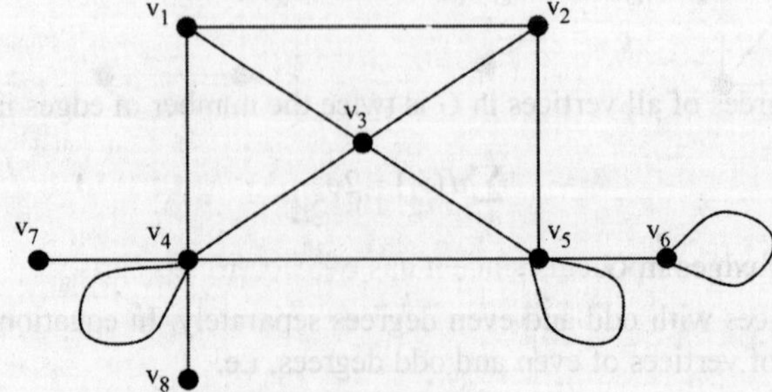

Fig. 10.12

10.8 Null Graph

A graph containing only isolated vertices is called a **null graph**. In other words, the set of edges in a null graph is empty. We denote the null graph on n vertices by N_n. Graph shown in Fig. 10.13 is N_3.

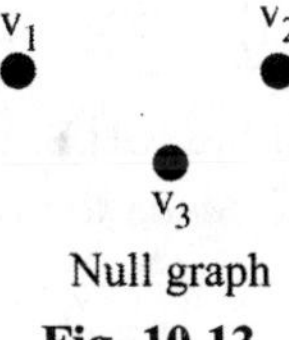

Null graph

Fig. 10.13

In a null graph G = (V, E) the edge set is empty but the vertex set V must not be empty, otherwise there will be no graph. Null graph is different from empty graph, which has no vertices and no edges. A graph of N6 class is shown in the Fig.10.14. In practice, an isolated vertex in a graph has very little importance.

Fig. 10.14

10.9 Connected Graph

If there exist at least one path between any two vertices of a graph G, then it is called a **connected graph**. If graph is not connected then it is called a **disconnected graph**. Each of the connected part of the disconnected graph is called **components**. Graphs shown in the Fig-10.15 (a) is connected while that of (b) and (c) are disconnected and disconnected null graphs respectively.

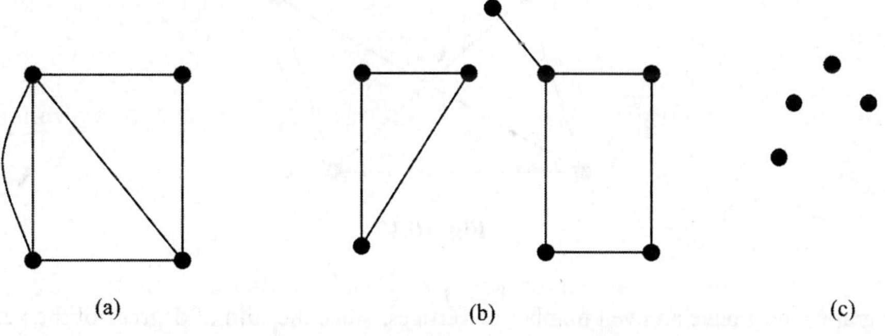

Fig. 10.15

The graph 10.15 (b) has two components since it has two connected parts.

10.10 Complete Graph

A graph $G(V, E)$ is said to be **complete** if every vertex in G is connected to all other verities of G. We denote the complete graph on n vertices by K_n. Fig. 10.16 shows K_1, K_2, K_3, K_4 and K_5. One can note that in a complete graph of n vertices, degree of every vertex is n-1 as every vertex is joined with every other vertex through one edge and the total number of edges in G is $\frac{n(n-1)}{2}$.

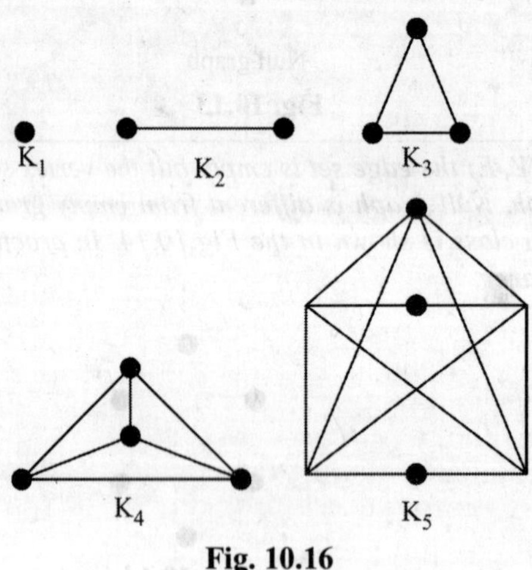

Fig. 10.16

10.11 Regular Graph

A graph in which all vertices are of equal degree is called a **regular graph** or simply a **regular**. A graph G is regular of degree k or ***k*-regular** if every vertex has degree k. The graph shown in Fig.10.17 is a regular of degree three. Usually a 3-regular graph is called a **cubic graph**. The regular graph of degree 0 and 1 are shown in the Fig.10.18. The 0-regular graph is the **trivial graph** with one vertex and no edge.

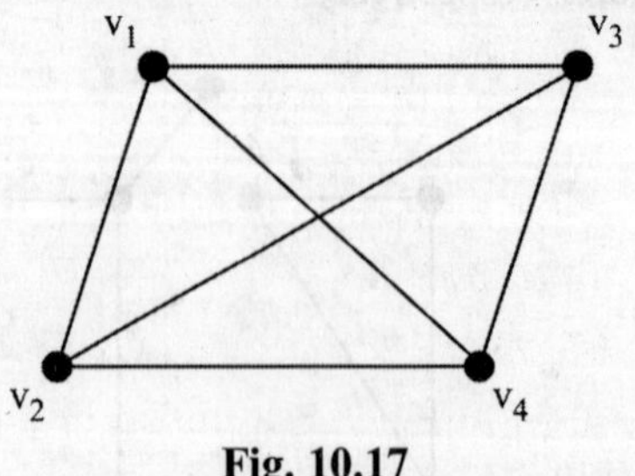

Fig. 10.17

REMARK

The 3-regular graphs must have an even number of vertices, since the sum of degrees of the vertices is an even number. The same can be seen in Fig.10.17 for a 3-regular graph with four vertices. Note that the complete graph with n vertices i.e., K_n is regular of degree $n-1$, null graph N_n is regular of degree 0.

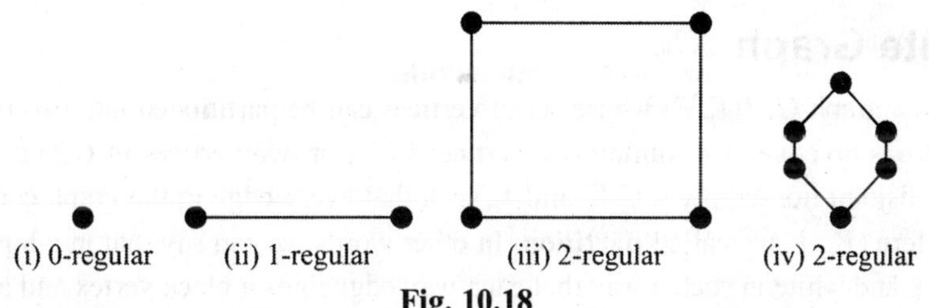

Fig. 10.18

10.12 Cycle Graph

A connected graph whose edges form a cycle of length n is called a **cycle graph** of order n. Here length n means graph consists of n edges. Cycle graphs of length n are denoted by C_n. For a cycle graph $n \geq 3$.

Graphs of class C_6 is shown in the Fig.10.19. Note that the cycle graph C_6 is regular graph of degree 2.

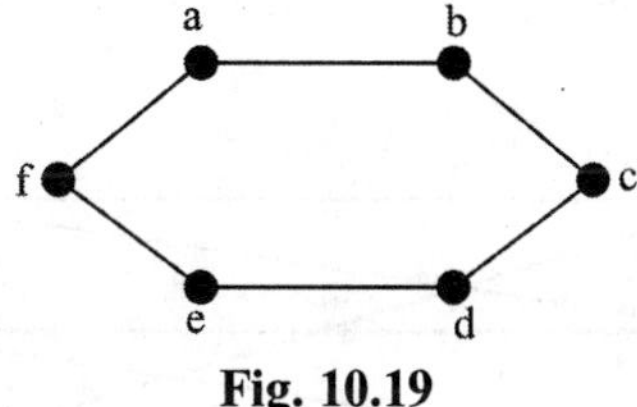

Fig. 10.19

10.13 Wheel Graph

A **wheel** of order n is a graph obtained by joining a single new vertex to each vertex of cycle graph of order (n-1). Wheels of order n are denoted by W_n. A graph of class W_5 is shown in the Fig.10.20. Here W_5 is obtained by adding vertex e to cycle graph C_4. Cycle graph C_4 consists of vertices a, b, c, d.

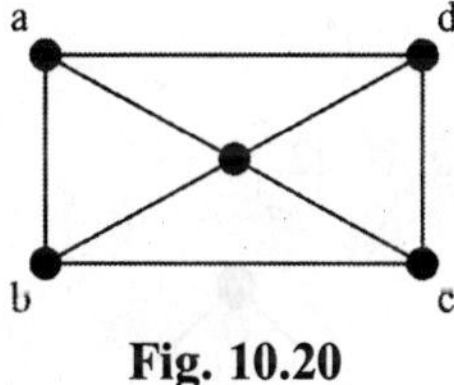

Fig. 10.20

10.14 Path Graph

Removing an edge from a C_n graph, a **path graph** of order n is obtained. A path graph of order n is denoted by P_n. A graph of class P_5 is shown in the following figure.

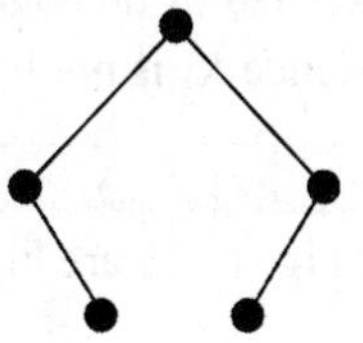

10.15 Bipartite Graph

A **bipartite graph** is a graph $G = (V, E)$ whose set of vertices can be partitioned into two nonempty subsets G_1 and G_2 such that there is no edge in E joining two vertices in G_1 or two vertices in G_2, i.e. its vertex set can be partitioned into two disjoint non-empty sets V_1 and V_2 such that every edge in the graph connects a vertex in V_1 and a vertex in V_2. Here (V_1, V_2) is called **partition**. In other words we can say that in a bipartite graph vertices can be coloured black and white in such a way that each way edge joins a black vertex and a white vertex.

Example 10.6

Show that C_6 is a bipartite graph.

Solution

C_6 is a bipartite graph can be shown by the fact that it can be partitioned into the two sets of vertices $V_1 = \{a, c, e\}$ and $V_2 = \{b, d, f\}$ such that every edge in the graph connects a vertex in V_1 and a vertex in V_2 so that no edge in G connects either two vertices in V_1 or two vertices in V_2 as shown in the Fig. 10.21.

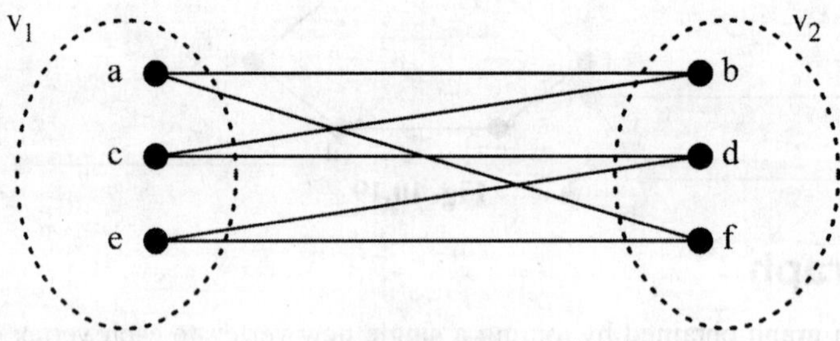

Fig. 10.21

Example 10.7

Show that K_3 is not a bipartite graph.

Solution

Let us take a K_3 graph as shown in the Fig. 10.22.

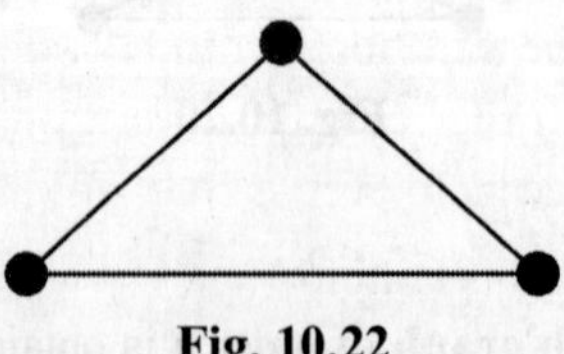

Fig. 10.22

Here, if we divide K_3 into two disjoint sets, one of the set will be containing two vertices. For this graph to be bipartite, these two vertices must not be connected by an edge. But this is not the case with K_3 as each vertex is connected to every other vertex by an edge. Hence K_3 is not bipartite.

Example 10.8

Are the graphs G_1, G_2, G_3 and G_4 shown in Fig. 10.23 are bipartite?

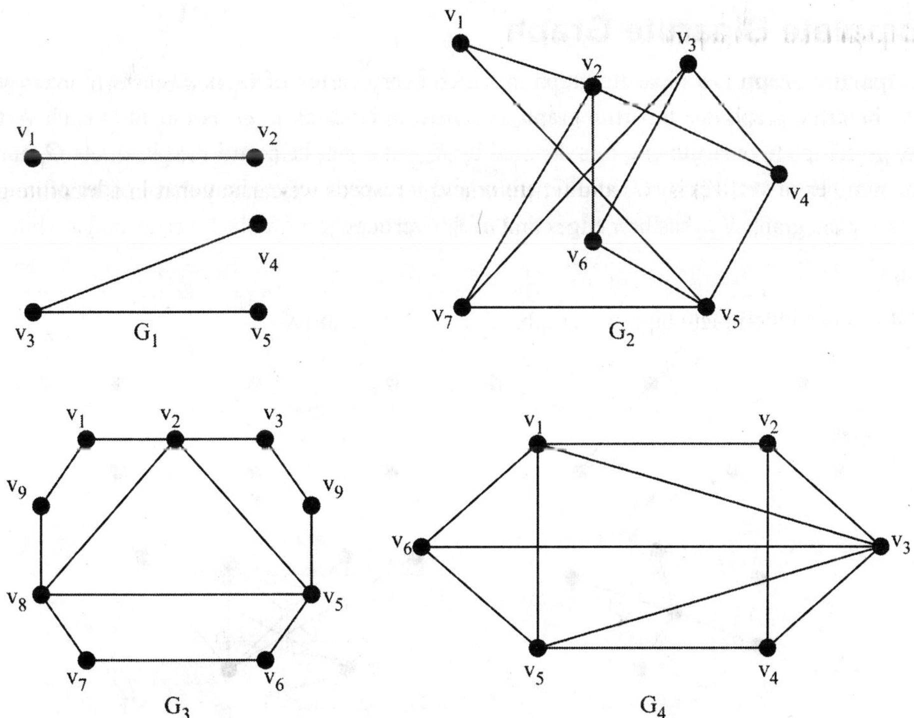

Fig. 10.23

Solution

i. If we refer to our definition, we find that when we partition the graph into two portions say vertex sets V_1 and V_2, then we find one edge is incident on one vertex in V_1 and one vertex in V_2. It is to be noted here that if v_i is a vertex in V_1 and v_2 is a vertex in V_2, it is not necessary that there is an edge between V_1 and V_2. If we see the graph G_1 of Fig.10.23, we find that if $V_1 = \{v_1, v_3\}$ and $V_2 = \{v_2, v_4, v_5\}$, then there is an edge between V_1 and V_2. However, not all edges between vertices in V_1 and V_2 are in graph. For instance $(v_2, v_3\}$ is not there. Hence G_1 is bipartite.

ii. Graph G_2 is bipartite as it can be partitioned into two disjoint sets of vertices $V_1 = \{v_2, v_3, v_5\}$ and $V_2 = \{v_1, v_4, v_6, v_7\}$ such that each edge connects a vertex of V_1 to vertex of V_2. Here again we can note that every vertex of V_1 is not adjacent to every vertex of V_2. For instance v_1 and v_3 are not adjacent.

iii. The Graph G_3 shown in Fig. 10.23 is not bipartite. Let us prove this arguing by contradiction. If graph G_3 had been bipartite it could be partitioned into two disjoint sets of vertices V_1 and V_2 such that $V_1 \cup V_2 = V$ (Refer 10.26-1) and each edge is incident on one vertex in V_1 and one vertex in V_2. Now consider the vertices v_2, v_5 and v_8. Since v_2 and v_8 are adjacent; one should be in V_1 and other should be in V_2. Again, v_2 and v_5 are adjacent and v_8 is in V_2, V_5 is in V_1. Further, v_2 and v_5 are adjacent and v_2 is in V_1, hence v_5 should be in V_2. But now we see that v_5 is both in V_1 and V_2, which is a contradiction since V_1 and V_2 are disjoint. Therefore, G_3 is not bipartite.

iv. Graph G_4 is not bipartite because it can not be partitioned into two disjoint sets V_1 and V_2 such that every edge in the graph connects a vertex in V_1 and a vertex in V_2. It can be verified by taking the vertices v_1, v_2 and v_3.

10.16 Complete Bipartite Graph

A **complete bipartite graph** is a bipartite graph in which every vertex of G_1 is adjacent to every vertex of G_2, i.e. a complete bipartite graph is a bipartite graph in which each vertex in G_1 is joined to each vertex in G_2 by just one edge. A complete bipartite graph is denoted by $K_{m,n}$ if it can be partitioned into sets G_1 and G_2 in such away that the number of vertices of G_1 and G_2 are m and n respectively. The normal order of m and n is such that $m \leq n$. Clearly the graph $K_{m,n}$ has $m \cdot n$ edges and $m + n$ vertices.

Example 10.9

Fig. 10.24 shows the complete bipartite graphs $K_{2,3}$, $K_{3,3}$, $K_{1,5}$ and $K_{2,4}$.

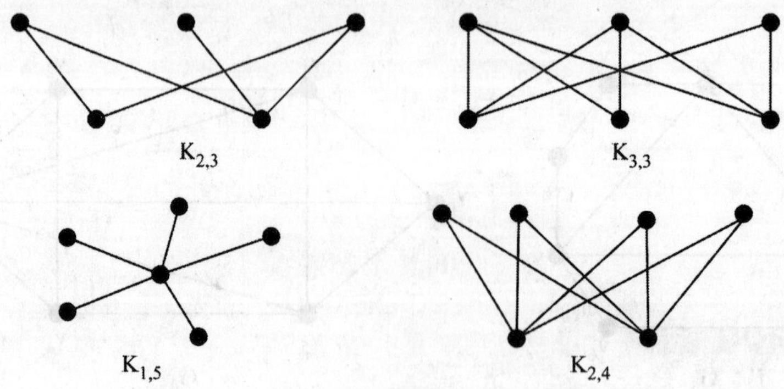

Fig. 10.24

Example 10.10

Draw the graph $K_{2,5}$.

Solution

Bipartite graph $K_{2,5}$ consists of seven vertices. These vertices are partitioned into a set G_1 of 2 vertices say a and b and a set G_2 of 5 vertices say p, q, r, s, and t, and all possible edges from vertices p, q, r, s, t. Thus there are 5×2=10 edges.

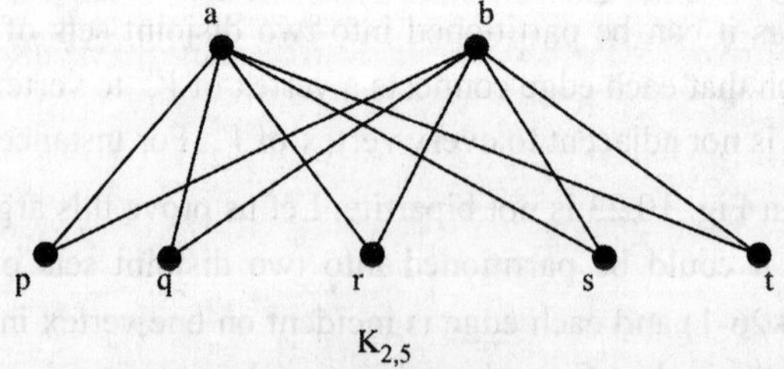

Fig. 10.25

Example 10.11

Which connected graphs can be both regular and bipartite?

Solution

Let us take a bipartite graph $K_{m,m}$. Since every complete bipartite graph is bipartite. This graph is regular of degree m since each vertex is connected to m other vertices and hence its degree is m. If m disjoint edges (edges connected to distinct pair of vertices) are deleted from $K_{m,m}$, then the subgraph (Refer 10.23) obtained in this way will also be regular. For example, the subgraph of $K_{4,4}$ shown in Fig.10.26 (a) is 3-regular. If we continue to delete m (1, 2, 3, …, m)disjoint edges and each time we obtain a regular graph of one less degree. It is to be observed here that continuously doing so at last we get a disconnected graph, but in any case their connected components have the desired properties. Graph in Fig. 10.26 (b) is a graph obtained after deleting 1- disjoint edge from $K_{4,4}$.

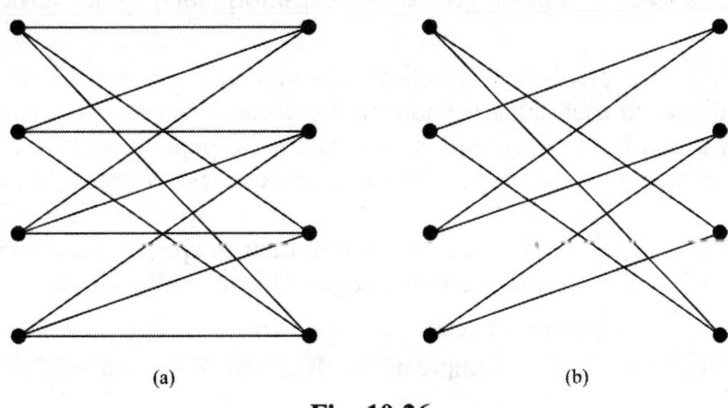

Fig. 10.26

10.17 Star Graph

Any graph that is $K_{1,n}$ is called a **star graph.**

10.18 Platonic Graph

Other interesting graphs are shown in Fig. 10.27. The graphs formed by the vertices and edges of the five regular platonic solids are shown in the Fig. 10.27.

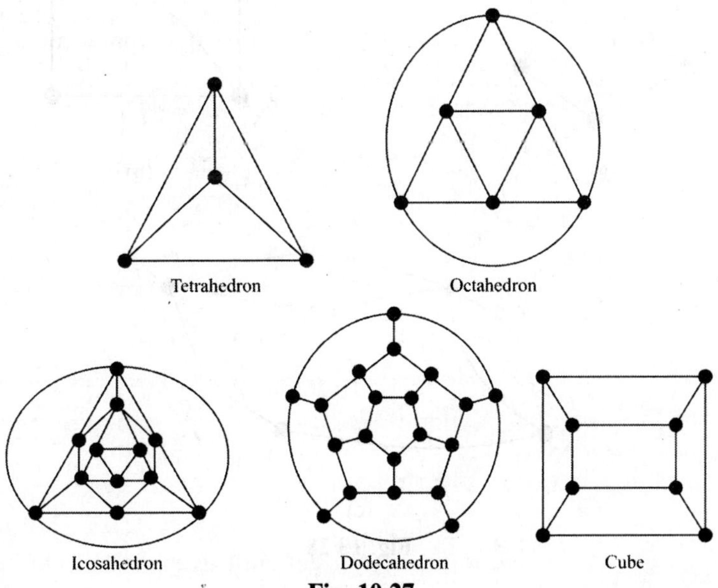

Fig. 10.27

10.19 Labelled Graph

A graph G is called a labelled graph if its edges or vertices are assigned of one kind or another. The difference between labelled and unlabelled graphs is more apparent when we try to count them.

10.20 Weighted Graph

There are many occasions where we want to give additional information to the vertices and/or the edges of a graph.

Let us take an example of a graph that represents the highway connections among different cities. We might wish to assign a number to each edge to indicate the distance between the two cities connected by the road. We can also assign a number to each vertex to indicate the population of a city. We can also take an example of a graph that represents the outcomes of the cricket matches. We may thus have graphs in which the number of edges shows the weights of the edges. A graph in which weights like capacities, distance, probabilities, etc. are assigned to every edge is called a **weighted graph**. The graphs in Fig.10.28 (a), (b) and (c) in which number on edges shows the weights of the edges. Ofcourse all weighted graphs are labelled graphs.

The weight between any two adjacent vertices v_i and v_j is written as $W(v_i, v_j)$ or if edge $e = v_i v_j$, then the weight is also represented by $W(e)$. For example in Fig. 10.28(b), $W(v_1, v_2) = 1.2$.

Example 10.12

A graph representing the traffic density of different roads in a city can be represented with the help of a weighted graph. Here weight may be assigned according to traffic density of each road. Graphs shown in Fig. 10.28 are weighted graphs.

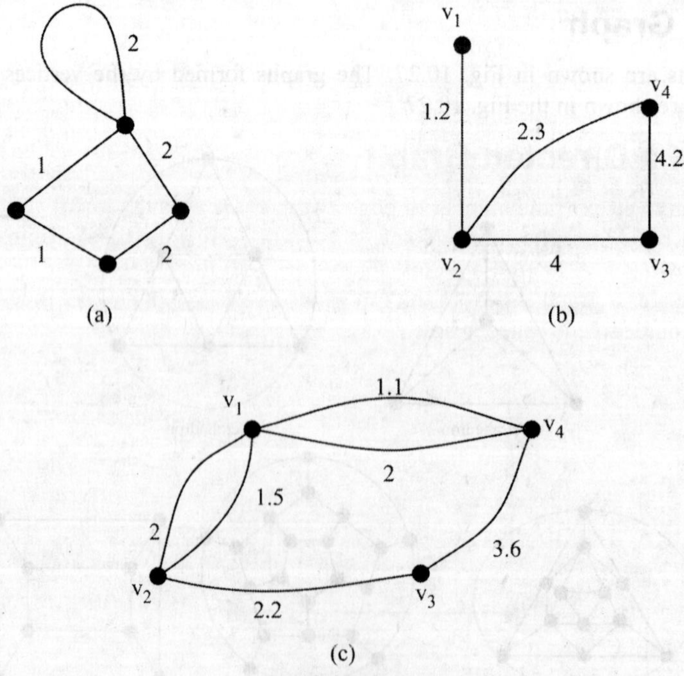

Fig. 10.28

10.21 Directed Graph

In a graph $G = (V, E)$, an edge which is associated with an ordered pair of $V \times V$ is called a **directed edge** of G, while an edge which is associated with an unordered pair of vertices is called an **undirected edge**. A graph in which every edge is directed is called a **directed graph** or **digraph**. A graph in which every edge is undirected is called an **undirected graph**. A graph in which

Some edges are directed and some are undirected, is called a **mixed graph**.

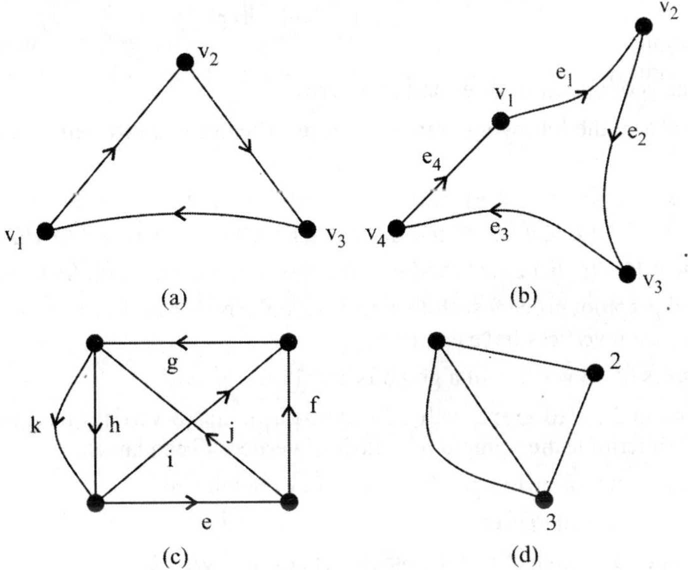

Fig. 10.29

10.22 Degree of a Directed Graph

In a directed graph for any vertex v_i the number of edges which have v as the initial vertex is called **outdegree** of the vertex v denoted by $d^+(v)$. The number of edges which have v as their terminal vertex is called the **indegree** of v denoted by $d^-(v)$. A loop at a vertex contributes 1 to the indegree and 1 to the outdegree of the vertex. For example consider a graph shown in the Fig.10.30. A vertex v with zero indegree is called a **source** and a vertex v with zero outdegree is called a **sink**.

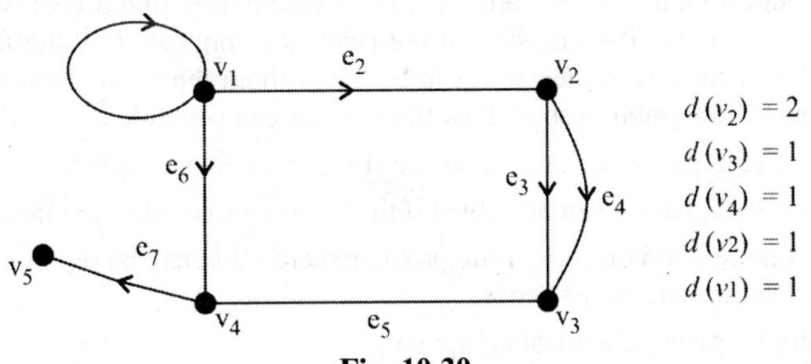

Fig. 10.30

Exercise 10.1

1. Define
   i. Simple graph
   ii. Multiple graph and give an example for each case.
2. Define
   i. Directed graph
   ii. Undirected graph.
3. Define
   i. Regular Graph
   ii. Null Graph
   iii. Complete Graph
4. Draw all simple graphs of one, two, three and four vertices.
5. Draw a picture of each of the following graphs and state whether it is directed or non directed and are they simple
   i. $G = (V, E)$, where $V = \{a, b, c, d, e\}$, and $E = \{\{a, b\}, \{b, c\}, \{a, c\}, \{a, d\}, \{d, e\}\}$.
   ii. $G = (V, E)$, where $V = \{a, b, c, d, e\}$, and $E = \{(a, a), (a, b), (b, c), (c, d), (e, d), (d, e)\}$.
   iii. $G = (V, E)$, where $V = \{a, b, c, d, e\}$, and $E = \{(a, b), (a, c), (a, d), (a, e), (e, c), (c, a)\}$.
6. Let G is nondirected graph of order 9 such that each vertex has degree 5 or 6. Prove that at least 5 vertices have degree 6 or at least 6 vertices have degree 5.
7. The sum of the degrees of all vertices of a graph is equal to ...... edges.
8. Suppose that G is an undirected graph with 12 edges. Let G has 6 vertices of degree 3 and the rest have degrees less than 3. Determine the minimum number of vertices G can have.
9. How many vertices will the following graphs have if they contain:
   i. 16 edges and all vertices of degree 2,
   ii. 21 edges, 3 vertices of degree 4, and the other vertices of degree 3,
   iii. 24 edges and all vertices of the same degree.
10. For any simple graph of n vertices, prove that the number of edges of G is less than or equal to $\frac{1}{2}n(n-1)$.
11. A sequence $d = (d_1, d_2, d_3, \ldots d_n)$ of nonnegative integers is said to be **graphical** if there exists a simple non directed graph with no self-loop with degree sequence d. Show that
    a. (2, 3, 3, 4, 4, 5) is not graphical
    b. (4, 3, 2, 2, 1) is graphical
    c. (2, 3, 3, 4, 5, 6, 7) is not graphical
    d. (2, 2, 4) is not graphical
12. i. Three married couples on a journey come to a river where they find a boat which can not carry more than two persons at a time. The crossing of the river is complicated by the fact that husbands are all very jealous and will not permit their wives to be left without them in a company where there are other men present. Construct a graph to show how the transfer can be made.
    ii. Prove that the puzzle in part (i) cannot be solved if there are four couples.
    iii. Prove that the puzzle in part (i) can be solved if there are four couples and the boat holds three persons.
13. Name five situations (games, activities, real life problems, etc.) that can be represented by means of graphs. Explain what the vertices and the edges denote.
14. Draw the graphs of the following chemical compounds
    a. Methane CH_4
    b. Ethane C_2H_6
    c. C_6H_6
    d. N_2O_3.

    [**Hint:** Represent atoms by vertices and chemical bonds between them by edges.]

15. Define cycle graph, bipartite graph, path graph and star graph. Give one example of each of the graph.
16. Consider the following graph.
    i. Describe formally the graph G in the diagram, that is, find the set $V(G)$ of vertices of G and the set $E(G)$ of edges of G.
    ii. Find the degree of each vertex.
17. Draw the following graphs:
    i. the complete graph K_6
    ii. the complete bipartite graph $K_{5,7}$
    iii. the complement of the cycle graph c_4.

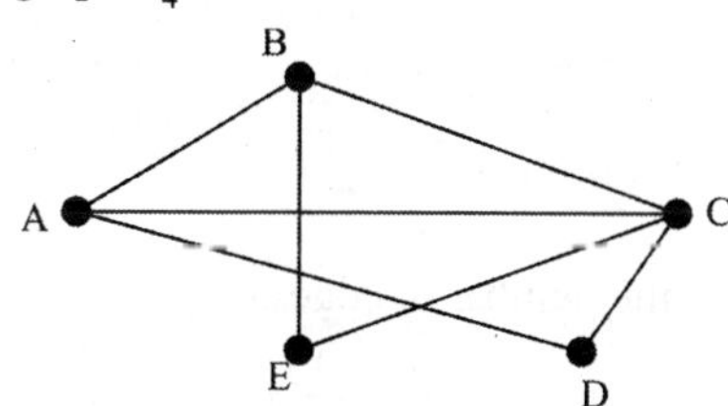

18. Determine whether the following graphs are bipartite?

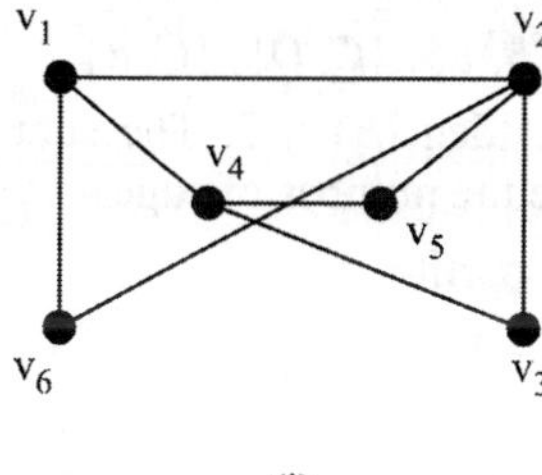

(i)

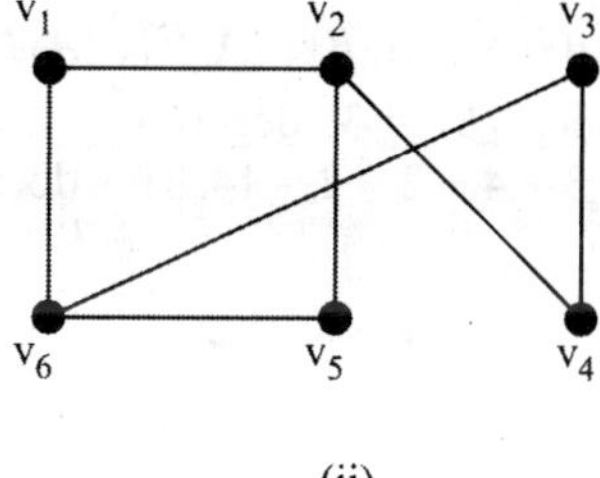

(ii)

19. Determine the value of m for which the following graphs are regular.
    i. C_m
    ii. W_m
    iii. K_m
    iv. $K_{m,n}$
20. How many vertices does a regular graph of degree 4 with 10 edges have?
21. Consider the complete graph K_n
    a. Find the number m of edges in K_n.
    b. Find the degree of each vertex in K_n.
    c. Find those values of n for which K_n is regular.
22. Show that a simple graph with $|V| = n$ is not bipartite if $|E| = \lfloor n^2/4 \rfloor$w.
23. A tripartite graph is a graph where the set of vertices may be partitioned into three subsets so that no edge has incident vertices in the same subset.
    i. Draw examples of $K_{2,3,3}$.
    ii. Give an example of a tripartite graph.
24. Determine the number of vertices and edges in (i) K_n (ii) C_n (iii) P_n.
25. Draw a connected graph that becomes disconnected when any edge is removed from it.
26. Consider the following multigraph G.
    i. Which of them are not connected?
    ii. If they are not connected, find the number of connected components.
    iii. Which are cycles free (without cycle)? If there are any, find the number of cycles.
    iv. Which are loops free (without loops)?

v. Which are (simple) graphs?

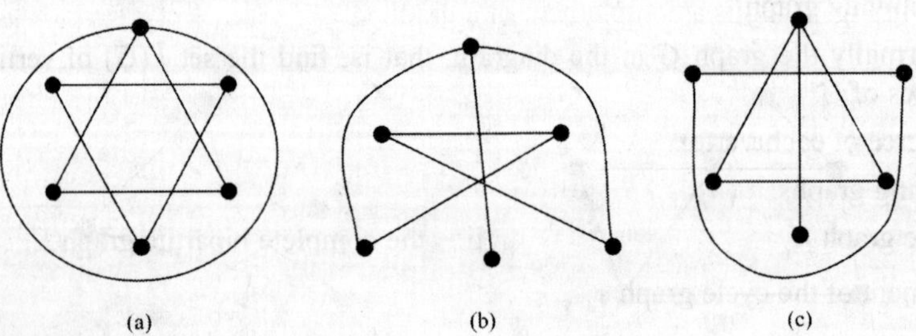

(a) (b) (c)

Answers

7. Twice the number of edges.
8. Equal to the sum of outdegrees of ½ the number of edges.
9. i. 16 ii. 13

   iii. solve $n \mid v \mid = 48$ for all divisors n of 48, where v denotes the number of vertices.
10. The number of edges $\leq C(n, 2)$ since each edge connects 2 vertices.
16. i. Seven pairs; $E(G) = [\{A, B\}, \{A, C\}, \{A, D\}, \{B, C\}, \{B, E\}, \{C, D\}, \{C, E\}$

    ii. deg $(A) = 3$, deg $(B) = 3$, deg $(C) = 4$, deg $(D) = 2$, deg $(E) = 2$, The sum of the degrees of the vertices is $3 + 3 + 4 + 2 + 2 = 14$. This does equal twice the number of edges.
18. i. Not bipartite ii. Not bipartite
19. i. for all $m \geq 3$ ii. for $m = 3$

    iii. for all $m \geq 1$
20. 5
21. a. m $Cn, 2)$ $\frac{n(n \quad 1)}{2}$ b. $n - 1$

    c. all n
24. i. n vertices and $\frac{n(n-1)}{2}$ edges. Ii. n vertices and n edges.

10.23 Complement of a Graph

The complement $\overline{G}$ of a simple graph $G = (V, E)$ is the simple graph with vertex set $V(G)$ defined by the edge v_i $v_j \in E(G)$ iff $v_I\, v_j \notin E(\ G)$. In other words we can say $\overline{G}$ is a simple graph with vertex set $V(G)$ in which two vertices are adjacent iff they are not adjacent in G. One can observe here that complement of a complete graph is null graph. Graphs shown in the Fig.10.31 are complement to each other.

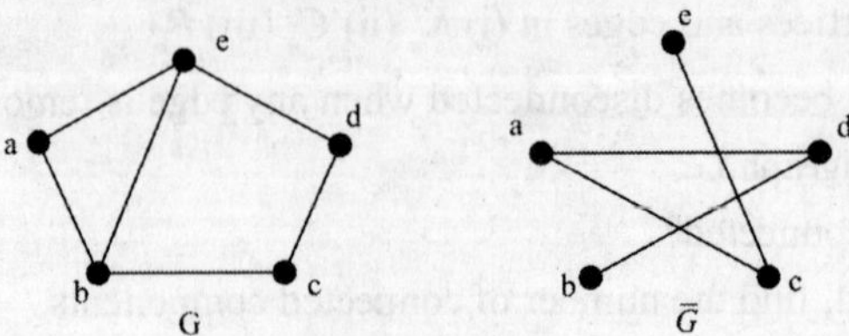

Fig. 10.31

Example 10.13

Graph $\bar{G}$ shown in Fig. 10.32 is a complement of a graph G.

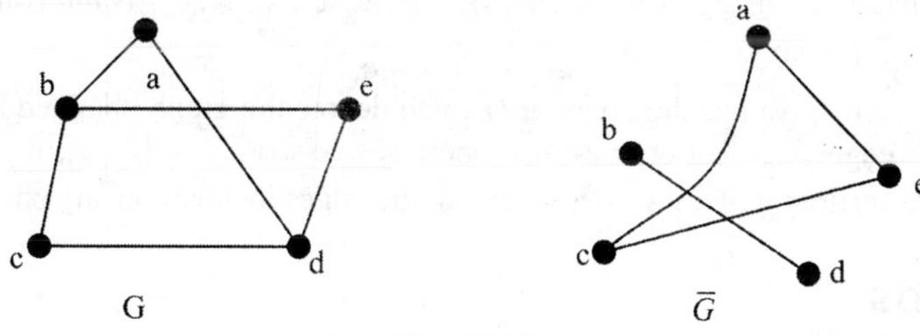

Fig. 10.32

Example 10.14

Graphs and their compliments are shown in Fig. 10.33.

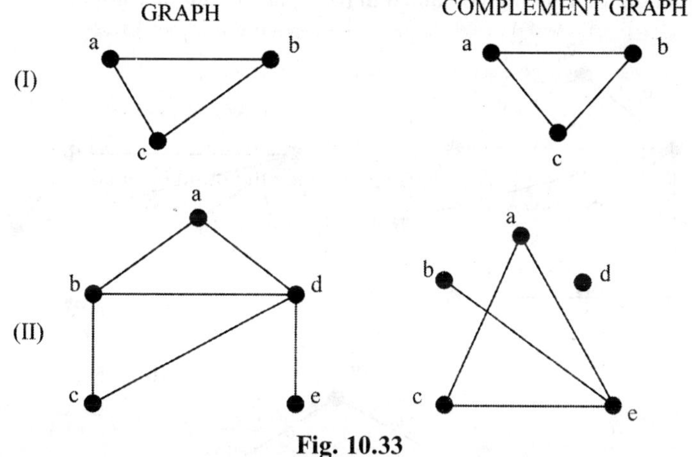

Fig. 10.33

Example 10.15

A graph and its complement is shown in the following graph.

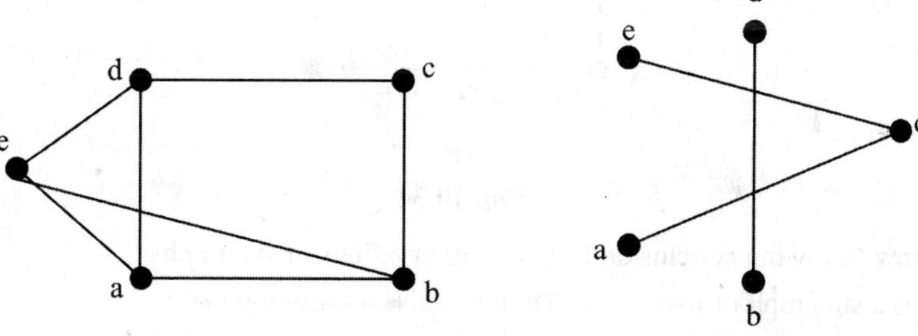

10.24 Graph Obtained by Deletion of Vertices and Edges

Consider a graph G which consists of many edges including edge e. The graph obtained by deletion of the edge e from the graph is denoted by $G - e$. More generally, $G - \{e_1, e_2, e_3, \ldots, e_n\}$ is the graph obtained by deleting the edges $\{e_1, e_2, e_3, \ldots, e_n\}$ from G.

Similarly, if v is the vertex, we use the notation $G - v$ to denote the graph obtained by deleting the vertex v from G, together with all edges incident on v. More generally, we write $G - \{v_1, v_2, v_3, \ldots, v_n\}$ for the graph obtained by removing the vertices $v_1, v_2, v_3, \ldots, v_n$ and all the edges incident on any one of them.

10.25 Subgraphs

Consider a graph $G = (V, E)$. A graph $g = (v, e)$ is said to be a subgraph of G if e is a subset of E and v is a subset of V such that edges in e are incident only with the vertices in v. For instance, the graph in Fig. 10.34(b) and (c) are a subgraph of Fig. 10.34(a).

The concept of subgraph is similar to the concept of subset in set theory. As a subset is contained in a set similarly a subgraph is thought of as being contained in (or a part of) another graph.

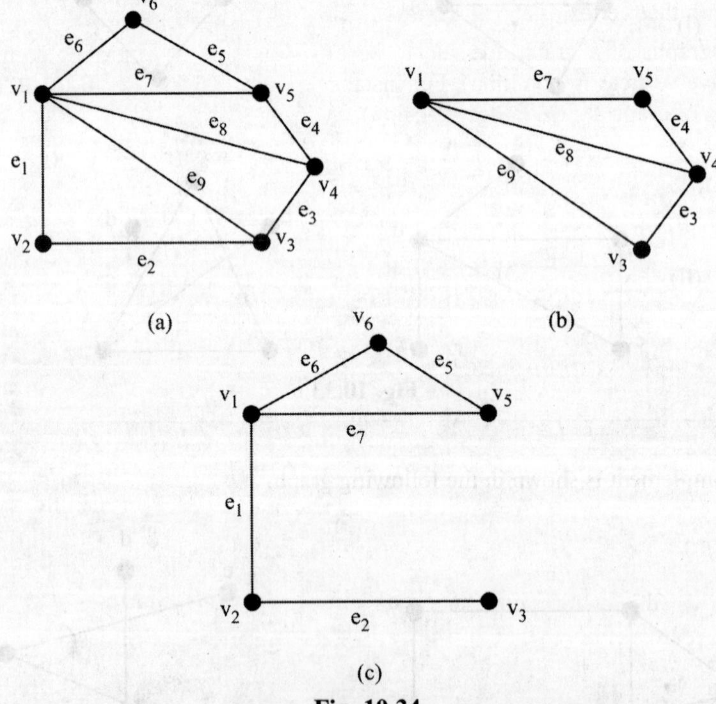

Fig. 10.34

Now one can draw following conclusion from the observations of subgraphs:

1. A graph G is a subgraph of itself. A null graph is also a subgraph of G.
2. A subgraph of a subgraph of G is a subgraph of G.
3. A single vertex in a graph G is a subgraph of G.
4. A single edge in G, together with its end vertices, is also a subgraph of G.

10.25-1 Spanning Subgraph

A subgraph of G is said to be a **spanning subgraph** if it contains all the vertices of G. Graph shown in Fig. 10.35(b) is a spanning subgraph of Fig.10.35 (a).

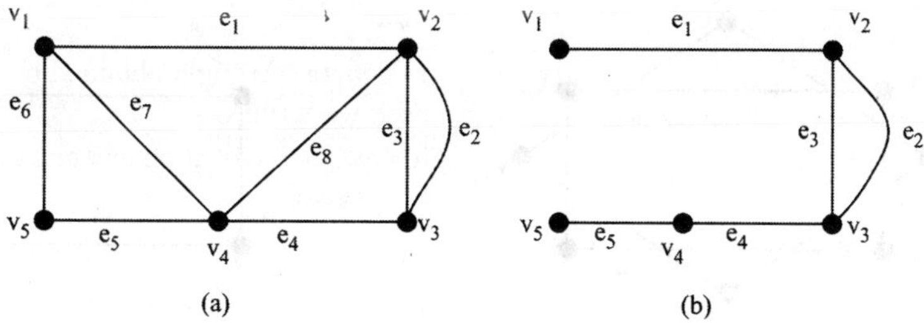

Fig. 10.35

10.25-2 Edge Disjoint and Vertex Disjoint Subgraphs

Any subgraph of a graph can be obtained by deleting certain edges and certain vertices. It is to be noted that by deletion of an edge we have vertices in places where they were, whereas deletion of vertex necessitates the deletion of any edge with that vertex as an end vertex.

Two or more subgraphs of a graph are said to be **edge disjoint** if subgraphs do not have any edge in common (they may have vertices in common). For instance graphs shown in Figs. 10.36 (b) and 10.36 (c) are edge disjoint subgraphs of graph shown in Fig. 10.36 (a).

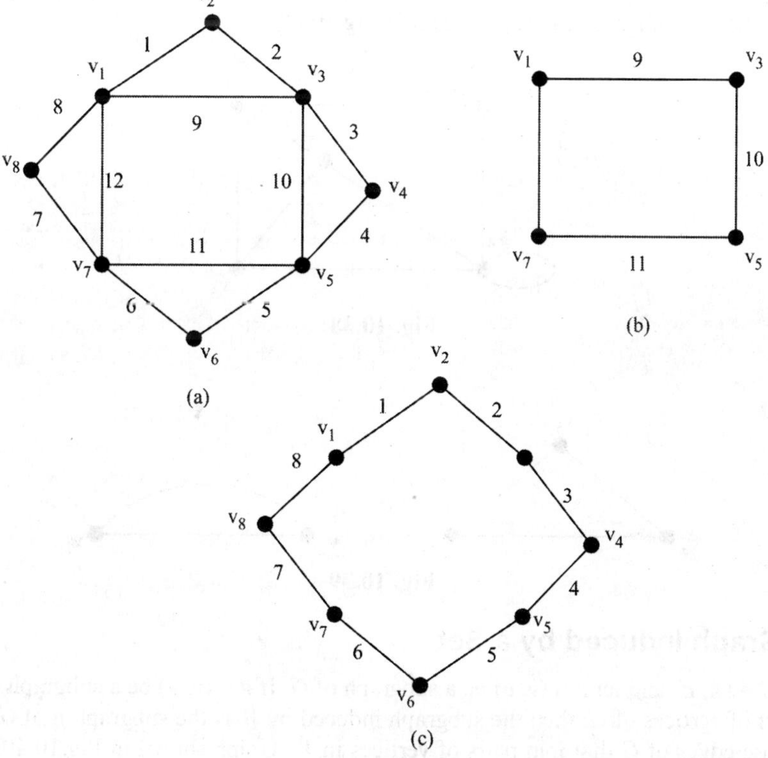

Fig. 10.36

Subgraphs that do not have vertices in common are said to be **vertex disjoint**. One can observe that the graphs that have no vertices in common cannot possibly have edges in common.

For instance graphs show in Fig. 10.37(b) and (c) are vertex disjoint graphs of graph Fig. 10.37(a).

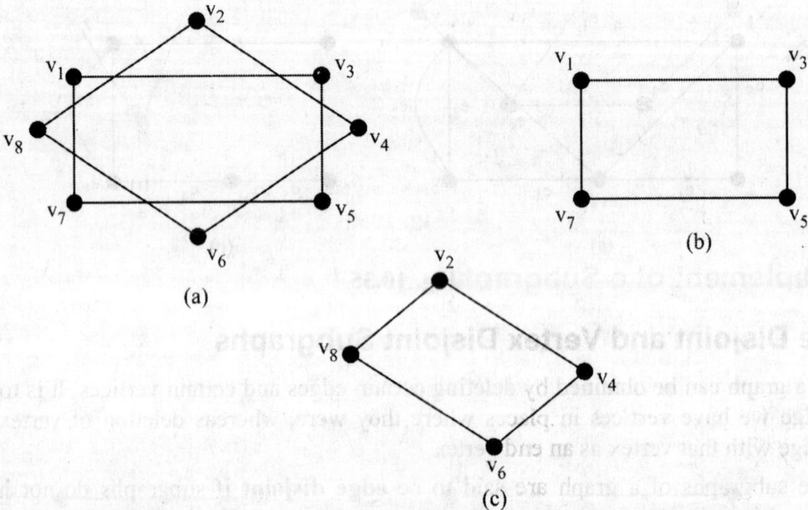

Fig. 10.37

Example 10.16

Find the vertex disjoint subgraph of graph shown in Fig.10.38.

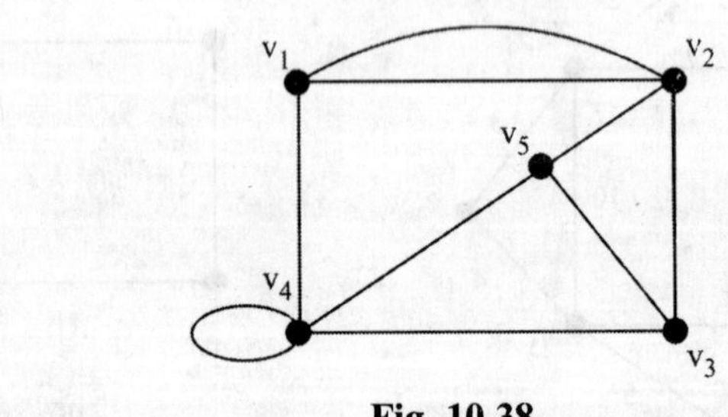

Fig. 10.38

Solution

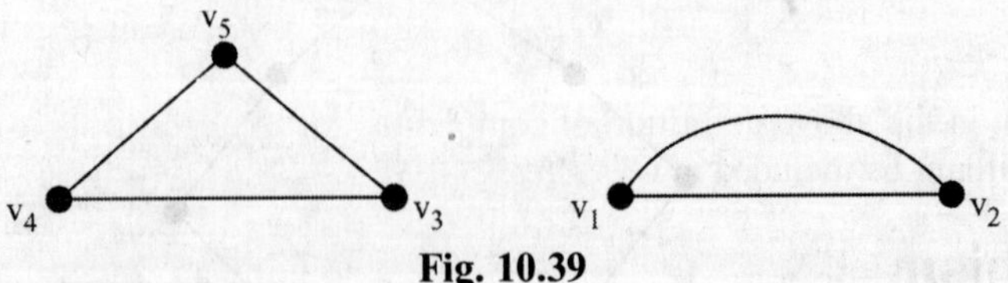

Fig. 10.39

10.25-3 Sub Graph Induced by a Set

Consider a graph $G = (V, E)$ and let $g = (v, e)$ be a subgraph of G. If $g = (v, e)$ be a subgraph of graph $G = (V, E)$ and W is any subset of vertices of G, then the subgraph induced by W is the subgraph g of G obtained by taking $v = W$ and e be those edges of G that join pairs of vertices in W. Graph shown in Fig.10.40 (b) is the subgraph induced by the set $W = \{v_1, v_2, v_3, v_4\}$ of the graph shown in Fig. 10.40 (a).

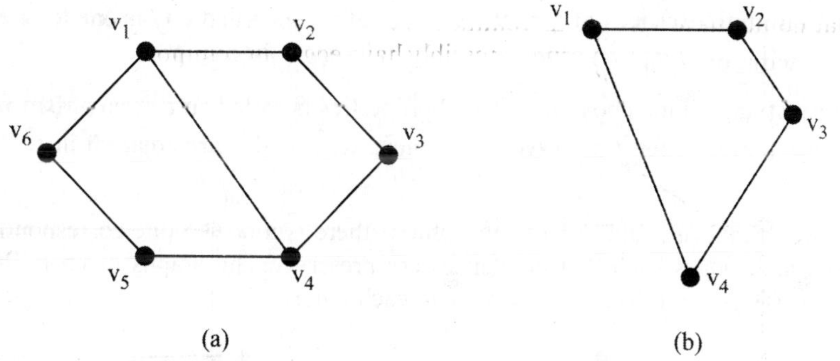

Fig. 10.40

10.25-4 Complement of a Subgraph

Consider a graph $G = (V, E)$. The complement of a subgraph $G' = (V', E')$ with respect to the graph G is another subgraph $G'' = (V'', E'')$ such that E'' is equal to $E - E'$ and V'' contains only the vertices with which the edges in E'' are incident. The graph shown in Fig. 10.41(C) is a complement of graph shown in Fig. 10.41(B), which is a subgraph of Fig.10.41 (A).

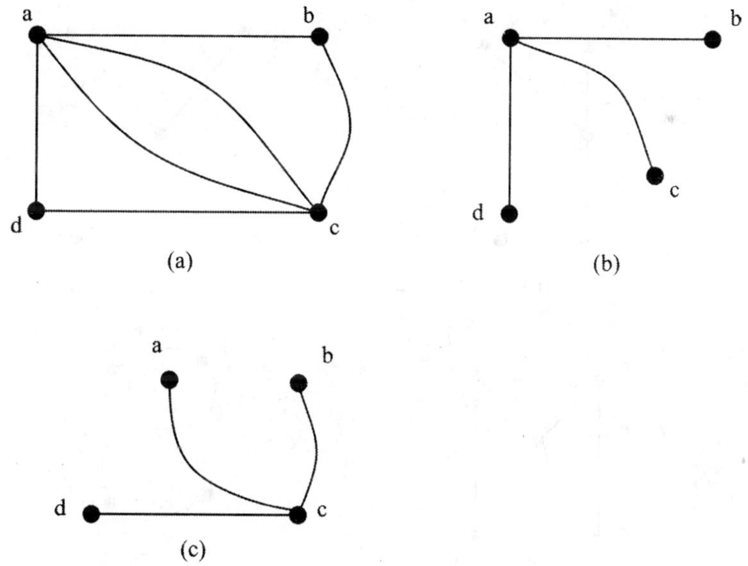

Fig. 10.41

It is to be noted that in the above definition of complement of a subgraph some isolated vertices in G that are not included in G' will not be included in G'' either.

10.26 Isomorphism

In geometry if two figures resemble in terms of geometric properties and have identical behaviour then the figures are called congruent. Likewise in graph theory two graphs are thought of as equivalent and called **isomorphic** if they have identical behaviour in terms of graph-theoretic properties. More precisely it can be defined as follows:

Two graphs $G = (V, E)$ and $G_1 = (V_1, E_1)$ are **isomorphic** if there exist a function

$f: V(G) \rightarrow V_1(G_1)$ from the vertices of G to the vertices of G_1 such that (i) f is one to one , (ii) f is onto, (iii) for each of vertices v_i and v_j of G; $\{v_i, v_j\} \in E$ if and only if $\{f(v_i), f(v_j)\} \in E_1$.

Any function f satisfying all the above mentioned properties is called an **isomorphism** from graph G to G_1. The condition (iii) says that function f preserves adjacency, i.e. v_i and v_j are adjacent in G iff $f(v_i)$ and $f(v_j)$ are adjacent in G_1.

In otherwords two graphs are said to be isomorphic if there is a one-to-one correspondence between their vertices and between their edges such that incidences are preserved.The graphs given in the Fig.10.42(a) and (b); (c) and (d); (e) and (f); (g) and (h) are isomorphic to each other.

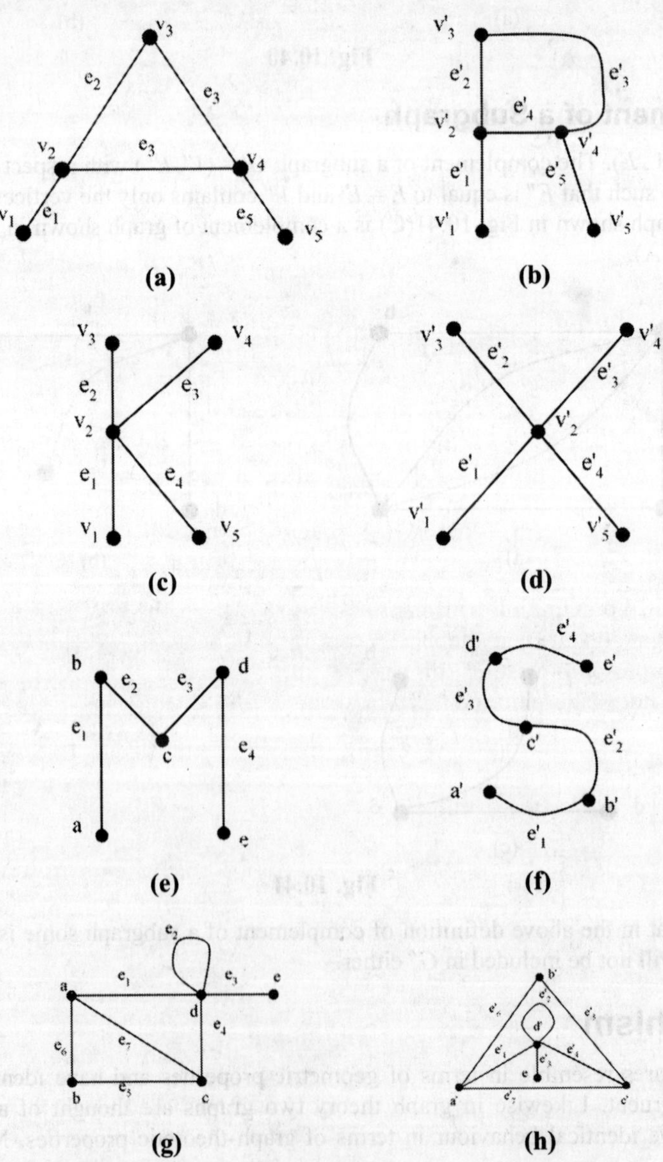

Fig. 10.42

From the Fig.10.42 one can see the correspondence between the graphs as: vertices v_is correspond to vertices v_i's and edges e_is correspond to edges e_i's of graphs G and G_1 respectively.

One can conclude from the above discussion that except for the labels (i.e. names) of their vertices and edges, isomorphic graphs are the same graph, may be drawn differently. It is not always easy to determine whether the given graphs are isomorphic or not. For example the graphs shown in the Fig. 10.43(*a*, *b*, *c*) are isomorphic, but in the first instance it doesn't seem like isomorphic.

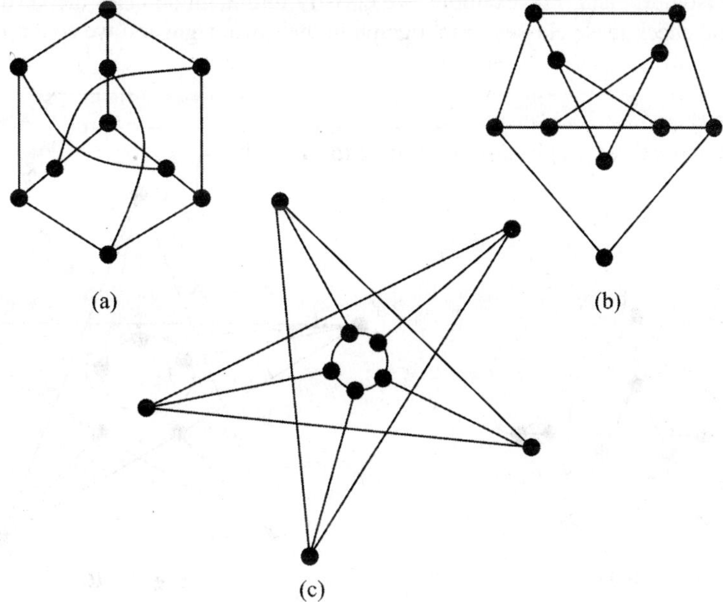

Fig. 10.43

It is apparent from the definition of isomorphism that two graphs which are isomorphic have the same number of vertices, edges and an equal number of vertices with a given degree; however this is not a sufficient condition for an isomorphism to exist, as can be seen from the graphs shown in Fig.10.44 satisfy all the three conditions yet they are not isomorphic and it can be verified from the following facts.

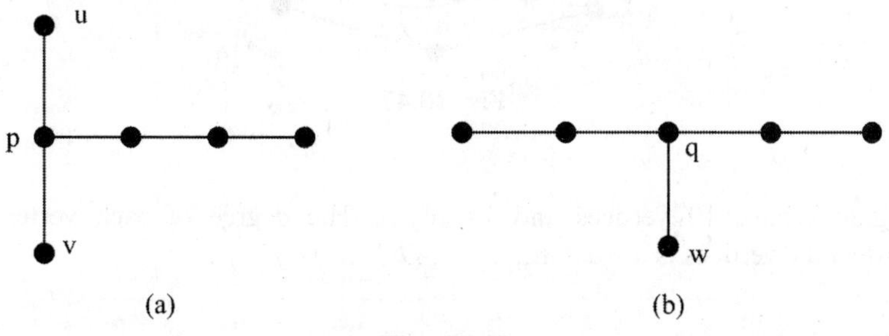

Fig. 10.44

i. If graph in Fig. 10.44 (a) is to be isomorphic to Fig. 10.44 (b) then vertex *p* must correspond to vertex *q* as there is no other vertex of degree three.

ii. In Fig. 10.44 (a) there are two pendant vertices *u* and *v* adjacent to *p* while in Fig. 10.44 (b) there is only one pendant vertex *w* adjacent to *q*.

10.26-1 Determining When Graphs are Not Isomorphic

Here are some tips, which may prove to be helpful in determining when two graphs are not isomorphic.

i. To find some property that isomorphic graphs must share but the two graphs in question are not sharing that. For example they have different number of vertices, edges, degrees (degree sequences).

ii. Some times two graphs may have the same degree (hence the same number of vertices and edges) and still not be isomorphic. For this, attempt to classify the vertices into classes according to some property preserved by isomorphism. For example, we classify into a single class the vertices of degree two in each graph and check these classes form a graph in their own right and we could ask if these graphs are isomorphic.

Example 10.17

Show that the following three graphs are isomorphic to each other.

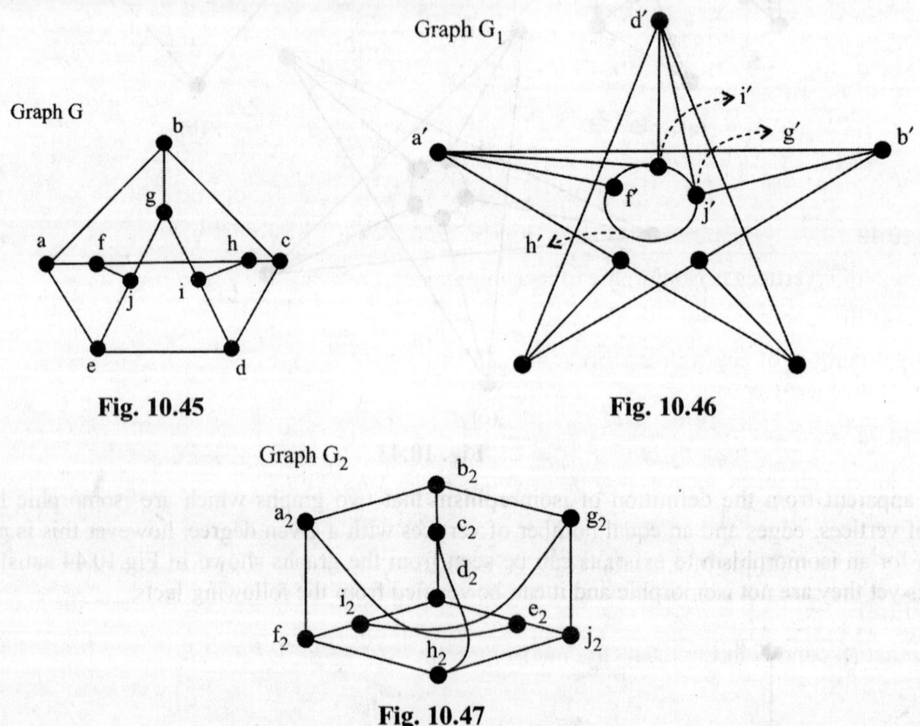

Fig. 10.45

Fig. 10.46

Fig. 10.47

Solution

All the three graphs have 10 vertices and 14 edges. The degree of each vertex is 3. Further, the correspondence between the vertices is $a - a' - a_2$, $b - b' - b_2$ ..., $j - j' - j_2$.

Example 10.18

Prove that two graphs are isomorphic if and only if their complements are isomorphic.

Solution

Let $\bar{G}_1$ and $\bar{G}_2$ be the complements of graphs of G_1 and G_2 respectively. Suppose G_1 is isomorphic to G_2. Then if u_1 and v_1 are 2 vertices in G_1 such that there exists an edge between them, then there exist u_2 and v_2 in G_2 such that there exists an edge between them. Hence, in $G_1 - \bar{G}_1$ there does not exist an edge between u_1 and

v_1 and similarly in $G_2 - \bar{G}_2$ there does not exist an edge between u_2 and v_2. Similarly, if there does not exist an edge between u_1 and v_1 in $\bar{G}_1$ there does not exist an edge between u_2 and v_2 in $\bar{G}_2$ while these pairs have an edge in G_1 G_2 and G_2 G_2 respectively. Thus, there exists a one to one correspondence between the vertices of $G_1 - G_2$ and $G_1 - G_2$, and hence in $(G_1 - G_2) \cup \bar{G}_1$ and in $(G_2 - G_2) \cup \bar{G}_2$. If G_1 is isomorphic to G_2, their complements $\bar{G}_1$ and $\bar{G}_2$ have one to one correspondence as indicated and are hence isomorphic. The converse is true similarly due to the fact that if G_1 and $\bar{G}_2$ are isomorphic, so are G_1 and G_2. Fig. 10.48 illustrates this.

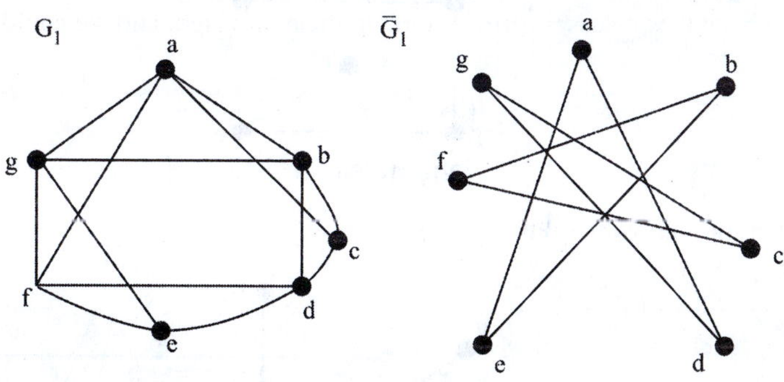

Fig. 10.48

Example 10.19

If a graph of n vertices is isomorphic to its complement how many vertices it must have?

Solution

The total number of edges in a complete graph is $n(n-1)/2$. Therefore, G and its complement must each have $n(n-1)/4$ edges. Hence n or $(n-1)$ be a multiple of 4.

Example 10.20

Can a graph with seven vertices be isomorphic to its complement.

Solution

Since 7 is not a multiple of 4, a graph of 7 vertices cannot be isomorphic to its complements.

Example 10.21

Show that $K_{3,3}$ and G in Fig. 10.49 are not isomorphic.

Solution

In graph $K_{3,3}$ and G, cycles of length 4 have different vertices, e.g. $(a\ e\ b\ f\ a)$, $(a'\ c'\ e'\ f'\ a')$. Further vertices in $K_{3,3}$ can be partitioned into 2 blocks (abc) (fed) such that there exists an edge from each vertex of one block to any vertex of other block and none within the block. G does not have such blocks.

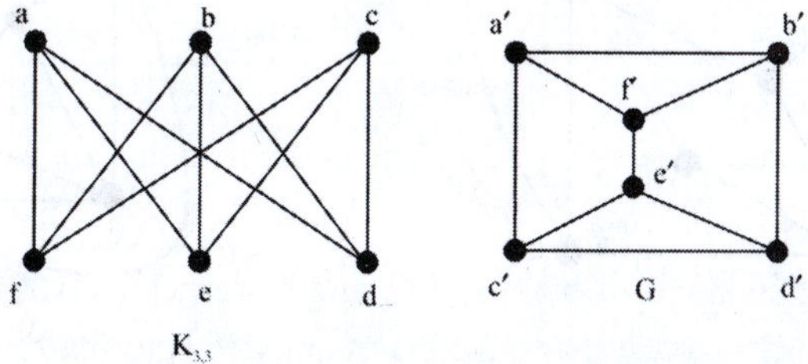

Fig. 10.49

Exercise 10.2

1. Find all connected subgraphs of the following graph.

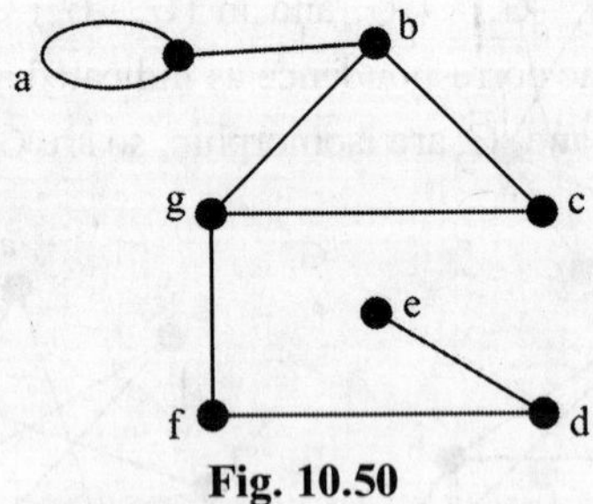

Fig. 10.50

2. Draw all subgraphs of the following graph.

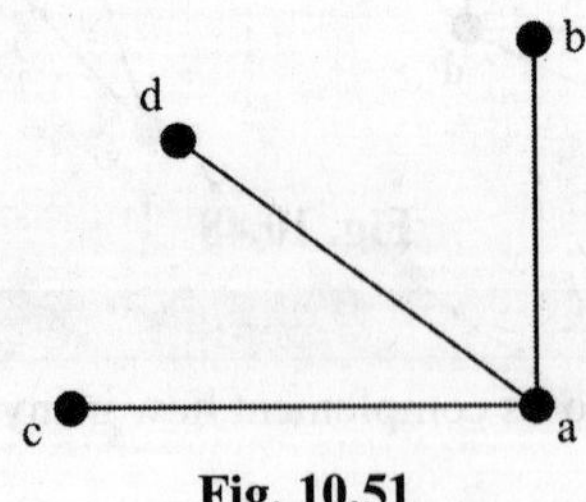

Fig. 10.51

3. How many subgraphs does K_3 have?
4. Which of the following pairs of non-directed graphs in following figures are isomorphic?

i.

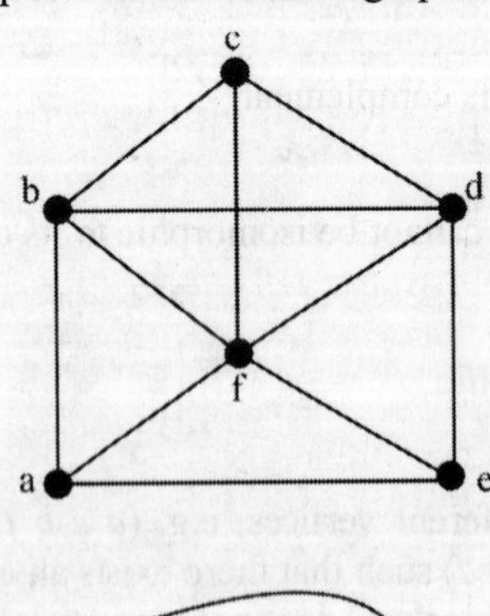

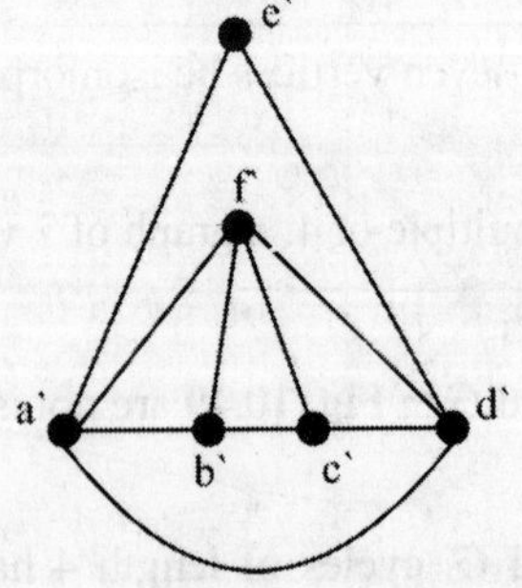

ii.

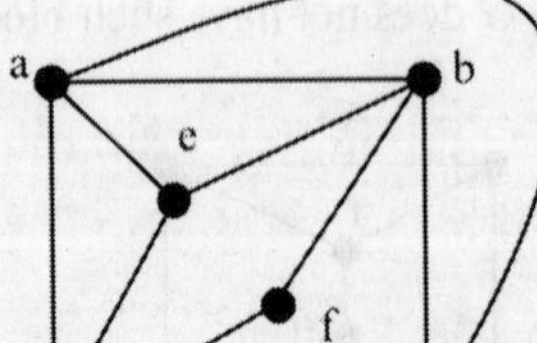

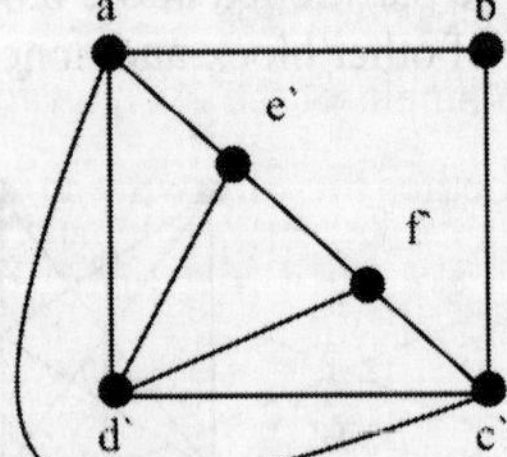

iii.

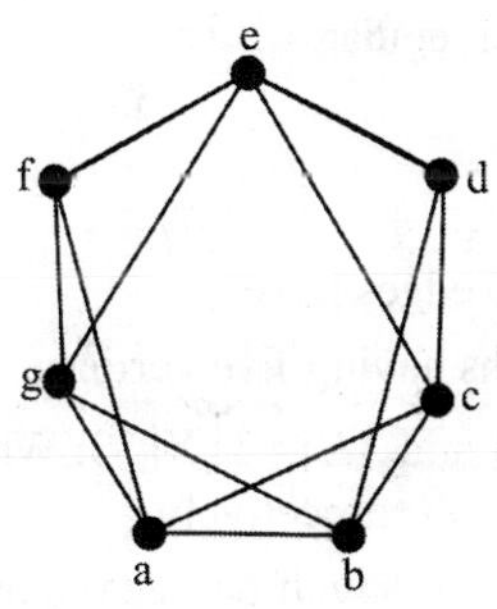

iv.

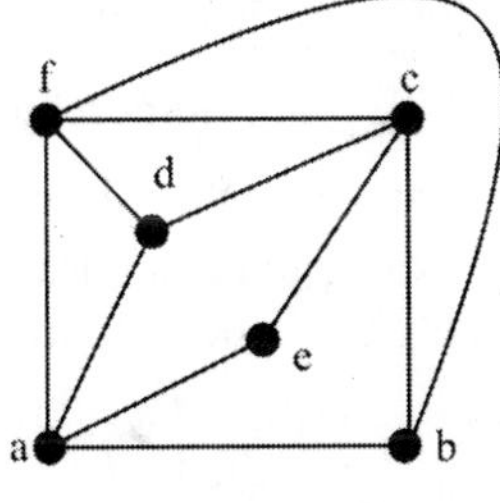

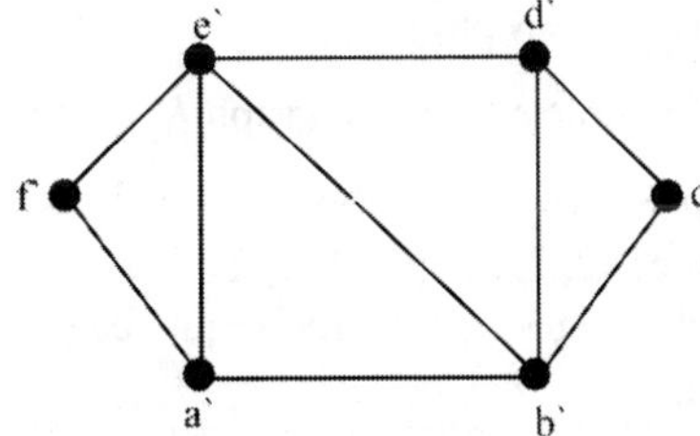

v.

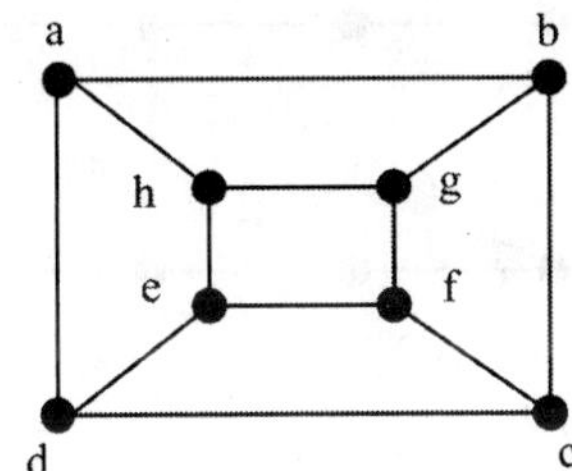

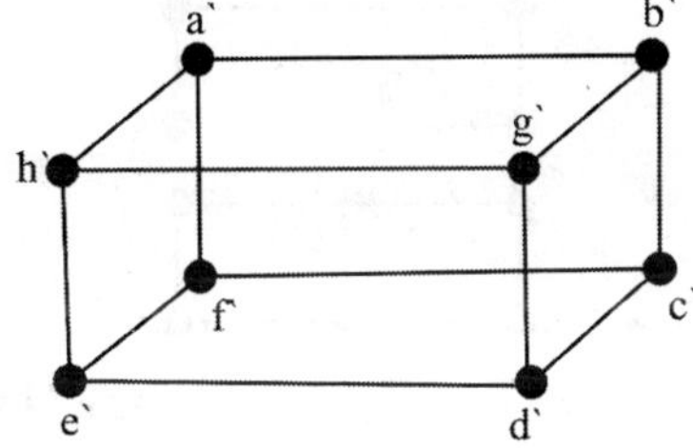

vi.

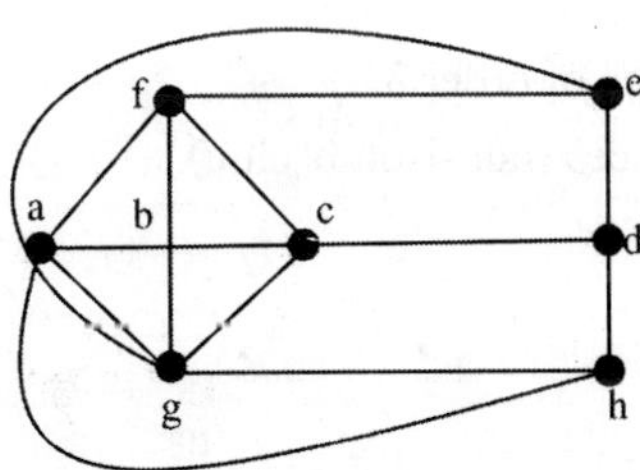

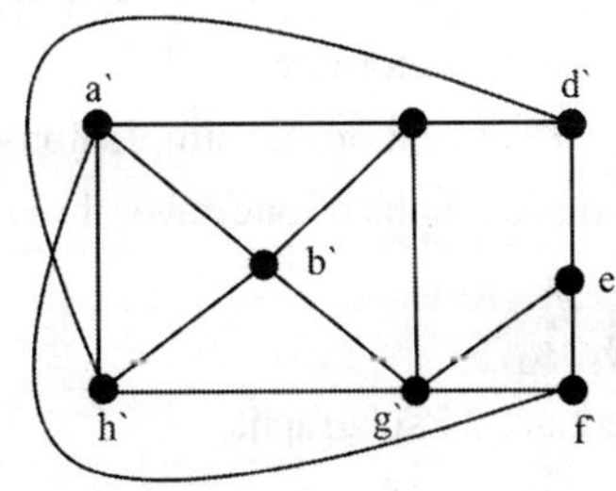

5. Draw the following graphs:

   i. the complete graph K_6

   ii. the complete bipartite graph $K_{5,7}$

   iii. the complement of the cycle graph c_4.

6. Determine whether the following graphs are bipartite?

   i.

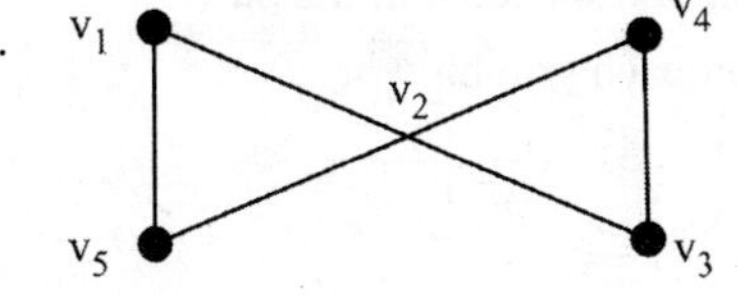

   ii.

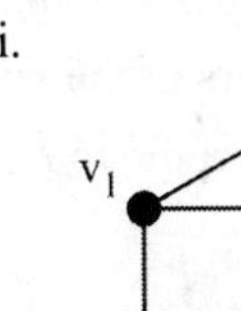

7. Determine the value of m for which the following graphs are regular.
   i. C_m
   ii. W_m
   iii. K_m
   iv. $K_{m,n}$
8. How many vertices does a regular graph of degree 4 with 10 edges have?
9. Draw all nonisomorphic, circuit free connected simple graphs having five vertices.
10. Let G be a simple graph all of whose vertices have degree 3 and $|E|=2|V|-3$. What can be said about G? Where $|V|$ and $|E|$ denote number of vertices and edges of G respectively.
11. Give an example of 2 nonisomorphic graphs of order 6 and size 6 with degree sequence (2, 2, 2, 2, 2, 2)
12. Is $k_{3,3}$ and k_6 are isomorphic.
13. Consider the complete bipartite graph $K_{m,n}$. Which of the graphs $K_{m,n}$ are isomorphic?
14. The two graphs G_1 and G_2 shown in Fig.10.52 have 8 vertices and 8 edges. Moreover, they have the same degree sequence (2, 2, 2, 2, 2, 2, 2, 2). Verify that they are not isomorphic. Is it true that every graph G with 8 vertices and 8 edges and degree sequence (2, 2, 2, 2, 2, 2, 2, 2) has to be isomorphic to one of G_1 and G_2? Explain

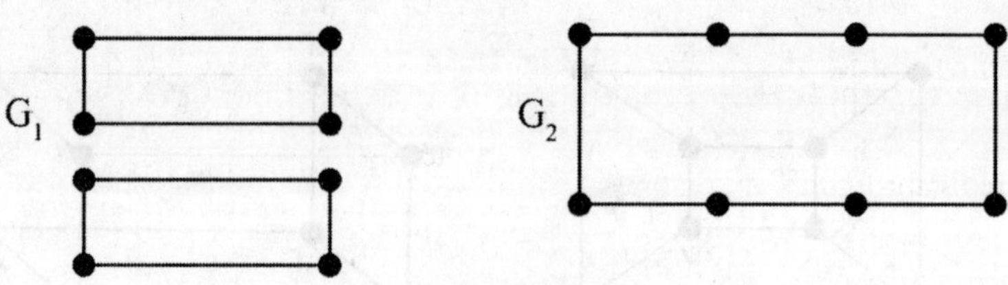

Fig. 10.52

15. Give an example of two non-directed graphs with 4 vertices and 2 edges that are not isomorphic. Verify that they are not isomorphic.
16. Determine all nonisomorphic simple non-directed graphs of order 3.
17. Draw all simple graphs of one, two, three and four vertices (non-isomorphic).

Answers

2. In total there are 34 subgraphs.
3. 17
4. i. Isomorphic; by removing the vertices of degree 5 and their incident edges from each graph, observe the isomorphism for remaining graph and extend.
   ii. Nonisomorphic, since vertices of degree 3 are adjacent in one graph and non-adjacent in the other.
   iii. Nonisomorphic, consider their complements.
   iv. Nonisomorphic, vertices of degree 3 are adjacent in one graph, non-adjacent in the other.
   v. Nonisomorphic; see the neighbours of the vertex of degree 5 in each graph.
10. For $|V|=6$, conclude that G is isomorphic to $K_{3,3}$ or the graph.
13. None are isomorphic.

10.27 Homeomorphic Graphs

We can obtain a new graph by dividing an edge of a given graph G with additional vertices. Two graphs G and G_1 are said to be **homeomorphic** if they can be obtained from the same graph or isomorphic graphs by this method. For instance, graphs shown in Fig. 10.53(b) and (c) are not isomorphic but they are homeomorphic since they can be obtained from the graph shown in Fig. 10.53(a) by adding appropriate vertices.

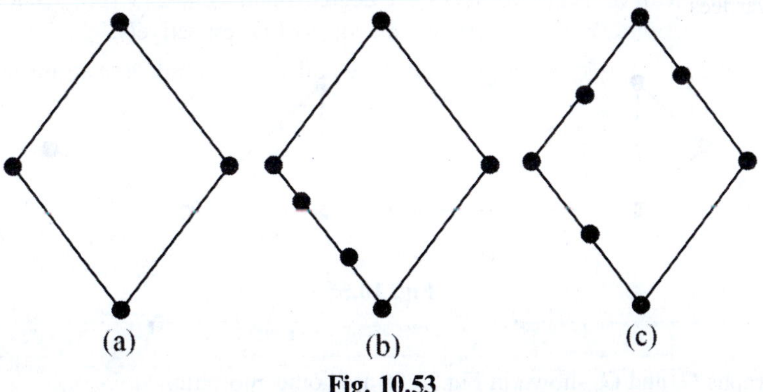

Fig. 10.53

Example 10.22

Show that the following figures are distinct, i.e. no two of them are isomorphic. Also show that G_2 and G_3 are homeomorphic.

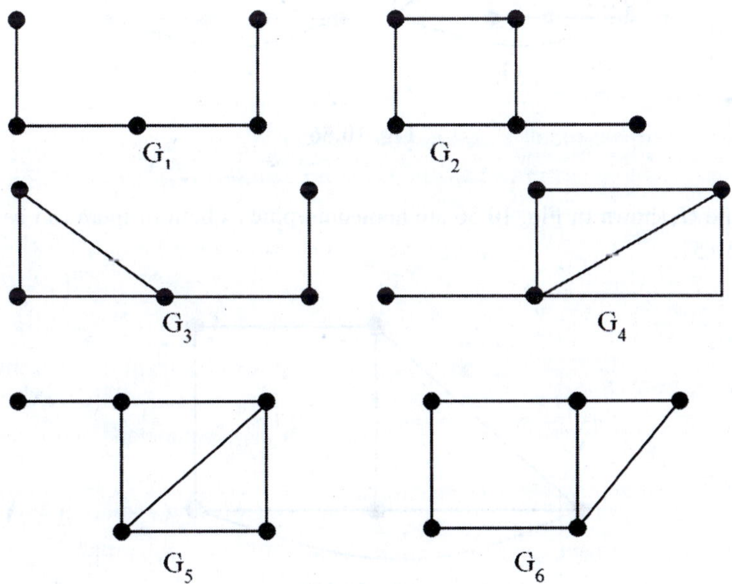

Fig. 10.54

Solution

In the given graphs the degrees of vertices of only G_2 and G_3 are same and the other graphs are having different degrees of vertices. Hence none of the graphs is isomorphic except possibly G_2 and G_3. But in G_2 and G_3 if we delete the vertices of degree 3, we obtain distinct subgraphs. Therefore G_2 and G_3 are also nonisomorphic. Hence all the graphs are distinct.

But graphs G_2 and G_3 are homeomorphic as it can be obtained from the following isomorphic graphs by adding appropriate vertices.

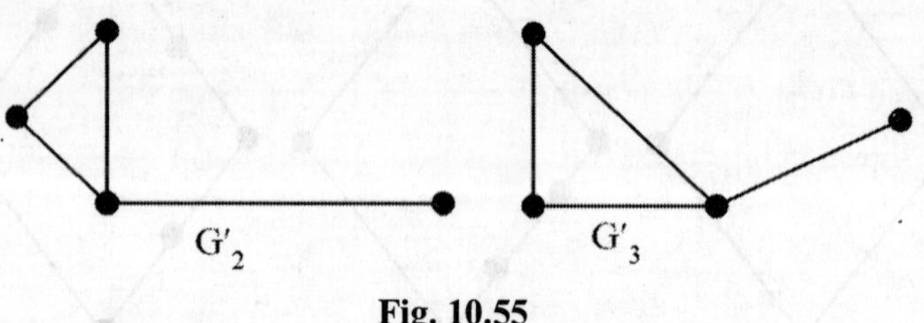

Fig. 10.55

Example 10.23

Show that the graphs G_1 and G_2 shown in Fig.10.56 are homeomorphic.

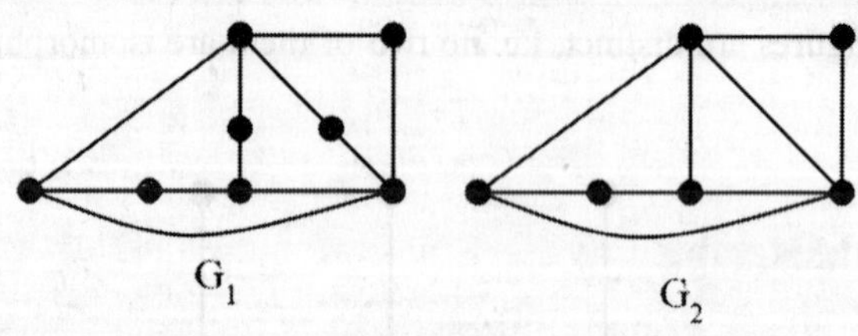

Fig. 10.56

Solution

The graphs G_1 and G_2 shown in Fig. 10.56 are homeomorphic as both of them can be reduced to the graph G_3 as shown in Fig. 10.57.

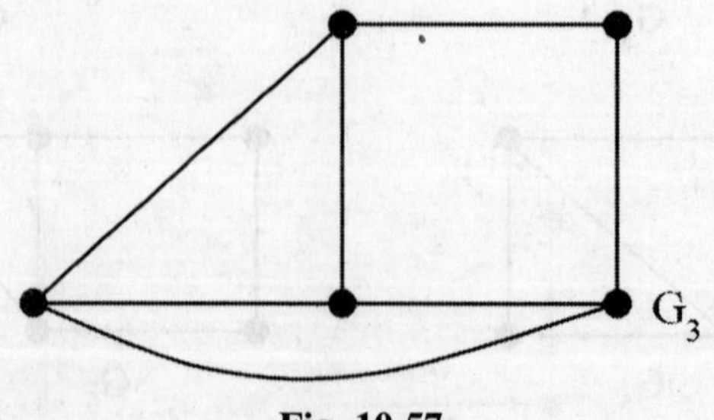

Fig. 10.57

If we let R be a relation defined on a set of graphs by the rule G_1RG_2 if G_1 and G_2 are homeomorphic then R is an equivalence relation.

10.28 Operations of Graphs

In this section we will discuss some operations on the graphs.

10.28-1 Union of Graphs

Consider two graphs G_1 and G_2. Let $V(G_1)$ and $V(G_2)$ denote the vertices of graph G_1 and G_2 and $E(G_1)$ and $E(G_2)$ denotes the edges of graph G_1 and G_2 respectively. The union of graph G_1 and G_2 be a graph such that

$$V(G_1 \cup G_2) = V(G_1) \cup V(G_2)$$

and
$$E(G_1 \cup G_2) = E(G_1) \cup E(G_2)$$

where $V(G_1 \cup G_2)$ and $E(G_1 \cup G_2)$ denote the vertices and edges of union of graphs G_1 and G_2.

For example graph shown in Fig.10.58, $G_1 \cup G_2$ is union of G_1 and G_2.

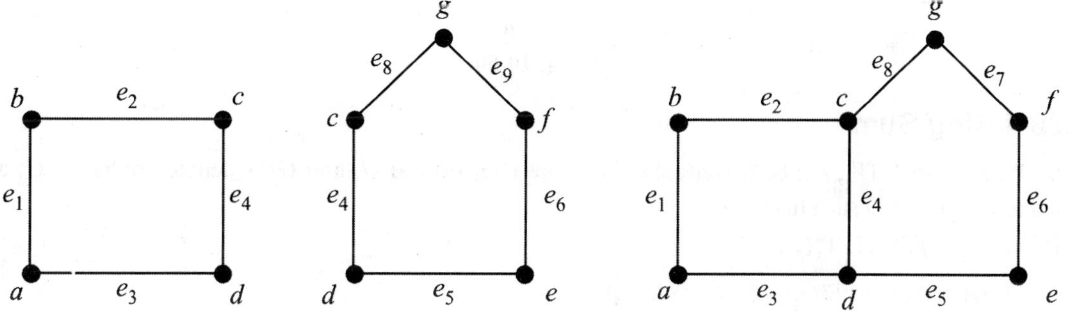

Fig. 10.58

10.28-2 Intersection of Graphs

Let G_1 and G_2 be two graphs with at least one vertex in common then the intersection of these two graphs will be a graph such that

$$V(G_1 \cap G_2) = V(G_1) \cap V(G_2)$$

and
$$E(G_1 \cap G_2) = E(G_1) \cap E(G_2)$$

For example graph shown in Fig.10.59 $G_1 \cap G_2$ represents the intersection of G_1 and G_2.

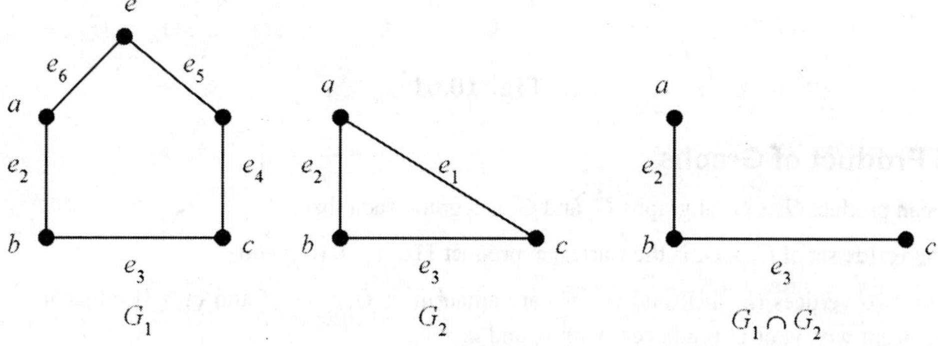

Fig. 10.59

10.28-3 Sum of Two Graphs

Consider two graphs G_1 and G_2. Let $V(G_1)$ and $V(G_2)$ denote vertices of graph G_1 and G_2 respectively. $V(G_1 \cap G_2) = \phi$, then the sum of graphs G_1 and G_2: $G_1 + G_2$ is defined as the graph whose vertex set is $V(G_1) + V(G_2)$ and the edge set of $G_1 + G_2$ consist of those edges which are in G_1 and G_2 and the edges obtained by joining each vertex G_1 to each vertex of G_2. For example see the graphs shown in Fig. 10.60.

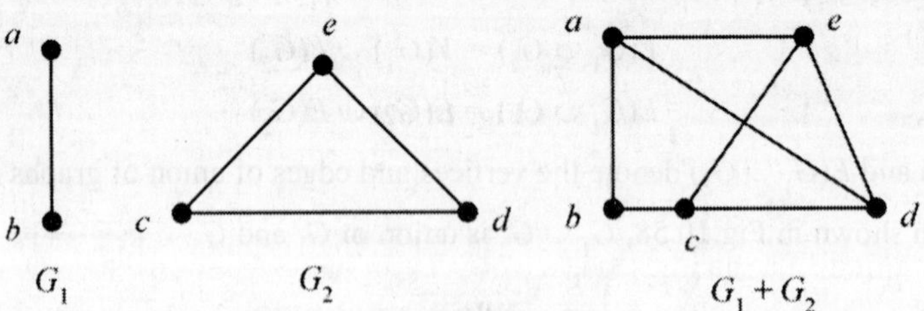

Fig. 10.60

10.28-4 Ring Sum

Let $G_1(V_1, E_1)$ and $G_2(V_2, E_2)$ be two graphs. Then the **ring sum** of G_1 and G_2 is denoted by $G_1 - G_2$ and is defined as the graphs G such that

i. $V(G) = V(G_1) \cup V(G_2)$

ii. $E(G) = E(G_1) \cup E(G_2) - E(G_1) \cap E(G_2)$.

The point (ii) says the graph G consists of those edges, which are either in G_1 or G_2 but not in both. The graphs shown in Fig.10.61 represent ring sum.

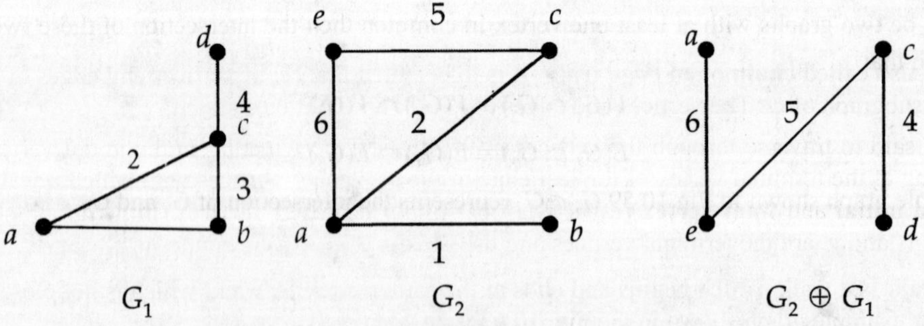

Fig. 10.61

10.28-5 Product of Graphs

The Cartesian product $G_1 \times G_2$ of graphs G_1 and G_2 is a graph such that

i. the vertex set of $G_1 \times G_2$ is the Cartesian product $V(G_1) \times V(G_2)$ and

ii. any two vertices (u_1, u_2) and (v_1, v_2) are adjacent in $G_1 \times G_2$ if and only if either $u_1 = v_1$ and u_2 is adjacent with v_2 or u_1 is adjacent with v_1 and $u_2 = v_2$.

For example:

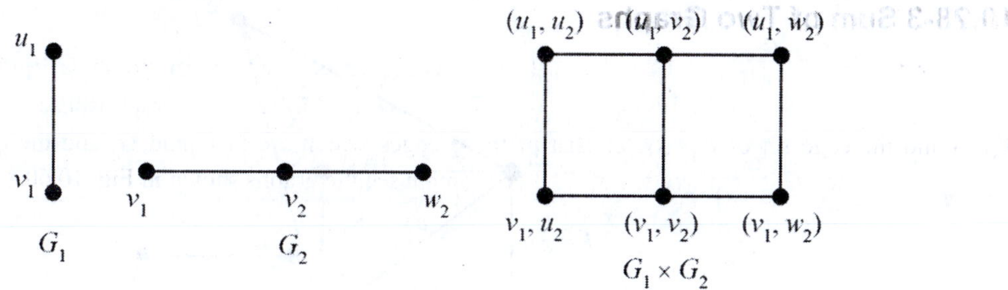

Fig. 10.62

10.29 Walks

A walk in a graph G is a finite alternating sequence of vertices and edges of the form $v_0, e_1, v_1, e_2, \ldots, e_{n-1}, v_{n-1}, e_n, v_n$, (starting and terminating with vertices) such that each edge is incident with the vertices preceding and following it.

For instance $v_1 e_1\ v_2\ e_2\ v_3\ e_3\ v_3\ e_4\ v_4\ e_5\ v_2\ e_6\ v_5$ is a walk in the Fig.10.63.

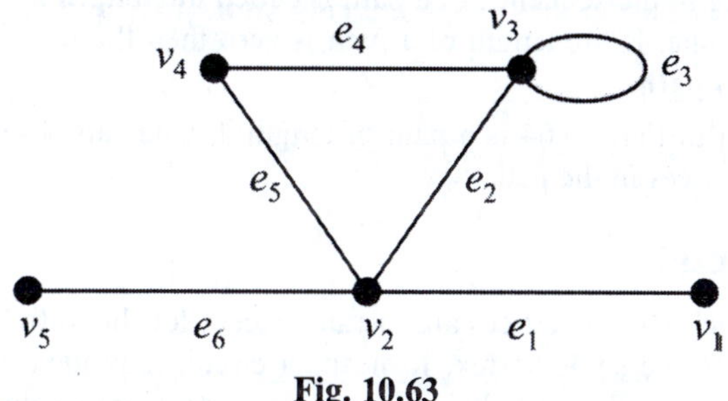

Fig. 10.63

A walk is also called **chain** or an **edge train**. It is to be noted that set of vertices and edges of a given walk in a graph is a subgraph of G. The vertices and edges in a walk need not be distinct.

A walk is said to traverse through the vertices in the sequence originating in the initial vertex of the first edge and ending at the terminal vertex of the last edge in the sequence. Vertices with which a walk begins and ends are called **initial** and **final vertex** respectively or simply **terminal vertices**. In the graph shown in Fig. 10.63, vertices v_1 and v_5 are the terminal vertices and the walk is $v_1\ e_1\ v_2\ e_2\ v_3\ e_3\ v_3\ e_4\ v_4\ e_5\ v_2\ e_6\ v_5$.

A **close walk** is a walk, which begins and ends at the same vertex. A walk, which is not close, is called an **open walk**. For example graph shown in the Fig. 10.63 is an open walk.

10.29-1 Trail

If the edges of a walk are distinct then we call it a **trail** or we can say a trail is a walk with no repeated edge. A trail may visit same vertex more than once, but only if it comes and goes from a different edge each time.

10.29-2 Path

A path is a walk that does not include any vertex twice except that its initial vertex may be same as final vertex. In other words if in a trail all the vertices are distinct except possibly terminal vertices, then it is called a **path**. Let us consider the graph shown in Fig. 10.64.

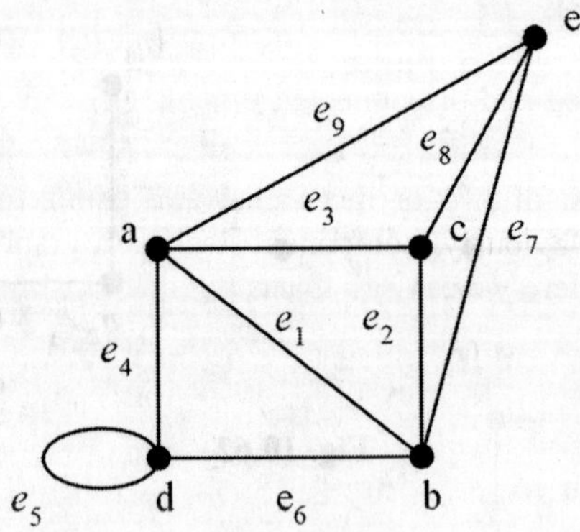

Fig. 10.64

Here, we see that $e\, e_8\, c\, e_2\, b\, e_1\, a\, e_4\, d\, e_5\, d\, e_6\, b$ is a trail, whereas $e\, e_8\, c\, e_2\, b\, e_1\, a\, e_4\, d$ is a path and $e\, e_8\, c\, e_3\, a\, e_9\, e\, e_7\, b\, e_2\, c\, e_3\, a$ is a walk.

10.29-3 Length of a Path

The number of edges appearing in the sequence of a path is called the **length of the path,** viz. an edge if its not a self loop is a path of length one. If the length of a path is zero then the path has no edge, it contains only a single vertex, is called a **trivial path**.

For example graph shown in Fig. 10.64 is a path of length 4. One can observe that the length of a path is one less than the number of vertices in the path.

10.29-4 Cycle and Circuit

A **cycle** is a closed path in which all the vertices are distinct except for the initial and final vertex. A **circuit** is a trail, which begins and ends on the same vertex. It means a circuit may have internal repeated vertices other than the end vertices. A cycle of length k is called a k-cycle. In a graph any cycle must have length two or more. Two different cycles are shown in the Fig.10.65.

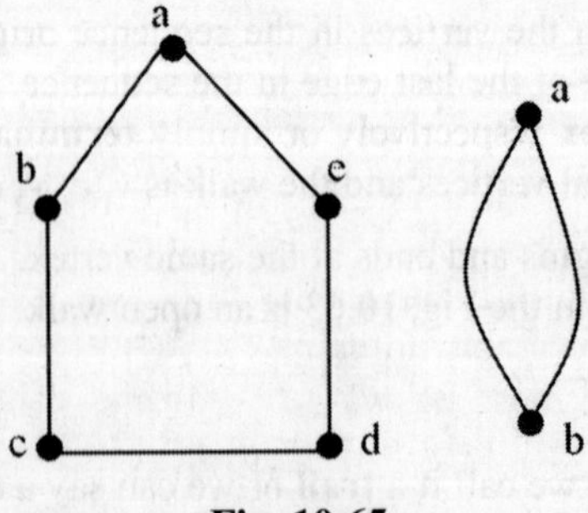

Fig. 10.65

For example, consider the graph shown in Fig. 10.64. Here

Cycle: $e\, e_7\, b\, e_2\, c\, e_3\, a\, e_9\, e$.

Circuit: $e\, e_9\, a\, e_3\, c\, e_2\, b\, e_1\, a\, e_4\, d\, e_6\, b\, e_7\, e$.

Caution! Many authors call walks as paths or simple paths and some of them call path chains.

10.30 Connectivity

We already discussed about connected and disconnected graphs. Here we take some theorems related to these graphs.

As a disconnected graph consist of two or more than two connected graphs. Each of these connected subgraphs is called a **component.** Obviously, component itself is a graph. For example graph shown in Fig. 10.66 consist of three components. Here we can say that the graph G is the union of three disjoint components A, B and C.

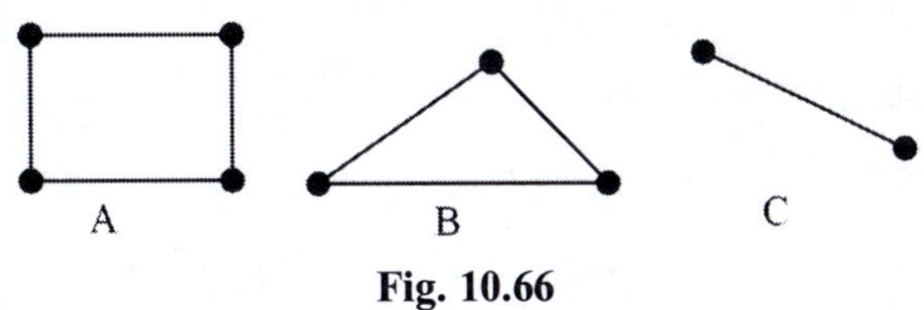

Fig. 10.66

Theorem 10.6

A graph G is disconnected iff its vertex set V can be partitioned into two nonempty, disjoint subsets S_1 and S_2 such that there exists no edge in G whose one end vertex is in subset S_1 and the other is in subset S_2.

Proof

To prove this theorem let us assume that such a partitioning exists. Consider two arbitrary vertices v_1 and v_2 of graph $G = (V, E)$ such that $v_1 \in S_1$ and $v_2 \in S_2$. As per our assumption no path can exist between vertices v_1 and v_2, otherwise there would be at least one edge whose one end vertex would be in S_1 and other in S_2. Hence, if a partition exists, graph G is not connected.

Conversely, let us assume G is a disconnected graph. Let v_1 be a vertex in G. Consider S_1 be the set of all vertices that are joined by paths to v_1. Since graph G is disconnected, S_1 does not include all vertices of G. And the remaining vertices will form a nonempty set S_2. Now it is clear that no vertex in S_1 is joined to any in S_2 by an edge. Hence the partition.

Theorem 10.7

If a graph, which may be connected or disconnected, has exactly two vertices of odd degree, there must be a path joining these two vertices.

Proof

Consider a graph $G = (V, E)$ with all even vertices (i.e. vertices with even degrees) except vertices v_1 and v_2, which are odd. From above theorem which is applicable for every graph (directed and undirected) and therefore for every component of a disconnected graph, no graph can have an odd number of odd vertices. This implies that vertices v_1 and v_2 in graph G must belong to the same component and hence must have a path between them.

Theorem 10.8

A graph G without parallel edges or self loop (i.e. simple graph) with n nodes and k components can have at most $(n - k)(n - k + 1)/2$ edges.

Proof

Let $n_1, n_2, \ldots, n_k$ be the number of vertices in each of the k components of a graph G.

$$n_1 + n_2 + \ldots, n_k = n, \quad \text{where } n_i \geq 1$$

Before proving the theorem let us take the inequality

$$\sum_{i=1}^{n}(n_i - 1) = n - k$$

$$(\sum_{i=1}^{n}(n_i - 1))^2 = n^2 + k^2 - 2nk$$

or $$\sum_{i=1}^{n}(n_i^2 - 2n_i) + k + \text{nonnegative cross terms} = n^2 + k^2 - 2nk$$

Therefore $(n_i - 1) \geq 0$, for all i

Squaring both sides, we have

$$\therefore \quad \sum_{i=1}^{n} n_i^2 \leq n^2 + k^2 - 2nk - k + 2n = n^2 - (k-1)(2n-k)$$

$$\Rightarrow \quad \sum_{i=1}^{n} n_i^2 \leq n^2 - (k-1)(2n-k) \qquad \dots (1)$$

As we know the maximum number of edges in the ith component of graph G is $(1/2)n_i(n_i-1)$. Therefore, the maximum number of edges in G is

$$\frac{1}{2}\sum_{i=1}^{k}(n_i - 1)n_i = \frac{1}{2}(\sum_{i=1}^{k} n_i^2) - \frac{n}{2}$$

$$\leq \frac{1}{2}\left[n^2 - (k-1)(2n-k)\right] - \frac{n}{2} \quad \text{(from 1)}$$

$$= \frac{1}{2}(n-k)(n-k+1)$$

10.31 Distance and Diameter

Consider a connected graph $G = (V, E)$. The distance between vertices v_i and v_j denoted by $d(v_i, v_j)$ or $d_G(v_i, v_j)$ is the minimum number of edges between any two points in G. For example in Fig.10.67 (a), $d(v_8, v_3)$ = 2. The **diameter** of the graph $G = (V, E)$ written as diam(G) orΔ, is the maximum distance between any two vertices, i.e. $\Delta = \underset{u,v \in V}{\text{Max}} \; d(u,v)$

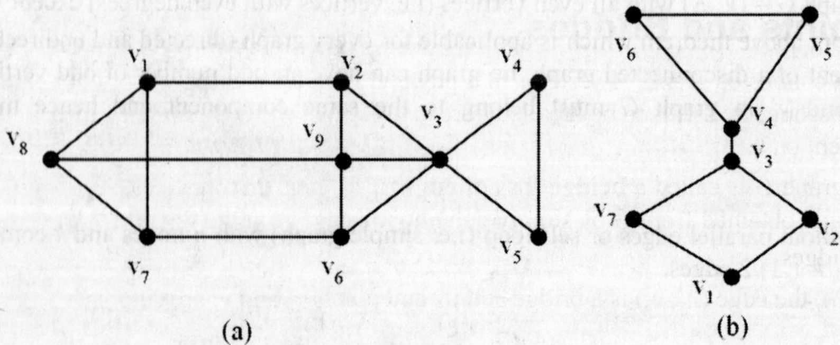

Fig. 10.67

Diameter of the graph shown in Fig. 10.67(b) is 4, i.e. $\Delta = 4$.

Example 10.24

The paths between the vertices u and v in the graph shown in the Fig. 10.68 are (a, e), (a, c, f), (b, c, e), (b, f), (b, g, h) and (b, g, i, k). Out of these (a, e) and (b, f) are two shortest paths, each of length 2, i.e. $d(u, v) = 2$.

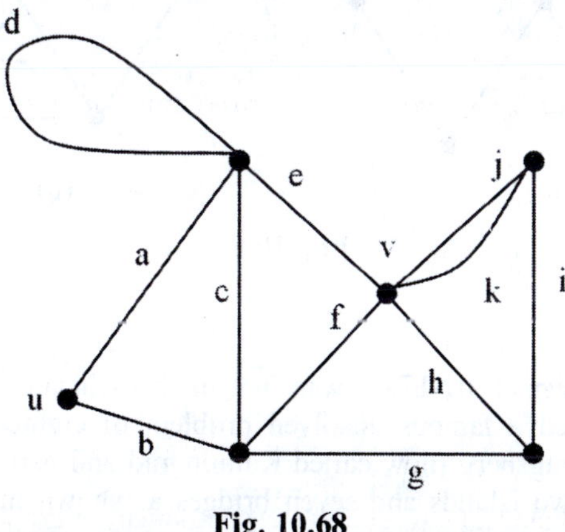

Fig. 10.68

Example 10.25

Under what condition is the equation $d(v_1, v_2) + d(v_2, v_3) = d(v_1, v_3)$ satisfied for $v_1, v_2, v_3 \in V$ in a simple graph $G = (V, E)$.

Solution

The given equation holds if there is a unique path of minimum length between the vertices v_1, v_2 and v_3.

10.32 Reachability

A vertex v_i of a simple graph is said to be **reachable** or **accessible** from v_j if there is a path from v_i to v_j. If a vertex v_i is reachable from a vertex v_j then the path of minimum length from v_i to v_j is called **geodesic**.

The length of geodesic from v_i to v_j is called the **distance** from v_i to v_j denoted by $d(v_i, v_j)$. It is assumed that $d(v_i, v_i) = 0$. We assume that every vertex is reachable from itself. Here, it is to be noted that reachability is a relation, which is reflexive, and transitive. Reachability need not be symmetric or anti-symmetric.

10.33 Cutpoints and Bridges

Let $G = (V, E)$ be a connected graph. A vertex v in a graph G is called a **cutpoint** (or **cut vertex** or **articulation point**) if $G - v$ is disconnected. It is to be recalled here that the $G - v$ is the graph obtained from G by deleting v and all edges incident on v.

An edge e of graph G is called a **bridge** (or **cut edge**) if G - e is disconnected. Again it is to be noted here that $G - e$ is the graph obtained from G by simply deleting the edge e. In the Fig.10.69 (a) vertex v_3 is a cutpoint and there are no bridges.

In Fig.10.69 (b), the edge (v_4, v_5) is a bridge and its end points v_4 and v_5 are cutpoints.

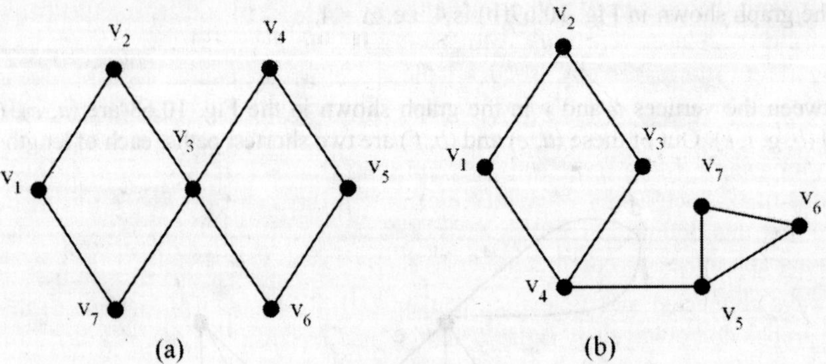

Fig. 10.69

10.34 Euler# Graphs

Swiss mathematician Leonard Euler (1707-1783) was first mathematician whose concept of using graph had been recorded. He in 1736 settled a famous unsolved problem of eighteenth century, which is known as Konigsberg Bridges problem. Konigsberg (now called Kaliningrad and part of the Russia) was located on the banks of Pregil River included two islands and seven bridges as shown in the Fig.10.70. The problem was beginning at the starting point and ending at the starting point can a person walk through town crossing all seven bridges but not crossing any bridge twice?

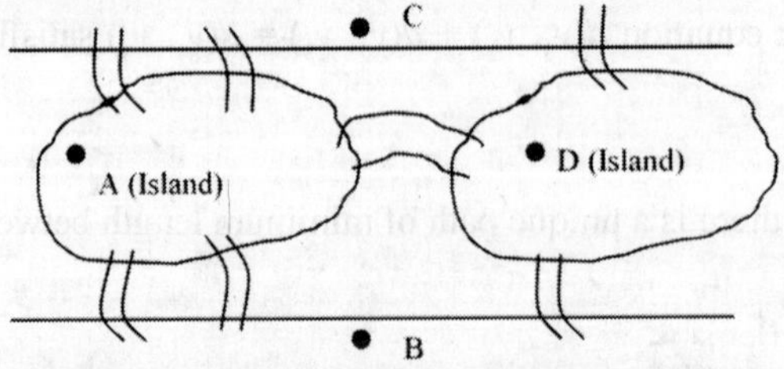

(a) The Konigsberg Bridges

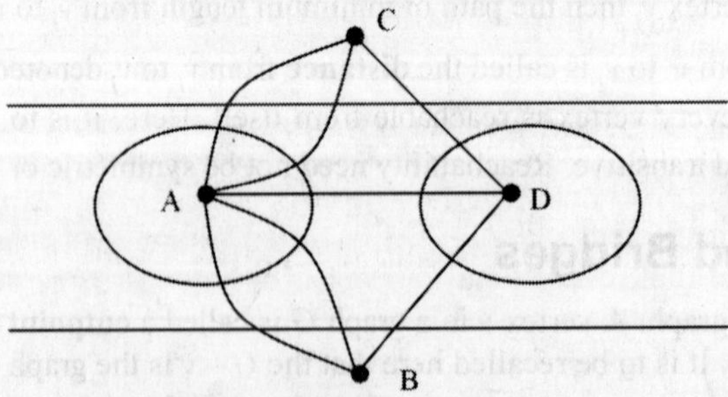

(b) Euler's graphical representation

Fig. 10.70

Leonhard Euler (1707-1783) was an enormously creative Swiss mathematician. He studied in Basel under Johann Bernoulli and in 1730 became a professor of physics and later of mathematics in St. Petersburg, Russia. He contributed to almost all branches of mathematics and its application to physical problems even after he became totally blind in 1766.

The problem of Konigsberg referred to the celebrated Swiss Mathematician L. Euler. To prove that such a walk is impossible, Euler replaced each of the areas A, B, C and D by vertices and each bridge by an edge joining the corresponding vertices, thereby producing the graph. On the pattern of this problem Euler discussed more general problems such as in what type of graph G is it possible to find a closed walk running through every edge of G exactly once? Later on such a walk is called **Euler line** and a graph that consists of an Euler line is called an **Euler graph**. More specifically a connected graph G is **Eulerian** if there exist a closed trail containing every edge of G. Such a trail is called **Eulerian trail** or **Eulerian line**. Some authors also refer this as Euler path. Note that this definition requires each edge to be traversed only once. Similarly, we define an **Eulerian circuit** in a graph to be a circuit that traverses each edge in a graph once.

As walk is always connected and Euler line contains all the edges of the graph. Hence, here after it is assumed that Euler graphs do not have any isolated vertices and are always connected.

Example 10.26

Which of the graphs shown in Fig.10.71 have an Euler circuit? If it does not contain Euler circuit, which have an Euler trail?

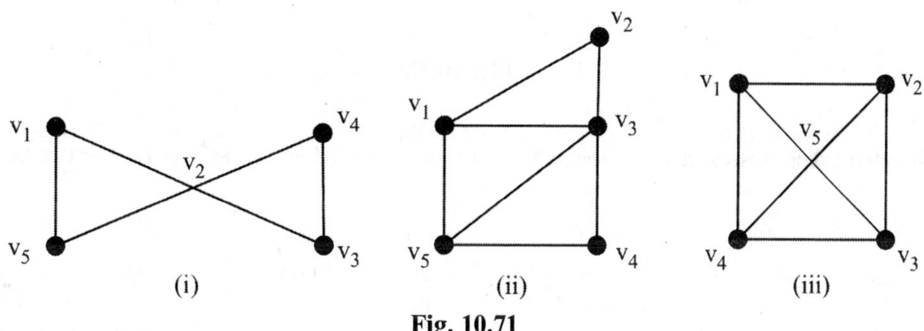

Fig. 10.71

Solution

The graph (i) has an Euler circuit v_1, v_2, v_3, v_4, v_2, v_5, v_1. We can observe here that all edges are traversed once. Readers can verify that neither (ii) nor (iii) has an Euler circuit. But (ii) has an Euler trail v_5, v_4, v_3, v_2, v_1, v_3, v_5 v_1. (iii) Doesn't have an Euler trail as well as Euler path.

Theorem 10.10

A connected graph G is an Euler graph iff all vertices of G are of even degree.

Proof

Let G be a Euler graph. Then G possesses an Eulerian trail, which begins and ends at vertex a, say. We observe that when we trace the trail, every time trail meets a vertex, it goes through two edges, which are incident with the vertex and have not been traced before. Since any trail (hence Eulerian trail) uses any edge once, each traversal of any vertex, say b represents a contribution of two to its degree. Hence deg (v) is even.

Conversely, suppose that graph G is connected and every vertex of G is even. Let us try to construct an Eulerian trail T_1 by beginning at an arbitrary edge e, say. We go on extending the trail T_1 by adding an edge after the other. During the tracing if the trail is not closed at any step (means if tracing started at vertex a and ended at b; where $a \neq b$) then last vertex b encountered must be of odd degree (i.e. odd number of edges incident b). Hence, T_1 can be incident on b. Continuing extensions of T_1 in this way we must have reached the initial vertex a, i.e. trail T_1 closed. If in this process T_1 includes all the edges of G then T_1 is Eulerian.

Let us assume that trail T_1 does not include all the edges of graph G. Then we remove those edges of G that have been traced by T_1 and for the remaining edges obtain a graph H (H is a subgraph of G). The degree of all vertices of H is even since T_1 contains vertices of even degree. Since G is connected let us assume that there exist an edge e' of H that has an end point a in T_1. Here, again we construct a trail T_2 in H starting with the vertex a' and using e'. As all the vertices of H are of even degree, this trail ultimately returns to the vertex where it started i.e. a' (see Fig. 10.72). We can join T_1 and T_1 to form a larger closed trail in G. We continue this process until all the edges of G are covered. In this way we obtain a Eulerian trail and hence G is Eulerian.

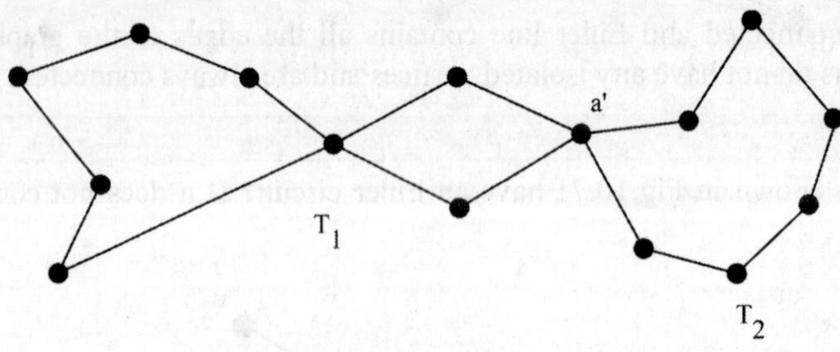

Fig. 10.72

Corollary 1

An undirected graph possesses an Eulerian trail iff it is connected and has either zero or two vertices of odd degree.

Proof

Let us assume that the graph possesses an Eulerian path. We know that this graph is connected. We can observe that when we trace the Eulerian trail, every time trail meets a vertex, it goes through two edges, which are incident with the vertex and have not been traced before. Now, one can conclude very easily that except for the two vertices at the two ends of the path, the degree of any vertex in the graph must be even. If there exist two distinct vertices at two ends of the Eulerian trail, then they are the only two vertices with odd degree. If they coincide, all vertices have even degree, and the Eulerian trail become an Eulerian circuit. This proves the necessary condition.

To prove the sufficient condition let us construct an Eulerian trail by starting at one of the two vertices that are of odd degree (if there is no vertex of odd degree we start at an arbitrary vertex) and going through the edges in such a way that no edge is traced more that once. And for a vertex of even degree whenever the trail enters through an edge, it will always have the vertex through another edge that is never been covered during tracing. Hence, when the tracing is finally ended we must have reached the other vertex of odd degree. Tracing all the edges of the graph in this way, we will have an Eulerian trail. If doing so some of the edges of the graph are left, we shall remove those edges that have been traced and for the remaining edges obtain a subgraph. The degrees of the vertices of this subgraph are all even. As the trail path is connected, this subgraph must touch the path that we have traced at least at one vertex. Here, again we can construct a path that passes through the edges beginning from one of these vertices. As vertices are of even degree, this trail ultimately returns to the vertex at which it started. We can join this trail with the trail we have already constructed to obtain one which begins and ends at the two vertices of odd degree.

Corollary 2

A connected multigraph has an Euler path but not an Euler circuit iff it has exactly two vertices of odd degree.

Corollary 3

If G is a connected graph and every vertex has even degree, then G has an Euler circuit.

Example 10.27

Show that the graph shown in Fig. 10.73 is not Eulerian.

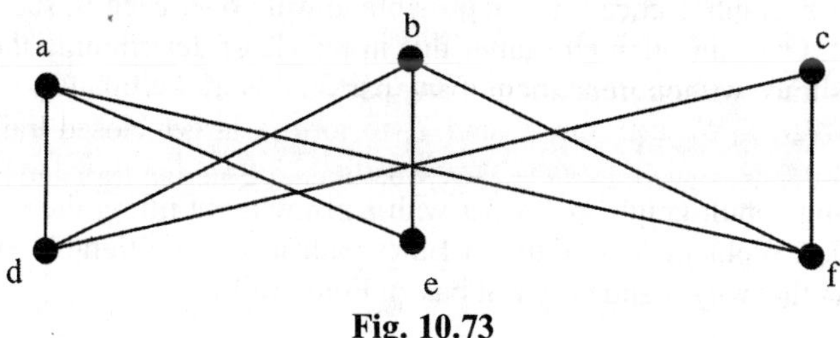

Fig. 10.73

Solution

Since all the vertices of graph shown in Fig. 10.73 are of degree 3 (i.e. odd). Therefore from corollary 2, the graph is not Eulerian.

Example 10.28

Consider a graph G shown in Fig. 10.74. Show that G has an Euler cycle. Find also an Euler circuit.

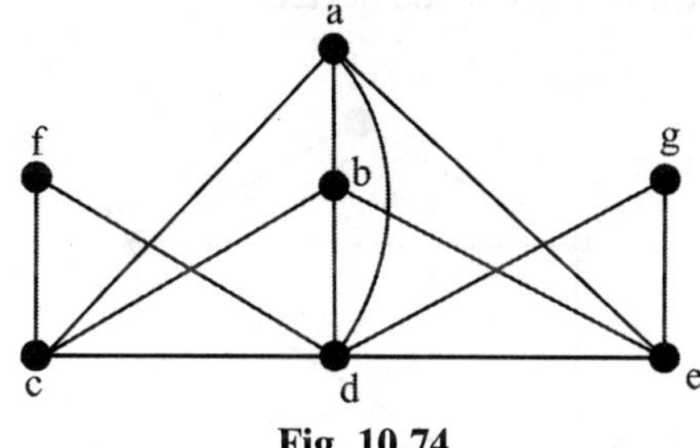

Fig. 10.74

Solution

Let us verify that the graph shown in Fig.10.74 has an Euler cycle with the help of corollary 3. We can note that the graph is connected and all the vertices are of even degree. Therefore G has an Euler cycle. By inspection, we find one Euler circuit as $(f, d, g, e, a, c, d, a, b, e, d, b, c, f)$.

Example 10.29

Is graph shown is Fig. 10.75 contains Euler circuit?

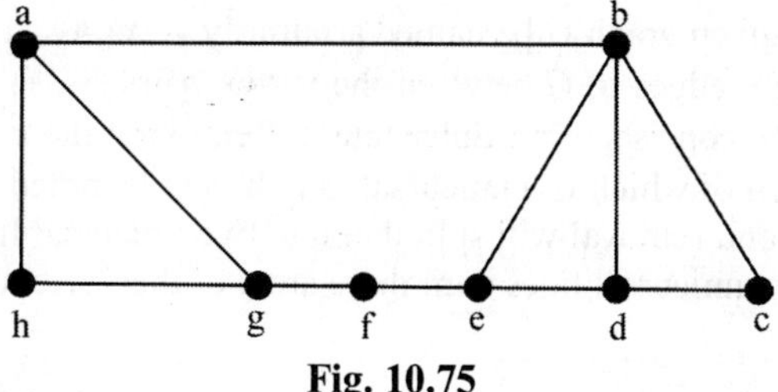

Fig. 10.75

It doesn't contain Euler circuit. Readers are advised to verify it.

10.34-1 Konigsberg Bridge Problem

Again looking back to the Konigsberg bridge problem we find that all the vertices are not of even degree. This implies that it is not an Euler graph. Hence it is not possible to walk over each of such bridges for one time and return to the starting point. One can often encounter this in puzzle of determining the possibility of drawing a given figure in a continuos trace without repetition of any part.

Example 10.30

The problem of drawing a multigraph on a paper with a pen without lifting the pencil or repeating any lines is clearly a matching with a problem of finding an Euler path in the multigraph. Here it is to be noted that multigraph can be drawn in this way if and only if it has an Euler trail.

10.34-2 Unicursal Line (Open Euler Line), Unicursal Graph

In Fig.10.76 the walk ($v_1\, v_3\, v_4\, v_1\, v_2\, v_4\, v_5\, v_2$) includes all the edges of the graph and doesn't pass through any edge more than once. This walk is not closed since the initial vertex is v_1 and final vertex is v_2. The open walk that covers all the edges of a graph without retracing any edge is called an **open Euler line** or **unicursal line**. A connected graph that has a unicursal line is called a **unicursal graph**. One can see here that by adding an edge between initial and final vertices of an open Euler line we shall get an Euler line. We conclude that a connected graph is unicursal iff it has exactly two vertices of odd degree.

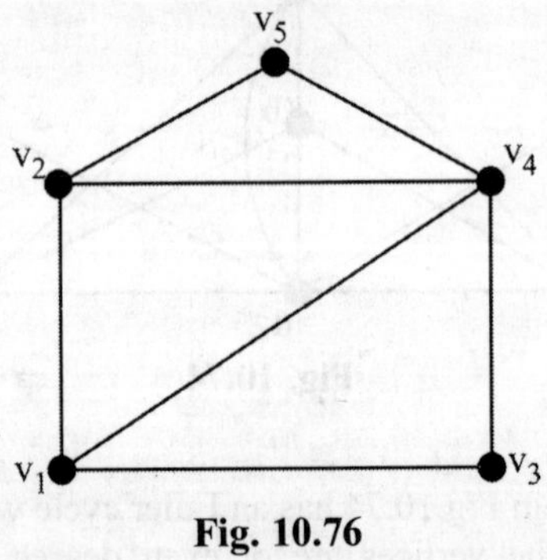

Fig. 10.76

Theorem 10.11

In a connected graph G with exactly $2n$ odd vertices, there exist n edge disjoint subgraphs such that they together contain all edges of G and that each is a unicursal graph.

Proof

Let the 'odd vertices of the given graph G be named arbitrarily $v_1, v_2, v_3 \ldots v_n, v_1', v_2', v_3', v_4', \ldots v_n'$. Let us form a new graph G' by adding n edges to G between the vertex pairs (v_1, v_1') (v_2, v_2'), …, (v_n, v_n'). As every vertex of G' is of even degree, G' consists of an Euler line λ. Removing the n edges just added from λ we find that λ will split into n walks, each of which is a unicursal line. It is to be noted here that first removal will leave a singular unicursal line, the second removal will split that into two unicursal lines and each successive removal will split a unicursal line into two unicursal lines until there are n of them. Hence the theorem.

Theorem 10.12

A connected graph G is an Euler graph if and only if it can be decomposed into circuits.

Proof

Let us decompose graph G into circuits i.e. G is a union of edge disjoint circuits. As we know degree of every vertex in a circuit is two, hence every vertex of G is of even degree. Thus G is an Euler graph.

Conversely, Suppose G is an Euler graph. Consider a vertex v_1. There are at least two edges incident at v_1. Out of these two edges let us consider that one of the edges be between v_1 and v_2. As vertex v_2 is also of even degree, it must have at least another edge, say between v_2 and v_3. Continuing doing so we arrive at a vertex from where we have started, thus forming a circuit C. Now, if we remove C from graph G. Graph of remaining vertices must also be of even degree. But this graph is not necessarily connected. Proceed in this fashion until no edge is left. Hence the theorem.

10.35 Arbitrarily Traceable Graphs

Consider a Euler graph shown in Fig. 10.77(a). Let us start from vertex v_1 and trace the path $v_1\ v_2\ v_3$. Here, at v_3 we have an option of going to v_1, v_3 or v_4. If we go for the first option then we would trace the circuit $v_1\ v_2\ v_3\ v_1$, which is not a Euler line. From here we conclude that starting from vertex v_1 it is not always that the entire Euler line can be traced by moving along any edge that has not traced earlier. Now following question arises:

Is there any property that a vertex has in an Euler graph such that an Euler line is always obtained when one follows any walk according to a particular rule that whenever one reaches at a vertex one should select any edge, which is not previously traversed?

Graph of this type is called **arbitrarily traceable graph** from vertex v. The graph shown in Fig. 10.77(a) is an arbitrarily traceable graph from vertex v_3 but not from any other vertex. The graph shown in the Fig. 10.77(b) is not arbitrarily traceable from any vertex.

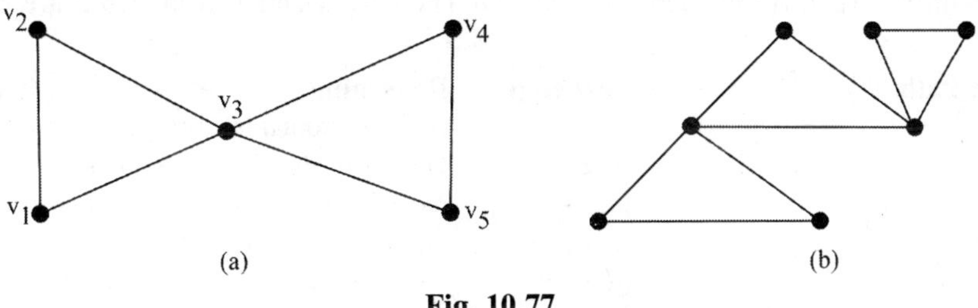

Fig. 10.77

Theorem 10.13

An Euler graph G is arbitrarily traceable from vertex v in G iff every circuit in G contains v.

Proof

Proof of the theorem is beyond the scope of this book however readers interested in the proof are advised to refer Ore, O. "A problem regarding the tracing of graphs," Rev. Elementary Math. Vol.6, 1961, 49-53.

10.36 Fleury's Algorithm

The algorithm constructs Euler tour (moving on a graph by covering all the edges of the graph and returning to the original vertex) in an Euler graph. Its approach is to construct a trail that grows to the desired Eulerian tour. Following steps construct Fleury's algorithm.

Step 1: Start at a vertex v_0 that has odd degree unless graph G is even in which case start at any vertex and set $W_0 = v_0$

Step 2: If the trail $W_i = v_0, e_1, v_1, \ldots, e_i, v_i$ has been chosen , choose an edge e_{i+1} different from $e_1, e_2, \ldots, e_i$ such that:

i. e_{i+1} is incident with v_i and

ii. If there is more than one such edge, select one that is not a bridge.

Step 3: Stop, if we contain every edge of G, otherwise repeat step 2.

Theorem 10.14

If G has one nontrivial component and at most two odd vertices, then Fleury's algorithm constructs an Eulerian trail.

Example 10.31

Use Fleury's algorithm to construct an Euler circuit for the graph in Fig. 10.78.

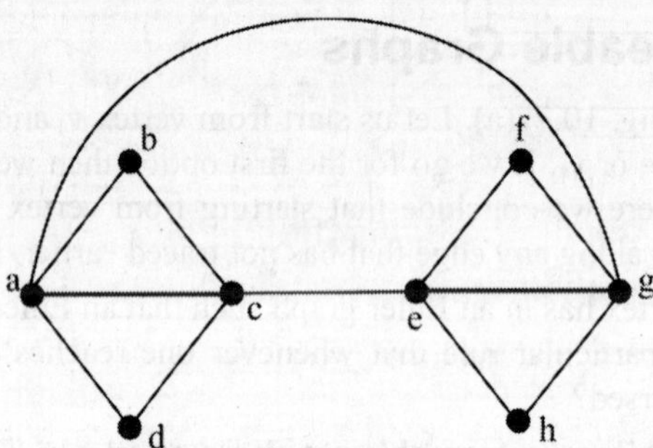

Fig. 10.78

Solution

Let us start with vertex a. Apply step 2 repeatedly. The results obtained in step 2 are generalized in the following table:

Current Path	Next edge	Reasoning
$p : a$	$\{a, b\}$	No edge from a is a bridge. Choose any one.
$p : a, b$	$\{b, c\}$	Only one edge from b remains.
$p : a, b, c$	$\{c, a\}$	No edge from c is a bridge. Choose any one.
$p : a, b, c, a,$	$\{a, d\}$	No edge from a is a bridge. Choose any one.
$p : a, b, c, a, d$	$\{d, c\}$	Only one edge from vertex d remains.
$p : a, b, c, a, d, c$	$\{c, e\}$	Only one edge from vertex c remains.
$p : a, b, c, a, d, c, e$	$\{e, g\}$	No edge from e is a bridge. Choose any one.
$p : a, b, c, a, d, c, e, g$	$\{g, f\}$	$\{a, g\}$ is a bridge. Choose $\{g, f\}$ or $\{g, h\}$.
$p : a, b, c, a, d, c, e, g, f$	$\{f, e\}$	Only one edge from f remains.
$p : a, b, c, a, d, c, e, g, f, e$	$\{e, h\}$	Only one edge from e remains.
$p : a, b, c, a, d, c, e, g, f, e, h$	$\{h, g\}$	Only one edge from h remains.
$p : a, b, c, a, d, c, e, g, f, e, h, g$	$\{g, a\}$	Only one edge from g remains.
$p : a, b, c, a, d, c, e, g, f, e, h, g, a$		

Here readers may note that at several vertices we could have other choices then what we have done here. In other words we can say that, if a graph has an Euler circuit, it is quite possible to have several different Euler circuits.

10.37 Hamiltonian Circuit and Path

A Hamiltonian* circuit is defined as a closed walk that traverses every vertex of a connected graph G exactly once except for the end vertices, which is also the terminal vertex. A circuit in a connected graph G is said to be **Hamiltonian** if it contains Hamiltonian circuit. It means a Hamiltonian circuit of a graph having n vertices consists of exactly n edges.

---

* The name "Hamiltonian" is derived from the Irish mathematician Sir William Rowan Hamilton, who invented a game in 1857 consisting of a solid regular dodecahedron.

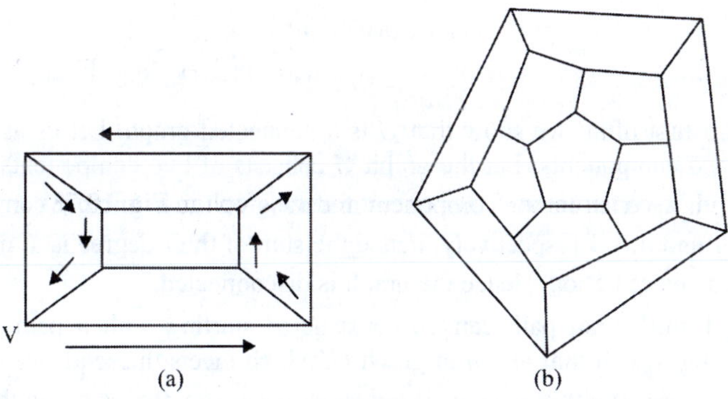

Fig. 10.79

It is to be noted here that an Eulerian circuit traverses every edge exactly once, but may repeat vertices while a Hamiltonian circuit traverses each vertex exactly once and hence miss some edges altogether. The graph shown in Fig.2.71 with heavy lines is a Hamiltonian circuit. Here when one traverses along the heavy lines he passes through each vertex exactly once.

Hamiltonian circuit is always a connected graph, but every connected graph does not have a Hamiltonian circuit. The graph shown in the Fig. 10.77(a) and (b) have a Hamiltonian circuit whereas the graph shown in Fig. 10.80 is non-Hamiltonian.

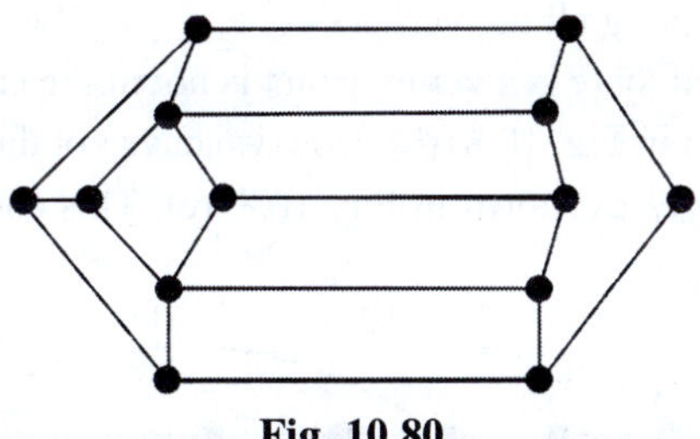

Fig. 10.80

Now from the definition of Hamiltonian circuit we can have following observations:

1. A complete graph with more than two vertices is Hamiltonian.
2. Every cycle graph is Hamiltonian.
3. Every platonic graph is Hamiltonian.

10.37-1 Hamiltonian Path

The path obtained by removing any one edge from a Hamiltonian circuit is called a **Hamiltonian path.** Hamiltonian path in a graph G visits every vertex of graph G. It is to be noted here that a Hamiltonian path is a subgraph of a Hamiltonian circuit, it implies that graph having a Hamiltonian circuit also has a Hamiltonian path. But the converse is not always true. The length of a Hamiltonian path if it exists in a connected graph of n vertices is $n - 1$.

For the existence of a Hamiltonian circuit (path) we need to consider only simple graphs because it traverses every vertex exactly once, it should not include self-loop or a set of parallel edges. If a general graph contains self-loop or parallel edges it can be made simple graph by removing cycle or parallel edges.

Theorem 10.15

A graph G with n vertices always has a Hamiltonian path if the sum of the degrees of every pair of vertex (v_i, v_j) in G satisfies the condition

$$d(v_i) + d(v_j) \geq n - 1.$$

Proof

To prove this theorem first of all we show that G is a connected graph. Let us assume that G has two or more than two disconnected components. Let the graph G consists of two components, one having n_1 vertices while other n_2. Suppose v_1 is a vertex in one component and v_2 is vertex in another component. Since degree of v_1 and v_2 be at most $n_1 - 1$ and $n_2 - 1$ respectively, hence the sum of their degree is at most $n_1 + n_2 - 2$ which is less than $n - 1$. But this is a contradiction. Hence the graph is disconnected.

Let us show how a Hamiltonian path can be constructed starting with a path containing single edge. Consider a path of $m - 1$ edges such that $m < n$ in graph G, which meets the sequence of vertices say $v_1, v_2, v_3, \ldots, v_m$. Obtain a path of m edges if either v_1 or v_m is adjacent to a vertex that is not in the path by extending the path to include this vertex. If this is not the case than both v_1 and v_m are adjacent only to the vertices that are in the path. Now we will show that in that case there is a circuit containing exactly the vertices $v_1, v_2, \ldots, v_m$. In case v_1 is adjacent to v_m then the circuit $(v_1, v_2, \ldots, v_m, v_1)$ will suffice, so let us assume here that v_1 is adjacent only to $v_{i_1}, v_{i_2}, \ldots v_{i_m}$ where $2 \leq i_j \leq m - 1$. And if v_m is adjacent to one of $v_{i_1 - 1}, v_{i_2 - 1}, \ldots, v_{i_m - 1}$, say v_{j-1}, then from Fig.10.81(a), one can see that the circuit $(v_1, v_2, v_3, \ldots, v_{j-1}, v_m, v_{p-1}, \ldots v_j, v_1)$ contains exactly the vertices $v_1, v_2, \ldots, v_m$. Suppose v_m is not adjacent to any one of $v_{i_2 - 1}, v_{i_2 - 1}, \ldots, v_{i_k - 1}$ then v_m is adjacent to at most $m - k - 1$ vertices. Hence, the sum of the degree of v_1 and v_m is at most $n - 2$ which is a contradiction.

Now we have got a circuit containing all the vertices $v_1, v_2, \ldots v_m$. Let us take one vertex, say v_z which is not in the circuit. Since G is connected there is a vertex v that is not in the circuit with an edge between v_j and v_k for some v_k in $\{v_1, v_2, \ldots v_m\}$ as shown in Fig. 10.81(b). Now we have got the path $(v_z, v_k, v_{k-1}, \ldots, v_m, v_{m-1}, \ldots, v_j, v_1, v_2, v_{k-1})$ which contains m edges as shown in Fig. 10.81(c), This construction is repeated until we get a path of $n - 1$ edges.

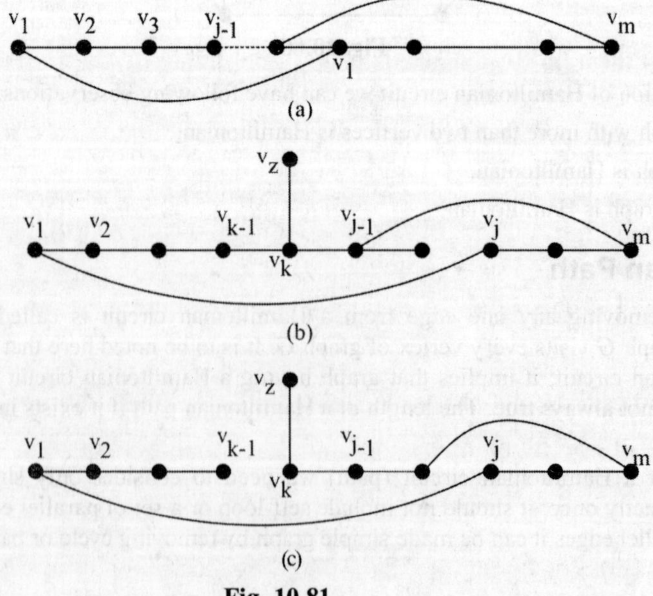

Fig. 10.81

Theorem 10.16 (Ore, 1960).

If G is a simple graph with n (≥ 3) vertices, and if

$$\deg(v) + \deg(w) \geq n$$

for each pair of non-adjacent vertices v and w, then G is Hamiltonian.

Proof

Let us assume that the theorem is not true and derive a contradiction. Let G be a non-Hamiltonian graph with n vertices, satisfying the given condition on the vertex degrees. Let us suppose that graph G is 'only just' non-Hamiltonian in the sense that addition of any further edge gives a Hamiltonian graph. It is to be noted here that addition of an extra edge does not violate the condition on the vertex degrees. From this we conclude that graph G contains a path $v_1, v_2, \ldots, v_n$ passing through every vertex. But we have assumed that graph G is non--Hamiltonian, hence vertices v_1 and v_n are not adjacent, and so $\deg(v_1) + \deg(v_n) \geq n$. This implies that there exist some vertex v_i adjacent to v_1 with the property that v_{i-1} is adjacent to v_n (See Fig. 10.82). But this gives us the required contradiction since $v_1, v_2, \ldots, v_{i-1}, v_n, v_{n-1}, v_{i+1}, v_i, v_1$ is then a Hamiltonian cycle.

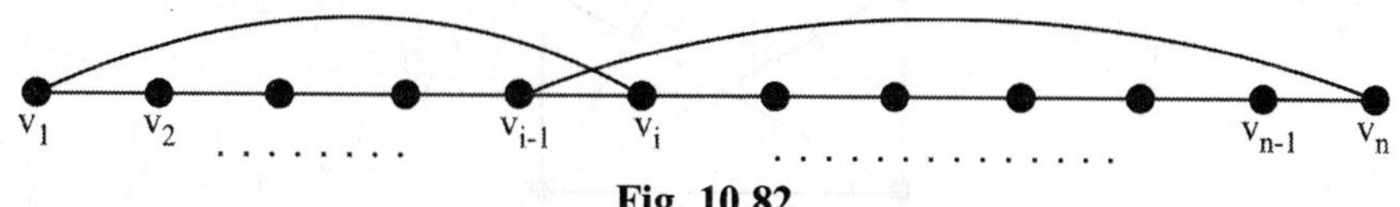

Fig. 10.82

Corollary (Dirac, 1962):

Let G be a simple graph with n (≥ 3) vertices then graph G is Hamiltonian if $\deg(v) > n/2$ for each vertex v.

Proof

This result follows immediately from theorem 10.16. As $\deg(v) + \deg(w) \geq n$ for each pair of vertices v and w irrespective of adjacency.

Example 10.32

Verify hypothesis of **theorem** 10.16 for the following graph.

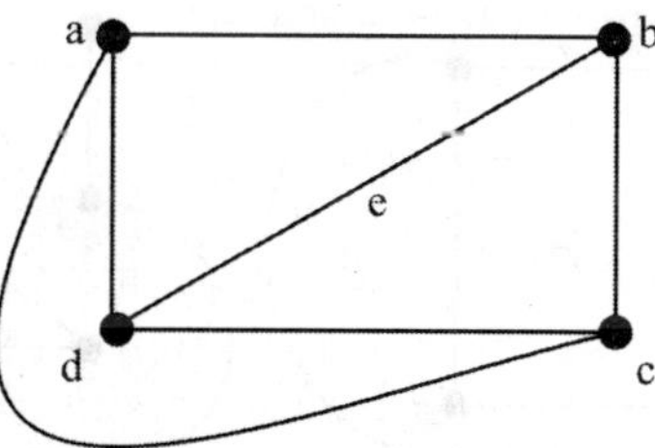

Solution

Here number of vertices is 5, i.e. $n = 5$. We can observe form the figure there exist three pair of non-adjacent vertices. Now for pair of vertices (b, d)

$$\deg(b) + \deg(d) = 3 + 3 = 6 > 5$$

For pair (a, e), $$\deg(a) + \deg(e) = 3 + 2 = 5$$

For pair (e, c), $$\deg(e) + \deg(c) = 2 + 3 = 5$$

In all the three cases sum of the degrees of nonadjacent vertices is either equal to or greater than number of vertices ($n = 5$).

Example 10.33

Show that K_n has a Hamiltonian circuit whenever $n \geq 3$.

Solution

In graph K_n, we can form a Hamiltonian circuit beginning at any vertex. Such a circuit can be formed by visiting vertices in any order, as long as initial and final vertices are same and visits each vertex exactly once. All this is possible because there exist edges in K_n between any two vertices.

Particular Case k_5:

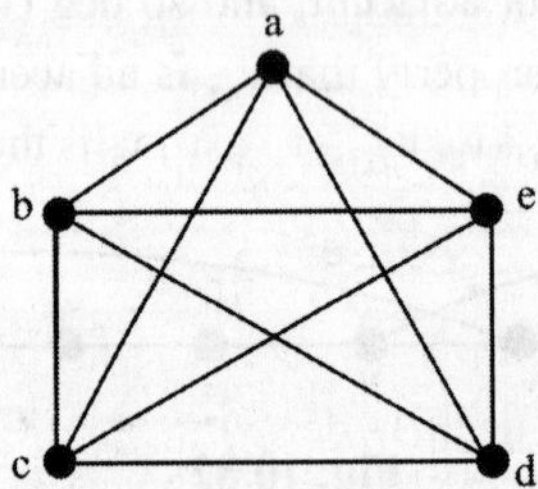

Solution

There exist a Hamiltonian circuit (a, b, c, d, e) in k_5 and k_5 in Hamiltonian. We can observe that each vertex of k_5 is of degree 4. And number of vertices of k_5 is 5, i.e. $n = 5$. Since $n = 5 \geq 3$ and $\deg(v) = 4 > \frac{n}{2} = \frac{5}{2}$ for each vertex. Hence according to hypothesis of corollary of theorem 2.17 it should be Hamiltonian.

Example 10.34

Which of the following graphs shown in Fig.10.84 have a Hamiltonian circuit? If not, why?

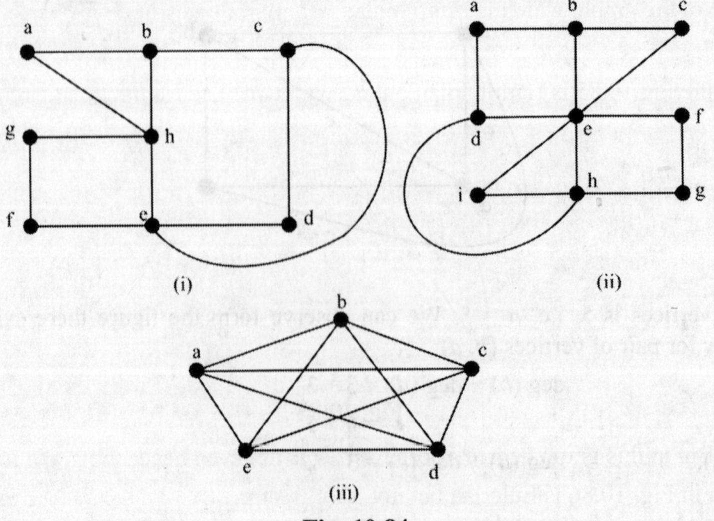

Fig. 10.84

Solutions

Graph (i) and (iii) have Hamiltonian circuits but (ii) has no Hamiltonian circuit (Readers are advised to verify it).

Theorem 10.17

Let the number of edges of graph G be m and the number of vertices be n respectively. Then G has Hamiltonian circuit if $m \geq \frac{1}{2}(n^2 - 3n + 6)$.

Proof

Let G be a graph and suppose u and v are any two non-adjacent vertices of G. Let H be a sub graph of G obtained by eliminating vertices u and v, and all the edges incident on u and v from G. Then H has $n - 2$ vertices and m – deg (u) – deg (v) edges (If u and v were adjacent then one fewer edge would have been removed). The graph H can have at the most $n \quad 2_{C_2}$ number of edges. This is possible when there is an edge connecting every pair of vertices. Thus the maximum possible number of edges of H is

$$n-2_{C_2} = \frac{(n-2)(n-3)}{2} = \frac{1}{2}(n^2 - 5n + 6)$$

$$\Rightarrow \quad m - \deg(u) - \deg(v) \leq \frac{1}{2}(n^2 - 5n + 6)$$

$$\Rightarrow \quad \deg(u) + \deg(v) \geq m - \frac{1}{2}(n^2 - 5n + 6)$$

By the hypotheses of the theorem

$$\deg(u) + \deg(v) \geq \frac{1}{2}(n^2 - 3n + 6) - \frac{1}{2}(n^2 - 5n + 6)$$

$\deg(u) + \deg(v) \geq n$

Example 10.35

Give an example of the graph, which is Hamiltonian but not Eulerian, and vice-versa.

Solution

Graph shown in the Fig.10.85 is Hamiltonian but not Eulerian. The Hamiltonian circuit is

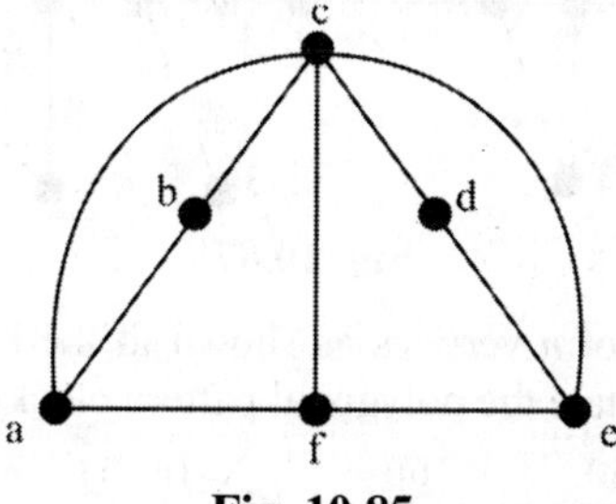

Fig. 10.85

a, b, c, d, e, f, a. But in this graph degree of each vertex is not even hence the graph is non Eulerian.

The graph shown in Fig. 10.86 is Eulerian but not Hamiltonian.

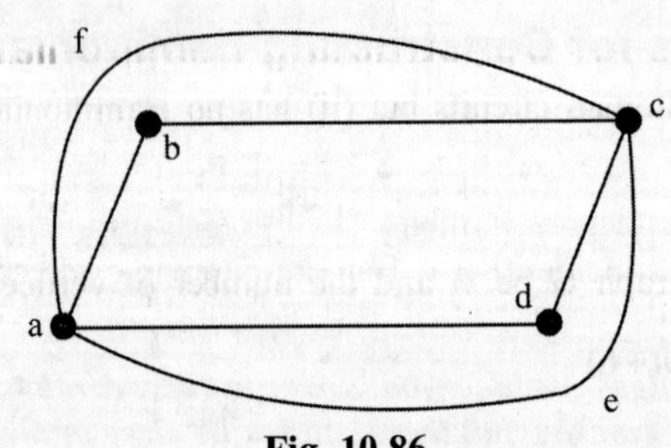

Fig. 10.86

In this graph we can observe that the degree of each vertex is even hence it is Eulerian. The graph is not Hamiltonian because we can't find any circuit in the graph, which visits all the vertices once except initial and final vertex.

10.37-2 Number of Hamiltonian Circuits in a Graph

A given graph may contain more than one Hamiltonian circuit. As far as all the edge disjoint Hamiltonian circuits in a graph is concerned, the exact number (of edge disjoint-Hamiltonian) is still an unsolved problem but the following theorem gives the number of edge-disjoint Hamiltonian circuit in a complete graph with odd number of vertices.

Theorem 10.18

In a complete graph with n vertices there are $(n-1)/2$ edge disjoint. Hamiltonian circuits, if n is an odd number ≥ 3.

Proof

Consider a complete graph G with n vertices. It has $n(n-1)/2$ edges and a Hamiltonian circuit in G consists of n edges.

$\therefore$ the number of edge-disjoint Hamiltonian circuit in G will not be greater than $(n-1)/2$.

There are $(n-1)/2$ edge disjoint Hamiltonian circuits, when n is odd. It can be shown as follows:

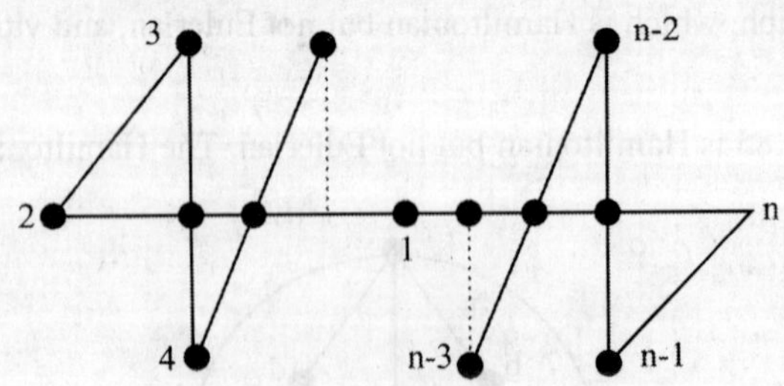

Fig. 10.87

The subgraph of a complete graph of n vertices as shown in the Fig.10.87 is a Hamiltonian circuit. If we keep the vertices fixed on a circle and rotate the polygonal pattern clockwise by

$$\frac{360}{(n-1)}, 2.\frac{360}{(n-1)}, 3.\frac{360}{(n-1)}, \ldots, \frac{(n-3)}{2}.\frac{360}{(n-1)} \text{ degrees.}$$

One can see here that in every rotation we are getting a new Hamiltonian circuit without any edge common to previous ones.

$\Rightarrow$ We have $(n-3)/2$ all edge disjoint new Hamiltonian circuits and also edge disjoint among themselves. Hence the theorem.

10.37-3 Some Basic Rules for Constructing Hamiltonian Paths and Cycles

1. If given graph G has n vertices, then Hamiltonian path and Hamiltonian circuit must contain $n - 1$ and n edges respectively.
2. If a vertex v in graph G has degree m, then a Hamiltonian path must contain at least one edge incident on v and at most two edges incident on v. But Hamiltonian cycle contains exactly two edges incident on v.

In particular, both edges incident on a vertex of degree two will be contained in every Hamiltonian cycle. We can say that there can not be three or more edges incident with one vertex in a Hamiltonian cycle.

1. *While building a Hamiltonian path or cycle no cycle that does not contain all the vertices of G can be formed.*
2. *During the construction of a Hamiltonian circuit if we have passed through a vertex v, then all other unused edges incident on v can be deleted because only two edges incident on v can be included in a Hamiltonian cycle.*

10.37-4 Nonexistence of a Hamiltonian path

In order to prove the nonexistence of a Hamiltonian path or cycle first of all we try to form parts of a Hamiltonian path and show that construction must always fail. Showing that we cannot visit all the vertices without visiting some vertices at least twice can show failure.

Example 10.36

Consider the problem of assigning duties to seven drivers in seven days so that two duties given by the same driver are not scheduled on consecutive days. If no driver serves more than four days. Show that it is always possible to schedule the duties.

Solution

Let G be a graph with seven vertices corresponding to the seven duties. The edge between any two vertices corresponds to two duties performed by different drivers. One can see here that degree of each vertex is at least 3, the sum of the degrees of any two vertices is at least 6. Hence, G always contains a Hamiltonian path, which corresponds to a suitable schedule for the seven duties.

Example 10.37

In the Fig.10.88 which is formed in the order of appearance of English alphabet. Does it form Hamiltonian path and Hamiltonian cycle.

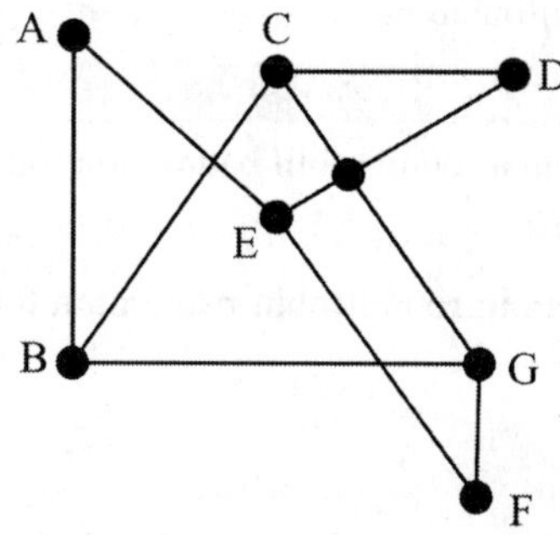

Solution

Going through the graph one can conclude that the path through the vertices of graph shown in the Fig.10.88 in the order of appearance in the English alphabet forms a Hamiltonian path. But graph has no Hamiltonian cycle.

Example 10.38

Show that there is no Hamiltonian cycle and Hamiltonian path for the graph shown in Fig.10.89.

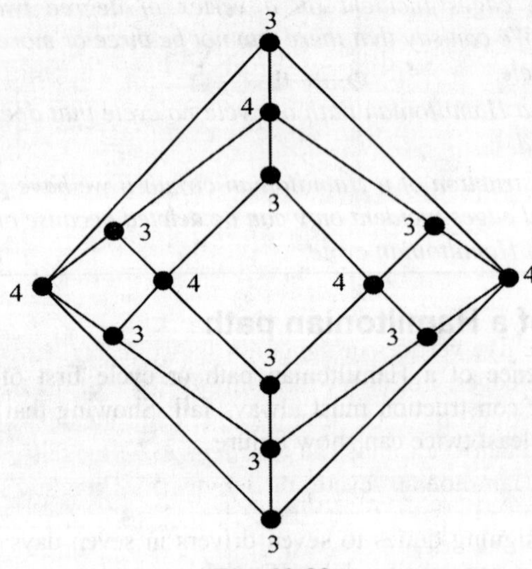

Fig. 10.89

Solution

From the Fig.10.89 it is clear that each vertex of the graph G has degree three or four. We can also observe that every edge connects a vertex of degree three with one of degree four. If we had a Hamiltonian cycle it would visit each vertex of the graph passing alternately through a 3-vertex and a 4-vertex. Hence in this cycle we will establish a one-to-one correspondence between the set of 3-vertices and the set of 4-vertices. But for the existence of Hamiltonian cycle in this graph there would be same number of 3-vertices and 4-vertices. Here in the Fig.10.89 we can see that total number of 3-vertices are eight and that of 4-vertices are six, which is not equal. Therefore, there can be no Hamiltonian cycle for the given graph.

Further, there is no Hamiltonian path, which can be verified with the help of the above reasoning. If there exit such a path, then the difference between the number of 3-vertices and 4-vertices must be one. But here difference is two, hence there is no Hamiltonian path.

Example 10.39

Is it always possible to rank players in a round robin badminton tournament?

Solution

Let us arrange the rank of the players in round-robin badminton tournament in following way s.t. player A will be ranked higher than player B if

1. A beats B or
2. A beats a player who beats B
3. A beats a player who beats another player who beats B and so on

Now as the outcome of the matches can be represented with the help of a directed graph which implies that existence of Hamiltonian path. Hence, it is always possible to rank the players.

Example 10.40

Show that the graph shown in Fig. 10.90 does not contain a Hamiltonian cycle.

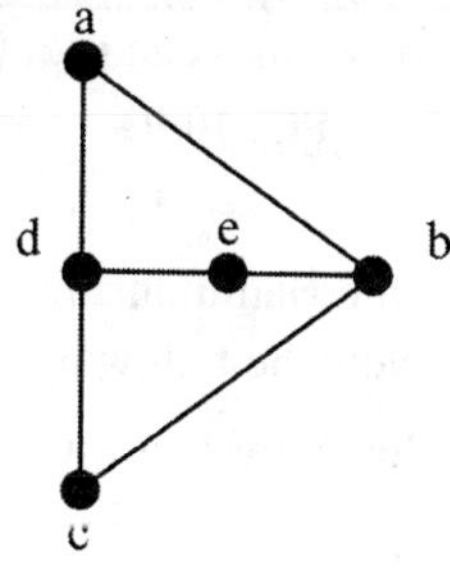

Fig. 10.90

Solution

We can observe that there are five vertices in the graph shown in Fig. 10.90. As it contains five vertices, a Hamiltonian cycle must have five edges. Let us try to eliminate some of the edges of the graph, leaving just a Hamiltonian cycle. For this, we would have to eliminate one edge incident at *b* and one edge incident at *d* because each vertex in a Hamiltonian cycle must have a degree 2. But this removal leaves only four edges, which is not sufficient for a Hamiltonian cycle of length 5. Therefore, this graph does not have a Hamiltonian cycle.

Example 10.41

Prove that there is no Hamiltonian cycle in each of the following graphs:

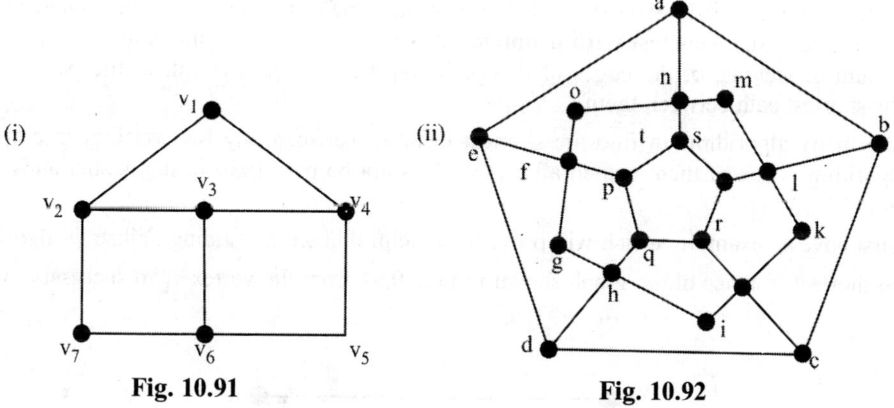

Fig. 10.91 **Fig. 10.92**

Solution

i. If we consider the vertices of degree 2, we see that the edges $\{v_1, v_2\}$, $\{v_1, v_4\}$, $\{v_2, v_7\}$, $\{v_7, v_6\}$, $\{v_5, v_6\}$ and $\{v_4, v_5\}$ must be included in a Hamiltonian cycle. But this forms a proper subcycle of *G*.

ii. Similar to the above proof (i), consider the vertices *o*, *g*, *i*, *k* and *m*.

Example 10.42

Show that any Hamiltonian cycle which contains $(v_3\ v_4)$ in the graph in Fig. 10.93 must also contain $(v_8\ v_3)$.

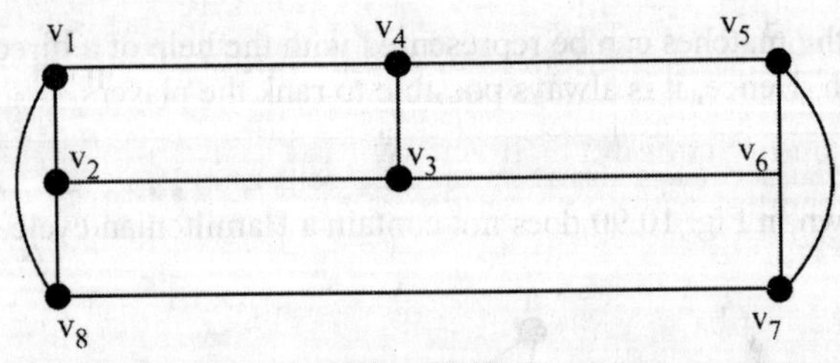

Fig. 10.93

Solution

To verify this let us first include $(v_3\ v_4)$ in the Hamiltonian cycle (H-cycle). Then, at v_4 we have to remove $(v_4\ v_1)$ or $(v_4\ v_5)$; at v_3 remove $(v_3\ v_2)$ or $(v_3\ v_6)$. Now the following cases are possible:

i. If we remove $(v_4\ v_1)$ and $(v_3\ v_2)$ then, the H- cycle must include $(v_1\ v_2)$ and $(v_2\ v_8)$. But here observe that unless $(v_8\ v_7)$ is included, the path $(v_1\ v_2\ v_8)$ gets disconnected with the rest of the graph. But if we include $(v_8\ v_7)$, we get H-path $(v_1\ v_2\ v_8\ v_7\ v_6\ v_3\ v_4\ v_5)$. But this is not a Hamiltonian cycle.

ii. If we remove $(v_4\ v_1)$ and $(v_3\ v_6)$ and include cycles $(v_4\ v_3)$ and $(v_2\ v_3)$, then, at v_2 we can remove $(v_2\ v_8)$, which will give the path $(v_4\ v_3\ v_2\ v_1\ v_8)$ and $(v_4\ v_5\ v_6\ v_7)$. But, unless $(v_7\ v_8)$ is included in H-cycle, these two paths get disconnected. But including $(v_7\ v_8)$, we get the H-cycle $(v_3\ v_4\ v_5\ v_6\ v_7\ v_8\ v_1\ v_2\ v_3)$.

10.38 Shortest Path Problem

Many real life problems can be modeled using graphs with weights assigned to their edges. For example finding shortest path in air distance between New Delhi and New York. What combination of flights has minimum total flight time between these two cities. What is the cheapest fare between these two cities. Similarly finding shortest distance between any two cities connected by highways.

All the above problems (Many move matching with this type of problems) can be modeled using weighted graph and is covered under **'shortest path problem'**. It is to be recalled that the length of a path in a weighted graph is the sum of weights of the edges of this path and the minimum length of the path between the two vertices is the shortest path between them.

There are many algorithms to find the shortest distance between any two vertices in a weighted graph. Dijkstra's algorithm is one of them named after E.W. Dijkstra born in 1930 in the Netherlands who has given this algorithm.

Let us first have an example, which will prove to be helpful in understanding Dijkstra's algorithm.

Find the shortest distance of the graph shown in Fig.10.94 from the vertex v_1 to successive vertices until v_6 is reached.

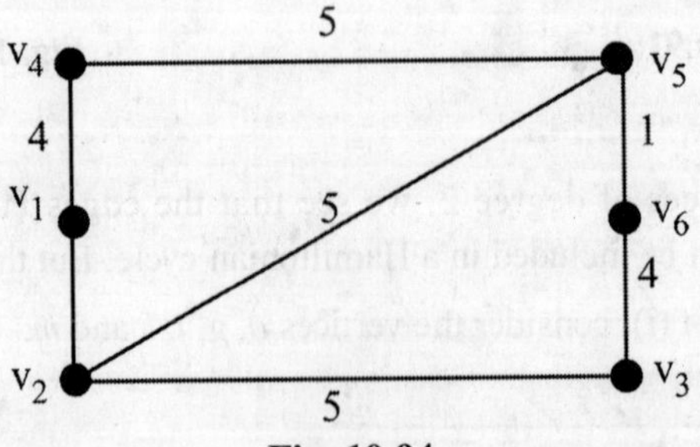

Fig. 10.94

From vertex v_1 we have two paths namely (v_1, v_2) and (v_1, v_4) until the find vertex v_6 is reached. Here the length of edge (v_1, v_2) and (v_1, v_4) is 6 and 4 respectively. Hence vertex v_4 is closest. Again we search for the paths that go through only v_1 and v_4 until the final vertex v_6 is reached. The shortest path to v_2 is still (v_1, v_2) having length 6 whereas the shortest path to v_4 is (v_1, v_4, v_5) having length 9. Therefore next closest vertex to v_1 is v_2.

Now to find the third closet vertex to v_1 we will examine only paths that go through v_1, v_2 and v_4 until the final vertex v_6 is reached. Here we can observe two paths (v_1, v_4, v_5, v_6) of path length 10 to v_6 and path (v_1, v_2, v_3) to v_3 of path length 11. Hence v_6 is the next closet vertex to v_1. Consequently (v_1, v_4, v_5, v_6) is the shortest path form v_1 to v_6 with length 11.

Now we enquire how to obtain a shortest path between v_1 to v_6 of an undirected connected simple weighted graph. Dijkstra's algorithm is based an finding the shortest length from vertex v_1 to first vertex and then length of the shortest path from v_1 to a second vertex and so on. This process in continued till we reach the final vertex v_6 with shortest path from v_1.

Dijkstra's algorithm is based on iterations. Adding one-vertex in each iteration forms distinguished set of vertices. Simultaneously, a labeling is done in each iteration. In the labeling procedure, assign a label to vertex v_i, with the length of the shortest path form v_1 to v_i that consist of only vertices already in the distinguished set. The new vertex added to the distinguish set should be that one which has minimal label among those vertices not already in the set.

Let v_1 and v_n be initial and final vertices between which the shortest path is to be determined.

10.38-1 Dijkstra's Algorithm

Following is the **Dijkstra's algorithm** for finding shortest path from one specified vertex v_1 to another specified vertex v_n.

Step 1: Assign a permanent label zero to the starting vertex v_1 and temporary labels ∞ to the remaining vertices.

Step 2: Each vertex v_j that is not got permanently labeled gets a new temporary label whose value is given by

$$\text{Min} \;\; \text{old label of } v_j, \text{ old label of } v_i \quad d(v_j, v_i)$$

where v_j is the latest vertex permanently labeled in the previous iteration and $d(v_i, v_j)$ is the direct distance between vertex v_i and vertex v_j. If direct distance is not given between vertex v_i and vertex v_j then assign $d(v_i, v_j) = \infty$.

Step 3: Select the smallest value among all the temporary labels so found and this becomes a permanent label of the corresponding vertex. In case of tie, select any one for permanent label.

Step 4: Repeat step (2) and (3) alternately until the destination vertex v_n gets permanent label. The following example illustrates this algorithm.

Example 10.43

Find the shortest path between vertex v_1 and v_7 with the help of Dijkstra's algorithm for the Fig.10.95.

Solution

By applying Step 1 of Dijkstra's algorithm assign permanent label zero to the starting vertex v_1 and ∞ to remaining all the vertices.

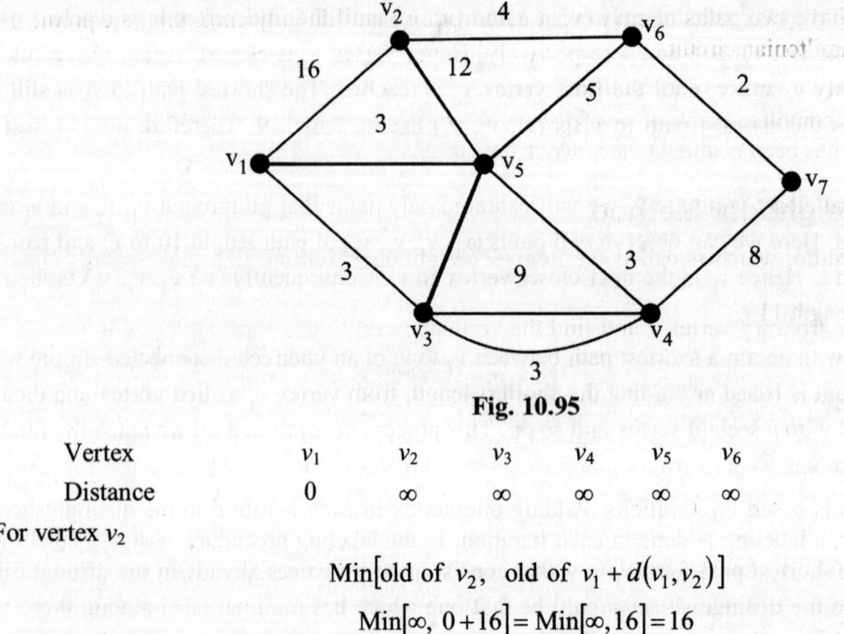

Fig. 10.95

Vertex	v_1	v_2	v_3	v_4	v_5	v_6
Distance	0	∞	∞	∞	∞	∞

For vertex v_2

$$\text{Min}[\text{old of } v_2,\ \text{old of } v_1 + d(v_1, v_2)]$$

$$\text{Min}[\infty,\ 0+16] = \text{Min}[\infty, 16] = 16$$

Applying Dijkstra's algorithm finally we have following table

Vertex	v_1	v_2	v_3	v_4	v_5	v_6	v_7
Distance	[0]	∞	∞	∞	∞	∞	∞
	0	16	∞	∞	[3]	∞	∞
	0	15	[5]	6	3	8	∞
	0	15	5	[6]	3	8	∞
	0	15	5	6	3	[8]	14
	0	15	5	6	3	8	[10]

Thus shortest path between vertex v_1 and v_7 is 10.

10.38-2 Travelling Salesperson Problem

Travelling salesman problem is one of the problems considered as puzzles by the mathematicians. Suppose a salesman wants to visit a certain number of cities allotted to him. He knows the distance (or cost or time) of journey between every pair of cities. His problem is to select such a route that starts from his home city, passes through each city once and only once and return to his home city in the shortest possible distance (or at the least cost or in least time).

This problem can closely be related to Hamiltonian circuits. If we represent the cities by vertices and road connecting two cities by edges we get a graph. Here, we can see with every edge e_i a number (distance or cost or time) is associated say w_i. This type of graph is called weighted graph with $w(e_i)$ being the weight of edge e_i.

Given a weighted graph, the problem of finding a circuit $(e_1, e_2, \ldots e_n)$ that visits every vertex exactly once and minimizes the sum of weights is called **travelling salesperson problem**, which is equivalent to determining Hamiltonian circuit for the graph. If the students are looking for a circuit which must visit each vertex at least once and if the task asked them to visit each vertex once and only once, then the problem would require them to find a Hamiltonian circuit.

In the graph if we have n vertices (cities) then there are $(n-1)!$ possible edges (routes). The total number of Hamiltonian circuits in a complete graph of n-vertices is $(n-1)!/2$ (As out of total $(n-1)!$ number of choices, each Hamiltonian circuit has been counted twice, hence the number is divided by 2).

10.38-3 Nearest Neighbour Method

Here we present a procedure, which is called the **nearest neighbour Method**, that gives results close to the shortest routes.

Step 1: Choose any arbitrary vertex v and find the vertex nearest to this starting vertex to form an initial path of one edge.

Step 2: Add new edges to the path already obtained by selecting new vertices in the following manner:

Let p denote the latest vertex that is most recently added to the path, then eliminate from consideration all edges incident with p that lead to the vertices that have already been chosen (including v). Now among the rest of the vertices, which have not been selected so far, pick the closest one to p and add to the path, the edge-connecting p and the new vertex. Go on doing so until all the vertices of graph G is included in the path.

Step 3: Join initial and the final vertex by an edge and form a circuit.

This algorithm produces a directed cycle in the complete graph but not necessarily the shortest one and different solution may be obtained with respect to different initial vertex.

Example 10.44

In the Fig.10.96 (a) starting with vertex v_1 construct a Hamiltonian circuit according to the nearest neighbour method.

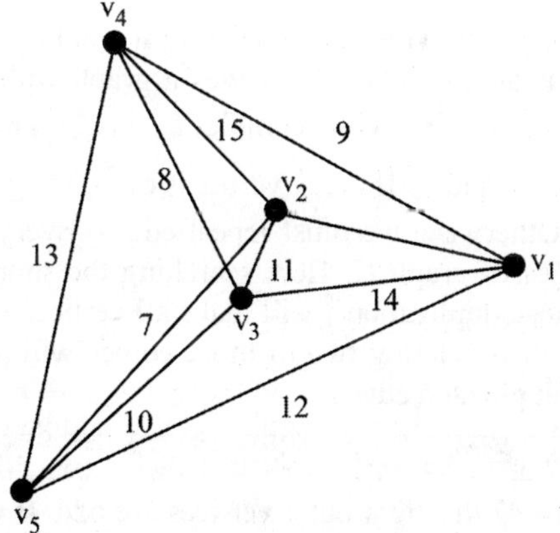

Fig. 10.96(a)

Solution

It has been asked to start with vertex v_1. Vertex nearest to v_1 is v_4 (because v_4 has the least weight of the four edges incident with v_1), hence select the edge $(v_1\ v_4)$. The next edge must have v_4 as an end vertex, and $(v_4\ v_1)$ is not allowed (one cannot return to v_1 , it has already been used), so smaller weight of the remaining edges is chosen namely $(v_4\ v_3)$. The weight with smallest edge originating at v_3 is $(v_3\ v_5)$. The next edge with smallest weight originating at v_5 is (v_5 v_2). The next edge with smallest weight originating at v_2 is (v_2 v_3). So algorithm finds route (v_1 v_4 v_3 v_5 v_2 v_1) as shown in Fig.10.96 (b). It is to be noted here that a different route may be obtained if one starts with different vertex.

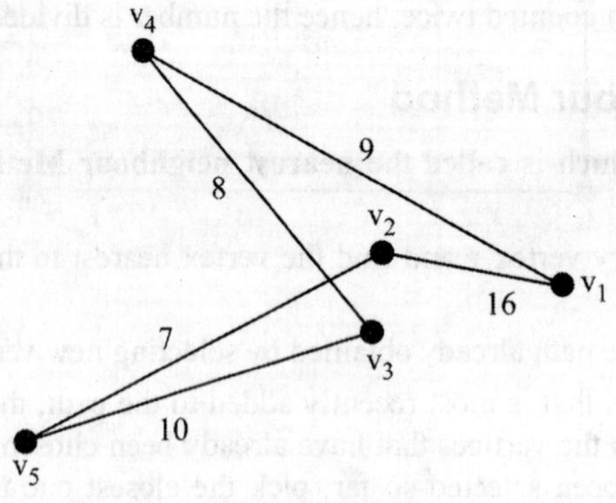

Fig. 10.96(b)

10.38-4 The Chinese Postman Problem

This problem is proposed by the Chinese mathematician **Guan Meigee** [1962]. Suppose a mail carrier traverse all edges in a road network starting and ending at the same vertex. If the edges have been assigned weights representing distance (or time). Our aim is to find a closed walk of minimum total length that uses all the edges. This is called the **chinese postman problem** in honour of **Guan Meigee**.

Proof

To prove this theorem let us use the principle of mathematical induction on size (number of edges) $e(G)$. The result is true for $e(G) = 1$. Consider the case when $e(G)>1$ and suppose that the claim holds for graphs with $e(G) - 1$ edges.

If graph G is even, then G has no cut-edge, since deleting it would leave subgraphs with one odd vertex*. Hence starting from some vertex v_i along an edge v_j leaves a graph with two odd vertices. By the induction principle the algorithm completes the circuit of G by completing a (v_i, v_j) trail in $G - v_iv_j$.

Suppose G has two odd vertices v_i and v_j. If every vertex is even, the graph G will be Eulerian and then sum the weights which is the answer. Otherwise, we must repeat edges. Every traversal is an Eulerian circuit of a graph obtained by duplicating edges of graph G. Here searching the shortest traversal is same to finding the minimum total weight of edges whose duplication** will make all vertices even. If we use an edge three or more times in making all vertices even, then deleting two of those copies will have all vertices even. There may be more than one ways to choose the duplicated edges.

Example 10.45

In the graph given in the Fig.10.97 the eight outer vertices are odd. If we match them around the outside to make the degrees even, the extra cost is $6 + 6 + 6 + 6 = 24$ or $1 + 11 + 11 + 1 = 24$.

---

* A graph with every vertex degree even has no cut edge.

** We say duplication because we need not use an edge more than twice.

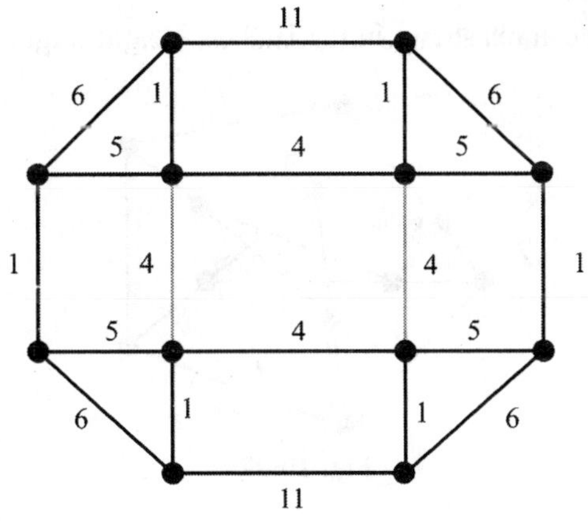

Fig. 10.97

We can do better by using all the vertical edges, which total only 18. Here it is to be observed that adding an edge from an odd vertex to an even makes the even vertex odd. We must continue adding edges until we complete a trail to an odd vertex. The duplicate edges must consist of collection of trails that pair up the odd vertices. We may restrict our attention to paths pairing up the odd vertices, though the path may need to intersect.

Exercise 10.3

1. Consider the graph shown in Fig.10.102. Find
   i. All cycles, if any
   ii. All cut points if any
   iii. All bridges, if any
   iv. The degree of each vertex
   v. All paths from *a* to *g*
   vi. All trails (distinct edges) from *b* to *c*
   vii. The distance from *a* to *c*, $d(a, c)$
   viii. diam (G), the diameter of G.
2. a. Consider a connected graph G. Prove that if G contains a cycle C, which contains an edge e; then G-e is still connected.

   b. Let G be a connected graph. Prove that if $e = \{u, v\}$ is an edge such that $G - e$ is disconnected, than u and v belong to different components of $G - e$.
3. List all the paths between vertices v_5 and v_6 in Fig.10.98. Also give the length of each of these paths.

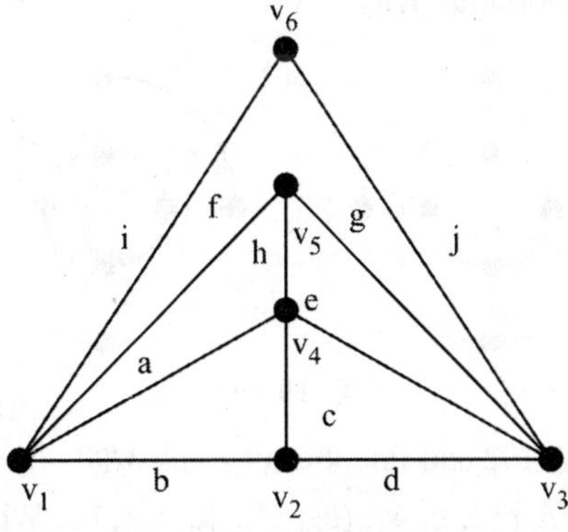

Fig. 10.98

4. i. What is the diameter of the graph shown in Fig.10.99.

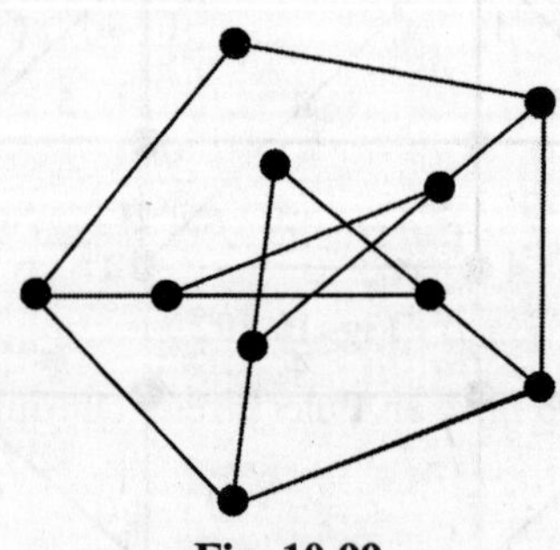

Fig. 10.99

ii. If the vertices of the graph represent computers and edges represent data communication links between computers. What is the physical significance of the graph in this case?

5. Prove that a simple graph with n vertices must be connected if it has more than $[(n-1)(n-2)]/2$ edges.
6. Prove that any two simple connected graphs with n vertices, all of degree two are isomorphic.
7. If the intersection of two paths is a disconnected graph. Show that the union of the two paths has at least one circuit.
8. In the graph shown in the Fig. 10.100 edges can be partitioned into two edge-disjoint paths.

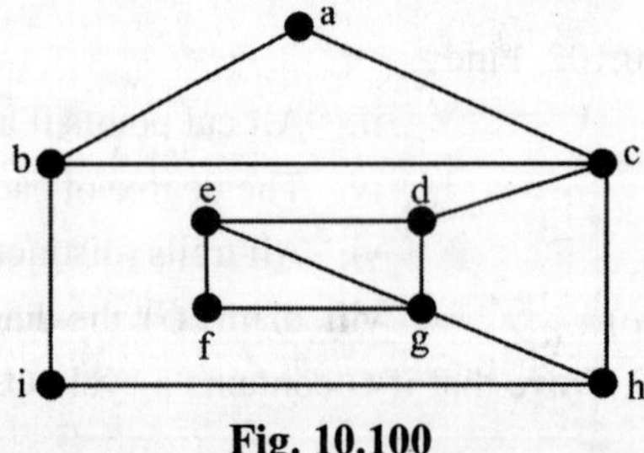

Fig. 10.100

a. Show one such partition.

b. Consider a connected graph G of odd number of vertices say m (m>0). Show that the edges in G can be partitioned into $m/2$ (edge-disjoint) paths.

9. Are two graphs in Fig.10.101 isomorphic? Why?

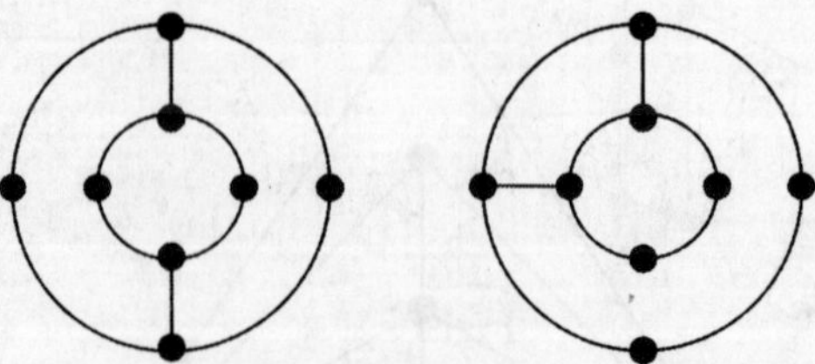

Fig. 10.101

10. Consider the graph given in Fig.10.102 and find the subgraph $M(V', E')$ generated by:

i. $V' = \{b, c, d, e, f\}$; (ii) $V' = \{a, c, e, g, h\}$; (iii) $V' = \{c, f, g, h\}$. Which of them are homeomorphic and which one isomorphic?

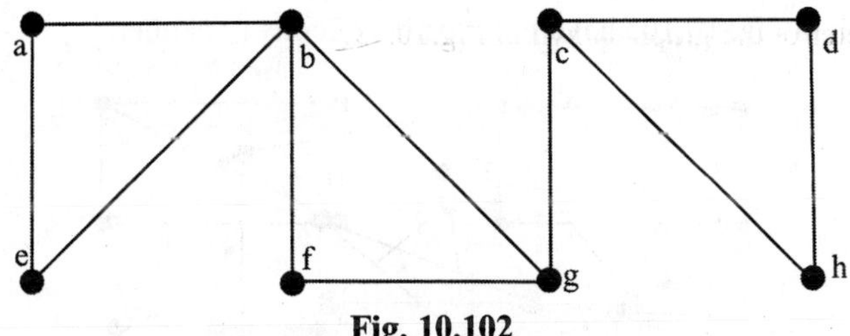

Fig. 10.102

11. Which of the graphs in the Fig.10.103 have an Euler circuit, an Euler path but not an Euler circuit or none of these?

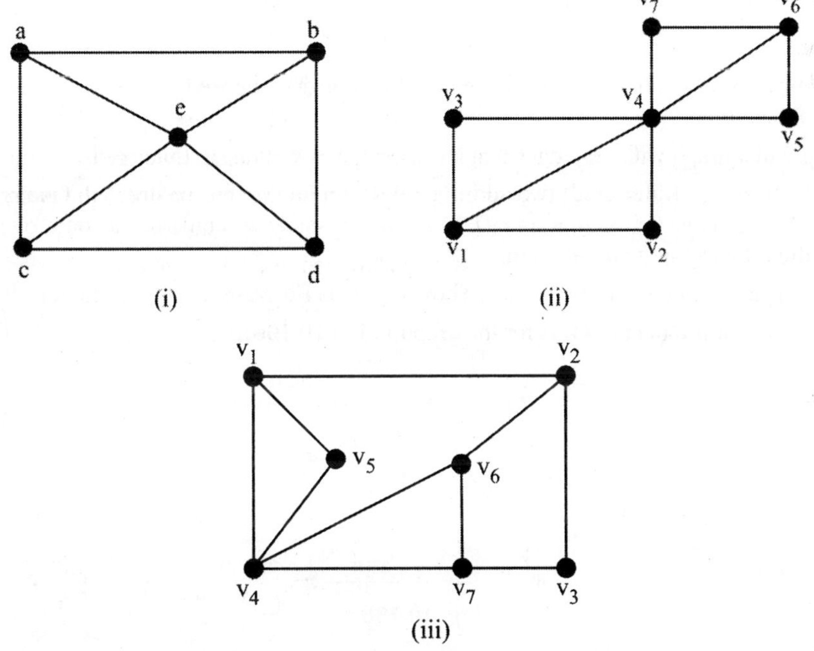

Fig. 10.103

12. Find an Euler circuit in Fig.10.104.

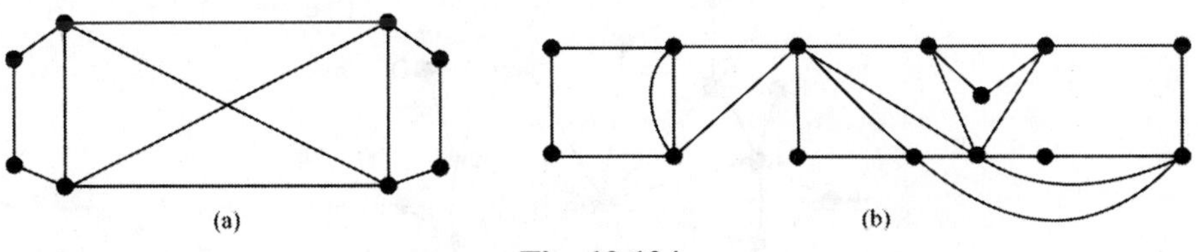

Fig. 10.104

13. Which of the multigraph in Fig. 10.105 have Euler paths, circuits or neither.

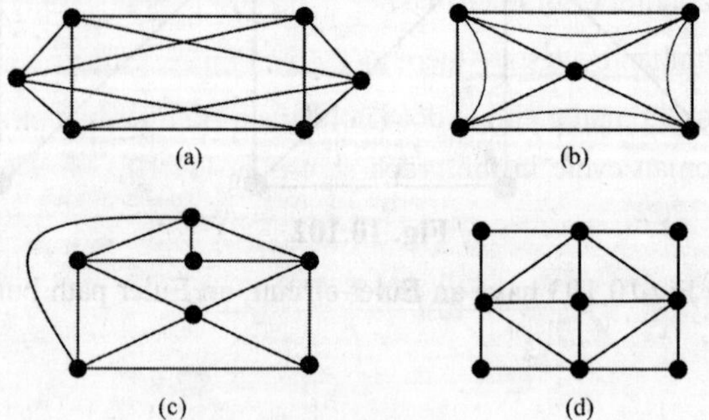

Fig. 10.105

14. Prove that a connected graph has an Euler circuit iff it can be decomposed to a set of elementary cycles that have no edges in common.
15. Give an example of a graph with ten edges that has a bridge as well as an Euler path.
16. At present in Konigsberg (Kalinigrad) two additional bridges have been constructed. One bridge is between regions B and C and other between B and D. Is it now possible to construct a route over all bridges of konigsberg without re-crossing any of them.
17. Suppose G and G' are Homeomorphic graphs. Show that G is Eulerian iff G' is Eulerian.
18. Find the number of Hamiltonian circuits for the graph in Fig.10.106.

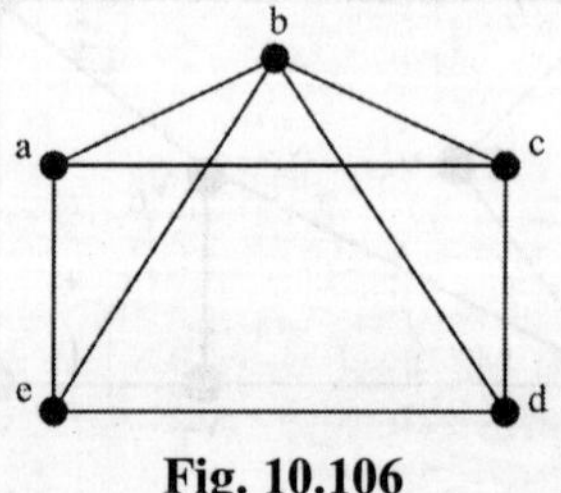

Fig. 10.106

19. Prove that there is no Hamiltonian cycle in the graph shown in Fig. 10.107.

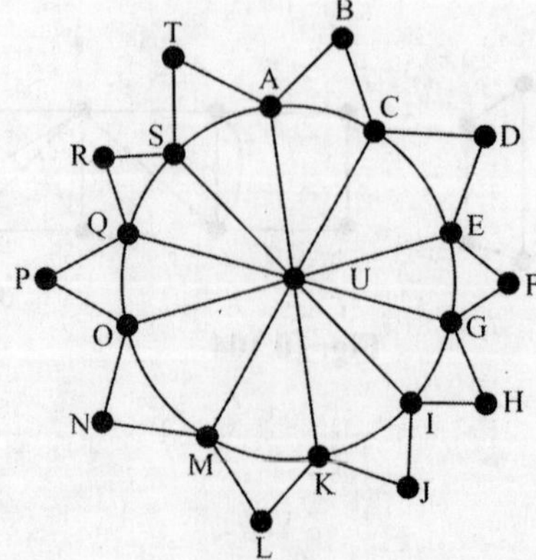

Fig. 10.107

20. a. Test whether there exist a Hamiltonian circuit and Hamiltonian path in $K_{4,4}$ and $K_{4,6}$?

b. What is the sufficient condition for the existence of a Hamiltonian circuit and Hamiltonian path in $K_{m,n}$.

21. How many different Hamiltonian cycles are there in K_n, a complete graph on n vertices?

22. Draw a graph that has a Hamiltonian path but does not have a Hamiltonian circuit.

23. Show that the any Hamiltonian cycle in the graph shown in Fig.10.108. Which contains the edge $\{c, d\}$ must also contain the edge $\{g, h\}$.

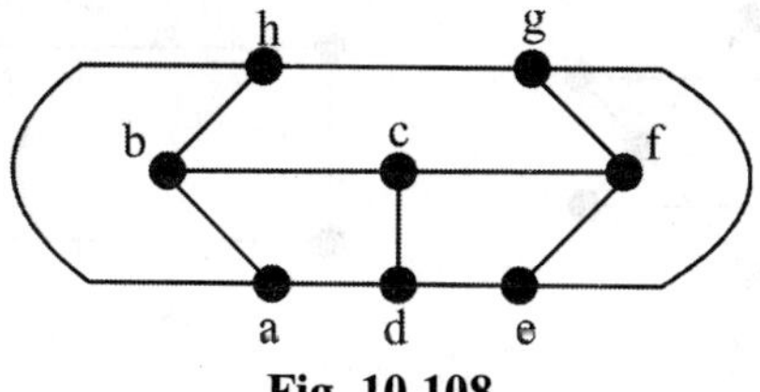

Fig. 10.108

24. Show that any Hamiltonian cycle in the graph shown in the Fig.10.109 must contain exactly two edges $\{v_1, v_8\}$, $\{v_3, v_4\}$, $\{v_9, v_{10}\}$. Further show that any Hamiltonian cycle that contains both edges $\{v_4, v_5\}$, $\{v_5, v_{10}\}$ cannot also contain $\{v_1, v_8\}$.

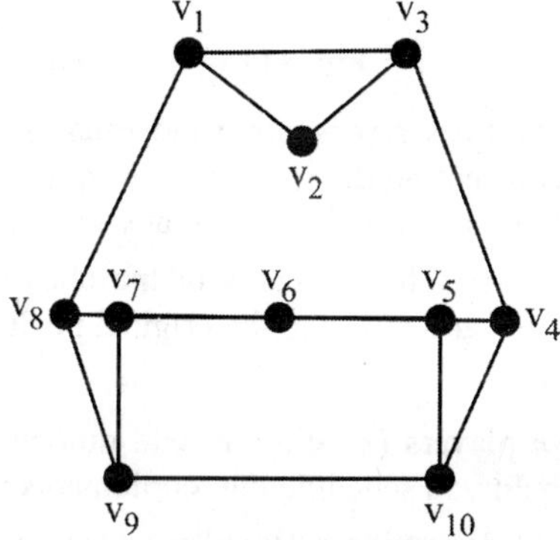

Fig. 10.109

25. Find a Hamiltonian cycle in each of the following graph.

i. The dodecahedron

ii. The octahedronz

iii.

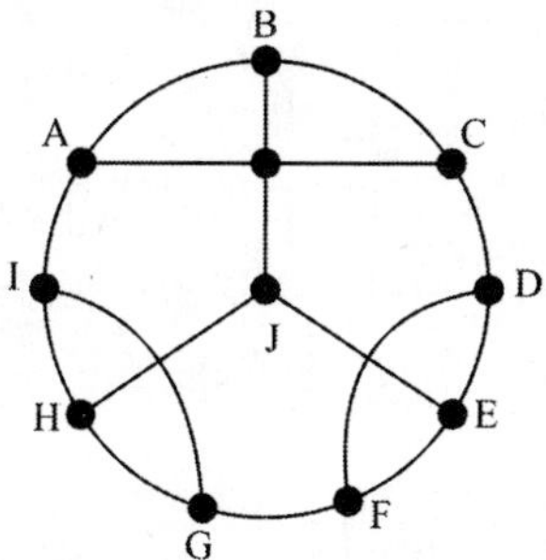

iv.

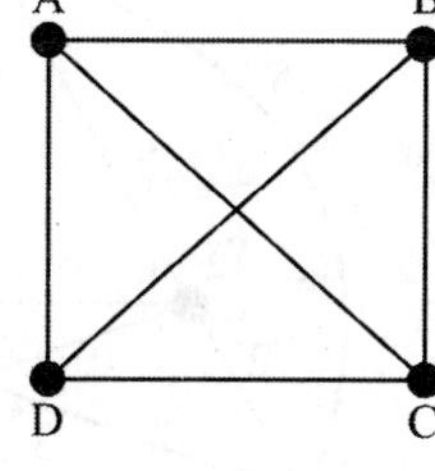

Fig. 10.110

26. Show that thee can be no path longer than Hamiltonian path if it exist in a graph.
27. a. Draw a graph that has both an Eulerian circuit and a Hamiltonian circuit.
    b. Draw a graph that has neither an Eulerian circuit nor a Hamiltonian circuit.
    c. Draw a graph that has an Eulerian circuit but has no Hamiltonian circuit.
    d. Draw a graph that has no Eulerian circuit but has a Hamiltonian circuit.
28. Which of the following graphs are (a) Eulerian (b) Hamiltonian. Determine the Eulerian path (or circuit) .

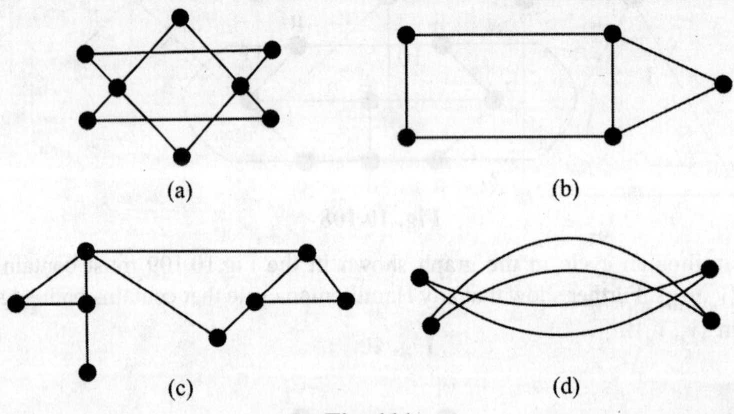

Fig. 111

29. a. Draw a graph with sixty-four vertices representing the squares of a chessboard. Join these vertices appropriately by edges, each representing the move of the knight. One can see that in this graph every vertex is of degree two, three, four, six or eight. How many vertices are of each type?
    b. Is it possible, starting from any sixty-four squares of the chessboard to move a knight such that it occupies every square exactly once and return to the original position? If so give one such tour?

    [**Hint:** Look for a Hamiltonian circuit]
30. A round-robin tournament among *n* players (*n* being an odd number) can be represented by a complete graph of *n* vertices. Discuss how would you schedule the tournaments to finish in the shortest possible time.
31. Use the nearest-neighbour method to determine a Hamiltonian circuit for the graph shown in Fig.10.112 (i) Starting at vertex v_1 (ii) Starting at vertex v_4 (iii) Determine the minimum Hamiltonian circuit.

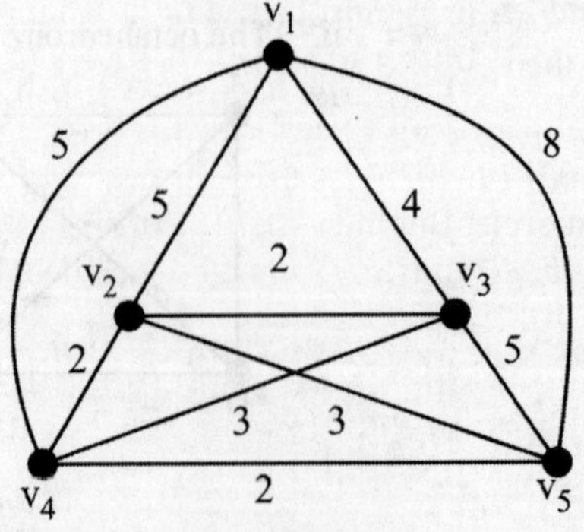

Fig. 10.112

32. Find the shortest path between the vertices A and P of the graphs shown in Figs.10.113 (a) and (b), where the numbers associated with the edges are the distance between the vertices.

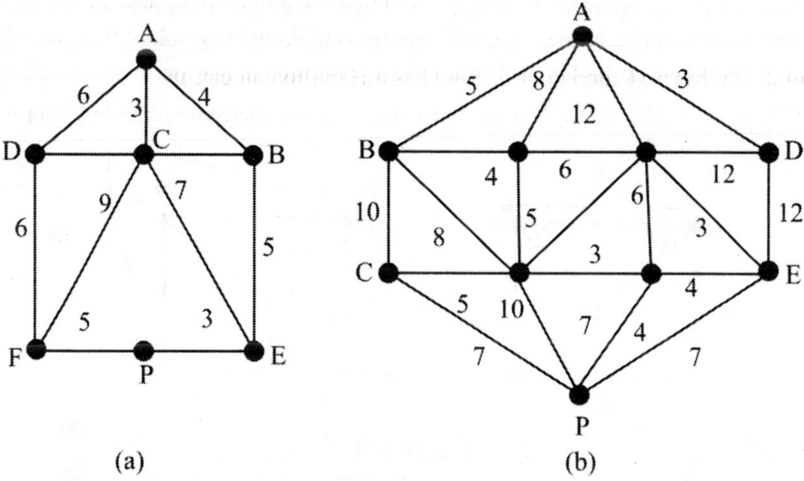

Fig. 10.113

Answers

1. i. *abea, bfgb, cdhc* ii. *b, c, g*
   iii. only $\{c, g\}$ iv. 2, 4, 3, 2, 2, 2, 3, 2
   v. *abg, abfg, aebg, aebfg* vi. *bge, bfgc, baebgc, baebfgc*
   vii. 3 viii. 4

10. i. $E' = \{be, bf, cd\}$, ii. $E' = \{ae, fg, gc\}$,
    iii. $E' = \{fg, gc, ch\}$. Further (i) and (ii) are isomorphic and (i) and (ii) are homeomorphic.

20. **Hint:** Adding a vertex by dividing an edge doesn't change the degree of the original vertices and simply adds a vertex of even degree.

24. In case two of $\{v_1, v_8\}$, $\{v_3, v_4\}$, and $\{v_9, v_{10}\}$, are not used (let us assume $\{v_3, v_4\}$, and $\{i, j\}$ are not used), then $\{v_4, v_5\}$, $\{v_5, v_{10}\}$ and $\{v_4, v_{10}\}$ must be used, giving a triangle subcycle, impossible. This implies that at least two of $\{v_1, v_8\}$, $\{v_3, v_4\}$ and $\{v_9, v_{10}\}$ must be used. But all three should not be used because in that case (i) Suppose $\{v_4, v_{10}\}$ is used, then $\{v_4, v_5\}$ and are $\{v_5, v_{10}\}$ not used. It means two of the three edges at e are not used; impossible.

28. Let us assume that $\{g, h\}$ is not used but $\{e, d\}$, is used. In this case edges $\{a, h\}$, $\{b, h\}$, $\{g, f\}$ and $\{e, g\}$ must be included in a Hamiltonian cycle. But then $\{a, b\}$ and $\{e, b\}$ have to be deleted otherwise a triangle subcycle will be formed. Hence, $\{b, c\}$ and $\{e, f\}$ must be included. But then there are three edges of the cycle incident on C. contradict.

10.39 Planar Graph

A graph G is said to be **planar** if it can be drawn on a plane in such a way that no two of its edges intersect (do not cross); otherwise G is said to be **nonplanar**. It is to be noted that if a graph G has been drawn with crossing edges, this doesn't imply that G is nonplanar, as there are different ways to draw the graph without crossovers*.

For example the graph shown in the Fig.10.114 (a) seems to be nonplanar, but it can be redrawn in Fig.10.114 (b) without crossovers.

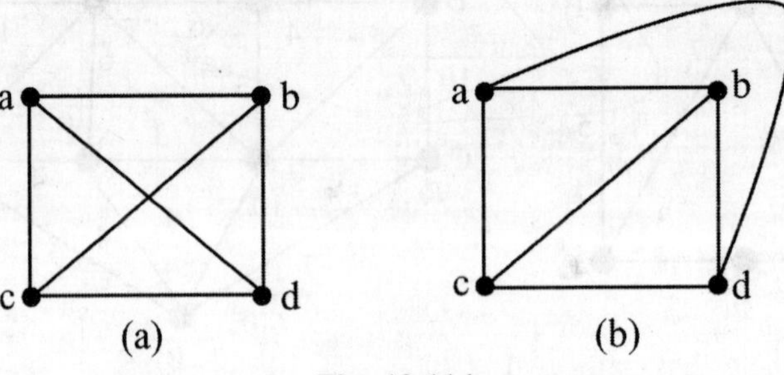

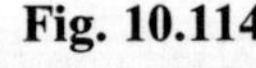

Fig. 10.114

The meeting of edges at a vertex is not considered as an intersection.

10.39-1 Embedding

Embedding is a drawing of a geometric representation of graph on any surface such that no edge intersect. An embedding of a planar graph G on a plane is called a plane representation of G. We say that a planar graph is a **plane graph** if it is drawn in a plane so that no two edges crossover.

Example 10.46

Draw a planar representation of both the graphs shown in Fig. 10.115, if possible.

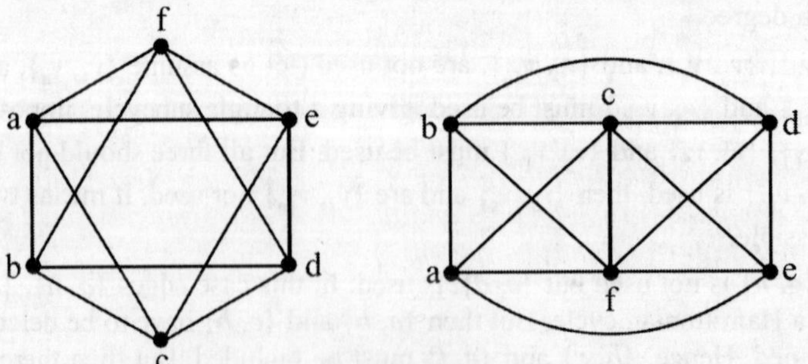

Fig. 10.115

* The point of intersection of edges are called **crossovers** and the crossing edges are called **cross-over** each other.

Solution (i)

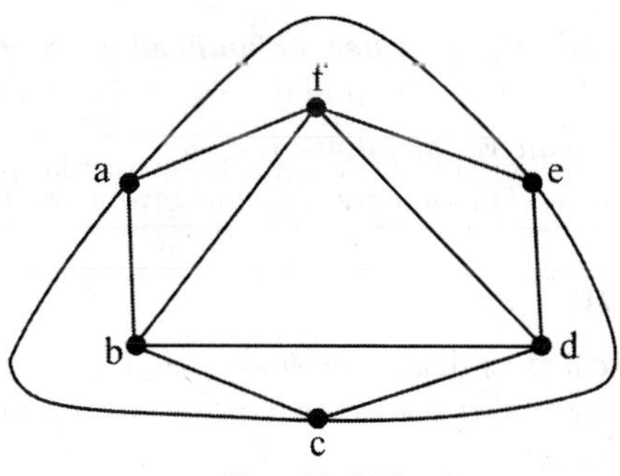

Fig. 10.116

10.39-2 Maps and Regions

A particular planar representation of a finite planar multigraph is called a **map**. If the underlying Multigraph is connected then we say that the map is **connected**.

A plane representation of a graph divides the plane into **regions** or in other words we can say a given map divides the plane into various **regions** (also called **windows**, **faces**, or **meshes**). For example, the map in Fig. 10.117 with six vertices and eleven edges divides the plane into six regions.

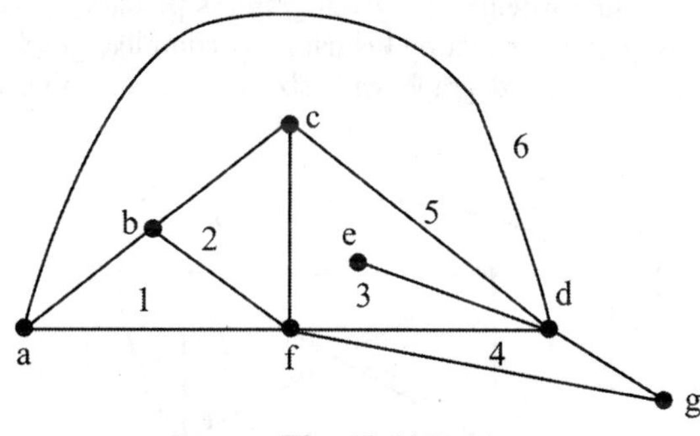

Fig. 10.117

Here, it is to be noted that five regions (1, 2, 3, 4, 5) are bounded while region 6 is not bounded.

A region is known by the set of edges or the set of vertices forming its boundary. Region is not defined in a nonplanar graph and a planar graph which is not embedded in a plane. For example graph shown in Fig. 10.118 has no region.

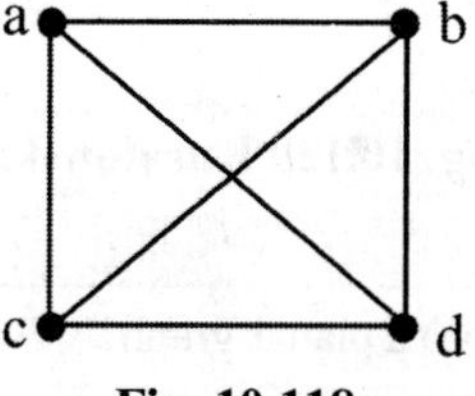

Fig. 10.118

i. *Border of each region of a map consists of edges. Edges not always form a cycle.*

ii.	*Region is a property of the specific plane representation of a graph.*
iii.	*The portion of the plane lying outside a graph embedded in a plane is infinite in extent. Such a region is said to be* ***infinite, outer, unbounded*** *or* ***exterior*** *region for that particular plane representation. The region 6 shown in the Fig.10.117 is unbounded.*
iv.	*A planar graph has exactly one infinite region.*
v.	*By changing the embedding of a given planar graph, the infinite region can be changed.*

10.39-3 Degree of a Region

The **degree** of any region r, represented as deg(r), mean the length of the cycle or closed walk which borders region r. We can observe that each edge either borders two regions or is contained in a region and will occur twice in any walk along the border of the region.

In the graph shown in the Fig.10.117 boundary of all the regions are cycles except for region 3. But, in case we move anticlockwise in region 3, starting at the vertex say c, then the closed path obtained is (c, d, e, d, f, c), where the edge $\{e, d\}$ occurs twice. The degrees of region 1, 2, 3, 4, 5, and 6 are respectively 3, 3, 5, 3, 4, 4.

10.39-4 Determining when Graphs are Planar

If none of geometric representation of G out of all possible geometric representations can be embedded in a plane then the graph G is nonplanar. But geometric graph G is planar if one can draw isomorphic graph of G that is embedded in the plane.

But the question how to determine whether the given graph is planar/nonplanar is answered by Kuratowski (Polish Mathematician) in the two graphs known by his name. A complete graph with five vertices K_5 is the first graph of Kuratowski, and regular connected graph with six vertices and nine edges $K_{3,3}$ is second graph of Kuratowski.

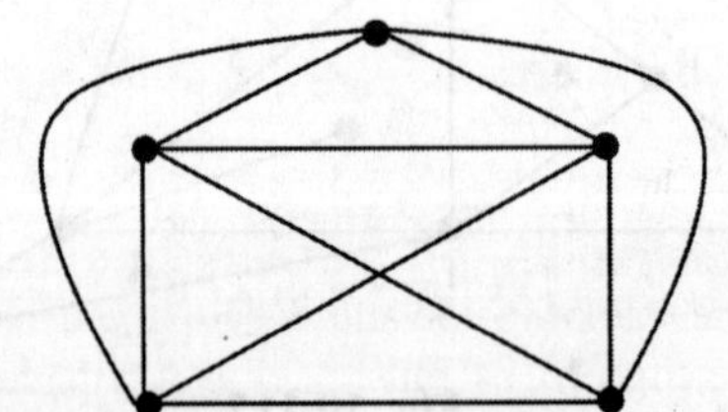

Fig. 10.119 Kuratowski First Graph K_5

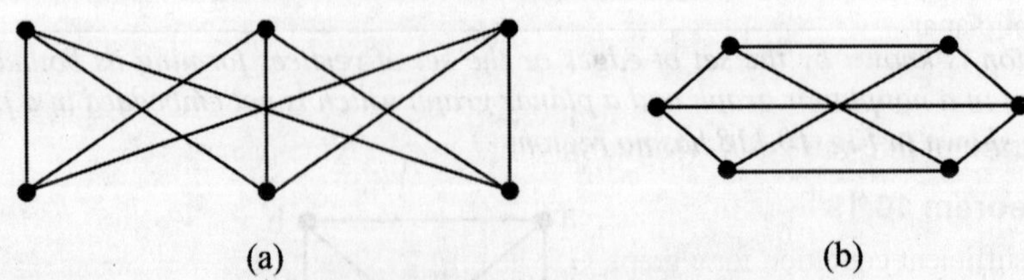

Fig. 10.120 Kuratowski Second Graph

Example 10.47

Is $K_{3,3}$ (Kuratowski's second graph) a planar graph?

Solution

Let us first try to find a systematic way to draw a graph in the plane without intersection of edges. The method we adopt involves following two ideas:

1. In geometric representation of a cycle in the plane, any edge not on the cycle must lie either inside the cycle or outside the cycle or the edge must intersect one of the edges of the cycle.

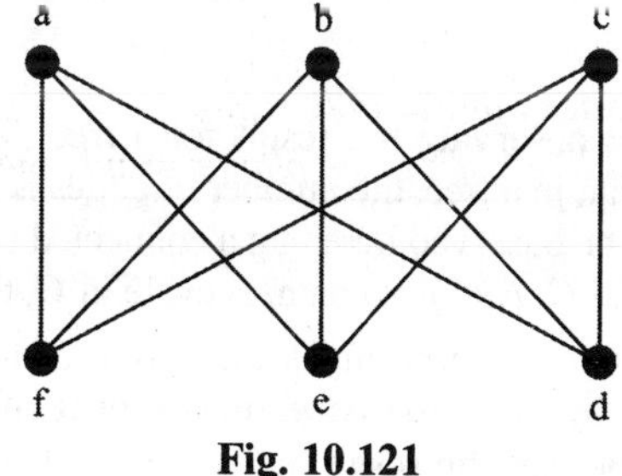

Fig. 10.121

2. The graph can be redrawn so that edges and vertices formerly outside the cycle are now inside the cycle and vice-versa i.e. the roles of being inside or outside the cycle are symmetric (equivalent). Consider the graph $K_{3,3}$ shown in Fig. 10.122.

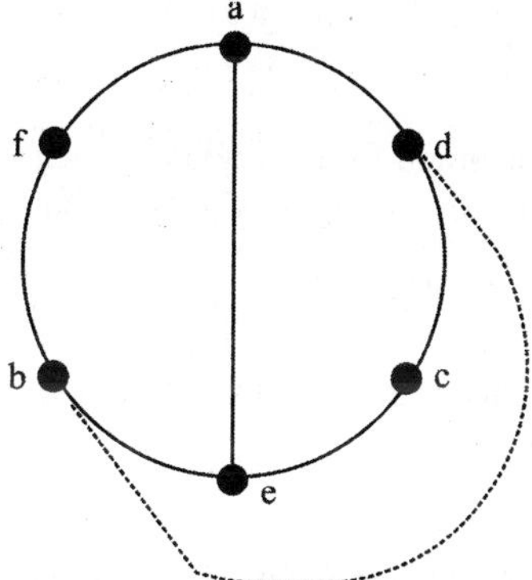

Fig. 10.122

Draw the cycle *a-d-c-e-b-f-a*. As the role of inside and outside the cycle is symmetric (equivalent). Let us take the edge $\{a, e\}$ inside the cycle then the edge $\{b, d\}$ must be drawn outside the cycle otherwise they will intersect. But we can observe that there is no way to draw the edge $\{c, f\}$. either inside or outside the cycle without crossing over either $\{b, d\}$ or $\{a, e\}$.

Hence $K_{3,3}$ is not planar.

Example 10.48

The complete graph of five vertices K_5 (Kuratowski first graph) is nonplanar.

Kuratowski's Theorem 10.19

Necessary and sufficient condition for a graph G to be planar is that G does not contain either of Kuratowski's two graphs or any graph homeomorphic to either of them i.e. K_5 and $K_{3,3}$.

10.40 Euler's Formula

A planar graph may be represented differently but the question is whether the number of regions obtained from each embedding is same or not? The answer is same, which can be obtained with the help of Euler's formula taken below.

Theorem 10.20 (Euler's Formula, discovered by Euler in 1752)

For any connected planar graph G, $v - e + r = 2$, where v, e, and r are the number of vertices, edges, and regions of the graph respectively.

Proof

Let us prove this formula by first observing the result for a tree. Since a tree determines only one region (by convention). Further, it is known that in a tree, the number of edges is one less than the number of vertices, i.e. $v - e + 1 = 2$ or $v - e = 1$ holds. It is to be observed here that a connected plane graph with only one region must be a tree otherwise it will contain a cycle in G. And if it contains cycle in G, there must exist an internal region and an external region.

The general result can be proved by induction on m, the number of regions determined by G. If $r = 1$, the result is already proved. Let us assume that the result holds for $r \geq 1$ and show that the result holds for $r + 1$ regions. Since G is connected and suppose it has $r + 1$ regions. Deleting an edge common to the boundary of two separate regions we obtain a plane graph G' with v vertices, e' edges and r' region having same number of vertices but reduces the number of edges and vertices by one. Applying the induction hypothesis, we find $v - e + r = v - e' + r'$

$$\Rightarrow v - e + r = 2.$$

REMARK

Above theorem actually holds for the graph for the regions of degree one; but through out this section we will assume that the graph is simple and number of edges is greater than one. This implies that the degree of each region is greater than or equal to three.

Observation

In a plane graph G, if the degree of each region $\geq k$ for $k \geq 1$, then $r \leq 2e$, here r is the number of regions and e is the number of edges.

Corollary

In a simple, connected planar graph G, with r regions, v vertices ($v \geq 3$) and e edges ($e > 1$), then

i. $e \geq (3/2)\, r$ ii. $e \leq 3v - 6$

iii. There is a vertex v of G such that $\deg(v) \leq 5$.

Proof

By Euler's Formula

$$v - e + r = 2.$$

As the sum of the degrees of the region equals $2e$. But each region has degree ≥ 3, hence

$$2e \geq 3r$$

$$\Rightarrow \quad e \geq \frac{3}{2} r$$

Substituting this in Euler's formula, we get

$$e \geq \frac{3}{2}(e - v + 2)$$

As for (iii) let us assume that each vertex has degree ≥ 6, then since degree of vertices of G is twice the number of edges, i.e. $\deg(v) = 2e$,

$$6v \leq 2e$$

and as we know from (i) or $$v \le \frac{1}{3}e$$

$$r \le \frac{2}{3}e$$

But, from Euler's formula $r + v = e + 2$, hence we have

$$\frac{2}{3}e + \frac{1}{3}e \ge r + v = 2 + e$$

or $$e \ge e + 2$$

or $$0 \ge 2$$

which is a contradiction. Hence degree (v) is less than 5.

Theorem 10.21

A complete graph K_n is planar if $n \le 4$.

Proof

It can be seen that K_n is planar for $n = 1, 2, 3, 4$ (By corollary (ii) of Euler's formula). Thus, we have only to show that K_n is nonplanar if $n > 5$. As K_5 is a complete graph of $v = 5$ vertices and has $e = 10$ edges. If the graph is planar then by $e \le 3v - 6$, we have

$$10 = e \le 3v - 6 = 3\times5 - 6 = 9$$

which is impossible. Thus K_5 (Kuratowski's second graph) is nonplanar.

Theorem 10.22

A complete bipartite graph $K_{m,n}$ is planar iff $m \le 2$ or $n \le 2$.

Proof

It can be seen easily that $K_{m,n}$ is planar if $m \le 2$ or $n \le 2$(in all these cases Euler's formula is satisfied). To prove for $m \ge 3$ and $n \ge 3$, it is sufficient to prove that $K_{3,3}$ is nonplanar. As we know that bipartite graph $K_{3,3}$ has six vertices and nine edges. If $K_{3,3}$ were planar by Euler's formula $r = e - v + 2 = 9 - 6 + 2 = 5$. But $K_{3,3}$ is a bipartite graph, it cannot have a cycle of odd length. This implies that each cycle has length ≥ 4. Therefore, degree of each region would have to be greater than or equal to 4. But for this we would have $4r \le 2e$ or $20 = 4\times5 \le 2\times9 = 18$, which is a contradiction. Hence $K_{m,n}$ is planar iff $m \le 2$ or $n \le 2$.

Exercise 10.4

1. Which of the following graphs shown in the Fig. 10.123 are planar? If planar draw.
2. Show that $K_5 - e$ and $K_{3,3} - e$ are planar for any edge e of K_5 and $K_{3,3}$ respectively.
3. Show that any graph with 4 or fewer vertices is planar.
4. Show that there are only five possible simple completely regular planar graphs, excluding the trivial graphs with degree ≤ 2. Sketch them.

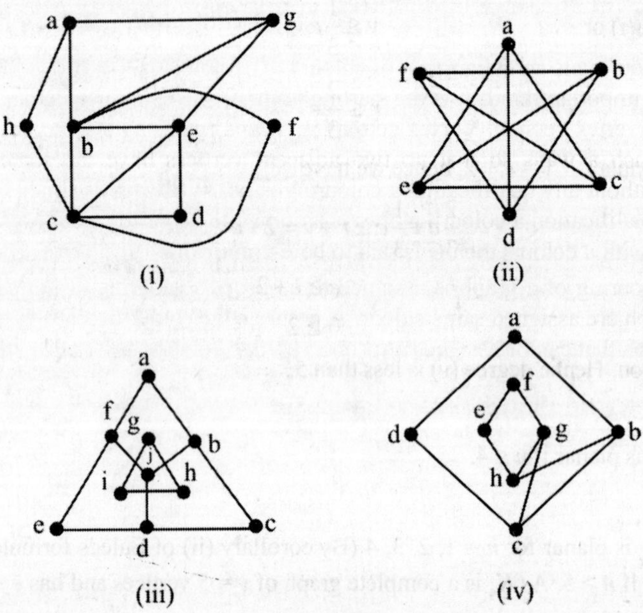

Fig. 10.123

5. Show that each of the graph shown in Fig. 10.124 is not planar by finding a subgraph homeomorphic to either K_5 or $K_{3,3}$.

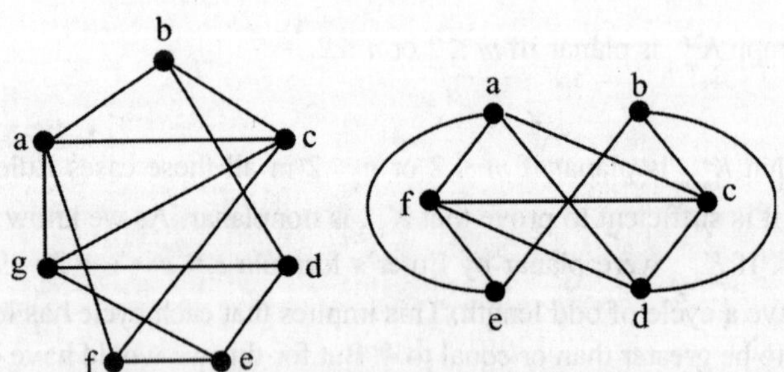

Fig. 10.124

6. Check whether the following graphs are planar. If it is, redraw it so that no edges cross, otherwise find a subgraph homeomorphic to either K_5 or $K_{3,3}$.

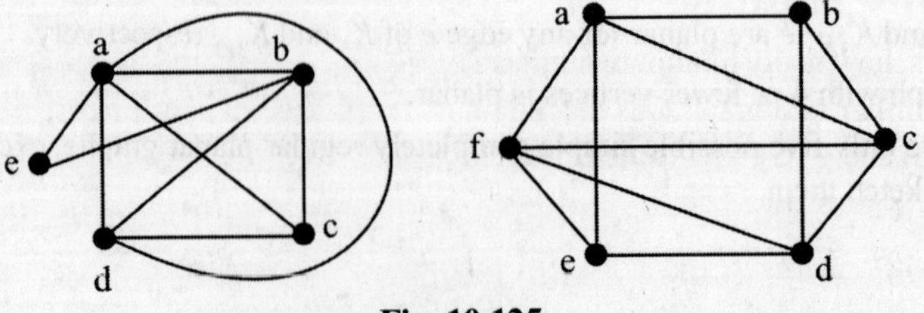

Fig. 10.125

10.41 Colouring

In graph theory graph colouring is an assignment of colours to certain object (vertex, edge, region etc.) in a graph. Vertex colouring is most important kind. It is the starting point of the entire subject and other colouring problem can be transformed into a vertex version. **Vertex colouring** means colouring the vertices of a graph. Colouring the vertices of a graph in such a ways so that no two adjacent vertices have the same colour is called **proper colouring**. When used without any qualification a colouring is nearly always assumed to be proper coluring. And when used without any qualification a colouring always assumed to be vertex colouring. If we consider a graph G with m vertices coloured with n colours then G is said to be ***n*-colourable**.

Similar to vertex colouring of a graph is assumed to be proper colouring of the edges which means no two adjacent edges of the graph are assigned same colour. A proper edge colouring with m-colours is called a (proper) ***m*-edge colouring**. A graph that can be assigned a (proper) k-edge colouring is *k*-edge colourable.

Example 10.49

a. Consider a complete graph G as shown in Fig. 10.126(a) with six vertices. Since every vertex is adjacent to another vertex, therefore, G requires six different colours for colouring. Hence G is 6-colourable.

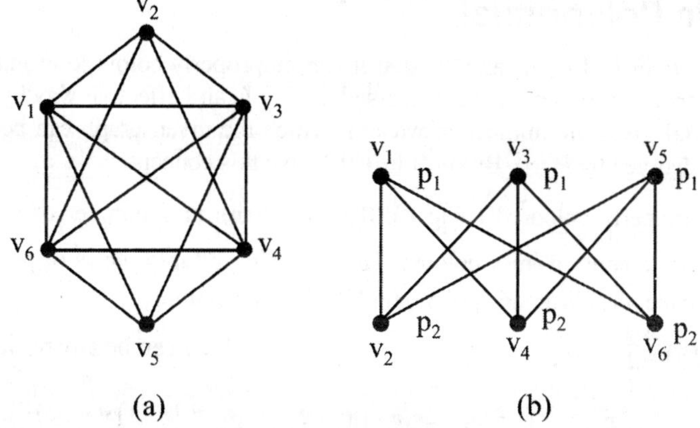

Fig. 10.126

b. Consider the graph G in Fig. 10.126 (b). One colour p_1, is assigned to vertices v_1, v_3, and v_5. The second colour p_2 is assigned to vertices v_2, v_4, and v_6 as they are adjacent to v_1, v_3, and v_5. All the vertices have been assigned with two colours in such a way that no two adjacent vertices have same colour. Hence G is 2-colourable.

10.42 Chromatic Number

Now the question arises, how many minimum numbers of colours are required for proper colouring.

The **chromatic number** of a graph G is the minimum number of colours n needed for colouring of the graph G. It is denoted by $\chi(G)$ and $\chi(G) = n$ means G is n-chromatic.

REMARK

A graph that can be assigned a n-colouring is n-colourable, and it is n- chromatic if its chromatic number is exactly n. For colouring problems, we need only simple connected graph because discarding loops and replacing all parallel edges with single edge without affecting adjacency does not change chromatic number.

Example 10.50

Consider the graphs shown in Fig. 10.126. Find the chromatic number.

Solution

a. In Fig. 10.126(a) the complete graph G requires six different colours for proper colouring so the chromatic number $\chi(G) = 6$.

b. In Fig. 6.1 (b), the graph G requires at least two colours for proper colouring. Thus $\chi(G) = 2$.

1. *A graph with only isolated vertices is 1-chromatic.*
2. *A graph is 1-chromatic if and only if it is totally disconnected.*

10.43 Chromatic Polynomial

Consider a graph of m vertices and let us assume that it can be properly coloured in many different ways using different colours. This property of graph can be expressed in simple and effective way by means of a polynomial. The **chromatic polynomial** gives the number of ways in which the given graph can be coloured by n or fewer number of colours and is denoted by $P_n(n)$. Birkhoff in 1912 gave this concept.

Let the graph G be properly coloured using i different colours in c_i number of ways. Since there are ${}^nC_i$ different ways of choosing i colours from n colours. There are ${}^nC_i$ different ways of properly coloring G using i colors out of n colours. Hence, the chromatic polynomial is given by sum of

$$P_m(n) = \sum_{i=1}^{n} c_i . {}^nC_i \qquad [\because i \text{ can be any positive integer from 1 to } n]$$

$$= c_1 . \frac{n}{1\,!} + c_2 \frac{n(n-1)}{2\,!} + c_3 \frac{n(n-1)(n-2)}{3\,!} + \ldots + c_n \frac{n\,(n-1)\,(n-2)\,\ldots.\,(n-m+1)}{n\,!}$$

Each c_i has to be evaluated individually for the given graph.

For example any graph of at least one edge requires two colours for proper colouring. So, c_1 the number of ways of properly colouring a non-trivial graph with one colour is 0 and with two colours is $c_2 = 2!$. Hence chromatic polynomial for a graph with one edge and two vertices is given by

$$P_2(n) = 0\frac{n}{1!} + 2\frac{n\,(n-1)}{2\,!}$$

$$= n\,(n-1)$$

1. *Other values of c_i depend on the nature of the graph.*
2. *For complete graph K_m with m vertices each vertex has (n -1) adjacent vertices, i.e. the graph can be coloured using m different colours in m! ways for proper colouring. Hence $c_m = m!$ and for $i < m$, $c_i = 0$.*

Theorem 10.23

The chromatic polynomial of a complete graph of m vertices is given by

$$P_m(n) = n(n-1)(n-2) \ldots (n-m+1) \text{ and vice versa.}$$

Proof

Consider a graph G with m vertices. There are n different ways of colouring of any vertex (first vertex) of graph. A second vertex can be coloured in $(n-1)$ ways, the third in $(n-2)$ ways and so on. Similarly the mth vertex can be coloured properly in $(n - m + 1)$ different ways iff every vertex is adjacent to each others. Therefore, chromatic polynomial is

$$P_m(n) = n(n-1)(n-2) \ldots (n-m+1).$$

Hence the graph is a complete graph and vice versa.

Example 10.51

Find the chromatic polynomial of K_4, complete graph of 4 vertices.

Solution

Consider a complete graph G with 4 vertices represented by K_4. Since minimum m colours required for proper coluring of complete graph of m vertices K_m. Therefore, $c_i = 0$, for $i = 1, 2, 3, \ldots, m - 1$. Hence for proper colouring of K_4, we require 4 colours. Thus $c_1 = c_2 = c_3 = 0$ and $c_4 = 4! = 4{\cdot}3{\cdot}2{\cdot}1 = 24$

$$\therefore P_4(n) = c_1\frac{n}{1} + c_2\frac{n(n-1)}{2!} + c_3\frac{n(n-1)(n-2)}{3!} + c_4\frac{n(n-1)(n-2)(n-3)}{4!}$$

$$= n(n-1)(n-2)(n-3)$$

$(\because c_1 = c_2 = c_3 = 0\ c_n = 4)$

Example 10.52

Find the chromatic polynomial of graph shown in Fig. 10.127.

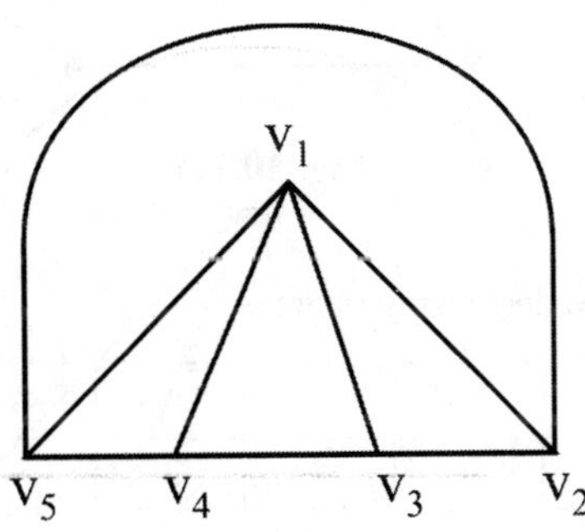

Fig. 10.127

Solution

The graph in Fig. 10.127 consists of triangles. Since a triangle requires minimum three colours for proper colouring. So $c_1 = c_2 = 0$. Now consider three colours p_1, p_2 and p_3 for colouring the vertices v_1, v_2, v_3 in $3! = 6$ differents ways. So $c_3 = 6$. Similarly, when we have four colours p_1, p_2, p_3 and p_4, then the vertices v_1, v_2, v_3 and v_4 can be properly coloured in $4! = 24$ ways and v_5 vertex can be coloured with the colour of either v_2 or v_3. Thus $c_4 = 2 \times 24 = 48$. And $c_5 = 5! = 120$. Hence,

$$P_5(n) = \sum_{i=1}^{n} c_i{}^nC_5$$

$$= c_1{}^nC_1 + c_2{}^nC_2 + c_3{}^nC_3 + c_4{}^nC_4 + c_5{}^nC_5$$

$$P_5(n) = 6\frac{n(n-1)(n-2)}{3!} + 48\frac{n(n-1)(n-2)(n-3)}{4!} + 120\frac{n(n-1)(n-2)(n-3)(n-4)}{5!} \quad [\because c_1 = c_2 = 0]$$

$$= n(n-1)(n-2) + 2n(n-1)(n-2)(n-3) + n(n-1)(n-2)(n-3)(n-4)$$

$$= n(n-1)(n-2)[1 + 2(n-3) + (n-3)(n-4)]$$

$$= n(n-1)(n-2)(n^2 - 5n + 7).$$

10.43-1 Wetch-Powell Algorithm for Vertex Colouring of a Graph

Wetch-powell algorithm gives upper bound of chromatic number, it does not yield the minimum number of colours required to paint the graph. Wetch-powell algorithm is described below:

Step 1: Order the vertices of G in decreasing degree. (Such an ordering may not be unique since some vertices may have the same degree).

Step 2: Use first colour c_1 to the first vertex and then in the requential order assign c_1.

Step 3: Repeat step 2 with a second colour c_2 and sequence of non-coloured vertices.

Step 4: Continue repeating with additional colours c_3, c_4 soon until all the vertices are coloured.

Example 10.53

Use Wetch-Powell algorithm to colour the graph shown in Fig. Find also the ψ (G).

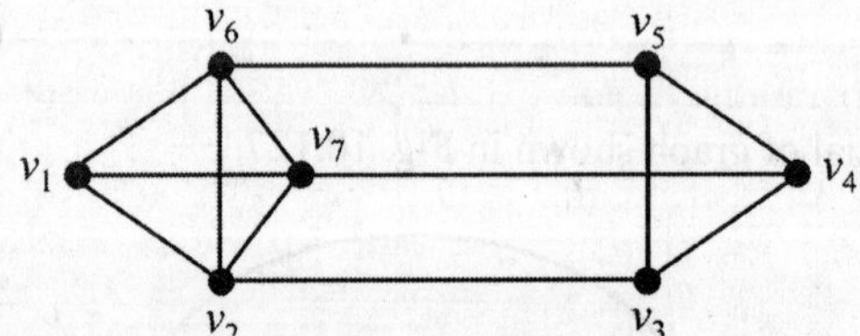

Fig. 10.128

Solution

First list the vertices of graphs G in decreasing degrees:

Vertex	v_6	v_2	v_7	v_1	v_3	v_4	v_5
Degree	4	4	4	3	3	3	3

Assign colour c_1 to the vertex v_6. Now find the vertices not adjacent to v_6 in the list. The next and only vertex on the list not adjacent to v_6 is v_3, and assign same colour c_1 to v_3. Now, move to next uncoloured vertex v_2 in the assign second colour c_2 to v_2. At this stage we observe, v_5 the only non-coloured vertex not adjacent to v_2 in this list. Colour this with G. Now we are lift with no uncoloured vertices which are non-adjacent to either vertex v_2 or v_5. Move to colour c_3. Assign colour c_3 to v_7. This is no uncoloured vertices adjacent to v_7 in the list. Now assign fourth colour c_4 to remaining uncoloured vertices.

Finally, we get the following colour:

Vertex	V_6	v_3	v_2	v_5	v_7	v_1	v_4
Degree	C_1	c_1	c_2	c_2	c_3	c_4	c_4

All the vertices are coloured and four colours were used. Hence $\psi(G) \le 4$. But the vertices v_1, v_2, v_6 and v_7 are adjacent to each other, hence at least four colours are required to colour these vertices. Thus $\psi(G) = 4$.

Example 10.54

An university offers seven courses coded as A, B, C, D, E, F, G. There are 12 students S1, S2, ..., S12. Who intend to lake these courses as given below:

S1: A, B, D; S2: A, B, G; S3: B, D; S4: A, C;
S5: C, D; S6: A, B; S7: D, F; S8: B, C;
S9: A, E; S10: A, E, G; S11: A, G; S12: A, E.

No student can appear for more than one course in a given time period, but courses can run in pair sessions. How many time periods required?

Solution

Period (1) Course A: S1, S2, S4, S6, S9, S10
Course D: S3, S5, S7

Period (2) Course B: S2, S3, S6, S8/
Course E: S1, S9, S10

Period (3) Course C: S1, S4, S5, S8/
Course G: S2, S10, S11

Period (4) Course F: S7, S11, S12.

Let the periods are shown as P_1, P_2, P_3 and $P4$ in the Fig.10.129. Here, to plane the teaching of the courses one must take care that when two or more course are to be tough in the same period, different class rooms are needed. The period should be programme in non-overlapping time intervals. No student can appear in two course at the same time.

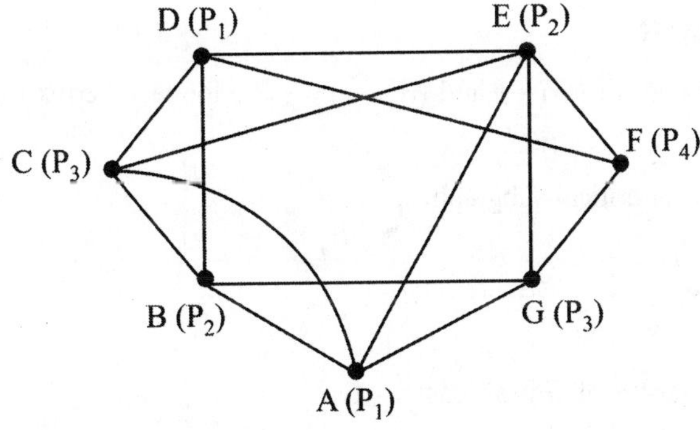

Fig. 10.129

10.44 Bichromatic Graph

If a graph G requires two colours for colouring then it is called **bichromatic graph.** In other words, if the graph can be partitioned into two non-empty subsets, such that each edge connects a vertex from each subset, then it can be painted with two different colours and called **bichromatic.**

Example 10.55

Consider famous three utilities puzzle problem. The objective of the puzzle is to supply three utilities gas, electricity and water to three houses x, y and z. Show that the problem is represented by bichromatic graph.

Solution

The vertices x, y and z show three houses and u_1, u_2 and u_3 represent the utilities gas, electricity and water respectively in graph G in Fig. 6.2. Consider that vertices x, y and z are painted by colour p_1 and utilities by p_2. No two houses and utilities are adjacent. Hence, we need only two colours to represent the puzzle. So the chromatic number of graph G is 2. i.e. $\chi(G) = 2$. Thus the puzzle problem is bichromatic.

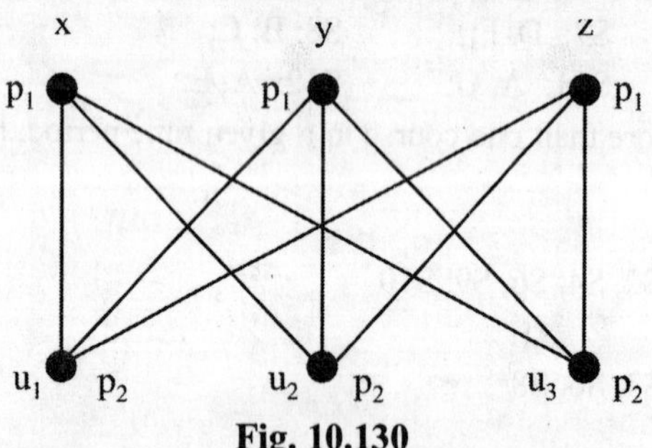

Fig. 10.130

1. *The chromatic number of graph G is less than or equal to total number of vertices in G. i.e. $\chi(G) \le m$ (total number of vertices).*
2. *A complete graph of m vertices requires m colours, i.e. χ (complete graph) = m.*
3. *If subgraph of G is n-chromatic then $\chi(G) \ge n$.*
4. *If degree of vertex is d then at the most d colours are required to paint the nodes adjacent to v.*
5. *$\chi(G)$ = maximum $\{\chi(K) / K$ is a connected component of $G\}$.*

10.45 Critical Graph

A graph G is said to be ***n*-critical** if $\chi(G) = n$ and $\chi(G - v) < \chi(G)$ for each vertex v of G.

REMARK

An n-chromatic graph has n-critical subgraph.

Theorem 10.24

If graph G is n-critical. Then

a. G is connected

b. The degree of each vertex of G is at least n - 1.

c. G can not be partitioned.

Proof

a. Let K be any component of disconnected graph G with chromatic number n. Suppose that v be any vertex of G not in K. Then $\chi(G - v) \ge n$. But this contradicts the fact of n-critical graph. Hence G is connected.

b. As given G is n-critical graph. For each vertex v of G, $\chi(G - v) \le n - 1$. If degree of vertex $d(v) < n - 1$, we need at the most $n - 2$ colours to paint the adjacent node of v. It implies that $(n - 1)$ colouring of G -

v can be extended to a $(n - 1)$ colouring of G by colouring v by the colour distinct from the colours adjacent of v. This fact contradict the $\chi(G) = k$. So degree of vertex $d(v) \geq n - 1$.

c. Let G_1 and G_2 are subgraphs of G then $G = G_1 \cup G_2$ and $G_1 \cap G_2 = K$, then $\chi(G_1) \leq k - 1$ and $\chi(G_2) \leq k - 1$. But, the vertices of $G_1 \cap G_2$ can be coloured with same colour in both sub-graphs. Hence, on combining they give $(k - 1)$ colouring of G. But this contradicts the fact $\chi(G) = k$. Hence, G can not be partitioned.

Lemma 10.1

Suppose G is a graph with $\chi(G) > n$ and can be partitioned in $G(x)$ and $G(y)$ in such a way that they are n-colourable, then the edge cut (x, y) has at least n edges.

Corollary

The degree of vertex is greater than or equal to $m - 1$ if m chromatic graph has at least m vertices.

Proof

Suppose K be an m-critical subgraph of m-chromatic graph G. According to theorem 6.3, degree of every vertex of K is at least $m - 1$ and so also in G. Clearly, K has at least m vertices because it is m-chromatic.

1. *For any graph G with maximum degree Δ of any vertex, chromatic number is defined as*

$$\chi(G) \leq 1 + \Delta(G).$$

2. *For any graph G, with minimum degree δ of any vertex, chromatic number is given by*

$$\chi(G) \geq m/(m - \delta(G)),$$

where m is the number of vertices of G.

Example 10.56

Find the chromatic number of graph G shown in Fig.10.131.

Solution

For graph G shown in Fig. 10.129, $d(v_7) = 6$ and degree of all other vertices is 3. The maximum degree of G, $\Delta(G) = 6$ and minimum degree, $\delta(G) = 3$. Also by above note 1, we have $\chi(G) \leq 1 + \Delta(G)$, i.e. in this case $\chi(G) \leq 7$.

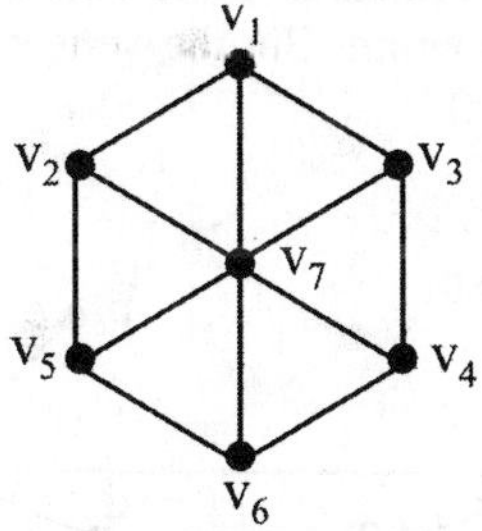

Fig. 10.129

Since there are simple circuits of length 3, therefore $\chi(G) \geq 3$. Now, if $\chi(G) \geq 6$, then by above corollary there will be at least 6 vertices of degree greater than or equal to 5. But this is a contradiction. Hence $\chi(G) \leq 5$.

It implies $3 \leq \chi(G) \leq 5$. But still we do not know the exact chromatic number. Let us start, painting the triangle v_4, v_3, v_7 with three different paints p_1, p_2, p_3 respectively. Now, since v_1 is adjacent to v_3 and v_7, we cannot paint it with, p_2 and p_3. So if we paint v_1 by colour p_1, then v_2 by p_2, v_5 by p_1 and v_6 by p_2. Thus the graph can be coloured by three different colours. So $\chi(G) = 3$.

10.46 Four Colour Problem

Till now, colouring of graph refers to proper colouring of its vertices and edges. Now, we consider a map in which the adjacent regions; not coloured with same paint will be termed as **proper colouring** of the map. If n colours are used to colour a map properly, it is called ***n*-colouring of map**. The proper colouring of regions is also called **map colouring**. Here we are not only interested in just proper colouring the regions of a given graph but our aim is also to use minimum number of colours while painting the regions. *Four colour conjecture** leads that every planar graph or map can be properly coloured with four colours.

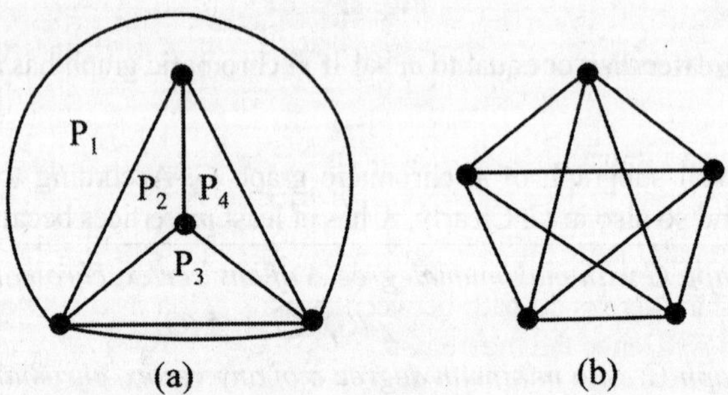

Fig. 10.130

10.47 The Five Colour Theorem

Theorem 10.25

Any finite planar graph G can be coloured using five colours (or fewer).

Proof

To prove the theorem, we use induction on the number of vertices of the graph. Let G be a planar graph with m vertices. Since G is a planar graph and connected, hence there exist at least one vertex of degree less then or equal to five. Let v be the vertex of degree ≤ 5, and consider the induced subgraph G' of G on $G - v$. According to induction hypothesis, induced subgraph G' requires no more than five colours. If the degree of vertex v is less than 4, then there is no difficulty for proper colouring. But the vertex v has exactly five neighbours and they are all differently coloured as show in Fig. 10.132(a).

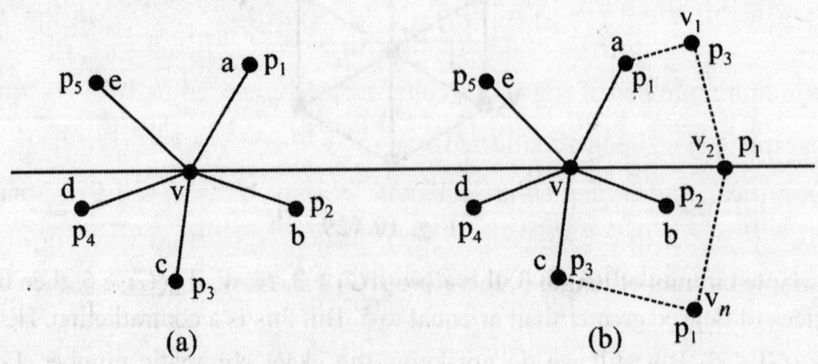

* The **four colour conjecture** states that four colours are sufficient for colouring any map on a plane.

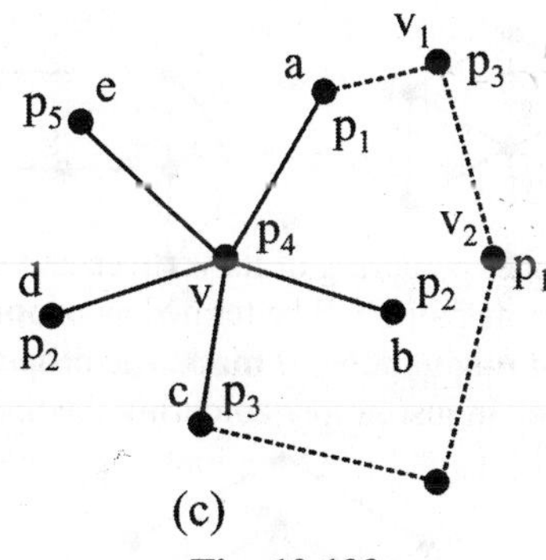

(c)

Fig. 10.132

Suppose there is a path between vertices a and c, say $a\ v_1\ v_2 \dots v_n\ c$ [see fig. 10.132(b)]. Colour alternatively a by p_1, v_1 by p_2, v_2 by p_1 and so on. Also, according to the property of planar graph there is no similar path between b and d (because it will intersect the path a and c). Hence there is no path of alternating colours p_2 and p_4 through vertices b to d. So, vertex d can be painted with colour p_2 and vertex b is still with colour p_2. Now we have colour p_4 left over with which we can paint vertex v.

Similarly, if we assume that there is no path between a and c and a path between b and d, then we can paint v with colour p_1 or p_3 instead of p_4. Hence the theorem.

Exercise 10.5

1. Define the following terms:

   a. chromatic number
   b. critical graph
   c. bichromatic graph
   d. n-colouring

2. Find the chromatic number of the "wheel" graph shown in following figure.

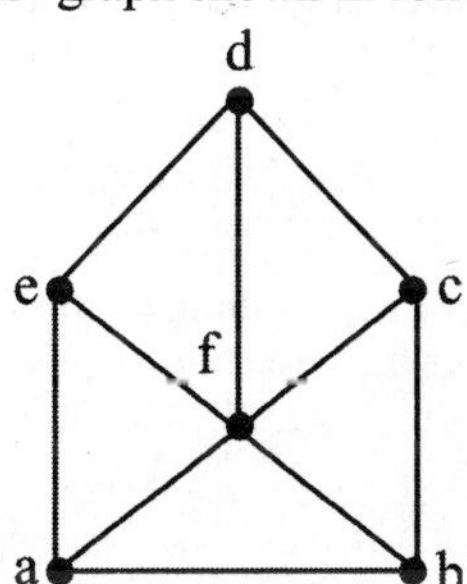

3. Determine the minimum number of edges in a connected m-vertex graph with chromatic number n.
4. Every n-critical graph is $\overline{n\ \ 1}$- edge-connected.
   [**Hint:** Let G is n-critical, and G_1 and G_2 are subgraph of G such that $\chi(G_1) \le n - 1$ and $\chi(G_2) \le n - 1$. Apply Lemma 10.1.]
3. Determine the chromatic number of each of the following graphs:

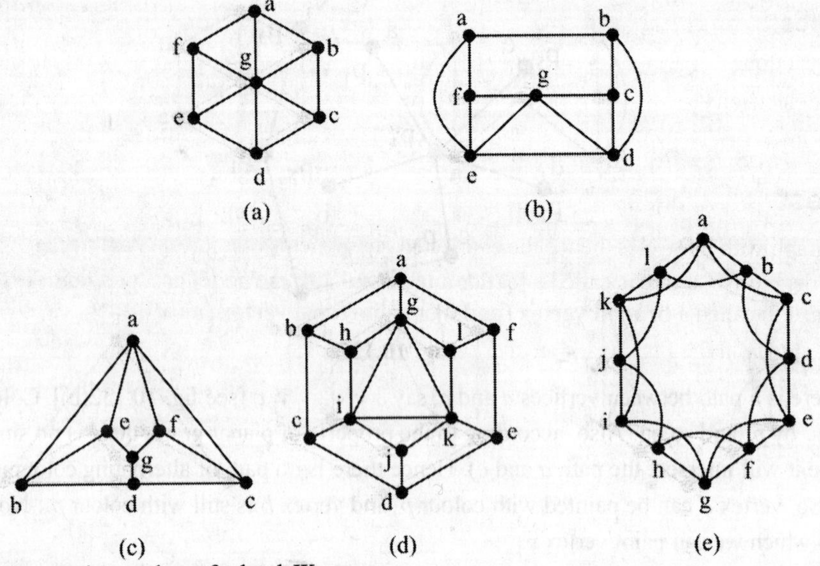

4. What is the chromatic number of wheel W_n.
5. Suppose the odd cycles in G are pair-wise intersecting meaning that every two odd cycles in G have a common vertex. Prove that $\chi(G) \leq 5$.
6. Sketch two different graphs that have the same chromatic polynomial.
7. Determine the chromatic polynomials of the following graphs:

   a. K_n b. K_n

   c. $K_{3,3}$
8. Show that any planar graph with 2 or fewer edges can be coloured by four colours.

Answer

5.. a. 3   b. 4

c. 4   d. 2

e. 4

10.48 Tree

A connected graph that contains no cycles (circuits) is called a **tree**. One can conclude from the definition that a tree is a simple graph, i.e. neither having self loop nor parallel edges as they both form cycles (circuits). Example of a tree is shown in the Fig.10.133. A tree with only one vertex is called a **trivial** or **degenerate** or **empty** tree, otherwise T is a **non-trivial tree**. The number of vertices in a tree is sometime called **order** of the tree.

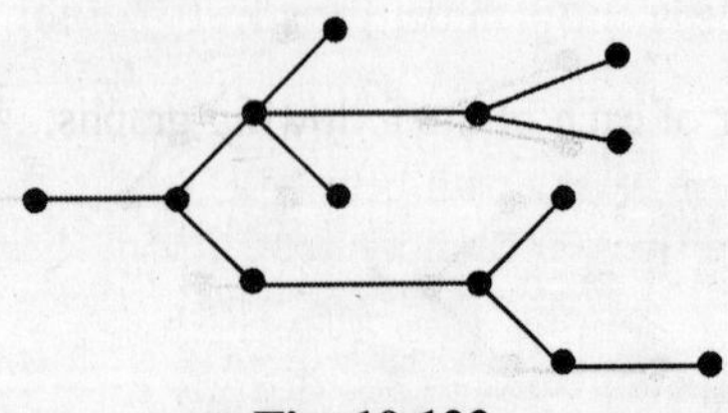

Fig. 10.133

Examples of tree

i. Chain reaction
ii. Administrative set up in an organization
iii. Flow of electricity
iv Water/ L P G /Petroleum supply pipeline
v. Postal Flow
vi. Decision Tree.

10.48-1 Forest

A **forest** *G* is a graph with no cycles, hence the connected components of a forest *G* are trees.

A vertex of degree 1 in a tree is called a **leaf** or a **terminal vertex (node)** or a **pendant vertex** and a vertex of degree larger than 1 is called a **branch vertex (node)** or an **internal vertex (node).**

In the graph shown in the Fig. 10.134 *a, c, d, f* and *i* are leaves and *b, e, h* and *g* are branch nodes.

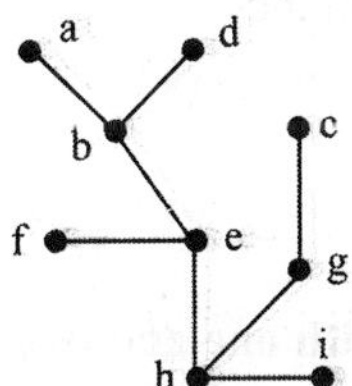

Fig. 10.134

Example 10.57

a. Graphs (i), (ii) and (iii) in Fig. 10.135 are trees while graphs (iv) and (v) are not tree

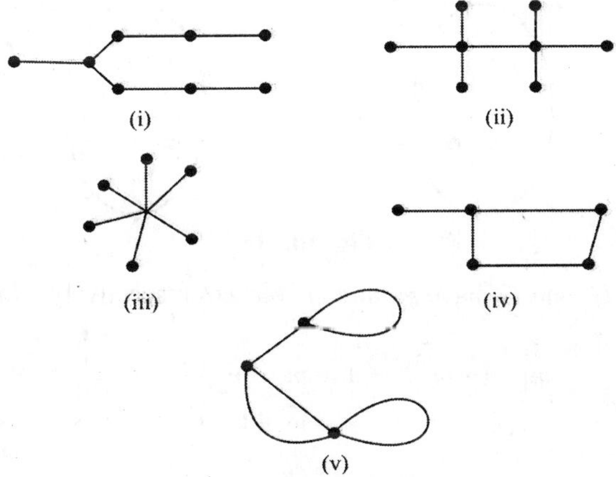

Fig. 10.135

Theorem 10.26

A simple graph *G* is a tree iff there is one path between every pair of vertices.

Proof

Graph *G* is connected because there exist a path between every pair of vertices. As it is given that there is one and only one path between a pair of vertices, *G* cannot have a circuit. This implies that *G* is a connected circuitless graph. Hence *G* is a tree.

Conversely, if *G* is a tree then *G* is connected.

⇒ There is a path between every pair of vertices

⇒ There must be only one path between every pair of vertices otherwise we can have a circuit

∴ G is a tree.

⇒ There is a unique simple path between every pair of vertices.

Theorem 10.27

If G is a tree with n vertices then it has exactly n–1 edges.

Proof

To prove this theorem we employ mathematical induction.

If $n = 1$, the number of vertices = 1 and the number of edges 0. Hence the result is true for $n = 1$.

If $n = 2$, the number of vertices = 2, and the number of edges = 1 (i.e. 2 – 1). See following figure.

Assume $n = m$, i.e. all trees with m number of vertices have exactly $m - 1$ edges.

Trees with one and two vertices

Let us assume that e_m be an edge connecting vertices v_i and v_j. As G is a tree, there exist only one path i.e. e_m between the vertices v_i and v_j. Hence, if we delete e_m, it will disconnect tree G into two components, say G_1 and G_2 as one can see in Fig. 10.136. It is easy to see that each such components has less than m+1 vertices and since there are no circuits in G, each of these components is a tree.

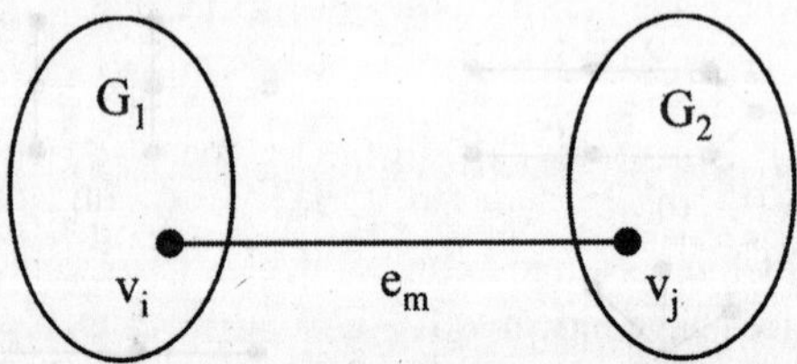

Fig. 10.136

Let the two components G_1 and G_2 have m_1 and m_2 vertices respectively. Now, the total number of edges will be

$$m_1 - 1 + m_2 - 1 + 1 = m_1 + m_2 - 1$$
$$= m + 1 - 1$$
$$= m.$$

Thus the theorem is true for $n = m + 1$ if it is true for $n = m$. Hence by mathematical induction tree G has exactly $n - 1$ edges.

Corollary

If G is nontrivial then G contains at least two vertices of degree 1.

Proof

Let the number of vertices of graph G be n. Then from the sum of degrees formula,

$$\sum_{i=1}^{n} \deg(v_i) = 2e = 2(n-1) = 2n - 2.$$

where e is the number of edges.

If we have only one vertex of degree 1, then

$$\deg(v_i) \geq 2 \quad \text{for } i = 2,3, \ldots ,n.$$

$$\sum_{i=1}^{n} \deg(v_i) = 1 + \sum_{i=2}^{n} \deg(v_i)$$

$$\geq 1 + 2n - 2$$

$$= 2n - 1$$

But then $2n - 2 \geq 2n - 1$ or $-2 \geq -1$

which is a contradiction. Hence, if G is a nontrivial tree then G contains at least 2 vertices of degree 1.

Theorem 10.27

Any connected graph G with n vertices and $(n - 1)$ edges is a tree.

Proof

Let us consider that the graph G has a circuit of length m, then there are m vertices and m edges on the circuit for each of the $n - m$ vertices not on the circuit [there is an incident edge on a geodesic to a vertices of the cycle]. As each such edge is different, the number of edges $n - m + m = n > n - 1$. This is a contradiction. Hence the theorem.

10.49 Distance in a Tree

The **distance** between two vertices u and v in a graph $G = (V, E)$, written $d_G(u, v)$ or simply $d(u, v)$ is defined to be the minimum number of edges between them (i.e. shortest one). There may be several shortest paths. If G has no such path, then $d(u, v) = \infty$.

10.50 Directed Tree

A **directed tree** is an cyclic digraph which is a tree as an undirected graph and which has one vertex called its **root** with indegree 0, while all other vertices have indegree 1. An isolated vertex is also a directed tree. In a directed tree any vertex which has out degree zero is called a **terminal vertex** or a **leaf**. All other vertices are called **branch vertices** (or nodes). The **level** of any vertex is the length of its path from the root.

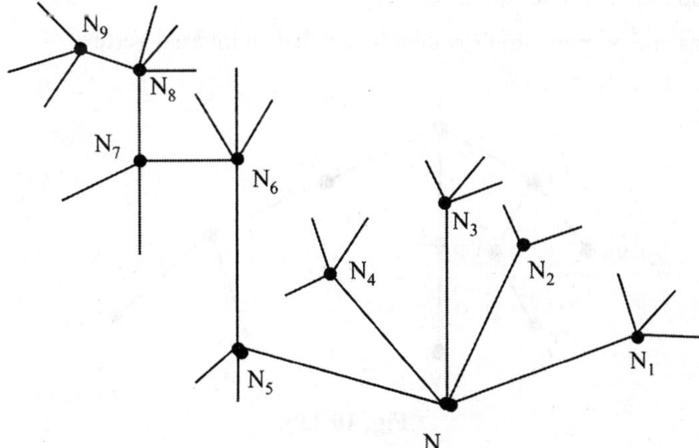

Fig. 10.137

Example 10.58

A directed tree $T = (V, E)$ is shown in Fig. 10.138, where $V = \{a, b, c, d, e, f, g, h\}$ and $E = \{(a, b), (a, c), (a, d), (b, e), (d, f), (e, g), (e, h)\}$. The root of T is the vertex a and the vertices at level 2 are e and f. It is to be noted that the directed trees are conventionally drawn with the root at the top and all edges going from the top of the page towards the bottom, so the direction of the edges sometimes not explicitly shown.

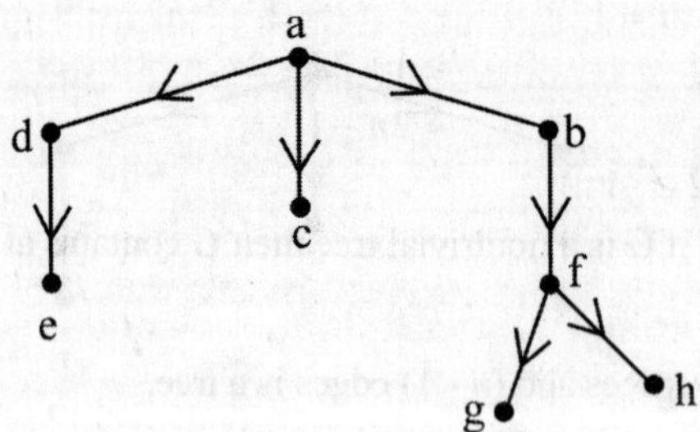

Fig. 10.138: A directed tree

10.51 Rooted and Binary Trees

A **rooted tree** is a tree in which there is one designated vertex called the **root**. The **level** of a vertex v in a rooted tree is the length of the path to v from the root and the maximum vertex level is called the **depth** or **height** of the tree. Those vertices with degree 1, other than root r are called **leaves** or **terminal vertices** of T and the directed path from a vertex to a leaf is called a **branch**. The vertices other than leaves are called branch vertices or nodes. Note that in a tree any vertex may be designated as a root. Generally, the term tree means trees without any root. However, for emphasis they are sometimes called **free trees** or **non-rooted trees** to differentiate them from the rooted kind. Tree shown in Fig.10.139 is a rooted tree with root r and 10 other vertices.

In a rooted tree T with root $r = v_o$ if $r = v_o, v_1, v_2, \ldots, v_{i-1}, v_i, v_{i+1}, \ldots, v_{n-1}, v_n$, then

i. v_{i-1} is the **parent** of v_i for $i = 1, 2, 3, \ldots, n-1$.

ii. $v_1, v_2, \ldots, v_{i-1}, v_i, v_{i+1}, \ldots, v_{n-1}$ are **ancestors** of v_n.

iii. v_i is a **child** of v_{i-1} for $i = 1, 2, 3, \ldots, n-1$. For example v_2 is a child of v_3.

iv. v_i is a **descendant** of v_{i-1}. For example v_3 is a descendant of v_2.

v. If v_i has no child then v_i is a terminal vertex.

vi. If a vertex v_i has one or more children then it is called an internal vertex.

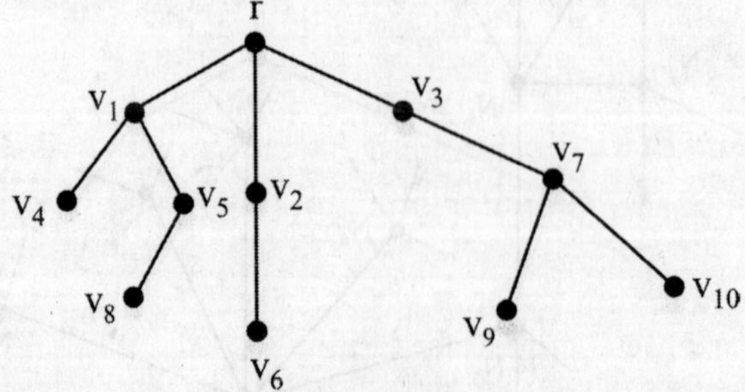

Fig. 10.139

The tree shown in Fig. 10.139 has five leaves v_4, v_6, v_8, v_9, v_{10}, level (v_6) = level (v_9) = 3, and depth of the tree is 3. In contrast to the natural trees, which have their root at the bottom, in graph theory rooted trees are typically drawn with their root at the top.

Example 10.59

Consider the tree in Fig. 10.140. Let vertex *a* has been designated as the root of the tree.

It is to be noted here that if we choose any other vertex say *m* in the tree, then the path from *a* to *m* will be unique. The vertices on this path (*a, d, k, m*) are described in genealogical terms as follows:

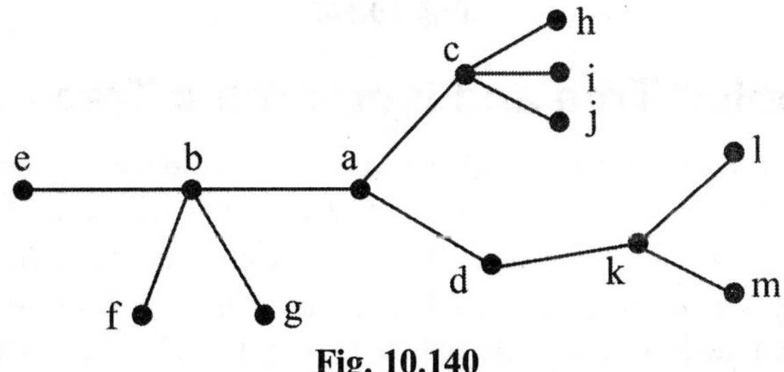

Fig. 10.140

i. *m* is a **child** (son or daughter) of *k* (so is *l*).

ii. *k* is *m*'s **parent** (father).

iii. *a, d* and *k* are *m*'s **ancestors**.

iv. *d, k* and *m* are **descendants** of *a*.

As from the Fig.10.140 one can see that genealogical relationships are often easier to visualize if we rewrite the tree in a way so that children are positioned below their parents.

From the Fig.10.141 one can think each vertex as the root of a tree that contains in addition to itself, all of its descendants, e.g. *c* is the root of a tree that contains *h, i* and *j*. And finally *h, i* and *j* are roots of the tree that contain only themselves.

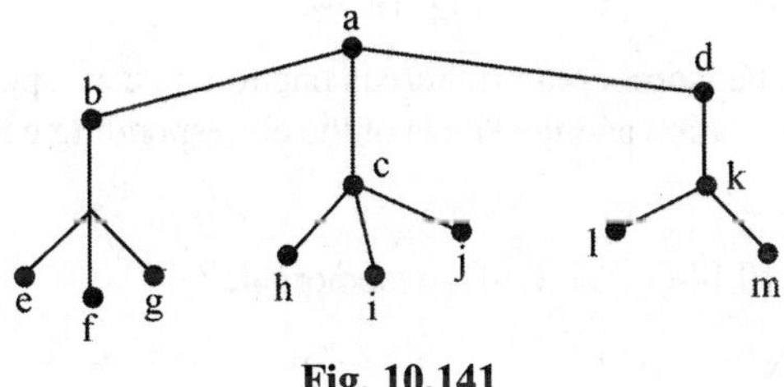

Fig. 10.141

Example 10.60

Fig. 10.137 is a rooted tree, with *N* as the root, which is an example of what is called a **decision tree**.

10.52 Subtree

Let *a* be a branch node in a tree *T*. Then a **subtree** with *v* as the root is a subgraph say $T' = (V', E')$ of *T* s.t. V' contains *v* and all its descendants, and E' contains the edges in all paths emanating from *v*. When we say a subtree of *v* we mean a subtree that has a son of *v* as root. For example graph shown in Fig. 10.142(a) in a subtree of graph shown in Fig. 10.141.

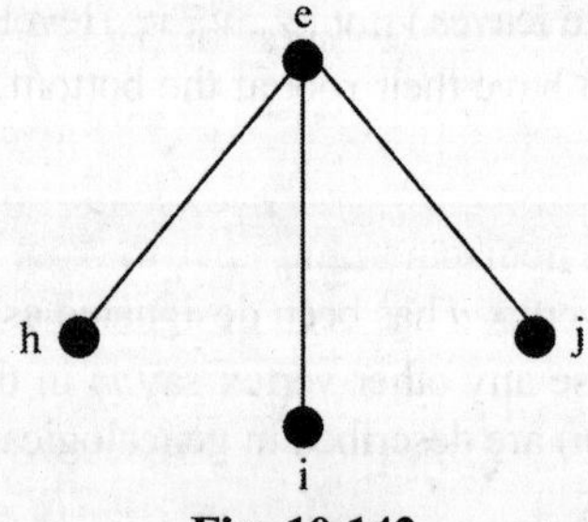

Fig. 10.142

10.53 Ordered Rooted Tree and isomorphic Tree

An **ordered rooted tree** is a rooted tree whose subtrees are put into a definite order and are themselves ordered trees. An empty tree (a tree with no vertices) and a single vertex are ordered rooted trees.

The trees shown in the Fig.10.143 (b) represents the same tree but they are of different order.

In an ordered tree the edges incident from each branch node labelled with integers 1, 2, 3, . . ., n. As a result of this the subtrees of a branch node can be referred to as the 1st, 2nd, 3rd,….. subtrees of the branch node corresponding to the labels of the edges incident from it. In an ordered tree it is not necessary that orders of tree are consecutive.

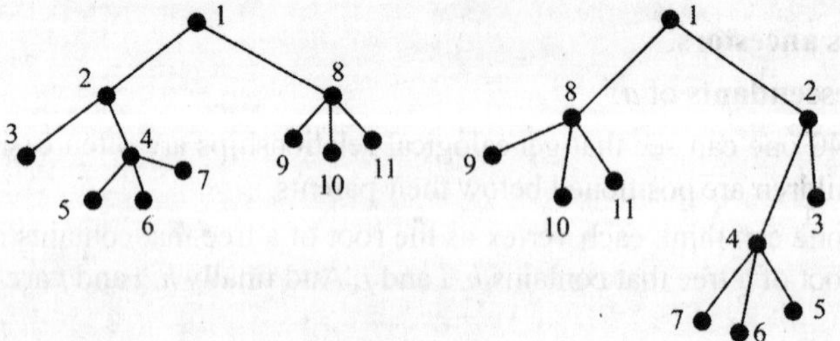

Fig. 10.143

Two ordered trees are said to be **isomorphic** if there is one to one correspondence between their vertices and edges that preserves the incidence relation and the labels of the corresponding edges match.

Example 10.61

The rooted tree shown in Fig 10.144 (a) and (b) are isomorphic.

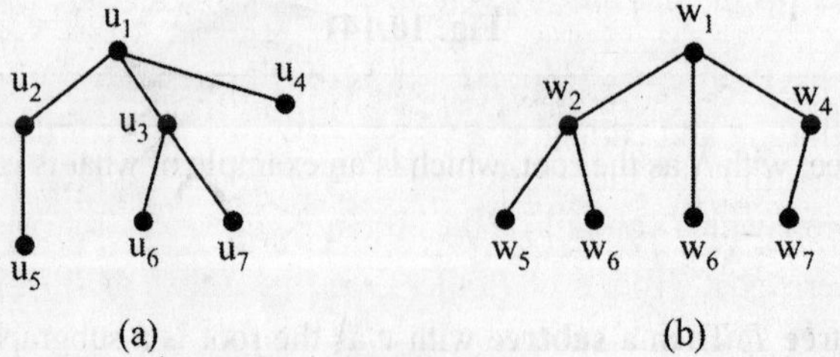

Fig. 10.144

Here the isomorphism is:

$$u_1 = w_1,\ u_2 = w_4,\ u_4 = w_3,\ u_3 = w_2,\ u_5 = w_7,\ u_6 = w_5,\ u_7 = w_6$$

It is to be noted that isomorphism in a tree need not be unique. Can you find another isomorphism of the tree shown in Fig. 10.144.

Example 10.62

The rooted tree shown in Fig. 10.145 is not isomorphic since the root of the tree T_1 is 3 but that of T_2 is 2.

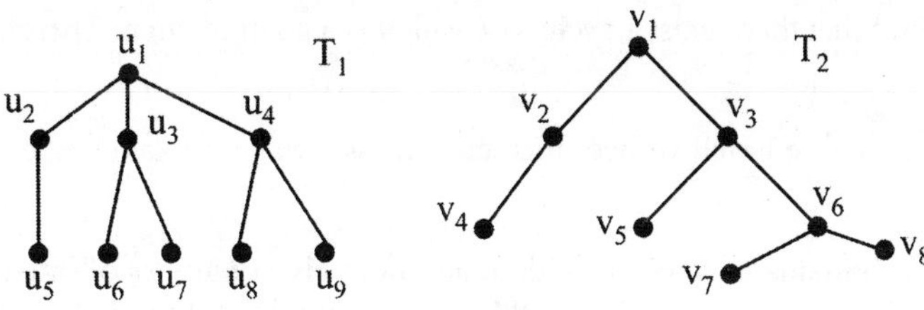

Fig. 10.145

Sometimes rooted trees are not isomorphic but if we take them as free tree, then they are isomorphic. For example trees shown in Fig. 10.146 are not isomorphic as rooted trees, but they are isomorphic as free trees.

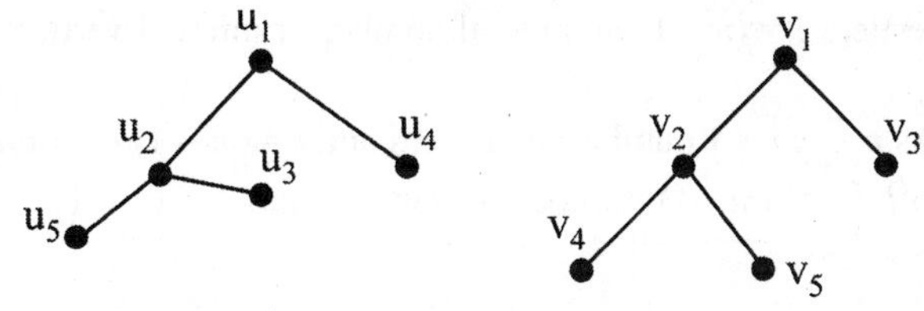

Fig. 10.146

10.54 *M*-ary Tree

A tree is said to be an ***m*-ary** tree if every parent (internal vertex) has at most *m*-children. An *m*-ary is said to be **complete** or **full *m*-ary** tree if every parent has exactly *m* children.. If $m = 2$, the trees are called **binary**. An *m*-ary tree is said to be regular if each of its branch nodes has exactly *m* children.

Theorem 10.28

Let $T = (V, E)$ be a rooted tree then:

i. There are no cycle in *T*.
ii. If *r* is the root of the tree then *r* is the only root of *T*.
iii. Each vertex in *T*, other than v_0 has indegree one.

Proof

i. Consider a cycle beginning and ending at a vertex *r*. We know from the definition of a tree that any vertex *v* cannot be equal to *r* and there must exist a path from *r* to *v*. But this implies that there is more than one path from *r* to *v* which contradicts the definition of a tree. Hence there are no cycles in tree *T*.
ii. Let us assume that there exist a root *r*' of *T* other then root *r*, then there must exist a path p_1 from *r* to *r*' and a path p_2 from *r*' to *r* since *r*' is a root.

   ⇒ There exist a cycle from *r* to *r*'. But this is not possible. Therefore the vertex *r* is unique.
iii. Consider a vertex *a* in *T* other then *r*. Then there exist a unique path from *r* to *w*. We can conclude from here that *a* has indegree at least one. In case the indegree of *a* is more than one then there must be

distinct vertices b and c such that the branch ba, ca are both in T. Let $a \neq r$, $c \neq r$, and there are paths p_2 from r to b and p_3 from r to c by definition of a tree.

$\Rightarrow$ There are two different paths from r to a. But this is a contradiction (by the definition of a tree with root r). Therefore, the indegree of a is one and r has indegree zero.

If possible, let the indegree of r is 1 and let branch ar be in T. There is a path from r to a by definition and a path from a to r. This implies that there exist a cycle in T which is a contradiction. Therefore, r has indegree zero.

Theorem 10.29

If T is a full m-ary tree with i internal vertices then there are $n = m.i + 1$ total number of vertices.

Proof

Since a full m-ary tree contains exactly m-children and here given number of internal vertices is i, therefore total number of vertices excluding root is $m.i$. Hence the tree contains $m.i +$ 1number of vertices including root.

REMARK

If number of leaves in a m-ary tree is l then $n = i + 1$. Now using above theorem and this equality we conclude the following: A full m-ary tee with has

a. total number of n vertices $i = (n - 1)/m$ internal number of internal vertices and $l = [(m - 1)\, n + 1]\, /m$ number of leaves.

b. i internal vertices has $n = m.i + 1$ number of vertices and $l = (m - 1)\, i + 1$ number of leaves.

c. l leaves has $n = (m.l - 1)/\, (m - 1)$ number of vertices and $i = (l - 1)\, /\, (m - 1)$ number of internal vertices.

Theorem 10.30

In an m-ary tree of height h has at most m^h number of leaves.

Proof

We use mathematical induction on height h to prove this theorem.

For $h = 1$: When height of m-ary tree is 1, then tree consists of a root with at most m-children which is leaf. Therefore there are at most $m^1 = m$ leaves in an m-ary tree of height 1.

For height less than h: Let us assume that the result is tree for all m-ary tree of height less than h.

For height equal to h: Now assume that T be an m-ary tree of height h. If we delete the edges from the root to each of the vertices of level 1 we obtain a subtree. The leave of T are the leaves of subtrees so obtained. Each of these subtrees has height less than or equal to h - 1. Hence by inductive hypothesis each of these rooted trees has at most m^{h-1} leaves. Since atmost m such subtrees are possible, each with a maximum of m^{h-1} leaves. Hence there are at most $m.m^{h-1} = m^h$ leaves. Hence the theorem

REMARK

If number of leaves in a m-ary tree is l then number of vertices $n = i + l$. Using above theorem and this equality we conclude the following:

A full m-ary tree with

a. total number of n vertices has $i = (n - 1)/m$ number of internal vertices and $l = [(m - 1)\, n + 1]\, /\, m$ number of leaves.

b. i internal vertices has $n = m\, i + 1$ number of vertices and $l = (m - 1)\, i + 1$ number of leaves.

c. l leaves has $n = m\, (\, l - 1)/(m - 1)$ number of vertices and $i = (l - 1)/(m - 1)$ number of internal vertices.

10.55 Binary Tree

A **binary tree** is a rooted tree in which every vertex has at most two children, and each child of a vertex is designated as its **left child** or **right child**. The subtrees rooted at the children of the root are the **left subtree** and the **right subtree** of the tree. In a more formal way, we can say a binary tree is a rooted tree in which each vertex has either one child or two children.

A binary tree *T* is frequently presented by a diagram in the plane called a **picture** of *T*. The difference between binary trees and ordered trees is that every vertex of a binary tree has exactly two subtrees (one or both of which may be empty), while a vertex of an ordered tree may have any number of subtrees. The Fig.10.147 represents a binary tree having following properties:

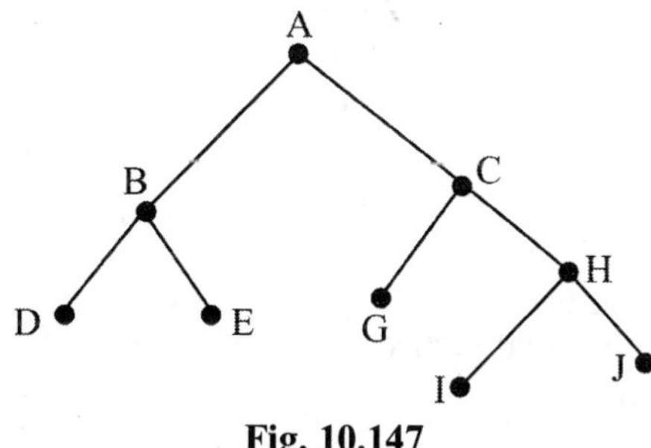

Fig. 10.147

i. Tree *T* consists of 9 vertices.

ii. The root of *T* is the vertex *A* at the top of the graph.

iii. A left downward slanted line at a vertex, say *v* indicates a left successor of *v* and a right downward slanted line *v* indicates a right successor of *v*.

Further in Fig. 10.147, we find that

a. *B* is a left successor (child) and *C* in a right successor (right child) of the root *A*.

b. the left subtree of the root *A* consists of the vertices *B*, *D* and *E*, and the right subtree of A consists of the nodes *C*, *G*, *H*, *I*, and *J*.

c. the vertices *A*, *B*, *C* and *H* have two successors and the vertices *D*, *E*, *G*, *I*, *J* are terminal vertices.

Example 10.63

Four distinct binary trees are shown in the Fig. 10.148.

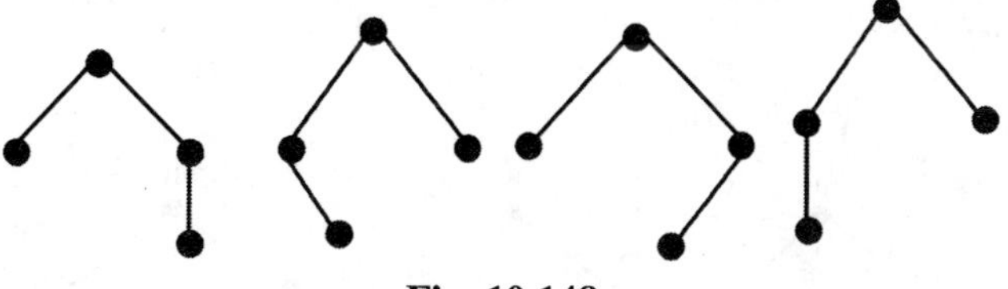

Fig. 10.148

10.55-1 Full Binary Tree

A tree in which every vertex has either zero or two subtrees is said to be a **full binary tree**. In other words, each vertex has either two or zero children. Tree shown in Fig. 10.147 is a full binary tree.

Note

i. *In a full binary tree there is exactly one vertex with degree 2 i.e. root and all other vertices have degree 1 or 3.*

ii. *A full binary tree will always have odd number of vertices, say n. Since there is exactly one vertex (root) of even degree, and the remaining n–1 vertices are of odd degrees. But the number of vertices of odd degree in a graph is always even, therefore n–1 is even. Hence n is odd.*

iii. *Let n be the total number of vertices and m be the number of pendant vertices in a full binary tree T. Then n – m –1 is the number of vertices of degree three.*

The number of edges in T

$$= \frac{1}{2}[m + 3(n - m - 1) + 2]$$

$$= n - 1$$

$$m = \frac{n+1}{2}$$

Theorem 10.31

If T is a full binary tree with m internal vertices, then there are $m + 1$ terminal vertices and $2m + 1$ total vertices.

Proof

Consider a full binary tree T with m internal vertices. Since each internal vertex contains two children hence m internal vertices will be having $2m$ children. Thus, the total number of vertices including the root is $2m + 1$ and

$$\text{Number of terminal vertices} = \text{Total vertices} - \text{Internal vertices}$$

$$= 2m + 1 - m = m + 1.$$

10.55-2 Similar Binary Trees

Two binary trees T and T' are said to be **similar** if they have the same structure or shape. The trees are said to be **copies** if they are similar and have the same contents as corresponding vertices.

Example 10.64

In the three trees T_1, T_2 and T_3 shown in Fig.10.149. Identify those that represent the same:

i. rooted tree

ii. ordered rooted tree

iii. binary tree

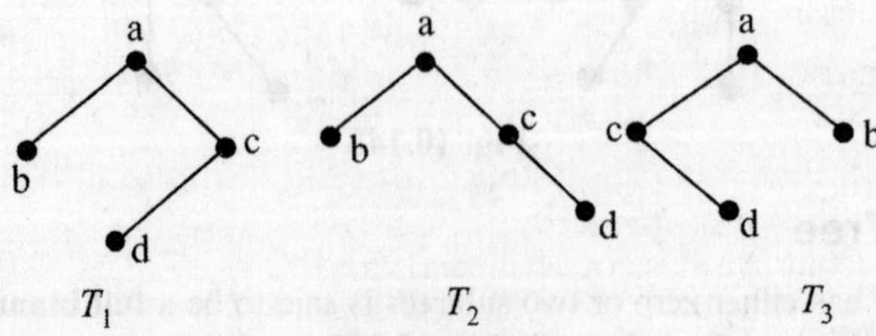

Fig. 10.149

Solution

i. As all the three trees have same root a, they represent same rooted tree. And root a is the root with children b and c, and c has the single child d.

ii. T_1 and T_2 are same ordered rooted tree but T_3 is different. As b is the first child of a in T_1 and T_2 but it is second child of a in T_3.

iii. All three trees, T_1, T_2 and T_3 represent different binary trees. Specifically T_1 and T_2 are different since we distinguish between left and right successors even if there is only one successor (which are not true for ordered rooted trees). d is the left successor of c in T_1 but a right successor of c in T_2.

10.56 Isomorphism in a Binary Tree

Consider two binary trees T_1, T_2 with their respective roots r_1 and r_2. The binary tree T_1 and T_2 are said to be **isomorphic** if there exists one-to-one, onto function f from the vertex set of T_1 to the vertex set of T_2 and satisfies the following:

i. $f(r_1) = r_2$

ii. If vertex v_i and v_j are adjacent in T_1 then $f(v_i)$ and $f(v_j)$ must be adjacent in T_2.

iii. If v is a left child (right child) of u in T_1 then $f(v)$ must be left child (right child) of $f(u)$ in T_2.

The function f is called **isomorphism**.

Example 10.65

i. The binary trees shown in the Fig.10.150 are isomorphic. The isomorphism is $f(u_i) = v_i$, where $i = 1, \ldots 5$.

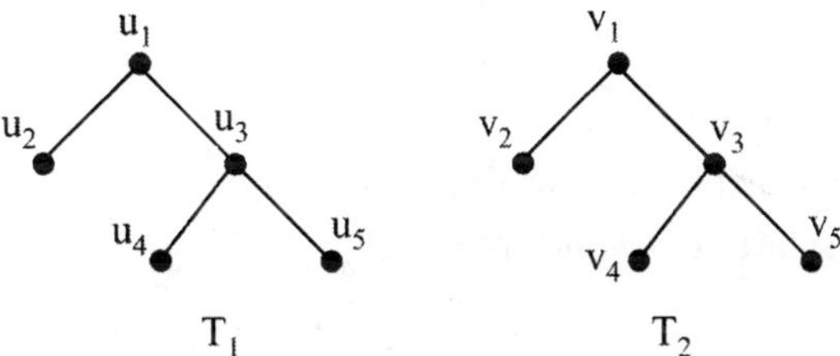

Fig. 10.150

ii. The binary trees in the Fig. 10.151 are not isomorphic.

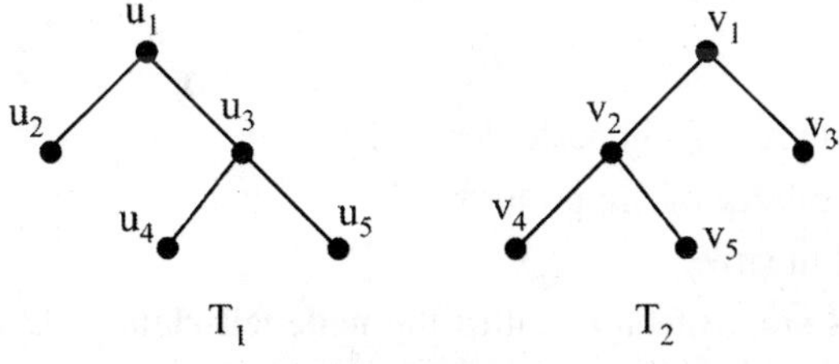

Fig. 10.151

Here, property (iii) of isomorphism is not satisfied. But, it is to be noted that the trees T_1 and T_2 are isomorphic as a rooted tree and as a free tree.

Example 10.66

Draw different possible non-isomorphic binary trees with three vertices.

Solution

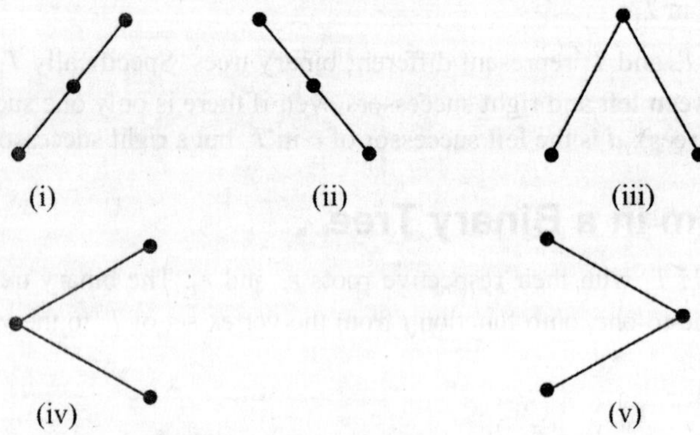

Fig. 10.152

There are five nonisomorphic binary trees with three vertices.

10.57 Traversing Binary Trees

The traversal of a binary tree consists of visiting each vertex of the tree in some defined order. The most common binary tree traversals are differentiated by the order in which the root and its subtrees are visited. The three standard traversals are as follows:

i. Preorder Traversal
   a. Process the root r of the tree.
   b. Traverse the left subtree of r in preorder.
   c. Traverse the right subtree of r in preorder.
ii. Inorder Traversal
   a. Traverse the left subtree of r in inorder.
   b. Visit the root r of the tree.
   c. Traverse the right subtree of r in order.

iii. Postorder Traversal
   a. Traverse the left subtree of r in postorder.
   b. Traverse the right subtree of r in postorder.
   c. Process the root of the tree.

The above three algorithms are sometimes called the **node-left-right** (*NLR*) traversals, the **left-node-right** (*LNR*) traversal and the **left-right-node** (*LRN*) traversal respectively.

Any traversal of an empty tree consists of doing nothing.

Example 10.67

Consider the tree shown in Fig.10.153. Here in the tree, A is the root, and its left subtree L_T consists of nodes B, D and E and its right subtree R_T consists of nodes C and F.

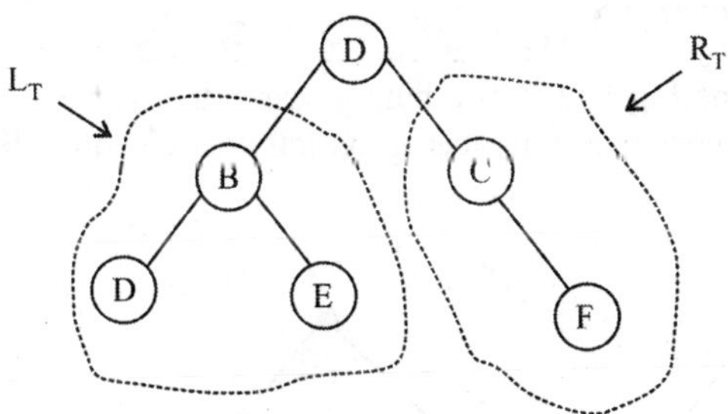

Fig. 10.153

i. **Preorder Traversal**

In the preorder traversal of *T*, first process root *A*, then traverse L_T and then traverse R_T. However, the preorder traversal of L_T visits the root *B* and then *D*, and *E*, and in a similar way the preorder traversal of R_T processes the root *C* and then *F*. Hence *A-B-D-E-C-F* is the preorder traversal of *T*.

ii. **Inorder Traversal**

In this traversal of *T*, first traverses L_T, visits *A* and then traverses R_T. Inorder traversal of L_T the order in which nodes visited are *D*, *B* and then *E*, and inorder traversal of R_T visits C and then *F*. Hence *D-B-E-A-C-F* is the inorder traversal of *T*.

iii. **Postorder Traversal**

In this traversal of *T*, first traverses L_T then traverses R_T, and then visits *A*. *D-E-B-F-C-A* is the postorder traversal of *T*.

Example 10.64

For the tree in Fig.10.154, the different orders in which the nodes processed are.

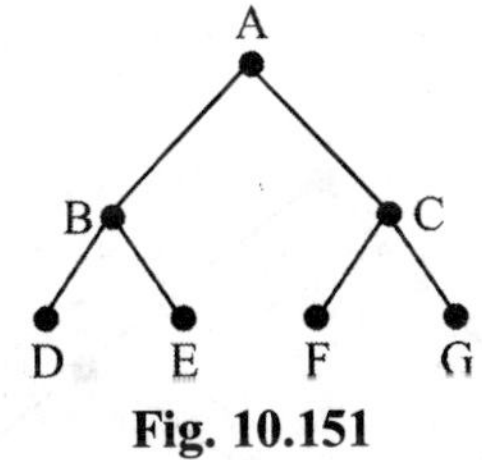

Fig. 10.151

i. Preorder Traversal	ABDECFG
ii. Inorder Traversal	DBEAFCG
iii. Post order Traversal	DEBFGCA.

10.58 Binary Search Trees

A **binary search tree** is a binary tree *T* in which each vertex *v* of *T* satisfies the following properties:

i. Each data item in the left subtree of *v* is less than the data item in *v*.

ii. Each data item in the right subtree of *v* is greater than the data item in *v*.

One can observe that the above property guarantees that the inorder traversals of *T* will give a sorted listing of the elements of *T*. See Fig. 10.155.

Example 10.68

The binary tree T shown in the Fig.10.155 is a binary search tree as it satisfies both the conditions of above article, i.e. each vertex v in T exceeds every number in its left subtree and is less than every number in its right subtree.

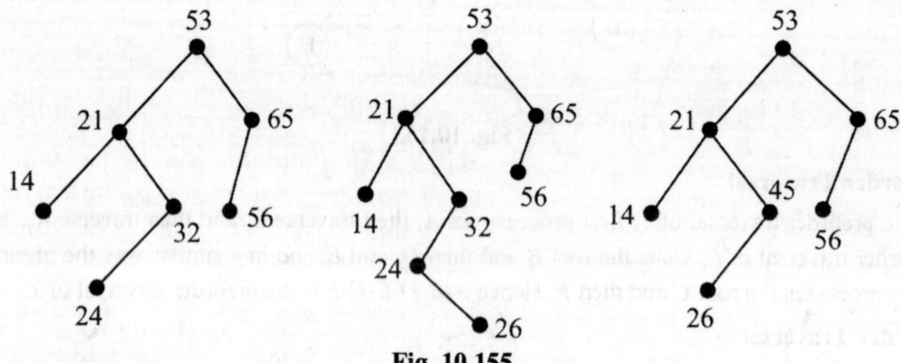

Fig. 10.155

Had we replaced 32 by 90 in (i) then T would not be binary search tree because 90 would be in the left subtree of 53 but 90 is greater than 53.

Example 10.69

Given a sequence of numbers 19, 25, 6, 9, 11, 21, 47, 8, 4, 39, 77. Create a binary search tree for the sequence.

Solution

First of all label the root with 19 and then compare it with 25, we see that 25 > 19, make a right child for the root and label it 25. Next as 6 < 19 (root), label a left child of the root with label 6. Continue this and go to next number 9. 9 < 19 (root) ⇒ 9 will appear as the label of some vertex in the left subtree of the root, but 9 > 6, hence 9 is the right child of the vertex. Continuing this we see that the binary search tree is as shown in the Fig.10.156.

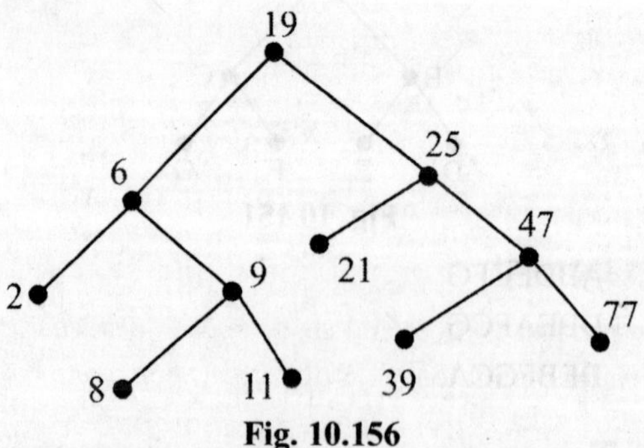

Fig. 10.156

One of the good features of binary search tree is that additional numbers appended to the original sequence can be incorporated into the binary search tree just as if these had been in the sequence all along.

Example 10.70

List the order in which the vertices are processed using preorder, inorder, and postorder interval.

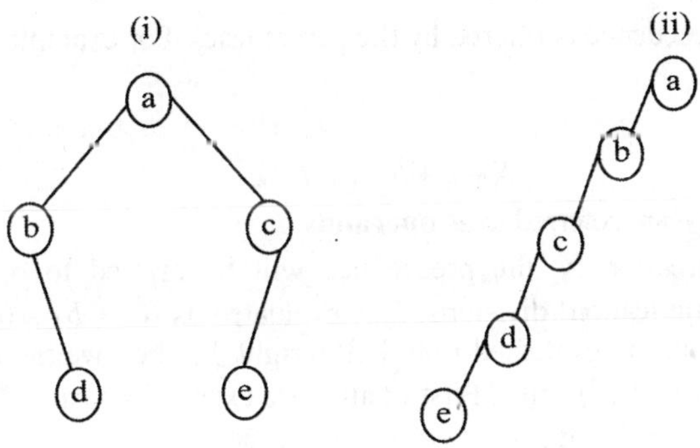

Fig. 10.157

Solution

i. Preorder a b d c e, Inorder b d a e c, Postorder d b e c a

ii. Preorder a b c d e, Inorder e d c b a, Postorder e d c b a

10.59 Path Length

The **path length** of a tree is defined as the sum of the path lengths from the root to all pendant vertices. The path length of the tree shown in the Fig.10.158 is 24 i.e. $1 + 3 + 4 + 4 + 4 + 4 + 4 = 24$.

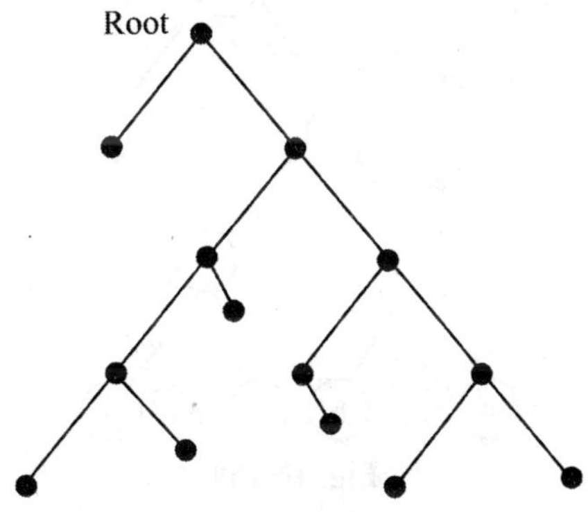

Fig. 10.158

10.59-1 Weighted Path Length

Consider a binary tree T with n pendant vertices and suppose each pendant vertices n_i is associated with positive real number w_i (weight). The **weighted path length** P of tree T is defined to be

$$P = \Sigma\, W_i l_i = W_1 l_1 + W_2 l_2 + \ldots + W_n l_n$$

where l_i is the length of the path from the root (level) to an external vertex N_i

10.60 Expression Trees

Algebraic expression can be represented by binary trees in which vertices of the trees are labeled with numbers, variables of operations that make up the expressions, For each vertex that contains an operators, its left subtree gives the left operand and its right subtree the right operand. The precedence of operators '*' (multiplication) and '/' (division) is considered to be of equal and higher value than that of '+' and '-'. If there is parentheses in and expression then the order of precedence is offered by the parentheses. For example in $a * (b + c)$, we first evaluate $(b + c)$ and then $a * (b + c)$.

Consider the expression

$$X = a * b - c / d + e \qquad \dots (1)$$

The variables a, b, c, d and e are referred to as **operands**.

Now to evaluate the expression X, the precedence will be offered to parentheses. Then according to convention precedence on multiplication/ divisions: X is evaluated as $((a * b) - (c/d)) + e$. Since multiplication/ divisions or addition / subtractions are evaluated from left to right. Further, we can analyse by noting that it is sum of two simple expressions $(a * b) - (c/d)$ and e. First of all these expressions can be decomposed down further into the difference of the expressions $a * b$ and c/d.

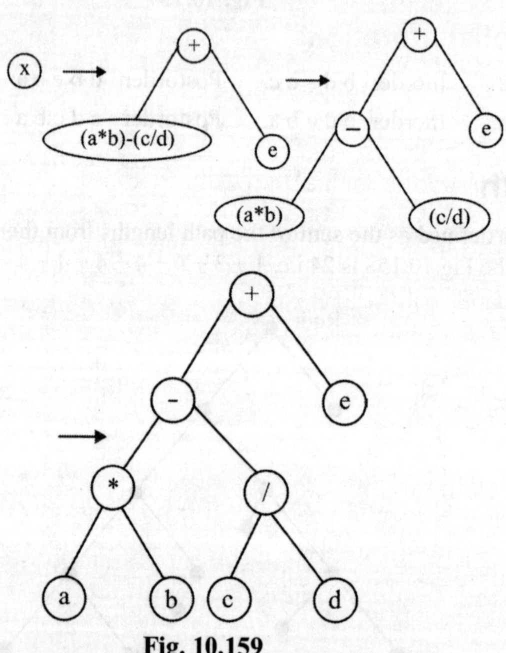

Fig. 10.159

Example 10.70

Draw expression tree for

a. $a * (b - c)/(d + e)$ b. $a^2 - b^2$

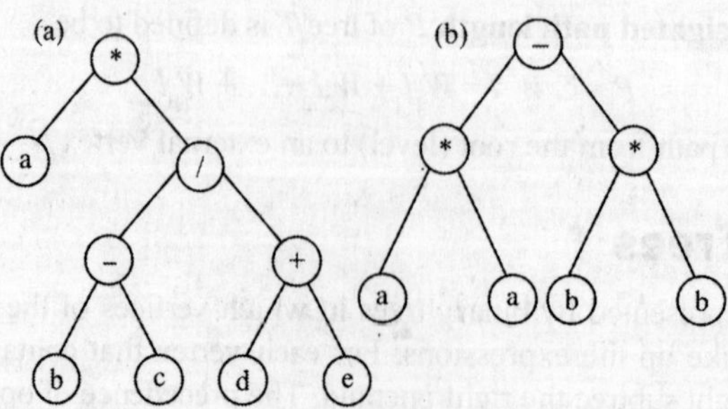

Fig. 10.160

When we decompose any expression into (left expression) (operation) (right expression), the expression tree of that expression is the binary tree whose root contains the operation and whose left and right subtrees are the trees of the left and right expressions, respectively. The evolution of the expression tree for the expression X is shown in the Fig.10.159.

A binary operation applied to a pair of numbers can be written in following three ways:

i. Infix

ii. Prefix

iii. Postfix.

10.60-1 Infix

In the infix form of an expression, an operator appears between its operands such as $a + b$ for the sum of a and b.

10.60-2 PREFIX (or Polish notation)

In prefix, operands follow the operators such as sum of a and b is written as $+ ab$. Expression $(a - b) * c + d/e$ in prefix is written as $+ * - abc/de$.

10.60-3 POSTFIX (or reverse Polish notation)

In postfix, the operator follows its operands such as sum of a and b is written as $ab+$. Expression $(a-b)*c + d/e$ in postfix is written as $ab-c*de +/$

The binary tree representation of the expression $((a + b) * c - (d/e)$ is shown in Fig.10.160.

The connection between traversals of an expression tree and these forms is simple:

The preorder traversal and postorder traversals of an expression tree will result in the prefix and postfix form of the expression respectively, however in order traversal of an operation tree will not in general yield the proper infix form of the expression.

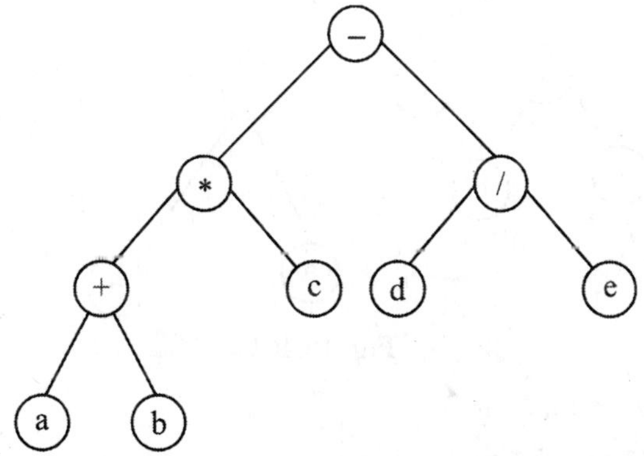

Fig. 10.161

Example 10.72

Represent the expression as a binary tree and write the prefix and postfix forms of the expression

i. $(a + b) * (c - d)$

ii. $(((a + b) * c + d) * e) - ((a + b) * c - d)$

Solution (i)

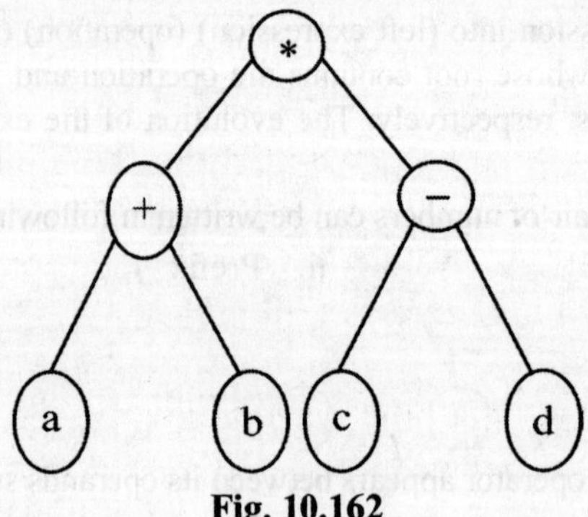

Fig. 10.162

Prefix: $* + ab - cd$

Postfix: $ab + cd - *$

(ii)

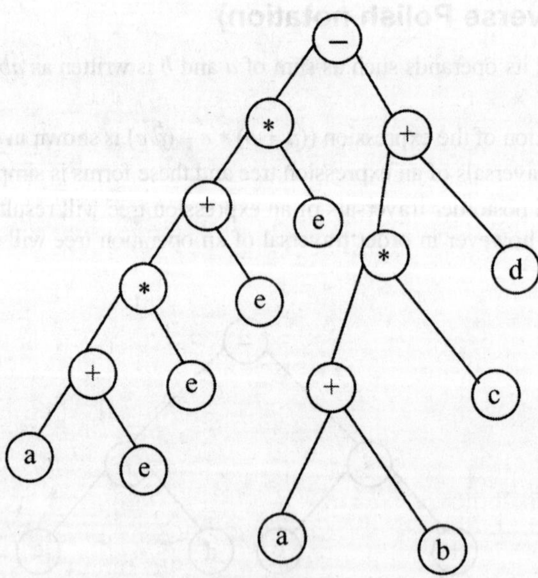

Fig. 10.163

Prefix: $- * + * + a\,b\,c\,d\,e + * + a\,b\,c\,d$

Postfix: $ab + c * d + e * ab + c * d + -$

Example 10.73

Represent the postfix expression as a binary tree and write the prefix form, the infix form and the fully parenthesised infix form of the expression:

i. $ab + c -$

Solution

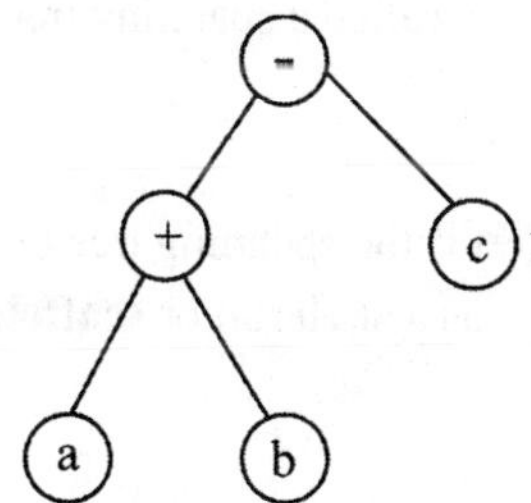

Fig. 10.164

Prefix: $-+abc$

Infix: $a+b-c$

Fully parenthesized infix form: $((a+b)-c)$.

ii. $abc**cde+/-$

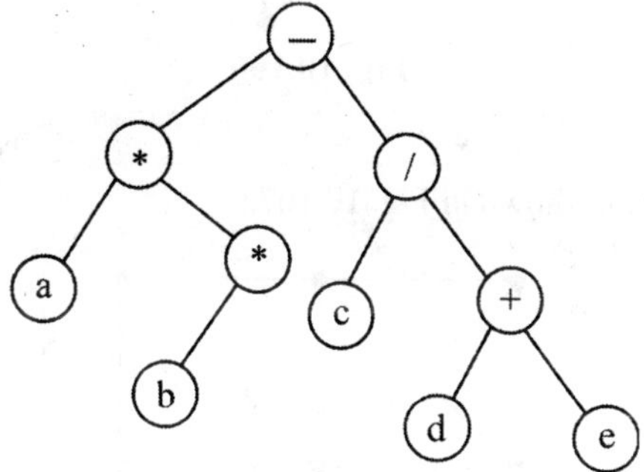

Fig. 10.165

Prefix: $-*a*c/c+de$

Infix: $a*b*c-c/(d+e)$

Fully parenthesised infix form: $((a*(b*c))-(c/(d+e)))$

Example 10.74

Find the value of the expression $abc+-$ if $a=1$, $b=c$, $c=3$.

Solution

–4, try yourself. 10.61 *Spanning Trees*

10.61 Spanning Trees

10.61-1 Spanning Set

Let $G=(V, E)$ is a connected undirected graph. A **Spanning set** for G is defined as a subset F of E such that (V, F) is connected.

10.61-2 Spanning Tree

A subgraph S of a connected graph $G = (V, E)$ is called a spanning tree of G if

i. S is a tree

ii. S contains all the vertices of G.

The subgraph in heavy lines in Fig. 10.166 is the spanning tree of the graph G.

Spanning trees some time also referred to as a **skeleton** or **scaffolding** of G. A spanning tree that is directed is called directed spanning tree of G.

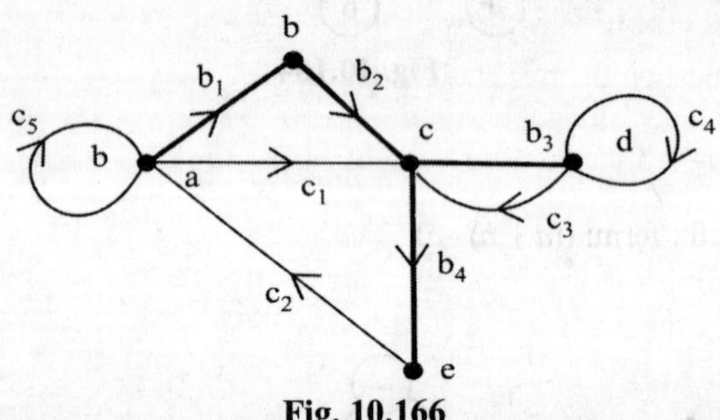

Fig. 10.166

Example 10.75

Draw three spanning trees of graph shown in Fig. 10.167.

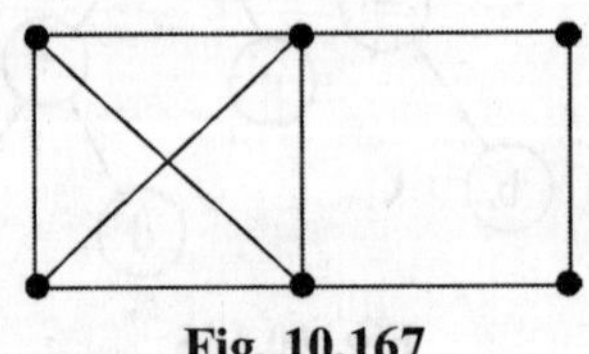

Fig. 10.167

Solution

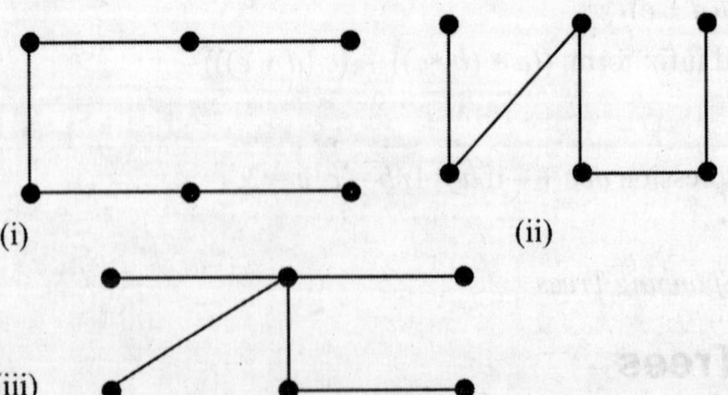

Example 10.76

Consider the graph shown in the Fig.10.167. Suppose that this graph represents a communication network in which the vertices correspond to communication links. What is the largest number of edges that can be deleted while still allowing the stations to communicate with each other?

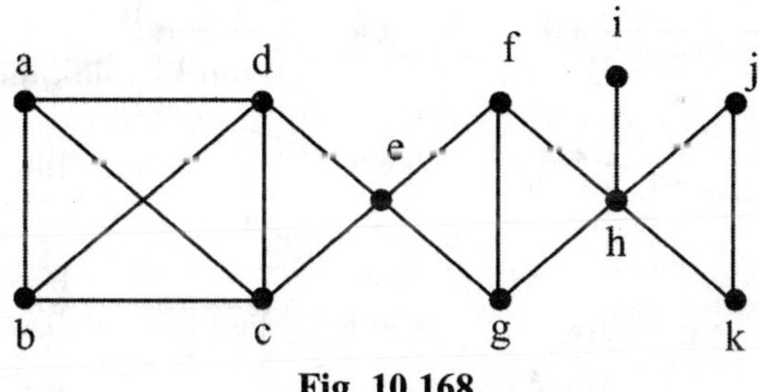

Fig. 10.168

Solution

As we know for communication the two stations are linked in one way. Here the cycle *dced* gives two ways for *d* and *e* to communicate, namely, along the path *dce* or directly from *d* to *e*. Even if the edge $\{d, e\}$ is deleted, *d* and *e* can still communicate via *c*. Likewise one edge of each cycle in *G* can be deleted. Continuing this we will get the spanning tree shown in Fig.10.169.

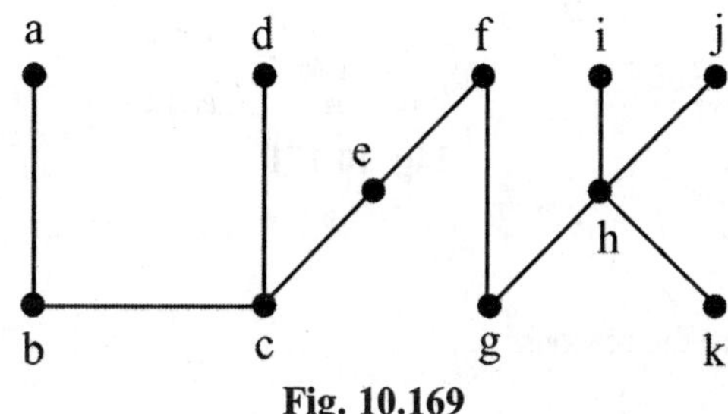

Fig. 10.169

Note that *G* has 15 edges and the spanning tree for *G* has 10 edges so 5 are to be deleted.

REMARK

If *G* is a connected graph with *m* edges, a spanning tree of *G* must have $n - 1$ edges (A tree with *n* vertices has exactly $n - 1$ edges). So, the number of edges that must be removed to obtain a spanning tree be $m - (n - 1) = m - n + 1$. This number is frequently called the **circuit rank** of *G*.

Example 10.77

Which of the following graph in Fig. 10.171(i) to (vi) are spanning tree of the graph *G* shown in Fig.10.170.

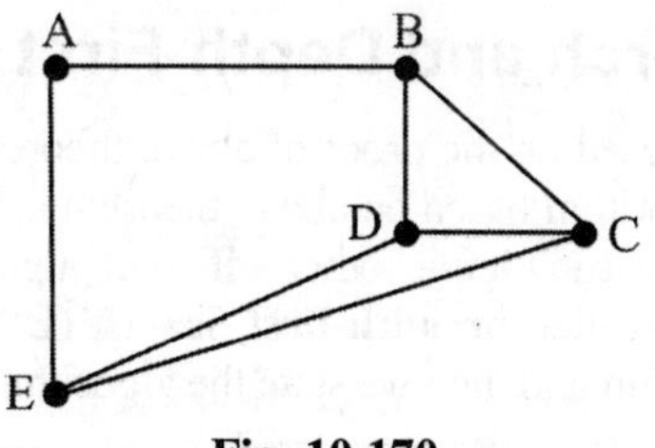

Fig. 10.170

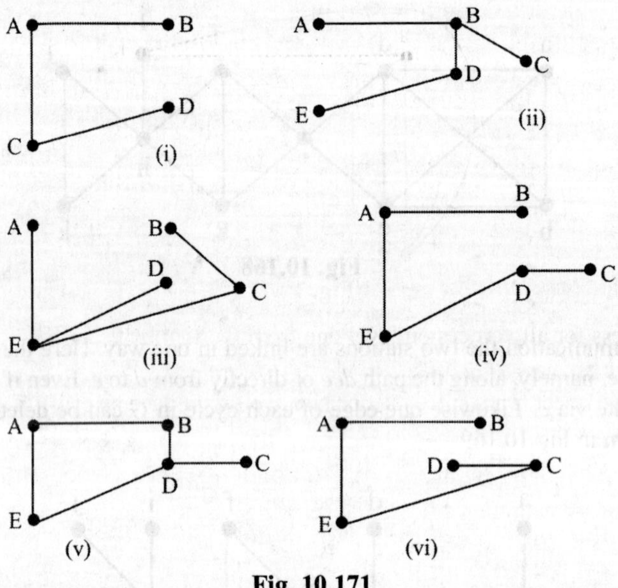

Fig. 10.171

Solution

(ii), (iii), (iv) and (vi).

Readers are advised to enquire the reasons.

Theorem 10.32

A graph *G* has a spanning tree if and only if *G* is connected.

Proof

If *G* has a spanning tree *T*, there is a path between any pair of vertices in *G* along the tree *T*. Thus *G* is connected. If the graph *G* is connected and acyclic, then *G* is a tree.

Suppose that *G* is having a cycle. If we delete an edge (but no vertex), the graph will remain connected. If after removal it becomes acyclic, we stop here. If it contains a cycle, we delete an edge from this cycle. Continuing this way, we finally produce an acyclic, connected graph *T* which is a tree. Since *T* contains all the vertices of *G*, *T* is a spanning tree of *G*.

10.62 Breadth-First Search and Depth-First Search

An algorithm to find a spanning tree based on the proof of above theorem could be designed to find the spanning tree for a connected graph. But the algorithm based on above theorem is not efficient and also time consuming due to process of finding cycles. Here we can define other efficient algorithms for finding a spanning tree of a connected graph. These algorithms are called **breadth-first Search** (BFS) and **depth-first Search** (DFS). Let us take an example to illustrate the algorithm and then we state the algorithm specifically.

Example 10.78

For illustration of breadth first search, see the graph shown in Fig.10.172. The basic idea of BFS is to process all the vertices on a particular level before going to next higher level.

Let us first select an ordering of the vertices say *abcdefgh* of graph *G* shown in Fig 10.172. Then we select *a* as first vertex and label it as the root. At this state *T* consists of the single vertex *a* and no edge. Add to *T* all edges (a, x) and vertices on which they are incident, for $x = b$ to h, that do not produce a cycle when added to *T*. Thus, we add edges (a, b), (a, c) and (a, g) to *T*. (We could use either of the parallel edges incident on *a* and *g*).

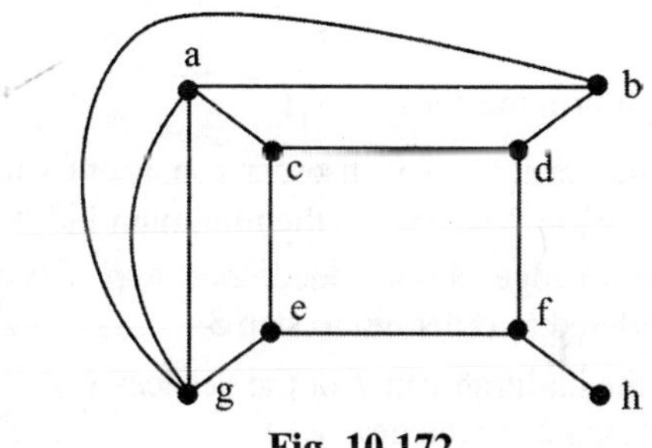

Fig. 10.172

Now repeat the process for all vertices on level one from the root by examining each vertex in the designated order.

b: Include (*b*, *d*). (we reject (*b*, *g*) since its inclusion would produced a cycle in *T*)

c: Include (*c*, *e*). (We reject (*c*, *d*) since its inclusion would produce a cycle in *G*)

g: None (Since inclusion of any of the edge at this stage would produce a cycle)

Repeat this procedure with the vertices on level 2.

d: Include (*d*, *f*)

e: none

Repeat this procedure with the vertices on level 3.

f: Include (*f*, *h*).

As no edges can be added to the single vertex *h* on level 4, the procedure ends. The spanning tree obtained is shown in Fig. 10.173.

The edges which are included in BFS are called **tree edges** and the edges that are rejected in the BFS are called **cross edges**. Hence, BFS partitions the edges of the graph *G* into two sets of of edges namely tree edges and cross edges. BFS can be used to test whether an arbitrary graph *G* with *n* vertices is connected or not. BFS can also be used to find minimum path lengths in an unweighted graph from a fixed vertex *v* to all other vertices. We note that the length of a shortest-path from *v* to a vertex on a level *i* of the spanning tree is *i*. **Dijkstras shortest path algorithm** for weighted graph can be considered as a generalization of BFS. Let us formalize the description of BFS.

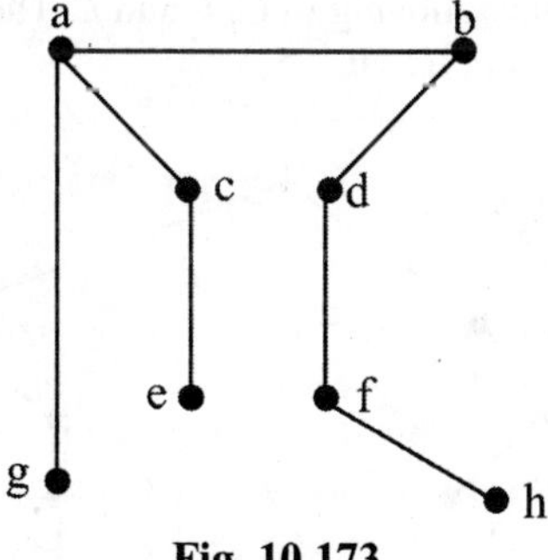

Fig. 10.173

10.62-1 Breadth-First Search for a Spanning Tree

Input: A connected graph *G* with vertices labelled $v_1, v_2, ..., v_n$.

Output: A spanning tree *T* for *G*.

Procedure

1. **Start:** Let v_1 be the root of T. Form the set $V = \{v_1\}$.
2. **Add new edges:** Consider the vertices of V in order consistent with original labeling. Then for each vertex $x \in V$, add the edge (x, v_i) to T where i is the minimum index such that adding the edge (x, v_i) to T does not produce a cycle. If no edge can be added, then stop; T is a spanning tree for G. When all the vertices of V have been considered in order, go to step 3.
3. **Update V:** Replace V by all the children v in T of the vertices x of V where the edges (x, v) were added in step 2. Go back and repeat step 2 for the new set V.

There is an alternative to BFS called **depth-first search**, which proceeds to successive levels in a tree at the earliest possible opportunity.

Example 10.79

Illustration of Depth-First search on the graph of Fig. 10.174.

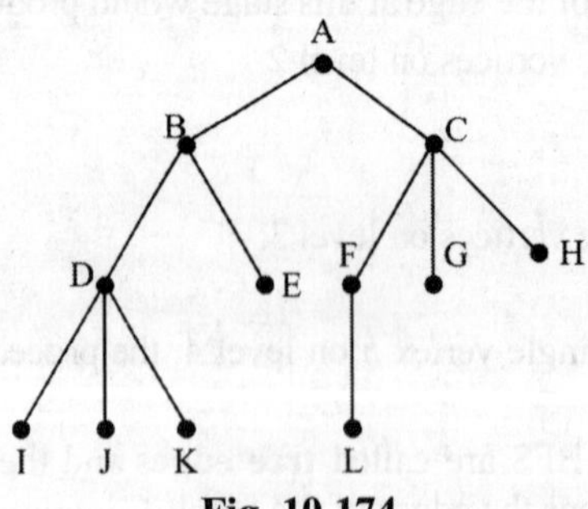

Fig. 10.174

Solution

In depth first Search basic concept is to penetrate as deeply as possible into a tree before fanning out to other vertices. For the illustration of this consider the tree of Fig. 10.174. In order to perform a depth first search, let us start at vertex A and move down to B, D and I. There is no scope for further penetration, we backtrack to D and then go down to j. Again we backtrack to D and go down to K. We now backtrack via D to B, from which we can go down to E. Again, backtracking to A then moving to C, F and L. Then to G and H via C and then backtrack to A. This gives us the labelled tree as shown in Fig. 10.175.

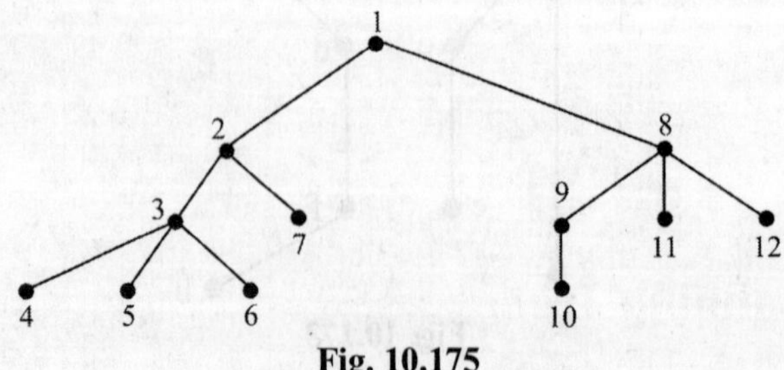

Fig. 10.175

DFS is also called **backtracking** since the algorithm returns to vertices previously visited to add paths.

10.62-2 Algorithm for Depth First Search For A Spanning Tree

Input: A connected graph G with vertices labeled $v_1, v_2, \ldots v_n$)

Output: A spanning tree T of G.

Procedure

1. **Visit a vertex:** Let v_1 be the root of T, and set $L = v_1$. (The name L stands for the vertex last visited)
2. **Find an unexamined edge and an unvisited vertex adjacent to L:** For all vertices adjacent to L, choose the edge (L, v_i) where i is the minimum index such that adding (L, v_i) to T doesn't create a cycle. If no such edge exists, go to step (iii); otherwise, add edge (L, v_i) to T and set $L = v_i$; repeat step (ii) at the new value for L.
3. **Backtrack or terminate:** If x is the parent of L in T, set $L = x$ and apply step (ii) at the new value of L. If L has no parent in T (so that $L = v_i$) then DFS terminates and T is a spanning tree for G.

Example 10.80

Use a DFS to find a spanning tree for the graph G shown in Fig. 10.176.

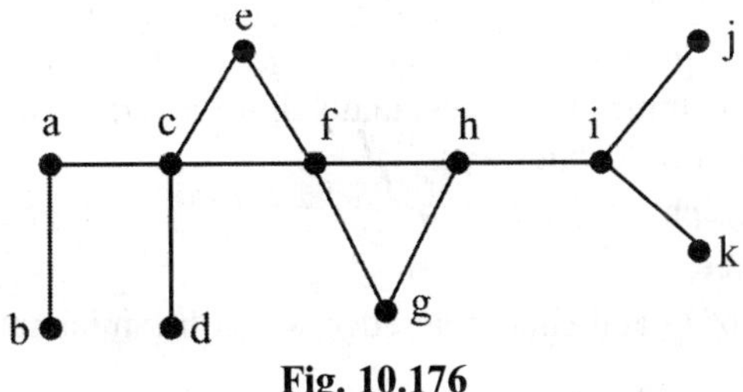

Fig. 10.176

Solution

Let us arbitrarily start with the vertex f (one can start with some other vertex also). Make a path by successively adding edges with vertices not already in the path, as long as possible. In this way, we are getting a path f, e, c, a, b (note that there is a possibility of building other paths). There is no further scope for penetration, we backtrack to a. But, there is no beginning at a containing vertices not in the path already visited. So backtrack to c. Now, form the path c, d. Again backtrack to e, and then to f. From f we can build a path f, g, h, i, j. Then backtrack to i and form the path i, k. This produces the spanning tree in Fig. 10.177.

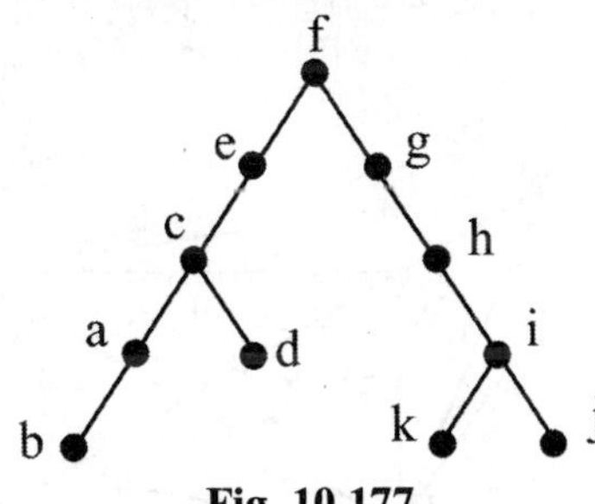

Fig. 10.177

10.63 Minimal Spanning Trees

10.63-1 Weight of a spanning Tree

Consider a connected weighted graph G. The **weight of a spanning tree** T of G is defined as the sum of weights of all the branches in T. Here, one can observe that the different spanning trees will have different weights.

10.63-2 Shortest Spanning Tree

A spanning tree in a weighted connected graph G with smallest weight is called a **shortest spanning tree** or **minimal spanning tree or shortest distance spanning tree.**

Applications

Suppose we have a collection of n cities and we want to construct a communication or transportation network connecting all the cities. If we know the cost of building the links between each pair of cities and our wish is to construct the network as cheaply as possible. Then with the help of shortest spanning tree in a connected weighted graph we can find such network.

10.63-3 Prim's Algorithm to Find Minimal Spanning Tree

Prim's algorithm does not require listing all the edges in order of nondecreasing weight or checking at each step if a newly selected edge from a circuit. This algorithm firstly chooses a vertex and then an edge with minimum weight incident on that vertex. Then iterate by choosing an edge which is not yet selected with minimum weight from the vertices obtained so far. Continue this process untill all the n vertices of graph G have been connected by $n-1$ edges. The algorithm consists of following steps:

Input: A connected weighted graph G.

Output: A minimal spanning tree.

Step 1: Select any vertex v_1 of G and choose an edge which has minimum weight among the edges of G incident with the selected vertex v_1.

Step 2: Now we have two vertices v_1 and v_2 (say). From these two vertices select the edge which is incident on these vertices and has minimum height. In case of tie of weights select any one of them.

Step 3: Continue step 2 with latest vertices in such a way that precisely one end of the new edge (to be selected) lie on the vertices already selected and other end should not lie on any of the selected vertex. Stop when all the n vertices are included.

Example 10.81

Find the minimal spanning tree of the following graph by Prim's algorithm.

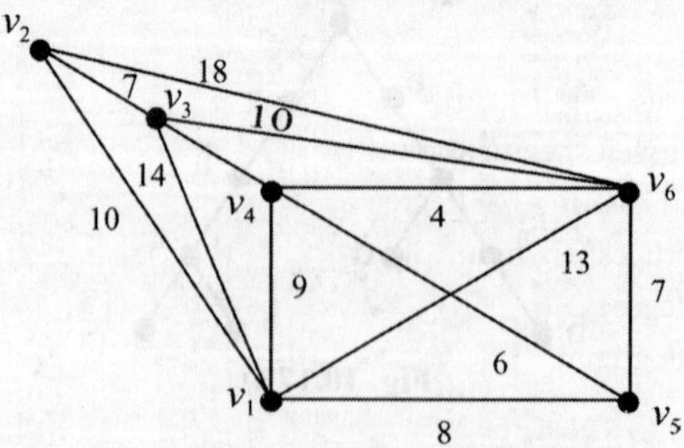

Solution

Step 1: Let us choose the vertex v_1 edge with smallest weight incident on v_1 is (v_1, v_5).

v_1 ———— v_5 (8)

Step 2: Now weights of different edges incident on v_1 and v_5 is given by:

$$w(v_1, v_6) = 13,\ w(v_1, v_4) = 9,\ w(v_1, v_3) = 14,\ w(v_1, v_2) = 10,\ w(v_5, v_6) = 7,\ w(v_5, v_4) = 6.$$

Out of these weights of different edge we choose edge (v_5, v_4) since it is minimum.

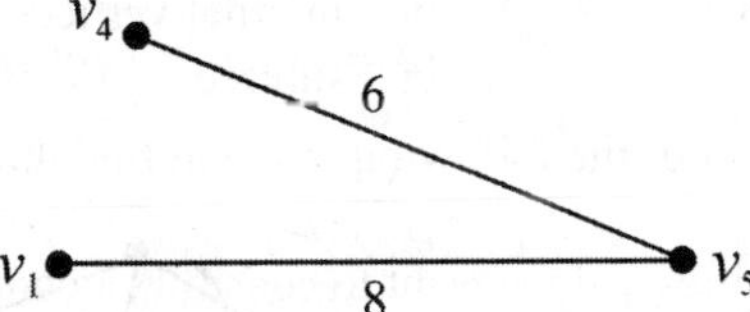

Step 3: Again weights of the all edges incident on v_1, v_4 and v_5.

(Except those whose both end lie on v_1, v_4 and v_5):

$w(v_1, v_2) = 10$, $w(v_1, v_3) = 14$, $w(v_1, v_6) = 13$, $w(v_4, v_3) = 4$, $w(v_4, v_6) = 4$, $w(v_5, v_6) = 7$.

Out of these weights of different edge we choose edge (v_4, v_6) since it is minimum.

It is to be noted here in place of edge (v_4, v_6) we could also have selected edge (v_4, v_3) since it weight is also 4. Selection of either of edge will give same minimal spanning tree. Proceeding similarly next sequences of edges of the resulting minimal tree will be (v_4, v_3), (v_3, v_2).

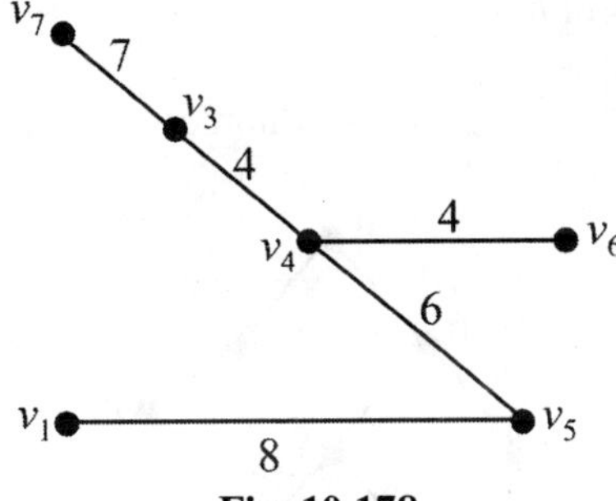

Fig. 10.178

Now all the (6) vertices are covered. One can observe that the minimal spanning tree as shown in Fig.10.178 contains 5 edges. The weight of minimal spanning tree is $8 + 6 + 4 + 4 + 7 = 29$.

Exercise 10.5

1. Find all the trees with six vertices.
2. Find the following graph with given specification if it exists.
   i. Tree with five vertices and having total degree 8.
   ii. Tree with four internal vertices, six terminal vertices.
   iii. Tree with all vertices of degree 2.
3. Show that every tree is bipartite graph.
4. For the following tree find:

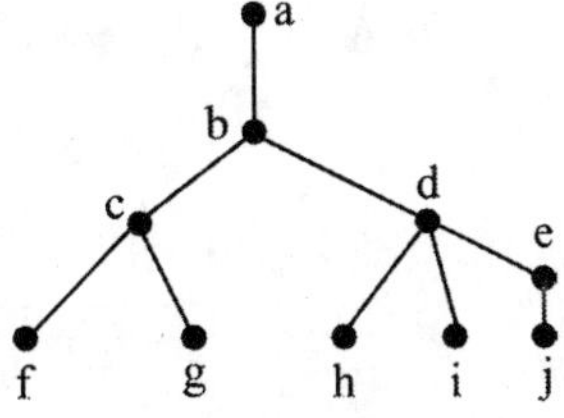

i. parent of d and g ii. children of c and e
iii. descendents of d and e. iv. internal vertices
v. leaves vi. subtree.

5. Find the path from root r to each leaf of the following tree and find the level of (i) h (ii) i.

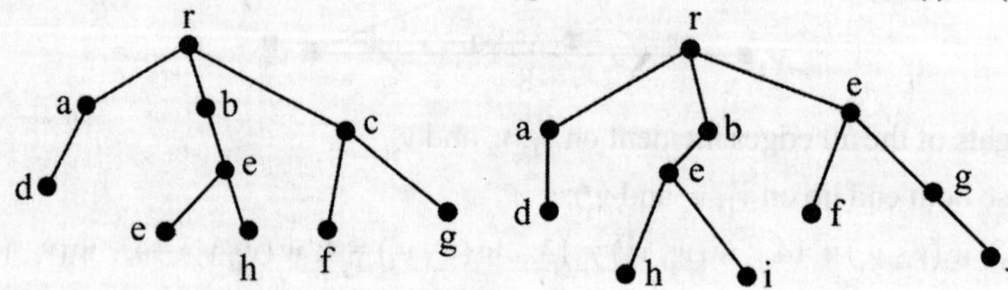

6. Draw a binary tree with six vertices and only one leaf.
7. Draw all distinct binary trees
   a. with three vertices b. with four vertices.
8. Consider the binary tree T in the Fig. 10.18
   Find (i) depth d (ii) the descendants of b
   List the vertices of T in
   i. preorder ii. Inorder
   iii. postorder traversals.

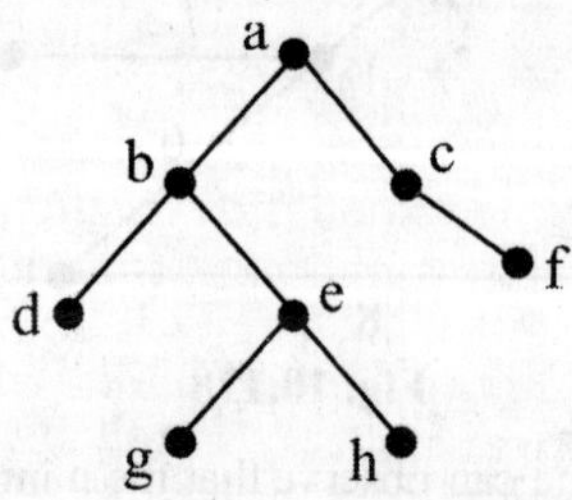

Fig. 10.179

9. Determine whether the following pair of free tree is isomorphic.

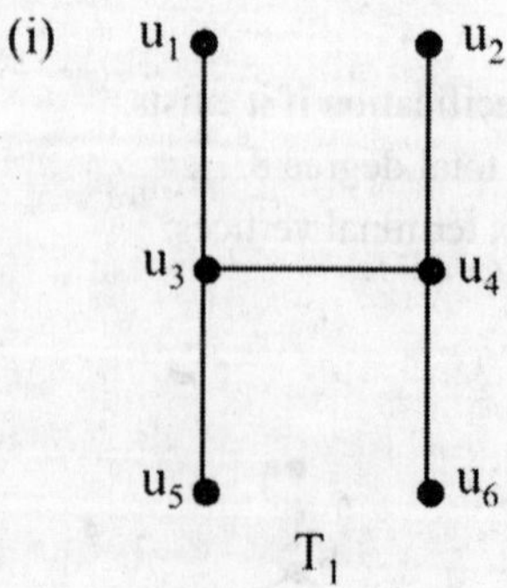

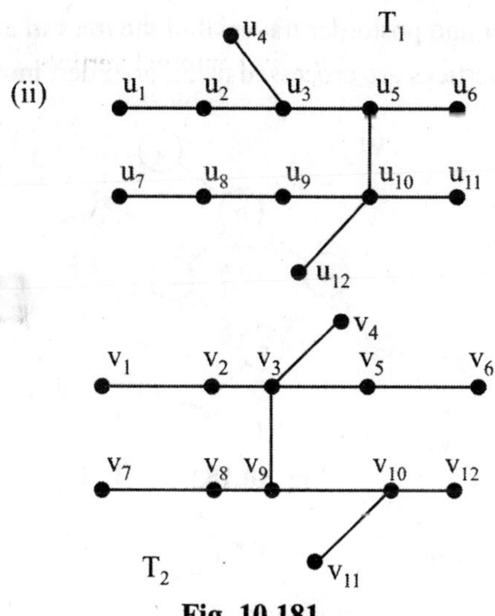

Fig. 10.181

10. List the order in which the vertices are processed using preorder, inorder and postorder traversal for the following:

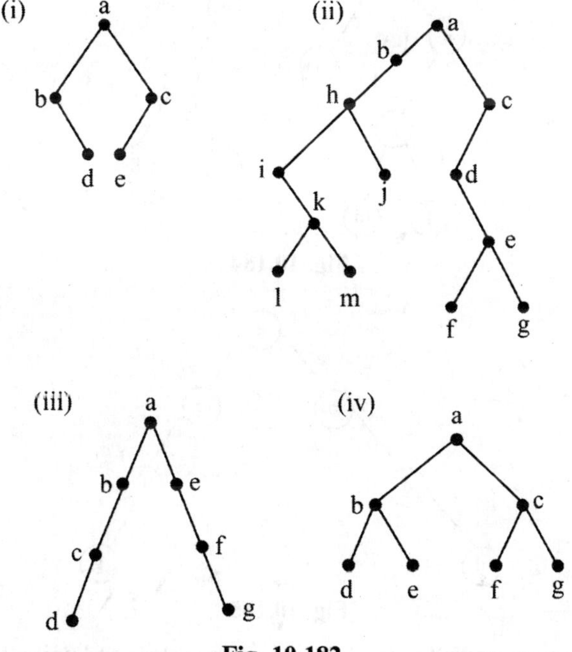

Fig. 10.182

11. Draw the expression tree for the following expressions:

i. $x(y + z)$

ii. $x^2 - 4yz$

iii. $\frac{a^3 \quad 1}{a^2 \quad 1}$

iv. $((dz + c)z + b)z + a$

12. a. List out the preorder, inorder and postorder traversal of the trees in question 11 above.
    b. List the order in which the vertices are processed using preorder, inorder and postorder traversal.

(i)

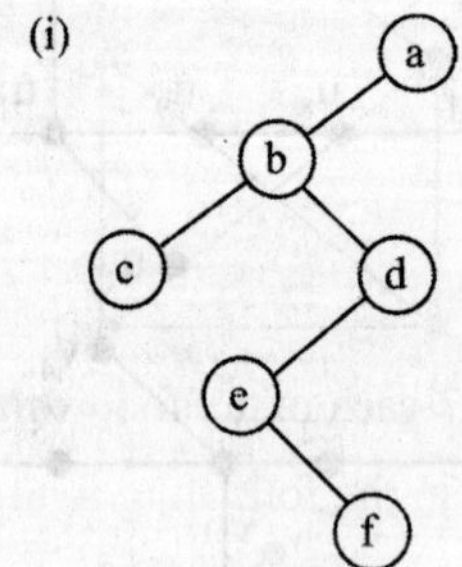

Fig. 10.183

(ii)

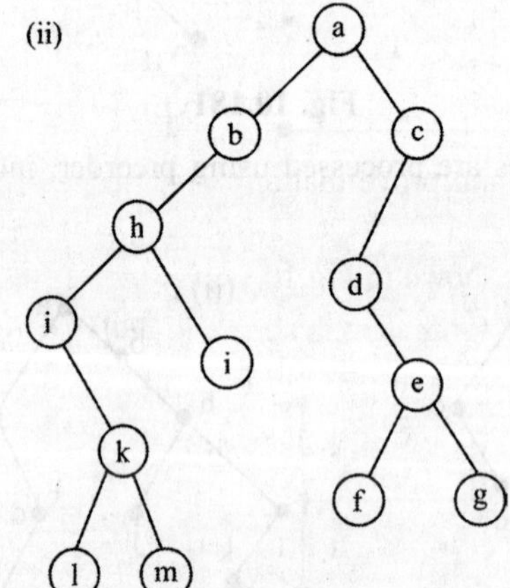

Fig. 10.184

(iii)

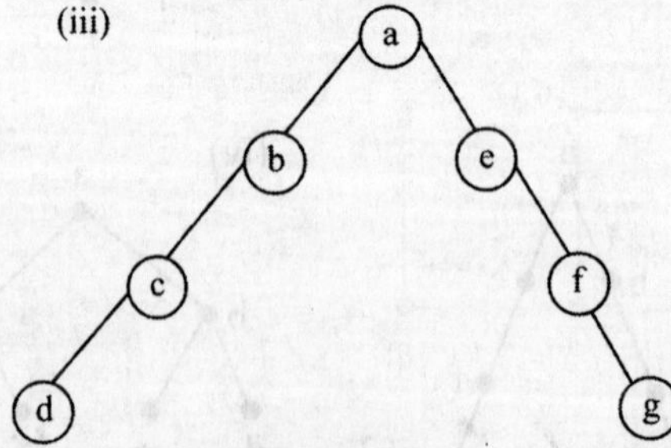

Fig. 10.185

13. a. Draw a binary tree corresponding to the following expressions and write the prefix and postfix form of the expression.

    i. $(a * b + c * d) - (a/d\text{-}(d + e))$ ii. $((a - c) *d)/ (a + (b + d))$.

    b. Represent following postfix expression as binary tree and write down the prefix form, usual infix form and the fully parenthesised infix form of the expression.

    i. $abcd + */e\text{-}$ ii. $ab + cd * ef/ - a *$.

14. i. Determine the different nonisomorphic spanning trees for the graph.

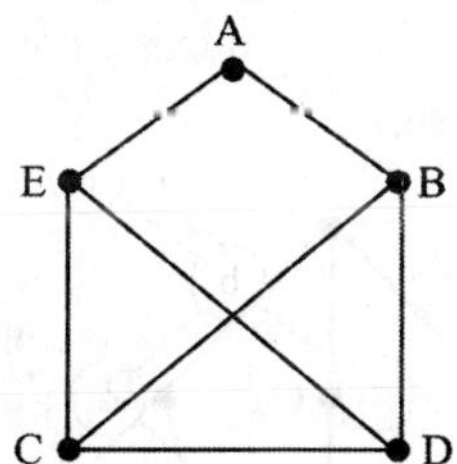

ii. How many different spanning trees does each of the following simple graphs have (a) K_3 (b) K_4 (c) $K_{2,2}$

15. Find all spanning trees of the graph shown in the following figure.

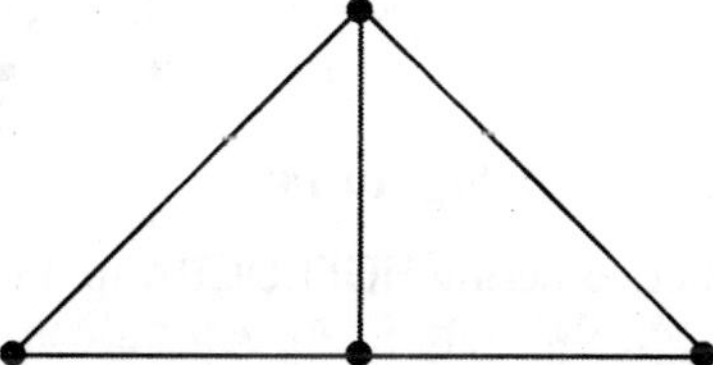

16. Define spanning tree of a graph G and prove that any given edge of a connected graph G is a branch of some spanning tree.
17. Find all spanning trees of the graphs shown in Fig.10.95.
18. Determine a minimum spanning tree for the graph shown in Fig.10.95.

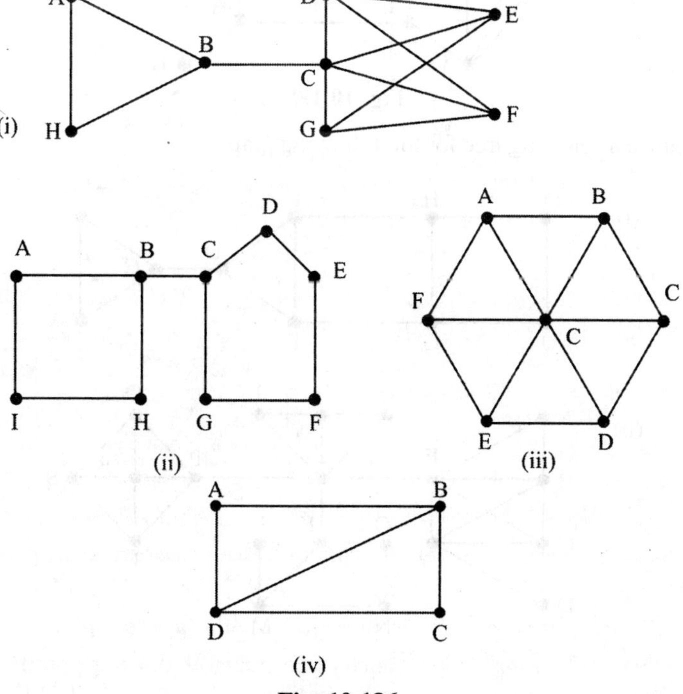

Fig. 10.186

v. The star graph $K_{1, n}$.

vi. C_n, the n-cycle graph.

19. Perform BFS and DFS on the following tree.

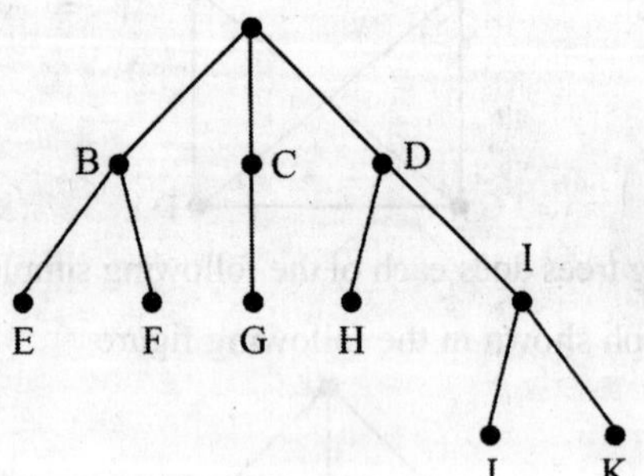

Fig. 10.187

20. Perform BFS and DFS with the vertices ordering HGFEDCBA to find a spanning tree for graph G.

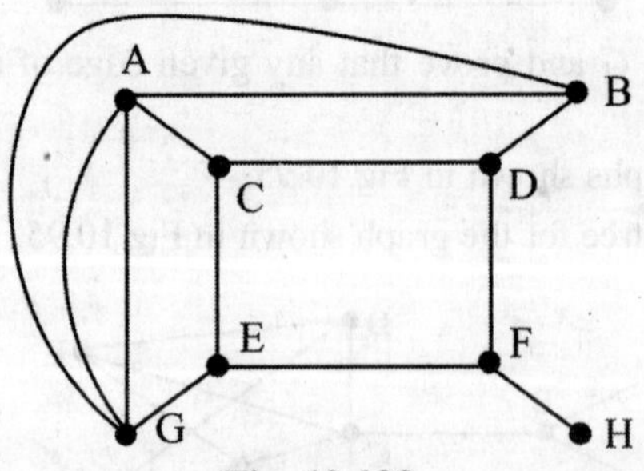

Fig. 10.188

21. Perform DFS to produce a spanning tree for the following graph.

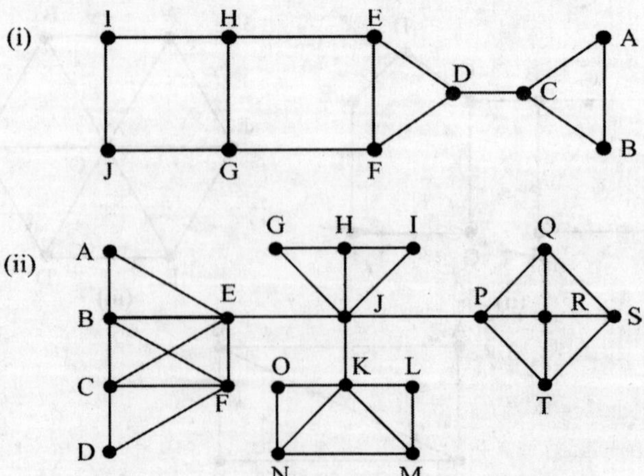

Answers

8. a. (i) $d = 4$ (ii) d, e, g, h

 h (i) *a b d c g h c f* (ii) *d b g e h a c f* (iii) *d g h e b f c*

11.

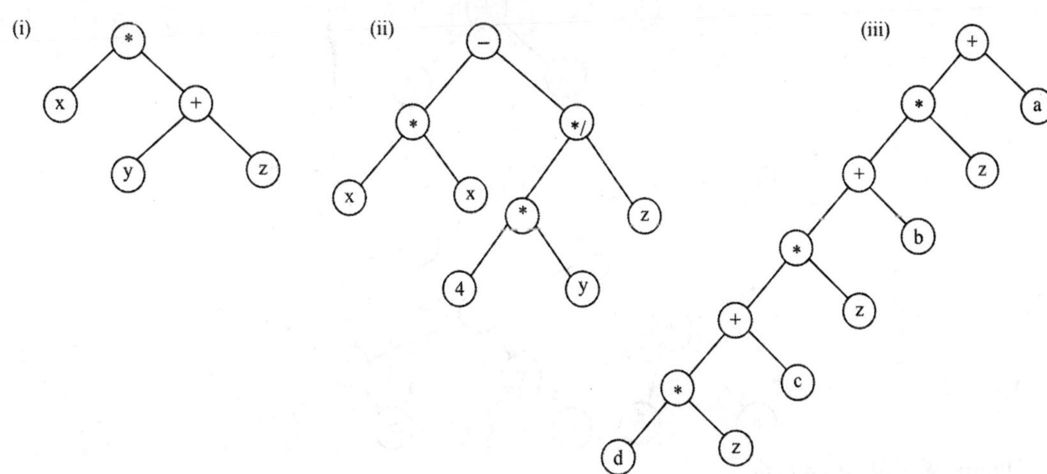

12. a.

 Preorder Inorder Postorder

 $*a + bc$ $a*b + c$ $abc +*$

 b. (iii) Preorder: *abcdefg*, Inorder: *dcbaefg*, Postorder: *dcbgfea*

13. a. (ii)

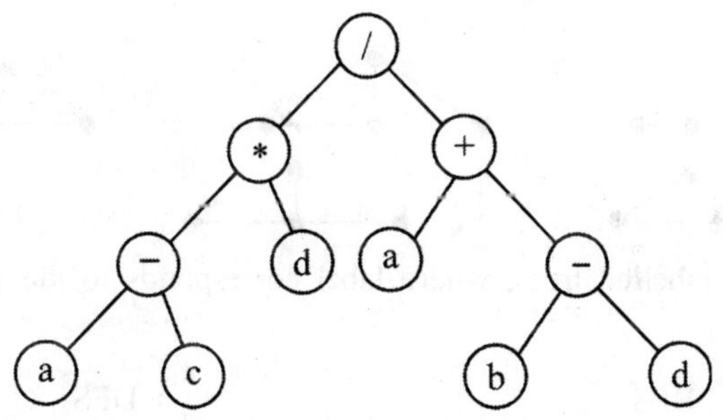

 Prefix form: $/* - acd + a - bd$

 Postfix form: $ac - d *abd - + /$

13. b. (i) Prefix: $- / a * b + cde$

 Infix: $a / (b * (c+d)) - e$

 Fully parenthesised infix: $((a/((b * (c + d))) - e)$

(i)

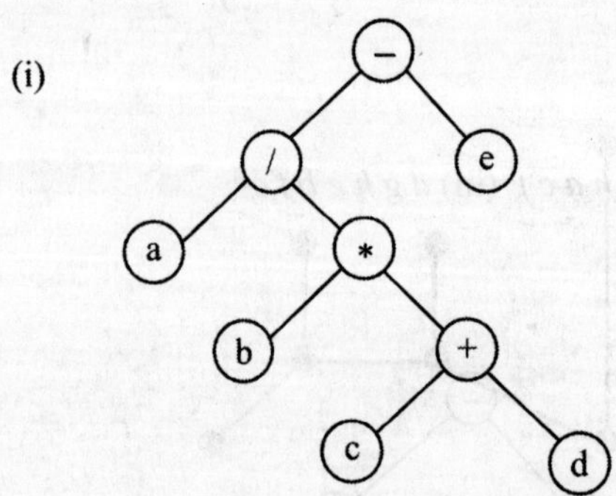

(ii)

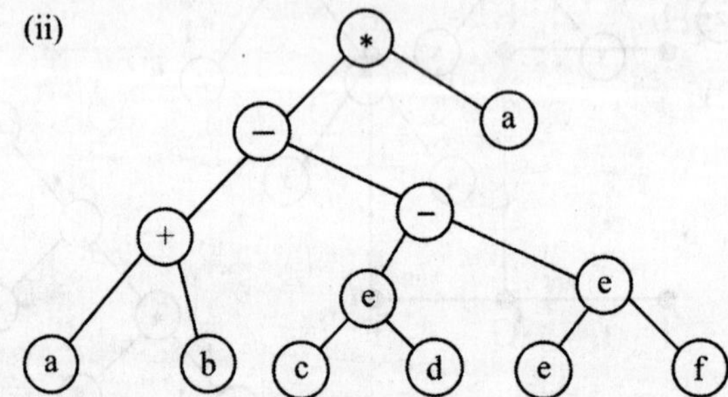

Prefix: *–+ *ab* –* *cd* / *efa*

Infix: $(a + b - (c * d - e/f)) * a$

Fully parenthesized infix form: $(((a + b - (c * d) - e/f))) * a)$.

15.

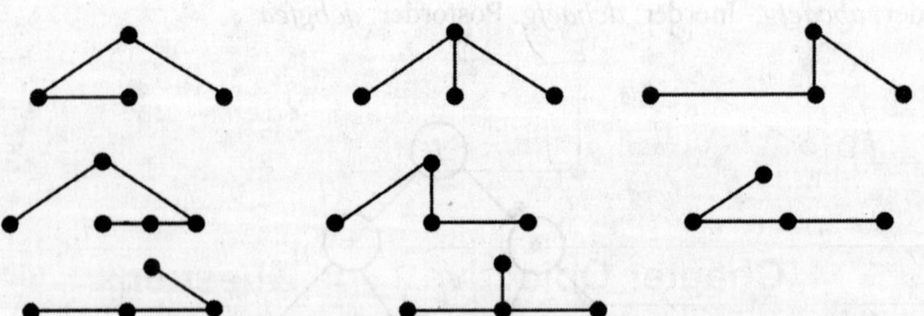

19. We obtain the following labelled trees, where label corresponds to the order in which the vertices are visited.

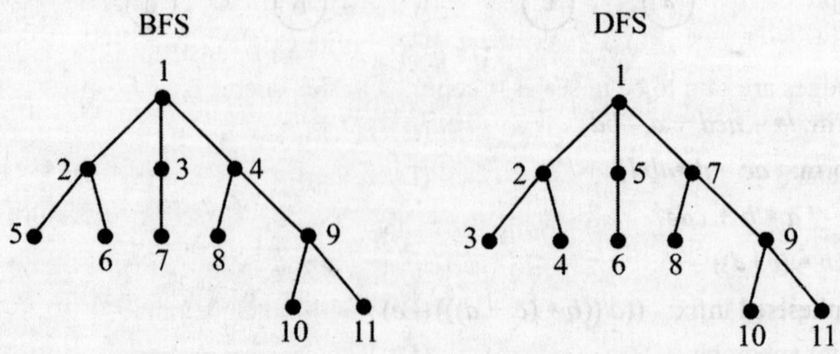

20. **DFS**: Path (H, F, E, G, B, D, C, A)
    BFS:

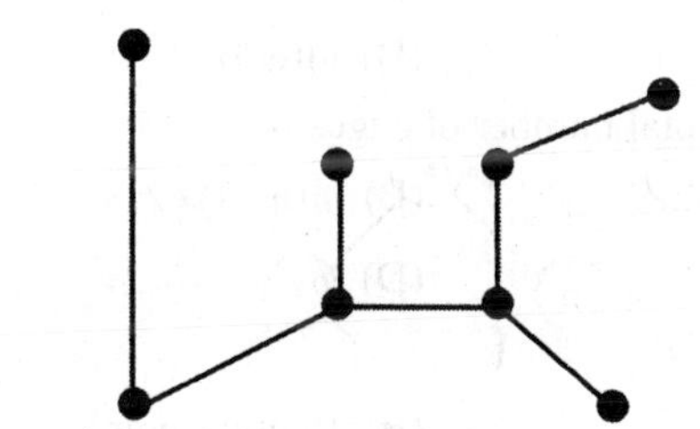

21. DFS (i)

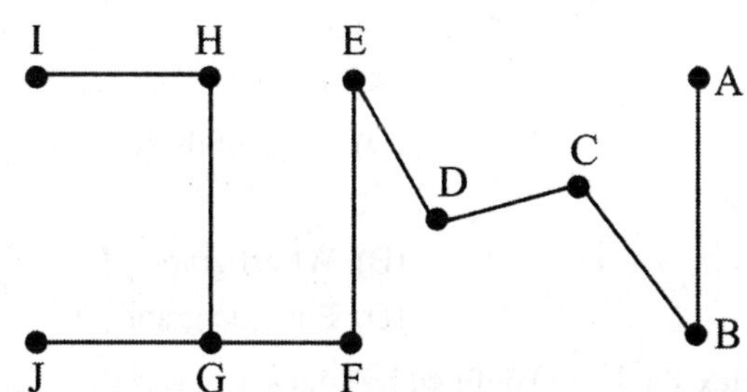

DFS (ii)

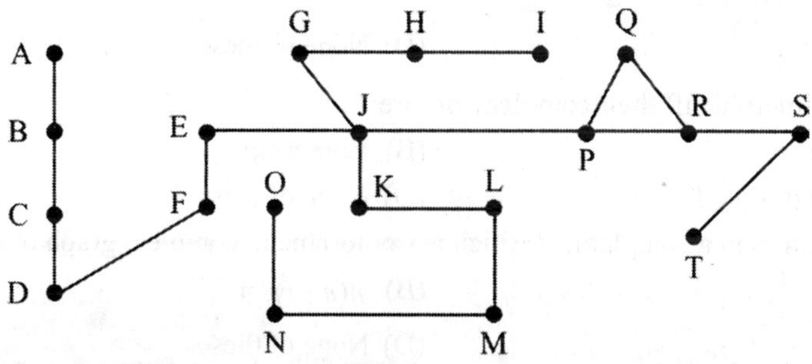

Chapter Objective Type Questions

A. Multiple choice questions.

1. A graph consists of loops and parallel edges is known as:
   (A) Trivial graph (B) Pseudo graph
   (C) Multigraph (D) Finite graph
2. Two adjacent edges are said to be in series if common vertex degree is:
   (A) is equal 2 (B) is less than 2
   (C) is greater than 3 (D) is zero
3. $\sum_{i=1}^{n} d(v_i) = 2e$ is:
   (A) Incidence property (B) Handshaking lemma
   (C) Edge disjoint subgraph (D) Isomorphism property

4. A component is the ____________ part of disconnected graph:
(A) isolated (B) exterior
(C) connected (D) interior

5. In a complete graph of n vertices, total number of edges is:
(A) $n(n-1)(n-2)/2$ (B) $n(n-1)/2$
(C) $(n-1)(n-2)$ (D) n

6. Cubic graph is an example of:
(A) Isolated graph (B) Regular graph
(C) Simple graph (D) Multigraph

7. In a cyclic graph:
(A) $n \le 2$ (B) $n \le 3$
(C) $n \ge 3$ (D) None of these

8. A graph $k_{1,n}$ represents:
(A) Bipartite graph (B) Wheel graph
(C) Star graph (D) Platonic graph

9. The $\bar{G}$ of $G(V, E)$ with vertex set $V(G)$ defined by edges $v_i v_j \in G$ iff:
(A) $v_i v_j \subseteq E(\bar{G})$ (B) $v_i v_j \notin E(\bar{G})$
(C) $v_i v_j \in E(\bar{G})$ (D) None of these

10. Two graphs are isomorphic iff their complements are:
(A) Homomorphic (B) Isomorphic
(C) Both ((A) and ((B) (D) None of these

11. The number of vertices in a complement which is isomorphic to complete graph of n vertices is:
(A) $n(n-2)$ (B) $n(n-1)/4$
(C) $n(n-1)$ (D) None of these

12. An undirected graph $G = (V, E)$ has:
(A) No self loops or no multi-edges (B) Self loops or multi-edges
(C) No self loops and multi-edges (D) Self loops and no multi-edges

13. A graph G without parallel edges or self loops with n nodes and k components have atmost ________ edges.
(A) $(n-k+1)/2$ (B) $(n-k)(n-k+1)/2$
(C) $((n-k)(n-k+1)+1)/2$ (D) None of these

14. $\Delta = \underset{u,v \in l}{\text{Max}}\, d(u, v)$ represents ____________ of a graph.
(A) area (B) diameter
(C) Both ((A) and ((B) (D) none of these

15. $\chi(G)$ represents the __________ of a graph.
   (A) Degree (B) Chromatic number
   (C) Pictorial (D) Edges number
16. Pre-order traversal of the following tree is:

A
B C
D E F G

   (A) ABDECFG (B) ADBEFCG
   (C) ABEDCFG (D) None of these
17. Which of the following is infix form of expression?
   (A) $+a*b$ (B) $c+a*b$
   (C) both ((A) and ((B) (D) None of these
18. A complete graph K_n will be planar if:
   (A) $n \leq 1$ (B) $n \geq 4$
   (C) $n \leq 4$ (D) All of these
19. A graph is said to be planar if it can be drawn on a plane in such a way that:
   (A) Two of its edges intersect (B) No two of its edges intersect
   (C) Its edges intersect and perpendicular (D) None of these
20. The four colour conjecture states that are sufficient for colouring any map on plane.
   (A) four colours (B) one of the four colour
   (C) both ((A) and ((B) (D) none of these

B. Fill in the blanks.

1. A ________ graph has one vertex and no edge.
2. An edge of a graph which joins a vertex to itself is called a __________.
3. A multigraph consists of __________ edges.
4. Even or odd property of a vertex depends upon its _________.
5. A pendant vertex has ____________ degree.
6. • • • • • • is an example of __________ graph.
7. The graphs formed by using five regular platonic solids are known as ___________.
8. In a ___________ every edge is directed.
9. A mixed graph consists of _____________ and ____________ edges.
10. An undirected graph consists of only ___________ edges.

11. A source is a vertex with __________.
12. A subgraph consisting of all the vertices of its graph is known as __________.
13. A trail is a __________ with no repeated edge.
14. A trail which begins and ends with same vertex known as __________.
15. A complete graph with more than two vertices is called __________.
16. The points of intersection of edges are called __________.
17. __________ colouring is known as the colouring the vertices of graph.
18. A connected graph that contains no cycles is called a __________.
19. A leaf is a vertex of degree __________.
20. A complete bipartite graph $K_{m,n}$ is __________ if $m \leq 2$ or $n \leq 2$.
21. __________ is the first graph of Kuratowski with five vertices.
22. An infinite region can be changed by changing __________ of a planar graph.
23. In any undirected graph the sum of degrees of all the nodes are __________ the number of edges.
24. The number of edges in an undirected graph with two vertices of degree 7, four vertices of degree 5, and the remaining four vertices of degree 6 is __________.

C. Which of the following is True (T) / False (F)?

1. A vertex of degree zero is called isolated vertex.
2. The number of vertices of odd degree in a graph G is always odd.
3. A null graph consists of only isolated vertices.
4. The sum of degrees of all vertices of a graph G is twice the number of edges in G.
5. All vertices of equal degree make a regular graph.
6. Edges of complete graph (C_n) is less than edges of a Path graph (P_n).
7. A weighted graph is a graph which uses weights like capacity, probability, distance etc.
8. A vertex with zero outdegree is called sink.
9. A subgraph of a subgraph of graph G is a subgraph of G
10. A graph with seven vertices is isomorphic to its complement.
11. A close walk has different start and end vertex
12. A path is a walk which includes every edge twice except the initial vertex.
13. The diameter of a graph $G = (V, E)$ is the maximum distance between any two edges.
14. A connected graph G is an Euler graph iff all vertices of G are of even degree.
15. Every platonic graph is Hamiltonian.
16. Region is a property of the specific plane representation of a graph.
17. A planar graph has exactly one infinite region.
18. In proper colouring, the two adjacent vertices has the same colour.
19. A biochromatic graph uses more than two colours for colouring
20. A complete graph of m vertices requires m colours.
21. The chromatic number of n graph is greater than or equal to total number of vertices in G.
22. A tree with only one vertex is called empty tree.
23. In a full binary tree every vertex has either zero or two subtrees

24. Two binary trees with same structure are known as similar binary trees.
25. Any finite planar graph can be coloured using five colours.
26. A particular planar representation of a finite planar multigraph is called a map.
27. Every graph is not its own subgraph.
28. The terminal vertex of a graph are of degree two.
29. A tree with n vertices has n edges.
30. A single vertex in a graph G is a subgraph of G.

Answers Objective Type

A.

1.	B	2.	A	3.	B	4.	C
5.	B	6.	B	7.	C	8.	C
9.	B	10.	B	11.	B	12.	B
13.	B	14.	B	15.	B	16.	A
17.	B	18.	C	19.	A	20.	A

B.

1.	Trivial	2.	Loop	3.	Parallel
4.	Degree	5.	One	6.	Null
7.	Platonic Graph	8.	Digraph\	9.	Directed, Undirected
10.	Undirected	11.	Zero	12.	Spanning subgraph
13.	Walk	14.	Circuit	15.	Hamiltonian
16.	Crossovers	17.	Vertex	18.	Tree
19.	One	20.	Planar	21.	K_5
22.	Embedding	23.	Twice	24.	29

C.

1.	T	2.	F	3.	T	4.	T
5.	T	6.	F	7.	T	8.	T
9	T	10.	F	11.	F	12.	F
13.	F	14.	T	15.	T	16.	T
17.	T	18.	F	19.	F	20.	T
21.	F	21.	T	22.	T	23.	T
24.	T	25.	T	26.	F	27.	F
28.	F	29.	T				

CHAPTER

11 Automata

11.0 Introduction

In this chapter we begin with the study of strings, alphabets, operations on strings and theory of automata. We survey the transition graph of automaton with table and also deal with the reachability of any string by automaton and discuss about the equivalence of deterministic and non-deterministic finite automaton. We will also focus on the equivalence of Moore and Mealy machines. In the last stage of this chapter, we give a brief idea about the minimization of deterministic finite automaton.

11.1 Basic Concepts of Automaton Theory

In this section we will discuss various terms used in automaton theory.

11.1-1 Alphabets

It is a finite non-empty set of characters (symbols), denoted by Σ (sigma). It contains:

i. Binary alphabet e.g. $\Sigma = \{0, 1\}$.

ii. The set of all lower-case English letters. e.g. $\Sigma = \{a, b, c\ldots\}$.

iii. It can be of the form $\Sigma = \{0, 1, a, b, \ldots\}$.

11.1-2 String over an alphabet

A string over an alphabet set Σ is a finite ordered list of characters chosen from the set Σ.

Example 11.1

i. 00110, 01, 00, 1 are all strings over an alphabet set $\Sigma = \{0, 1\}$.

ii. *abab*, *aaabbb*, *ab*, *a* are all strings over an alphabet set $\Sigma = \{a, b\}$.

The lower-case letters u, v, w, ... denotes string names.

Example 11.2

Let $w = ababa$. Here w is a string having the specific value *ababa*.

11.1-3 Length of String

Let u be a string, then the **length** of u is denoted by $|u|$ and defined as "the number of positions for characters that it contains".

Example 11.3

i. If $u = 010011$, then $|u| = 6$.

ii. If $u = abaa$, then $|u| = 4$.

In general, we say that the length of string is "the number of characters" in the string which means that "the number of positions of symbols in the string".

11.1-4 Empty String

The string having no symbol is called **empty string**. In other words "If the length of string is zero then it is called empty string or **null string**". This is denoted by $\in$. Here it is to be noted that $|\in| = 0$.

11.1-5 Reverse of a String

It is obtained by writing the symbols in reverse order. If u is a string then u^R denotes its reverse.

Example 11.4

i. If u = abcbca then u^R = acbcba.

ii. If $w = a_1\, a_2 \ldots a_{n-1}\, a_n$ then $w^R = a_n\, a_{n-1} \ldots a_2\, a_1$.

11.1-6 Powers of an alphabet

Let Σ be the set of alphabet, than Σ^m is the set of all strings of length m with symbols from Σ.

Example 11.5

If $\Sigma = \{a, b\}$, then $\Sigma^0 = \{\in\}$ because only empty string is of zero length.

$\Sigma^1 = \{a, b\}$, set of all strings over $\Sigma = \{a, b\}$ of length one.

$\Sigma^2 = \{aa, ab, ba, bb\}$, set of all strings of length 2.

$\Sigma^3 = \{aaa, aab, aba, baa, abb, bab, bba, bbb\}$ set of strings of length 3and so on.

In example 11.5, we have $\Sigma = \{a, b\}$ and $\Sigma^1 = \{a, b\}$. But one should not get confused between these two. As it is clear from the example that $\Sigma = \{a, b\}$ is the set of alphabet whose members are symbols a and b while $\Sigma^1 = \{a, b\}$ is a set of strings of length one.

11.1-7 The Kleene Star (closure)

Let Σ be the set of alphabet. The kleene star Σ^*, called the **closure** of the alphabet set Σ, is the set of all strings over Σ.

Example 11.6

i. If $\Sigma = \{0\}$, then $\Sigma^* = \{\in, 0, 00, 000, \ldots\}$.

ii. If $\Sigma = \{a, b\}$, then $\Sigma^* = \{\in, a, b, aa, ab, ba, bb, \ldots\}$

iii. If $\Sigma = \{a, b, c\}$, then $\Sigma^* = \{\in, a, b, c, aa, ab, ac, ba, bb, bc, \ldots\}$

iv. If $\Sigma = \varnothing$ (empty set), then $\Sigma^* = \{\in\}$.

In another way it can be written as $\Sigma^* = \Sigma^0 \cup \Sigma^1 \cup \Sigma^2 \cup \ldots$

Σ^+ is the set of all non-empty strings from alphabet Σ, i.e. $\Sigma^+ = \Sigma^* - \{\in\}$.

Example 11.7

If $\Sigma = \{x\}$, then $\Sigma^* = \{\in, x, xx, xxx \ldots\}$ and $\Sigma^+ = \{x, xx, xxx \ldots\}$.

11.1-8 Sub string

Any string *u* is called a **substring** of string *w* if it appears within *w*.

Example 11.8

If *w* = *abbaba* be a string over an alphabet set $\Sigma = \{a, b\}$ then *a*, *ab*, *abb*, . . . are all substrings of *w*.

11.1-9 Concatenation

Let *u* and *v* be any two strings of alphabet Σ. The concatenation of *u* and *v* denoted by *uv* is a new string obtained by *u* followed by *v*.

Let $u = a_1 a_2 \ldots a_i$ and $v = b_1 b_2 \ldots b_j$ then $uv = a_1 a_2 \ldots a_i b_1 b_2 \ldots b_j$

Since length of *u* = *i* and length of *v* = *j*, therefore length of *uv* = *i* + *j*. *i.e.* the length of concatenation of two strings is the sum of the individual length. i.e. $|uv| = |u| + |v|$. So concatenation with empty string does not change the string i.e., $\in$ is called **identity** for concatenation, e.g. $u\in = u = \in u$.

If *w* = *u.v*, then *u* is called **Prefix** of *w* and *v* is a **Suffix** of *w*. Generally, if *w* = *uv* then *u* and *v* are substrings of *w*. Also $u\in = u = \in u$ therefore every string is a substring of itself, is a prefix of itself and is a suffix of itself.

Example 11.9

i. Let *u* = *abaa*, *v* = 0110 then *uv* = *abaa*0110 and *vu* = 0110*abaa*.

ii. If *w* = 0111, then the

   Set of prefixes = {$\in$, 0, 01, 011, 0111}.

   Set of suffixes = {$\in$, 1, 11, 111, 0111}.

11.1-10 Palindrome

The string of the form *abcba*, *abaaba* etc. are called **palindrome**. The same can be said in other words as the string which is same whether written forward or backward.

If the length of palindrome is even then it can be obtained by the concatenation of string and its reverse.

11.1-11 Languages

Let Σ be a set of alphabet and Σ^* is the set of all strings over Σ. Consider a subset of Σ^*. This subset is called **language** over Σ and denoted by *L* i.e. $L \subseteq \Sigma^*$.

Some important facts about languages.

1. It can be empty i.e. L = $\varnothing$. The language with this property is called **empty language**. The empty language $\varnothing$ is the language over any alphabet Σ.
2. If a language consists of only empty string, it will be denoted by $L = \{\in\}$ and it is language over any alphabet.

3. Languages of all strings having *n a*'s followed by *n b*'s or $n \geq 0$ are given by the set {∈, *ab*, *aabb*, *aaabbb*, ...}.
4. Set of all strings over Σ = {*a*, *b*} with equal numbers of *a*'s and *b*'s is {∈, *ab*, *ba*, *abab*, ...}.
5. Σ* is a language for any alphabet Σ.

One must not get confused with ∅ and {∈}. ∅ has no string i.e. empty language whereas {∈} is a language having one string namely ∈.

Example 11.10

If Σ = {*a*, *b*} then Σ* = {∈, *a*, *b*, *aa*, *ab*, *ba*, *bb*, ...}. Then the set { *ab*, *abb*, *aabb* } is a language over Σ. Since its length is finite so it is a finite language.

11.1-12 Regular language

A language is called **regular language** if some finite automaton recognizes it.

11.2 Deterministic Finite Automaton

The first type of automaton we study in detail are finite acceptors which are deterministic in their operation. The term 'deterministic' means that on each input there is one and only one state to which the automaton can transist from its current state. A finite automaton has several parts. It has set of states and rules for going from one state to another, depending upon the input symbol. Further, it has an input alphabet that indicates the allowed input symbol. It also has an initial state (or start state) and a set of **final states** or (**accepting states**). Finite automaton can be defined as follows:

A **deterministic finite automaton** (in short **DFA**) is a list of five objects (5-tuple) $A = (Q, \Sigma, \delta, q_0, F)$

Where A is the name of DFA

Q is a finite set of states

Σ is a finite set of input symbols

q_0 is a start state or initial state and $q_0 \in Q$

F is a set of final states and $F \subseteq Q$.

δ is a transition function defined from $Q \times \Sigma$ to Q i.e. it maps a state and an input symbol to a state.

Now we will describe the operation of finite automaton in detail. DFA is a simple language recognition device. The block diagram of a DFA is given in Fig. 11.1.

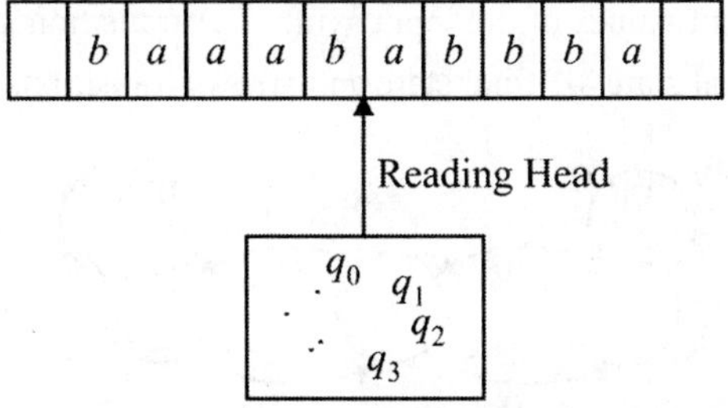

Fig. 11.1: Block diagram of a Finite Automaton

Main parts of a finite automaton are as follows:

11.2-1 Input tape

A single dimension array of squares constructs an input tape. In this each square will contain a single symbol from input alphabet Σ.

11.2-2 Reading head

It is a movable reading head (pointer) which can move either to the left or to the right and observe the position of a symbol on the input tape. Initially, it is placed at the leftmost square of the tape and the movement of reading head to the right side only.

11.2-3 Finite control

It is main part of a finite automaton, also called **black box** and contains all the transition states either **deterministic** or **non-deterministic** of set Q, where deterministic means a state not moves towards more than one state at a time whereas in case of non-deterministic one state may move to different states.

11.3 Transition Function

A function usually denoted by 'δ' whose input is a pair of state and alphabet symbol and whose output is a single state, i.e. it maps a state and an input string of alphabet into a state. Sometimes it is called as **next state function** because it causes the change in the state.

Mathematically, if a finite automaton A is in state $q_i \in Q$ and the symbol read from the input tape is $x_j \in \Sigma$, then $\delta(q_i, x_j) = q_k$ and $q_k \in Q$, which is an unique state to which A passes.

As we know that a finite automaton having a set of finite states and a set of input alphabet Σ. A transition function takes one state and one input symbol returns to a state. This notion gives us a rule for transitions in a DFA. For each input symbol there is exactly one transition out of each state. Initially, there is one state from which the automaton starts, is called **initial** or **start** state denoted by q_0 is indicated by the arrow labeled "Start". A state with a double circle is called **final state**.

A DFA operates in the following manner:

When DFA receives an input string, the processing begins with A's initial state. The automaton receives the symbols from the input string one by one from left to right. After reading each symbol, A moves from one state to another along the transition (the arrows going from one state to another) taking the labelled value. After reading the last symbol, A gives its output. Now if A is in the final state, then the output is accepted otherwise rejected.

Now we can define, a **transition graph** as a graphical representation of a finite automaton in which the vertices represent states and the edges represent transitions. The labels of vertices are the names of states while the labels of the edges are the current values of the input symbol. A transition graph is shown in Fig. 11.2.

In this graph q_0 represents initial state, q_2 final state and arrows transitions.

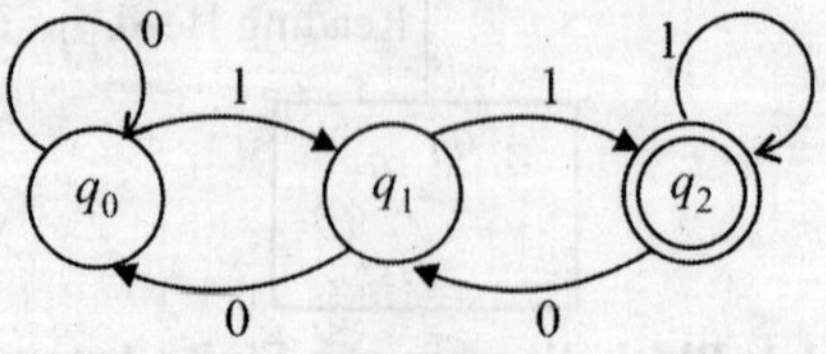

Fig. 11.2

Example 11.11

In the graph shown in Fig. 11.2 represents a DFA,

$$A = (Q, \Sigma, \delta, q_0, F),$$

where $Q = \{q_0, q_1, q_2\}$, $\Sigma = \{0, 1\}$, $F = \{q_2\}$

and δ is given by

$\delta(q_0, 0) = q_0$, $\delta(q_0, 1) = q_1$

$\delta(q_1, 0) = q_0$, $\delta(q_1, 1) = q_2$

$\delta(q_2, 0) = q_1$, $\delta(q_2, 1) = q_2$.

11.4 Acceptability of a String by a Deterministic Finite Automaton (DFA)

A finite automaton $A = (Q, \Sigma, \delta, q_0, F)$ accept a string w if, $\delta(q_0, w) = q_f$, where $q_f \in F$.

Example 11.12

One can see in the Fig. 11.2 that DFA accepts the string 011. The sequence of states is $q_0 \xrightarrow{0} q_0 \xrightarrow{1} q_1 \xrightarrow{1} q_2$ i.e. starting in state q_0, the symbol 0 read first. Since there is a loop (*the transition from the state to the state itself*) the automaton remains in the same state q_0. By reading next input symbol 1, the automaton goes into state q_1. Now by reading 1, it goes into q_2 state. Now since there is no symbol left in our string and the automaton is in final state q_2, therefore the string 011 is accepted by the given automaton.

11.5 Transition Table

The table which represents the list of transition rules (functions) of a finite automaton is called **transition table**. With the help of transition table one can easily observe the different transitions of DFA.

Example 11.13

Consider a DFA, $A = (Q, \Sigma, \delta, q_1, F)$

where, $Q = \{q_1, q_2, q_3\}$, $\Sigma = \{0, 1\}$, $F = \{q_2\}$

δ is given by

$\delta(q_1, 0) = q_1$, $\delta(q_1, 1) = q_2$

$\delta(q_2, 0) = q_3$, $\delta(q_2, 1) = q_2$

$\delta(q_3, 0) = q_2$, $\delta(q_3, 1) = q_2$.

The transition table for given δ is

Table 11.1

State / Σ	Input	
	0	1
$\rightarrow q_1$	q_1	q_2
(q_2)	q_3	q_2
q_3	q_2	q_2

11.6 Construction of a Transition Table from Transition Graph

Let us explain it with the help of following example.

Example 11.14

Consider a transition graph shown in Fig. 11.3.

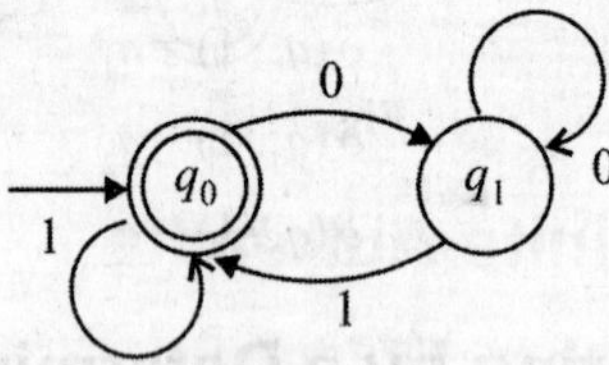

Fig. 11.3

which shows a finite automaton $A = (Q, \Sigma, \delta, q_0, F)$ in which $Q = \{q_0, q_1\}$, $\Sigma = \{0, 1\}$. Here q_0 is initial state and q_0 is a final state. The transition function δ is given by

$\delta(q_0, 1) = q_0$, $\delta(q_0, 0) = q_1$

$\delta(q_1, 0) = q_1$, $\delta(q_1, 1) = q_0$

corresponding to the transition graph. The transition table is given in Table 11.2.

Table 11.2

State / Σ	Input	
	0	**1**
→ (q_0)	q_1	q_0
q_1	q_1	q_0

11.7 Construction of a Transition Graph Corresponding to a Transition Table

Example 11.15

Consider a transition table given in Table 11.3

Table 11.3

State / Σ	Input	
	0	**1**
→ q_0	q_0	q_1
q_1	q_2	q_1
(q_2)	q_0	q_1

According to Table 11.4, we have a DFA

$A = (Q, \Sigma, \delta, q_0, F)$, where $Q = \{q_0, q_1, q_2\}$, $\Sigma = \{0, 1\}$, q_0 = initial state, $F = \{q_2\}$ and δ is given in Table 11.3. The transition graph corresponding to Table 11.3 is given in Fig. 11.4.

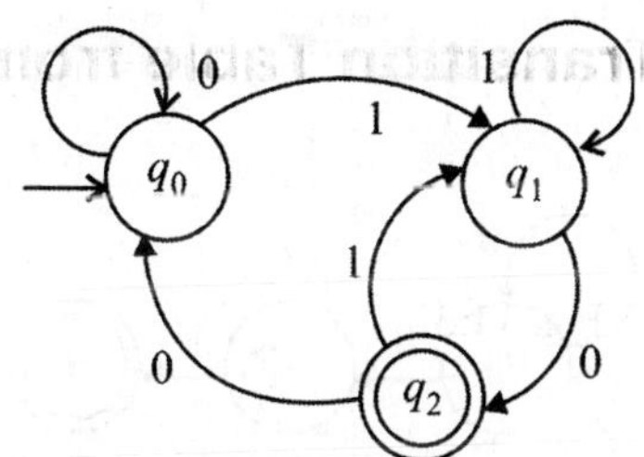

Fig. 11.4

11.8 Non Deterministic Finite Automata

In all previous examples of finite automaton we have seen that there is exactly one transition out of each state for every possible input symbol. These all are cases of the deterministic computation. Let us consider an automaton A given in Fig. 11.5.

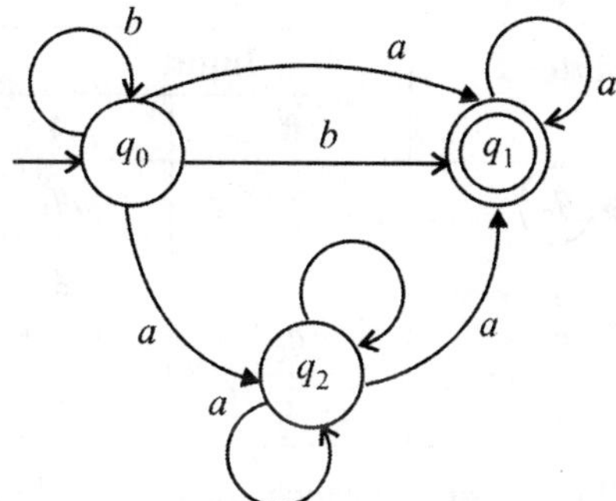

Fig. 11.5

Here it is clear that there are three states namely q_0, q_1 and q_2. q_0 is its **initial state** and q_1 is its **final state**. The input alphabet are 'a' and 'b' and the transition function is defined in Fig. 11.5. If the automaton is in state q_0 and input symbol is 'a' then the question comes into mind is "what will be the next state?" From Fig.11.5 it is clear that the next state will be either q_1 or q_2. Thus, the movement of automaton 'A' is not unique i.e. A is not deterministic. Such type of automaton is called **non deterministic finite automaton (NDFA)**. Now we can say that NDFA, $N = (Q, \Sigma, \delta, q_1, F)$ also has five parts like DFA,

Where

Q = Non empty finite set of states.

Σ = Set of alphabet

q_0 = Initial state

$F \subseteq Q$ is the set of final states

δ is a transition function defined as

$\delta: Q \times \Sigma_\epsilon \rightarrow P(Q)$

where

$\Sigma_\epsilon = \Sigma \cup \{\epsilon\}$ and $P(Q)$ is the power set of states.

Example 11. 16

Consider an NDFA given in Fig. 11.6.

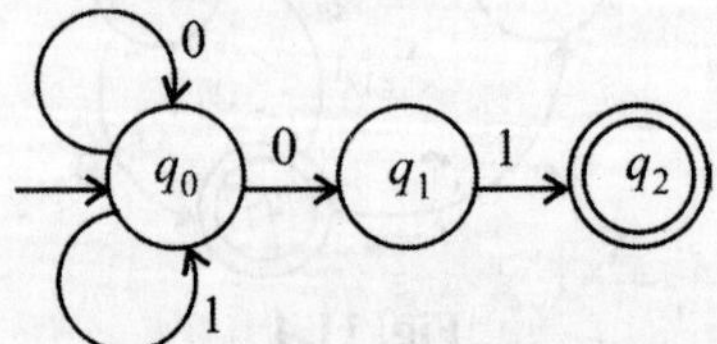

Fig. 11.6

This NDFA accepts the strings 001, 011101, 001101 and rejects 01110, 00, 010.

Example 11.17

For the DFA given in Table 11.4

Table 11.4

State / Σ	Input	
	0	**1**
→ (q_0)	q_2	q_1
q_1	q_3	q_0
q_2	q_0	q_3
q_3	q_1	q_2

Find out the strings among the following strings which are accepted by DFA.

i. 101101 ii. 11111

iii. 000000

Solution

i.

$$\begin{aligned}\delta(q_0, 101101) &= \delta(q_1, 01101)\\ &= \delta(q_3, 1101)\\ &= \delta(q_2, 101)\\ &= \delta(q_3, 01)\\ &= \delta(q_1, 1)\\ &= \delta(q_0, \in) = q_0.\end{aligned}$$

As q_0 is its final state, the string is accepted.

ii. and

iii. do yourself.

Example 11.18

For the transition graph of a DFA, '*A*' in Fig. 11.7 obtain the sequence of states for the input sequence 000101. Also find an input sequence not accepted by *A*.

Solution

The sequence of states are given by

$$\delta(q_0, 000101) = \delta(q_0, 00101)$$
$$= \delta(q_3, 0101)$$
$$= \delta(q_4, 101)$$
$$= \delta(q_4, 01)$$
$$= \delta(q_4, 1)$$
$$= \delta(q_4, \in) = q_4$$

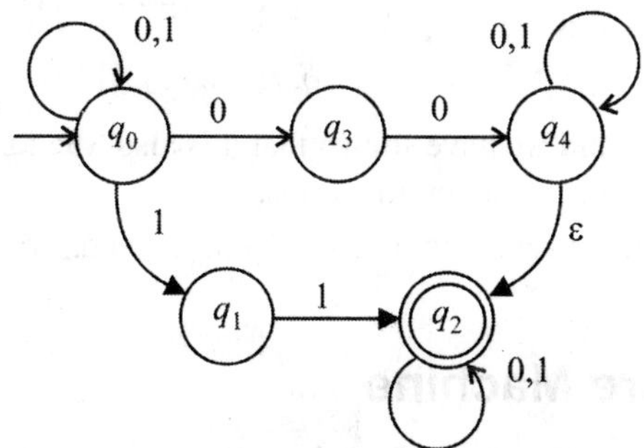

Fig. 11.7

11.9 Equivalence of Deterministic and Non-Deterministic Finite Automaton

We know that deterministic and non-deterministic automaton recognize the same languages.

As NDFA have more power than DFA's, so we expect that NDFA recognize more languages.

Two finite automaton A_1 and A_2 are said to be equivalent if $L(A_1) = L(A_2)$. i.e. if both accept the same language. It means if a language is recognized by an NDFA then there exists a DFA that also recognize the same language. Let us now take the most important theorem related to the equivalence of NDFA and DFA.

Theorem 11.1

For every non-deterministic finite automaton, there exists an equivalent deterministic finite automaton i.e. one that accept the same language.

Proof

Let $A = (Q_1, \Sigma, \delta_1, \{q_0\} F_1)$ be a DFA constructed from NDFA $N = (Q_2, \Sigma, \delta_2, \{q_0\} F_2)$ by the subset construction. Here we wants to show that $L(A) = L(N)$. Let w be any string over Σ, then we have to show that $\delta_1(\{q_0, w) = \delta_2(q_0, w)$. Let us prove it by mathematical induction on the string w.

Let w be an empty string i.e. $|w| = 0$. Obviously by the definition of DFA and NDFA both $\delta_1(\{q_0\}, x)$ and $\delta_2(\{q_0\}, x)$ are in state $\{q_0\}$. Hence, the result. Let this result is true for every string of length n. Now we will show that it is also true for the strings of length $(n + 1)$.

Let $w = xa$ be a string of length $(n + 1)$, where x is a string of length n and 'a' is final symbol of w. Since $|x| = n$, therefore by induction

$$\delta_1, (\{q_0\}\ x) = \delta_2(q_0, x).$$

If $\{P_1, P_2, \dots, P_k\}$ be the set of states for N then

$$_2(q_0, w) \quad \bigcup_{i\ 1}^{k} {}_2(P_i, a) \qquad \ldots(1)$$

Now $$\delta_1(\{P_1, P_2, \ldots P_k\}, a) = \bigcup_{i\ 1}^{k} {}_2(P_i, a) \qquad \ldots(2)$$

Also $$\delta 1(\{q0\}, x) = \{P_1, P_2, \ldots, P_k\} \qquad \ldots(3)$$

By using equations (1), (2) and (3), we get

$$\delta_1(\{q_0\}, w) = \delta_1(\delta_1(\{q_0\}, x)\, a)$$
$$= \delta_1(\{P_1, P_2, \ldots, P_k\}, a) = \bigcup_{i\ 1}^{k} {}_2(P_i, a)$$
$$= \delta_2(q_0, w)$$

As the result is true for $|w| = 0$ and we have shown that it is also true for $|w| = n+1$ when the result is true for $|w| = n$. Hence the given statement is true for all strings.

Therefore, A and N both accept the same string w iff $\delta_1(\{q_0\}, w)$ or $\delta_2(q_0, w)$ respectively contain a state in F_2. Therefore $L(A) = L(N)$.

11.10 Mealy and Moore Machine

11.10-1 Finite Automaton with Outputs

Earlier we have considered the finite automaton with the property that either they accept the string or reject the string (they have binary output). Now consider the automaton in which the output is chosen from some other alphabet. If the output function depends only on the present state, the automaton is called **Moore machine** and if the output function depends on both present state and present input than the automaton is called **Mealy machine**. The most general definitions of the two machines are as follows:

11.10-2 Moore Machine

A Moore machine is a collection of six objects $(Q, \Sigma, \Gamma, \delta, \gamma, q_0)$ where,

- Q represent the finite set of states.
- Σ is the input alphabet
- Γ is the output alphabet
- δ is a transition function mapping a state and an input into a state i.e. $\delta : Q \times \Sigma \rightarrow Q$.
- γ is the output function mapping a state into output alphabet, i.e. $\gamma : Q \rightarrow \Gamma$
- q_0 is the initial state.

11.10-3 Mealy Machine

A Mealy machine is a collection of six-objects $(Q, \Sigma, \Gamma, \delta, \gamma, q_0)$, where all the symbols except γ have the same meaning as in the Moore machine. Here γ is the output function mapping $\Sigma \times Q$ into Γ, i.e. $\gamma : \Sigma \times Q \rightarrow \Gamma$.

i. *In a Mealy machine the number of characters in output string is same to the number of letters in input string. But in Moore machine the length of output string is one more than the length of input strings.*

ii. *As Moore and Mealy machines do not accept or reject an input string, therefore they do not define a language of accepted words, hence they have no final states.*

Now we discuss the procedures for conversion of a Moore machine into a Mealy machine and vice-versa.

11.11 Minimization of Finite Automaton

As we know that a DFA recognize a unique language, but the converse does not hold. So for a given language there exists more than one DFA that accepts it. All these DFAs are called **equivalent automaton**. But it is to be noted that two or more equivalent automata will differ only in the number of possible states other things will remains same. We also know that in a finite automaton a state may be final or non-final state. Two states p and q of a DFA are equivalent if both $\delta(p, w)$ and $\delta(q, w)$ are final states or both are non final states for all $w \in \Sigma^*$. As Σ^*contains infinite number of strings so it is very difficult to construct $\delta(p, w)$ or $\delta(q, w)$ for all $w \in \Sigma^*$. Now two states p and q are n-equivalent ($n \geq 0$) if both $\delta(p, w)$ and $\delta(q, w)$ are either final states or both non final states for all w such that
$|w| \leq n$.

The relations which we have defined above are **equivalence relations** on the set of states Q. This equivalence relation on Q partitions Q into disjoint equivalence classes. Let S and S_n be the partitions of Q, here elements of S_n are n-equivalent classes. If p and q are n-equivalent for all $n \geq 0$, then we say that they are equivalent. Further if p and q are $(n+1)$-equivalent then also they are n-equivalent. Continuing in this way at last we get $S_n = S_{n+1}$ for some n, where S_n is the set of n-equivalence classes.

11.11-1 Procedure for Minimization of DFA

Before taking the procedure, let us take the statement of a result whose proof is obvious and is the key for the construction of minimum state automaton.

Result

Two state p and q are said to be $(n+1)$- equivalent if, firstly they are n-equivalent and secondly $\delta(\text{p, b})$ and $\delta(\text{q, b})$ are also n-equivalent for all $b \in \Sigma$.

Now, if we know about n-equivalence classes then by using above result we can construct $(n+1)$-equivalence classes. The procedure for minimization of a DFA is given by:

Step 1: First construct the set of all final states of given DFA say it is Q_1^0. Then set $Q_2^0 = Q - Q_1^0$, where Q is the set of all states. Finally the 0-equivalence S_0 is given by $S_0 = \{Q_1^0, Q_2^0\}$.

Step 2: Consider any subset Q_t^n of S_n let q_1, q_2 are in Q_t^n. q_1 and q_2 will be $(n+1)$-equivalent if $\delta(q_1, b)$ and $\delta(q_2, b)$ are in the same equivalence class in S_n, for all $b \in \Sigma$.

Now divide Q_t^n again into $(n+1)$-equivalence classes. By continuing this procedure for every Q_t^n in S_n we get all the elements of S_{n+1}.

Repeat this procedure of construction of S_n for $n = 1, 2, \ldots$ until $S_n = S_{n+1}$.

At last we get the automaton with minimum number of states in which the states are the equivalence classes i.e. the elements of S_n and whose transition table is obtained by replacing the state by its corresponding equivalence class.

Example 11.19

Construct a minimum state automaton equivalent to a given automaton A, whose transition graph is given in Fig. 11.8.

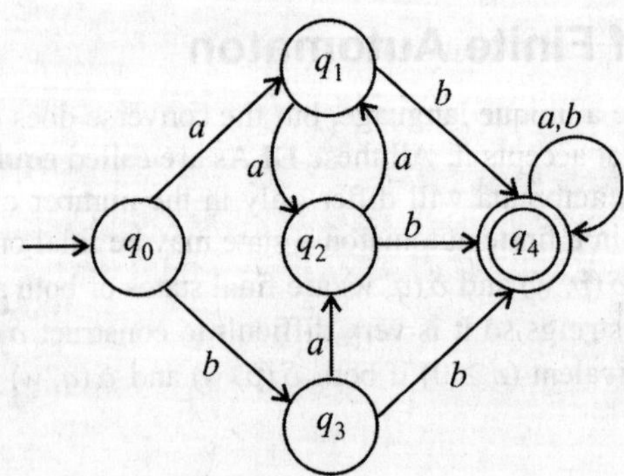

Fig. 11.8

Solution

The transition table corresponding to transition graph in Fig. 11.8 is given in Table 11.5.

Table 11.5

State/Σ	a	B
$\rightarrow q_0$	q_1	q_3
q_1	q_2	q_4
q_2	q_1	q_4
q_3	q_2	q_4
(q_4)	q_4	q_4

By step 1, we have $Q_1^0 = F \quad \{q_4\},\ Q_2^0 = Q \quad Q_1^0$

$= \{q_0, q_1, q_2, q_3\}$

Therefore, $S_0 = \{Q_1^0, Q_2^0\}$

$= \{\{q_4\}, \{q_0, q_1, q_2, q_3\}\}$

Since $\{q_4\}$ has only one element so it can not be divided again. So $Q_1 \quad q_4$

Consider $q_0, q_1 \in Q_2^0$. One can observe from Table 11.5 that under '*a*' column, the entries corresponding to q_0 and q_1 are q_1 and q_2, which lie in Q_2^0, i.e. they lie in the same set and under '*b*' column, the entries are q_3 and q_4. Since $q_3 \in Q_2^0$ and $q_4 \in Q_1^0$, therefore q_0 and q_1 are not 1-equivalent. Similarly q_0 is not 1-equivalent to q_2 and q_3.

Now consider q_1 and q_2 and observe that the entries under '*a*' column are q_2 and q_1, both are in Q_2^0. Similarly the entries under '*b*' column are q_4 and q_4, which is in Q_1^0 hence, q_1 and q_2 are 1-equivalent. Similarly q_1 and q_3 are 1-equivalent.

Hence $S_1 = \{\{q_4\}, \{q_0\}, \{q_1, q_2, q_3\}\}$

Now $\{q_4\}$ and $\{q_0\}$ are also in S_2 as they cannot be divided again. Consider q_1 and q_3. The entries under '*a*' column are q_2 and q_1 which are in the same set in S_1 and the entries under '*b*' column are q_4 and q_4 which is also in the same set in S_1. Therefore q_1 and q_2 are 2-equivalent.

Transition table for minimum state automaton is given in Table 11.6.

The transition graph for minimum state automaton is given by the Fig. 11.9.

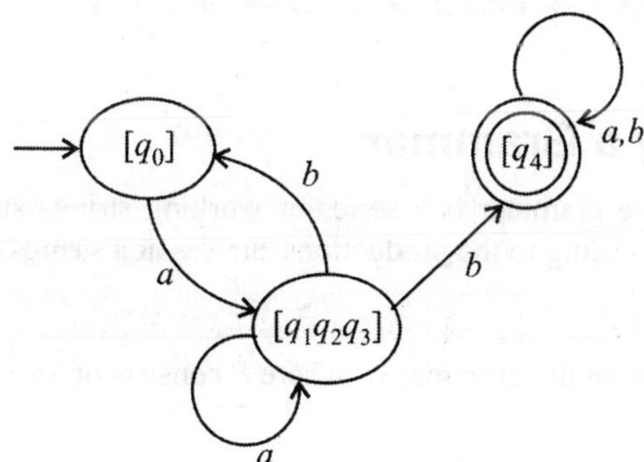

Fig. 11.9

Consider q_1 and q_3. The entries under 'a' column are q_2 and q_2 which are in same set in S_1. Also entries under 'b' column are q_4 and q_4, which are in the same set in S_1. Therefore q_1 and q_3 are 2-equivalent. Thus

$$S_2 = \{ \{q_4\}, \{q_0\}, \{q_1, q_2, q_3\} \} = S_1$$

Hence S_1 gives the equivalence classes. The minimum state automaton is

$$A' = \{ Q', \{ a, b\}, \delta', q_0, F')$$

Where, $Q' = \{ [q_4], [q_0], [q_1, q_2, q_3]\}$

$q'_0 = [q_0]$

$F' = [q_4]$ and δ' is given by the Table 11.6.

Table 11.6

State/Σ	a	b
$\rightarrow [q_0]$	$[q_1\, q_2\, q_3]$	$[q_1\, q_2\, q_3]$
$[q_4]$	$[q_4]$	$[q_4]$
$[q_1\, q_2\, q_3]$	$[q_1\, q_2\, q_3]$	$[q_4]$

11.12 Grammar

A phrase-structure grammar $G = (V, \Sigma, P, S)$ consists of

i. a finite non-empty set of variables V.

ii. a finite non-empty set of terminal elements Σ.

iii. a finite set of productions P.

iv. a start symbol S.

Every production in P is the form $\alpha \rightarrow \beta$, where α, β are strings on $V \cup \Sigma$. α must have at least one non terminal on its left side.

Example 11.20

Let $G = (V, \Sigma, P, S)$ is a phrase structure grammar where-

i. $V = \{ A, B, S\}$

ii. $\Sigma = \{a, b\}$

iii. P is the set of productions $S \to ABa,\ A \to BB,\ B \to ab,\ AB \to b$.

iv. S is the start symbol.

11.13 Derivations by a Grammar

A derivation in a phrase-structure grammar is a series of working strings starting from the initial symbol S, which by making substitution according to the productions, arrives at a string of all terminals.

Example 11.21

The string *abababa* is derivable from the grammar G, where P consists of $S \to ABa$, $A \to BB$, $B \to ab$, $AB \to b$.

Solution

We know that
$$\begin{aligned} S &\to ABa \\ &\to Aaba && [\because B \to ab] \\ &\to BBaba && [\because A \to BB] \\ &\to Bababa && [\because B \to ab] \\ &\to abababa && [\because B \to ab] \end{aligned}$$

11.13-1 Language Generated by a Grammar

Let $G = (V, \Sigma, P, S)$ be a phrase-structure grammar. The language generated by G, denoted by $L(G)$, is the set of all strings of terminals that are derivable from the starting state S. i.e.

$$L(G) = \{u \in \Sigma^* : S \Rightarrow u\}.$$

Example 11.22

Let $G = (V, \Sigma, P, S)$ be a grammar where $V = \{S, A\}$, $\Sigma = \{a, b\}$ and P is the set of productions $S \to aA \,/\, b$, $A \to aa$. Find $L(G)$.

Solution

From $S \to b$, we get $S \Rightarrow b$. Hence $b \in L(G)$.

Also, from $S \to aA \to aaa$ $\quad [\because A \to aa]$

we get, $S \Rightarrow aaa$.

No additional words can be drived. Hence $L(G) = \{b, aaa\}$.

Example 11.23

Let $G = (V, \Sigma, P, S)$, where $V = \{S\}$, $\Sigma = \{0, 1\}$ and S is the set of productions $S \to 0S1/\wedge$. Find $L(G)$.

Solution

As $S \to \wedge$ is in P. So $S \Rightarrow \wedge$. Hence $\wedge \in L(G)$.

Also for $n \geq 1$,

$$\begin{aligned} S &\to 0S1 \\ &\to 00S11 && [\because S \to 0S1] \\ &\to 000S111 && [\because S \to 0S1] \\ &\quad \vdots \\ &\to 000\ldots0S1\ldots111 \end{aligned}$$

or $$S \Rightarrow 0^n S1^n \quad \text{[using } S \to 0S1 \text{ } n \text{ times]}$$

But $S \to \wedge \quad \therefore S \Rightarrow 0^n 1^n$

Hence $$0^n 1^n \in L(G) \quad \forall n \geq 0.$$

Thus $$0^n 1^n \subseteq L(G) \text{ for } n \geq 0 \qquad \dots (1)$$

Now we have to show that $L(G) \subseteq 0^n 1^n$. For this, let $u \in L(G)$. The derivation of u starts with S. If $S \to \wedge$, we get $u = \wedge$.

If $S \to 0S1$ and at any stage we use $S \to \wedge$ we get u. Also the terminal string u is obtained by applying $S \to \wedge$ only. Thus

$$S \Rightarrow 0^n S1^n \Rightarrow 0^n 1^n \qquad [\because S \to \wedge]$$

Hence $$L(G) \subseteq 0^n 1^n \qquad \dots (2)$$

From (1) and (2), we can say that

$$L(G) = 0^n 1^n : n \geq 0.$$

11.14 Types of Phrase-Structure Grammars

A phrase-structure grammar can be classified according to the types of productions that are allowed. Chomsky classified the grammars into four types in terms of productions.

i. **Type-0:** A type-0 grammar has no restriction on its productions.

ii. **Type-1:** A type-1 grammar can have productions only of the form $\alpha \to \beta$ such that $|\alpha| \leq |\beta|$ or of the form $\alpha \to \wedge$. It is also known as **context-sensitive grammar**. Every type-1 grammar is a type-0 grammar. Language generated by this grammar is known as context-sensitive language.

iii. **Type-2**: A type-2 grammar can have productions only of the form $\alpha \to \beta$, where α is a single symbol from V. It is also know as **context-free grammar.** Every type-2 grammar is a type-1 grammar. Long age generated by this grammar is context-free language.

iv. **Type-3:** A type-3 grammar can have productions only of the form $\alpha \to \beta$ with α = variable and either β = concatenation of terminal and variable or a terminal. This grammar is know as **regular grammar.** Every type-3 grammar is a type-2 grammar. Language generated by this grammar is regular language.

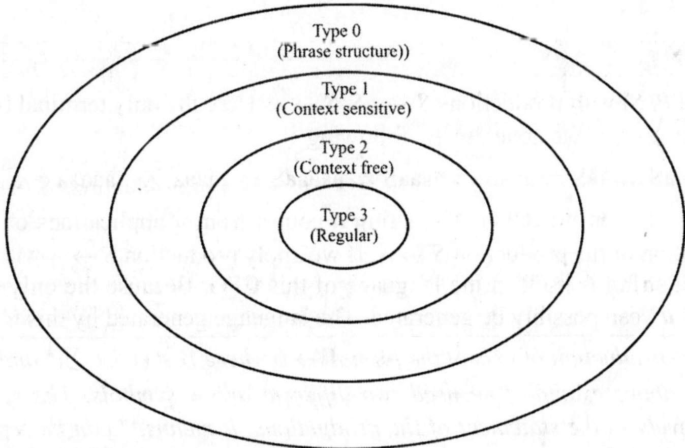

Fig. 11.10: Types of grammars (Chomsky Hierarchy)

Example 11.24

i. Type-1 grammars are of the type
$S \to A,\ \ S \to \wedge,\ A \to aBB,\ \ bB \to 00,\ \ aA \to a$ etc.
ii. Type-2 grammars are of the type
$S \to \wedge,\ \ S \to A,\ A \to ab,\ \ A \to aaB$ etc.
iii. Type-3 grammars are of the type
$S \to \wedge,\ \ S \to a,\ S \to aA,\ \ A \to a,\ A \to aB$ etc.

11.15 Context-Free Grammar

The concept of context free grammar or CFG was invented by the linguist Noam Chomsky in 1956. Chomsky gave several mathematical models for languages. Context free languages are also known as non-regular languages.

11.15-1 Definition

A context free grammar (CFG), is a collection of three things.

i. An alphabet Σ of a letters called terminals.
ii. A set of symbols called nonterminals, one of which is the symbol S, standing for "start symbol".
iii. A finite set of productions of the form
One nonterminal $\to$ Finite string of terminals/non-terminals

Example 11.25

The grammar $G = (\{S\}, \{a\}, P, S)$ where P, the set of productions is given by $S \to SS$, $S \to a$, $S \to \wedge$, is called context free grammar.

Example 11.26

The grammar G whose productions are $S \to aAS \mid a$, $A \to SbA \mid SS \mid ba$ is called a context free grammar.

11.15-2 Context Free Language (CFL)

The language generated by a CFG is the set of all strings of terminals that can be produced from the start symbol S using the productions as substitutions. A language generated by a CFG is called a context-free language or CFL.

Example 11.27

Consider $G = (\{S\}, \{a\}, P, S)$ with productions $S \to aS$, $S \to \wedge$. Here the only terminal be a. If we apply $S \to aS$ five times and then apply $S \to \wedge$, we generate the following

$$\mathrm{S} \Rightarrow \mathrm{aS} \Rightarrow \mathrm{aaS} \Rightarrow \mathrm{aaaS} \Rightarrow \mathrm{aaaaS} \Rightarrow \mathrm{aaaaaS} \Rightarrow \mathrm{aaaaa}\ \wedge = \mathrm{aaaaa} = \mathrm{a}^5$$

This is a derivation of a^5 in this CFG. The string a^n comes from n applications of the production $S \to aS$ followed by one application of the production $S \to \wedge$. If we apply production $S \to \wedge$ without the production $S \to aS$, we find that the null string is itself in the language of this CFG. Because the only terminal is a, it is clear that no words outside of $\boldsymbol{a}^*$ can possibly be generated. The language generated by this CFG is exactly $\boldsymbol{a}^*$.

i. *If every production of G is of the form $A \to \beta$ where $\beta \in (V \cup \Sigma)^*$ and $A \in V$.*
ii. *In the above example, we used two different arrow symbols. The symbol "$\to$" we employ exclusively in the statement of the productions. It means, " can be replaced by ", as in $S \to aS$. Another arrow symbol " $\Rightarrow$" we employ between the unfinished stages in the generation of our word. It means, "can develop into" as in $aaS \Rightarrow aaaS$. These "unfinished stages" are strings of terminals and non-terminals called as working strings.*

11.15-3 Left Most Derivation

If we replace the 'left most' variable only by its productions then the derivation is said to be the left most derivation.

11.15-4 Right Most Derivation

If we replace the 'right most' variable only by its productions then the derivation is said to be the right most derivation.

11.15-5 Derivation Trees

The sequence of applications of the rules or productions that produces the finished string of terminals from the starting symbol is called a derivation of the word. This derivation in CFG can be represented by a tree known as **derivation tree** or **parse tree** or **generation tree** or **production tree**. To draw this tree, we start with the symbol S. Every time we use a production to replace a non-terminal by a string. We draw downward lines from the non-terminals to each character in the string.

11.15-6 Ambiguity in CFG

A CFG is called **ambiguous** if for at least one string in the language that it generates there are more than one possible derivations of the string that correspond to different derivation trees. If a CFG is not ambiguous, it is called **unambiguous.**

Example 11.28

Show that the grammar $S \rightarrow XaX,\ X \rightarrow aX \mid bX \mid \wedge$ is ambiguous.

Solution

Consider a string *aabb*.

The two left most derivations are given in fig. 11.11 & 11.12

$S \Rightarrow XaX$	
$\Rightarrow aXaX$	$[X \rightarrow aX]$
$\Rightarrow a\wedge aX$	$[X \rightarrow \wedge]$
$\Rightarrow aaX$	
$\Rightarrow aabX$	$[X \rightarrow bX]$
$\Rightarrow aabbX$	$[X \rightarrow bX]$
$\Rightarrow aabb\wedge$	$[X \rightarrow \wedge]$
$\Rightarrow aabb$	

$S \Rightarrow XaX$	
$\Rightarrow \wedge aX$	$[X \rightarrow \wedge]$
$\Rightarrow aX$	
$\Rightarrow aaX$	$[X \rightarrow aX]$
$\Rightarrow aabX$	$[X \rightarrow bX]$
$\Rightarrow aabbX$	$[X \rightarrow bX]$
$\Rightarrow aabb\wedge$	$[X \rightarrow \wedge]$
$\Rightarrow aabb$	

Therefore, grammar is ambiguous.

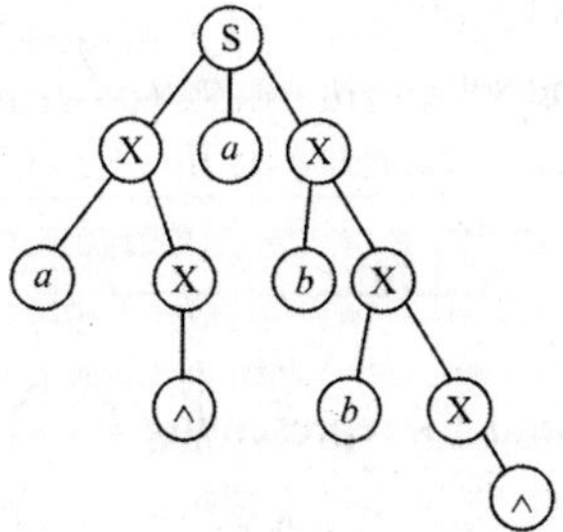

Fig. 11.11

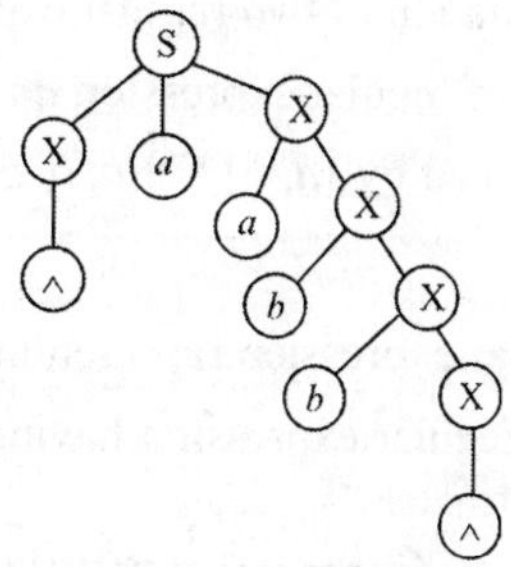

Fig. 11.12

11.16 Regular Expression

Regular expressions play an important role in the study of finite automaton and its applications. The regular expression gives us an algebraic description of deterministic and non-deterministic finite automaton. Therefore, the regular expressions can define exactly in the same languages as that of the finite automata and these languages are called **regular languages**. Regular expressions play an important role in computer science applications, such as text-formatting systems, in UNIX, lexical analysis etc. These systems use a regular expression like notation for describing patterns that the user wants to find in a file. Different search systems convert the regular expression into either a DFA or an NDFA, and simulate that automaton on the file being searched.

To construct an arithmetic expression we use binary operations addition and multiplication like (4+5)×2. Similarly, to construct a regular expression, we have union, concatenation and closure like $(0 \cup 1)\ 1^*$. As we know that the value of any arithmetic expression is a number and the value of a regular expression is a language. Therefore, the value of arithmetic expression given above is 18 while the value of regular expression is the language consisting of all strings starting with a '0' or a '1' followed by any number of 1's.

11.17 Formal Definition of a Regular Expression

As we know that a regular expression gives us an algebraic relation between certain sets of strings. Actually these expressions give a description of the languages accepted by finite automaton.

Now, we give a formal definition of regular expression over an alphabet Σ recursively as follows:

1. $\in$, $\varnothing$ and any element of Σ (terminal symbol) are regular expressions. These are called **primary regular expressions**.
2. If P_1 and P_2 are any two regular expressions, then
   i. Union $(P_1 + P_2)$ is a regular expression.
   ii. Concatenation $(P_1 . P_2)$ is a regular expression.
3. If P is a regular expression, then P^* is a regular expression.
4. Any expression obtained recursively by a finite number of applications of the rules 1-4 is a regular expression.

11.18 Regular Set

The set represented by a regular expression is called a **regular set**.

Notations

1. The regular expression 'a' denotes the set $\{a\}$.
2. The union of two regular expressions $a + b$ denoted by $\{a, b\}$.
3. The concatenation of two regular expression ab denoted by {ab}.
4. The closure of regular expression a i.e. a^* denoted by the set $\{\in, a, aa, aaa, \ldots\}$.
5. $(a + b)^*$ denoted by $\{a, b\}^*\ \forall\ a, b \in \Sigma$.

Example 11.29

i. 00 is a regular expression representing {00}.

ii. (0+1)* is a regular expression having all strings of 0's and 1's representing $\{0, 1\}^* = \{\in, 0, 1, 00, 01, 10, 11, \ldots\}$.

iii. The set {01, 10} is the union of {01} and {10} represented by 01+10.

iv. The set $\{abba\}$ is the concatenation of a, b, b and a. Therefore, $\{abba\}$ is represented by $abba$.

v. The set $\{a, aa, aaa, \ldots\}$ is the concatenation of a and $\{a\}^*$. Therefore it can be written as $a(a)^*$.

11.19 Algebraic Laws for Regular Expressions

Like arithmetic expressions, the regular expressions have a number of laws that work for them. Many of these are similar to the laws for arithmetic expressions if we think of union as addition and concatenation as multiplication. However, there are a few places where the analogy breaks-down, and there are also some laws that apply to regular expressions but have no analogy for arithmetic expression specially when the closure operator is involved. In this section we give a catalog of that laws.

I_1 $\varnothing$ is the identity for union
$\varnothing + P = P$, where P is a regular expression.

I_2 $\varnothing$ is annihilator for concatenation
$\varnothing P = P \varnothing = \varnothing$

I_3 $\in$ is the identity for concatenation
$\in P = P \in = P$

I_4 $\in^* = \in$ and $\varnothing^* = \in$.

I_5 Idempotent law for union
$P + P = P$

I_6 $P^* P^* = P^*$

I_7 $P P^* = P^* P$

I_8 $(P^*)^* = P^*$

I_9 $\in + P P^* = P^* = \in + P^* P$

I_{10} $(PQ)^* P = P(QP)^*$

I_{11} $(P + Q)^* = (P^* Q^*)^* = (P^* + Q^*)^*$

I_{12} Right distributive law of concatenation over union
$(P + Q)R = PR + QR$

Left distributive law of concatenation over union
$P(Q + R) = PQ + PR$

Example 11.30

Give a regular expression for the set of all strings having at least one pair of consecutive zeros.

Solution

Every string in the language must contain 00 somewhere. Before and after 00 the string contains arbitrary string over $\{0, 1\}$ which is $(0+1)^*$. Therefore, the regular expression is given by $(0+1)^*\ 00\ (0+1)^*$.

Example 11.31

Give a regular expression for the set of all strings on $\{a, b\}$ end by either an 'a' or 'bb'.

Solution

The first part of this example is the set of all strings on $\{a, b\}$ which is $(a+b)^*$. The second part gives the union of 'a' and 'bb' i.e. $(a+bb)$. The final result is the concatenation of $(a+b)^*$ and $(a+bb)$. Therefore, the regular expression is given by $(a+b)^*\ (a+bb)$.

11.20 Transition Graph and Regular Expressions

In this section we will show that the languages accepted by finite automaton are precisely the languages denoted by regular expressions. For it, we study a theorem, which gives the relation between transition systems and regular expressions.

It is to be noted that if R is a regular expression, then L (R) denote regular language.

Theorem 11.2

Let R is a regular expression then there exists NDFA with $\in$-moves that accepts $L(R)$.

Proof

Let us prove this by the principle of induction on the number of operators in the regular expression R in such a way that there exists NDFA N with $\in$-moves, having one final state and no transitions out of this final state, such that $L(N) = L(R)$.

Basic Step

If the number of operators in R is zero, then the expression R must be $\in$, $\varnothing$ or a_i for some $a_i \in \Sigma$. The transition systems (NDFA) given in Fig.11.13, clearly satisfy the above conditions.

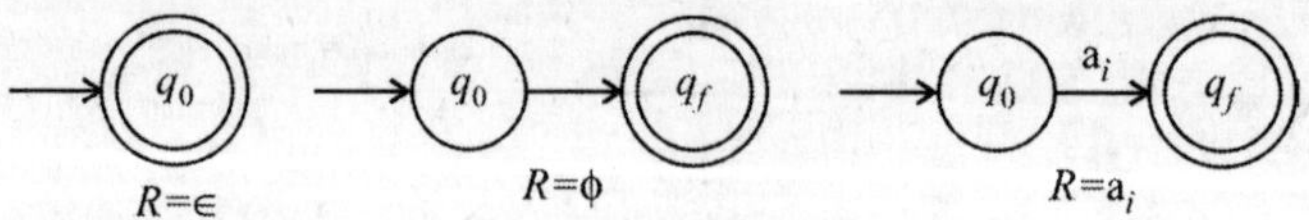

Fig. 11.13

Induction Step

First, we assume that the theorem is true for regular expressions with less than k operators such that $k \geq 1$. Let R have k operators, then there are three different cases depending on the form of R, like $R = R_1 + R_2$ (union), $R = R_1 R_2$ (concatenation), $R = R_1{}^*$ (closure), where R_1 and R_2 are regular expressions, each having k operators or less.

By hypothesis, R_1 and R_2 can be represented by the transition graphs $N(R_1)$ and $N(R_2)$ respectively. The transition graph for $R = R_1 + R_2$ is given by the Fig. 11.14. We introduce a new initial state which is connected to the initial states of $N(R_1)$ and $N(R_2)$ by $\in$-moves and a new unique final state. The final states of $N(R_1)$ and $N(R_2)$ are connected to this final state by $\in$-moves.

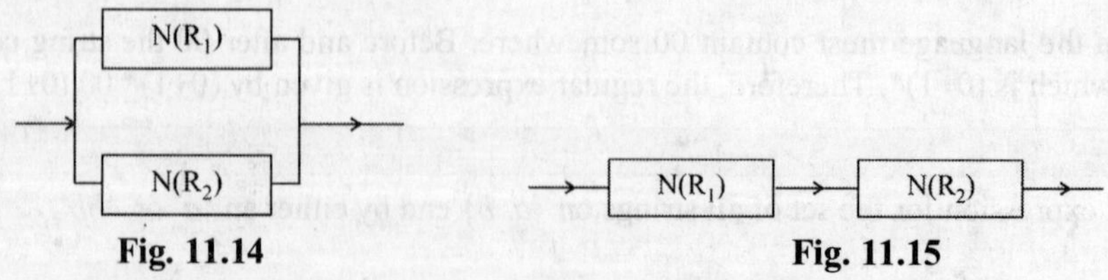

Fig. 11.14 **Fig. 11.15**

Similar construction for $R_1 R_2$ and $R_1{}^*$ is given by Fig. 11.15 and Fig. 11.16 respectively.

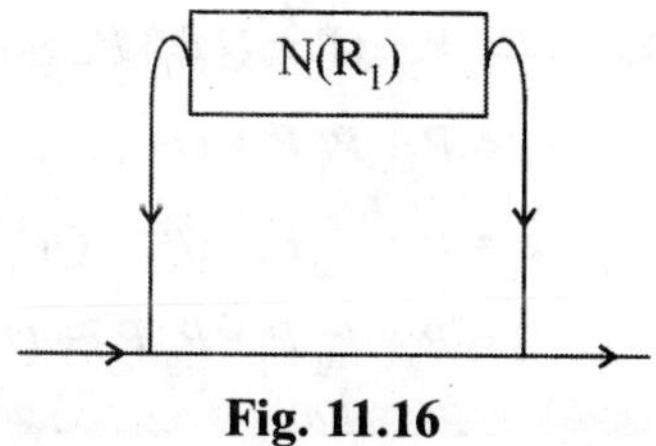

Fig. 11.16

Example 11.32

Construct a transition system for the regular expression $(a + b)^*$ over $\Sigma = \{a, b\}$.

Solution

Let $R = (a + b)^*$ be a regular expression over $\Sigma = \{a, b\}$. Then the transition graph for R is given by Fig. 11.7.

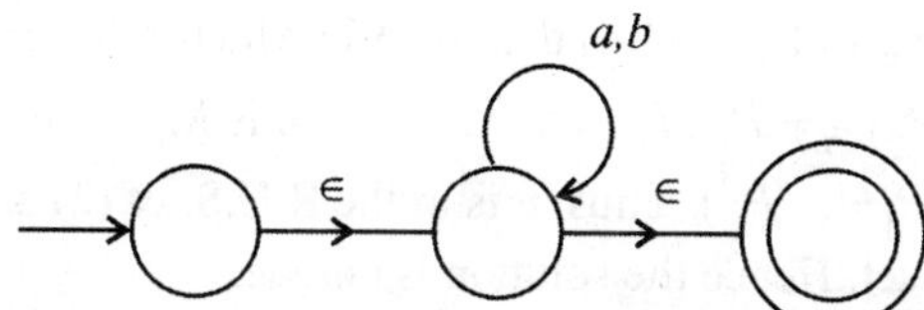

Fig. 11.7: Transition system for $(a+b)^*$

11.21 Arden's Theorem

When we simplify a regular expression, we try to replace it by a simpler regular expression equivalent to the given expression. The following theorem known as Arden's theorem is a useful tool to simplify such type of regular expressions.

Theorem 11.3

Let P_1 and P_2 be two regular expressions over alphabet Σ. If $\in$ does not belongs to P_1, then the equation $Q = P_2 + Q P_1$ has a unique solution given by $Q = P_2 P_1^*$.

Proof

We have $\qquad Q = P_2 + Q P_1 \qquad \ldots(1)$

Then $Q = P_2 P_1^*$ will be a solution of (1) if it satisfies (1)

$$
\begin{aligned}
\text{R.H.S.} \quad &= P_2 + Q P_1 \\
&= P_2 + (P_2 P_1^*) P_1. \\
&= P_2 + P_2 (P_1^* P_1) && \text{[Associativity]} \\
&= P_2 [\in + P_1^* P_1) && [\text{by } I_{12}] \\
&= P_2 P_1^* && [\text{by } I_9] \\
&= Q = \text{L.H.S.}
\end{aligned}
$$

Therefore, (1) is satisfied by $Q = P_2 P_1^*$ which means that $Q = P_2 P_1^*$ is a solution of (1).

Now, we have to show that this solution is unique. Replace Q by $P_2 + QP_1$ on the R.H.S. of (1), we get

$$Q = P_2 + QP_1.$$

$$= P_2 + (P_2 + Q P_1) P_1$$

$$= P_2 + P_2 P_1 + Q P_1^2 \qquad \text{[by } I_{12}]$$

$$= P_2 + P_2 P_1 + (P_2 + QP_1) P_1^2$$

$$= P_2 + P_2 P_1 + P_2 P_1^2 + Q P_1^3 \qquad \text{[by } I_{12}]$$

- - - - - - - - - - - - -

- - - - - - - - - - - - -

$$= P_2 + P_2 P_1 + P_2 P_1^2 + ... + P_2 P_1^n + Q P_1^{n+1} \quad \text{for } n \geq 0$$

$$Q = P_2 (\in + P_1 + P_1^2 + ... + P_1^n) + Q P_1^{n+1} \qquad ...(2)$$

Let Q satisfies (1), then it will satisfy (2)

Consider a string $u \in Q$ of length n. Then u lies in the set $P_2 (\in + P_1 + P_1^2 + ... + P_1^n) + Q P_1^{n+1}$. Since P_1 does not has $\in$, therefore there is no string of length less than $(n + 1)$ which belongs to $Q P_1^{n+1}$. Therefore u does not belongs to $Q P_1^{n+1}$. Hence, u is in $P_2 (\in + P_1 + P_1^2 + ... + P_1^n)$ i.e. u is M $\in + P_2 P_1^*$. This implies $u \in P_2 P_1^i$, for some $i \geq 0$ and hence $u \in P_2 (\in + P_1 + ... P_1^i)$. Thus, u is in the R.H.S. of (2) and hence in L.H.S. of (2) i.e. $u \in Q$. Thus Q and $P_2 P_1^*$ are the same set. Hence the solution is unique.

11.21-1 Applications of Arden's Theorem

Arden's theorem is used to obtain a regular expression recognized by a transition system. To obtain the regular expression the following assumptions are used

i. There should be no $\in$-moves in the transition graph.
ii. The initial state should be unique.
iii. If q_k is a final state then q_k represents the regular expression accepted by the system.

β_{ij} is the regular expression representing the set of edge label from q_i to q_j. In case there is no such edge then $\beta_{ij} = \varnothing$. Therefore, we get the following equations in $q_1 .. q_n$.

$$q_1 = q_1 \beta_{11} + q_2 \beta_{21} + ... + q_n \beta_{n1} + \in$$

$$q_2 = q_1 \beta_{12} + q_2 \beta_{22} + ... + q_n \beta_{n2}$$

. . . .

$$q_n = q_1 \beta_{1n} + q_2 \beta_{2n} + ... + q_n \beta_{nn}$$

Now, by using the substitutions several times and using Arden's theorem one can get q_k in terms of β_{ij}'s.

Example 11.33

Consider the transition system given in Fig. 11.18. Prove that the strings recognized are (0 + 0(1 + 00)*1) 0 (1 + 00)* 0.

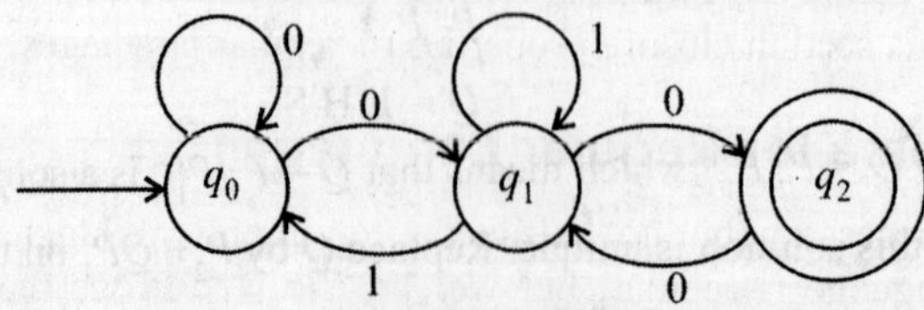

Fig. 11.18

Solution

Since the transition system given in Fig.11.18 have only one initial state and there is no $\in$-moves, so we can apply Arden's theorem. The basic equations for q_0, q_1 and q_2 are:

$$q_0 = q_0 0 + q_1 1 + \in \quad ...(1)$$
$$q_1 = q_0 0 + q_1 1 + q_2 0 \quad ...(2)$$
$$q_2 = q_1 0 \quad ...(3)$$

By substitution, from (2) we have

$$q_1 = q_0 0 + q_1 1 + q_1 00 = q_0 0 + q_1 (1 + 00) \quad \text{[by } I_{12}\text{]}$$

By Arden's theorem, we get

$$q_1 = q_0(0) + q_1(1 + 00)^* \quad ...(4)$$

Again by substituting (4) in (1), we get

$$q_0 = q_0 0 + q_0 0\,(1 + 00)^* 1 + \in$$
$$= q_0\,(0 + 0\,(1 + 00)^* 1) + \in \quad \text{[by } I_{12}\text{]}$$

Again by Arden's theorem

$$q_0 = \in (0 + 0\,(1 + 00)^* 1)^*$$

Since $\in$ is an identity, then

$$q_0 = (0 + 0\,(1 + 00)^* 1)^* \quad ...(5)$$

Now, from (4)

$$q_1 = (0 + 0\,(1 + 00)^* 1)^*\, 0\,(1 + 00)^* \quad ...(6)$$

From (3),

$$q_2 = (0 + 0\,(1 + 00)^* 1)^*\, 0(1 + 00)^* 0 \quad ...(7)$$

Since q_2 is a final state, the set of strings recognised by the graph is

$$(0 + 0\,(1 + 00)^* 1)^*\, 0(1 + 00)^* 0.$$

11.22 Regular Expressions and Finite Automaton

In the construction of a finite automaton equivalent to a given regular expression, we first construct a transition graph with $\in$-moves by using theorem 1 which is equivalent to the given regular expression. Then construct a transition table for the transition graph obtained after removing $\in$-moves.

Construct equivalent DFA and reduce the number of states if possible.

11.23 Properties of Regular Languages

As we know that a language which can be defined by a regular expression is called a **regular language.** There are several questions related to regular languages, one can ask. The important one is "are all languages regular?". The answer is no. The languages, which are not regular, are called **non regular languages.** In this section, we give a basic result called the **Pumping lemma** which is used to prove whether certain languages are not regular. We also give in this section closure properties of regular languages.

11.23-1 Pumping Lemma for Regular Languages

The pumping lemma gives a necessary (but not sufficient) condition for a language to be regular. We cannot use this lemma to set up that language is regular, but we can use it to prove that a language is not regular, by showing that the language does not obey the lemma.

The pumping lemma uses the Pigeonhole Principle (if P pigeons are placed into less than P holes, some hole has to have more than one pigeon in it) in another form. The proof of pumping lemma is based on the fact that in a transition graph with n vertices, any walk of length greater than or equal to n must repeat some vertex.

Theorem 11.4

Let L be an infinite regular language. Then there exists a constant n (the number of states of the smallest automaton that accepts L) such that, for every string $w \in L$ with $|w| \geq n$, there exists $w_1, w_2, w_3 \in \Sigma^*$ with $w = w_1 w_2 w_3$, $|w_2| \geq 1$, $|w_1 w_2| \leq n$, $\forall i = 0, 1, 2, \ldots, w_1 w_2^i w_3 \in L$. That is we can find a non–empty string w_2 not too far from the beginning of w which can be "pumped". i.e. repeating w_2 any number of times or deleting it when $i = 0$, keeps the resultant string in the language L.

Proof

Since L is regular, it is accepted by a DFA. $A = (Q, \Sigma, \delta, q_0, F)$. Let A has n-states say $q_0, q_1 \ldots q_{n-1}$. Since L is infinite therefore, consider any string w with $|w| \geq n$, say $w = a_1, a_2 \ldots a_m$, where $m \geq n$ and each a_k is an input symbol.

For $i = 1, 2, \ldots, m$ let $\delta(q_0, a_1 \ldots a_i) = q_i$.

As we have only n different states, so it is impossible for each of the $(n + 1)$ states $q_0, q_1 \ldots q_n$ to be distinct. Thus, by the Pigeonhole Principle there exists two integers s and t, $0 \leq s < t \leq n$ such that $q_s = q_t$.

Now let
$$w = w_1 w_2 w_3$$
where
$$w_1 = a_1, \ldots a_s$$
$$w_2 = a_{s+1} \ldots a_t$$
$$w_3 = a_{t+1} \ldots a_m$$

Such that $|w_1 w_2| \leq n$. The path with path value w in the transition graph of A is shown in Fig. 11.19.

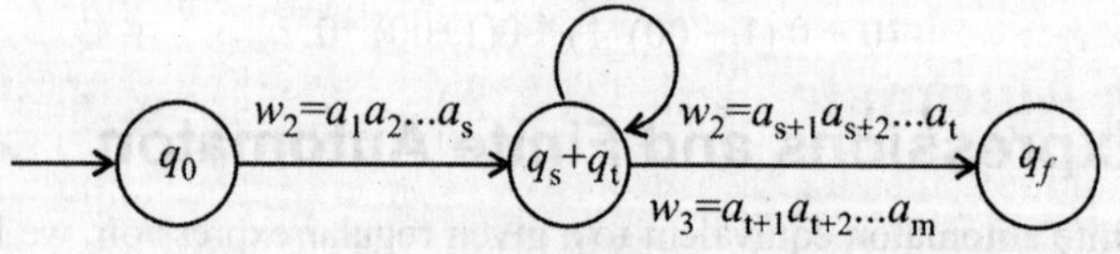

Fig. 11.19

Since $s < t$, the string w_2 is of length at least one. And since $t \leq n$, so the length of this path is not more than n. If $q_f \in F$ i.e. if $w \in L(A)$, then $a_1 a_2 \ldots a_s a_{t+1} \ldots a_m$ is also in $L(A)$, since we have a path from q_0 to q_f which passes through q_s but not the loop w_2.

$$\therefore \delta(q_0, a_1 a_2 \ldots a_s a_{t+1} \ldots a_m) = \delta(\delta(q_0, a_1 a_2 \ldots a_s), a_{t+1} \ldots a_m)$$
$$= \delta(q_s, a_{t+1} \ldots a_m) = q_f$$

Similarly, we can go around the loop of the Fig.11.19 as many times as we like.

Therefore, $a_1 a_2 \ldots a_s (a_{s+1} a_{s+2} \ldots a_t)^i a_{t+1} a_{t+2}, \ldots a_m$ is in $L(A)$ for an $i \geq 0$.

i.e. $\quad w_1 w_2^i w_3 \in L(A) \; \forall i \geq 0.$

11.23-2 Applications of Pumping Lemma

Let us see some examples of "how the pumping lemma is used"?. In the following examples we shall propose a language and use the pumping lemma to prove that the language is not regular.

To summarize, the steps needed to prove that a language L is not regular, are as follows:

Step 1: Assume that L is regular.

Step 2: Let n be the number of states in the corresponding DFA.

Step 3: Pick a 'suitable' string w with $|w| \geq n$

Step 4: Show that, for every decomposition of w into $w_1\ w_2\ w_3$ such that $|w_2| \geq 1$ and $|w_1\ w_2| \leq n$, there exists $i \geq 0$ such that $w_1\ w_2^i\ w_3 \notin L$.

Step 5: Conclude that assumption (1) is false.

Failure to proceed through these steps have no effect on the proof that L is not regular but it does prove that L is regular. If the language is finite then the pumping lemma is useless, because all finite languages are regular.

Example 11.34

Show that the language $L = \{0^m\ 1^m : m \geq 0\}$ is not regular.

Solution

Consider the language L is regular. Let n be the constant of pumping lemma. Let $w = 0^n\ 1^n$, then $|w| \geq n$. Fig. 11.20 shows how we can decompose $w = w_1\ w_2\ w_3$ with $|\ w_1\ w_2| \leq n$ and $|w_2| \geq 1$.

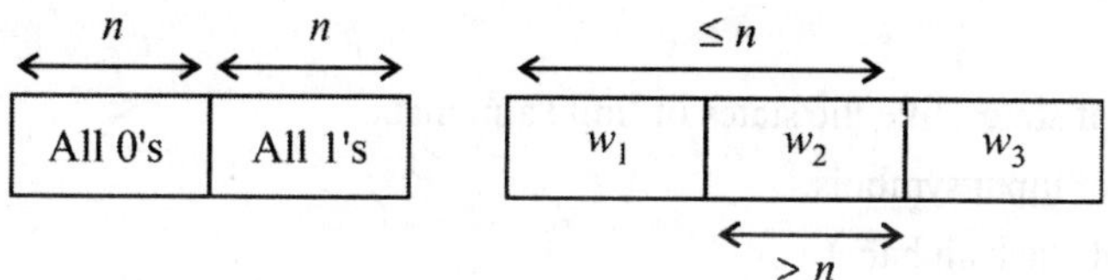

Fig. 11.20: Decomposing the string w into possible choices for w_1, w_2 and w_3

$w_1\ w_2$ must be a string of 0's, so pumping w_2 will give more 0's than 1's. It shows that the pumped string is not in L, which is a contradiction. Therefore L is not regular.

11.24 Pushdown Automata

Earlier we have seen that a finite automaton recognize the regular languages. Finite automata cannot recognize the context free languages because a finite automaton has strictly finite memories, whereas the recognition of a context-free language may require storing an unbounded amount of information. Consider $L = \{0^m\ 1^m : m \geq 0\}$. We know that this is not a regular language. When we scan a string from this language, we must not only check that all 0's precede the first 1, we must also count the number of 0's. Since m is unbounded, this counting cannot be done with a finite memory.

This problem can be solved by adding an auxiliary memory in the form of a '**stack**'. A stack is a "first come last out" list i.e. the most recently added element is removed first. For example, the set of plates in a marriage parties. We can remove plates only at the top of the set not from the bottom. In similar manner, we can add plates only at the top of the set not in the bottom. This type of arrangement where a finite automata has a stack gives us a class of machine called **pushdown automata** (PDA).

From this discussion we can say that the pushdown automata is a nondeterministic finite automata having three parts such as an input tape, a finite control and a stack shown in Fig.11.21.

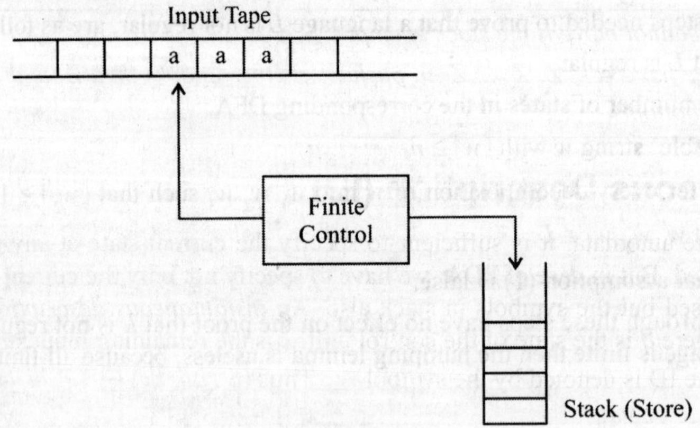

Fig. 11.21

11.25 Definition

A pushdown automata consists of seven components

$$P = (Q, \Sigma, \Gamma, \delta, q_0, Z_0, F)$$

where

i. Q is a finite set of states, like the states of finite automata.

ii. Σ is a finite set of input symbols.

iii. Γ is a finite set of stack alphabet.

iv. δ is the transition function which takes as argument a triple $\delta(q, a, x)$, where :

1. q is a state in Q.
2. a is either an input symbol in Σ or $a = \wedge$, the empty string.
3. x is a stack symbol i.e. $x \in \Gamma$.

The output of δ is a finite set of ordered pairs (p, r), where 'p' is the new state and r is the string of stack symbols that replaces x at the top of the stack.

v. *q_0 is the initial state in Q.*

vi. *Z_0 is the stack start symbol.*

vii. $F \subseteq Q$ is the set of final states.

Example 11.35

Let $P = (Q, \Sigma, \Gamma, \delta, q_0, Z_0, F)$

where $Q = \{q_0, q_1, q_2\}$, $\Sigma = \{0, 1\}$, $\Gamma = \{0, 1, Z_0\}$, $F = \{q_2\}$ and δ is given by

$\delta(q_0, 0, Z_0) = \{(q_0, 0Z_0)\}$, $\quad \delta(q_0, 1, Z_0) = \{(q_0, 1Z_0)\}$

$\delta(q_0, 0, 0) = \{(q_0, 00)\}$, $\quad \delta(q_0, 0, 1) = \{(q_0, 01)\}$

$\delta(q_0, 1, 0) = \{(q_0, 10)\}$, $\quad \delta(q_0, 1, 1) = \{(q_0, 11)\}$

$\delta(q_0, \wedge, Z_0) = \{(q_0, Z_0)\}$, $\quad \delta(q_0, \wedge, 0) = \{(q_1, 0)\}$

$\delta(q_0, \wedge, 1) = \{(q_1, 1)\}$, $\quad \delta(q_1, 0, 0) = \{(q_1, \wedge)\}$

$\delta(q_1, 1, 1) = \{(q_1, \wedge)\}$, $\quad \delta(q_1, \wedge, Z_0) = \{(q_2, Z_0)\}$.

i. *A PDA cannot move if the stack is empty.*

ii. *When we have* $\alpha = Z_1 Z_s \ldots Z_n$ *in pushdown store (stack) then* Z_1 *is the topmost element and* Z_n *is at bottom.*

11.26 Instantaneous Description (ID)

As we know that in finite automata, it is sufficient to specify the current state at any time and the remaining input string to be processed. But in case of PDA, we have to specify not only the current state and the remaining input string to be processed but the symbols in stack also. An *instantaneous description* (ID) of a PDA is the ordered tried (q, x, r), where q is the state of the control unit, x is the remaining input string to be processed and r is the stack variable. The ID is denoted by the symbol $\vdash$: Thus $(q_1, au, bv) \vdash (q_2, u, cv)$ is possible iff $(q_2, c) \in \delta(q_1, a, b)$.

11.27 Reflexive-Transitive Closure of $\vdash$

The reflexive-transitive closure of $\vdash$ is denoted by $\vdash^*$ defined as a definite sequence of n moves, where n is non-negative integer.

11.28 Acceptance by a Pushdown Automata

An input string is said to be accepted by a PDA if the automata is in a final state at the end of the string.

Example 11.36

Consider $P = (Q, \Sigma, \Gamma, \delta, q_0, Z_0, F)$, where $Q = \{q_0, q_1\}, \Sigma = \{a, b\}, \Gamma = \{a, b, Z_0\}, F = \{q_1\}$ and δ is given by

$$\delta(q_0, a, Z_0) = \{(q_0, a Z_0)\}, \quad \delta(q_0, b, Z_0) = \{(q_0, b Z_0)\}$$

$$\delta(q_0, a, a) = \{(q_0, aa)\}, \quad \delta(q_0, b, b) = \{(q_0, bb)\}$$

$$\delta(q_0, a, b) = \{(q_1, ab)\}, \quad \delta(q_0, b, a) = \{(q_1, ba)\}$$

$$\delta(q_1, a, a) = \delta(q_1, b, b) = \{(q_1, \wedge)\}, \quad \delta(q_1, b, a) = \{(q_1, \wedge)\}$$

Suppose the input string is *aaabbb*. We get the following moves:

$$\begin{aligned} \delta(q_0, aaabbb, Z_0) &\vdash \delta(q_0, aabbb, a Z_0) \\ &\vdash \delta(q_0, abbb, aa Z_0) \\ &\vdash \delta(q_0, bbb, aaa Z_0) \\ &\vdash \delta(q_1, bb, baaa Z_0) \\ &\vdash \delta(q_1, b, aaa Z_0) \\ &\vdash \delta(q_1, \wedge, aa Z_0) \end{aligned}$$

At the end of input string, the PDA is in final state $\{q_1\}$. Therefore the input string *aaabbb* is accepted by the given PDA.

11.29 Pushdown Automata and CFL

Let L be a context free language generated by a grammar G in Greibach normal form. Then we can construct a PDA. P accepting L using the following formulas.

i. F_1: $\delta(q_1, \wedge, A) = \{(q, x) : A \to x \text{ is in productions}\}$

ii. F_2: $\delta(q, b, b) = \{(q, \wedge)\}\ \forall\ b \in \Sigma$.

Example 11.37

Construct pushdown automata for the grammar $S \to aAA$, $A \to aS \mid bS \mid a$.

Solution

Let P be the PDA given by

$$P = (\{q\}, \{a, b\}, \{S, A, a, b\}, \delta, q, S, \varphi)$$

and δ is given by the rules F_1 and F_2 as follows :

i. $\delta(q, \wedge, S) = \{(q, aAA)\}$

ii. $\delta(q, \wedge, A) = \{(q, aS), (q, bS), (q, a)\}$

iii. $\delta(q, a, a) = \{(q, \wedge)\}$

iv. $\delta(q, b, b) = \{(q, \wedge)\}$

Let us apply this on the string *abaaaa*.

$$\begin{aligned} \delta(q, abaaaa, S) &\vdash \delta(q, abaaaa, aAA) \\ &\vdash \delta(q, baaaa, AA) \\ &\vdash \delta(q, baaaa, bSA) \\ &\vdash \delta(q, aaaa, SA) \\ &\vdash (q, aaaa, aAAA) \\ &\vdash \delta(q, aaa, AAA) \\ &\vdash \delta(q, aaa, aaa) \\ &\vdash \delta(q, \wedge). \end{aligned}$$

11.30 Context Free Grammars for PDA

If $P = (Q, \Sigma, \Gamma, \delta, q_0, Z_0, F)$ be a PDA, there must exists a context-free grammar $G = (V, \Sigma, P, S)$, where

$$V = \{S\} \cup \{[q, Z, q'] : q, q' \in Q, Z \in \Gamma\}$$

And the productions P are obtained using the following rules.

i. S - productions are given by $S \to [q_0, Z_0, q]\ \forall\ q \in Q$.

ii. Each move of the type $\delta(q, a, Z) \to (q', \wedge)$ gives the production of the form $[q, Z, q'] \to a$.

iii. Each move of the type $\delta(q, a, Z) \to (q_1, Z_1 Z_2 \ldots Z_n)$ gives the productions of the form

$$[q, Z, q'] \to a\,[q_1, Z_1, q_2]\ [q_2, Z_2, q_3] \ldots [q_n, Z_n, q'].$$

Example 11.38

Construct a context free grammar G accepted by the pushdown automata.

$$P = (\{q_0, q_1\}, \{a, b\}, \{Z_0, Z_1\}\ \delta, q_0, Z_0, \{q_1\})$$

Where δ is given by

$$\delta(q_0, a, Z_1) = \{(q_0, Z_0 Z_1)\}$$
$$\delta(q_0, b, Z_0) = \{(q_0, Z_0 Z_0)\}$$

$$\delta(q_0, a, Z_0) = \{(q_1, \wedge)\}.$$

Solution

Let (V, Σ, P, S) be the required grammar.

Where

$V = \{S, [q_0, Z_0, q_0], [q_0, Z_0, q_1], [q_0, Z_1, q_0], [q_0, Z_1, q_1], [q_1, Z_0, q_0], [q_1, Z_0, q_1], [q_1, Z_1, q_0], [q_1, Z_1, q_1]\}$

The set of productions are

i. $S \to [q_0, Z_0, q_0]$

ii. $S \to [q_0, Z_0, q_1]$

iii. $\delta(q_0, a, Z_0) = \{(q_0, Z_0 Z_1)\}$ gives

$$[q_0, Z_1, q_0] \to a\,[q_0, Z_0, q_0]\,[q_0, Z_1, q_0]$$
$$[q_0, Z_1, q_0] \to a\,[q_0, Z_0, q_1]\,[q_1, Z_1, q_0]$$
$$[q_0, Z_1, q_1] \to a\,[q_0, Z_0, q_0]\,[q_0, Z_1, q_1]$$
$$[q_0, Z_1, q_1] \to a\,[q_0, Z_0, q_1]\,[q_1, Z_1, q_1]$$

iv. $\delta(q_0, b, Z_0) = \{(q_0, Z_0 Z_0)\}$ gives

$$[q_0, Z_0, q_0] \to b\,[q_0, Z_0, q_0]\,[q_0, Z_0, q_0]$$
$$[q_0, Z_0, q_0] \to b\,[q_0, Z_0, q_1]\,[q_1, Z_0, q_0]$$
$$[q_0, Z_0, q_1] \to b\,[q_0, Z_0, q_0]\,[q_0, Z_0, q_1]$$
$$[q_0, Z_0, q_1] \to b\,[q_0, Z_0, q_1]\,[q_1, Z_0, q_1]$$

v. $\delta(q_0, a, Z_0) = \{(q_1, \wedge)\}$ gives

$$[q_0, z_0, q_1] \to a.$$

11.31 Pumping Lemma for Context-Free Languages

Pumping lemma is a tool for showing that certain languages are not context-free. This theorem is in the same spirit as pumping lemma for regular languages. The pumping lemma for regular languages can always find one small string to pump while in pumping lemma for context free languages we can always find at most two short, nearby substrings that we can pump i.e., we may repeat both of the strings 'i' times, for any integer i, and the resulting string will still be in the language.

11.31-1 Pumping Lemma

Theorem 11.5

Let L be a context-free language. Then there exists a positive integer n such that

i. Every $w \in L$ with $|w| \geq n$ can be decomposed as $w = uvxyz$ for some strings u, v, x, y, z.

ii. $|vxy| \leq n$.

iii. $|vy| \geq 1$.

Then $uv^i x y^i z \in L$ for all $i \geq 0$.

Proof

Let L be a context free language generated by a context free grammar G. If $\wedge \in L$ then consider L - $\{\wedge\}$ and construct a grammar G in CNF generating L - $\{\wedge\}$. If $\wedge \notin L$ then construct a grammar G in CNF generating L.

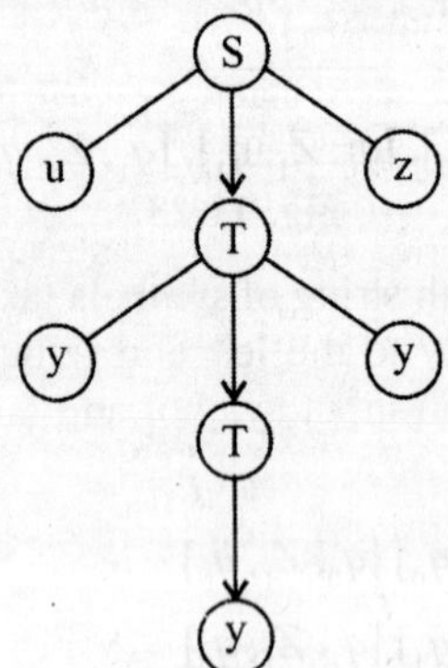

Fig. 11.22

Now, starting with a CNF grammar G such that $L(G) = L$ - $\{\wedge\}$. Since there are no unit productions and no Null productions, v and y cannot both be empty. Thus $|vy| \geq 1$. Since the length of any string on the right side of productions is bounded by l (say) then the length of the derivation of w in L must be at least $|w| / l$. But here L is of infinite length so there exists arbitrarily long derivations and corresponding derivation trees of arbitrary height. Now, consider such high derivation tree and some sufficiently long path from root to leaf of the tree. But G have a finite number of variables, therefore, there must be some variable that repeats on this path as shown in Fig.11.22 .

From this derivation tree, we have

$$S \overset{*}{\Rightarrow} u\,T\,z \overset{*}{\Rightarrow} uv\;T\,yz \overset{*}{\Rightarrow} uvxyz$$

Where u, v, x, y and z are all strings of terminals only. From Fig. it is clear that $T \overset{*}{\Rightarrow} v\,T\,y$ and $T \overset{*}{\Rightarrow} x$. So all the strings $uv^i\,xy^i\,z$, $i \geq 0$ can be generated by the grammar G and finally in L.

Also in the derivations $T \overset{*}{\Rightarrow} v\,T\,y$ and $T \overset{*}{\Rightarrow} x$, we can assume that there is not repetition of variables. Thus, the lengths of v, x and y depends only on the productions of the grammar. Therefore, $|vxy| \leq n$.

11.32 Turing Machines

In 1936, A.M. Turing proposed the Turing machine as a model of "any possible computation". In other words, the Turing machine is capable of performing any calculation which can be performed by any computing machine.

The basic model of Turing machine is given in Fig.11.23

The machine consists of a **finite control**, which can be in any of a finite set of states. There is a tape divided into squares or cell; each cell can hold any one of a finite number of symbols.

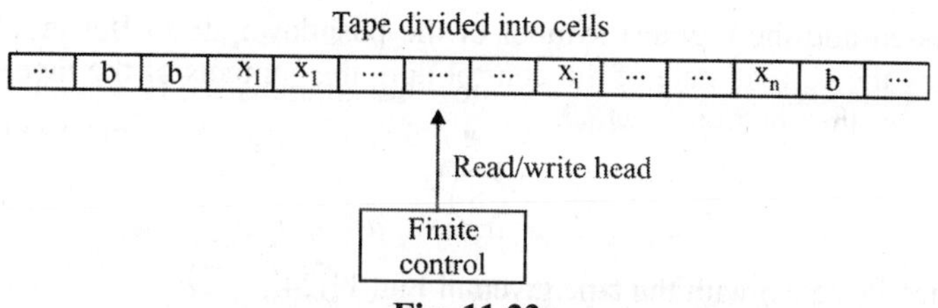

Fig. 11.23

Initially, the **input**, which is a finite-length string of symbols taken from the input alphabet, is placed on the tape. All other tape cells, extending infinitely to the left and right, initially hold a special symbol called the **blank**. The blank is a tape symbol, but not an input symbol and there may be other tape symbols besides the input symbols and the blank.

There is a **tape head** that is always positioned at one of the tape cells. The Turing machine is said to be scanning that cell. Initially, the tape head is at the leftmost all that hold the input.

A **move** of the Turing machine is a function of the state of the finite control and the tape symbol scanned.

In one move, the Turing machine will:

i. Change state
ii. Write a tape symbol in the cell scanned
iii. Move the tape head left or right.

11.33 Definition

A Turing machine (TM) is a seven-tuple

$$M = (Q, \Sigma, \Gamma, \delta, q_0, b, F),$$

where

i. Q is a finite non-empty set of states.
ii. Σ is the finite non-empty set of input symbols.
iii. Γ is the finite set of tape symbols.
iv. δ is the transition function mapping the states of finite automata and tape symbols to states, tape symbols and movement of the head. Mathematically.
v. $\delta(Q, \Gamma) = (Q, \Gamma, \{L, R\})$.
vi. q_0 is the initial (start) state and $q_0 \in Q$.
vii. $b \in \Gamma$ is the blank symbol. b is not an input symbol.
viii. $F \subseteq Q$ is the set of final or accepting states.

11.34 Representation of Turing Machines

We can represent a Turing machine in the following three ways:

i. Using instantaneous descriptions with move-relations.
ii. Using transition table.
iii. Using transition graph.

11.34-1 Instantaneous Descriptions for Turing Machines

In order to describe formally what a TM does, we need to develop a notation for instantaneous descriptions (ID), like the notation we developed for PDA's. We have defined ID of a PDA in terms of the current state, input string to be processed and the topmost symbol of the pushdown store. But in TM any configuration is completely determined by the current state of the control unit, the contents of the tape, and the position of the read-write head. We will use the notation in which

$$x_1 q x_2$$

or

$$a_1\, a_2 \,\ldots\, a_{k-1}\, q\, a_k\, a_{k+1} \ldots a_n$$

is the ID of a machine in state q with the tape given in Fig. 11.24.

The symbols $a_1, a_2, \ldots, a_n$ show the tape contents, while q defines the state of finite control unit.

The convention is taken so that the position of read-write head is over the cell containing the symbol immediately following q.

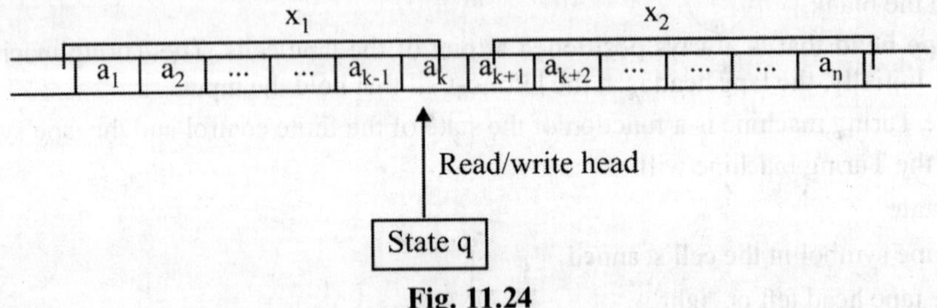

Fig. 11.24

11.34-2 Representation by Transition Table

We can represent the transitions of a Turing machine graphically. A transition diagram consists of a set of vertices corresponding to the state of Turing machine. Directed edges are used to represent transition of states. The edges are labeled by one or more triples of the form (x, y, d), where $x, y \in \Gamma$ and 'd' is a direction, either L (left) or R (right). Thus,

$\delta(q, x) = (p, y, d)$ then the label of edge will be (x, y, d) from q to p.

As for other kinds of diagrams (transition), we represent the initial state by an arrow entering that state. Final states are indicated by double circles. From transition diagram, we cannot read the symbol 'b' directly.

Example 11.39

M is a Turing machine given by the transition diagram in Fig. 11.25. Find a computation sequence of *M* for processing the input string 000111.

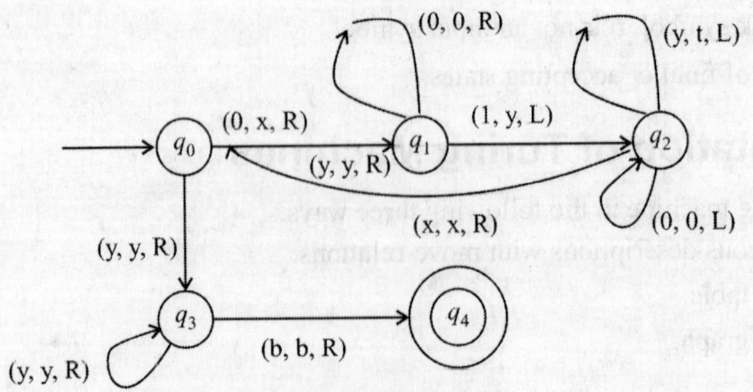

Fig. 11.25

Solution

The initial tape input is b000111b. As the initial state is q_0 so the computation sequence is

[0, q_0] 00111 $\xrightarrow{(0,x,R)}$ x [0, q_1] 0111 $\xrightarrow{(0,0,R)}$ x0 [0, q_1] 111

$\xrightarrow{(0,0,R)}$ x00 [1, q_1] 11 $\xrightarrow{(1,y,L)}$ x0 [0, q_2] y11

$\xrightarrow{(0,0,L)}$ x [0, q_2] 0y11 $\xrightarrow{(0,0,L)}$ [x, q_2] 00y11

$\xrightarrow{(x,x,R)}$ x [0, q_0] 0y11 $\xrightarrow{(0,x,R)}$ xx [0, q_1] y11

$\xrightarrow{(0,0,R)}$ xx0 [y, q_1] 11 $\xrightarrow{(y,y,R)}$ xx0y [1, q_1] 1

$\xrightarrow{(1,y,L)}$ xx0 [y, q_2] y1 $\xrightarrow{(y,y,L)}$ xx [0, q_2] yy1 $\xrightarrow{(0,0,L)}$ x [x, q_2] 0yy1

$\xrightarrow{(x,x,R)}$ xx [0, q_0] yy1 $\xrightarrow{(0,x,R)}$ xxx [y, q_1] y1 $\xrightarrow{(y,y,R)}$ xxxy [y, q_1] 1

$\xrightarrow{(y,y,R)}$ xxxyy [1, q_1] $\xrightarrow{(1,y,L)}$ xxxy [y, q_2] y $\xrightarrow{(y,y,L)}$ xxx [y, q_2] yy

$\xrightarrow{(y,y,R)}$ xx [x, q_2] yyy $\xrightarrow{(x,x,R)}$ xxx [y, q_0] yy $\xrightarrow{(y,y,R)}$ xxxy [y, q_3] y

$\xrightarrow{(y,y,R)}$ xxxyy [y, q_3] $\xrightarrow{(y,y,R)}$ xxxyyy [b, q_3] $\xrightarrow{(b,b,R)}$ xxxyyy [b, q_4]

This Turing machine accepts strings of the form $\{0^n 1^n : n \geq 0\}$.

11.35 Transition Table for Turing Machine

If the transitions 'δ' of a TM is in the form of a table then this table is known as transition table for TM.

Example 11.40

Table below gives the rules of the transition function δ of a TM having seven state q_0, q_1, ... q_6. The tape symbols are 0, 1 and *b*. q_0 is initial state while q_6 is the only final state.

Table 11.7

State	Tape Symbols		
	0	1	b
$\rightarrow q_0$	(q_1, b, R)	(q_5, b, R)	-
q_1	$(q_1, 0, R)$	$(q_2, 1, R)$	-
q_2	$(q_3, 1, L)$	$(q_2, 1, R)$	(q_4, b, L)
q_3	$(q_3, 0, L)$	$(q_3, 1, L)$	(q_0, b, R)
q_4	$(q_4, 0, L)$	(q_4, b, L)	$(q_6, 0, R)$
(q_5)	(q_5, b, R)	(q_5, b, R)	(q_6, b, R)
q_6	-	-	-

We can construct a transition table from a transition diagram and vice-versa.

11.36 Turing Machines as Language Accepters

Consider a TM $M = (Q, \Sigma, \Gamma, \delta, q_0, b, F)$. A string *w* is written on the tape and the tape head begins at the left most symbol. If, after a sequence of moves, the TM enters a final state, then the string *w* is accepted by *M* otherwise not. For example, the string 000111 is accepted by the TM given in example.

Example 11.41

Construct a TM that can accept the set of all even palindromes over $\Sigma = \{0, 1\}$.

Solution

We have the following steps for processing even length plindromes.

i. The TM scans the first symbol of the input tape either 0 or 1, erases it and changes state q_1 or q_2.
ii. TM scans the remaining part without changing the tape symbol until it encounters *b*.
iii. The read-write head moves to the left. If the right most symbol tallies with the leftmost symbol, the right most symbol is erased. Otherwise TM halts.
iv. The read-write head moves to the left until b is encountered.

The required TM is given in the Table 11.8.

Table 11.8

State	Tape Symbol		
	0	1	b
$\rightarrow q_0$	(q_1, b, R)	(q_2, b, R)	(q_7, b, R)
q_1	$(q_1, 0, R)$	$(q_1, 1, R)$	(q_3, b, L)

q_2	$(q_1, 0, R)$	$(q_2, 1, R)$	(q_4, b, L)
q_3	(q_5, b, L)	-	-
q_4	-	(q_6, b, L)	-
q_5	$(q_5, 0, L)$	$(q_5, 1, L)$	(q_0, b, R)
q_6	$(q_6, 0, L)$	$(q_6, 1, L)$	(q_0, b, R)
q_7 (circled)	-	-	-

Exercise 11.1

1. Determine whether the strings 101, 110001, 1001, 1111 are accepted or rejected by the transition graph given in Fig. 11.2.
2. Let $L_1 = \{ a, b^2 \}$, $L_2 = \{ a^2, ab, b^3 \}$, $L_3 = \{ a^2, a^4, a^6 \ldots\}$ denote the languages over an alphabet $\Sigma = \{a, b\}$. Then find
   i. L_1L_2
   ii. L_1L_3
   iii. L_1L_1.
3. Let $\Sigma = \{a, b\}$. Construct an automaton A which will accept those words from Σ which end in two b's.
4. Consider the words $u = a^3 b^2 a^2 b$, $v = b^2 a b$. Find
   i. uv
   ii. vu
   iii. u^2.
5. Find $|u|$, $|v|$, $|uv|$, $|vu|$ for the words u and v in exercise 3.
6. Construct an automaton A over $\Sigma = \{a, b\}$ which will accept precisely those words from Σ which have an even number of a's.
7. Let $\Sigma = \{0, 1\}$ construct an automaton A which will accept those words from Σ which begin with a '0' followed by (zero or more) 1's.
8. Describe the string w in the language L accepted by the automaton A in the following figure.

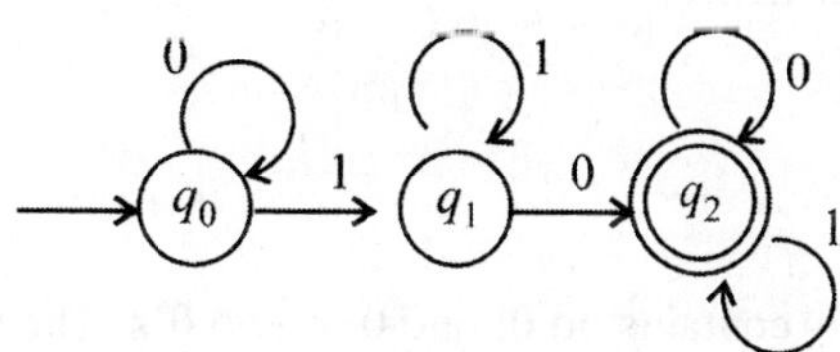

9. Let $G = (\{A, B, S\}, \{0, 1\}, P, S)$, where P consists of $S \to 0AB$, $A0 \to S0B$, $A1 \to SB1$, $B \to SA$, $B \to 01$. Show that $L(G) = \varphi$.
10. Show that $G_1 = (\{S\}, \{a, b\}, P_1, S)$ where $P_1 = \{S \to aSb \mid ab\}$ is equivalent to $G_2 = (\{S, A, B, C\}, \{a, b\}, P_2, S)$, where P_2 consists of $S \to AC$, $C \to SB$, $S \to AB$, $A \to a$, $B \to b$.
11. Find the language generated by the grammar $S \to 0S1 \mid 0A1$, $A \to 1A \mid 1$.
12. Let $G = (\{S, C\}, \{a, b\}, P, S)$, where P is $S \to aCa$, $C \to aCa \mid b$. Find $L(G)$.
13. If G is $S \to aS \mid bS \mid a \mid b$. Find $L(G)$.

14. Represent the following sets by regular expression.
    i. The set of all strings over $\{0, 1\}$ which has at most two zeros.
    ii. $\{a^n \mid n$ is divisible by 2 or 3 or $n = 5\}$.
    iii. The set of all strings over $\{a, b\}$ beginning and ending with a.
15. Write regular expressions for the following languages.
    i. The set of strings over alphabet $\{a, b, c\}$ containing at least one a and at least one b.
    ii. The set of strings of 0's and 1's with at most one pair of consecutive 1's.
    iii. The set of all strings of 0's and 1's such that every pair of adjacent 0's appears before any pair of adjacent 1's.
    iv. The set of strings of 0's and 1's whose number of 0s is divisible by five.
16. Describe, in the English languages, the sets represented by the following regular expressions.
    i. $a\,(a + b)^*\,ab$
    ii. $a^* \, b + b^* \, a$
    iii. $(aa + b)^* \, (bb + a)^*$
    iv. $a^* + (ab + a)^*$.
17. Show that the language $a^{n^2} : n \quad 1$ is not context free.
18. Prove that the language $L = \{ww\!: w \in \{a, b\}^*\}$ is not context free.
19. Prove that the language $\{0^i 1^j 2^i 3^j : i \geq 1$ and $j \geq 1\}$ is not context free.
20. Show that the language $\{a^m b^m c^n : m \leq n \leq 2m\}$ is not context free.

Answers

1. No, No, No, Yes.
9. We start with $S \to 0AB \to 0ASA \to 0A0ABA$ or $S \to 0A1 \to 0A0B \to 0S0B1$. If we apply $S \to 0AB$ we get A in the sentential form. If we try to eliminate A using $A0 \to S0B$ or $A1 \to SB1$. Here we cannot eliminate A or S. So $L(G) = \varphi$.
10. Here $L\,(G) = \{a^n b^n \mid n \geq 1\} = L(G_2)$.
11. $L\,(G) = \{0^m 1^n : n > m \geq 1\}$.
12. $L\,(G) = \{a^n ba^n : n \geq 1\}$.
13. $L\,(G) = \{a, b\}^+$.
14. i. Here we have three cases: w contains no 0; one 0 or two 0's. The required regular expression is $1^* + 1^*01^* + 1^*01^*01^*$
    ii. $(aa)^*+(aaa)^*+(aaaaa)^*$
    iii. $a\,(a + b)^*\,a$.
16. i. The set of all strings with a and ending in ab.
    ii. The strings are either strings of a's followed by one b or string of b's followed by one a.
    iii. The set of all strings of the form uv where a's occur in pairs of u and b's occur in pairs of v.
    iv. The strings are either strings of a's or strings of $ab + a$.

Chapter Objective Type Questions

A. Multiple choice questions

1. When $x = 010$ and $y = 1$, then the value of xy and yx will be-
   (A) $xy = 0101, yx = 1010$ (B) $xy = 0111, yx = 1010$
   (C) $xy = 0101, yx = 1110$ (D) $xy = 0101, yx = 1011$
2. Finite Automata is a collection of tuples:
   (A) $(Q, \Sigma, \Gamma, \delta, q_0, z_0, F\}$ (B) (V, T, P, S)
   (C) $(Q, \Sigma, \delta, q_0, F\}$ (D) $(Q, \Sigma, \delta, \lambda, q_0)$
3. Finite Automata is divided by:
   (A) Four parts (B) One part
   (C) Three parts (D) Two parts
4. DFA stands for:
   (A) Duplicate Finite Automata (B) Determine Finite Automata
   (C) Deterministic Finite Automata (D) Definition Finite Automata
5. Transition table is represented in:
   (A) Graphical form (B) Tabular form
   (C) Hierarchical form (D) None of these
6. Finite Automata is a collection of 5 tuples $(Q, \Sigma, \delta, q_0, F)$ in which δ is a:
   (A) Set of state (B) Finial state
   (C) Transition Function (D) Input set
7. A finite Automata which accepts more than one input is known as:
   (A) RE (B) DFA
   (C) CFG (D) NDFA
8. My-Hill Nerode theorem was proposed by:
   (A) John Nerode (B) Amion Nerode
   (C) John Myhill (D) Amion Myhill
9. Multiple final states consist of:
   (A) PDA (B) NFA
   (C) RE (D) DFA
10 The finite state system is a:
   (A) Control mechanism (B) Very good design tool
   (C) Mathematical model (D) All of these
11. $r^*$is equivalent to L_1^* , i.e. closure. Here $r^*$is known as:
   (A) Positive closure (B) Kleen closure
   (C) Negative closure (D) None of these
12. The regular expression for the language accepting all combination of a's over the set $\Sigma = \{a\}$ is:
   (A) $R = a^+$ (B) $R = a^*$
   (C) $R = aa^+$ (D) $R = aa^+$

13. What is the right arrangement of the various operators based on their precedence?
(A) Kleen closure, Associative, Union operator (B) Associative, Kleen, Union operator
(C) Union operator, Associative, Kleen closure (D) Associative, Union operator, Kleen closure
14. $\wedge + RR^*$ is equivalent to
(A) R (B) $+$
(C) R^* (D) Ø
15. Arden's theorem states that
(A) $R = P + QR$ (B) $R = R + PQ$
(C) $R = Q + RP$ (D) $R = Q + RP^*$
16. $\{a^{2n}/n \geq 1\}$ is represented by the regular expression:
(A) $(aa)^*$ (B) a^*
(C) aa^*a (D) a^*a
17. Mapping function can be written as:
(A) $(q_0, q_2, \ldots, q_i)$ (B) $\delta(q_0, q_1, q_2, q_3, \ldots q_i)$
(C) $\epsilon(q_0, q_1, \ldots q_j)$ (D) $\wedge(q_0, q_1, \ldots q_j)$
18. If L_1 and L_2 are two languages then $L_1 \cup L_2$ is:
(A) Regular (B) Not Regular
(C) Context free grammar (D) Finite automata
19. The finite automata is a collection of $(Q, \Sigma, \delta, q_0, F)$, where Q is a:
(A) Output alphabet (B) Transition Function
(C) Set of states (D) Set of input symbol
20 The compiler uses the program of lexical analyzer in the process of:
(A) Tokens (B) Compilations
(C) Transitions (D) UNIX
21. A machine in which output symbol depends upon the present input symbol and present state of the machine is:
(A) Moore machine (B) Mealy machine
(C) RE to DFA (D) CFG
22. A finite state machine is a machine in which the next state is decided by current state and current input symbol:
(A) Mealy machine (B) Moore machine
(C) Mealy to Moore (D) Both (A) and (B)
23. The context free grammar can formally be defined as a set denoted by $G =$
(A) (Q, S, q_0) (B) $(q_0, S, \delta, \Delta, F)$
(C) (V, T, P, S) (D) (q_0, δ, Δ, F)
24. In context free grammar V and T are set of:
(A) Terminal (B) Non-terminals
(C) Start Symbol (D) Terminals and Non-terminals

25. The language generated by G is denoted by $L(G)$. The language L is called:
   (A) Regular expression
   (B) Context Free Language
   (C) Finite automata
   (D) PDA
26. If $G = (V, T, P, S)$ some context free grammar then generation of some language using specific rules is called
   (A) Sentential Form
   (B) FA
   (C) Derivation
   (D) PDA
27. The derivation tree is also known as:
   (A) Right most derivation
   (B) Left most derivation
   (C) Parse tree
   (D) Positive closure
28. A leaf nodes is always be a:
   (A) Non-terminal node
   (B) Terminal node
   (C) Start symbol
   (D) Derivation tree
29. The grammar can be derived in either left most derivation or rightmost derivation. One can draw a derivation tree known as:
   (A) Parse tree
   (B) Production
   (C) Terminal
   (D) Non- terminal
30. If there can be more than one left most or right most derivation possible, then that grammar is said to be:
   (A) Context free grammar
   (B) Ambiguous grammar
   (C) Regular expression
   (D) PDA
31. The Chomsky's normal form (CNF) can be defined as:
   (A) Non terminal → non terminal.
   (B) Terminal → Terminal Non terminal
   (C) Terminal
   (D) Only non terminal
32. The rule for GNF is:
   (A) Non terminal → one terminal
   (B) Non terminal → one terminal. Any number of non-terminals
   (C) Terminal → any number of terminals.
   (D) Terminal → one non terminals.
33. The context free grammar or languages are closed under:
   (A) Intersection
   (B) complement
   (C) Kleen closure
   (D) Intersection or complement
34. The decision properties of CFLs are:
   (A) Emptiness
   (B) Finiteness
   (C) Membership
   (D) All of these
35. PDA consists of:
   (A) Input tape
   (B) Finite control
   (C) stack
   (D) all of these
36. Finite control is pointed by-
   (A) Input symbol
   (B) Current symbol
   (C) Stack symbol
   (D) All of these

37. PDA can be defined as a collection of components which is-
(A) Seven (B) Five
(C) Four (D) Six
38. Which symbols are used while drawing the PDA?
(A) Start (B) Accept
(C) Reject (D) All of these
39. ID stands for:
(A) Intension description (B) Instantaneous description
(C) Integrity description (D) None of these
40 How many types of operation are used on stack in a PDA?
(A) One (B) Four
(C) Two (D) Three
41. A PDA accepting a language $\{L= a^n b^n \mid n \geq 1\}$:
(A) (q_1, ϵ) (B) (q_2, ϵ)
(C) (q_0, ϵ) (D) (q_1, ϵ) (q_2, ϵ)
42. PDA can be collection of components which is Γ:
(A) Input set (B) Mapping function
(C) Start symbol (D) Stack alphabet
43. PDA can be collection of components which is z_0:
(A) Mapping function (B) Stack alphabet
(C) Start symbol (D) Final state
44. Multiple path for the same input is:
(A) PDA (B) DPDA
(C) NPDA (D) All of these

B. Fill in the blanks.

1. The difference of two languages L_1 and L_2 is denoted by ______.
2. Let $\Sigma = \{a, b\}$, then Σ^* equals to ______.
3. Full form of NDFA is ______.
4. QST is a substring of ______.
5. The union of two regular expressions R_1 and R_2 is written as ______.
6. (0* 1*)* is equals to ______.
7. In Moore machine ______ number of tuples are used.
8. The formula $\lambda'(q, a) = \lambda(\delta(q, a))$ is used for the conversion of ______.
9. $a^*(a+b)^*$ is equivalent to ______.
10. The derivation of w from start symbol S can be written as $S \rightarrow W$ which is known as ______.
11. A root node is always a node indicating ______.
12. A derivation is read from ______.
13. GNF stands for ______.

14. The context free grammar or languages are not closed under _____ .
15. PDA stands for ____ .
16. Full form of DPDA is_____ .
17. The language can be accepted by PDA using _____approaches.
18. Non- deterministic pushdown automata is similar to____.

C. Which of the following statements is True (T)/ False (F)?

1. DFA is a Subset of NFA.
2. If string x=110, then x^2=110100.
3. The Kleen closures are denoted by *.
4. The concatenation of two regular expression R_1 and R_2 is written as R_1 - R_2.
5. $\emptyset+R$ is equivalent to R^*.
6. Mealy machine is five tuple.
7. The regular expression can be modeled by the FA.
8. In the context free grammar $G=(V, T, P, S)$, the P represents set of production.
9. $S\rightarrow xyx$, $S\rightarrow ayx$, $S\rightarrow abx$, $S\rightarrow aba$, is an example of right most derivation.
10. A context free grammar is called inherently ambiguous if all the production rules in this grammar are ambiguous.
11. The productions in which one non terminal gives another non terminal are known as E production.
12. PDA is less powerful than FA.
13. NDPDA stands for Non-distinct Push Down Automata.
14. NPDA is more powerful than DPDA.
15. If x=01, y=101 and z=011, then $xyzy$ is 01101011101.
16. Let R={(1, 2), (2, 3), (2, 4)} be relation in {1, 2, 3, 4}.Then R^+={(1, 1), (2, 2), (3, 3), (4, 4), (1, 3), (1, 2), (2, 3), (3, 4)}.
17. If R={(a, b), (b, c), (c, a)} is a relation in {a, b, c}. Then R^2={(a, b), (b, c), (c, a), (a, c), (b, a), (c, b), (a, a), (b, b), (c, c)}.

Answers (Objective Type Questions)

A.

1. A	2. C	3. D	4. C
5. B	6. C	7. D	8. C
9. D	10. D	11. B	12. B
13. A	14. C	15. C	16. A
17. B	18. A	19. C	20. B
21. B	22. B	23. C	24. D
25. B	26. C	27. C	28. B
29. A	30. B	31. A	32. B

33. C	34. D	35. D	36. B
37. A	38. D	39. B	40. C
41. B	42. D	43. C	44. C

B.

1. $L_1 - L_2$
2. {Σ, *a*, *b*, *aa*, *ab*, *ba*, *bb* …}
3. Non-deterministic finite automata
4. QQSSTT
5. $R_1 + R_2$
6. (0+1)*
7. Six
8. Moore to mealy conversion
9. (*ab*)*
10. Sentential form
11. Start symbol
12. Left to right
13. Greibach normal form
14. Intersection
15. push down automata
16. Deterministic Pushdown Automata
17. Two
18. NDA

C.

1. T	2. F	3. T	4. T
5. F	6. F	7. T	8. T
9. F	10. T	11. F	12. F
13. F	14. T	15. T	16. T
17. T			

Conversion of a Moore Machine into a Mealy Machine

Let M_1 and M_2 are Moore machine and Mealy machine respectively. Let 'a' be the output for Moore machine for its initial state. The two machines are said to be equivalent if for all input string 'u',

$$a\, Z_{M_1}(u) = Z_{M_2}(u).$$

The procedure for the conversion of a Moore machine into a Mealy machine is given below:

1. We will define the output function γ_1 for Mealy machine as a function of present state and input symbol, therefore

$$\gamma_1(q_i, a) = \gamma(\delta(\Sigma_i, a))$$

for all states q_i and input symbol 'a'.

2. The transition function of a Mealy machine will be same as that of given Moore machine.

Example

Construct a Mealy machine equivalent to the Moore machine given below:

Present State	Next state		Output
	0	1	
$\rightarrow q_0$	q_1	q_2	1
q_1	q_3	q_2	0
q_2	q_2	q_1	1
q_3	q_0	q_3	1

Solution

For each input symbol in a Moore machine, construct the pair of next state and its corresponding output and we get the following table.

Present State	Next State			
	0		1	
	State	Output	State	Output
$\rightarrow q_0$	q_1	0	q_2	1
q_1	q_3	1	q_2	1
q_2	q_2	1	q_1	0
q_3	q_0	1	q_3	1

which is the Mealy machine for given Moore machine.

Conversion of a Mealy Machine into a Moore Machine

Example

Construct a Moore machine equivalent to a Mealy machine given in the following table.

Present State	Next State			
	0		1	
	State	Output	State	Output
$\to q_1$	q_1	1	q_2	0
q_2	q_4	1	q_4	1
q_3	q_2	1	q_3	1
q_4	q_3	0	q_1	1

Solution

For the construction of a Moore machine that accept the same set of input sequences let us see the next state column for any state, say q_t and find out the different outputs related with q_t in that column. Now divide q_t in such a manner that the number of different states being equal to the number of different outputs related with q_t. For example, q_1 is related with one output namely 1 and q_2 is related with two outputs 0 and 1. q_3 is related with two outputs 0 and 1. Also q_4 is related with one output 1. So we divide q_2 into q_{20} and q_{21}. Similarly we divide q_3 as q_{30} and q_{31}. The new table obtained is given described below.

Present State	Next State			
	0		1	
	State	Output	State	Output
$\to q_1$	q_1	1	q_{20}	0
q_{20}	q_4	1	q_4	1
q_{21}	q_4	1	q_4	1
q_{30}	q_{21}	1	q_{31}	1
q_{31}	q_{21}	1	q_{31}	1
q_4	q_{30}	0	q_1	1

Rearranging the pair of states and outputs, we get the following table.

Present State	Next State		
	0	1	Output
$\to q_1$	q_1	q_{20}	1
q_{20}	q_4	q_4	0
q_{21}	q_4	q_4	1
q_{30}	q_{21}	q_{31}	0
q_{31}	q_{21}	q_{31}	1
q_4	q_{30}	q_1	1

From above table, we can see that the initial state q_1 is related with output 1, means if a machine starts at state q_1 with input $\in$ then we get an output 1. It means that the Moore machine accept a null sequence which is rejected by Mealy machine. To remove this problem we add a new starting state q_0, having the same state transitions as q_1 have with output '0'.

The modified table is given in below.

Present State	Next State		
	0	1	Output
$\rightarrow q_0$	q_1	q_{20}	0
q_1	q_1	q_{20}	1
q_{20}	q_4	q_4	0
q_{21}	q_4	q_4	1
q_{30}	q_{21}	q_{31}	0
q_{31}	q_{21}	q_{31}	1
q_4	q_{30}	q_1	1

BIBLIOGRAPHY

1. Deo Narsingh, Graph Theory with Application to Engineering and Computer Science. Prentice Hall of India, New Delhi (1986).
2. Fraligh J.B., A first Course in Abstract Algebra, Narosa Publishing House, New Delhi (1982).
3. Gerstring J. L., Mathematical Structure for Computer Science, W. H. Freeman and Company (2004).
4. Grimaldi Ralph P., Discrete and Combinatorial Mathematics, Pearson Education Pvt. Ltd., New Delhi (2003).
5. Harary F., Graph Theory, Addison Wisley(1969).
6. Herstein I. N., Topics in Algebra, John Wiley & Sons , New York (1999).
7. Hopcroft J. E. and Ullman J. D., Introduction to Automata Theory, Languages and Computation, Narosa Publishing House, New Delhi (1979).
8. Johnsonbaugh Richard, Discrete Mathematics, Pearson Education (Singapore) Pvt. Ltd., New Delhi (2002).
9. Joshi K. D., Foundations of Discrete Mathematics., John Wiley & Sons(1989).
10. Kenneth A. Ross and Charles R.B. Wright, Discrete Mathematics, Prentice Hall Inc., New Jersy (1988).
11. Kolman Bernard, Busby Robert C. and Ross Sharon Cutler, Discrete Mathematical Structures, Pearson Education (Singapore) Pvt. Ltd., New Delhi (2001).
12. Linz, P. An Introduction to Formal languages and Automata, Narosa Publishing House, New Delhi (2002)..
13. Liu C. L., Elements of Discrete Structures, Tata McGraw Hill Pub. Co. Ltd., New Delhi (2000).
14. Mishra K.L.P. and Chanderasekaran N.,Theory of Computer Science, Prentice Hall of India, New Delhi (1998).
15. Mott. Joe L., Kandel Abraham, Baker P. Theodore, Discrete Mathematics for Computers Scientists and Mathematicians, Prentice Hall of India, New Delhi (2001).
16. Rao Shankar G., Discrete Mathematical Structure, New Age International Pub., New Delhi (2002).
17. Rosen Kenneth H., Discrete Mathematics and its application, Tata McGraw Hill Pub. Co. Ltd., New Delhi
18. Seymour Lipschutz, Marc Lipson , Theory and Problems of Discrete Mathematics, Tata McGraw Hill Pub. Co. Ltd., New Delhi.
19. Sooryanarayana B. And ranganath G. K., A Text Book of Graph Theory and its Application, S. Chand & Company Ltd., New Delhi (2001).
20. Trembley J. P. and Manohar R., Discrete Mathematical Structures With Application To Computer Science., Tata McGraw Hill Pub. Co. Ltd., New Delhi (1997).
21. Truss John, Discrete Mathematics for Computer Science, Addison Wesley (2000).
22. Wilson Robinson J., Graph Theory, Pearson Education (Singapore) Pvt. Ltd., New Delhi (2002).

INDEX

D

E

Q

R

S

T